MW00837367

Periodic Chart of the Elements

Periods

Group																	
1 **IA**	**2** **IIA**	**3** **IIIB**	**4** **IVB**	**5** **VB**	**6** **VIB**	**7** **VIIB**	**8**	**9** **VIIIB**	**10**	**11** **IB**	**12** **IIB**	**13** **IIIA**	**14** **IVA**	**15** **VA**	**16** **VIA**	**17** **VIIA**	**18** **VIIIA**

1 | 1 **H** 1.0079 | | | | | | | | | | | | | | | | | 2 **He** 4.0026

1 **H** 1.0079 / **VIIIA** 17

Period 1: 1 **H** 1.0079

2 | 3 **Li** 6.941 | 4 **Be** 9.0122 | | | | | | | | | | | 5 **B** 10.811 | 6 **C** 12.011 | 7 **N** 14.007 | 8 **O** 15.999 | 9 **F** 18.998 | 10 **Ne** 20.180

3 | 11 **Na** 22.990 | 12 **Mg** 24.305 | | | | | | | | | | | 13 **Al** 26.982 | 14 **Si** 28.086 | 15 **P** 30.974 | 16 **S** 32.066 | 17 **Cl** 35.453 | 18 **Ar** 39.948

4 | 19 **K** 39.098 | 20 **Ca** 40.078 | 21 **Sc** 44.956 | 22 **Ti** 47.88 | 23 **V** 50.942 | 24 **Cr** 51.996 | 25 **Mn** 54.938 | 26 **Fe** 55.847 | 27 **Co** 58.933 | 28 **Ni** 58.693 | 29 **Cu** 63.546 | 30 **Zn** 65.39 | 31 **Ga** 69.723 | 32 **Ge** 72.61 | 33 **As** 74.922 | 34 **Se** 78.96 | 35 **Br** 79.904 | 36 **Kr** 83.80

5 | 37 **Rb** 85.468 | 38 **Sr** 87.62 | 39 **Y** 88.906 | 40 **Zr** 91.224 | 41 **Nb** 92.906 | 42 **Mo** 95.94 | 43 **Tc** 98.906[a] | 44 **Ru** 101.07 | 45 **Rh** 102.91 | 46 **Pd** 106.42 | 47 **Ag** 107.87 | 48 **Cd** 112.41 | 49 **In** 114.82 | 50 **Sn** 118.71 | 51 **Sb** 121.76 | 52 **Te** 127.60 | 53 **I** 126.90 | 54 **Xe** 131.29

6 | 55 **Cs** 132.91 | 56 **Ba** 137.33 | 57 ***La** 138.91 | 72 **Hf** 178.49 | 73 **Ta** 180.95 | 74 **W** 183.85 | 75 **Re** 186.21 | 76 **Os** 190.2 | 77 **Ir** 192.22 | 78 **Pt** 195.08 | 79 **Au** 196.97 | 80 **Hg** 200.59 | 81 **Tl** 204.38 | 82 **Pb** 207.2 | 83 **Bi** 208.98 | 84 **Po** 209.98[a] | 85 **At** 209.99[a] | 86 **Rn** 222.02[a]

7 | 87 **Fr** 223.02[a] | 88 **Ra** 226.03[a] | 89 **†Ac** 227.03[a] | 104 **Rf** 261.11[a] | 105 **Db** 262.11[a] | 106 **Sg** 263.12[a] | 107 **Bh** 262.12[a] | 108 **Hs** (265)[b] | 109 **Mt** (266)[b]

* | 58 **Ce** 140.12 | 59 **Pr** 140.91 | 60 **Nd** 144.24 | 61 **Pm** 146.92[a] | 62 **Sm** 150.36 | 63 **Eu** 151.97 | 64 **Gd** 157.25 | 65 **Tb** 158.93 | 66 **Dy** 162.50 | 67 **Ho** 164.93 | 68 **Er** 167.26 | 69 **Tm** 168.93 | 70 **Yb** 173.04 | 71 **Lu** 174.97

† | 90 **Th** 232.04 | 91 **Pa** 231.04 | 92 **U** 238.03 | 93 **Np** 237.05[a] | 94 **Pu** 239.05[a] | 95 **Am** 241.06[a] | 96 **Cm** 244.06[a] | 97 **Bk** 249.08[a] | 98 **Cf** 252.08[a] | 99 **Es** 252.08[a] | 100 **Fm** 257.10[a] | 101 **Md** 258.10[a] | 102 **No** 259.10[a] | 103 **Lr** 262.11[a]

Source: These data are based on the definition that the mass of the ^{12}C isotope of carbon is exactly 12 amu. The atomic weight of an element is the weighted average of the masses of the isotopes of that element. Data from "Atomic Weights of the Elements," *Pure and Applied Chemistry*, **63**, *975* (1991).

[a]Radioactive element has no stable isotope. The mass is for one of the element's common radioisotopes.
[b]Mass number of the most stable isotope of this radioactive element.

Experimental
Organic Chemistry

EXPERIMENTAL ORGANIC CHEMISTRY

DANIEL R. PALLEROS

University of California, Santa Cruz

JOHN WILEY & SONS, INC.

New York ■ Chichester ■ Weinheim ■ Brisbane ■ Toronto ■ Singapore

ACQUISITIONS EDITOR   Jennifer Yee

SENIOR PRODUCTION EDITOR   Elizabeth Swain

SENIOR MARKETING MANAGER   Charity Robey

SENIOR DESIGNER   Karin Gerdes Kincheloe

ILLUSTRATION EDITORS   Sandra Rigby and Edward Starr

PHOTO EDITOR   Lisa Gee

COVER PHOTO   Paul Schraub

COVER CONCEPT   Daniel R. Palleros

ILLUSTRATIONS   Fine Line Illustrations, Inc.

This book was set in 10/12 Palatino Light by Laser Words and printed and bound by Courier/Westford. The cover was printed by Lehigh.

This book is printed on acid-free paper. ∞

To order books or for customer service, call 1(800)-CALL-WILEY (225-5945).

Library of Congress Cataloging in Publication Data:

Palleros, Daniel R.
        Experimental organic chemistry / Daniel R. Palleros.
                p.   cm.
        Includes bibliographical references and index.
        ISBN 0-471-28250-2 (cloth : alk. paper)
        1. Chemistry, Organic Laboratory manuals.   I. Title.
QD261.P335   1999   99-35417
547'.078–dc21   CIP

Printed in the United States of America

10 9 8 7 6 5 4 3 2

To my students

Contents

Preface

To the Students

You are about to embark in what may be a life-changing experience. After taking organic chemistry, a stroll in the forest will never be the same. You will perceive the world around you with renewed senses as if a new dimension had been added. This dimension is knowledge.

Organic chemistry is everywhere, from the delicate smell of violets to the paper these words are printed on. In the organic chemistry laboratory you will learn to decode some of nature's secrets, and in doing so, you will learn to use tools that will extend your senses to the molecular level. You will also learn a new language that will enable you to describe what you see through the magnifying glass of experimental organic chemistry.

In the laboratory you will learn to observe, interpret, and predict organic chemistry. You will develop your problem-solving skills and will gain a sense of self-confidence and self-reliance. By the end of this course you will be proficient in the most important aspects of laboratory work and will be able to design and interpret your own experiments.

In developing this laboratory course I tried to be loyal to my conviction that we learn only what we enjoy. Whenever possible, I made an effort to relate the experiments to everyday life and to provide social and historical notes. Many of the experiments have biological applications or are connected to medicine. For example, you will prepare diiodotyrosine, a compound related to the thyroid hormones that control metabolism and development. You will transform a pain-relief ointment, such as Bengay, into aspirin. You will isolate caffeine from tea. You will analyze vitamin tablets. You will isolate the components of milk. You will prepare fragrances that mimic the smell of fruits and flowers. You will obtain volatile oils from plants. You will make an ant pheromone (pheromones are compounds used by animals to communicate with members of the same species). You will explore the chemistry of steroids. You will make plastics and dye fabrics.

One of the first lessons that you will learn in the laboratory is to be patient. In this fast-paced society it seems that patience is a virtue that has fallen by the wayside. As a scientist in-training you will learn to follow nature at its own speed and to respect its timing. Very often, there is little we can do to hasten the course of an experiment. In the laboratory you will learn to budget your time efficiently while waiting for an experiment to conclude. You will also learn to take frustration in stride. An experiment may not work as planned and must be repeated. But a failed experiment gives us an opportunity to learn from our mistakes. Another lesson you will learn in the laboratory is to be faithful to your data. You will learn to accept your results and treat them with respect, avoiding embellishment and intentional omissions. Remember that the building of science is made of truth and its foundations are rooted in trust. It is in the laboratory where the advances of science are made. Without laboratory work, science would be just a poetic fabrication.

You are not alone in the endeavor of discovering experimental organic chemistry. All of us working in the laboratory, from the beginner to the most seasoned chemist, are always learning new aspects of this science, art, and craft.

The Subject of Experimental Organic Chemistry

In this course you will learn to work with organic chemicals by obtaining them, identifying them, and transforming them. Organic chemicals can either be obtained from natural sources, by a technique called *extraction*, or can be prepared from other chemicals by a technique called *synthesis*. When a compound is extracted from nature or synthesized in the laboratory, it is hardly ever obtained in a pure form. Normally the chemical of interest is in a mixture with other components from which it must be separated. After the *separation*, the chemist must make sure that residual amounts of the other components, called impurities, are eliminated from the compound of interest. This process is called *purification*. Separation and purification are carried out by similar techniques (see diagram below), and the differences between the two procedures are more conceptual than operational. In the separation of a mixture, the chemist usually recovers all of its components in separate containers. In the purification of a chemical, on the other hand, the chemist is only interested in that chemical and tosses away the impurities. In this course you will learn many separation and purification techniques such as recrystallization, distillation, chromatography in all of its different forms (gas chromatography, thin-layer chromatography, column chromatography, and high-performance liquid chromatography), sublimation, liquid-liquid extraction, and acid-base extraction. Units 4–10 and 14 are devoted to these subjects.

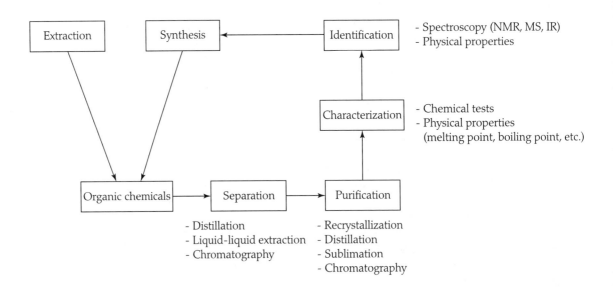

After a compound has been purified, it must be characterized and identified. *Characterization* and *identification* go hand in hand and are performed by similar techniques. Characterization, which usually precedes identification, is the study of the physical and chemical properties of a compound. Charaterization is done by chemical tests, which by a simple color change in a test tube can give valuable information about the chemical, and also by the determination of physical properties such as melting point, boiling point, refractive index, and density. Identification, on the other hand, is a more involved procedure and implies the elucidation of the structure of the compound. Identification is normally accomplished by a battery of chemical and spectroscopic methods. Nuclear magnetic resonance (NMR), infrared spectroscopy (IR), and mass spectrometry (MS) are among the most valuable tools to this end. You will find numerous spectroscopic applications at the end of the experiments beginning in Unit 11. Units 30–34 are fully devoted to spectroscopy.

Organization

This book is organized into three sections. **Section 1** (The Basics) contains three units: Laboratory Safety (Unit 1), Basic Concepts (Unit 2), and Basic Operations (Unit 3). Read Unit 1 and familiarize yourself with the safe practices in the laboratory *before* you start your experimental work. You can delay reading Unit 2 and Unit 3 until a later time if you want. These units contain basic information about common operations and procedures in the laboratory. Some of this information is a review of what you have already learned in general chemistry; other information will be new to you. You will find references to Units 2 and 3 throughout the book.

Section 2 (Units 4–29) contains the experiments. Each unit is composed of two parts: the background information and the experiments themselves. There are exercises at the end of each background part with answers to selected problems at the end of the book. Try to work out these problems; they will give you a deeper understanding of the subject and will prepare you to do and enjoy the experiments. Interlaced in these exercises you will find the story of Synton X. Plosiff, a "slightly vain, somewhat absent-minded, but totally dedicated graduate student" who is our fictional chemistry hero. I hope that you like him and his friends.

Each experiment begins with an overview where you will find a brief description of the aims and organization of the experiment. Read this part carefully because it will provide you with the "big picture" of the experiment. At the end of each experimental part you will find guidelines, in the form of **pre-lab** and **in-lab questions**, to help you prepare for the experiment and write a lab report. Answer the pre-lab questions before you start your experimental work. They will help you understand and take full advantage of the experiment. The in-lab questions will assist you in interpreting and analyzing your data.

Section 3 (Units 30–34) deals with spectroscopy: infrared (Unit 31), ultraviolet-visible (UV-vis, Unit 32), nuclear magnetic resonance (Unit 33), and mass spectrometry (Unit 34). All of the spectroscopic techniques are grouped together in one section because they are routinely used in most of the experiments. You should read the spectroscopy units as the course progresses according to the needs of the experiments or as indicated by your instructor.

To the Instructors

One of my main goals in writing this book was to acknowledge that chemistry, like music, can only be learned in a progressive manner. Basic laboratory techniques should be learned and practiced before they can be properly put to use to explore more advanced problems. The experiments in Units 4–11 and 14 introduce the most important techniques in the organic chemistry laboratory through extensive practice. Recrystallization, distillation, extraction, and chromatography are among the techniques presented in these units. The treatment of these techniques is thorough from both a theoretical and a practical standpoint.

The units in this book are modular and do not need to be introduced in the order in which they are presented. However, I have found that students understand the concept of chromatography better if the subject is presented after liquid-liquid extraction and if a connection between partitioning between two liquid phases and partition chromatography is made. Also, high-performance liquid chromatography is better appreciated by the students if they have a previous knowledge of gas chromatography.

In general, two experiments, or one experiment with several independent parts, are offered in each unit. The teachers can select the parts that suit their needs. To help focus the students, the suggested background reading is listed at the beginning of each experiment.

Beginning in Unit 12, experiments on the chemistry of functional groups are introduced in an order similar to that in which most organic chemistry textbooks present them. However, these units are also modular and teachers can alter this order without important consequences. The background information on the chemistry of functional groups summarizes only the most important aspects of their chemistry, with emphasis on experimental issues.

Unit 18 covers the subject of chemical kinetics and reaction mechanisms. The principles of kinetics are also illustrated in Units 19 and 29. In these experiments the students will go beyond the determination of rate constants and will learn to use chemical kinetics as a tool for the elucidation of reaction mechanisms. Unit 21 is about oxidation and reduction reactions. Grouping these subjects in one unit enables students to compare different methods and see their applications and limitations. Unit 23 deals with multistep synthesis. With a few exceptions, organic synthesis as a subject in its own right is not discussed in introductory lecture textbooks. Given the growing importance of this subject, a discussion of the concepts of retrosynthesis and synthons is presented in this unit. Units 26 and 27 deal with the chemistry of polymers and dyes. Both units provide a wide range of experiments to choose from. Unit 28, on bioorganic chemistry, introduces students to the use of microorganisms to perform stereospecific transformations.

The units on spectroscopy (Units 30–34) are intended to be used as the course progresses. From Experiment 11 on, most experiments include spectroscopic analysis. If the instrumentation is available, the students should obtain their own spectra. State-of-the-art IR, ^{1}H-NMR, and ^{13}C-NMR, and sometimes the mass spectra, are provided in this book for comparison and interpretation. It is up to the teacher to decide how much use they want to make of spectroscopic techniques. The treatment of IR and NMR includes a discussion of the technique of Fourier transform. Despite its growing use and importance, this subject is only tangentially treated in most textbooks. I tried to remedy this situation by providing a discussion that can be followed by students without much background in physics and math. Extensive sets of correlation tables are offered for UV-visible, IR, MS, and, especially, NMR.

There has been a major change in the teaching of experimental organic chemistry in the last 10 years. The widespread use of microscale experiments brought about by the work of D. Mayo, R. Pike, and S. Butcher in 1986 (*Microscale Organic Laboratory*, Wiley, New York) has revolutionized the way in which experimental organic chemistry is taught. Although the advantages offered by the microscale approach are many, I feel that students should also learn the macroscale approach because, out of necessity, in academia as well as in industry, some chemical operations are performed at the macroscale level. This book offers both approaches, especially when general laboratory techniques such as distillation, recrystallization, and extraction are introduced. Most of the syntheses, however, are at the micro or semi-micro level. By the end of the course the students should be in a position to decide at which scale an experiment would be best performed.

The experiments presented in this book emphasize the importance of critical thinking through extensive practice in problem solving and chemical analysis. The pre-lab and in-lab questions are designed to make the students actively participate in the experiments. In many of the experiments, the students will analyze and identify unknown samples. To this end they will make use of different techniques such as melting point, gas chromatography, thin-layer chromatography, polarimetry, refractive index, preparation of derivatives, and infrared and nuclear magnetic resonance spectroscopies. Many experiments have open-ended questions; others are puzzles for the students to solve. For example, Experiment 12 covers an investigation of the main product of a dehydration reaction; Experiment 16B involves a puzzle in electrophilic aromatic substitution; Experiment 18 covers the study of the kinetics and mechanism of a nucleophilic aromatic substitution reaction; Experiment 19B involves the study of the hydrophobic effect in the Diels-Alder reaction; Experiment 20A explores the identification of carbonyl compounds; Experiment 21 includes the identification of a reduction product; Experiment 23B shows the synthesis and identification of ionones; and Experiment 28 involves the study of a reduction pathway mediated by yeast, which also includes NMR conformational analysis.

The experiments have been designed to be performed by each student individually. I believe that in the organic chemistry laboratory the students should develop a sense of self-confidence in their practical skills that can only be obtained by working independently. I also believe that students, as scientists in-training, should learn to share results. Some experiments are designed so students can team up to investigate different parts of a larger problem. They still work individually, but they need to consult each other along the way and share their results in order to bring the experiment to a conclusion.

"Safety First" boxes are incorporated at the beginning of each experiment. They highlight the most important hazards in dealing with the chemicals. However, they are not a substitute for the Material Safety Data Sheets, which should be made available to the students in the laboratory. **"Cleaning up"** guidelines are offered at the end of each experiment. They will help in the collection, segregation, and disposal of the waste. Because waste disposal procedures vary locally, these guidelines should be checked by laboratory personnel against local and state regulations before disposing of any chemical, or before instructing the students to do so.

All the experiments presented in this book have been tried and optimized in our teaching laboratories. They work as planned under the conditions described here. Suggestions and comments are welcome and will help us in future editions.

Acknowledgments

No book is the work of just one person. Many people helped me at different stages in the production of this book, and to all of them I am grateful. I cannot list the thousands of students who used this book in its preliminary format and who were instrumental in refining the experiments presented here, but I am indebted to all of them, and to all of them this work is dedicated. I am also grateful to my teaching assistants who were always on the lookout for ways to improve the experimental procedures. Particular thanks go to Kris Kessler for her constant help and technical support, to Keith Oberg who carefully read the early stages of the manuscript, to Monika Hennig and Alvin Go for their patience in proofreading the text, to Jim Loo for his enthusiastic technical support in obtaining the NMR spectra, to Mary Howe for her help with the mass spectra, to Lori Nicholson, Beau Willis, Barbara Klipfel, Buddy Morris, Dan Blunk, and Bob Bailey for their assistance with waste disposal and safety matters, to Linda Locatelli for sharing her expertise in plants and seeds with me, and to Micheline Markey for her artistic talents in obtaining some of the photographs shown here.

I am especially indebted to the people at Wiley: Jennifer Yee, my Editor, who patiently and skillfully directed the entire project, Nedah Rose who believed in this project from the beginning, Elizabeth Swain, Senior Production Editor, who carefully managed the metamorphosis of the manuscript into this book, Lisa Gee who researched the sources of many of the photographs, Sandra Rigby who coordinated the rendering of the artwork, and Karin Gerdes Kincheloe who designed the cover. I would also like to thank the reviewers of the manuscript for their comments and suggestions, among them Jared A. Butcher (Ohio University), Clifford W. J. Chang (University of West Florida), Craig Fryhle (Pacific Lutheran University), Ronald Lawler (Brown University), Michael F. McCormick (Emory University), J. Ty Redd (Southern Utah University), L. Kraig Steffen (Fairfield University), and Robert D. Walkup (Texas Tech University). I would also like to thank all of my colleagues at UCSC for their constant support. I am particularly grateful to Professors Joseph Bunnett and Anthony Fink for their encouragement. Finally, I would like to acknowledge my friends from the University of Buenos Aires, School of Exact and Natural Sciences, for their inspiration, especially my dear friend Susi Socolovsky for her unfailing support over the years.

DANIEL R. PALLEROS
Santa Cruz, California

SECTION 1

The Basics

Laboratory Safety

A laboratory should be a safe and comfortable place to work, but no environment is safe without the cooperation of its inhabitants. Although the experiments in this manual do not involve extremely dangerous substances, the use of hazardous chemicals cannot be entirely eliminated. Before engaging in experimental work, read the following guidelines and observe them at all times in the laboratory.

1.1 GENERAL SAFETY GUIDELINES

Clothing

- Wear **safety goggles** at all times no matter what you are doing or where you are in the lab. For maximum protection against splashes, safety goggles **should not have open holes on the sides**; the venting apertures, necessary to minimize fogging, should have splashguards Figure 1.1. Goggles with antifog lenses are commercially available. If fogging occurs, clean the lens with a piece of tissue paper and antifog lotion, if available. Repeat the cleaning as necessary. This operation should be performed outside the lab. Do not store your goggles in your glassware drawer because they may inadvertently get contaminated. If you need prescription glasses, wear them beneath the safety goggles.

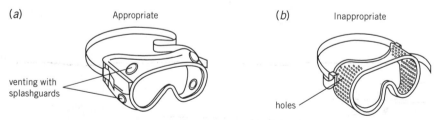

(a) Appropriate   (b) Inappropriate

venting with splashguards

holes

Figure 1.1 Safety goggles: a) with splashguards (indirect venting), appropriate for chemistry lab; b) with open holes (direct venting), inappropriate for chemistry lab.

- Always wear **protective clothing** such as long pants, long skirts, and shoes that completely cover your feet. Sandals, short pants, mini-skirts, and tank tops are not allowed in the lab. Aprons and lab-coats, although recommended, are no substitute for long pants because they do not fully protect the legs.

- Confine **long hair, loose-fitting sleeves,** and **scarves.** If they are not confined, they may cause accidents or catch fire.
- **Do not** wear **obtrusive rings, bracelets,** and **chains.** They can catch on lab equipment, and cause spillage and accidents; if they come in contact with electrical sources, an electric shock may occur.

Good Laboratory Habits

- **Use common sense.**
- **Do not eat, drink, smoke,** or **apply makeup.**
- Take a **short break** if necessary. Go outside and breathe fresh air. A 10-minute break during a 4-hour lab period is refreshing.
- **Never leave an ongoing experiment unattended.** If you have to leave the lab for a few minutes, notify your instructor.
- **Do not keep** your **books, backpacks,** and **coats on the bench top.** Instead, keep them in the designated area.
- **Do not sit on the benches.**
- Keep **exits** and **aisles free of obstructions** at all times.
- Keep your **bench space clean** and **tidy** while working.
- **Never work alone** in the laboratory.
- **Do not perform unauthorized experiments.** Such experiments are strictly prohibited.
- **Do not invite** or **receive visitors** in the laboratory.
- **Do not rush. Do not run. Do not push.**

Precautions

- Locate the **safety-showers, eye-wash fountains, fire extinguishers, first-aid kit,** and **emergency phone** before beginning your lab work.
- Before each experiment, **familiarize** yourself with the **hazards** (flammability, reactivity, stability, and toxicity) of the compounds involved. Such information can be found in the Material Safety Data Sheets (MSDS) provided by the manufacturer (discussed later). Also consult the Merck Index and the Aldrich Catalog for additional information. Record this information in your lab book.
- **Be aware of your surroundings.** Know what your neighbors are doing. If somebody near you appears to be performing an unsafe operation, point out the hazard immediately. Prevention is always the best medicine.
- **Never** use **open flames** in the laboratory without your instructor's permission.
- In case of an **emergency evacuation** of the building, all the students and instructors should **meet outside** the building at a **spot designated in advance by the instructor.**
- If you are **pregnant** or planning a pregnancy while taking the organic chemistry laboratory, contact your instructor and your doctor. They will provide you with information regarding potential risks to you and the embryo.

Before Leaving the Lab

- **Clean the bench.** Other students will appreciate a spotless work area.
- **Turn off gas, water, air, steam,** and **vacuum** valves at or near your bench space.
- **Unplug all electrical appliances** (hot plates, heating mantles, water pumps, etc.).
- **Wash your hands** thoroughly with soap and water.

Handling Chemicals

- Consider all chemicals **poisonous**.
- Use the **fume hood** when handling **organic solvents** and volatile compounds such as toluene, hexane, diethyl ether, petroleum ether, and methylene chloride.
- **Never use your mouth** to carry out a chemical operation (fill a pipet, start a siphon, etc.).
- **Dispose of chemical waste in the containers provided for that purpose**. They should be clearly labeled and are usually placed in the fumehood. Follow proper procedures as indicated by your instructor.
- **Do not contaminate reagents**. Most organic chemicals are very expensive. Use **clean** and **dry** pipets to dispense them. Take just the amount you need, do not waste them. Never pour unused reagents back into stock bottles.
- **Keep stock bottles in their designated spots** so everyone can find them easily.
- **Label all containers**. Do not use chemicals from unlabeled containers.
- **Read labels** carefully.
- **Do not inhale, smell**, or **taste chemicals**. Do not touch your face without washing your hands.
- **Do not wear gloves** unless instructed to do otherwise. There are several types of gloves offering different levels of chemical protection. Always check their chemical resistance and recommended usage. If they are not worn properly, they will only give you a false sense of security.
- **Do not reuse** contaminated **gloves**.

Handling Glassware

- **Check your glassware before use**. It is important to check round-bottom flasks and condensers carefully. Star cracks, like the one shown in Figure 1.2, may cause glassware to break. Discard any piece of glassware with a star crack.
- When **inserting or removing** a piece of **glass tubing** from a stopper or hose, always **wrap the tubing** with paper towels or a cloth to avoid cutting yourself. Use water, glycerol, or grease as lubricants as necessary.
- Place **broken glassware** in the **appropriate disposal container**.
- Place broken or disposable **contaminated glassware** in the appropriate **hazardous waste container**.

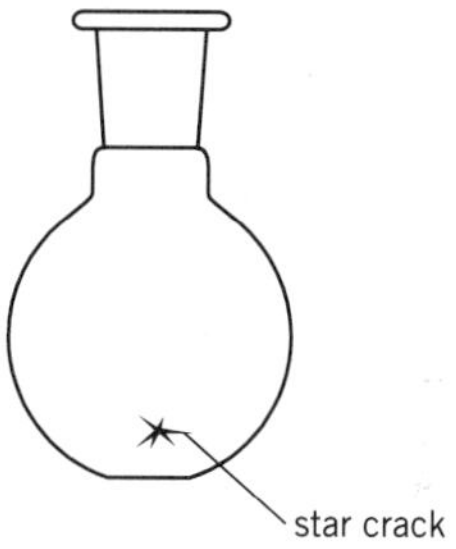

Figure 1.2 Round-bottom flask with star crack.

In Case of Accident

- If you **injure** yourself, **notify your instructor immediately**. Check your local emergency number.
- **Small cuts** can be treated by **washing** the wound, removing any pieces of glass, and applying **pressure** with a **sterile pad**.
- Do **not move injured persons** unless they are in further danger.
- If **chemicals get in your eyes**, immediately flush them with water for at least 15 minutes using an **eye-wash fountain** or **eye-wash cup**. Seek medical attention.
- In case of **ingestion of a hazardous chemical** contact the local **Poison Control Center** immediately. The phone number should be posted by the emergency phone. Meanwhile follow the **first aid instructions** shown in the **MSDS**. Never give anything by mouth to someone who is unconscious.
- In case of a small **fire**, get your instructor's attention immediately and follow instructions. In case of a large fire, evacuate the building immediately. See instructions under "**In Case of Natural Disaster or Fire Alarm**."

- If your **clothing is on fire, do not run**. Use <u>water, a blanket, or a coat</u> to put out the fire. If necessary, **roll on the floor**. Get prompt medical attention.

Spills

- If you spill a chemical on your skin, **remove all contaminated clothing** and immediately **flush with cold water** for at least 15 minutes. Notify your instructor. Check the MSDS for delayed effects. Check for a reaction during the next 24 hours. See a physician.

- If you have a **chemical spill on a large area of your body**, use the **safety-shower** immediately as you remove any contaminated clothing. Avoid spreading the chemical on the skin. See a physician.

- **Mercury is a highly toxic element**. If you break a mercury thermometer, notify your instructor immediately. Do not try to collect the mercury yourself. Dispose of the broken thermometer in the appropriate waste container.

- If you spill a **nonhazardous** solid or liquid, immediately **wipe it up** using paper towels or an absorbent material such as vermiculite. If the spill is large (>100 mL) or if it poses a danger to you or someone else, notify your instructor. Do not attempt to clean up large spills.

- If you spill a **hazardous** material **notify your instructor** immediately.

- Dispose of **material used to clean up a spill** in the proper **waste containers**.

In Case of Natural Disaster or Fire Alarm

- If the **fire alarm goes off**, turn off gas, water, vacuum, steam, and any source of heat that you may be using (do this only if it does not pose further danger to you) and **exit the building as soon as possible**. Do not run.

- **Do not use the elevators to evacuate** the building.

- **Gather** outside the building in the **pre-arranged spot**.

- In case of **natural disaster follow the instructor's orders**. In case of **earthquake find shelter** under a desk or a door frame. When the tremor has subsided turn off gas, water, steam, vacuum, and any source of heat that you may be using (do this only if it does not pose further danger to you) and **exit the building as soon as possible**. Do not run.

1.2 CHEMICAL TOXICITY

All chemicals are toxic. Paracelsus, a Renaissance physician, said in one of the most widely cited quotes in toxicology: "All substances are poisons.... The right dose differentiates a poison and a remedy." Toxic chemicals are often divided into two main categories: **acutely toxic** and **chronically toxic substances**. While acutely toxic substances cause immediate damage after a single exposure (for example, hydrogen cyanide), a chronically toxic substance has noticeable toxic effects after repeated exposure. Substances in this category usually have a time delay between the exposure and the onset of the effects (examples include lead and mercury). These two categories are not mutually exclusive, and most compounds exhibit both acute and chronic effects depending on the dose.

Toxic chemicals are often classified according to their effects. For example, **carcinogenic** compounds are malignant-tumor causing agents, **mutagenic** compounds cause heritable genetic changes, and **teratogenic** compounds produce malformation of the developing embryo. A very crude way of assessing acute chemical toxicity is through animal experimentation. **LD_{50} (lethal dose 50)** is the amount that ingested, injected, or applied to the skin in a single dose causes the death of 50% of the test animals. LD_{50} is expressed in mg, g, kg, and so on per kg of body weight. LD_{50} depends, of course, on the test animals and on the conditions used for the experiment. **LC_{50} (lethal concentration 50)** is used for airborne and water pollutants, and is defined as the concentration of a chemical in the air or water needed to kill 50% of the test animals. LC_{50} is given in mg/L, mg/m^3, ppm, and so on. When trying to determine the potential hazards of a chemical to humans, LD_{50} and LC_{50} should be regarded only as rough guides because the extrapolation from animal studies to human subjects is not direct. Chemicals that are relatively safe for animals may be toxic to humans and vice versa. LD_{50} and LC_{50} studies are of only limited value for understanding the chronic effects of toxic chemicals. More significant to human subjects are studies using cultured human tissues rather than live animals. These types of studies are slowly becoming more widespread.

In the laboratory, inhalation of chemicals is unavoidable and it may result in chronic or acute intoxication if the level of chemical in the air is above a given limit. Several indices have been devised to indicate the toxicity of volatile compounds. The **P.E.L.** or **Permissible Exposure Limit**, is the maximum permitted average concentration of airborne chemical in volume parts per million (ppm) for a worker exposed 8 hours daily. For relatively nontoxic vapors (for example, ethanol, acetone, and ethyl acetate) values in the range of 400–1000 ppm are common; on the other hand, very toxic chemicals, for example benzene and chloroform, have P.E.L. values in the range of 1–2 ppm.

Through the action of the Occupational Safety and Health Administration (**OSHA**) it is now required that **Material Safety Data Sheets** (**MSDS**) accompany any shipment of hazardous materials. The MSDS contain updated information on physical data, reactivity, fire and explosion hazards, toxicity and health hazards, and spill and waste disposal procedures. The MSDS should be read before using any chemical. A sample of the MSDS for aspirin, obtained from the Aldrich Chemical Company, is shown in Figure 1.3 (on pp. 8–11). For information about online resources see the section at the end of this unit.

1.3   DEALING WITH CHEMICALS AND WASTE DISPOSAL

Labeling of Chemicals

In 1966, the National Fire Protection Association introduced a pictorial label to indicate chemical hazards (Fig. 1.4). Four diamonds contain information about **health** (left, blue diamond), **fire** (top, red diamond), **instability** (right, yellow diamond), and **special information** (bottom, white diamond) such as reactivity with water and oxidation power. Each diamond has a number from 0 to 4 indicating the chemical hazard in that category; 4 is very hazardous and 0 is not hazardous. The health hazard depicted in these labels refers to acute toxicity rather than chronic effects.

When preparing chemicals in the laboratory, labeling them properly is the first step to good laboratory practices. Unlabeled chemicals are very costly to dispose of and can create potentially dangerous situations. You should write the name of the compound clearly on the label with waterproof ink and secure it perfectly. You may also provide the chemical structure, although **it should not substitute for the chemical name**. Hazard information should be included if available. The student's name, the amount of material, melting or boiling points if known, the date, and the lab book page where the synthesis, isolation, or purification is described should also be included (Fig. 1.5). In the case of potential explosives such as peroxides and peroxide-forming compounds (see section on "**Chemicals Worth Knowing**"), the dates when the chemical was received and opened (if the chemical was purchased) or the date when it was prepared (if it was synthesized in the laboratory) should be included.

MATERIAL SAFETY DATA SHEET

```
SECTION 1. - - - - - - - - - CHEMICAL IDENTIFICATION- - - - - - - - - -
    CATALOG #:          239631
    NAME:              ACETYLSALICYLIC ACID, 99+%
SECTION 2. - - - - - COMPOSITION/INFORMATION ON INGREDIENTS - - - - - -
    CAS #:    50-78-2
    MF: C9H8O4
    EC NO:  200-064-1
  SYNONYMS
    ACENTERINE * ACESAL * ACETICYL * ACETILSALICILICO * ACETILUM
    ACIDULATUM * ACETISAL * ACETOL * ACETONYL * ACETOPHEN * ACETOSAL *
    ACETOSALIC ACID * ACETOSALIN * O-ACETOXYBENZOIC ACID * 2-
    ACETOXYBENZOIC ACID * ACETYLIN * 2-(ACETYLOXY)BENZOIC ACID *
    ACETYLSAL * ACETYLSALICYLIC ACID (ACGIH) * ACETYLSALICYLSAURE (GERMAN)
    * ACIDO ACETILSALICILICO (ITALIAN) * ACIMETTEN * ACIDE
    ACETYLSALICYLIQUE (FRENCH) * ACIDO O-ACETIL-BENZOICO (ITALIAN) *
    ACIDUM ACETYLSALICYLICUM * ACISAL * ACYLPYRIN * ASA * A.S.A. * A.S.A.
    EMPIRIN * ASAGRAN * ASPERGUM * ASPIRDROPS * ASPIRIN * ASPIRINA 03 *
    ASPRO CLEAR * ASTERIC * AC 5230 * BENASPIR * BENZOIC ACID, 2-
    (ACETYLOXY) - (9CI) * BIALPIRINIA * CAPRIN * O-CARBOXYPHENYL ACETATE *
    COLFARIT * CONTRHEUMA RETARD * DELGESIC * DOLEAN PH 8 * DURAMAX * ECM
    * ECOTRIN * EMPIRIN * ENDYDOL * ENTERICIN * ENTEROSARINE * ENTROPHEN *
    GLOBOID * HELICON * IDRAGIN * ISTOPIRIN * KAPSAZAL * KYSELINA 2-
    ACETOXYBENZOOVA (CZECH) * KYSELINA ACETYLSALICYLOVA (CZECH) *
    MEASURIN * MEDISYL * MICRISTIN * NEURONIKA * NOVID * POLOPIRYNA *
    RHEUMIN TABLETTEN * RHONAL * SALACETIN * SALCETOGEN * SALETIN *
    SOLPYRON * TRIPLE-SAL * XAXA * YASTA * ZORPRIN *
SECTION 3. - - - - - - - - - - HAZARDS IDENTIFICATION - - - - - - - - -
  LABEL PRECAUTIONARY STATEMENTS
    TOXIC
    TOXIC BY INHALATION, IN CONTACT WITH SKIN AND IF SWALLOWED.
    IRRITATING TO EYES, RESPIRATORY SYSTEM AND SKIN.
    POSSIBLE RISK OF HARM TO THE UNBORN CHILD.
    CALIF. PROP. 65 REPRODUCTIVE HAZARD.
    POSSIBLE SENSITIZER.
    TARGET ORGAN(S):
    BLOOD
    DO NOT BREATHE DUST.
    IN CASE OF ACCIDENT OR IF YOU FEEL UNWELL, SEEK MEDICAL ADVICE
    IMMEDIATELY (SHOW THE LABEL WHERE POSSIBLE).
    IN CASE OF CONTACT WITH EYES, RINSE IMMEDIATELY WITH PLENTY OF
    WATER AND SEEK MEDICAL ADVICE.
    AFTER CONTACT WITH SKIN, WASH IMMEDIATELY WITH PLENTY OF WATER.
    TAKE OFF IMMEDIATELY ALL CONTAMINATED CLOTHING.
    WEAR SUITABLE PROTECTIVE CLOTHING, GLOVES AND EYE/FACE
    PROTECTION.
SECTION 4. - - - - - - - - - FIRST-AID MEASURES- - - - - - - - - - -
    IN CASE OF CONTACT, IMMEDIATELY FLUSH EYES WITH COPIOUS AMOUNTS OF
    WATER FOR AT LEAST 15 MINUTES.
    IN CASE OF CONTACT, IMMEDIATELY WASH SKIN WITH SOAP AND COPIOUS
    AMOUNTS OF WATER.
    IF INHALED, REMOVE TO FRESH AIR. IF NOT BREATHING GIVE ARTIFICIAL
    RESPIRATION. IF BREATHING IS DIFFICULT, GIVE OXYGEN.
    IF SWALLOWED, WASH OUT MOUTH WITH WATER PROVIDED PERSON IS CONSCIOUS.
    CALL A PHYSICIAN.
    WASH CONTAMINATED CLOTHING BEFORE REUSE.
SECTION 5. - - - - - - - - - FIRE FIGHTING MEASURES - - - - - - - - -
  EXTINGUISHING MEDIA
    WATER SPRAY.
    CARBON DIOXIDE, DRY CHEMICAL POWDER OR APPROPRIATE FOAM.
  SPECIAL FIREFIGHTING PROCEDURES
    WEAR SELF-CONTAINED BREATHING APPARATUS AND PROTECTIVE CLOTHING TO
    PREVENT CONTACT WITH SKIN AND EYES.
SECTION 6. - - - - - - - - ACCIDENTAL RELEASE MEASURES- - - - - - - -
    WEAR SELF-CONTAINED BREATHING APPARATUS, RUBBER BOOTS AND HEAVY
    RUBBER GLOVES.
    SWEEP UP, PLACE IN A BAG AND HOLD FOR WASTE DISPOSAL.
    AVOID RAISING DUST.
    VENTILATE AREA AND WASH SPILL SITE AFTER MATERIAL PICKUP IS COMPLETE.
SECTION. 7. - - - - - - - - - HANDLING AND STORAGE- - - - - - - - - -
    REFER TO SECTION 8.
```

Figure 1.3 Material Safety Data Sheet for aspirin.

SECTION 8. - - - - - EXPOSURE CONTROLS/PERSONAL PROTECTION- - - - - -
    CHEMICAL SAFETY GOGGLES.
    RUBBER GLOVES.
    NIOSH/MSHA-APPROVED RESPIRATOR.
    SAFETY SHOWER AND EYE BATH.
    USE ONLY IN A CHEMICAL FUME HOOD.
    DO NOT BREATHE DUST.
    DO NOT GET IN EYES, ON SKIN, ON CLOTHING.
    AVOID PROLONGED OR REPEATED EXPOSURE.
    WASH THOROUGHLY AFTER HANDLING.
    KEEP TIGHTLY CLOSED.
    STORE IN A COOL DRY PLACE.
SECTION 9. - - - - - - PHYSICAL AND CHEMICAL PROPERTIES - - - - - - -
  APPEARANCE AND ODOR
    WHITE POWDER
  PHYSICAL PROPERTIES
    MELTING POINT: 138 C TO 140 C
SECTION 10. - - - - - - - -STABILITY AND REACTIVITY - - - - - - - - -
  INCOMPATIBILITIES
    STRONG OXIDIZING AGENTS
    STRONG ACIDS
    STRONG BASES
  HAZARDOUS COMBUSTION OR DECOMPOSITION PRODUCTS
    TOXIC FUMES OF:
    CARBON MONOXIDE, CARBON DIOXIDE
SECTION 11. - - - - - - - - TOXICOLOGICAL INFORMATION - - - - - - - -
  ACUTE EFFECTS
    HARMFUL IF SWALLOWED, INHALED, OR ABSORBED THROUGH SKIN.
    CAUSES EYE AND SKIN IRRITATION.
    MATERIAL IS IRRITATING TO MUCOUS MEMBRANES AND UPPER
    RESPIRATORY TRACT.
    PROLONGED OR REPEATED EXPOSURE MAY CAUSE ALLERGIC REACTIONS IN CERTAIN
    SENSITIVE INDIVIDUALS.
  CHRONIC EFFECTS
    OVEREXPOSURE MAY CAUSE REPRODUCTIVE DISORDER(S) BASED ON TESTS WITH
    LABORATORY ANIMALS.
    TARGET ORGAN(S):
    BLOOD
  RTECS #: VO0700000
    SALICYLIC ACID, ACETATE
  TOXICITY DATA
    ORL-CHD LDLO: 104 MG/KG   LANCAO 2,809,1952
    UNR-MAN LDLO: 294 MG/KG   85DCAI 2,73,1970
    ORL-RAT LD50: 200 MG/KG   34ZIAG -,67,1969
    IPR-RAT LD50: 340 MG/KG   NYKZAU 62,11,1966
    REC-RAT LD50: 790 MG/KG   34ZIAG -,67,1969
    ORL-MUS LD50: 250 MG/KG   ARZNAD 5,572,1955
    IPR-MUS LD50: 167 MG/KG   USXXAM # 4376771
    SCU-MUS LD50: 1020 MG/KG   CPBTAL 28,1237,1980
    ORL-DOG LD50: 700 MG/KG   ARZNAD 21,719,1971
    IVN-DOG LD50: 681 MG/KG   AIPTAK 149,571,1964
    ORL-RBT LD50: 1010 MG/KG   GTPZAB 24 (3),43,1980
    ORL-GPG LD50: 1075 MG/KG   JAPMA8 47,479,1958
    ORL-HAM LD50: 3500 MG/KG   ATSUDG 7,365,1984
    ORL-MAM LD50: 1750 MG/KG   IJMRAQ 81,621,1985
  TARGET ORGAN DATA
    SENSE ORGANS AND SPECIAL SENSES (TINNITUS)
    BEHAVIORAL (ALTERED SLEEP TIME)
    BEHAVIORAL (SOMNOLENCE)
    BEHAVIORAL (TREMOR)
    BEHAVIORAL (CHANGE IN MOTOR ACTIVITY)
    BEHAVIORAL (COMA)
    BEHAVIORAL (ANALGESIA)
    LUNGS, THORAX OR RESPIRATION (ACUTE PULMONARY EDEMA)
    LUNGS, THORAX OR RESPIRATION (RESPIRATORY DEPRESSION)
    LUNGS, THORAX OR RESPIRATION (RESPIRATORY STIMULATION)
    GASTROINTESTINAL (NAUSEA OR VOMITING)
    GASTROINTESTINAL (OTHER CHANGES)
    LIVER (HEPATITIS: HEPATOCELLULAR NECROSIS, DIFFUSE)
    LIVER (LIVER FUNCTION TESTS IMPAIRED)
    KIDNEY, URETER, BLADDER (CHANGES IN TUBULES)
    KIDNEY, URETER, BLADDER (URINE VOLUME DECREASED OR ANURIA)
    KIDNEY, URETER, BLADDER (HEMATURIA)

Figure 1.3 (*Continued*)

```
                    BLOOD (HEMORRHAGE)
                    MUSCULO-SKELETAL (OTHER CHANGES)
                    PATERNAL EFFECTS (SPERMATOGENESIS)
                    PATERNAL EFFECTS (TESTES, EPIDIDYMIS, SPERM DUCT)
                    MATERNAL EFFECTS (MENSTRUAL CYCLE CHANGES OR DISORDERS)

                    MATERNAL EFFECTS (PARTURITION)
                    EFFECTS ON FERTILITY (PRE-IMPLANTATION MORTALITY)
                    EFFECTS ON FERTILITY (POST-IMPLANTATION MORTALITY)
                    EFFECTS ON FERTILITY (OTHER MEASURES OF FERTILITY)
                    EFFECTS ON EMBRYO OR FETUS (EXTRA EMBRYONIC STRUCTURES)
                    EFFECTS ON EMBRYO OR FETUS (FETOTOXICITY)
                    EFFECTS ON EMBRYO OR FETUS (FETAL DEATH)
                    EFFECTS ON EMBRYO OR FETUS (OTHER EFFECTS TO EMBRYO OR FETUS)
                    SPECIFIC DEVELOPMENTAL ABNORMALITIES (CENTRAL NERVOUS SYSTEM)
                    SPECIFIC DEVELOPMENTAL ABNORMALITIES (EYE, EAR)
                    SPECIFIC DEVELOPMENTAL ABNORMALITIES (CRANIOFACIAL)
                    SPECIFIC DEVELOPMENTAL ABNORMALITIES (BODY WALL)
                    SPECIFIC DEVELOPMENTAL ABNORMALITIES (MUSCULOSKELETAL SYSTEM)
                    SPECIFIC DEVELOPMENTAL ABNORMALITIES (CARDIOVASCULAR SYSTEM)
                    SPECIFIC DEVELOPMENTAL ABNORMALITIES (BLOOD AND LYMPHATIC SYSTEMS)
                    SPECIFIC DEVELOPMENTAL ABNORMALITIES (OTHER DEVELOPMENTAL ABNORMALITIES)
                    EFFECTS ON NEWBORN (STILLBIRTH)
                    EFFECTS ON NEWBORN (LIVE BIRTH INDEX)
                    EFFECTS ON NEWBORN (OTHER NEONATAL MEASURES OR EFFECTS)
                    EFFECTS ON NEWBORN (BIOCHEMICAL AND METABOLIC)
                    EFFECTS ON NEWBORN (OTHER POSTNATAL MEASURES OR EFFECTS)
                    NUTRITIONAL AND GROSS METABOLIC (DEHYDRATION)
                    NUTRITIONAL AND GROSS METABOLIC (BODY TEMPERATURE INCREASE)
                    BIOCHEMICAL EFFECTS (OTHER)
                    ONLY SELECTED REGISTRY OF TOXIC EFFECTS OF CHEMICAL SUBSTANCES
                    (RTECS) DATA IS PRESENTED HERE. SEE ACTUAL ENTRY IN RTECS FOR
                    COMPLETE INFORMATION.
           SECTION 12. - - - - - - - - ECOLOGICAL INFORMATION - - - - - - - - -
                    DATA NOT YET AVAILABLE.
           SECTION 13. - - - - - - - - DISPOSAL CONSIDERATIONS - - - - - - - -
                    DISSOLVE OR MIX THE MATERIAL WITH A COMBUSTIBLE SOLVENT AND BURN IN A
                    CHEMICAL INCINERATOR EQUIPPED WITH AN AFTERBURNER AND SCRUBBER.
                    OBSERVE ALL FEDERAL, STATE AND LOCAL ENVIRONMENTAL REGULATIONS.
           SECTION 14. - - - - - - - - - TRANSPORT INFORMATION - - - - - - - -
                    CONTACT ALDRICH CHEMICAL COMPANY FOR TRANSPORTATION INFORMATION.
           SECTION 15. - - - - - - - - REGULATORY INFORMATION - - - - - - - - -
                  EUROPEAN INFORMATION
                    TOXIC
                    R 23/24/25
                    TOXIC BY INHALATION, IN CONTACT WITH SKIN AND IF SWALLOWED.
                    R 36/37/38
                    IRRITATING TO EYES, RESPIRATORY SYSTEM AND SKIN.
                    R 63
                    POSSIBLE RISK OF HARM TO THE UNBORN CHILD.
                    S 22
                    DO NOT BREATHE DUST.
                    S 45
                    IN CASE OF ACCIDENT OR IF YOU FEEL UNWELL, SEEK MEDICAL ADVICE
                    IMMEDIATELY (SHOW THE LABEL WHERE POSSIBLE).
                    S 26
                    IN CASE OF CONTACT WITH EYES, RINSE IMMEDIATELY WITH PLENTY OF
                    WATER AND SEEK MEDICAL ADVICE.
                    S 28
                    AFTER CONTACT WITH SKIN, WASH IMMEDIATELY WITH PLENTY OF WATER.
                    S 27
                    TAKE OFF IMMEDIATELY ALL CONTAMINATED CLOTHING.
                    S 36/37/39
                    WEAR SUITABLE PROTECTIVE CLOTHING, GLOVES AND EYE/FACE
                    PROTECTION.
                  REVIEWS, STANDARDS, AND REGULATIONS
                    OEL = MAK
                    ACGIH TLV-TWA 5 MG/M3                            DTLVS* TLV/BEI, 1997
                    OEL-AUSTALIA:TWA 5 MG/M3 JAN 1993
                    OEL-BELGIUM:TWA 5 MG/M3 JAN 1993
                    OEL-DENMARK:TWA 5 MG/M3 JAN 1993
                    OEL-THE NETHERLANDS:TWA 5 MG/M3 JAN 1993
                    OEL-RUSSIA:STEL 0.5 MG/M3 JAN 1993
```

Figure 1.3 *(Continued)*

```
OEL-SWITZERLAND:TWA 5 MG/M3 JAN 1993
OEL-UNITED KINGDOM:TWA 5 MG/M3 JAN 1993
OEL IN BULGARIA, COLOMBIA, JORDAN, KOREA CHECK ACGIH TLV
OEL IN NEW ZEALAND, SINGAPORE, VIETNAM CHECK ACGIH TLV
NIOSH REL TO ACETYLSALICYLIC ACID-AIR: 10H TWA 5 MG/M3
 NIOSH* DHHS # 92-100, 1992
NOHS 1974: HZD 84517; NIS 12; TNF 1764; NOS 16; TNE 10452
NOES 1983: HZD X1189; NIS 1; TNF 27; NOS 1; TNE 191; TFE 82
NOES 1983: HZD 84517; NIS 9; TNF 491; NOS 22; TNE 10776; TFE 6083
EPA GENETOX PROGRAM 1988, NEGATIVE: SPERM MORPHOLOGY-MOUSE
EPA GENETOX PROGRAM 1988, INCONCLUSIVE: MAMMALIAN MICRONUCLEUS
EPA TSCA SECTION 8 (B) CHEMICAL INVENTORY
EPA TSCA TEST SUBMISSION (TSCATS) DATA BASE, JUNE 1998
 U.S. INFORMATION
CALIFORNIA PROPOSITION 65:
THIS PRODUCTS IS OR CONTAINS CHEMICAL(S) KNOWN TO THE STATE OF
CALIFORNIA TO CAUSE FEMALE REPRODUCTIVE TOXICITY.
SECTION 16. - - - - - - - - - OTHER INFORMATION- - - - - - - - - - - -
THE ABOVE INFORMATION IS BELIEVED TO BE CORRECT BUT DOES NOT PURPORT TO
BE ALL INCLUSIVE AND SHALL BE USED ONLY AS A GUIDE. SIGMA, ALDRICH,
FLUKA SHALL NOT BE HELD LIABLE FOR ANY DAMAGE RESULTING FROM HANDLING
OR FROM CONTACT WITH THE ABOVE PRODUCT. SEE REVERSE SIDE OF INVOICE OR
PACKING SLIP FOR ADDITIONAL TERMS AND CONDITIONS OF SALE.
COPYRIGHT 1998 SIGMA-ALDRICH CO.
```

Figure 1.3   (*Continued*)

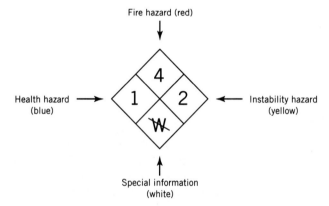

Figure 1.4   A label indicating health, fire, and instability hazards. The symbol **W̶** in the Special Information diamond means "reactive with water" or, in other words, "avoid contact with water."

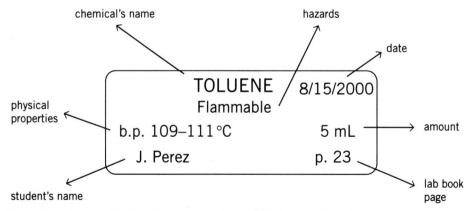

Figure 1.5   Sample label for chemicals made or purified in the laboratory.

Segregation and Storage of Chemicals

Only minimum amounts of chemicals should be stored in the laboratory and they should be segregated according to their **compatibility**. Although there is no general agreement regarding how chemicals should be segregated, the purpose of the segregation is always to avoid dangerous reactions that may occur when chemicals mix by accidental spillage. For storage purposes, chemicals can be classified into **radioactive, explosive, peroxides** and **peroxide-forming compounds, flammables, corrosives** (acids and bases), **oxidizers**, and **highly toxic**. A few examples of chemical segregation are: acids should be separated from bases and metals such as sodium, potassium, magnesium. They should also be separated from chemicals such as cyanide salts that may react with them to give hydrogen cyanide (a lethally dangerous gas). Oxidizing compounds (permanganates, perchlorates, nitrates, etc.) should not be stored with organic acids or flammable materials. For more information on the subject see Reference 3.

In general, chemicals should be stored in a cool, dry place away from direct sunlight and heat. **Corrosive** chemicals such as acids should be stored using secondary containment such as trays that would contain an accidental spillage. **Flammable** compounds are substances that catch fire very easily. Flammability depends on the vapor pressure and the propensity of the compound to form combustible mixtures with air. Flammable chemicals such as ethanol, acetone, toluene, and most organic solvents except chlorinated ones, should be stored in fire-resistant cabinets.

Gases used in the laboratory are contained in gas cylinders. Gas cylinders should be secured to a wall or a bench to prevent them from falling over. They should be stored away from flames, heat sources, spark-producing machinery, and corrosive chemicals. They should always be transported using a cart and with their valve cap on.

Waste Disposal

Chemical operations always generate unwanted chemicals. While some chemicals can be recycled or put to some use, others must be disposed of. The best way of dealing with chemical waste is to *minimize* it in the first place. This is possible by reducing the scale of the experiment. Whenever possible, the use of small-scale experiments is encouraged.

Only **nonhazardous** (nonradioactive, nontoxic, nonflammable, noncorrosive) waste can be disposed of in the drain or can go to a sanitary landfill. Only a handful of substances can generally be disposed of in this way. The criteria to determine whether a waste is nonhazardous *change locally*. **State and local regulations should be checked before disposing of chemicals.** For example, in California, a compound must have an LD_{50} of at least 5000 mg/kg to be considered nontoxic. According to this, sodium chloride (table salt) with a $LD_{50} = 3000$ mg/kg is considered toxic and should be disposed of as hazardous waste. Nonhazardous solutions can be flushed down the drain if the pH is near neutrality; however, *dilution* of hazardous wastes to render nonhazardous solutions is prohibited by law. Remember this environmentalist saying: "Dilution is not the solution to pollution." **Students should refrain from pouring chemicals in the sink or in the trash can unless they are *specifically instructed to do so by their instructor.***

Once a substance (or mixture of substances) has been labeled "hazardous waste," disposal must be in accordance with federal, state, and local regulations. The Resource Conservation and Recovery Act of 1977 (RCRA) and later modifications regulate (at the federal level) the handling of hazardous waste. However, more stringent requirements are often set by local agencies. Familiarize yourself with the federal, state, and local regulations and dispose of waste properly. Most schools provide the teaching laboratories with instructions regarding waste disposal. Follow them. **The cleanup procedures offered in this book at the end of each experiment should**

be considered as *guidelines*, **and should not be followed without previously checking your local regulations**.

Most of the chemicals generated in university laboratories are disposed of by incineration. Only a small amount is sent to secure landfills (for example, asbestos). Incineration is preferred over secure landfill because it does not imply a long-term containment that may eventually result in hazardous malfunction such as a leakage into the ground. Nonchlorinated organic solvents (for example, hexane and acetone) are used as fuel to heat cement manufacturing kilns. Chlorinated organic solvents, because of their lower combustibility, must be mixed with other organic chemicals before they can be used for fuel blending, making the process less efficient. Aqueous solutions are not used for fuel blending and must be chemically treated or incinerated by appropriately mixing them with combustible materials. These processes make the disposal of aqueous waste very expensive. Mercury waste is also very expensive to dispose of. If mercury waste is added to a container of liquid waste it increases the disposal cost of that waste at least ten times.

The disposal of chemicals generated in the laboratory should be done in such a way as to avoid the mixing of incompatible reagents. Among the undesired consequences of mixing two or more incompatible chemicals are fire, heat, and gas evolution, and even explosions. To avoid this, tables of chemical compatibility have been generated; they are very useful in deciding if two types of chemicals can be mixed in the same waste container. For more information on this subject see Reference 6. In the teaching laboratories, students must carefully read the waste container labels before disposing of any chemical.

Waste containers should be properly labeled. If a bottle that used to contain another chemical is used as a waste container, all residues of the old chemical should be removed before the bottle is used to collect waste. The old label should be defaced, and a tag with the names and proportions of the constituents of the waste and the date should be secured around the neck. A waste bottle should never be filled to the rim; instead, it should have an empty head space of about 1/10 of the total volume. Waste bottles should be properly capped.

Broken glassware should never be disposed of in the trash can because it may injure the person emptying the trash container. Place broken glass in especially designed cardboard boxes with plastic lining (Fig. 1.6). Used Pasteur pipets and broken glass contaminated with hazardous materials should be disposed of in a separate broken-glass box and labeled as hazardous waste.

In dispensing chemicals to carry out the indicated reactions, students should be aware that any unused portion cannot be returned to the original stock container and

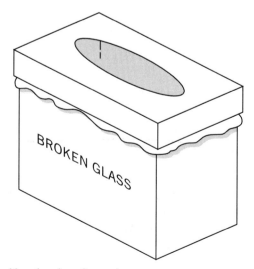

Figure 1.6 Cardboard box for glass disposal.

therefore, should be disposed of as waste. To decrease the amount of waste, only the necessary amount of chemicals should be taken.

Chemicals Worth Knowing

Some hazardous chemicals deserve special attention by virtue of their potentially devastating effects or their widespread use. Peroxides, peroxide-forming compounds, cyanides, perchloric, and picric acids are among them.

Peroxides and Peroxide-Forming Compounds   Peroxides are compounds with general structure: R—O—O—R, where R is an alkyl or aryl group. Peroxides are explosives with sensitivity to shock, sparks, heat, friction, impact, and light. Moreover, organic peroxides are highly flammable substances. Ethers and some hydrocarbons are particularly prone to form peroxides when exposed to air for long periods of time (Fig. 1.7). Cyclic ethers, and ethers derived from primary and secondary alcohols, are particularly susceptible to form peroxides during storage. Peroxides and peroxide-forming compounds should be labeled with the dates when the bottle was received and opened and its expected shelf life as stated by the manufacturer. There are several tests that can be used to determine the level of peroxides in a liquid; however, if you suspect that a chemical may be contaminated with peroxides, you should take the safest approach and dispose of the chemical following appropriate procedure instead of trying to remove the peroxide. **Never distill a liquid suspected of containing peroxides**. Grinding, friction, and impact with solid peroxides should be avoided. Since metals can catalyze the decomposition of peroxides, the use of metal spatulas should be avoided. Peroxides and peroxide-forming compounds should be stored in air-tight containers to protect them from light and heat. They should be kept away from spark sources.

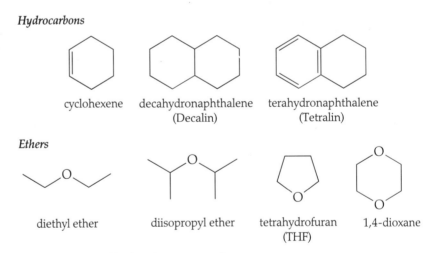

Figure 1.7   Some peroxide-forming compounds.

Cyanides   Hydrogen cyanide, HCN, is one of the most potent poisons known to humans. A few inhalations of this gas are enough to kill a human being. Hydrogen cyanide is produced by mixing sodium or potassium cyanide (which also are highly toxic) with acids. Inhalation and ingestion of, or skin contact with cyanides can be fatal. Cyanide poisons the body by replacing oxygen in the blood by binding preferentially to the hemoglobin molecules in the red blood cells. When working with cyanides, strict supervision is essential. Under no circumstance should a person working with cyanides be alone in the laboratory. Ampules of amyl nitrite (an antidote) should be readily available (in a prominent place in the hood or pinned to the worker's lab coat) any time an experiment with cyanide is performed. The person performing the experiment, as well as coworkers, should be familiar with the first aid practices for cyanide poisoning (see, for example, Ref. 2). If poisoning occurs, medical help should be sought immediately.

Perchloric Acid   Perchloric acid ($HClO_4$) is a powerful oxidizing agent. Contact with organic materials and reducing agents should be avoided because it may result in explosions. Transition metal perchlorates and perchlorate esters are also explosive. Special handling is necessary (see Ref. 8).

Picric Acid   When dry, picric acid (a relative of TNT) is highly explosive. Commercial picric acid contains 10–15% water, and does not present a serious hazard; however, if the amount of water is less than 10%, it becomes shock, heat, friction, and spark sensitive and should be handled accordingly. Picrate salts are also explosive. When handling these chemicals you should be fully aware of safe storage, handling, and disposal (see Refs. 1 and 3).

piciric acid
(2,4,6-trinitrophenol)

1.4 ONLINE SAFETY RESOURCES

Information about laboratory safety and chemical toxicity is available on the Internet, but as the number of sites on the Web increases day by day, so does the complexity of the search to locate relevant addresses. Listed below are a few sites with valuable information.

For a concise discussion of toxicity, flammability, and reactivity of some selected chemicals commonly used in the laboratory, you can check the safety page of the Howard Hughes Medical Institute:

http://www.hhmi.org/science/labsafe/lcss/start.htm

Websites with rapid access to MSDS can be found below. The first site gives access to MSDS from several providers, the second site is kept by Oxford University, and the third by the Sigma-Aldrich Company.

http://www.chem.plu.edu/cirrus.html
http://physchem.ox.ac.uk/MSDS/
http://www.sigma-aldrich.com

Toxicity and safety information about pesticides can be found at the following site, which is a cooperative effort of the University of California, Davis; Oregon State University; Michigan State University; and Cornell University:

http://ace.ace.orst.edu/info/extoxnet/

The Website for the Occupational Safety and Health Administration, (OSHA) where information about toxic chemicals can be found, is:

http://www.osha.gov/index.html

Information about carcinogens can be found in the Website for the International Agency for Research on Cancer (IARC), which depends on the World Health Organization and is located in France:

http://www.iarc.fr

BIBLIOGRAPHY

1. Prudent Practices in the Laboratory. Handling and Disposal of Chemicals. Assembly of Mathematical and Physical Sciences. National Research Council. National Academy Press, Washington, D.C., 1995.

2. Safety in Academic Chemistry Laboratories. American Chemical Society. 5th ed., Washington, D.C., 1990.

3. Safe Storage of Laboratory Chemicals. D.A. Pipitone, Ed. 2nd ed. Wiley, New York, 1991.

4. Hazardous Materials Handbook. R.P. Pohanish and S.A. Greene. Van Nostrand Reinhold, New York, 1996.

5. Calculated Risks. Understanding the Toxicity and Human Health Risks of Chemicals in Our Environment. J.V. Rodricks. Cambridge University Press, Cambridge, 1992.

6. A Method for Determining the Compatibility of Hazardous Wastes. H.K. Hatayama et al. EPA 600/2-80-076.

7. CRC Handbook of Laboratory Safety, A.K. Furr, Ed. 3rd ed., CRC Press, Boca Raton, Fl, 1990.

8. Perchloric Acid and Perchlorates. A.A. Schilt. G. Frederick Smith Chemical Company, Columbus, Ohio, 1979.

9. Hazards in the Chemical Laboratory. S.G. Luxon, Ed. 5th ed., Royal Society of Chemistry, Cambridge, UK, 1992.

10. The Sigma–Aldrich Library of Chemical Safety Data. R.E. Lenga, Ed. 2nd ed., Sigma-Aldrich Corporation, Milwaukee, 1988.

11. Bruegel. W.S. Gibson. Thames and Hudson, New York, 1977.

EXERCISES

Problem 1

If we want to take a peek at everyday life in a northern European village in the 16th century, all we have to do is go to a library and open a book on Bruegel. Pieter Bruegel the Elder (ca. 1525–1569), a painter of the Flemish Renaissance, bequeathed us a legacy of anecdotal vignettes on common activities such as hunting, harvesting, dancing, and playing. His paintings show, with encyclopedic detail and no lack of humor, ordinary people going about their daily routines. In Figure 1.8 "The Alchemist," an engraving dated 1558, Bruegel shows us an alchemist, his wife, and their three children. A helper pumping the bellows and a scholarly man directing the operation from his desk are also depicted. The laboratory is in a state of total disarray. The wife showing an empty pocket and the children playing with equally empty pots are telling us that there is no food in the household. The book on the desk reads "Alghe mist," a pun on "alchemist" and "algemist," which in Dutch means "all is lost." This picture shows the low esteem that society had for alchemists. In his usual moralistic style, Bruegel shows, through the window, the family finally going to the poorhouse (for an excellent book on Bruegel's work see Ref. 11).

(a) Study the picture carefully and make a list of all the violations of modern safety standards.

(b) Make a list of instruments and apparatus in use and operations being performed.

Problem 2

Synton X. Plosiff, a slightly vain, somewhat absent-minded but totally dedicated grad student, loves organic synthesis, classical music, and working out in the gym. On an unusually hot California morning, Synton is wearing his favorite shorts and a white tank top printed with the legend "Survival of the Fittest" in faded red ink. After about 1 hour of heavy training in the gym, he heads to his lab where he has left a reaction running overnight. The next step in the synthesis is the addition of sodium cyanide. It is 9 in the morning and the lab is deserted.

Figure 1.8 "The Alchemist," engraving by P. Galle after a drawing by Pieter Bruegel, 1558.

Synton goes first to his office, opens his bag, and takes a six-pack of tomato juice out; he mechanically pulls a can out of the plastic rings and puts the rest in his tiny refrigerator. He's thinking about his reaction and the many ways to increase what until now has been a miserable percentage yield. Sipping warm tomato juice, he walks into his lab while electrons are jumping, bonds are breaking, and nucleophiles are dancing in his mind. He checks the reaction mixture that, to his chagrin, has mysteriously turned into a heavy brown syrup. He turns on the radio, opens his drawer, takes out a few spatulas and a piece of weighing paper and starts wandering about the lab searching for sodium cyanide. After checking every single cabinet (from "c" to "s"), he finally finds a jar with sodium cyanide inside a plastic bag next to a container of sodium hydroxide in the fumehood. He unwraps the sodium cyanide and places the jar on the bench top next to a small bottle of dilute hydrochloric acid, when the telephone in his office rings. He rushes to answer it, thinking it may be his girlfriend, Ester. But he is wrong. It is his boss who wants to know how the reaction is going. He isn't pleased with the news of the heavy brown syrup and says that he is coming to the lab right away. A minute later he enters the lab and finds Synton lying unconscious on the floor; his broken glasses are lying at his side.

(a) What may have happened to Synton?

(b) Enumerate the things that Synton did that morning in violation of safety rules.

Problem 3

In each case indicate the correct option:

1. If you spill a small amount of a nonhazardous solid on the lab bench, you should:

(a) call emergency services; (b) notify the instructor; (c) clean the spill yourself; (d) evacuate the building.

2. If you spill a hazardous material on the lab bench, you should:

(a) call emergency services; (b) notify the instructor; (c) clean the spill yourself; (d) evacuate the building.

3. If you have a chemical spill on a large area of your body, the first thing you should do is:

(a) roll on the floor; (b) remove any contaminated clothing as you use the safety shower; (c) call emergency services; (d) notify your instructor.

4. If someone next to you ingests a hazardous chemical, you must:

(a) give the poisoned person a glass of milk; (b) immediately call the local Poison Control Center; (c) immediately try to induce vomiting; (d) encourage the victim to cough.

5. If you break a mercury thermometer, you must:

(a) collect the mercury yourself; (b) leave the mercury on the floor; (c) immediately notify your instructor; (d) evacuate the building at once.

6. If the fire alarm goes off, you must:

(a) exit the building as soon as possible; (b) run outside as fast as possible; (c) call emergency services; (d) ask your instructor for instructions.

7. Hazardous chemicals must be disposed of:

(a) in the trash can if they are solid; (b) in any empty container near the sink; (c) in containers labeled for that purpose; (d) down the drain.

8. In the lab you can use your mouth to:

(a) pipet water; (b) start a siphon; (c) moisten a label; (d) smile.

Problem 4

Some of the activities listed below are allowed and others are prohibited in the laboratory. Circle the ones that are allowed, and box those that are prohibited:

tasting chemicals; putting broken glassware in the trash can; keeping your lab notebook on your bench; using open flames; talking to your neighbor; keeping your backpack on your lab bench; inhaling chemicals; wearing long pants; wearing a lab coat; wearing sandals; drinking water; working alone; leaving experiments unattended; leaving the lab for a short while; using your mouth to pipet a liquid; wearing a wedding band; pouring unused chemicals back into the stock bottles; wearing loose, long hair; running; wearing safety goggles; wearing shorts; keeping a water bottle on your bench.

Problem 5

(a) Give full names for the following acronyms:

P.E.L.

MSDS

LD_{50}

LC_{50}.

(b) Link the two columns:

Carcinogenic	• heritable genetic changes
Teratogenic	• malignant tumors
Mutagenic	• malformation of embryo

Basic Concepts

In this unit we will discuss important concepts that are extensively used in the rest of the book. Some of them have been already introduced in general chemistry; some are new. In reviewing the old subjects (which include polarity, H-bonds, balancing chemical equations, and calculation of percentage yield) we will focus on their applications to organic chemistry. Among the new material you will find a description of organic solvents, searching for physical data, and the sources of chemical information including online services.

All these subjects are gathered here for easy reference. This unit is like a keepsake box. Take what you need, when you need it, and leave the rest. You will find references to this unit throughout the book.

2.1 POLARITY AND H-BONDS

The interaction between molecules, called **intermolecular interactions**, are responsible for their physical properties such as boiling point, melting point, and solubility. These intermolecular interactions are the result of the **polarity** of molecules and their tendency to form **hydrogen bonds**. In this section we will briefly review these two important concepts.

Dipole Moment and Dielectric Constant

Unlike density or melting point, which are easy to quantify, there is no universal agreement on how to measure the polarity of molecules. The **dipole moment** and the **dielectric constant** are two properties often used to describe polarity. We will discuss both of them here. Other polarity indices have also been defined in connection with chromatography, and are relevant where this separation technique is concerned (Units 7–10).

Molecules can be regarded as geometric entities with a core of positive charges surrounded by a negative electron cloud. For most molecules, the center of positive charge does not coincide with the negative charge center, and as a result, the molecule is a permanent dipole. If the charges, q, are separated by a distance r, the **dipole moment**, μ, is defined as the product between q and r. The dipole moment is a vector and, by convention, is represented as an arrow going from the positive to the negative end ($\leftrightarrow$). Two charges, one positive and one negative, equivalent to the electron charge (1.60×10^{-19} C) and separated by 1 Å (10^{-10} m) have a dipole moment of 1.60×10^{-29} C $\times$ m. A common unit for dipole moment is the debye (D); $1\,D = 3.336 \times 10^{-30}$ C $\times$ m. Thus, the dipole moment of an electron and a proton separated by 1 Å is:

$$\mu = 1.60 \times 10^{-29}\ \text{C} \times \text{m} \frac{1\ \text{D}}{3.336 \times 10^{-30}\ \text{C} \times \text{m}} = 4.80\ \text{D}$$

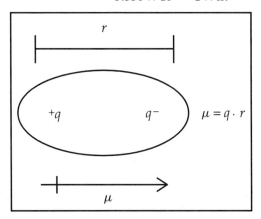

The dipole moment originates in the difference in electronegativity between atoms. For example, the HCl molecule has a permanent dipole moment because chlorine is more electronegative than hydrogen, and therefore it pulls the bond electrons toward itself creating a negative charge density, δ^-, on the chlorine atom, and leaving a positive charge density, δ^+, on the hydrogen. In contrast, a chlorine molecule does not have a permanent dipole moment because both chlorine atoms have the same electronegativity.

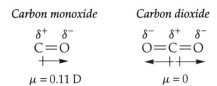

The dipole moment depends also on the molecular geometry. The dipole moment of a molecule is the vector addition of the dipole moments for individual bonds. For example, the dipole moment of carbon monoxide is 0.11 D, but the dipole moment of carbon dioxide is zero because the two opposite vectors cancel each other out.

$$
\begin{array}{cc}
\textit{Carbon monoxide} & \textit{Carbon dioxide} \\
\overset{\delta^+\ \ \delta^-}{\text{C}=\text{O}} & \overset{\delta^-\ \ \delta^+\ \ \delta^-}{\text{O}=\text{C}=\text{O}} \\
\mu = 0.11\ \text{D} & \mu = 0
\end{array}
$$

By a similar argument it can be concluded that trichloromethane ($CHCl_3$, most commonly known as chloroform) must have a finite dipole moment, whereas the dipole moment of tetrachloromethane (CCl_4, carbon tetrachloride) is zero.

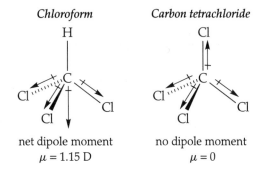

Chloroform

net dipole moment
$\mu = 1.15$ D

Carbon tetrachloride

no dipole moment
$\mu = 0$

Molecules with no dipole moment because of symmetry, such as carbon dioxide and carbon tetrachloride, are not necessarily considered nonpolar. These molecules are made up of individual dipoles that are able to interact with other dipoles, and thus the molecule is considered polar. Let's consider yet another example: 1,4-dioxane. The molecule has no net dipole moment because of symmetry; however, each individual oxygen atom with its lone electron pairs can interact with the positive end of other dipoles, and thus 1,4-dioxane is a polar compound.

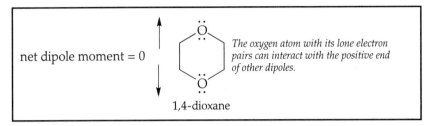

net dipole moment = 0

The oxygen atom with its lone electron pairs can interact with the positive end of other dipoles.

1,4-dioxane

To describe the polarity of solvents the **dielectric constant** is often used. The dielectric constant of a substance, generally represented by ε, measures the effect of the substance on the transmission of an electric field. If two particles with charges, q_1 and q_2, are separated by a distance, r, the force between them in a vacuum, F, is given by Coulomb's law:

$$F = k\frac{q_1 \times q_2}{r^2} \tag{1}$$

where k is a proportionality constant. If instead of a vacuum, the two charged particles are immersed in a liquid or solid, the force between them, F', is diminished because the medium molecules can orient themselves and shield the net charge felt by the particles. The dielectric constant of the medium measures the decrease in this force.

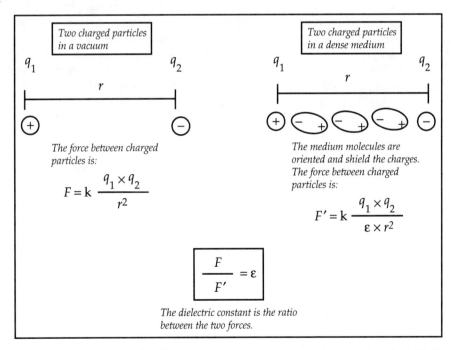

Two charged particles in a vacuum

q_1 q_2

r

The force between charged particles is:

$$F = k\ \frac{q_1 \times q_2}{r^2}$$

Two charged particles in a dense medium

q_1 q_2

r

The medium molecules are oriented and shield the charges. The force between charged particles is:

$$F' = k\ \frac{q_1 \times q_2}{\varepsilon \times r^2}$$

$$\frac{F}{F'} = \varepsilon$$

The dielectric constant is the ratio between the two forces.

Polar compounds, with a high tendency to get oriented in an electric field, have large dielectric constants, whereas nonpolar compounds with little orientation in an electric field have very low dielectric constants. Water has one of the highest dielectric constant values ($\varepsilon = 80.1$), while hydrocarbons (pentane, hexane, etc.) have the lowest value ($\varepsilon = 1.89$). Because the dielectric constant is the ratio of two forces, it has no units. The dipole moment and the dielectric constant are related and, in general, compounds with high dipole moments also have high dielectric constants.

Hydrogen Bonding

The covalent bond between a hydrogen and an electronegative element such as O or N is polarized with the positive charge density on the hydrogen.

$$\underset{\delta^- \ \ \ \delta^+}{O-H} \qquad \underset{\delta^- \ \ \ \delta^+}{N-H}$$

The positive end of the O—H and N—H dipoles can interact with the negative end of other dipoles, especially with oxygen and nitrogen atoms, forming what is called a **hydrogen bond** or simply an **H-bond**.

Four important H-bonds (dashed lines)

$$\underset{\text{donor} \qquad \text{acceptor}}{\overset{\delta^- \quad \delta^+ \quad \delta^-}{O-H\text{-}\text{-}\text{-}O}} \qquad \underset{\text{donor} \qquad \text{acceptor}}{\overset{\delta^- \quad \delta^+ \quad \delta^-}{N-H\text{-}\text{-}\text{-}O}}$$

$$\underset{\text{donor} \qquad \text{acceptor}}{\overset{\delta^- \quad \delta^+ \quad \delta^-}{O-H\text{-}\text{-}\text{-}N}} \qquad \underset{\text{donor} \qquad \text{acceptor}}{\overset{\delta^- \quad \delta^+ \quad \delta^-}{N-H\text{-}\text{-}\text{-}N}}$$

The molecule that provides the hydrogen is called the **H-bond donor**, and the one that accepts the hydrogen is called the **acceptor**. Good hydrogen bond donors are compounds with O—H and N—H bonds such as water, alcohols, phenols, amines, carboxylic acids, and amides. Good hydrogen bond acceptors are compounds with oxygen or trivalent nitrogen. *All compounds that are good H-bond donors are also good H-bond acceptors* (with the exception of hydronium and ammonium cations, which are only H-bond donors). *However, not all H-bond acceptors are donors.* For example, there are molecules with oxygen or nitrogen but without O—H and N—H bonds, such as aldehydes, ketones, ethers, esters, nitrocompounds, and so on, which can accept H-bonds but cannot donate them. Hydrogens attached to carbon do not participate in H-bonds.

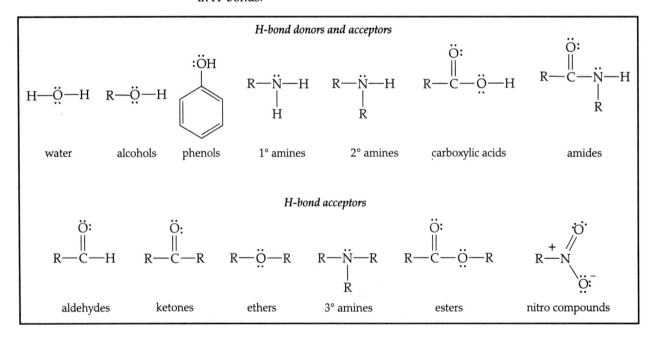

The formation of H-bonds has far-reaching consequences in organic chemistry. Compounds that can donate and accept H-bonds (compounds with O—H or N—H bonds) have higher boiling and melting points than compounds of similar molecular mass but without these bonds. Also, compounds with N—H and O—H bonds are, in general, more soluble in water than compounds lacking such groups.

We will come back to the subject of polarity and H-bonds when discussing specific topics such as boiling point (Unit 6), and chromatography (Units 7–10).

2.2 PHYSICAL DATA

It is very important to collect physical data about the compounds used in the laboratory *before* the beginning of the experiment. Knowing the density and boiling point of liquids, the melting point of solids, and the toxicity and hazards of all the chemicals is crucial to the success of the experiment. We can find all this information, and more, in handbooks available in libraries and laboratories, and also in the Material Safety Data Sheets (MSDS) supplied by the manufacturer. The most common and useful handbooks are the **Merck Index**, published by the Merck Company, the **CRC Handbook of Chemistry and Physics** published by the CRC Press, and the **Aldrich Catalog**, a list of chemicals sold by the Aldrich Company. The Merck Index is also available on CD ROM. In these handbooks, the chemicals are listed in alphabetical order, but they are often found under different names. For example, aspirin is found under the name *Aspirin* in the Merck Index, under the name *Acetylsalicylic acid* in the *Aldrich Catalog*, and under *Benzoic acid, 2-(acetyloxy)* in the *CRC Handbook* (76th edition). These handbooks also offer a list of synonyms that is useful in identifying the compounds. They also have useful molecular formula indexes. The molecular formula index has the names of all the compounds in the handbook with a given formula and gives the page where the compound can be found. For example the molecular formula for aspirin is $C_9H_8O_4$. The *CRC Handbook* (76th edition) has four compounds listed with this molecular formula, one of them under the name *Benzoic acid, 2-(acetyloxy)*, which corresponds to aspirin. When you are not sure under which name a chemical may be listed, first check the molecular formula index to find the name used in the handbook.

The information provided by these handbooks includes the molecular mass (MM), boiling point (bp), melting point (mp), density (d), refractive index (n_D), solubility, toxicity, and flammability. Depending on the scope of the experiment, different pieces of information must be gathered. For example, if we are performing a distillation experiment, we must collect the boiling points of the compounds, and may omit their solubility properties and refractive indexes. If we are doing a recrystallization instead, we would like to have solubility and melting point information.

Molecular Mass (MM), also called the Molecular Weight (MW) or Formula Weight (FW), is always given in grams per mole (g/mol).

Density (d or δ) is useful for liquids but of little use for solids except for plastics. It is given in grams per milliliter (g/mL), which for all practical purposes is equivalent to g/cm^3. For example, the density listing for methanol in the Merck Index includes:

$$d_4^0 0.8100 \quad \text{and} \quad d_4^{20} 0.7915$$

The superscript indicates the temperature at which the density was measured, and the subscript 4 indicates that the density of water at 4°C (1.000 g/mL) was used as a reference.

The **boiling point** (bp) is the temperature at which the vapor pressure of a liquid equals the applied pressure. The boiling point depends strongly on the applied pressure; the **normal boiling point** is determined at one atmosphere (760 Torr or 760 mm Hg). The pressure value at which the boiling point was determined is usually given as a subscript. For methanol, the Merck Index includes:

$$bp_{760} 64.7°C \quad \text{and} \quad bp_{200} 34.8°C$$

aspirin

Table 2.1   **Physical Data of Chemicals**

Name	Structure	Molecular mass (g/mol)	Boiling point (°C)	Melting point (°C)	Density (g/mL)	Refractive index	Solubility	Toxicity and hazards
aspirin	OCOCH$_3$ / COOH	180.15	—	138–140	—	—	1 g in 300 mL H$_2$O; 5 mL ethanol	LD$_{50}$ 200 mg/kg (oral rat) toxic; irritant
methanol	CH$_3$OH	32.04	64.7	−98	d$_4^{20}$0.7915	n^{20}$_D$ 1.3292	∞H$_2$O ∞ ethan ∞ ether	highly toxic / flammable

If the pressure is not indicated, it is understood that the number refers to the normal boiling point.

The **melting point** (mp) is the temperature at which solid coexists in equilibrium with liquid at normal atmospheric pressure. The melting point value is sometimes followed or preceded by the legend d. or dec., meaning that the solid decomposes during melting. Sometimes there are two melting points listed for the same compound. This may arise from different crystal forms, different recrystallization solvents, or just experimental error. Usually the melting point considered less reliable is given in parentheses. The recrystallization solvent is sometimes mentioned in parentheses next to the melting point.

The **refractive index** (n_D) is a useful property for the identification of liquids (discussed in Unit 11). The subscript D refers to the light source used (the D line of sodium). The temperature is normally indicated by a superscript. For example, the refractive index of methanol at 20°C is:

$$n_D^{20}1.3292$$

The **solubility** of a chemical (called the **solute**) in a solvent is the maximum possible concentration of the chemical in that solvent at any given temperature. The solubility of organic and inorganic chemicals in water and organic solvents is a very important piece of information when reactions in solution are performed. The Merck Index usually expresses the solubility as the minimum volume of solvent necessary to dissolve 1 gram of compound. Ethanol, methanol, ether (diethyl ether), acetone, and chloroform are the solvents most frequently listed. The symbol ∞ means that the compound is soluble in all proportions in the solvent in question. When two liquids are soluble in all proportions they are said to be **miscible**.

You should always research the **toxicity** and **hazards** of the chemicals before you use them. This information may prevent complications and headaches (real and figurative). The LD$_{50}$ can be used as a rough guide for the acute toxicity of chemicals and can be found in the Merck Index and the MSDS (Unit 1). A detailed hazards profile of the chemical is found in the MSDS.

The best way to organize the physical data information is in a table format. Table 2.1 gathers the physical data for two of the chemicals that we have considered in this section: aspirin and methanol.

2.3   SOLVENTS

Solvents play a pivotal role in organic chemistry. They provide the environment where reactions occur. They are used to extract chemicals from complex mixtures, and also to separate and purify all types of compounds. In this section we will discuss some of the most frequently used solvents and stress their important characteristics. In order of decreasing polarity they are:

Water. When possible, it is the solvent of choice. It is cheap, readily available, nonflammable, and nontoxic. An H-bond donor and acceptor. Very polar. Dissolves polar compounds such as inorganic and organic salts (especially salts of sodium and potassium), organic molecules with five carbon atoms or fewer, small sugars, and proteins. Does not dissolve nonpolar compounds. Difficult to evaporate.

Dimethyl sulfoxide. (CH_3SOCH_3; DMSO) Very polar. Excellent solvent for nonpolar and polar compounds including salts. An H-bond acceptor. Useful in organic synthesis. Totally miscible with water. Very difficult to evaporate. Irritant. Low flammability.

Methanol. (CH_3OH; MeOH) Less polar than water and DMSO but more polar than most organic solvents. Because of the -OH group, it is a good H-bond donor and acceptor. Dissolves organic compounds of medium polarity, especially at high temperatures. It is a good recrystallization solvent (Unit 4). Totally miscible with water. Toxic and flammable. Easy to evaporate.

Ethanol. (CH_3CH_2OH; EtOH) Also called simply "alcohol." Less polar than methanol. Because of the -OH group, it is a good H-bond donor and acceptor. Dissolves organic compounds of medium polarity, especially at high temperatures. It is a good recrystallization solvent (Unit 4). Totally miscible with water. It is available as a 95% solution in water (190 proof) and as absolute ethanol (100%; 200 proof). Less toxic than methanol. It is often rendered unfit for human consumption, and called **denatured alcohol**, by addition of small amounts of methanol, acetone, or toluene. Flammable. Easy to evaporate.

Acetone. (CH_3COCH_3) An excellent solvent for polar and nonpolar compounds alike. It does not dissolve salts. Because of the carbonyl group it is a good H-bond acceptor. Totally miscible with water. Toxic and flammable. Easy to evaporate.

Acetic acid. (CH_3COOH; AcOH) Polar. Acidic. Dissolves polar and medium polarity compounds. Totally miscible with water. An H-bond donor and acceptor. Difficult to evaporate. Corrosive. Flammable.

Ethyl acetate. ($CH_3COOCH_2CH_3$; AcOEt) Medium polarity. Partially miscible with water. Lighter than water. Because of the carbonyl group it is an H-bond acceptor. Moderately easy to evaporate. Irritant. Flammable.

Methylene chloride. (CH_2Cl_2) Also known as dichloromethane. Medium polarity. Dissolves nonpolar and medium polarity compounds. Immiscible with water. Denser than water. Easy to evaporate. Nonflammable. A possible carcinogen. When possible it should be replaced by other solvents. Toxic. Irritant.

Tetrahydrofuran. (THF) Dissolves nonpolar and medium polarity compounds. Miscible with water. Very useful in organic synthesis. Limited shelf life because it forms explosive peroxides on exposure to air. Irritant. Flammable.

Chloroform. ($CHCl_3$) Dissolves nonpolar and medium polarity compounds. Immiscible with water. Denser than water. A possible carcinogen with very low P.E.L. Used for specific applications (for example, the deuterated form ($CDCl_3$) is used as an NMR solvent) but largely replaced by other solvents with higher P.E.L.

Diethyl ether. ($CH_3CH_2OCH_2CH_3$; Et_2O) Also called simply "ether." Excellent solvent for nonpolar and low polarity compounds. Less polar than THF. Immiscible with water. Limited shelf life because it forms explosive peroxides on exposure to air. Lighter than water. Very easy to evaporate. Toxic. Flammable.

***tert*-Butyl methyl ether.** ((CH_3)$_3$C-OCH_3; BME) Same uses as diethyl ether but with less tendency to form explosive peroxides. Immiscible with water. Lighter than water. Flammable. Very easy to evaporate. Excellent solvent for liquid-liquid extractions. Irritant.

Toluene. ($C_6H_5CH_3$) Low polarity. A good solvent for nonpolar compounds. Does not form H-bonds. Immiscible with water. Lighter than water. Moderately easy to evaporate. Toxic. Flammable.

Carbon tetrachloride. (CCl_4) Dissolves nonpolar compounds. Immiscible with water. Denser than water. A possible carcinogen with very low P.E.L. It has been largely replaced by other solvents with higher P.E.L.

Benzene. (C_6H_6) Low polarity. Dissolves nonpolar compounds. Does not form H-bonds. Immiscible with water. Lighter than water. A known carcinogen. It should be avoided in the laboratory. It has been largely replaced by toluene and other solvents with higher P.E.L. Flammable.

Hexane(s). A mixture of isomeric hexanes. Among the least polar organic solvents. Does not form H-bonds. Dissolves only nonpolar compounds. Immiscible with water. Lighter than water. Easy to evaporate. Flammable. Irritant.

Petroleum ether. It is not an ether in the chemical sense of the word. It is a mixture of low-boiling point hydrocarbons. Available in different boiling point ranges (40–60°C; 60–80°C; and so on; fractions with boiling points higher than 60°C are often called **ligroin**). Among the least polar organic solvents. Does not form H-bonds. Dissolves only nonpolar compounds. Immiscible in water. Lighter than water. Very easy to evaporate. Toxic. Flammable.

Table 2.2 summarizes the most important properties of some common solvents.

2.4 BALANCING CHEMICAL EQUATIONS

The communication of chemical information is essentially visual; chemical ideas are conveyed in the form of detailed equations showing the structures of reactants and products. Without a pictorial description of chemical transformations there would be no organic chemistry. Often, the chemical equation includes not only the reactants and products, but also the reaction conditions such as solvent, temperature, and catalyst. The latter are generally written on the arrow, for example:

$$\tag{2}$$

Here, CH_2Cl_2 is the solvent and the reaction is performed at 25°C.

It is also common to write on the arrow the reagents that do not contribute with carbon atoms to the final product. This includes inorganic compounds such as Br_2, HBr, HCl, H_2, and so on, and organic reagents acting as acids or bases. For example, the elimination reaction of *tert*-butyl bromide with sodium ethoxide (EtONa) in ethanol as a solvent (EtOH) can be represented as:

$$\tag{3}$$

In this representation, the other reaction products, which in this case are ethanol and NaBr, are omitted. A balanced equation showing all the products is given below:

$$\tag{4}$$

Table 2.2 Useful Solvents

Solvent	M.M.	b.p.	dens.	f.p.	flam.	ε	P.I.	μ	n	S.W.	W.S.	Az.bp	%W	P.E.L.	Structure
acetic acid	60.05	118.0	1.049	16.7	x	6.15	6.2	1.70	1.3718	M	M	N	N	10	CH_3COOH
acetone	58.08	56.3	0.790	−94.7	xx	20.7[a]	5.1	2.69[b]	1.3587	M	M	N	N	750	CH_3COCH_3
acetonitrile	41.05	81.6	0.782	−43.8	xx	37.5	5.8	3.44[b]	1.3441	M	M	76.0	14.2	40	CH_3CN
benzene	78.11	80.1	0.879	5.5	xxx	2.28	2.7	0[b]	1.5011	0.18[a]	0.063[a]	69.3	8.8	1	C_6H_6
n-butanol	74.12	117.7	0.810	−88.6	x	15.8	3.9	1.75	1.3993	7.81	20.07	92.7	42.5	50	$CH_3(CH_2)_2CH_2OH$
butanone	72.10	80.1	0.805	−86.7	xx	18.51	4.7	2.76	1.3788	24.0	10.0	73.4	11.3	200	$CH_3COCH_2CH_3$
t-butyl methyl ether	88.15	55.2	0.741	−108.6	xxx	—	2.5	1.32	1.3689	4.8	1.5	52.6	4.0	40[d]	$CH_3OC(CH_3)_3$
carbon tetrachloride	153.84	76.8	1.594	−23.0	0	2.24	1.6	0[b]	1.4601	0.08	0.008	66.0	4.1	2[e]	CCl_4
chloroform	119.39	61.2	1.489	−63.6	0	4.81	4.1	1.15	1.4458	0.82	0.056	56.1	2.8	2[f]	$CHCl_3$
cyclohexane	84.16	80.7	0.779	6.5	xxx	2.02	0.2	0	1.4262	0.01	0.01	69.0	9.0	300	C_6H_{12}
diethyl ether	74.12	34.6	0.713	−117.4	xxx	4.34	2.8	1.15	1.3524	6.89	1.26	34.1	1.26	400	$(CH_3CH_2)_2O$
dimethylformamide (DMF)	73.09	153.0	0.949	−60.4	x/0	36.7	6.4	3.86	1.4305	M	M	—	—	10	$HCON(CH_3)_2$
dimethylsulfoxide (DMSO)	78.13	189.0	1.100	18.5	x/0	46.7	7.2	3.9	1.4783	M	M	—	—	N.L.	CH_3SOCH_3
1,4-dioxane	88.10	101.3	1.034	11.8	x	2.25	4.8	0.45	1.4224	M	M	87.8	18.0	25	$C_4H_8O_2$
ethanol	46.07	78.5	0.789	−114.1	x	26.0	5.2	1.70	1.3610	M	M	78.2	4.0	1000	CH_3CH_2OH
ethyl acetate	88.10	77.1	0.901	−84.0	xx	6.02[a]	4.4	1.88	1.3724	8.7	3.3	70.4	8.5	400	$CH_3CH_2OCOCH_3$
n-hexane	86.17	68.7	0.659	−95.3	xxx	1.89	0.1	0.08	1.3749	0.001[a]	0.01	61.5	—	50	$CH_3(CH_2)_4CH_3$
isopropanol	60.09	82.3	0.785	−88.0	x	19.92[a]	3.9	1.66[c]	1.3772	M	M	80.3	12.6	400	$CH_3CH(OH)CH_3$
methanol	32.04	64.7	0.791	−98.0	x	32.7[a]	5.1	1.70	1.3290	M	M	N	N	200	CH_3OH
methylene chloride	84.94	39.8	1.326	−95.1	0	8.93[a]	3.1	1.14	1.4241	1.6	0.24	38.1	1.5	50	CH_2Cl_2
petroleum ether	U	30–60	0.64–0.67	−49	xxx	—	0.1	—	1.36–1.38	—	—	—	—	N.L.	5–6-Carbon Hydroc.
tetrahydrofuran (THF)	72.10	66.0	0.888	−108.5	xx	7.6	4.0	1.75	1.4072	M	M	—	—	200	C_4H_8O
toluene	92.13	110.6	0.867	−95.0	xx	2.38[a]	2.4	0.31[b]	1.4969	0.074	0.03[a]	84.1	13.5	50	$Ph-CH_3$
water	18.02	100.0	0.998	0.0	0	80.1	10.2	1.87	1.3330	—	—	—	—	N.L.	H_2O

M.M.: molecular mass (g/mol); b.p.: boiling point (°C); dens.: density (g/mL) at 20°C referenced to water at 4°C; f.p.: freezing point (°C); flam.: flammability, 0: nonflammable; x/0: nonflammable but combustible; x flammable; xx: very flammable; xxx: highly flammable; ε: dielectric constant at 20°C; P.I.: polarity index (L.R. Snyder, *J. Chromat.* **92**, 223 (1974)); μ: dipole moment (D) at 25°C; n: index of refraction at 20°C; S.W.: solubility in water (%w/w) at 20°C, M: miscible in all proportions (L.R. Snyder, *J. Chromat. Sci.*, **16**, 223 (1978)); W.S.: water solubility in solvent (%w/w) at 20°C, M: miscible in all proportions; Az. bp: boiling point of azeotrope with water (°C), N: nonazeotropic mixture; %W: % water in azeotrope; P.E.L.: permissible exposure limit: the maximum permitted average concentration of airborne solvent in ppm (volume) for workers exposed 8 hours daily, according to the California Code of Regulations, Title 8, Section 5155, 1998; these are also the threshold-limit values, time-weighted average (TLV-TWA), recommended by the American Conference of Governmental Industrial Hygienists (ACGIH), except where noted; N.L.: not listed; U: undetermined. [a] at 30°C; [b] at 20°C; [c] at 30°C; [d] not listed, the given value is the TLV-TWA suggested by ACGIH; [e] TLV-TWA (ACGIH): 5 ppm; [f] TLV-TWA (ACGIH): 10 ppm.

In general, we adopt a minimalist approach to the representation of chemical reactions, and we usually prefer Equation 3 to Equation 4. However, there are instances when we need as detailed a picture of the reaction as possible. This is particularly true in the laboratory where we need to mix the reactants in the right molar amounts in order to carry out the reaction successfully. To figure out the amounts of material needed we must have a full description of the reaction including all reactants and products with their stoichiometric coefficients. In other words, we need a **balanced equation**.

The first step in balancing a chemical equation is to find out all the chemicals used or produced in the reaction. If one of them is missing, our final balanced equation would be incorrect. Let us consider, for example, the synthesis of ethyl chloride by the reaction of ethanol with thionyl chloride ($SOCl_2$). SO_2 and HCl are also reaction products. A balanced equation is shown below:

$$CH_3CH_2OH + SOCl_2 \longrightarrow CH_3CH_2Cl + SO_2 + HCl \qquad (5)$$

If we did not know, for example, that HCl was a byproduct, we would have balanced the equation as:

$$2CH_3CH_2OH + SOCl_2 \longrightarrow 2CH_3CH_2Cl + SO_2 + H_2O \qquad (6)$$
(wrong reaction)

Equation 6 implies that two molecules of ethanol react with one molecule of thionyl chloride, which is not the case experimentally. The correct stoichiometry for the reaction is given by Equation 5.

A question frequently asked by students is how to balance a chemical reaction that affords two or more products with the same molecular formula but different structures (these compounds are called isomers). For example, the dehydration of 2-methylcyclohexanol with phosphoric acid as a catalyst gives a mixture of 1-methyl and 3-methylcyclohexene. A balanced equation for the transformation is given below:

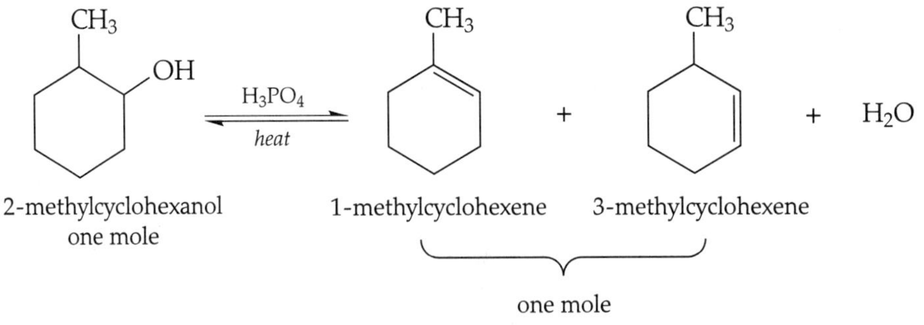

Notice that there are two rings on the right side of the equation (representing the two products), and only one on the left side (the starting material). However, the equation is correctly balanced because it is understood that only *one mole of the mixture* of methylcyclohexenes forms per *each mole* of 2-methylcyclohexanol that reacted.

2.5   CONCENTRATION UNITS

There are many ways to express the concentration of solutions. Molarity, mass per volume, and percentage are commonly used in organic chemistry.

The concentration expressed in molarity units gives the *number of moles of solute per liter of solution*. To calculate the molarity of a solution the number of moles of solute is divided by the volume of the solution (in liters) in which it is contained.

$$\text{molarity} = \frac{\text{number of moles solute}}{\text{volume of solution(L)}} \qquad (8)$$

Another useful way to express concentration is as the mass of solute (in grams or milligrams) per volume of solution (usually 1 mL, 100 mL, or 1 L). For example, if 50 g of sodium chloride is dissolved in water and the final volume of the solution is 1000 mL, the concentration in different mass/volume units is:

$$[\text{NaCl}] = \frac{50\text{ g}}{1000\text{ mL}} = 50\frac{\text{g}}{\text{L}} = 0.050\frac{\text{g}}{\text{mL}} = 50\frac{\text{mg}}{\text{mL}} \qquad (9)$$

Another way of calculating concentrations is as the percentage of solute in solution. This can be expressed as a weight or volume percentage.

The *weight percentage expresses the mass of solute as a percentage of the mass of the solution*. It is represented by the symbol w/w, which means **weight in weight**. For example, if 100 grams of sodium chloride is dissolved in water to make 1000 g of solution, the concentration of sodium chloride is:

$$[\text{NaCl}] = \frac{100\text{ g}}{1000\text{ g}} \times 100 = 10\%(\text{w/w}) \qquad (10)$$

The *volume percentage expresses the volume of solute as a percentage of the volume of the solution*. For instance, if 25 mL of methanol (MeOH) is mixed with water and the final solution volume is 100 mL, the concentration of methanol is:

$$[\text{MeOH}] = \frac{25\text{ mL}}{100\text{ mL}} \times 100 = 25\%(\text{v/v}) \qquad (11)$$

The **v/v** symbol means **volume in volume**. This concentration notation is particularly useful for liquid solutes.

Sometimes the concentration is expressed as the *mass of solute per 100 mL of solution*. This is called **weight in volume (w/v)** percentage. For example, if 12 g of sodium chloride is dissolved in water to make 200 mL of solution, the concentration as weight in volume (w/v) percentage is:

$$[\text{NaCl}] = \frac{12\text{ g}}{200\text{ mL}} \times 100 = 6\%(\text{w/v}) \qquad (12)$$

It should be noted that in all the cases discussed here including w/w, v/v, and w/v, the solute percentage is always *with respect to the whole solution and not just the solvent*.

Finally, there is another concentration notation that is particularly useful for mixtures of liquids. It describes a solution of liquids by the volume of each individual component in the mixture. For example, a mixture of 30 mL of water, 20 mL of methanol, and 10 mL of acetone is described as water: methanol: acetone (30 : 20 : 10 or 3 : 2 : 1).

2.6 MOLES AND MILLIMOLES

When performing a synthesis the reagents must be mixed in the correct molar proportions. Failure to do so may result in side reactions, unwanted products, and loss of time and money. To calculate the number of moles of a given compound, the mass (grams) is divided by the molecular mass (g/mol).

$$\boxed{\#\text{moles} = \frac{\text{mass(g)}}{\text{MM(g/mol)}}} \qquad (13)$$

For example, 0.25 g of aspirin (MM 180.15, Table 2.1) is equivalent to 0.00139 mol, Equation 14:

$$\#\text{moles} = \frac{\text{mass(g)}}{\text{MM(g/mol)}} = \frac{0.25\ \text{g}}{180.15\ \text{g/mol}} = 0.00139\ \text{mol} \tag{14}$$

In organic chemistry we frequently handle amounts much smaller than a mole. This is especially true in microscale experiments, where only milligram amounts of material are used. In such cases, instead of moles a smaller unit, the **millimole**, is preferred. One millimole, represented by **mmol**, is one thousandth of the mole, or in other words there are 1000 mmol in one mol:

$$1\ \text{mmol} = 10^{-3}\ \text{mol};\quad 1000\ \text{mmol} = 1\ \text{mol} \tag{15}$$

To convert moles into mmoles the number is multiplied by 1000. Thus, for the previous calculation, 0.00139 moles of aspirin is equivalent to 1.39 mmol:

$$0.00139\ \text{mol} \times 1000\frac{\text{mmol}}{\text{mol}} = 1.39\ \text{mmol} \tag{16}$$

To calculate the number of moles in a *volume of pure liquid*, the volume is first converted into mass using the density, d, Equation 17.

$$\text{mass(g)} = \text{d(g/mL)} \times \text{Volume(mL)} \tag{17}$$

The number of moles is then calculated using Equation 13. For example, 2.5 mL of methanol (d 0.7915 g/mL; MM 32.04 g/mol, Table 2.1) corresponds to 1.98 g or 0.0618 mol:

$$\text{mass(g)} = \text{d(g/mL)} \times \text{Volume(mL)} = 0.7915\frac{\text{g}}{\text{mL}} \times 2.5\ \text{mL} = 1.98\ \text{g} \tag{18}$$

$$\#\text{moles} = \frac{\text{mass(g)}}{\text{MM(g/mol)}} = \frac{1.98\ \text{g}}{32.04\ \text{g/mol}} = 0.0618\ \text{mol} = 61.8\ \text{mmol} \tag{19}$$

In using these equations you should be aware of the units and make sure that they cancel out properly.

To calculate the number of moles of a chemical in solution, the concentration and volume are needed. If the concentration is given in g/mL, the mass of solute is obtained by multiplying the concentration by the volume of the solution:

$$\text{mass(g)} = \text{concentration (g/mL)} \times \text{Volume (mL)} \tag{20}$$

For example, 1.3 mL of a solution of sodium chloride in water with a concentration of 0.5 g/mL has 0.65 g of NaCl:

$$\text{mass} = 0.5\frac{\text{g}}{\text{mL}} \times 1.3\ \text{mL} = 0.65\ \text{g} \tag{21}$$

The mass of sodium chloride can then be converted into moles with the help of Equation 13 and the molecular mass for sodium chloride.

If the concentration of the solution is given in molarity, the calculation of the number of moles is even simpler. Multiplication of the solution volume (in L) by the molarity gives the number of moles, Equation 22:

$$\#\text{moles} = \text{concentration (mol/L)} \times \text{Volume(L)} \tag{22}$$

If the concentration is given as weight percentage, as is often the case with concentrated inorganic acids, also called **mineral acids**, the calculation of number of moles involves the density of the acid solution, and is explained in the next section.

2.7 MINERAL ACIDS

In the organic chemistry laboratory we often use concentrated (c) mineral acids such as hydrochloric acid, HCl (c), hydrobromic acid, HBr (c), nitric acid, HNO_3 (c), sulfuric acid, H_2SO_4 (c), and phosphoric acid, H_3PO_4 (c). All these concentrated acids are solutions of the acid in water. The concentration of commercial concentrated mineral acids is usually shown on the bottle's label as **% weight in weight** (w/w). As you may recall from section 2.5, % weight in weight means the mass of solute, pure acid in this case, per 100 g of solution. For example, the concentration of HCl (c) is typically 37% w/w, which means that there are 37 g of HCl per 100 g of solution.

To calculate the number of moles of acid in a given volume of the concentrated solution, you need its concentration (in % w/w), the density of the concentrated solution (also shown on the bottle's label), and the molecular mass of the acid. Let's calculate as an example the number of moles of HCl in one liter of its concentrated solution. All the information needed is gathered in Table 2.3, which also contains the typical concentration and density values for other mineral acids.

Table 2.3 **Typical Concentrated Mineral Acids**

Acid	Concentration (%w/w)	Density (g/mL)	Molecular Mass (g/mol)	Concentration (mol /L)
HBr	48	1.490	80.92	8.84
HCl	37	1.200	36.46	12.2
HNO_3	70	1.420	63.01	15.8
H_2SO_4	98	1.840	98.08	18.4
H_3PO_4	85	1.685	98.00	14.6

The first step in the calculation is to figure out the mass of the solution. To do this the solution density is used. For concentrated HCl, the density is 1.200 g/mL, thus the mass in one liter is:

$$\text{mass solution (g)} = d\ (g/mL) \times \text{Volume (mL)} = 1.200\frac{g}{mL} \times 1000\ mL = 1,200\ g \quad (23)$$

The next step is to calculate how much of this total mass is pure HCl. Since the concentrated solution has 37 g of pure HCl per 100 g of solution, multiplying the mass of the solution by the concentration of HCl (37 g HCl/100 g solution) the mass of HCl is obtained:

$$\text{mass HCl (g)} = \text{mass solution} \times \%HCl\ (w/w)$$

$$= 1,200\ g\ sol. \times 37\ g\ HCl/100\ g\ sol. = 444\ g\ HCl$$

Finally, to calculate the number of moles, the mass of HCl is divided by its molecular mass:

$$\#moles = \frac{mass(g)}{MM(g/mol)} = \frac{444\ g}{36.46\ g/mol} = 12.2\ mol \quad (24)$$

This number corresponds to the molarity of the acid given in the last column of Table 2.3.

2.8   CALCULATION OF YIELDS

To advance in the laboratory we must assess the progress of our experimental work at every step. Whether we are isolating a chemical from a natural source or synthesizing it from scratch, the calculation of the percentage yield is a simple way of measuring our achievements.

Yield of an Isolation

In the isolation of a product from a complex mixture, the **percentage yield** is calculated as the ratio of the mass of the product over the mass of the whole mixture multiplied by 100. For example, if we extract the fat from milk cream and obtain 2.1 g of fat from 10 g of cream, the percentage yield is:

$$\% \text{yield} = \frac{\text{mass of product}}{\text{mass of mixture}} \times 100 \qquad (25)$$

$$\% \text{yield} = \frac{2.1 \text{ g}}{10 \text{ g}} \times 100 = 21\%$$

Yield of an Organic Synthesis

The percentage yield for an organic synthesis has a different meaning. To calculate the percentage yield of a chemical reaction you must first calculate the number of moles of reactants and products, and use the balanced equation to determine what is *the maximum amount of product that the reaction can yield*. This amount is called the **theoretical yield**.

When chemicals are mixed to carry out a reaction, the mixture is usually nonstoichiometric. This means that there is often a molar deficit of one of the reactants and an excess of others. At the end of the reaction, the chemicals in excess will still be present, while the one present in a molar deficit would be completely consumed. This compound is called the **limiting reagent** because it is used up in the reaction. The limiting reagent determines the theoretical yield. Let's further clarify these concepts with the calculation of percentage yield for a chemical reaction.

In the transformation shown below, acetic acid reacts with methanol in the presence of sulfuric acid, a catalyst, to give methyl acetate (an ester) and water. This type of reaction, called esterification, is very useful for making esters in the laboratory. We will postpone discussing its chemistry until Unit 22; for now we will consider it as a prototype for any reaction of the form: $A + B \longrightarrow C + D$:

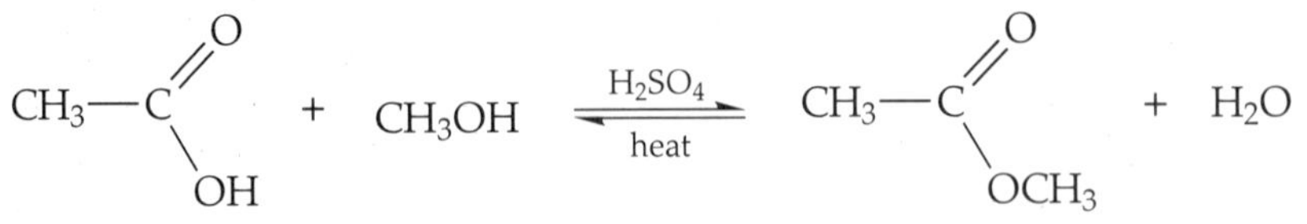

$$CH_3-\overset{\displaystyle O}{\overset{\|}{C}}\diagdown_{OH} + CH_3OH \underset{\text{heat}}{\overset{H_2SO_4}{\rightleftharpoons}} CH_3-\overset{\displaystyle O}{\overset{\|}{C}}\diagdown_{OCH_3} + H_2O \qquad (26)$$

acetic acid   methanol   methyl acetate

Let's assume that in the laboratory we mix 1.0 mL of acetic acid with 5 mL methanol in the presence of a catalytic amount of sulfuric acid (100 μL). After heating for about one hour we obtain 770 mg of methyl acetate. What is the percentage yield in the synthesis of the ester?

To answer this question we must first calculate number of millimoles of reagents and products. The volumes of reactants used and mass of product obtained are summarized in Table 2.4. This table also gathers the density and molecular mass of reactants and products that are needed to calculate the number of mmoles. The figures above the double line are the data, and the question marks indicate those quantities that must be calculated.

Table 2.4 **Table of Reagents for the Synthesis of Methyl Acetate**

	Acetic acid	Methanol	Sulfuric acid	Methyl acetate
volume	1.0 mL	5.0 mL	100 μL	—
density (g/mL)	1.049	0.7915	1.84 for 98% w/w	0.9342
MM (g/mol)	60.05	32.04	98.08	74.08
mass (g)	?	?	?	0.770
mmol	?	?	?	?

The mass for each chemical is calculated using Equation 17, and then the number of mmoles is calculated with Equation 13. The volume of sulfuric acid is given in **microliters** (μL), which is a useful unit for small volumes. The conversion between microliters and milliliters is as follows: $1\,\mu L = 10^{-6}\,L = 10^{-3}\,mL$. Thus, $100\,\mu L = 0.100\,mL$. (To give you an idea of the size of a microliter, remember that a drop of water contains about 30 μL).

The last two rows of Table 2.4, with the calculated figures, are shown below:

	Acetic acid	Methanol	Sulfuric acid	Methyl acetate
mass (g)	1.05	3.96	0.180	0.770
mmol	17.5	123.6	1.83	10.4

Our calculations indicate that 17.5 mmol of acetic acid, 123.6 mmol of methanol, and 1.83 mmol of sulfuric acid were mixed. According to the balanced equation, Equation 26, the reaction between acetic acid and methanol is 1 to 1, or in other words, one mole of methanol is required per each mole of acetic acid. Since the number of mmol for methanol is larger than the number of mmol for acetic acid, an excess of methanol remains unreacted at the end of the reaction, while all the acetic acid reacts. Therefore, **acetic acid is the limiting reagent**.

If acetic acid is totally consumed and transformed into methyl acetate, the maximum number of mmol of methyl acetate that can be obtained must be equal to the number of mmol of acetic acid, because according to the reaction stoichiometry, one mole of ester is formed per each mole of acetic acid reacted. Thus, the theoretical yield for this reaction is 17.5 mmol of methyl acetate. *The **theoretical yield** is the maximum amount of product that a reaction can afford, assuming that it goes to completion and ignoring experimental losses.*

$$\text{theoretical yield} = 17.5 \text{ mmoles}$$

The esterification reaction is an equilibrium reaction. This means that at the end of the reaction there is always some reactant left. However, for the calculation of the theoretical yield it is normally assumed that the reaction is irreversible and goes to completion.

The ***actual yield*** *of the reaction is the amount obtained experimentally,* in this case, 10.4 mmol. The **percentage yield** is the ratio between the actual and theoretical yields, times 100:

$$\boxed{\%\text{yield} = \frac{\text{actual yield}}{\text{theoretical yield}} \times 100} \tag{27}$$

In our example the percentage yield is:

$$\%\text{yield} = \frac{10.4 \text{ mmol}}{17.5 \text{ mmol}} \times 100 = 59.4\%$$

The student may wonder why in this reaction sulfuric acid isn't the limiting reagent if it contributes the lowest number of mmol of all three "reagents" (1.83 mmol). The answer is simple. Sulfuric acid is the *catalyst* of the reaction and not a true reagent. It is *consumed* and *regenerated* during the reaction where it is needed just to speed up the reaction rate.

2.9 SCALING UP, SCALING DOWN

Often in the organic chemistry lab you will need to scale up or down experimental procedures regarding the synthesis or isolation of chemicals. If you want to scale up a synthesis by a factor of 2, masses and volumes should be multiplied by this factor, but concentrations, temperatures, and times should remain the same. Glassware should be changed accordingly. As an example, the following procedure is shown at two different scales.

In a 25-mL Erlenmeyer flask mix 1.0 g of A with 3.0 mL of a 5% solution of B in water. Let the system react for 15 minutes at 35°C. Collect the precipitate by vacuum filtration and recrystallize it from about 10 mL of ethanol using a 25-mL Erlenmeyer flask.

Here is the same procedure scaled up by a factor of 2.5:

In a 25-mL Erlenmeyer flask mix 2.5 g of A with 7.5 mL of a 5% solution of B in water. Let the system react for 15 minutes at 35°C. Collect the precipitate by vacuum filtration and recrystallize it from about 25 mL of ethanol using a 50-mL Erlenmeyer flask.

2.10 KEEPING LAB BOOKS AND WRITING LAB REPORTS

Documentation of scientific information is the keystone in the building of science. You should keep a record of your experiments in a concise, precise, and thorough manner. To this end you should always use a laboratory notebook and never loose sheets of paper that can be easily misplaced. The lab book should be bound and have numbered pages. Spiral notebooks do not make good lab books because the pages can be easily removed.

You should prepare before each experiment by reviewing the theory and finding information about the chemicals involved. You should document this work in your lab book. If your lab book is well organized before you start the experiment, you will save time and effort. Having an organized lab book will also help you write a good lab report. To assist you in this task, pre-lab and in-lab questions are included at the end of each experiment.

Laboratory work is training in meaningful observation. When you are performing an experiment you should always write all the data and pertinent information in your lab book. This includes casual observations such as color changes, formation of a precipitate, evolution of a gas, or the malfunctioning of an instrument. To help you collect and analyze the data, in-lab questions are provided at the end of the experiments. Refer to them as you are performing the experiment.

It is very important that you do a preliminary analysis of your results as they are generated. Do not wait until you get home or until the next day to interpret your data. It may be too late. By then, some results may be meaningless and the experiment, or part of it, must be repeated. This is easy to do on the spot, but it takes more time and effort after completion of the experiment when chemicals and instrumentation have been put away.

At the end of the experiment you will normally be asked to write a lab report. Consult with your instructor regarding the format and the extent of the lab report. The following guidelines will help you in this endeavor.

The lab report should include the following parts:

1. Experiment # and title.
2. Your name, your instructor's name, the lab section.
3. The date.
4. An **INTRODUCTION** including:
   - Purpose of the experiment. Mention the techniques introduced by the experiment. If there is a hypothesis to prove, it should be included here. For example, "Does the dehydration of alcohols follow Zaitsev's rule?"
   - Theoretical background. A brief discussion of the theory behind the experiment. Answering the pre-lab questions will help you build up this part. If the experiment involves a chemical reaction, balanced chemical equations should be shown in this part. If requested, a reaction mechanism should be included.
5. An **EXPERIMENTAL PART** including:
   - A table with the physical properties of reagents, solvents, and products. The table should contain: molecular mass, structure, melting point, boiling point, density, solubility, flammability (for solvents), toxicity and hazards, as required. For synthesis experiments, the table should also include mass, volume, and mmoles used in the reaction.
   - A flowchart outlining the experimental procedure.
   - Observations such as color change, precipitation, solubilization of a solid, gas evolution, and so on.
   - Equipment used. For example, if you are performing a gas chromatography experiment, mention the instrument model, the type of column used, and the experimental conditions. Some of the in-lab questions will help you write this part.
   - Data. This includes melting points, boiling points, retention times, gas chromatography traces, IR spectra, and so on. To avoid repetition, present the data in a table format whenever possible.
   - If the experiment involves the isolation of a compound you should include the yield. In synthesis experiments, the theoretical, actual, and %yields should be included.
6. A **DISCUSSION** including:
   - Interpretation of your data, comparison with previously reported results, and brief comment on sources of error in your determinations. Some of the in-lab questions will help you write the discussion. Avoid colloquial comments such as "overall, the experiment went smoothly." Instead, let your data and your conclusions speak for themselves.
7. Include **REFERENCES** (*Merck Index, CRC Handbook,* etc.) at the end.

A sample of a lab report for selected parts of Experiment 1 (recrystallization of acetanilide and the identification of an unknown) is shown in Figure 2.1.

2.11   CHEMICAL LITERATURE

Before we engage in experimental work we must research the chemical literature to make sure, on one hand, that we have all the pertinent information to make our work successful, and on the other hand, that we are not repeating work and mistakes made by somebody else. A few hours spent in the library may save us days or weeks of frustration.

Student: Synton X. Plosiff   Date: 1/5/2000

Instructor: Krystal de Rocca   Lab section: Tue 1-5; room 271

Experiment 1

Recrystallization of acetanilide and identification of an unknown by melting point

Introduction

<u>Learning objectives</u>: (In-lab #1) In this experiment we will purify acetanilide by recrystallization from water. We will also identify an unknown solid by the mixture melting point technique.

<u>Recrystallization</u>: (Pre-lab #1-3; 5c) Recrystallization is a purification technique for solids. It consists of the following steps: 1) selection of the solvent; 2) dissolving the sample in the minimum volume of boiling solvent; 3) doing a hot filtration (if necessary) to remove insoluble impurities; 4) allowing the system to cool down to crystallize the solid; 5) separating the solid from the liquid (mother liquor) by vacuum filtration (this removes soluble impurities); 6) washing the solid on the filter with a small volume of cold solvent; 7) drying the solid.

A good recrystallization solvent (Pre-lab #4; 9d) should dissolve the sample very well at high temperatures, and sparingly at low temperatures. It should dissolve impurities very well or not at all (the nature of the impurities is usually unknown). The bp of the solvent should be in the range 50-120°C. The solvent should not react with the sample. Other things being equal, we should choose the solvent with the lowest toxicity. The bp of the solvent should be lower than the mp of the solid.

If the solubilities in the hot (S^H) and cold (S^C) solvent are known, the theoretical % Recovery (% R) of the recrystallization can be calculated (Pre-lab #10):

$$\% \ R \ = \ 100 \times (S^H - S^C)/S^H$$

In principle, theoretical % R doesn't depend on the total amount of material to be recrystallized. If S^H = 1 g/100 mL and S^C = 0.03 g/100 mL, the % R is:

$$\% \ R = 100 \times (1 - 0.03)/1 = 97 \ \%, \ \text{regardless of the amount of solid.}$$

<u>Melting point</u>. (Pre-lab #5a,b; 6; 9a-c) The melting of a pure organic compound occurs in a very narrow temperature range, called the melting point range, which is normally less than 2°C. Most impurities have a dual effect on the mp: They lower the mp and extend the mp range. If the melting range is narrow it doesn't necessarily mean that the compound is pure. For example, the melting point of eutectic mixtures is sharp. Also, some impurities, especially inorganic compounds (NaCl, chalk, sand, etc.) do not change the mp of organic solids.

Figure 2.1   A sample lab report.

Mixture melting point is a technique for the identification of organic solids. We mix two compounds (A and B) with the same mp. If the mp of the mixture is identical to the mp of the individual components, then A and B are the same. If the mp is different, then A and B are different compounds.

We should heat the sample slowly in the vicinity of the mp to make sure that the thermometer and the sample are at the same temperature (Pre-lab #7). A resolidified sample should not be reused to take a second mp because the sample may contain decomposition products that act as impurities and lower the mp. (Pre-lab # 8)

Experimental

Table of reagents (Pre-lab #11)

compound	MM (g/mol)	structure	mp (°C)	bp (°C)	density (g/mL)	flamm.	solubility	toxicity/ hazards
acetanilide	135.17		113– 115	304– 305	–	–	1 g/185 mL H$_2$O (room temp.) 1 g/20 mL H$_2$O (100°C)	LD$_{50}$ 800 mg/kg (oral rat) Toxic Irritant
water	18.02	H$_2$O	0	100	1.00	0	–	nontoxic

Flowchart: (Pre-lab # 12) Experimental procedure according to "Experimental Organic Chemistry"

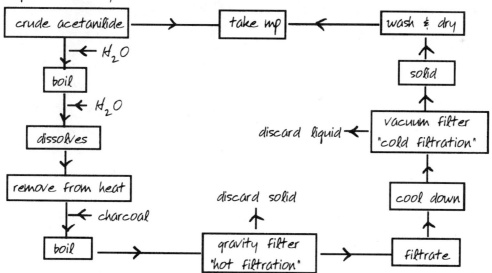

Glassware and equipment : (Pre-lab #13) spatula; 2 Erlenmeyer flasks (125 mL); glass rod; hot plate; short-stem funnel; filter paper; boiling chips; ice-water bath; filter flask (125 mL); Buchner funnel; watch-glass; oven; mp apparatus; capillary tubes.

Figure 2.1 (*Continued*)

<u>Data</u> (In-lab #2; 3; 10-13)

Recrystallization of acetanilide

	mass (g)	mp (°C)
before recrystallization	3.0	105-109
after recrystallization	1.3	110-112

Recrystallization solvent: water

<u>Observations:</u> Some acetanilide crystallized on the filter paper during the hot filtration and was lost. Crude acetanilide was beige. Acetanilide recrystallized as white flakes. The recrystallized product was dried in an oven (70°C) for 1 hour before weighing and determining the mp.

% Recovery:

$$\%R = 100 \times \text{mass after recryst/mass before recryst}$$
$$\% R = 100 \times 1.3 \text{ g}/3.0\text{ g} = 43\%$$

Identification of an unknown

Unknown number: 501; white solid.

mp (fast): 150-152°C; mp (slow): 155-157°C.

According to the table of unknowns, the possible compounds are: m-bromobenzoic acid (mp 155-157°C) and salicylic acid (mp 156-158°C).

<u>Mixture mp</u>

unknown + m-bromobenzoic acid: 120-135°C

unknown + salicylic acid: 155-157°C

Discussion and conclusions

The recrystallization of acetanilide gave a % Recovery of 43%. Loss of product during the hot filtration accounts, in part, for this low recovery. The melting point of the recrystallized product (110-112°C) is higher than the mp of the crude material (105-109°C). This indicates that the recrystallization was effective in removing impurities. However, the mp of the recrystallized product is lower than the literature value (113-115°C). Presence of residual water (recrystallization solvent) and error in determining the mp may account for this.

Based on the mixture mp, I can conclude that the unknown # 501 was salicylic acid, because there was no depression in the mp of the mixture.

Bibliography

Experimental Organic Chemistry, D. Palleros.

Merck Index, 12th edition. Aldrich Catalog, 1998-1999.

Figure 2.1 (*Continued*)

The wealth of chemical information could be overwhelming. More than 700,000 articles related to chemistry were published in 1996 alone. However, with the widespread use of computers the search for chemical information is becoming easier and faster every day. The primary source of chemical information is peer-reviewed scientific journals. Some journals publish original results in all fields of chemistry and others offer comprehensive reviews on specific subjects. The reviews are written by authorities in the field and comment on the most important and recent bibliography. Physical and spectroscopic data can be found in handbooks, catalogs, and special collections. The information contained in journals is compiled, sorted out, and catalogued by different services, among them the *Chemical Abstracts*, the *Beilstein Handbook of Organic Chemistry*, and the *Current Contents*.

Let's consider the most important sources of chemical information separately.

Journals

Some general journals for all areas of chemistry include:

- *Journal of the American Chemical Society*, also known as JACS, is arguably the most important and prestigious journal in chemistry. There is free Internet access to the table of contents of the current issue (http://www.ChemCenter.org).

- *Journal of Chemical Education*. Published by the Division of Chemical Education of the American Chemical Society. Articles with emphasis on chemistry education. Many good ideas for teachers and students. There is substantial free access on the Internet. Some articles and abstracts are available online free of charge. Index also available online (http://jchemed.chem.wisc.edu/).

- *Chemical Reviews*. Also published by the American Chemical Society. Comprehensive reviews on different subjects written by authorities in the field. The Internet access is http://www.ChemCenter.org.

- *Angewandte Chemie. International Edition in English*. A German journal translated into English. It publishes original articles and reviews. There is free Internet access to the table of contents and the abstracts to some "hot" papers (http://www.vchgroup.de/home/angewandte).

Important journals in the field of organic chemistry include:

- *The Journal of Organic Chemistry*. Published by the American Chemical Society. Covers all areas of organic chemistry. There is free Internet access to the table of contents of current issue (http://www.ChemCenter.org).

- *Tetrahedron* and *Tetrahedron Letters*. The first publishes full articles in all areas of organic chemistry, and the second is for the fast publication of short communications, most of them dealing with organic synthesis.

- *Synthesis*. A journal devoted to synthetic organic chemistry. There is free Internet access to the table of contents of recent issues (http://www.thieme.com/chem.htm).

- *Journal of the Chemical Society. Perkin Transactions I*. Published by the Royal Society of Chemistry (UK). Mainly devoted to organic synthesis. There is free Internet access to the table of contents (http://chemistry.rsc.org/rsc/).

- *Journal of the Chemical Society. Perkin Transactions II*. Published by the Royal Society of Chemistry (UK). Specializes in physical organic chemistry (reaction mechanisms, kinetics, spectroscopy, quantum mechanics). There is free Internet access to table of contents (http://chemistry.rsc.org/rsc/).

Other publications important in organic chemistry are:

- *Organic Syntheses*. Published annually by Wiley (New York). It describes the synthesis of diverse organic compounds. The syntheses are checked by independent chemists for accuracy. There are eight **Collective Volumes**.

- *Fieser and Fieser's Reagents for Organic Synthesis*. Published by Wiley (New York). Written by the legendary couple Louis and Mary Fieser from Harvard University. Describes the uses and applications of the most important reagents in organic chemistry. Started in 1967 and is periodically upgraded.
- *Encyclopedia of Reagents for Organic Synthesis*. Leo A. Paquette, editor. Wiley (Chichester, New York; 1995). A description of 3000 reagents with examples of use and references.
- *Advanced Organic Chemistry. Reactions, Mechanisms and Structure*. J. March. 4th ed. Wiley, New York, 1992. An excellent compilation of the most important reactions in organic chemistry with references.
- *Vogel's Textbook of Practical Organic Chemistry*. A. I. Vogel, B. S. Furniss, A. J. Hannaford, P. W. G. Smith, and A. R. Tatchell. Longman, Harlow, 5th ed., 1989. A classic laboratory textbook with many macroscale syntheses. Complete and thorough.

Laboratory Handbooks

- *Aldrich Catalog of Fine Chemicals*. Published by the Aldrich Company. Updated annually. It contains more than 27,000 chemical products. It cross-references to Beilstein (see below) and the Merck Index. Lists physical properties, and references spectroscopic information. Gives the Chemical Abstract Service (CAS) registry number (see below). Has molecular formula index. Also available on CD-ROM.
- *Merck Index. An Encyclopedia of Chemicals, Drugs, and Biologicals*. Published by the Merck Company (Whitehouse Station, NJ). S. Budavari, editor. Listing of about 10,000 chemicals; includes physical data, uses, toxicity, and hazards. References synthesis. Gives CAS registry number. Has molecular formula index. Upgraded periodically. Also available on CD ROM.
- *CRC Handbook of Chemistry and Physics*. Published by the CRC Press (Boca Raton, FL). D.R. Lide, editor. Contains a table of organic compounds with physical properties; cross-references Beilstein. Upgraded annually. Has molecular formula index.
- *Lange's Handbook of Chemistry*. Published by McGraw-Hill (New York). J. A. Dean, editor. Physical data for most important organic chemicals. Upgraded periodically.

Other Reference Handbooks

- *Dictionary of Organic Compounds*. Published by Chapman and Hall, 6th ed., 1996 (nine volumes). It contains physical data, references on synthesis, derivatives, applications, toxicity, and hazards of about 150,000 compounds. Has molecular formula index. Gives CAS registry number. Also available on CD ROM and online. For more information: http://www.chapmanhall.com.
- *CRC Handbook of Data on Organic Compounds*. Published by the CRC Press (Boca Raton, FL), 3rd ed., 1994. Physical data, extensive spectral data information. Molecular formula index. CAS registry numbers. Beilstein references. Also available on CD ROM.
- *Handbook of Natural Products Data*. Atta-ur-Rahman and Viqar Uddin Ahmad. Elsevier (Amsterdam, 1994). Contains spectral data. Has molecular formula and molecular mass indexes.
- *Physicians' Desk Reference*. Published annually by the Medical Economics Co. (Montvale, NJ), contains information about prescription and nonprescription drugs.

Toxicity and Hazards Handbooks

- *Hazardous Materials Handbook.* R.P. Pohanish and S.A. Greene. Published by Van Nostrand Reinhold (New York, 1996). It contains a brief description of fire, exposure, and health hazards, chemical reactivity, water pollution, and emergency response for most important chemicals.
- *The Sigma-Aldrich Library of Regulatory and Safety Data.* R.E. Lenga, and K.L. Votoupal, 1st ed., 1992. A compilation on the toxicity, safety, and hazards of more than 20,000 chemicals.

Spectral Collections

- *Aldrich Library of ^{13}C and ^1H FT-NMR Spectra.* C.J. Pouchert, and J. Behnke. 1st Edition, 1993. A collection of ^1H and ^{13}C-NMR spectra. The ^1H- and the ^{13}C-NMR spectra were obtained at 300 and 75 MHz, respectively.
- *Aldrich Library of NMR Spectra.* C.J. Pouchert. 2nd ed., 1983. A collection of ^1H-NMR spectra obtained with a 60 MHz instrument.
- *Aldrich Library of FT-IR Spectra.* C.J. Pouchert. 1st ed., 1985. A collection of more than 17,000 IR spectra.
- *Sadtler Standard Spectra.* Published by Sadtler Research Laboratories, Inc. (Philadelphia, PA). There are different collections containing IR, UV, ^1H- and ^{13}C-NMR spectra.

Compilation Services

The journals listed here are only a small fraction of all the chemistry journals published worldwide. The information contained in them is compiled, sorted out, and catalogued by different services, among them the *Chemical Abstracts*, the *Beilstein Handbook of Organic Chemistry*, and the *Current Contents*.

The *Chemical Abstracts*, published by the American Chemical Society, is the most complete chemical information resource. It has more than 16 million compounds on record. It gives a short abstract in English of articles on chemistry published anywhere in the world and in any language beginning in 1907. There are thirteen cumulative indexes, one every 10 years until 1957 (1917; 1927; 1937; 1947 and 1957), and one every 5 years thereafter (1957–61; 1962–66; 1967–71; 1972–76; 1977–81; 1982–86; 1987–91 and 1992–97). The cumulative indexes include formula, general subjects, chemical substances, authors, and patents. There is also an index by the **registry number**, which is a unique number that the Chemical Abstracts Service (CAS) assigns to every compound. For example, the CAS registry number for aspirin is 50-78-2. This number is extremely useful for carrying out searches, especially by computer. Searches on the *Chemical Abstract* can also be performed (for a fee) online. Information about the CAS can be obtained at http://www.cas.org.

The *Beilstein Handbook of Organic Chemistry* is everything but a handbook. Its latest edition (1995) has 480 volumes. It is the most complete database for organic chemistry. It contains more than 6 million compounds. It is also the oldest compilation, starting in 1881 and reaching as far back as 1779. It was originally published in German, but it is now published in English. More than 350 properties are listed for some chemicals. The search can also be done online. Its Internet site is http://www.beilstein.com.

The *Current Contents* article database is very useful to find titles and authors of current articles published by the most important journals. Most college libraries subscribe to this service. The online database contains almost 7 million citations from 6500 journals in many different fields (not only chemistry) and goes back as far as July 1989. Retrieval of information is done by title words or authors.

Online Resources

Other online resources include **MEDLINE** and **MEDLINE PLUS**, which are databases with citations and abstracts from more than 8,000 journals in medicine, life science, and health administration. The database contains almost 2 million citations from 1992. Medline backfiles from 1966 to 1991 are also available as separate databases. Many libraries subscribe to this service.

EXERCISES

Problem 1

Circle the H-bond donors, and box the H-bond acceptors.

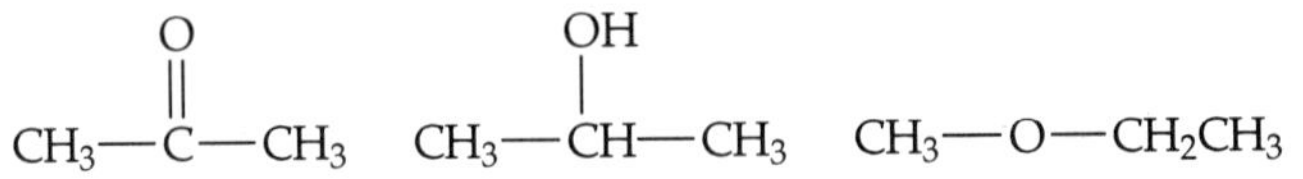

acetone   isopropanol   ethyl methyl ether

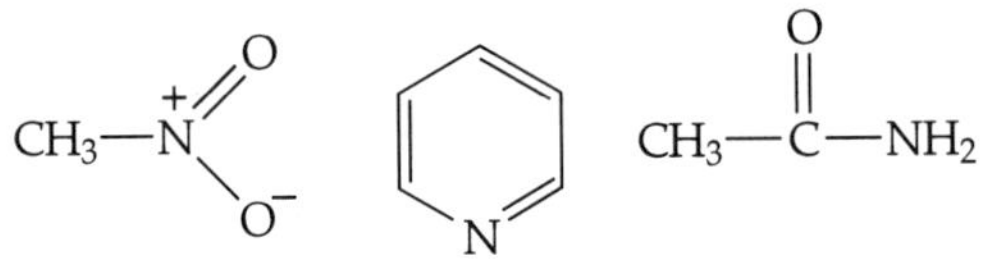

nitromethane   pyridine   acetamide

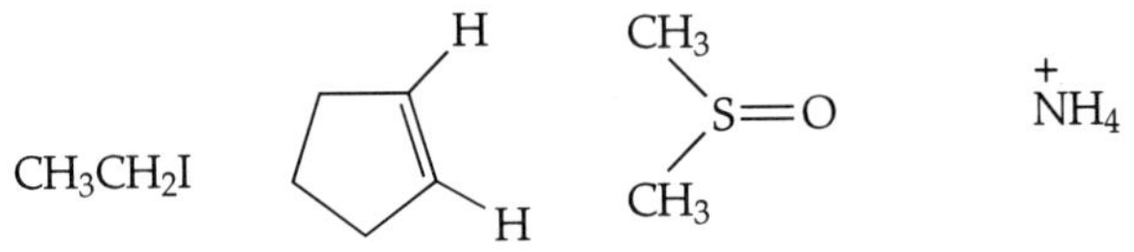

CH_3CH_2I

iodoethane   cyclopentene   dimethyl sulfoxide   ammonium cation

Problem 2

Based on geometric considerations, predict which of the following molecules have a permanent dipole moment.

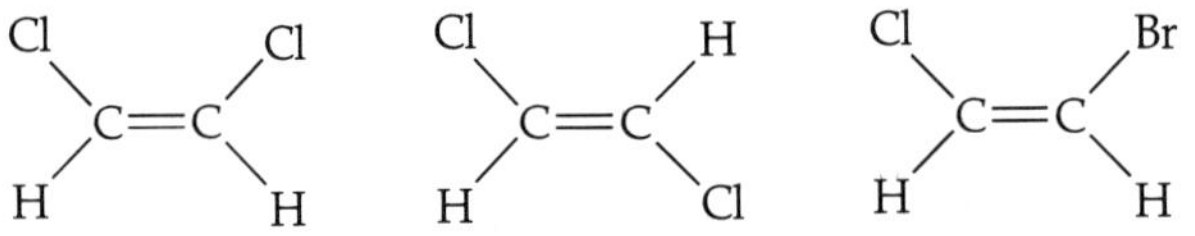

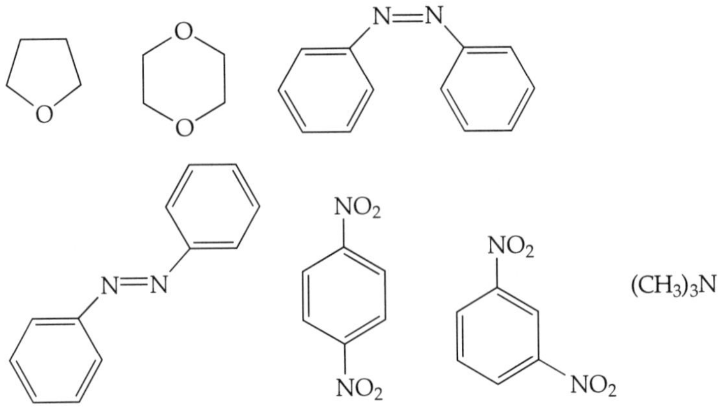

$(CH_3)_3N$

Problem 3

Balance the following equations by replacing the question mark with the appropriate stoichiometric coefficient. In each case determine the limiting reagent and the theoretical yield for each product when the molar amounts indicated under the reagents are mixed.

(a)

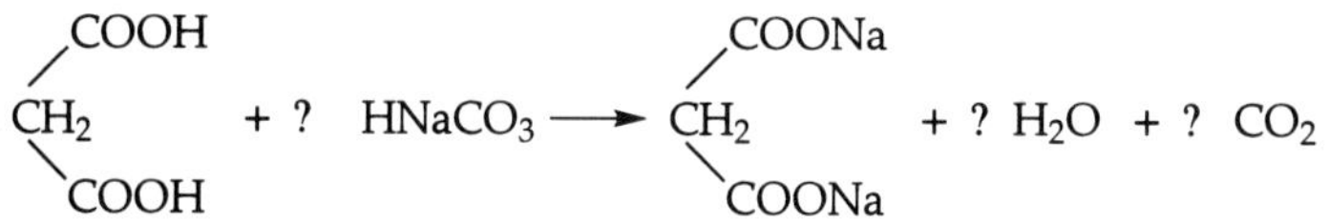

molar amounts used:

1 mmol   1.8 mmol

(b)

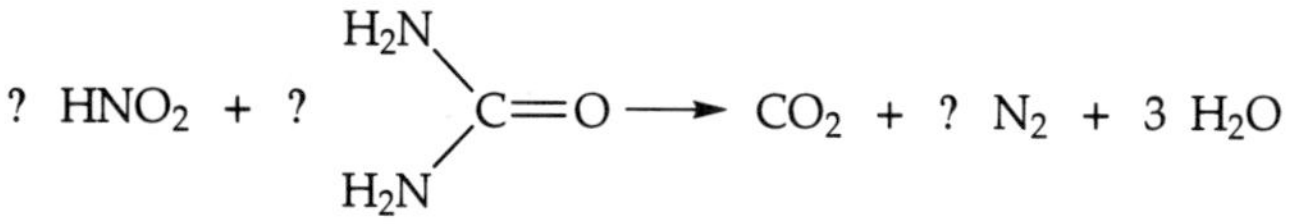

molar amounts used:

3.7 mmol   7.4 mmol

(c)

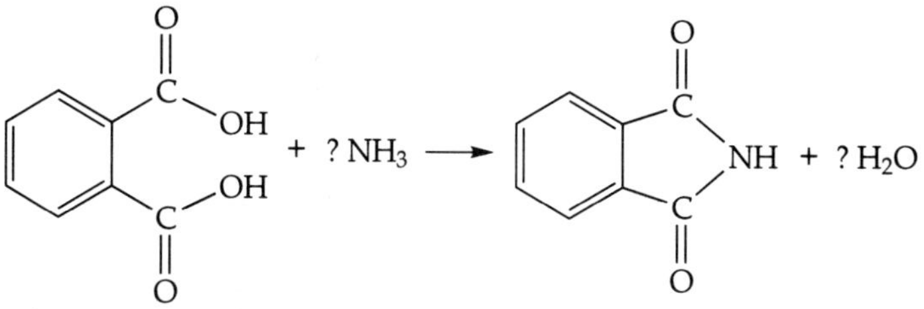

molar amounts used:

2.3 mmol   5.2 mmol

(d)

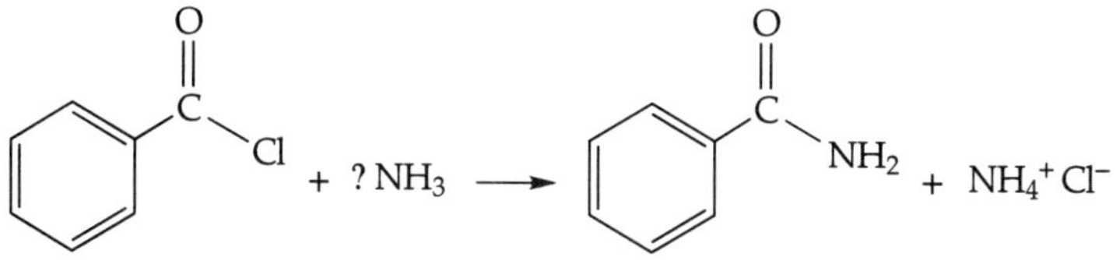

molar amounts used:

5.7 mmol   10.3 mmol

(e)

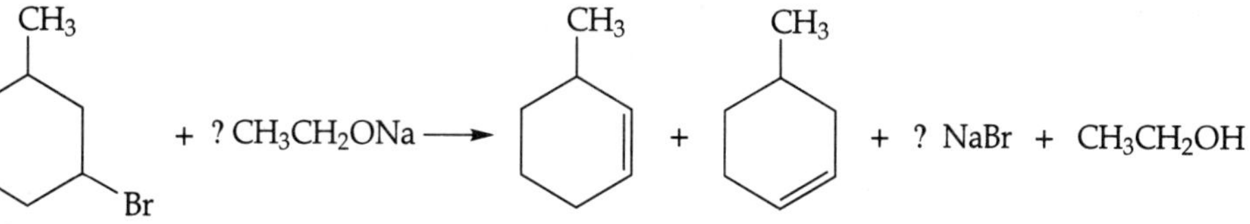

molar amounts used:

2.3 mmol   15.0 mmol

Problem 4

Calculate the concentration of an aqueous solution of hydroiodic acid (56% w/w, density = 1.70 g/mL; MM HI = 127.91 g/mol) in molarity, g/L and %w/v.

Problem 5

When 970 μL of 1-methylcyclohexene (MM = 96.17 g/mol; d = 0.809 g/mL) was reacted with 3.0 mL of concentrated HBr (48% w/w; d = 1.49 g/mL; MM HBr = 80.92 g/mol) it gave 1.10 g of 1-bromo-1-methylcyclohexane (MM = 177.09). Calculate the number of mmoles for each compound. Determine the limiting reagent. Calculate the theoretical yield of the product and the percentage yield. The balanced equation is shown below.

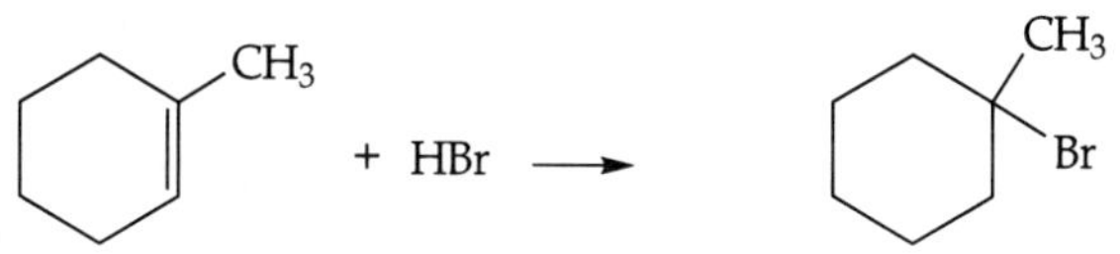

1-methylcyclohexene   1-bromo-1-methylcyclohexane

Problem 6

When benzil reacts with periodic acid, benzoic acid is obtained. The balanced equation is shown below. Calculate the percentage yield of the synthesis if 1.05 g of benzoic acid (MM = 122.12 g/mol) was obtained from 1.05 g of benzil (MM = 210.23 g/mol) and 3.3 mL of a 1.6 M solution of periodic acid was used.

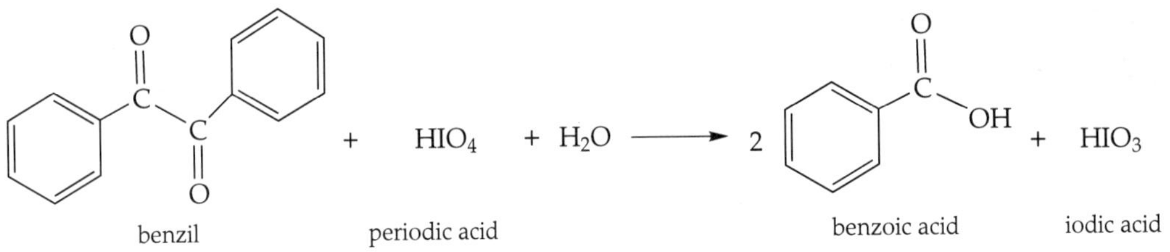

benzil   periodic acid   benzoic acid   iodic acid

UNIT 3

Basic Operations

Only the simplest and most common operations used in the organic chemistry laboratory are discussed in this unit. You will find references to this unit throughout the book. Complex operations such as recrystallization and distillation are discussed in special units.

3.1 HANDLING LIQUIDS

Small amounts of liquid are more difficult to handle than similar amounts of solid. At the microscale level, liquids should never be poured. They should be transferred with the aid of pipets and syringes.

Pasteur Pipet

A Pasteur pipet (Fig. 3.1) is a small piece of glass tubing with a capillary end. It can hold up to 2 mL of liquid and is filled using a small rubber bulb or a pluringe (see below). Liquids with low boiling points (such as acetone and methylene chloride) have a tendency to squirt out of the pipet tip. They are better handled using a pluringe instead of a rubber bulb.

Pluringe

A pluringe is a simple device to draw small volumes (0.1–2 mL) with a Pasteur pipet. It consists of a 1-mL or 3-mL plastic syringe connected to a piece of latex or Tygon (registered mark of Norton Co.) tubing by a plastic adaptor (Fig. 3.2). Pluringes are

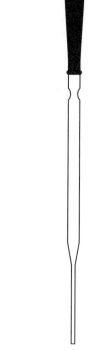

Figure 3.1 A Pasteur pipet.

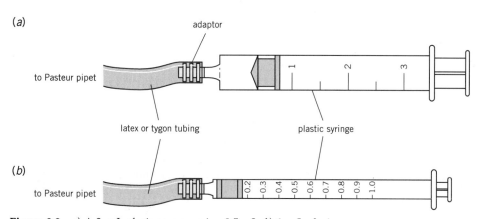

Figure 3.2 *a*) A 3-mL pluringe measuring 0.5 mL; *b*) 1-mL pluringe.

45

ideal for transferring volatile liquids, which tend to squirt out of the end of the pipet when ordinary rubber bulbs are used. Pluringes are not recommended when a high accuracy in volume is required; automatic pipets (see below) should be used instead. *Pluringes should always be used while connected to a Pasteur pipet.*

Operation Connect the latex or Tygon tubing of the pluringe to the end of the Pasteur pipet. With the plunger all the way in, immerse the tip of the pipet in the liquid. Move the plunger up to draw the desired volume. No air bubbles should be drawn with the liquid. Deliver the liquid by pushing the plunger slowly.

Automatic Pipet

These pipets are very useful for accurate measurement of small volumes, from 1 µl to 1000 µL (1000 µL = 1 mL), Fig. 3.3. They are available from several manufacturers and come in different levels of sophistication. There are basically two types: a) **single volume**, which are set to deliver a predetermined volume, and b) **adjustable**, which can be set to deliver different volumes.

Operation See Figure 3.3. Select the pipet with the appropriate volume range 1–20 µL; 10–200 µL; 100–1000 µL, and so forth. Adjust the volume by turning the volume control button (1). Select the appropriate disposable tip (2) and push it into the end of the barrel (3). Push the filling button (4) to the first stop position (5). Immerse the tip into the liquid and slowly release the filling button to the release position (6). Remove the tip from the liquid. Holding the tip against the walls of the receiving vessel, slowly push the filling button to the second stop position (7). Discard the tip.

Automatic pipets are very expensive and should be treated with absolute care and respect. *Always keep a loaded pipet vertical and never load a pipet without a disposable tip.* The liquid should never wet the barrel, but if by accident the liquid comes in contact with the barrel, clean it immediately with a piece of tissue paper.

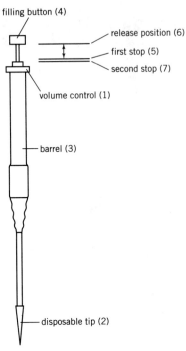

Figure 3.3 An automatic pipet.

3.2 HEATING ——————————————————————

In the modern organic chemistry laboratory the most important heat sources are the heating mantle and the hot plate. Because most organic solvents are flammable, Bunsen burners should be avoided; they should be restricted only to specific applications.

Heating Mantle

Heating mantles are useful for heating round-bottom flasks (Fig. 3.4). The mantle should be connected to a rheostat, which in turn is plugged into the wall outlet. The rheostat controls the voltage received by the mantle and regulates the heat. Never plug a heating mantle directly into the wall outlet because it may burn the heating element. Heating mantles are available in different sizes (25 mL–10 L). The round-bottom flask should fit into it snugly.

Hot Plates

These are used to heat flat-bottom glassware such as beakers and Erlenmeyer flasks. Some are equipped with a rotating magnet under the hot surface that allows stirring if a magnetic stir bar is placed in the reaction mixture. The stir bars are small magnets coated with Teflon (registered mark of the DuPont Co.) to make them inert (Fig. 3.5).

Hot plates are also useful for heating water baths and sand baths. These baths are used to heat reaction mixtures at predetermined temperatures. To monitor the

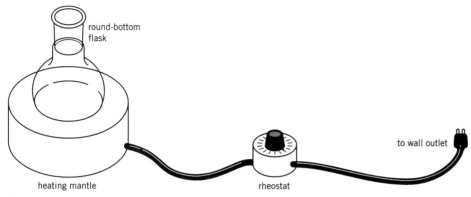

Figure 3.4 A heating mantle with rheostat.

(*a*)

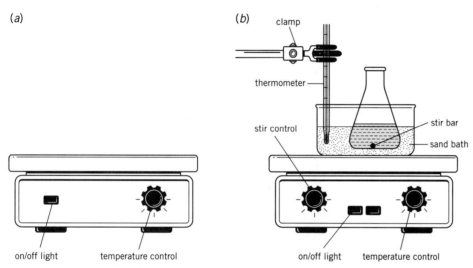

(*b*)

Figure 3.5 *a*) Hot plate; *b*) hot plate with magnetic stirrer and sand bath.

temperature, a thermometer is clamped to a ringstand and immersed in the bath near the reaction vessel. The thermometer should be placed in the bath when the bath is cold. If the thermometer is put in a hot bath, the bulb may break because of thermal shock. Water baths can be used for temperatures up to the boiling point of water. Sand baths are useful up to 250°C.

Bunsen Burner

Open flames should be avoided in the laboratory because most organic solvents are flammable. Never use a Bunsen burner in the laboratory without approval of your instructor. They should be used only when very high temperatures are necessary (Fig. 3.6).

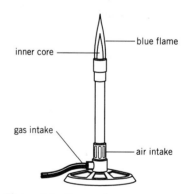

Figure 3.6 Bunsen burner.

3.3 FILTRATION

Filtration is an operation for separating a liquid from a solid. There are two types of filtrations in the organic chemistry laboratory: **gravity filtration** and **vacuum filtration**.

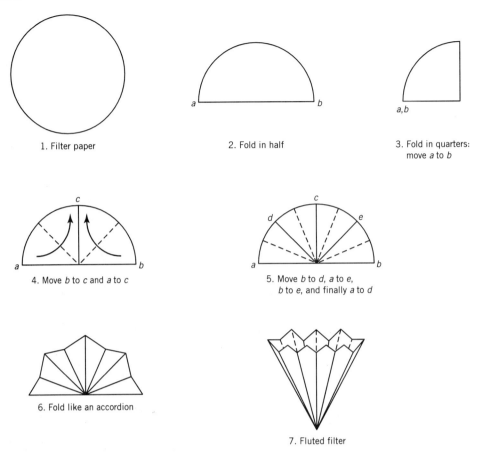

1. Filter paper

2. Fold in half

3. Fold in quarters: move *a* to *b*

4. Move *b* to *c* and *a* to *c*

5. Move *b* to *d*, *a* to *e*, *b* to *e*, and finally *a* to *d*

6. Fold like an accordion

7. Fluted filter

Figure 3.7 Folding filter paper to make a fluted cone for gravity filtration.

Gravity Filtration

Gravity filtration is the filtration of choice when one is interested in the liquid rather than the solid, or when the solid particles are very fine. There are two ways of carrying out a gravity filtration: with the help of a glass funnel and fluted filter paper (Fig. 3.7 and 3.8), and with the help of Pasteur pipets and cotton (Fig. 3.9).

Glass-Funnel Filtration This uses a cone of fluted filter paper on top of a glass funnel. Folding the filter paper accordion-like, as indicated in Figure 3.7, maximizes the filter surface area and results in faster filtrations. Filter papers with different pore sizes, or **porosity**, are available. The porosity determines the flow rate of the filtration. Very rapid filter papers have coarse porosity and are recommended for filtering solids with large particle size (larger than 20 μm = 0.02 mm). An example of this type of paper is Whatman #4. Slow papers, such as Whatman #3, have fine pores and are recommended only when the solid has very fine particle size. Most filtrations are successfully performed using paper of medium-coarse porosity such as Whatman #1 or Fisher P8[1].

Operation See Figure 3.8. Choose a suitable glass funnel and place it in the neck of an Erlenmeyer flask. The setup should not be top heavy. If the funnel is too big for the receiving flask, support the funnel on a ring and secure it to a ringstand. The funnel should not fit too tightly in the neck of the flask because this will prevent air flow and the filtration will stop. A paper clip placed as indicated in Figure 3.8*a* allows air flow. Choose the filter paper of the right porosity and appropriate diameter. The diameter of the disk should be about twice the diameter of the funnel. Make a cone of fluted filter paper as indicated in Figure 3.7 and place it in the funnel. The edge of the paper should be about 5 mm below the rim of the funnel. Secure the filter paper in place by adding a small aliquot of the same solvent to be filtered. Discard the filtrate. Pour the

(a)

(b)

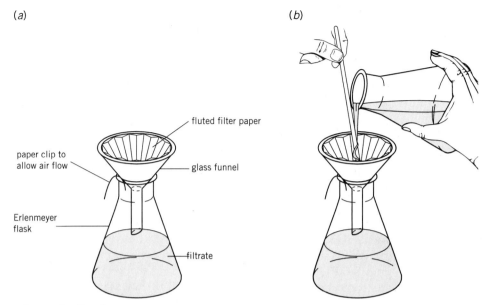

fluted filter paper

paper clip to
allow air flow

glass funnel

Erlenmeyer
flask

filtrate

Figure 3.8 Setup for gravity filtration.

(a)

(b)

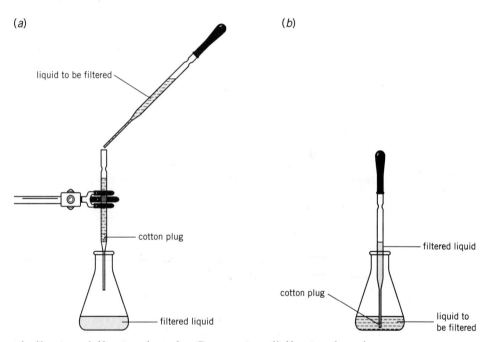

liquid to be filtered

cotton plug

filtered liquid

filtered liquid

cotton plug

liquid to
be filtered

Figure 3.9 Microscale filtration: *a*) filtration through a Pasteur pipet. *b*) filtration through a filter-Pasteur pipet.

suspension to be filtered with the help of a glass rod to direct the liquid into the funnel (Fig. 3.8*b*). To prevent puncturing the filter paper, avoid touching the center of the cone with the glass rod. If the particles are very fine, let them settle at the bottom of the flask before the filtration; pour most of the liquid first without disturbing the solid, and then pour the rest of the liquid mixed with the solid. Let the system drain well and wash out residual liquid in the solid by adding a small aliquot of fresh solvent.

Filtration Using a Pasteur Pipet   Pasteur pipets make excellent filtration devices, especially at the microscale level. There are two different ways of using a Pasteur pipet to filter suspensions. The most common way is illustrated in Figure 3.9*a*. A small cotton plug is positioned, with the help of a microspatula or copper wire, just above the capillary end. The suspension is transferred with the help of a second Pasteur pipet and the filtrate is collected in a receiving flask.

Another way of using Pasteur pipets as filtering devices is illustrated in Figure 3.9*b* with the so-called **filter-Pasteur pipet**. This pipet allows the transferring and filtration of a liquid in *one step*. A very small cotton plug, approximately the size of half a grain of rice, is introduced into a Pasteur pipet with the help of a fine wire. *The cotton plug should be positioned at the very end of the capillary tube*. It should be neither very tight nor loose. The liquid is drawn using a rubber bulb or pluringe. The liquid passes through the cotton, leaving behind solid particles. The liquid is then delivered into the receiving flask. Filter-Pasteur pipets are ideal for removing solids from minute amounts of liquid and also for transferring liquids with very low boiling points, which have a tendency to squirt out the end of regular Pasteur pipets.

Vacuum Filtration

Vacuum filtration is the operation of choice when one is interested in the solid rather than the liquid. Funnels made of porcelain with a perforated plate are used for this purpose. They are available in two different shapes: cylindrical or **Buchner funnels**, and conical or **Hirsch funnels** (Fig. 3.10). They are always used with a flat disk of filter paper covering the holes of the perforated plate.

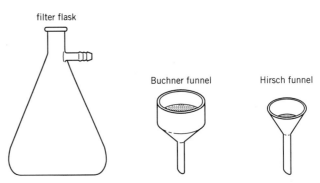

Figure 3.10   Filter flask and funnels for vacuum filtration.

With the aid of a rubber stopper, the funnel is attached to the neck of a filter flask, and the sidearm of the flask is connected to a vacuum line. As the suspension to be filtered is poured into the funnel, the liquid passes through the filter pushed by the higher external pressure. Vacuum can be generated by a **water aspirator**, an oil pump, or a house vacuum line. A water aspirator (Fig. 3.11) is the most inexpensive way of creating a vacuum in the laboratory. It consists of a plastic or metal tube, with a hole on the side, connected to a water faucet. Its operation is based on the Bernoulli principle. When turbulent water is forced to pass at high speed through a pipe with a hole on the side, air is drawn into the pipe through the hole. If a closed container is connected to the hole, air is evacuated from the container and a vacuum is created.

It is customary to interpose a trap between the filter flask and the water aspirator (Fig. 3.12). The trap prevents water from being sucked from the aspirator into the filter flask if a sudden drop in water pressure occurs. The maximum vacuum that can be obtained with a water aspirator is approximately 25 Torr (the vapor pressure of water at room temperature).

Operation   See Figure 3.12. Clamp the filter flask to a ringstand. Insert the funnel in the neck of the flask by using a snug-fitting rubber stopper or rubber disk. Using thick-walled rubber tubing, connect the sidearm of the filter flask to the vacuum trap. Select the right filter paper type for the filtration in hand. Whatman #1 is appropriate for most separations. Position the disk of filter paper flat on the perforated plate. The paper should cover the holes but it should not touch the walls (Fig. 3.13). If

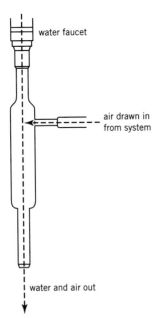

Figure 3.11   Water aspirator.

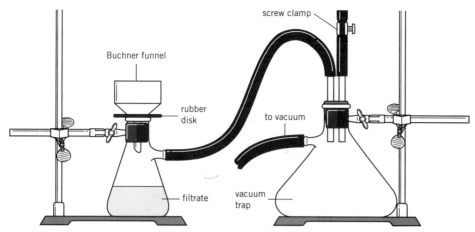

Figure 3.12 Vacuum filtration apparatus with trap.

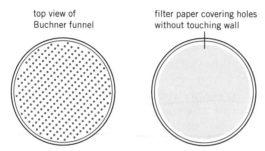

Figure 3.13 Correct positioning of filter paper for vacuum filtration.

you are using a water aspirator, turn the water on full blast and then close the screw clamp of the vacuum trap. Wet the filter paper with a few drops of the same solvent used in the filtration. This will firmly secure the paper to the funnel and prevent the solid from slipping under the paper. With the aid of a glass rod pour the suspension into the funnel. The liquid will pass through and the solid will be collected on the filter. To minimize mechanical losses, rinse the walls of the vessel that contained the suspension with a small aliquot of fresh solvent and transfer the rinse to the funnel. Let the liquid drain completely. Open the screw clamp in the vacuum trap. Add a small aliquot of fresh solvent to the solid on the filter and let the solid soak for a few seconds. Close the screw clamp. This step will wash the solid, removing traces of the liquid. Let the solid dry on the filter with the vacuum on for a few minutes. Open the screw clamp in the vacuum trap, and then turn off the aspirator (never do this in the reverse order because water would be sucked from the aspirator into the trap). With the help of a spatula remove the solid from the filter.

When you are working at the semimicro level, the vacuum filtration apparatus shown in Figure 3.14 can be used. It consists of a thick-walled test tube with sidearm and is equipped with a Hirsch funnel. A small test tube is placed inside to collect the filtrate.

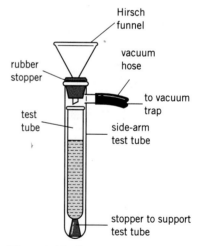

Figure 3.14 Semimicro vacuum filtration apparatus.

3.4 REFLUXING

Temperature is one of the variables that one can easily control in the laboratory to speed up or slow down a reaction. Most organic reactions are performed in the liquid phase using a suitable solvent. A convenient way of carrying out a reaction

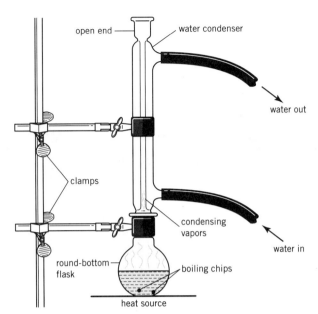

Figure 3.15   A reflux setup.

at high temperature, and keeping it constant, is with a **reflux apparatus** (Fig. 3.15). A reflux apparatus consists of a round-bottom flask attached to a condenser in a vertical position. The round-bottom flask is heated to make the solvent boil. The solvent boils at a particular temperature (Table 2.2) that remains constant as long as the heat supply is on. The solvent vaporizes and ascends through the flask into the condenser. As the vapors reach the cool surface of the condenser, they condense and drip down back to the round-bottom flask. This **refluxing** operation makes possible the heating of reaction mixtures at a constant temperature for long periods of time. No solvent escapes to the atmosphere during the boiling process.

Operation   Choose an appropriate heat supply (heating mantle, sand or water bath) for your system. Clamp the round-bottom flask to a ringstand and add a few pieces of porous material, called **boiling chips**, or a magnetic stir bar. The boiling chips allow the smooth formation of bubbles during boiling, and in doing so they prevent boil-over and mix the liquid at the same time. Attach a water-cooled condenser and clamp it to a ringstand. Attach a length of latex or Tygon tubing to the lower nipple of the condenser and connect it to the cold water faucet. Attach another length of tubing to the upper nipple and run it to a sink. This connection will force water to run up against gravity and makes the cooling efficient (if water flows down, the condenser will never get filled). To facilitate the insertion of the nipples into the tubing, wet the nipples with a few drops of water. Turn on the water slowly and make sure that the outlet tubing stays in the sink. *Do not turn the water on full blast!* A flow rate of a few drops per second is enough to cool down most systems.

Turn on the magnetic stirrer (if you are using one) and then turn on the heat. Make sure that the top of the condenser is open to the atmosphere. Never heat a closed system. As the liquid boils and vapors ascend through the condenser you will observe a reflux ring above which the condenser stays dry. The ring should be positioned no higher than the middle of the condenser. If the ring is too close to the open end of the condenser vapors may escape to the atmosphere. After the desired period of time has elapsed, turn off the source of heat and, if necessary, remove it from contact with the flask. Keep the water running through the condenser a little longer until the system has completely cooled down.

3.5 EVAPORATION

It is sometimes necessary to remove the solvent at the end of a reaction. This can be done by boiling off the solvent, preferably under a partial vacuum. This operation is called **evaporation**. The boiling point of liquids is a function of the applied external pressure (Fig. 3.16). As the applied pressure increases, the boiling point also increases. When the applied pressure is 1 atm = 760 Torr, the liquid boils at its *normal boiling point.* If the pressure is decreased by connecting the system to a vacuum line, the liquid boils at a lower temperature. This reduction in the boiling point results in easier and cleaner evaporations: easier because it takes less heat, and cleaner because decomposition reactions, which are accelerated by higher temperatures, are minimized.

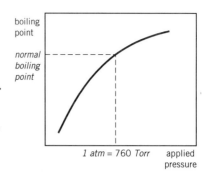

Figure 3.16 Boiling point of a typical liquid as a function of the external pressure.

Operation   The most convenient way of carrying out an evaporation is with the help of a **rotary evaporator** or **rota-vap**. A very simple rota-vap is shown in Figure 3.17. Place the liquid in a round-bottom flask with a ground-glass joint compatible in size with that of the rota-vap. Make sure that the joint is perfectly clean and dry. The rota-vap should be connected to the vacuum line through a vacuum cold trap (Fig. 3.18). The cold trap collects the evaporated solvent. With the vacuum-release valve of the cold trap open (see Fig. 3.18) turn on the vacuum and attach the flask to the neck of the rota-vap. The partial vacuum will keep the flask in place. If available, a clip can be used to secure the flask to the rota-vap (see section 3.7, Fig. 3.22). Turn on the rota-vap motor to make the flask spin. Heat the flask in a water bath and close the screw clamp in the vacuum trap to obtain maximum vacuum. If bubbling in the flask becomes too intense, tilt the whole assembly so that the flask moves up and less of it is immersed in the water bath. The spinning of the flask distributes the liquid in a large area and makes the evaporation fast and smooth. Boiling chips should not be added to the liquid. Once the solvent has evaporated, remove the flask from the water bath by raising or tilting the whole assembly. Stop the motor and, *while holding the flask with your hand*, release the vacuum by opening the screw clamp in the vacuum trap and then turning off the aspirator or vacuum line. Remove the flask from the rota-vap. Remove the solvent from the cold trap. More sophisticated rota-vaps are equipped with a condenser and a receiving flask directly attached to their bodies (Fig. 3.19).

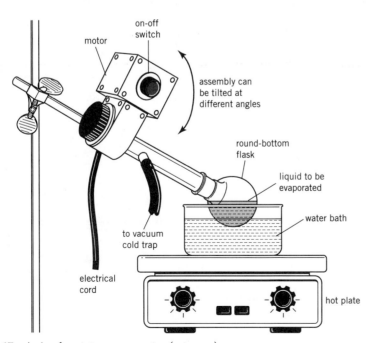

Figure 3.17 A simple rotary evaporator (rota-vap).

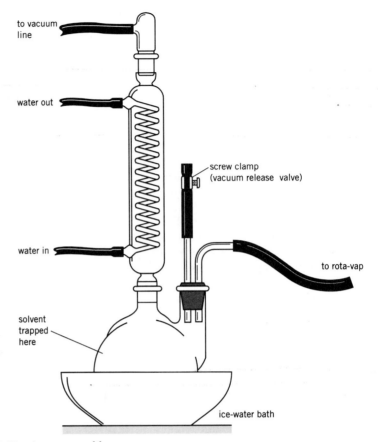

to vacuum line

water out

screw clamp (vacuum release valve)

water in

to rota-vap

solvent trapped here

ice-water bath

Figure 3.18   A vacuum cold trap.

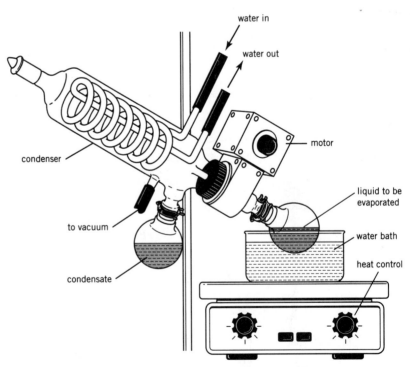

water in

water out

condenser

motor

to vacuum

liquid to be evaporated

water bath

heat control

condensate

Figure 3.19   A rotatory evaporator with condenser and receiving flask.

3.6   CENTRIFUGATION

Centrifugation is a convenient method for separating two liquid phases or a finely divided solid from a liquid. It is used in the organic chemistry laboratory to carry out microscale recrystallizations (Unit 4) and to break emulsions after liquid–liquid extractions (Unit 5). Simple centrifuges have an on-off switch and work at constant speed, usually between 2000 and 3500 rpm. Some have a timer that will automatically turn the motor off; more sophisticated ones have a variable speed control.

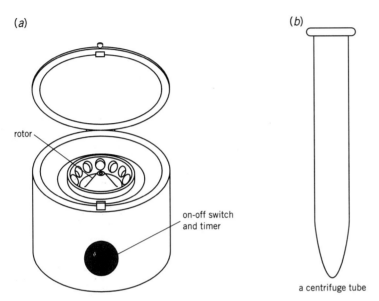

Figure 3.20   *a*) A bench top centrifuge; *b*) a centrifuge tube.

Operation   See Figure 3.20. Place the sample in a centrifuge tube. Centrifuge tubes are made of plastic or thick glass to resist the pull of the centrifugal forces. Place the tube in the centrifuge rotor and counterbalance the rotor with another tube of exactly the same weight and positioned opposite to the first tube. Never counterbalance a test tube by eyeballing its contents. Two tubes that look alike may have very different weights; if used as a pair, breakage would inevitably result. Some rotors have a screw-in lid that secures the rotor in place during centrifugation. Tighten up the rotor lid, if there is one, lower the centrifuge lid, and turn on the power. Let the system spin for the desired length of time. If the centrifuge is properly balanced, it should make a steady and uniform noise after full speed is reached. Rattle is not a noise that you would like to hear from a centrifuge. It is an indication that the centrifuge is not properly balanced. If you hear a rattling noise, stop the centrifuge at once and balance the tubes. Allow the rotor to stop completely before opening the lid. Never stop the rotor with your hands.

3.7   CARING FOR GLASSWARE

Before using any piece of glassware, check for cracks and fissures, especially for star cracks in round-bottom flasks (Fig. 1.2). Do not use glassware if it is cracked because it may suddenly break, especially if it is heated. Always wash your glassware at the end of the experiment and dispose of the chemicals properly according to local regulations. Soap, warm water, and a brush are appropriate to remove most organic chemicals adhered to the glass walls. If this fails, use a little ethanol or a little acetone to rinse the glassware and dispose of the rinses properly. Do not use acids or bases to clean glassware unless you are specifically instructed to do so.

For most applications in the organic chemistry lab, the glassware should be dry. It can be dried in an oven if time permits, but it is usually more convenient to dry the inside with pieces of rolled-up paper towel. This method of drying is enough for most applications. If scrupulously dry glassware is required, as in some water-sensitive reactions, the glassware can be heated with a heating gun (hair dryer) or flamed with a Bunsen burner (making sure that no flammable solvents are around!).

Ground-glass joints deserve a special mention. They come in different sizes specified by two numbers such as ⵣ 24/40, ⵣ 14/20, and ⵣ 14/10. The symbol ⵣ stands for "standard taper" and refers to ground-glass joints. The first number gives the maximum inside diameter (in mm) of the outer joint, or the maximum outside diameter of the inner joint, and the second number gives the length of the joint (also in mm; Fig. 3.21).

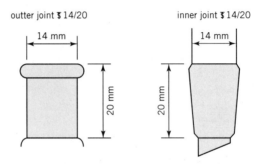

Figure 3.21   Ground-glass joints ⵣ 14/20.

Ground-glass joints should always be thoroughly cleaned before assembling an apparatus. Grease is not usually recommended to lubricate these joints, especially at the microscale level, because organic solvents from the reaction mixtures may come in contact with the lubricant and dissolve it, which would cause contamination.

Take special precautions when working with strongly alkaline solutions such as NaOH and KOH solutions, which are known to fuse ground-glass joints together even after a short exposure. Make sure that the joints are always alkali free. As a further precaution when working with alkaline solutions, lubrication with a small amount of grease (the size of a grain of rice for a ⵣ 24/40 joint) is recommended.

Clips and clamps are often used to keep ground-glass joints together and prevent glassware from falling down. One of the most commonly used clips is the Keck clip, which is made of plastic and comes in different sizes (Fig. 3.22a). The clip fits snugly

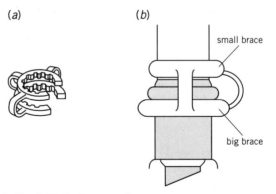

Figure 3.22   a) Keck clip; b) Keck clip around ground-glass joints.

into the joint. The small brace surrounds the inner joint and the big brace the outer joint (Fig. 3.22b). Keck clips are particularly useful to attach round-bottom flasks to rota-vaps (section 3.5) and condensers in distillation apparatus (Unit 6).

When inserting glass tubing, a thermometer, or a glass nipple into a stopper or rubber hose, always hold the glass wrapped up in a cloth or several layers of paper towel to protect your hand in case of breakage. Lubricate the glass with a few drops of water and apply a rotational motion as you insert it. Don't apply strenuous force because the glass may break.

3.8 MEASURING PRESSURE

When performing vacuum distillations (section 6.10) it is important to measure the pressure remaining in the system. To this end a manometer similar to that depicted in Figure 3.23 is used. Figure 3.23a shows a typical manometer with a U-tube filled with mercury. The manometer is enclosed in a plastic chamber to prevent mercury spills if the U-tube breaks. The manometer is connected to the vacuum line; the stopcock allows the isolation of the manometer from the vacuum line. Once the system has been evacuated, the stopcock is slowly opened. The vacuum pulls the mercury up the right arm. The difference in the mercury level between both arms is read on the scale, which is calibrated in millimeters (1 mm Hg = 1 Torr) as shown in Figure 3.23b. This type of manometer is useful to measure pressures in the range 5–120 Torr.

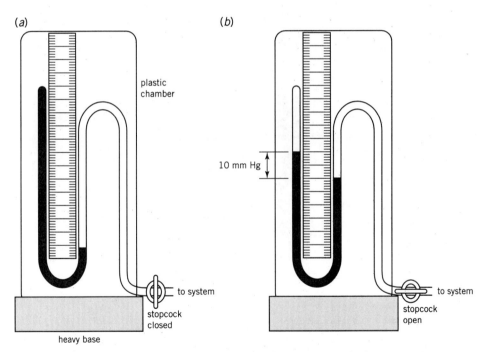

(a)   (b)

plastic chamber

10 mm Hg

to system

stopcock closed

to system

stopcock open

heavy base

Figure 3.23   A manometer; a) isolated from system (stopcock closed); b) connected to system.

The vacuum should never be released with the stopcock open because the mercury will rise in the left arm very quickly and may break the glass. Instead, before the vacuum is released, the manometer should be isolated by closing the stopcock. Once the system has reached atmospheric pressure, the stopcock is slowly opened.

SECTION 2

The Experiments

Recrystallization and Melting Point

4.1 OVERVIEW

One of the major tasks in the organic chemistry laboratory is purifying chemicals. Solid organic compounds can be purified successfully by **recrystallization**. Recrystallization involves several steps:

1. choosing a good recrystallization solvent;
2. dissolving the sample in the minimum amount of boiling solvent;
3. filtering the hot solution to get rid of insoluble impurities (**hot filtration**);
4. cooling down the solution to induce crystallization;
5. separating the solid from the solution (called the **mother liquor**) by vacuum filtration (**cold filtration**);
6. washing the solid on the filter with a small amount of cold solvent to remove traces of the mother liquor;
7. drying the solid to remove traces of solvent.

If the hot solution is clear and transparent without insoluble impurities the hot filtration can be skipped.

The purity of the recrystallized solid is usually checked by melting point determinations. The **melting point** of a substance is the temperature at which the liquid and the solid phases coexist in equilibrium at atmospheric pressure. Experimentally, a **melting point range** rather than a melting point is determined; the melting point range is the temperature interval between the beginning and ending of the melting process. Pure organic solids have a very narrow melting point range, usually 0.5–2°C. For example, a sample of pure cholesterol melts at 148–149°C. Most impurities have a dual effect on the melting point of organic compounds: they depress the melting point and extend the melting point range; for example, a contaminated sample of cholesterol may melt between 100 and 125°C, rather than at 148–149°C. Melting points are also used to *identify* organic compounds by the technique of **mixture-melting point**. The following sections discuss the principles of recrystallization and melting point determination.

4.2 RECRYSTALLIZATION

How Recrystallization Works

The process of recrystallization relies on the fact that the solubility of most organic compounds increases as the temperature of the solvent rises. Figure 4.1 displays the solubility of benzoic acid (see structure below) in water as a function of the temperature. The upper limit of temperature is dictated by the boiling point of the solvent.

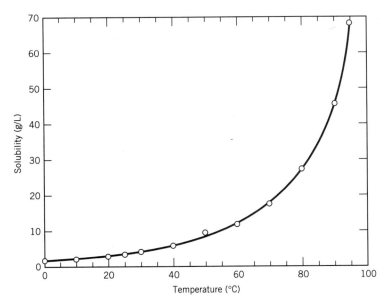

Figure 4.1 Solubility of benzoic acid in water.

To understand the process of recrystallization, consider the following example. Let's say we want to purify 10 g of benzoic acid that is contaminated with 0.2 g of salicylic acid. Let's also assume that the solvent of choice is water. Table 4.1 provides the solubilities of both acids at 20°C and near the boiling point of water.

benzoic acid   salicylic acid

The first question that we should ask is: What volume of boiling water is needed to dissolve the mass of benzoic acid to be purified ($m = 10$ g)? We calculate this volume, V_{hot}, using the solubility value of benzoic acid in hot water, S^H.

$$V_{hot} = \frac{m}{S^H} = \frac{10 \text{ g}}{6.8 \text{ g}/100 \text{ mL}} = 147 \text{ mL} \tag{1}$$

In such a volume of boiling water *all* of the salicylic acid present in the sample (0.2 g) also dissolves, as can be deduced from its solubility value near 100°C.

Table 4.1 Solubility of Benzoic and Salicylic Acids

| Compound | Solubility (g/100 mL) | |
	at 20°C (S^C)	near 100°C (S^H)
benzoic acid	0.29	6.80
salicylic acid	0.22	6.67

The second question is: How much benzoic acid precipitates out when the system is cooled down to 20°C? After cooling down, the amount of benzoic acid that remains in solution, $m_{solution}$, is determined by its solubility in the cold solvent, S^C, and the volume of the solution:

$$m_{solution} = V_{hot} \times S^C = 147 \text{ mL} \times 0.29 \text{ g}/100 \text{ mL} = 0.43 \text{ g} \qquad (2)$$

The amount of benzoic acid that precipitates out, $m_{recovered}$, is the difference between the initial amount, 10 g, and the mass that remains in the cold solution:

$$m_{recovered} = m - m_{solution} = (10 - 0.43) \text{ g} = 9.57 \text{ g} \qquad (3)$$

A third question to be answered is: How much salicylic acid coprecipitates with benzoic acid, if any?

Because the solubility of salicylic acid at 20°C is 0.22 g/100 mL, 147 mL of water can dissolve up to 0.32 g of this acid (Eq. 4):

$$147 \text{ mL} \times 0.22 \text{ g}/100 \text{ mL} = 0.32 \text{ g} \qquad (4)$$

This amount of salicylic acid is larger than the amount present in the sample (0.2 g), and therefore, all of the salicylic acid stays in solution and is separated from the recrystallized benzoic acid in the cold filtration.

Finally, we can calculate the **percentage recovery of the recrystallization**. Because we started with 10 g of benzoic acid, and 9.57 g precipitated from the cold solution, the % Recovery of benzoic acid is 95.7%:

$$\% \text{ Recovery} = \frac{m_{recovered}}{m} \times 100 = \frac{9.57 \text{ g}}{10 \text{ g}} \times 100 = 95.7\% \qquad (5)$$

Substituting Equations 3, 2, and 1 into Equation 5, in that order, the expression of % Recovery as a function of the solubilities is obtained:

$$\% \text{ Recovery} = \frac{S^H - S^C}{S^H} \times 100 = \left(1 - \frac{S^C}{S^H}\right) \times 100 \qquad (6)$$

As can be observed from Equation 6, the % Recovery of a recrystallization is independent of the total amount of sample to be purified; it depends only on the solubilities in the cold and hot solvent. The smaller the ratio S^C/S^H or, in other words, the higher the solubility in the hot solvent and the lower the solubility in the cold solvent, the larger the % Recovery. It is important to note that the % Recovery calculated using Equation 6 is theoretical rather than experimental, because the losses of product associated with handling the solid and the solutions have not been considered. The recrystallization of the sample of benzoic acid is illustrated in Figure 4.2.

The example above is useful for understanding how recrystallization works, but it does not provide a realistic picture of how to approach recrystallization in the laboratory. Most often, the solubilities of the compounds to be purified are unknown

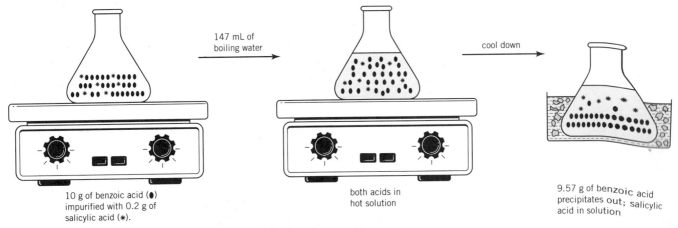

147 mL of boiling water

cool down

10 g of benzoic acid (●) impurified with 0.2 g of salicylic acid (✱).

both acids in hot solution

9.57 g of benzoic acid precipitates out; salicylic acid in solution

Figure 4.2 Recrystallization of benzoic acid impurified with salicylic acid.

and, thus, it is not possible to calculate how much solvent is needed to dissolve the sample. The nature and amount of impurities are also often unknown. Despite this lack of information, recrystallization is still one of the most powerful techniques to purify solid compounds. The art of recrystallization is rather empirical; its success depends strongly on experimentation as opposed to prediction. The following discussion is a detailed description of the steps involved in recrystallization.

Recrystallization Steps

1. Selection of the Solvent   Before beginning a recrystallization you must select an appropriate solvent. Knowing the structure of the compound may help you choose a good solvent. However, solvent selection is an experimental problem solved by trial and error. Let's examine the criteria for choosing a good solvent.

> 1.   *The compound to purify should be very soluble in the hot solvent and have very limited solubility in the cold solvent.*

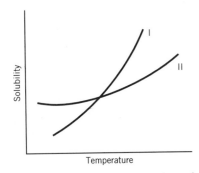

Figure 4.3 Solubility of hypothetical compound A in solvents I and II as a function of temperature.

Figure 4.3 demonstrates this point. The solubility of compound A in solvent I is more sensitive to temperature than in solvent II. Solvent I is a better recrystallization solvent than solvent II because it yields a larger percentage recovery of sample upon cooling. As already shown in Equation 6, the % Recovery depends on the ratio of the solubilities in the hot and cold solvent S^C/S^H; when this ratio is small due to a limited solubility in the cold solvent or a high solubility in the hot solvent, the % Recovery is large. Because information on the variation of solubility with temperature is usually unavailable, experimentation is the way to choose a solvent with good solubility behavior.

Solvents such as water, methanol, ethanol, toluene, and hexane are good choices to begin the search. It is not trivial to predict which solvent would make a good recrystallization solvent. The rule of thumb "like-dissolves-like" can help guide your search. Polar compounds are more likely to be dissolved by polar solvents and nonpolar compounds by nonpolar solvents. For a wider selection of useful solvents see Table 2.2.

Small amounts (30–50 mg) of the sample are placed in different test tubes. Each tube is labeled with the name of the solvent to be tested; about 0.5 mL of the cold solvent is added; if the sample dissolves completely, the solubility in the cold solvent is probably too high and a new solvent should be tried. If the sample does not dissolve in the cold solvent, the test tube is then heated to reach the boiling point while stirring with a glass rod. If the sample does not completely dissolve at this point, more boiling solvent is added drop-wise while the tube is immersed in a sand bath to keep up the boiling. More solvent is added until all the solid dissolves. If it takes more than 3 mL to dissolve the sample in the hot solvent, the solubility in this solvent is probably too low to make it a good recrystallization solvent. A new recrystallization solvent should be tried. If it takes less than 3 mL to dissolve the sample in the boiling solvent, then the sample is cooled down in an ice-water bath while the walls of the tube are gently scratched with the glass rod to help crystallization. If the sample does not crystallize back, a new solvent should be tried. The best solvent is the one that dissolves the sample (at the boiling point) in the smallest volume and affords the largest amount of solid after cooling down. When no single solvent seems to work, mixtures of solvents can be used as discussed in section 4.4.

> 2. *Ideally, the impurities should either be soluble in the cold solvent or very insoluble in the hot solvent.*

Very soluble impurities are removed easily during the cold filtration. They stay in solution while the solid is collected on the filter. On the other hand, impurities scarcely soluble in the hot solvent are separated during the hot filtration; they are retained on the filter. Figure 4.4 shows the solubility as a function of temperature for compound A and two hypothetical impurities (X and Y). Impurity X can be removed in the hot filtration because of its low solubility, while impurity Y, displaying a high solubility, can be separated in the cold filtration.

Sometimes, impurities with similar solubility behavior to the compound to be purified can be removed from the sample provided they are present in small amounts. For example, salicylic and benzoic acids have similar solubilities in water (Table 4.1). However, as you may recall, a sample of benzoic acid can be purified successfully by recrystallization from water, as long as the amount of salicylic acid is small enough to stay completely in solution upon cooling.

Since the nature and amount of impurities are often unknown, the condition that the impurities should be either very soluble or very insoluble is difficult to test beforehand and must be checked by the end result; in other words, the success of the recrystallization is assessed by comparing the purity of the final product with that of the starting material. An easy way to do this is by melting point determinations.

> 3. *The boiling point of the solvent should be in the range 50–120°C.*

Solvents with boiling points lower than 50°C are too volatile to be useful for recrystallization. They evaporate very quickly at room temperature and are difficult to handle, especially during filtration. A solvent like diethyl ether, b.p. 35°C, is often avoided in recrystallization. Not only is it a highly volatile liquid, but it is also very flammable and has a tendency to form explosive peroxides.

On the other hand, solvents with high boiling points are difficult to remove from the recrystallized product. The last step of a recrystallization involves removing traces of solvent. If the boiling point is too high, evaporation of the solvent at room temperature is difficult. The evaporation can be sped up by subjecting the solid to high temperatures, but this may result in partial decomposition of the product.

> 4. *The boiling point of the solvent should be lower than the melting point of the product.*

Although this condition is not necessary, it is highly recommended. If the sample melts at a temperature lower than the boiling point of the solvent, the solid may melt during the heating process, before the solvent boils. The melted product separates as oily droplets that are difficult to get into solution.

> 5. *The solvent should not react chemically with the product to be purified.*

Although most organic solvents are quite inert, some display reactivity toward certain chemicals. For example, do not use acetone for recrystallizing amines because

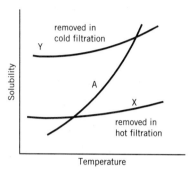

Figure 4.4 Solubility of compound A and impurities X and Y as a function of temperature.

condensation products may result. Avoid ethyl acetate in recrystallizing alcohols especially if you suspect traces of acid present in the sample. Do not use acetic acid with acid labile substances.

> 6. *Given several choices, the solvent with the lowest toxicity should be chosen.*

This criterion was often overlooked in the past. Solvents such as benzene, chloroform, carbon tetrachloride, dioxane, and pyridine, which are suspected or proven human carcinogens or mutagens, should be avoided as much as possible. When they cannot be replaced with other solvents, they should be handled under a fume hood with good ventilation. For most applications, toluene is a good substitute for benzene. Ethyl acetate and *t*-butyl methyl ether can sometimes replace dioxane. The permissible exposure limit (P.E.L.) of airborne solvents for workers exposed 8 hours daily, as given in Table 2.2, can be used as a rough guide to estimate the toxicity of the solvent. Solvents with low P.E.L. numbers should be avoided whenever possible.

2. Dissolving the Sample Once a good recrystallization solvent has been chosen, the bulk of the sample is dissolved in an Erlenmeyer flask. A beaker should never be used to carry out a recrystallization because the solvent evaporates too quickly through its large mouth; an Erlenmeyer flask, because of its conical shape, provides a glass surface where the ascending vapors of the hot solvent condense and fall back into the flask, minimizing the losses due to evaporation.

Small portions of hot solvent are added and the suspension of solid and solvent is brought to a boil by heating on a hot plate. A glass rod is used to stir the system and avoid boil-over (Fig. 4.5). Also, small pieces of porous material (**boiling chips**) are added while the liquid is still cold. The porous material favors the formation of bubbles and prevents boil-over by making the boiling smooth. More hot solvent is added in small portions until all the solid dissolves. If part of the solid does not seem to dissolve even after more hot solvent has been added, it is most likely due to the presence of insoluble impurities. The addition of the solvent should be stopped at this point; further addition of solvent will decrease the yield.

When very volatile solvents are used, it is recommended to use a 10 to 20% excess of solvent after the sample is dissolved; this extra volume would make up for evaporation during the filtration. If there is insoluble material to be filtered, a 10 to 20% excess of hot solvent is also recommended to avoid premature crystallization of the sample during the hot filtration.

3. Hot Filtration Insoluble impurities present in the hot solution (and any boiling chips added) are removed in this step (Fig. 4.6). Hot filtration is usually performed with a short-stem funnel (or stemless funnel, if available) and fluted filter paper (Fig. 3.7); a small piece of wire or a paper clip on the mouth of the Erlenmeyer flask allows proper venting of the system (see Fig. 4.6*a* and Fig. 3.8). The hot filtration should be carried out very carefully to avoid crystallization of the solid in the filter. To this end, it is convenient to preheat the funnel with boiling solvent.

A small portion of fresh recrystallization solvent is placed in an Erlenmeyer flask (where the filtrate from the hot filtration will be collected). One or two boiling chips are added and the funnel and fluted filter paper are placed on top. (If a stemless funnel is used, Fig. 4.6*b*, a beaker is preferable to an Erlenmeyer flask because it offers a better support for the funnel.) The whole assembly is placed on a hot plate and the funnel is heated by the ascending solvent vapors. The solvent is rapidly removed from the Erlenmeyer flask (but not the boiling chips) and the hot solution is filtered by pouring a small aliquot at a time. In the meantime, both Erlenmeyer flasks, the one that receives the filtrate and the one that contains the sample, should be kept on the hot plate. If despite all these precautions substantial crystallization occurs during the hot filtration, the filtration should be stopped, the crystals transferred back to the original Erlenmeyer flask, more solvent added if necessary to redissolve the sample, and the solution filtered again using a new piece of filter paper.

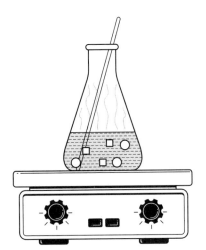

Figure 4.5 Dissolving the sample in the hot solvent.

(a)

(b)

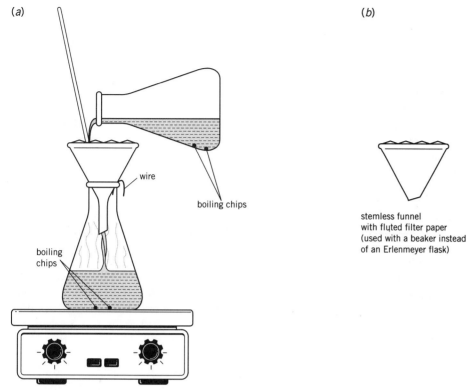

stemless funnel
with fluted filter paper
(used with a beaker instead
of an Erlenmeyer flask)

Figure 4.6 *a*) Hot filtration; *b*) stemless funnel.

Filter papers of fine porosity should be avoided during hot filtration because they result in very slow filtration, which leads to premature crystallization. When the particle size of the impurity is such that it either passes through the filter paper or clogs its pores, the use of diatomaceous earth or filter aid is recommended (discussed in section 4.3). The hot filtration is unavoidable when decolorizing charcoal is used to remove colored impurities (section 4.3).

4. Cooling Down   After the solution has been hot-filtered it is left to crystallize at room temperature first, and then placed in a cold bath (Fig. 4.7). If crystallization is too rapid, as happens when the hot solution is directly placed in a cold bath, small crystals with a large surface area form. These crystals can **adsorb** large amounts of impurities and should be avoided. (**Adsorption** is strictly a surface process that should not be confused with **absorption**, which implies penetration of a gas or a liquid into a solid. Adsorption is treated in Unit 8.) On the other hand, very slow crystallization results in the formation of big crystals that sometimes may occlude impurities. An intermediate rate of crystallization is recommended; to this end, the hot solution is allowed to cool down on the bench top and then in an ice-water bath. An ice-water bath should not be used with solvents such as acetic acid, cyclohexane, and dimethylsufoxide, which have freezing points between 20 and 0°C and solidify upon cooling.

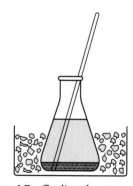

Figure 4.7   Cooling down.

Sometimes the compound fails to crystallize even though the solution is saturated. In such cases it is usually helpful to gently scratch the walls of the flask with a glass rod with a vertical motion; this process is believed to loosen small particles of glass that act as nucleation points where crystals grow. Another technique often used to induce crystallization is called **seeding**; it consists of adding a few crystals of the compound under consideration, which act as true nucleation particles. Crystals to be used as seeds can be obtained by evaporating a few drops of the solution on a watch-glass while rubbing it at the same time with a glass rod.

Some compounds, especially those with very low melting points, sometimes do not crystallize but rather separate as oils. Oils can be solidified by continuously stirring on an ice-water bath with a glass rod while the walls of the flask are frequently scratched. If this fails to crystallize the oil, the cold solution should be heated again with an extra 10–20% volume of solvent and then cooled down. If this addition of solvent fails to prevent oiling, a very fast cooling may be tried. The hot solution is rapidly cooled in an ice-water bath while stirring and scratching with the help of a glass rod. If this approach fails to induce crystallization, adding charcoal (section 4.3) and performing a hot filtration may help eliminate some of the impurities that prevent crystallization. Also, letting the oil and solution stand undisturbed in the refrigerator for several hours (or even days) may eventually result in the slow crystallization of the compound. If all these methods fail, a different recrystallization solvent should be used. It should be noted that solidified oils are amorphous (noncrystalline) solids that adsorb and occlude impurities very easily.

5. Cold Filtration   Once the compound has completely precipitated out of solution, the suspension is vacuum-filtered using a Buchner or Hirsch funnel depending on the total amount of solid. An assembly similar to that illustrated in Figure 4.8a and Figure 3.12 is used. A disk of filter paper that *covers the holes of the funnel without touching the walls* is positioned on the funnel (Fig. 3.13), the vacuum is turned on, and the filter paper is wetted with a small portion of the cold solvent to adhere it to the funnel and prevent it from moving when the bulk of the solution is added. The crystals are then collected by pouring the cooled suspension through the funnel. The filtrate, called the **mother liquor**, is saturated with the compound of interest and contains the soluble impurities. The mother liquor is sometimes saved to recover the solid in solution by evaporation of the solvent, to increase the yield of the overall process. The impure solid so obtained is subjected to a second recrystallization to afford a **second crop** of material. Usually, the purity of the second crop is less than that of the first crop.

6. Washing and Drying the Solid   The solid on the filter paper, called the **cake**, contains traces of the mother liquor with soluble impurities; all traces of mother liquor should be eliminated from the cake to increase the purity of the final product. To this end, a small portion of fresh cold solvent is added to the funnel with the vacuum off. The cake is carefully stirred with a glass rod or spatula (without moving the filter paper) and the vacuum is turned on again. With the aid of a spatula the cake is pressed against the filter to remove traces of the solvent (Fig. 4.8b). The drying of the solid starts at this point. Air forced to pass through the solid by the effect of vacuum

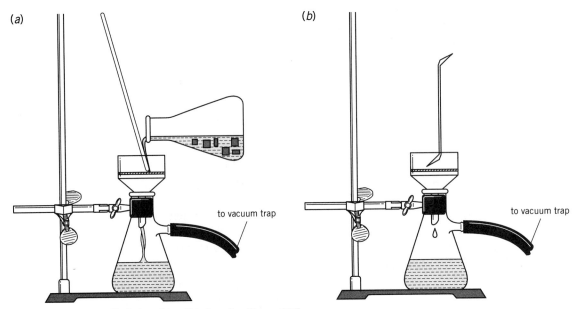

(a)   (b)

to vacuum trap   to vacuum trap

Figure 4.8   *a*) Cold filtration; *b*) drying the solid. See also Figure 3.12.

accelerates the evaporation of the solvent. If time permits, the solid can be left on the filter with the vacuum on for several hours; to avoid the deposition of dust it should be covered with a piece of paper perforated with a few small holes. After the initial drying on the filter, the cake is transferred to a plate made of porous material, to absorb the remaining solvent. A spatula is used to press the solid against the absorbing material. By repeatedly spreading and scraping the solid with the spatula, the porous plate absorbs the remaining solvent. The solid, on a piece of filter paper or on a watch glass, can also be dried in an oven at a temperature below its melting point or in a desiccator. (A desiccator is a glass chamber that can be closed tightly and contains a moisture-absorbing material such as silica.)

4.3   DECOLORIZING CHARCOAL

Very often solid organic chemicals contain small amounts of colored impurities. These impurities either arise from oxidation and decomposition reactions due to inappropriate storage or they consist of byproducts from the synthesis or isolation. These colored contaminants are usually very polar because they have double or triple bonds and chromophore groups (chromophore, from Greek; *chroma*, color; *phoros*, producer) such as $-NO_2$; $-CN$; $-N=N-$, and so on. They can be removed by adding activated charcoal to the hot solution. **Activated charcoal**, also known by the trade name Norit (registered mark of Norit Americas, Inc.), has a very large surface area on which polar compounds are selectively **adsorbed**. Because the compound to be purified can also be adsorbed on the charcoal particles, the amount of carbon should not be excessive. About 10–20 mg of charcoal per gram of sample is enough for most purposes.

The carbon is added to the hot (not boiling) solution of the sample. If the charcoal is added to the boiling solution, the tiny particles of carbon will cause effervescence and undesirable boil-over. After adding charcoal the solution is brought to a gentle boil with continuous stirring. The chemist should be careful because charcoal suspensions have a great tendency to bump. The hot suspension is filtered by gravity filtration by using a preheated short-stem funnel equipped with fluted paper. This operation should be carried out carefully in order to avoid premature crystallization on the filter.

Charcoal should not be separated by ordinary vacuum filtration because tiny carbon particles may pass through the paper pores drawn by the vacuum. As described in the previous paragraph, a gravity filtration gets rid of charcoal effectively. An alternative method, however, uses vacuum filtration and a **filter aid**. A filter aid is a material that prevents charcoal from passing through the paper. The most commonly used filter aid is **diatomaceous earth**, known as Celite (registered mark of the Celite Corp.).

When a filter aid is used, the charcoal suspension is prepared as described above and brought to a boil. In the meantime, fresh recrystallization solvent is added to a small amount of filter aid contained in an Erlenmeyer flask. The suspension of solvent and filter aid is brought to a boil over a hot plate with *continuous* stirring, and then poured over the Buchner funnel (with a disk of filter paper) on a vacuum filtration apparatus (Fig. 8a). The filtrate should be clear and the filter aid should form a thin and uniform layer over the filter paper. The filtrate is discarded and the filtration apparatus quickly reassembled. The charcoal suspension is vacuum-filtered using the Buchner funnel with the layer of filter aid. It is important not to let the filter aid dry and crack on the funnel because charcoal particles may pass through the cracks. About 0.5 g of filter aid is enough for a Buchner funnel 4 cm in diameter. Filter aid is not only recommended to remove charcoal by vacuum filtration, it is also useful to filter any suspension with very fine particles that clog the filter paper.

4.4 RECRYSTALLIZATION FROM MIXED SOLVENTS

Often the compound to be purified is such that no single solvent can be found to carry out the recrystallization successfully. Some compounds have such a solubility behavior that most solvents either dissolve them very well at low and high temperatures or do not dissolve them at all. In such cases mixtures of solvents should be used.

The sample is first dissolved using the minimum amount of a boiling solvent in which it is readily soluble (**good solvent**). Next, while the solution is kept near boiling with continuous stirring, the solvent in which the sample is sparingly soluble (**bad solvent**) is added drop-wise; this addition results in the formation of cloudiness where the drops of the bad solvent hit the solution (the compound precipitates), but the cloudiness rapidly disappears as the bulk of the solution is stirred; the drop-wise addition of the bad solvent is continued until the cloudiness persists. At this point, a few drops of hot good solvent are added just to make the cloudiness disappear in the hot solution. The solution is removed from the heat source, hot-filtered if necessary, and allowed to crystallize at low temperature. It is vacuum-filtered and washed on the filter using a cold aliquot of the mixture of the solvents. This mixture should have the same proportion of good and bad solvents as used in the recrystallization. If such a mixture cannot be reproduced it is advisable to wash the crystals with a cold aliquot of the bad solvent. When the proportion of solvents needed to carry out a recrystallization from a mixed solvent is already known, the solvents can be mixed from the beginning and the mixture used as if it were a single solvent.

Solvents used as a pair for recrystallization should be **miscible** (soluble in each other) and have relatively close boiling points. See the solvent miscibility chart at the beginning of Unit 5. If the boiling points are too different, the composition of the mixture will change during boiling. Good pairs of recrystallization solvents are ethanol-water; methanol-water; acetone-petroleum ether; toluene-petroleum ether; ethanol-toluene, and acetic acid-water. In each of these pairs, the first solvent listed is, in most applications, the good solvent and the second one is the bad one.

4.5 MICROSCALE RECRYSTALLIZATION

Purification of small amounts of solids (20–200 mg) can be successfully achieved by microscale recrystallization. Macro and microscale recrystallizations are based on the same principles; they differ only in some operational aspects. *Avoiding unnecessary transfers is the key to handling small amounts of chemicals.* There is a special type of glassware called the **Craig tube** to carry out microscale recrystallizations (Fig. 4.9*a*).

In the microscale recrystallization, **dissolution** of the sample and **recovery** of the crystals after cooling are performed in the Craig tube. The crystals are recovered by centrifugation instead of filtration. The Craig tube consists of a lower and an upper part. The lower part is used to dissolve the sample. The upper part, which resembles a plunger, is where the solid collects after the liquid is removed by centrifugation. The head of the plunger and the mouth of the tube are made of ground glass.

The sample is dissolved in the hot solvent in the lower section of the Craig tube while stirring constantly by rolling a microspatula between the thumb and the index finger. If no insoluble impurities are present, the system is allowed to cool down and the crystals are recovered by *centrifugation*. To recover the solid, the plunger is inserted in the mouth of the tube. A piece of thin copper wire or a length of thread is tied around the stem of the plunger as shown in Figure 4.9*b*, and the whole assembly is inverted and placed in a centrifuge tube. The system is spun down in a centrifuge for about 3 minutes (section 3.6). During centrifugation, the liquid passes through the ground glass joint and the solid is retained on the head of the plunger (Fig. 4.9*b*). The assembly is removed from the centrifuge and the Craig tube is retrieved with the help of the hanger.

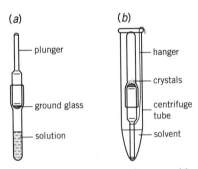

Figure 4.9 The Craig tube assembly for microscale recrystallization.

If insoluble impurities are present in the hot solution, a microscale hot filtration should be performed. A Pasteur pipet in which a small cotton plug has been inserted just above the stem is a convenient filter; Pasteur pipets with short stems are recommended for this step. A slight excess of solvent is added to the solution to be filtered to prevent crystallization in the pipet during the filtration. If decolorizing charcoal is added to remove colored impurities, the use of pelleted activated carbon is recommended (for example, Norit RO 0.8); the large carbon pellets (0.8 mm diameter) are easily removed during the hot filtration. The hot solution is transferred to the Pasteur pipet (with the aid of another Pasteur pipet) and collected in the bottom part of a Craig tube (Fig. 4.10). The solution is concentrated to saturation by boiling in a sand bath to remove the excess of solvent; stirring with a microspatula during the evaporation of the solvent avoids boil-over. When the solid starts to crystallize a few drops of hot solvent are added to completely dissolve the sample. The Craig tube is removed from the heat source and crystallization is allowed to proceed at room temperature first, and later in an ice-water bath. The solid is recovered by centrifugation as already described.

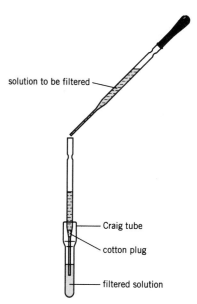

Figure 4.10 Microscale hot filtration.

4.6 MELTING POINT

Most organic compounds have well-defined melting points. The melting point is the temperature at which the solid coexists in equilibrium with the liquid at atmospheric pressure. Under equilibrium conditions, the melting point of a solid is identical to the **freezing point** of the liquid. Pure organic compounds have sharp melting points; the process of melting occurs in a very narrow range of temperature or **melting point range** (0.5–2°C). For example, the melting points reported in the *Merck Index* for benzoic and salicylic acids are shown in Table 4.2.

Table 4.2 **Melting Points**

Compound	m.p. (°C)
benzoic acid	122–123
salicylic acid	157–159

When a pure solid is heated its temperature rises steadily as long as heat is supplied (Fig. 4.11). When the temperature of the system reaches the melting point, droplets of liquid form. During the melting process, the temperature remains constant. The system absorbs heat, which is used to disrupt the crystal structure by loosening the interactions that hold the crystal together. The energy transferred to the system during the melting process is called **heat of fusion**. After the last particle of solid melts, the temperature rises again as long as the heat source is on.

Impurities have a dramatic influence on the melting points of organic compounds. They lower the melting point and widen the melting point range. For example, while pure benzoic acid melts sharply at 122–123°C, a sample of wet benzoic acid (water is the impurity) may melt in the range 110–117°C. A melting point range larger than 3°C is an indication that the sample is either impure or wet. The effects of impurities on melting points are also illustrated in Figure 4.11.

Eutectic Mixtures

To better understand the effects of impurities on the melting point, let us consider the following case. Compounds A and B melt at 130°C and 150°C, respectively. If we mix A with increasing amounts of B, B behaves as an impurity for A and it lowers its

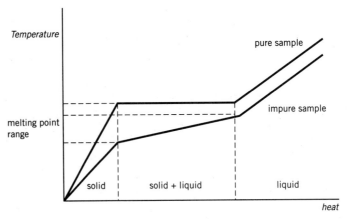

Figure 4.11 Temperature variation during a melting point determination as a function of the heat supply.

melting point. For example, a mixture containing 80% A and 20% B (*x* in Fig. 4.12*a*) melts in the range 110–125°C; whereas another mixture having 60% A and 40% B (*y* in Fig. 4.12*a*) melts in the range 110–120°C. In general, the melting points of mixtures with increasing amounts of B change as shown in the left-side curve of Figure 4.12*a*. On the other hand, increasing amounts of A depress the melting point of B as illustrated on the right-side curve of Figure 4.12*a*. The point where both curves intersect is called the **eutectic point**.

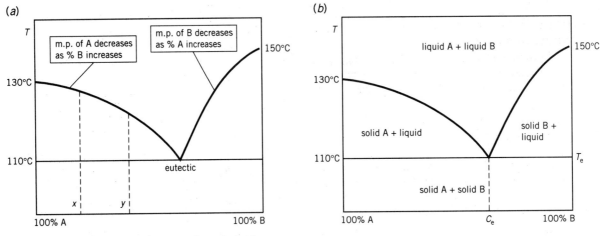

Figure 4.12 Melting point phase diagram of a typical mixture.

The lowest melting point for any mixture of A and B is T_e, the **eutectic temperature**, 110°C in this example. C_e represents the **eutectic composition**, which in this example is approximately 35% A and 65% B. T_e and C_e are characteristic for any given pair of compounds. Figure 4.12*a* has been redrawn in Figure 4.12*b*, which represents a **phase diagram** for mixtures of A and B. Points above the curves represent liquid states of the system and points below T_e correspond to mixtures of solid A + solid B. Points in the circular sections represent solid in equilibrium with liquid.

One thing is remarkable from this phase diagram: when a solid mixture of composition C_e is heated, it melts sharply at T_e. Eutectic mixtures represent a case of impure samples with sharp melting points. Phase diagrams like Figure 4.12 explain the behavior of many binary mixtures such as camphor and naphthalene, benzoic and *trans*-cinnamic acids, and *o*-nitrophenol and *o*-toluidine.

Impurities that do not dissolve in the melted sample *do not affect the melting point.* In general, inorganic salts are not soluble in organic liquids, and thus when a mixture of a solid organic compound and an inorganic salt is heated, the organic compound melts sharply at its melting point and the inorganic material, which usually has a very high melting point, remains as a solid.

Why Do Impurities Lower the Melting Point?

The molecules inside a crystal are not fixed in space; rather they are constantly moving around an equilibrium position. When a pure crystalline compound is heated, the energy absorbed by the system makes the molecules move faster and farther around the equilibrium position. This increase in molecular motion is reflected by an increase in the temperature of the solid. When enough heat is absorbed and the temperature reaches the melting point, the molecules finally break loose and the solid melts.

When two solids are mixed and heated together, the molecules in the crystals try to break loose as heat is absorbed. If the compounds do not dissolve in one another in the liquid state, or in other words, if they have no affinity for one another (as in the case of an organic chemical and an inorganic salt), one compound will not feel the presence of the other and each one will melt independently and at the same melting point as in the pure form. On the other hand, if the two compounds "like" each other and dissolve in one another in the liquid state, as heat is absorbed and molecules try to break loose, the molecules of one compound find the molecules of the other at the interface between the crystals, and together separate as a liquid solution *before the melting point of either compound is reached.* This is the beginning of melting and takes place at a temperature lower than the melting points of the pure compounds. As more heat is absorbed by the system, the temperature increases and more solid dissolves in the liquid solution until all the solid disappears.

Mixture Melting Point

A melting point determination is not only an easy way to check the purity of a solid sample but it can also be used to identify organic compounds. Let us pretend that we have two samples, A and B, with the same melting point of 150–151°C, and we want to determine whether they are the same compound. To answer this question we need to determine the *mixture-melting point*. In other words, we need to determine the melting point of a mixture of A and B. If A and B are identical, the mixture should have the same melting point as A and B alone (150–151°C). In contrast, if A and B are different compounds, one behaves as an impurity to the other and, therefore, any mixture of the two should have a lower melting point and a wider melting range than either substance alone.

Melting Point Determination

Melting points can be easily determined in the laboratory. The **capillary tube method** is the most common and reliable. Small amounts of sample are introduced into a capillary tube with one end sealed. The capillary tube is placed in a heating device near a thermometer. Heat is supplied to the system and the temperature is read when the solid melts. Different kinds of melting point apparatus are available (Fig. 4.13).

The Melt–Temp apparatus is in widespread use. The thermometer is placed in a metallic block, which has room for three capillary tubes. Heat supply is controlled by the resistor knob on top of the apparatus. A small magnifying glass allows the observation of the melting process.

The Thomas–Hoover apparatus consists of an oil bath in a glass vessel that is heated by an electric resistor. A dial on the front controls the heating rate. The capillary tubes are immersed in the oil near the thermometer bulb and the melting process is observed through a magnifying glass that is mounted on the front of the apparatus.

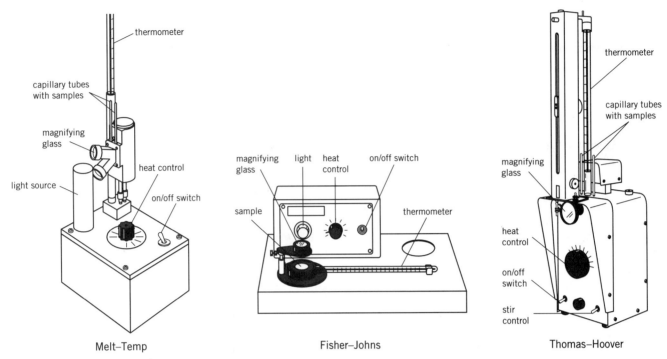

Figure 4.13   Melting point devices.

The Fisher–Johns apparatus is ideal for the determination of melting points when only small amounts of material are available. The sample, a few crystals, is placed between two microscope cover slips and observed to melt through the magnifying glass mounted on top of the hot block. The temperature is read on a thermometer inserted horizontally into the block.

For a precise determination of melting points several factors should be considered. The capillary tube should contain no more than 1–2 mm of sample. Large samples result in misleadingly wide melting point ranges. Heat should be applied slowly. Heating the system too rapidly may cause the thermometer to record a temperature different from that of the sample. In reading the temperature, the eyes should always be in the same plane as the top of the mercury column to avoid parallax error. If a sample vaporizes from the solid state during heating (this process is called sublimation and is discussed is section 4.7) the melting point should be determined using a capillary tube that has both ends sealed. If a compound decomposes with heat, the melting point is usually lower and unsharp because the decomposition products act as impurities to the sample. This problem can be minimized by preheating the apparatus without the sample to a temperature 20–15°C below the actual melting point, and then introducing the capillary into the heating block and continuing heating

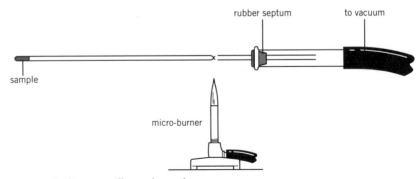

Figure 4.14   Sealing a capillary tube under vacuum.

at a moderate rate until the sample melts. Because decomposition of samples with heat usually involves oxidation by atmospheric oxygen, a more precise melting point of solids that decompose during melting can be obtained using an evacuated capillary tube. The capillary tube is filled with the solid in the customary fashion and the open end connected to a vacuum line through a rubber septum. The capillary tube is sealed with the aid of a micro-burner (Fig. 4.14).

4.7 SUBLIMATION

Sublimation involves purifying solids with relatively high vapor pressure by direct evaporation and condensation without passing through the liquid state. This process is very useful to purify solids contaminated with nonvolatile impurities. Sublimation is simpler than recrystallization because it does not require the use of solvents; however, its applications are limited because it can be used only with volatile solids.

Figure 4.15 shows a phase diagram for a hypothetical compound. Each point on the diagram corresponds to a state of the system characterized by a temperature (T) and applied pressure (P). Line AO represents equilibrium between solid and vapor and gives the sublimation temperature of the solid at different applied pressures. Line OB represents the boiling point of the liquid as a function of the external pressure. Line OC shows the change in the melting point with the applied pressure. Any point (T,P) that falls in the vapor region (right of line AOB) describes a system that exists in the vapor state. If the point (T,P) is in the liquid region (above COB), it represents a system in its liquid state. If the point (T,P) is in the solid region (left of AOC), the system is in the solid state. The points on the lines AO, OB, and OC describe solid-vapor, liquid-vapor, and solid-liquid equilibria.

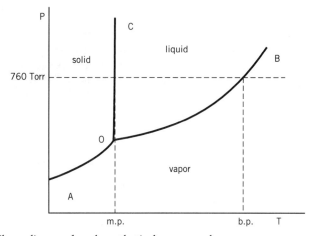

Figure 4.15   Phase diagram for a hypothetical compound.

Some solids have relatively high vapor pressures at temperatures below their melting points and they can be sublimed before they melt. The sublimation temperature is the temperature at which the vapor pressure of the solid equals the applied pressure. Only a few organic solids reach a vapor pressure of 760 mm Hg (760 Torr) before they melt. However, many organic chemicals have high vapor pressure near their melting points, which makes sublimation possible; technically, these compounds do not sublime but rather evaporate. Caffeine, benzoic acid, and naphthalene are a few examples of compounds that can be purified by sublimation at atmospheric pressure in the vicinity of their melting points. Other compounds with much lower vapor pressure can be sublimed only at reduced pressure. Figure 4.15 shows that reducing the applied pressure decreases the sublimation temperature (line AO).

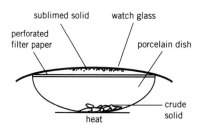

Figure 4.16   Atmospheric-pressure sublimator.

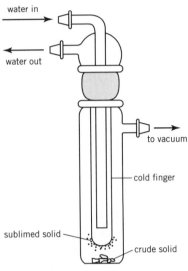

Figure 4.17 Reduced-pressure sublimator.

Sublimation at normal pressure can be performed in a porcelain dish covered with a watch glass with the convex side up (Fig. 4.16); a piece of filter paper with a number of small holes should be placed between the watch glass and the porcelain dish. The filter paper prevents the sublimed solid from falling back into the dish. This type of sublimator is useful for purifying 1–20 g of solid. For small samples (100–500 mg) a sublimator made with half of a Petri dish covered by a disk of filter paper and a beaker is very efficient (Fig. 5.24, Exp. 5). Figure 4.17 shows an apparatus for sublimation under reduced pressure. This sublimator has a water circulating system that cools the surface where the solid deposits (cold finger). The sublimator is heated (hot plate, sand or oil bath) at a temperature about 30°C below the melting point of the solid. If necessary, the sublimator shown in Figure 4.17 can be connected to a vacuum line. The crystals are collected on the cold surface.

Sublimation of water (ice) is a common operation to remove water from samples that cannot be heated at high temperatures (for example, sugars, proteins, and amino acids). This process, called **freeze drying**, has many applications in the food industry such as the production of instant tea and coffee. The water solution is frozen and water is sublimed at very low pressure (10^{-3} mm Hg), leaving behind the nonvolatile solid materials.

BIBLIOGRAPHY

1. Vogel's Textbook of Practical Organic Chemistry. A.I. Vogel, B.S. Furniss, A.J. Hannaford, P.W.G. Smith, and A.R. Tatchell. 5th ed. Longman, Harlow, UK. 1989.

2. Micro and Semimicro Methods, N.D. Cheronis, in Technique of Organic Chemistry, Vol. VI. A. Weissberger, ed. Interscience, New York, 1954.

3. Determination of Melting and Freezing Temperatures. E.L. Skau, J.C. Arthur, Jr., and H. Wakeham, in Technique of Organic Chemistry, Vol. I, Part I, A. Weissberger ed. Interscience, New York, 1960.

EXERCISES

Problem 1

The figure shows the solubility of a compound A in three different solvents (I–III) as a function of temperature. Which solvent would you choose to recrystallize A? Why?

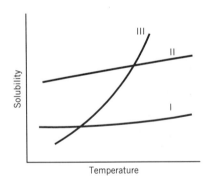

Problem 2

The table below provides the solubility values of an organic compound in three different solvents:

Solvent	at 10°C	at b.p. of solvent
toluene	0.5 g/100 mL	10.0 g/100 mL
cyclohexane	0.3 g/100 mL	0.7 g/100 mL
ethyl acetate	3.0 g/100 mL	15.0 g/100 mL

(a) Calculate the % Recovery upon recrystallization from each solvent disregarding experimental losses.

(b) Which solvent would you choose to carry out the recrystallization?

Problem 3

The figure below shows the solubilities (S) of compounds A, B, and C in different solvents as a function of temperature (T). Choose the best solvent (s) to recrystallize:

(a) A with impurities B and C

(b) B with impurities A and C

(c) C with impurities A and B

(d) B with impurity C

Briefly justify your answer.

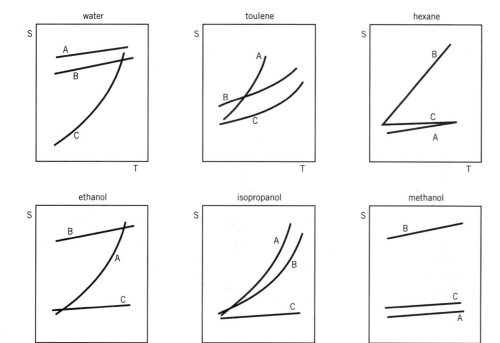

Problem 4

The table below provides the solubilities of compounds A and B in ethanol.

	Solubility (g/100 mL)	
	at 0°C	at the boiling point
A	0.1	7.0
B	0.9	1.3

You are purifying 10 g of compound A contaminated with 1 g of compound B.

(a) Calculate the minimum volume of ethanol that you would need to recrystallize the sample.

(b) Calculate the % Recovery of A upon recrystallization, disregarding experimental losses.

(c) Is A pure after recrystallization?

(d) Is A pure after recrystallization if A (10 g) is contaminated with 1.5 g of B? What is the % of A in the recrystallized product?

Problem 5

A 7-g sample of benzoic acid was carefully recrystallized using the minimum amount of boiling water (103 mL). Upon crystallization at 20°C, 95.7% of benzoic acid was recovered. Calculate the solubility of benzoic acid in water at 100°C and 20°C.

Problem 6

Circle the correct option:

Compound B is contaminated with charcoal. The solubility of B in ethanol is 20 g/L at 75°C and 0.2 g/L at 0°C. In order to recrystallize 15 g of B from ethanol you will need **75 liters/1 liter/750 mL/75 mL** of solvent. Charcoal will be separated during the **cold/hot filtration**. After cooling **0.15 g/1.5 g/14.85 g** will precipitate. The ideal % Recovery for this recrystallization is **99%/19.8%/97.5%**. The solubility of compound B in water is 100 g/L at 100°C and 33 g/L at 0°C; therefore, water is a **better/worse** recrystallization solvent since the percent recovery from water is **100%/33%/67%**.

Problem 7

A sample of aspirin is contaminated with table salt and sand. Aspirin can be recrystallized from water. In which steps are the salt and sand removed?

Problem 8

You suspect an unknown sample is one of the following compounds (m.p.): urea (133°C), 2,4-dimethylacetanilide (133°C), phenacetin (134°C), 3-chlorobenzamide (134°C). A list of the mixture melting points of the unknown with each one of the possible compounds is in the table. Identify the unknown and briefly explain your reasoning.

Unknown +	m.p. range (°C)
urea	120–130
2,4-dimethylacetanilide	105–120
phenacetin	133–134
3-chlorobenzamide	115–128

Problem 9

Three samples named X, Y, and Z have identical melting points, 148–149°C. When a 2:1 mixture of X and Y is prepared its melting point is 105–130°C. The melting point of a 1:1 mixture of Y and Z is 148–149°C. Discuss the identity of X, Y, and Z. Predict the melting point range of a 2:1 mixture of X and Z.

Problem 10

(*Previous Synton: Problem 2, Unit 1.*)   Whereas ordinary people fall in love over a cup of coffee, Synton and Ester fell in love over a melting point apparatus. They met in their sophomore year in college in the O-chem lab. The anxiety of the first day coupled with their love at first sight made them a little bit sloppier than usual. They were sharing the melting point apparatus when they noticed that their capillaries for mixture-melting point had been mixed up. They had previously determined the melting points of their unknowns, and (what a coincidence!) they both were 186–187°C. Synton remembered preparing two samples by mixing his unknown with succinic acid and with hippuric acid. Ester mixed hers with hippuric acid and camphoric acid. All these acids have m.p. 186–187°C. They decided to take the melting points of the four mixtures anyway, and they found that two mixtures melted at 186–187°C, one mixture at 120–140°C, and the other at 155–170°C. Could Synton and Ester both have had camphoric acid? Could both have had hippuric acid?

Recrystallization of Acetanilide and Urea

In this experiment you will learn to:

1. Perform recrystallization at the macroscale level by purifying acetanilide using water as a solvent and activated charcoal as a decolorizing agent.

2. Select a good recrystallization solvent for urea.

3. Perform recrystallization at the microscale level by purifying a few milligrams of urea.

4. Assess the efficiency of the purification process by determining the melting point before and after recrystallization.

5. Determine the melting point of mixtures. You will study mixtures of benzoic and *trans*-cinnamic acids. Since the two compounds melt sharply when they are in a pure state, you will determine the effect that mixing them has on their melting points.

6. Identify an unknown compound by mixture-melting point.

Check with your instructor to find out which parts of the experiment you will perform. Before you begin this experiment familiarize yourself with the structures below.

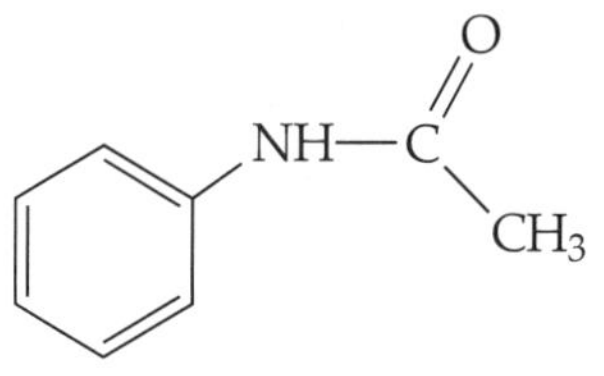

acetanilide

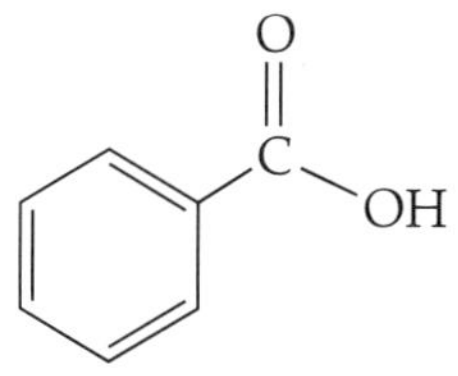

benzoic acid

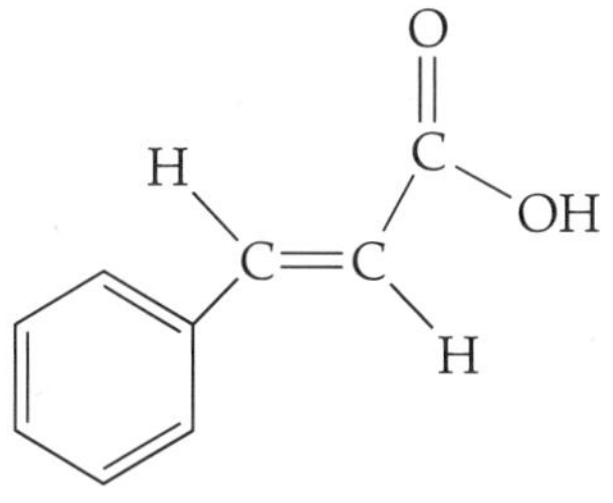

trans-cinnamic acid

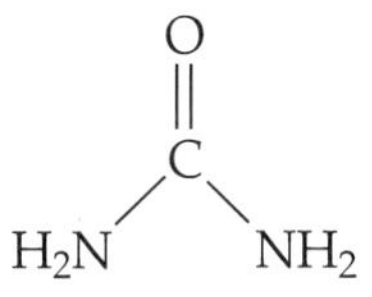

urea

PROCEDURE

E4.1 RECRYSTALLIZATION OF ACETANILIDE

> **Background reading:** 4.1–3; 3.3
> **Estimated time:** 2 hours

Important Note. Recrystallization is one of the most challenging operations in introductory organic chemistry laboratory. It requires organization, attention, and promptness. Read the sections on gravity and vacuum filtration (Unit 3). Have all the glassware required for this operation in readiness before you start the purification.

Obtain a sample of impure acetanilide; save a few milligrams for melting point determination. Review the flowchart below, which summarizes the steps for recrystallizing acetanilide (Fig. 4.18).

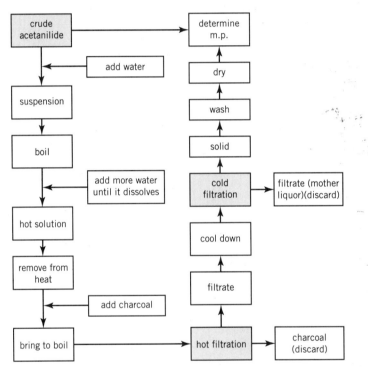

Figure 4.18 Flowchart for the recrystallization of acetanilide.

Dissolving the Sample Place 2 grams of impure acetanilide in a 125-mL Erlenmeyer flask. Add 30 mL of water and one or two boiling chips and bring the suspension to a boil by heating it on a hot plate. Stir the system frequently with a glass rod (with round ends). As soon as boiling begins, add more water drop-wise until all the solid (which may have just turned into oily droplets of melted acetanilide) dissolves (up to 10 more mL of water may be required to completely dissolve the acetanilide). When all the acetanilide dissolves, remove the flask from the hot plate. If crystals easily deposit from solution as it cools, add about 3 mL of water. Now slowly add a dash of activated charcoal (approximately 100 mg; never add charcoal to a solution on the verge of boiling, it will cause boil-over). Place the flask on the hot plate again and bring the system to a second boil. Due to the large surface area of charcoal, the suspension has a tendency to bump; avoid this by stirring while heating. Hot-filter the suspension using the procedure described below.

Hot Filtration After adding charcoal and before bringing the suspension to a second boil, set up the following apparatus. Place one or two boiling chips and approximately 10 mL of water into a 125-mL Erlenmeyer flask. Cut a small piece of copper wire, about an inch in length, bend it into a U and place it over the lip of the flask (see Fig. 3.8). Place a short-stem funnel equipped with a fluted piece of filter paper (see Fig. 3.7; for 5-cm and 6-cm diameter funnels use 9-cm and 12.5-cm diameter filter paper, respectively) on top of the flask, and place the complete apparatus onto the hot plate. When the funnel has been heated by the steam generated by the 10 mL of water, **quickly** discard the remaining water (but keep the boiling chips) and begin the hot filtration by pouring the acetanilide-charcoal suspension into the fluted filter paper with the aid of the glass rod to direct the solution into the funnel (see Fig. 4.6). Use a paper towel or a hand cloth to wrap the flask to avoid burning your hand. Fill the funnel no more than half full at any given time and maintain the temperature of both solutions to prevent premature crystallization in the funnel. To avoid making a hole in the filter, make sure that the glass rod does not touch its center. If substantial crystallization of acetanilide occurs in the filter, stop the hot filtration and transfer as much acetanilide as possible from the filter to the Erlenmeyer flask. Bring the system to a boil again, adding more water, if necessary, to dissolve the acetanilide completely. Filter again using a new piece of fluted filter paper.

Cooling Down After completing the hot filtration, allow the flask to cool to room temperature before placing it in an ice-water bath. If the system cools down too quickly, small crystals form and adsorb a large amount of impurities from the mother liquor.

Cold Filtration After crystallization is complete, collect the crystals by vacuum filtration. Set up the filtration apparatus using a 125-mL filter flask and a Buchner funnel (Fig. 3.12). Make sure that the filter paper covers all the holes on the Buchner funnel without touching the walls (Fig. 3.13). If the filter paper touches the walls of the funnel, small creases are formed, allowing the passage of solid and decreasing the overall yield of the process. If necessary, trim the filter paper with scissors. Position the filter paper on the funnel, then turn the vacuum on and wet the filter paper with 5–10 mL of cold water. This will adhere the paper to the funnel and prevent it from moving during the cold filtration. Discard the water in the filter flask. Filter your cold acetanilide-water mixture by pouring it into the Buchner funnel with the aid of a glass rod (Fig. 4.8a). To help the transfer of the solid acetanilide, gently swirl the contents of the Erlenmeyer flask before pouring the suspension into the funnel.

Washing and Drying the Solid Once you finish filtering the whole solution, and the liquid stops dripping from the funnel, turn off the vacuum and add a few milliliters of ice-cold water to the funnel to wash the crystals. Stir the solid very gently with a glass rod or a spatula without touching the filter paper. Turn the vacuum on and press the crystals with a spatula (tip should be slightly bent) to remove as much water as possible (Fig. 4.8b). Let the solid dry on the filter with the vacuum on for 5–10 minutes. Transfer the solid to a pre-tared watch-glass and spread it in a thin layer. With the help of tweezers, carefully remove any boiling chip from the solid. Dry the solid in an oven at approximately 70°C. Weigh the dried system after it reaches room temperature and calculate the mass of pure acetanilide by difference. Alternatively, spread the solid acetanilide over a piece of dry filter paper on a watch-glass, (pre-tare the watch-glass with the filter paper), cover the acetanilide with another watch-glass, and let it dry it in your drawer until the next section. Calculate the % Recovery of pure acetanilide. Determine the melting points of crude and purified material. Before you determine the melting point of the recrystallized acetanilide, spread a small amount over a porous plate. By spreading and scraping the solid repeatedly with a spatula, the plate absorbs the residual water.

Safety First

- When handling hot Erlenmeyer flasks, use paper towels or a piece of cloth to avoid burning your hand. Do not use clamps as handles. They make top-heavy systems prone to tip over.

E4.2 MICROSCALE RECRYSTALLIZATION OF UREA

Background reading: 4.5; 3.6
Estimated time: 2 hours

Safety First

- Methanol, ethanol, isopropanol, and toluene are flammable solvents.

Choosing the Recrystallization Solvent Place about 30 mg of finely crushed urea in each of five different small test tubes; the test tubes should be **perfectly dry**. Label each tube with the name of the solvent to be tested, for example, water, methanol, ethanol, isopropanol, and toluene. Add about 0.3 mL of the appropriate solvent at room temperature to each tube and observe whether the urea dissolves or not. If the solid dissolves, this solvent is probably not a good recrystallization solvent. If the urea does not dissolve, place the tube in a sand bath and bring the liquid to a boil. Stir the system by rolling a microspatula between your fingers. If the solid does not dissolve, keep adding solvent while boiling and stirring. Observe how much solvent is necessary to dissolve the solid (if the crystals do not dissolve after adding 1 mL of hot solvent, consider this solvent inadequate for the recrystallization of urea). Place the solution in an ice-water bath. Scratch the walls of the tubes with a microspatula and observe the recovery of solid. Choose as the recrystallization solvent the one that, after dissolving the urea when hot, affords the largest amount of solid upon cooling.

Doing the Recrystallization Weigh approximately 100 mg of urea directly in the lower section of a Craig tube. Deliver 3–5 mL of the appropriate solvent in a small test tube. Heat the solvent until it boils in a sand bath. Transfer about 1 mL of the hot solvent into the Craig tube with the aid of a Pasteur pipet and a pluringe (Fig. 3.2). Place the tube in the sand bath while holding it with your fingers. Stir continuously with a microspatula by rolling it between your fingers until the liquid boils. Add more boiling solvent from the test tube until all the solid dissolves. When the dissolution is complete place the Craig tube in a 50 mL beaker and let it cool down undisturbed on the bench top. Finally, cool the system in an ice-water bath. Place the plunger on top of the mouth of the Craig tube. Carefully place the system in a centrifuge tube and invert it (Fig. 4.9). Weigh the whole assembly. Balance the centrifuge by placing another centrifuge tube filled with sand or water of identical total weight in the 180°C position with respect to the hole where the Craig tube is placed. Spin it down for 3 minutes. The mother liquor will be collected in the centrifuge tube and the crystals will be trapped around the mouth of the Craig tube. Remove the Craig tube from the centrifuge. Disassemble it. Scrape the crystalline product from the head of the plunger and the mouth of the tube into a piece of preweighed weighing paper. Let the solid air dry. Weigh the paper with the crystals. Calculate the percent recovery of the process. Transfer the crystals to a clean porous plate for final drying. Determine the melting point.

E4.3 DETERMINATION OF MELTING POINTS

Safety First

- Care of thermometer. Thermometers are very fragile. Mercury is a toxic element and its spillage is difficult to clean up. Never cool a thermometer down by placing it in ice or cold water; the thermal shock may break it. Never shake a thermometer.

Background reading: 4.6
Estimated time: 1.5 hours

Using a spatula, crush about 100 mg (a spatulaful) of the sample on a watch-glass or small beaker. Thrust the crystals into the capillary tube by tapping the open end of the capillary against the watch-glass or beaker (Fig. 4.19). Invert the capillary tube and tap the closed end against the bench top to make the solid descend. If the crystals are stuck in the capillary, apply a rotatory motion to it by gently pulling a file across the capillary while holding it loosely between your thumb and your index finger. Alternatively, drop the capillary tube through a glass tube (about 40 to 60 cm long and 5–8 mm diameter) onto the bench top. Repeat this process until 1–2 mm of well-packed solid accumulates at the end of the capillary (Fig. 4.19). Place the capillary tube inside one of the wells of the melting-point apparatus and turn on the heat. At the beginning the temperature should increase at a rate of 10–15°C per minute, until a temperature approximately 20°C below the expected melting point is reached. At this point reduce the heat rate so that the temperature increases at a rate of 2–3°C per minute. Using this procedure, determine the melting point of acetanilide and urea before and after recrystallization.

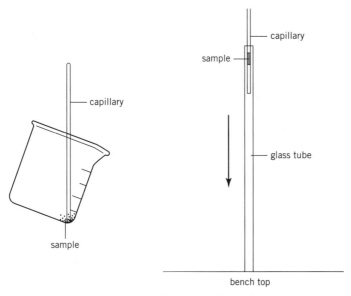

Figure 4.19 Filling the capillary tube for melting point determination.

Melting Point of a Mixture: Benzoic and *trans*-Cinnamic Acids

This part of the experiment illustrates the effect of impurities on the melting point of pure organic chemicals. Benzoic and *trans*-cinnamic acids are two compounds with melting points 121–122°C and 133–134°C, respectively. Determine the melting point of pure benzoic acid, *trans*-cinnamic acid, and a mixture of both as assigned by your instructor. Mixtures with different ratios of benzoic and *trans*-cinnamic acids will be analyzed by the whole class. Prepare approximately 100 mg of a mixture of both acids with a percentage of *trans*-cinnamic acid (20, 40, 50, 60, or 80%, by weight) as assigned by your instructor and determine its melting range. Share your results with the rest of the class. From a plot of melting range versus composition, estimate the eutectic temperature and eutectic composition of the mixture.

Identifying an Unknown by the Mixture-Melting Point Technique

Obtain an unknown sample and write down its identification number. The table below lists the names of the possible unknowns. Determine its approximate melting point by heating the melting point apparatus at a rate of 10°C per minute. In most cases this rough melting point is enough to narrow the possibilities to two compounds from the table below. Obtain a sample of the suspected compounds (about 50 mg) and mix them separately with an equal amount of your unknown. Determine the melting point of each mixture along with that of a new sample of your unknown; do these determinations simultaneously and carefully (at a rate of 2–3°C/minute near the melting point). Watch the process closely, and compare the melting of the three samples; pay attention to differences and similarities during melting. Identify your unknown. If necessary repeat the determination.

Notes

- To **save time**, you can determine up to three melting points simultaneously by using the three capillary wells of the melting point apparatus.

- **Never reuse** a capillary tube to determine the melting point for a second time.

Compound	mp (°C)	Compound	mp (°C)
benzophenone	49–51	*m*-toluic acid	111–113
4′-bromoacetophenone	50–52	acetanilide	113–114
biphenyl	70–71	benzoic acid	121–122
trans-crotonic acid	70–71	*trans*-stilbene	122–124
o-anisic acid	99–101	*m*-bromobenzoic acid	155–157
dimethyl fumarate	101–102	salicylic acid	156–158

EXPERIMENT 4 REPORT

Pre-lab

As part of your pre-lab preparation answer the following questions in your lab book. Be brief and concise.

1. Outline the recrystallization steps.
2. What type of impurity is removed during the hot filtration?
3. What type of impurity is removed during the cold filtration?
4. Enumerate the properties of a good recrystallization solvent.
5. Briefly explain the following terms:
   (a) melting point range
   (b) mixture-melting point
   (c) mother liquor
6. What effects do most impurities have on the melting point of organic compounds?
7. Why should the temperature of the melting point apparatus increase slowly in the vicinity of the melting point?
8. Explain why a second determination of the melting point should not be performed on the resolidified sample used for the first determination.
9. Discuss the validity of the following statements:
   (a) If the melting point of a mixture of A and B is lower than the melting points of A and B, A and B cannot be the same compound.
   (b) If an organic sample has a sharp melting point, it must be a pure single compound.
   (c) Inorganic materials such as carbon (charcoal), sodium chloride, and sand do not depress the melting point of organic compounds.
   (d) The nature of the impurities should be known before attempting a recrystallization.
10. Disregarding experimental losses, calculate the % Recovery of 1 g, 2 g, and 3 g of a compound A after recrystallization if the solubilities of A in the hot and cold solvent are 1 g/100 mL and 0.03 g/100 mL, respectively. Does the % Recovery depend on the total mass to be recrystallized?
11. Make a table showing the physical properties of acetanilide, urea, *trans*-cinnamic acid, benzoic acid, water, ethanol, methanol, isopropanol, and toluene (molecular mass, structure, m.p., b.p., density, solubility in water and organic solvents, flammability [for solvents only], and toxicity/hazards). Use Table 2.2, the *Merck Index*, and/or the *CRC Handbook of Chemistry and Physics*, and/or the *Aldrich Catalog*. Synonym of acetanilide: *N*-phenylacetamide.
12. Make a flowchart similar to that shown in Figure 4.18, for the recrystallization of: a) acetanilide, and b) urea.

Cleaning Up

- Discard used filter papers in the "Recrystallization–Solid Waste" container.
- Flush the mother liquors from the recrystallization of acetanilide down the drain.
- Discard the organic solvents from the recrystallization of urea in the container labeled "Recrystallization–Organic Solvents Waste."
- Dispose of capillaries and Pasteur pipets in the "Recrystallization–Solid Waste" container.
- Turn in the recrystallized products in labeled screw-capped vials.
- Discard unused portions of solid samples in the "Recrystallization–Solid Waste" container.
- Return unused unknowns to your instructor.

13. Make a list of the glassware necessary for the recrystallization of: a) acetanilide, and b) urea.

In-lab

1. Summarize the learning objectives of Experiment 4. Mention the names of the new techniques introduced.

E4.1 Recrystallization of Acetanilide

2. Report the melting points of crude and recrystallized acetanilide. Compare them with the literature value (the one from the table in pre-lab 11). Discuss your results.

3. Calculate the experimental % Recovery of acetanilide after recrystallization. Discuss any source of error.

$$\%\text{Recovery} = (\text{mass pure acetanilide}/\text{mass crude acetanilide}) \times 100$$

E4.2 Recrystallization of Urea

4. Report the solubility behavior of urea in the hot and cold solvents. Use a table format.

5. Which solvent did you choose for the recrystallization? Why?

6. Calculate the experimental % Recovery of purified urea and discuss any source of error.

7. Report the melting points of crude and purified urea. Discuss your results.

E4.3 Melting Point Determination

Mixture Melting Point: Benzoic and trans-Cinnamic Acids

8. Plot the melting point ranges (start and finish points) of benzoic acid, *trans*-cinnamic acid, and their mixtures as a function of the % composition of the mixtures (use the data generated by the entire lab section). Estimate the composition and the melting point of the eutectic mixture.

9. Compare your results with literature values. The eutectic composition of mixtures of benzoic and *trans*-cinnamic acids is reported to be 53% benzoic acid and 47% *trans*-cinnamic acid and the eutectic temperature is 82°C (Ref. 1).

Identifying an Unknown by the Mixture-Melting Point Technique

10. Write the identity number of your unknown.

11. Report the approximate and accurate melting points of your unknown.

12. Give the names of the suspected compounds.

13. Report the mixture-melting points and the identity of your unknown. Discuss your results.

---

1. Determination of Melting and Freezing Temperatures. E.L. Skau, J.C. Arthur, Jr., and H. Wakeham, in *Technique of Organic Chemistry. Physical Methods of Organic Chemistry*, Ed. A. Weissberger, 3rd ed. Vol. I, Part I, Chapter VII, p. 300; Interscience, New York, 1959.   **BIBLIOGRAPHY**

Extraction

5.1 INTRODUCTION

If we were to make home-made Italian dressing by mixing vinegar, oil, and salt we would obtain a two-layer system, "oil and vinegar"; but in which layer is the salt? Our intuition (or taste) might tell us that the salt is in the lower (vinegar) layer. Why? Is there any way to predict the distribution of a compound between two layers of liquids? How can the concentrations in each layer be calculated? In order to answer these questions let's attack the problem of extraction.

In the most general sense of the word, **extraction** is a physical process by which a compound (or a mixture of compounds) is transferred from one phase to another. Every time we make coffee or tea, an extraction process takes place. Water-soluble components in the tea leaves or in the coffee beans are transferred from a solid phase, the leaves or beans, into a liquid phase, the boiling water. This process is known as **solid-liquid extraction**. On the other hand, the oil-vinegar-salt system exemplifies **liquid-liquid extraction**.

Liquid-liquid extraction is a basic operation that should be mastered in the organic chemistry laboratory. By liquid-liquid extraction we can isolate single components from a mixture. For example, if we want to determine the presence of an organic pollutant, such as DDT, in a stream of water, the first step would be its isolation from the water by liquid-liquid extraction with an organic solvent, as we will explain in this section. The physical process that rules liquid-liquid extraction is known as **solvent-solvent partitioning**, or the distribution of solutes between a pair of solvents.

5.2 SOLVENT-SOLVENT PARTITIONING

Solvents such as diethyl ether, *tert*-butyl methyl ether, methylene chloride, and toluene have a very limited solubility in water. The solubility of water in such solvents is also very low. We say that these solvents are **immiscible** with water (see Fig. 5.1 and Table 5.1). For example, if toluene is mixed with water, a two-layer (or two-phase) system is obtained (Fig. 5.2). The upper layer contains the less-dense solvent, which in this case is toluene (density: 0.867 g/mL) and the lower layer contains the denser solvent, water in this case (density: 1.00 g/mL). Each layer is saturated with the other solvent. Saturation values between water and organic solvents are gathered in Table 5.1. The organic solvent layer is called the **organic layer** or **organic phase**, and the water layer, the **aqueous layer** or **aqueous phase** (from Latin: *aqua*, water).

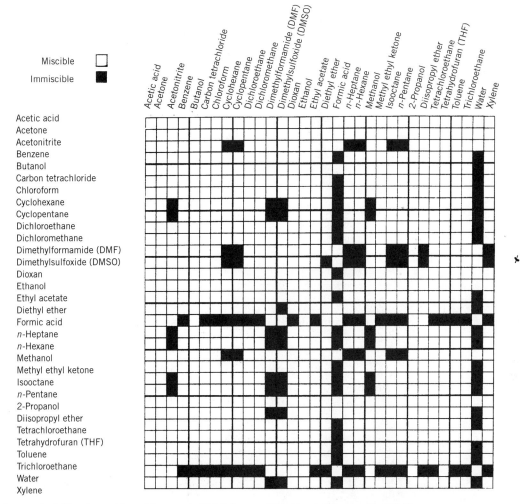

Figure 5.1 Solvent miscibility chart. (From Pharmacia FPLC System Handbook, Pharmacia, 1986.)

Useful Extraction Solvents
(for their physical properties including toxicity, see Unit 2)

toluene	methylene chloride or dichloromethane	chloroform	diethyl ether	*tert*-butyl methyl ether

CH₃ (toluene)
CH₂Cl₂ (methylene chloride or dichloromethane)
CHCl₃ (chloroform) — *hydrophobic i.e. non-polar*
CH₃CH₂—O—CH₂CH₃ (diethyl ether)
CH₃C(CH₃)₂—O—CH₃ (*tert*-butyl methyl ether)

Let's suppose that we add a solute A to a mixture of water and toluene, shake the system thoroughly to attain equilibrium, and allow it to settle. Solute A will be present in both layers but we expect to find it in a larger amount in the solvent for which it has a higher affinity. The distribution of solute A between both solvents is dictated by the **partition coefficient** of A between the solvents. The *partition coefficient is the equilibrium constant for the distribution of a solute between two immiscible layers.* For the system of a solute A distributed between toluene (T) and water (W), the partition coefficient, $K_{T/W}$ is the ratio of the concentrations of A in toluene and water

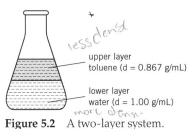

less dense

upper layer toluene (d = 0.867 g/mL)

lower layer water (d = 1.00 g/mL)

more dense

Figure 5.2 A two-layer system.

Table 5.1 Miscibility of Water and Organic Solvents at Room Temperature

Solvent	Solubility of water in solvent (g/100 mL solution)	Solubility of solvent in water (g/100 mL solution)
benzene	0.047	0.062
n-butanol	17.0	6.8
butanone	9.8	16.1
carbon tetrachloride	0.018	0.080
chloroform	0.12	0.74
cyclohexane	0.004	0.01
diethyl ether	1.24	5.6
ethyl acetate	2.92	8.20
n-hexane	0.007	0.001
methylene choloride	0.32	1.6
t-butyl methyl ether	1.1	4.5
n-octanol	4.14	—
toluene	0.046	0.074

at equilibrium (Eq. 2).

$$A_{water} \underset{}{\overset{K_{T/W}}{\rightleftharpoons}} A_{toluene} \qquad (1)$$

where

$$K_{T/W} = \frac{\text{concentration A in toluene}}{\text{concentration A in water}} \qquad (2)$$

The partition coefficient can also be expressed as:

$$K_{W/T} = \frac{\text{concentration A in water}}{\text{concentration A in toluene}} \qquad (3)$$

Both definitions for the partition coefficient (Eqs. 2 and 3), are correct and there is a reciprocal relationship between them (Eq. 4):

$$K_{T/W} = (K_{W/T})^{-1} \qquad (4)$$

Which partition coefficient, $K_{T/W}$ or $K_{W/T}$, we choose to use is not important as long as we carry out the calculations consistently.

The partitioning of A between both layers depends on the affinity of A for toluene and water. If A "likes" toluene better than water, or in other words, if toluene is a better solvent for A than water, the concentration of A in the toluene layer will be larger than the concentration of A in water and, thus, the partition coefficient $K_{T/W} > 1$. On the other hand, if A has higher affinity for water than for toluene, the partition coefficient $K_{T/W}$ will be less than unity. In a few words: *the partition coefficient measures the relative affinity of a given compound for two solvents.*

The partition coefficient is a constant that, in principle, depends neither on the total amount of solute distributed between the two layers, nor on their volumes. The partition coefficient depends only on the temperature. Being a ratio of concentrations, the partition coefficient has no units.

In the first approximation, the partition coefficient *can be estimated as the ratio of the solubilities* of the compound in both solvents. For example, let's suppose that we dissolve 1.1 g of caffeine in 100 mL of water and then add 100 mL of chloroform, a solvent of very limited miscibility with water, and shake the system. The solubility of caffeine in water is approximately 1.8 g/100 mL, while the solubility in chloroform is ten times larger (about 18.0 g/100 mL); therefore, we can expect that most of the caffeine originally in the water layer will be transferred to the organic layer after mixing. The partition coefficient for caffeine between chloroform and water, $K_{C/W}$, can be *estimated* as:

$$K_{C/W} \approx \frac{\text{solubility in chloroform}}{\text{solubility in water}} = \frac{18.0 \text{ g/100 mL}}{1.8 \text{ g/100 mL}} = 10 \qquad (5)$$

We can use the partition coefficient $K_{C/W}$ to calculate the concentrations of caffeine in the water-chloroform system. In our example, the total amount of caffeine distributed between both layers is 1.1 g, and since caffeine prefers chloroform 10 times over water, after equilibrium is attained the amount in the chloroform layer will be 1.0 g and the amount left in the water will be 0.1 g.

$$K_{C/W} = \frac{[\text{caffeine}]_{\text{chloroform}}}{[\text{caffeine}]_{\text{water}}} \approx 10 = \frac{1 \text{ g/100 mL}}{0.1 \text{ g/100 mL}} \qquad (6)$$

In the first approximation, no matter how much caffeine is present in the system it will be always partitioned between these two layers so that the ratio of the concentrations is 10.

Caffeine can be recovered from the organic phase by evaporation of the solvent. Disregarding mechanical loses, the evaporation will yield 1 g of caffeine. The recovery of caffeine from the organic phase, expressed as a percentage of the original amount, is the % Recovery. It measures the efficiency of the extraction (Eq. 7):

$$\% \text{ Recovery} = \frac{\text{amount recovered}}{\text{original amount}} \times 100 = \frac{1 \text{ g}}{1.1 \text{ g}} \times 100 = 91\% \qquad (7)$$

In general, the efficiency of a liquid–liquid extraction depends on the partition coefficient, the volume of solvent used, and the number of extractions. To extract an organic compound from an aqueous phase into an organic solvent, it is desirable to find an organic solvent in which the partition coefficient $K_{\text{org/water}}$ is as high as possible. Increasing the volume of organic solvent also increases the % Recovery as more solute is extracted into the larger organic phase. Performing several consecutive extractions also results in a higher efficiency for the extraction process. We will come back to these ideas in the following sections.

Applications of Liquid–Liquid Extraction

Let's recall the golden rule of solubility: **like-dissolves-like**. Highly polar and ionic compounds are readily soluble in water but have very low solubility in most organic solvents, which are less polar than water (polarity is discussed in section 2.1). On the other hand, organic compounds of medium and low polarity are more soluble in organic solvents than in water. This selective solubility behavior can be advantageously used to separate compounds by liquid-liquid extraction.

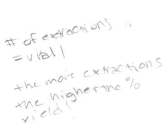

Examples of very polar molecules with higher affinity for water

NaCl

sodium chloride
(a salt)

glucose
(a sugar)

$H_3\overset{+}{N}CH_2COO^-$

glycine
(an amino acid)

Examples of medium and low-polarity molecules with higher affinity for organic solvents

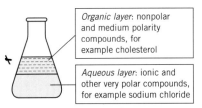

DDT
(low polarity)

cholesterol
(low polarity)

aspirin
(medium polarity)

hydrophobic low solubility w/ H₂O nonpolar

Figure 5.3 Partitioning between an organic phase and an aqueous phase. In this particular example the organic layer is the upper layer.

Organic layer: nonpolar and medium polarity compounds, for example cholesterol

Aqueous layer: ionic and other very polar compounds, for example sodium chloride

Due to their high solubility in water and low solubility in organic solvents, very polar solutes, such as ionic compounds and small sugars, display very low partition coefficients between most organic solvents and water ($K_{org/water} \ll 1$). Therefore, these solutes remain in the water phase even after several extractions with organic solvents. Conversely, organic compounds of medium and low polarity, for example, cholesterol, DDT, and aspirin, have larger partition coefficients ($K_{org/water} > 1$) and, thus, are easily extracted into organic solvents (Fig. 5.3).

In the isolation and synthesis of organic chemicals, as we will see in the experiments ahead, more often than not, the compound of interest is contaminated with ionic compounds, particularly salts such as NaCl, KCl, or Na_2SO_4. Extraction using water and organic solvents such as ethyl ether, methylene chloride, or toluene separates such mixtures very effectively. The salts and other very polar compounds go to the water layer while organic compounds of low and medium polarity are extracted into the organic phase. Once the compound of interest is in the organic layer, it can be easily recovered by evaporation of the solvent.

A mixture of two or more organic compounds of medium polarity can also be separated by liquid-liquid extraction provided the compounds have different acid-base properties. In such cases selective separation can be achieved by adjusting the pH of the aqueous layer to ionize one of the species. This process is called **acid-base extraction** and is treated in Unit 14.

Liquid-liquid extraction is often used, among many other applications, to avoid evaporating of water. An organic compound in an aqueous solution can be recovered, in principle, by evaporation of the water. However, the high boiling point of water requires extensive heating, which may result in decomposition of the compound. A more efficient way of recovering the compound from the aqueous solution is to carry out a liquid-liquid extraction with an appropriate organic solvent. The solute is extracted into the organic solvent, which is then evaporated. Evaporation of organic solvents is relatively easy because they usually have lower boiling points than water (for example, methylene chloride: 39.8°C and diethyl ether: 34.6°C).

In general, to ensure complete extraction of the solute into the organic solvent, several consecutive extractions are carried out. For example, if we want to determine how much DDT is present in a given volume of water, we would like to extract 100% of the DDT into the organic solvent. To this end, the water is mixed with the organic solvent, the layers separated (as explained in section 5.3), and the aqueous layer remaining after this first extraction is extracted again with another volume of fresh organic solvent. Eventually, a third extraction may be necessary. All the organic layers are combined and evaporated.

A fundamental question about liquid-liquid extraction arises when we have a limited amount of solvent at our disposal to carry out the extraction. What is more efficient: to use all the solvent at once or to divide it into smaller portions and carry out several consecutive extractions? It can be shown that the second option is always more efficient. In the organic chemistry laboratory, it is customary to carry out no more than two or three consecutive extractions. Beyond that point, there is little gain in the total amount of material extracted. For a more detailed account on the subject the reader is referred to Reference 3. *I went to the woods because I wished to live deliberately, to front only the essential facts of life, and see if I could not learn what it had to teach, and not, when I came to die, discover that I had not lived. Henry David Thoreau*

Partitioning: A Closer Look

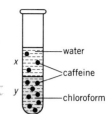

Figure 5.4 Partitioning of caffeine between water and chloroform.

How can we calculate in a systematic way the concentration of a solute partitioned between two layers? To carry out these calculations we need the partition coefficient and the mass balance for the system. Let's go back to our caffeine example where 1.1 g was partitioned between equal volumes of chloroform and water (Fig. 5.4). Calling y the mass of caffeine in chloroform, and x the mass in water at equilibrium, the mass balance is:

$$x + y = 1.1 \text{ g} \tag{8}$$

Rewriting Equation 6 we obtain:

$$K_{C/W} = \frac{[\text{caffeine}]_{\text{chloroform}}}{[\text{caffeine}]_{\text{water}}} = 10 \tag{9}$$

The concentration in each layer is given by the ratio between the mass of caffeine and the layer's volume:

$$[\text{caffeine}]_{\text{chloroform}} = \frac{y}{V_{\text{chloroform}}} \tag{10}$$

$$[\text{caffeine}]_{\text{water}} = \frac{x}{V_{\text{water}}} \tag{11}$$

In our example $V_{\text{chloroform}} = V_{\text{water}}$. Substituting Equations 10 and 11 into Equation 9, the volumes cancel out and Equation 12 is obtained:

$$K_{C/W} = \frac{y}{x} = 10 \tag{12}$$

Solving for x and y with the help of Equation 8,

$$x = 1.1 \text{ g} - y \implies \frac{y}{1.1 \text{ g} - y} = 10 \implies y = 1.0 \text{ g} \quad \text{and} \quad x = 0.1 \text{ g}$$

These figures agree with the amounts previously calculated in a more intuitive way.

Let's generalize these calculations for a solute partitioned between equal volumes of a solvent 1 and a solvent 2 (Fig. 5.5). If the total mass of solute is m, and the partition coefficient between solvent 2 and solvent 1 is $K_{2/1}$, then the mass of solute in solvent 2 (x) and the mass in solvent 1 (y), at equilibrium, are given by Equations 13 and 14.

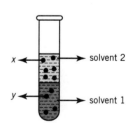

Figure 5.5 Partitioning between two layers.

$$x = m\frac{K_{2/1}}{1 + K_{2/1}} \tag{13}$$

$$y = m\frac{1}{1 + K_{2/1}} \tag{14}$$

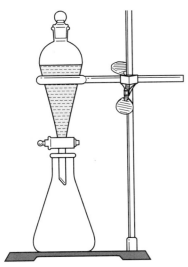

Figure 5.6 Separatory funnel in ringstand.

5.3 MACROSCALE LIQUID-LIQUID EXTRACTION

At the macroscale level (10 mL and up) solvent-solvent extractions are carried out with a **separatory funnel** (Fig. 5.6). Separatory funnels are available in different sizes, normally from 25 mL total capacity up to 2000 mL.

Experimental Technique

In this discussion we will assume that one of the solvents is water. The solution to be extracted is placed in an appropriate separatory funnel supported on an iron ring (Fig. 5.6), and with the stopcock closed. The total capacity of the separatory funnel should be at least twice the volume of the solution. The extraction solvent is added and the funnel is tightly stoppered. Usually the amount of extracting solvent is between 25 and 50% of the volume of the solution to be extracted.

The funnel is grasped with the palm of the right hand (left hand for lefties) around the body and the thumb and the first two fingers around the stopper (Fig. 5.7). The left hand is placed around the stem and stopcock (Fig. 5.8). The funnel is vigorously shaken to allow the phases to mix well. **Never point the stem toward your face or anybody nearby**. With the funnel inverted the stopcock is carefully opened with the left hand to release the built-up pressure. The stopcock is now closed and the funnel is shaken again for several seconds. The funnel is ''vented'' again. This process is repeated several times until the noise of the released pressure subsides. If there is no venting noise to start with (some solvents do not develop pressure) a total of a 1 to 2-minute shaking period is enough to attain equilibrium.

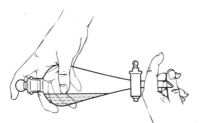

Figure 5.7 Shaking the separatory funnel.

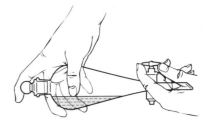

Figure 5.8 Venting the separatory funnel.

The funnel is returned to the ring and the layers allowed to separate forming a well-defined interface (Fig. 5.9). The lower layer is collected in an Erlenmeyer flask by opening the stopcock and making sure that the stopper is off, otherwise the liquid will not drain. If the interface between the layers is not clearly defined, swirling the funnel may help to settle the boundary between layers. This rotational motion also helps to collect scum and small particles of insoluble material in the neck of the funnel (Fig. 5.10). These materials are usually unwanted and should be drained along with the undesired layer.

One of the most common and frustrating mistakes made by beginners to liquid-liquid extraction is to keep the wrong phase and discard the one with the desired product. In principle, the identity of the layers can be predicted by the density of the solvents. However, a high solute concentration may change the density of the solution to the point that the layer with the less-dense solvent becomes the denser and lower layer. As a matter of precaution, **all phases should be saved (perfectly labeled) until the end of the experiment**. If there is any doubt concerning the identity of the layers the following test should be performed. Place about 1 mL of water in a test tube and add a few drops of the upper layer with a Pasteur pipet. If the drops dissolve in the water, it means they come from the aqueous phase. If the drops are immiscible with the water in the test tube, they belong to the organic phase.

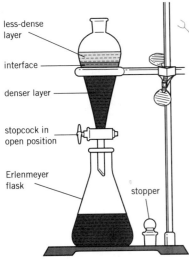

less-dense layer

interface

denser layer

stopcock in open position

Erlenmeyer flask

stopper

Figure 5.9 Draining the separatory funnel.

Breaking Emulsions

After shaking, the two phases may fail to separate sharply and may instead form a suspension of one liquid in the other. This is called an **emulsion** and although it is desirable in our Italian dressing, it is unwelcome in the laboratory. There is no foolproof trick to "break" the emulsion and get a clean separation between layers, but the following suggestions may help. Swirl the funnel very gently (funnel in the upright position Fig. 5.10). Let the system stand still in the ring for a minute or so and repeat the process. If the emulsion persists, add a small amount of a saturated solution of sodium chloride in water. The sodium chloride increases the ionic strength of the water phase and decreases its miscibility with the organic solvent. This process is called **salting-out** and may help break the emulsion. If this does not work and you are not interested in the aqueous layer, separate as much of the water layer as possible and place the organic phase plus the emulsion in an Erlenmeyer flask containing some magnesium sulfate (or other drying agent; see below) covering the bottom of the flask. Swirl the flask occasionally over a period of 5–10 minutes and then gravity-filter the system using fluted filter paper (Fig. 3.8). Rinse the magnesium sulfate on the filter with a few milliliters of the same organic solvent. Discard the magnesium sulfate. The solution should be clear now. If it is not, repeat the process. Another way of breaking emulsions is to centrifuge the layers for a few minutes. Glass tubes are preferred for this operation because some organic solvents may dissolve plastic centrifuge tubes. If all these suggestions fail to break the emulsion, a different solvent should be used to perform the extraction.

Figure 5.10 Swirling the separatory funnel.

5.4 MICROSCALE LIQUID-LIQUID EXTRACTION

When small volumes (0.5–10 mL) are to be extracted, a conical vial or a screw-capped test tube (Fig. 5.11) are the glassware of choice. Conical vials are available in different sizes (0.5–5 mL); test tubes are suitable when the volume to be extracted is between 2 and 10 mL.

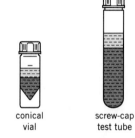

conical vial   screw-cap test tube

Figure 5.11 Conical vial and screw-cap test tube for microscale liquid-liquid extraction.

Experimental Technique

The solution of the desired compound and the extracting solvent are placed in the vial (or tube); the total volume should not be larger than three-fourths of the maximum capacity of the vial. The vial is tightly capped and shaken vigorously for a few seconds. The pressure is released by carefully unscrewing the cap. This process is repeated four to six times. The lower layer is withdrawn with the aid of a Pasteur pipet and placed in a clean test tube (Fig. 5.12). In doing this operation squeeze the rubber bulb before immersing the Pasteur pipet in the extraction vial. This will prevent air bubbles from disrupting the interface between the phases. Always retrieve the lower layer first, regardless of the layer of interest; *it is always much easier to remove the lower than the upper layer*.

Very volatile solvents such as diethyl ether and methylene chloride have a tendency to squirt out the tip of the pipet. There are several tricks to avoid this. One way is to saturate the inside of the pipet with the solvent's vapors by drawing *fresh solvent* several times before attempting the transfer of the organic layer. Another way is to put a small cotton plug in the capillary tip of the Pasteur pipet (Fig. 3.9b). Alternatively, the use of a pluringe instead of a rubber bulb minimizes the squirting problem.

After removal of the lower layer, if the upper layer is the one being extracted, a volume of fresh extracting solvent is added to the original vial and the extraction repeated. If the lower layer is the one being extracted, the upper layer remaining in the extraction vial is transferred to a clean test tube, and the lower layer is returned to the vial to repeat the extraction with a few more milliliters of extraction solvent.

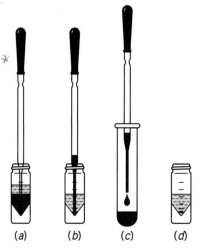

(a)   (b)   (c)   (d)

Figure 5.12 Microscale separation of layers using a Pasteur pipet: a and b) denser layer taken with Pasteur pipet; c) denser layer transferred to test tube; d) less-dense layer left in vial.

5.5 DRYING THE ORGANIC LAYER

After contact with an aqueous phase, any organic solvent will be saturated with water; although the amount of water is usually a small percentage of the total volume (for example, *tert*-butyl methyl ether saturated with water contains 1.1% H_2O at 20°C; Table 5.1), the **total amount of water** could be considerable, especially when large volumes of organic solvents are used. This poses a problem in the recovery of the solute from the organic layer, because evaporation of the solvent renders a product mixed with water. To avoid these wet residues, water should be removed before evaporation.

Most organic solvents are easily dried with the aid of anhydrous salts or other **drying agents**. Anhydrous salts remove water from the organic solvent by forming **hydrates** in which the water molecules are bound as **water of hydration**, (Eq. 15).

$$\text{salt} + n\text{H}_2\text{O} \rightleftharpoons \text{salt} \cdot (\text{H}_2\text{O})_n \qquad (15)$$

In Equation 15, n is the number of water molecules retained per molecule of salt; n varies from salt to salt ($n = 1/2$ to 10). The larger the number of water molecules in the hydrate the larger the **water capacity** of the drying agent. Another factor to take into account in choosing a drying agent is the **efficiency of the hydration process** determined by the hydration equilibrium constant. This equilibrium is very sensitive to temperature. At temperatures higher than room temperature the equilibrium is shifted to the left. To maximize the efficiency of the process, drying should be conducted at temperatures below 40°C.

Different drying agents are recommended depending on the nature of the solvent and the compound to be isolated. They are listed below.

Anhydrous magnesium sulfate. Its drying action is due to heptahydrate $MgSO_4 \cdot 7H_2O$. It is a fast and efficient drying agent and can be used for most compounds.

Anhydrous sodium sulfate. The hydrated form is $Na_2SO_4 \cdot$ this desiccant are commercially available: powder and granular. for microscale operations because its removal from the liquid p filtration. Although this drying agent has a high water capacity It is chemically inert and can be used with most compounds.

Anhydrous calcium sulfate. This is a rapid and chemicall has poor water capacity (it forms the hemihydrate $2CaSO_4 \cdot H_2$ limited to the removal of small amounts of water. This drying known as Drierite (registered mark of W.A. Hammond Drierit

Anhydrous calcium chloride. It forms the hexahydrate drying agent with high water capacity. Slow in its action. Its app reduced due to its reactivity toward alcohols, phenols, amines, it forms coordination complexes.

Anhydrous potassium carbonate. It forms $K_2CO_3 \cdot 2H$ basic drying agent, it must not be used for drying acids, ph compounds. It has a moderate drying capacity.

Sodium and potassium hydroxides. Due to their basic agents are only recommended to dry amines (basic organic co be avoided when drying acids, phenols, esters, and amides. I for drying chloroform solutions, because chloroform react

Sodium. A very powerful drying agent; removes trace such as toluene, benzene and diethyl ether and aliphatic ar It reacts violently with water. It should be handled with people only.

Table 5.1 Miscibility of Water and Organic Solvents at Room Temperature

Solvent	Solubility of water in solvent (g/100 mL solution)	Solubility of solvent in w (g/100 mL solution)
benzene	0.047	0.062
n-butanol	17.0	6.8
butanone	9.8	16.1
carbon tetrachloride	0.018	0.080
chloroform	0.12	0.74
cyclohexane	0.004	

Molecular sieves (3 Å and 4 Å). Small spheres (2–3 mm diameter) made of porous aluminosilicates. The average pore size is 3 and 4 Å, respectively. Water molecules are small enough to pass into the sieve where they are trapped as water of hydration. It is a rapid and efficient drying agent. Molecular sieves can be reactivated by heating in an oven at 300°C.

Procedure

A small amount (a spatulaful) of the desired drying agent is added to the organic phase; the system is stoppered (avoid rubber stoppers because some solvents may corrode them) and swirled occasionally for a period of 2–5 minutes (Fig. 5.13). If the drying agent is clumped together at the bottom of the flask, it means that water is still present in the solvent. More drying agent should be added in small portions until the solid runs as free particles in suspension when the flask is swirled. The drying agent is removed by gravity filtration using fluted filter paper (Fig. 3.8). The drying agent on the filter is rinsed with a small portion of fresh solvent to remove the last traces of organic solution from the filter. The filtrate may be directly collected in a pre-tared, round-bottom flask and the solvent evaporated using a rotatory-evaporator (Fig. 3.17). For microscale drying, the drying agent may be separated from the solution by removing the liquid with the aid of Pasteur pipets as shown in Figure 3.9. Drying agents with large particle size (granular form) are sometimes used, especially at the microscale level, because they are easy to separate. However, they are not as efficient as in their finely ground powder form.

Figure 5.13 Organic phase with drying agent.

5.6 SOLID-LIQUID EXTRACTION

Isolation of natural products is an ancient art; plant extracts have been used for centuries for diverse purposes, from medicinal preparations to love potions. The chemistry of natural products is a well-defined area of organic chemistry with far-reaching applications in biochemistry, pharmacology, and medicine.

The components in natural materials, such as plant and animal tissues, have a wide range of polarities and thus their selective extraction can be performed by choosing solvents of the right polarity. Nonpolar compounds such as fats, waxes, terpenes, and some steroids can be extracted by nonpolar solvents, such as petroleum ether (a mixture of low molecular mass aliphatic hydrocarbons). Methanol, a solvent of medium-high polarity, is recommended for the extraction of pigments, alkaloids, tannins, flavonoids, and other polar compounds. Finally, water extracts very polar compounds such as salts, small sugars, and proteins. The extraction of different families of compounds from a natural source can be performed sequentially (see Fig. 5.14). The specimen to be analyzed (dried and finely ground) is first extracted with petroleum ether, the extract is separated from the residue, and the residue is subjected to a second extraction with methanol. The residue from the second extraction is finally extracted with water. Each extract contains several families of compounds that can be further separated and analyzed by selective extraction with other solvents, acid-base extraction, chromatography, or distillation. The whole process is called **screening**.

There are two major techniques to perform solid-liquid extractions: **batch** and **continuous extraction**. In batch solid-liquid extraction, the solid specimen is mixed with the desired solvent and heated. High temperatures and long extraction times increase the yield of the process. However, heat should be used only if the compounds to be extracted are stable at high temperatures. Recommended extraction times range from a few minutes to several hours depending on the nature of the solute and the desired efficiency of the process.

Batch extraction is especially useful when the specimen is very rich in the component to be extracted or when the yield of the process is a matter of little

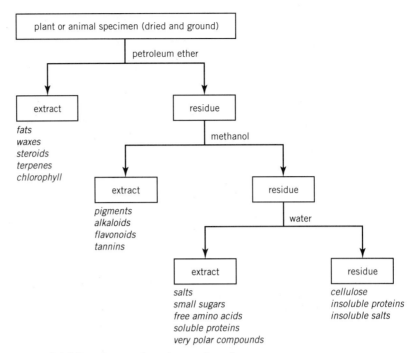

Figure 5.14   Solid-liquid extraction of natural products.

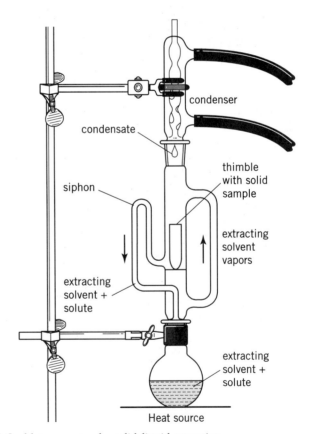

Figure 5.15   A Soxhlet apparatus for solid-liquid extraction.

consequence. If the desired component is only slightly soluble in the extraction solvent, or if it is present in very low amounts, or if the yield of the extraction is an important concern, batch solid-liquid extraction is inefficient and should be replaced by continuous extraction.

A Soxhlet extraction apparatus is used in continuous extraction (Fig. 5.15). The extraction solvent is placed in a round-bottom flask on a heating mantle or other heating device. Heat is supplied to vaporize the solvent; the vapors ascend through the connecting tube, condense on the surface of the condenser, and drip down in the extraction chamber. The extraction chamber contains the specimen placed in a porous thimble (usually made of filter paper). The extraction takes place as the solvent comes in contact with the sample. The solution accumulates in the chamber until it reaches the top of the siphon, and then it is siphoned to the distillation flask. As a result of this continuous process, the solution in the distillation flask is enriched in the desired compounds. This operation is usually carried out for several hours to ensure complete extraction. If the compounds to be isolated are thermolabile, Soxhlet extraction may lead to decomposition because the solution is continuously boiled. In such cases, low-boiling point solvents are recommended.

5.7   COUNTERCURRENT DISTRIBUTION

Can two solutes with similar partition coefficients be separated by liquid-liquid extraction? If the partition coefficients are similar, no separation is possible by simple liquid-liquid extraction. However, separation of compounds with similar K values can be successfully performed by **countercurrent distribution** (Fig. 5.16). The principles that govern countercurrent distribution are also fundamental to **partition chromatography**. Partition chromatography is one of the most powerful separation techniques used in the organic chemistry laboratory. Thus, understanding the principles outlined in this section will help you comprehend the physical process behind chromatography.

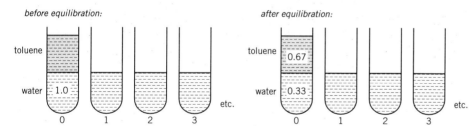

Figure 5.16   Countercurrent distribution.

Let us suppose that we have 11 tubes with 100 mL of water each (they are numbered from 0 to 10). We dissolve 1 g of a compound A in the 100 mL of water contained in tube 0, add 100 mL of toluene, and shake the system to equilibrate. Let's also assume that the partition coefficient for compound A between toluene and water is $K_{T/W}^{A} = 2$.

Using Equations 13 and 14, the amount of A in the toluene layer (x) and the water layer (y) can be calculated:

$$\text{mass in toluene:} \quad x = 1 \, \text{g} \frac{2}{1+2} = 0.67 \, \text{g}$$

$$\text{mass in water:} \quad y = 1 \, \text{g} \frac{1}{1+2} = 0.33 \, \text{g}$$

Both phases are now separated, the toluene layer is transferred from tube 0 to the next tube to the right, tube 1, and 100 mL of fresh toluene is added to tube 0. Tubes 0 and 1 are shaken to equilibrate. The amount of A in each layer after equilibration is

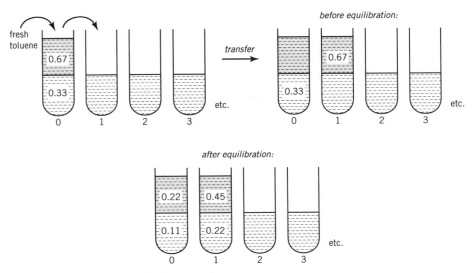

Figure 5.17 Distribution after first transfer.

shown in Figure 5.17. Notice that in each tube the ratio of the concentrations is 2 : 1 (as expected from the partition coefficient).

In the next step the toluene layer from tube 1 is transferred to tube 2, the toluene layer from tube 0 is transferred to tube 1, and 100 mL of fresh toluene is added to tube 0. The tubes are shaken to equilibrate. The amount of A in each layer after equilibration is shown in Figure 5.18.

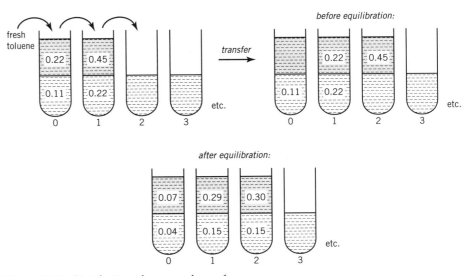

Figure 5.18 Distribution after second transfer.

This process is repeated several times; *in every transfer the upper layer is moved to the next tube to the right and fresh toluene is added to tube 0*. The amount of A in each tube after 10 transfers is shown in Figure 5.19.

In Figure 5.20, the total amount of A in each tube is plotted against the tube number. As can be observed, solute A is distributed among 11 tubes with maximum concentration in tube 7.

Let's suppose now that another solute B with partition coefficient $K^B_{T/W} = 0.5$ is subjected to the same process. After ten transfers, the amount of B in each layer is also shown in Figure 5.19. As can be observed, the maximum concentration for B is located in tube 3.

If compounds A and B are both present in tube 0 at the beginning of the whole process, then a reasonably good separation of the two will be achieved after ten

Distribution of A

0	0	0	0.010	0.037	0.089	0.151	0.175	0.133	0.060	0.012	*Toluene*
0	0	0	0.005	0.018	0.044	0.074	0.086	0.066	0.030	0.006	*Water*

Total amount:

0   0   0   0.015   0.055   0.133   0.225   0.261   0.199   0.090   0.018

Distribution of B

0.006	0.030	0.066	0.086	0.074	0.044	0.018	0.005	0	0	0	*Toluene*
0.012	0.060	0.133	0.175	0.151	0.089	0.037	0.010	0	0	0	*Water*

Total amount:

0.018   0.090   0.199   0.261   0.225   0.133   0.055   0.015   0   0   0

Tube number:

0   1   2   3   4   5   6   7   8   9   10

Figure 5.19 Distribution of A ($K_{T/W} = 2$) and B ($K_{T/W} = 0.5$) between toluene (T) and water (W) after ten transfers.

transfers. Tubes 0–3 will contain mainly compound B (the amount of A in tube 3 is less than 6% of that of B). Tubes 4–6 would contain a mixture of A and B. Tubes 7–10 will contain mainly A. This can be visualized in Figure 5.21. Complete separation (where virtually all of A is totally separated from all of B) can be accomplished by increasing the number of transfers.

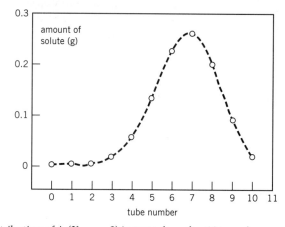

Figure 5.20 Distribution of A ($K_{T/W} = 2$) in test tubes after 10 transfers.

Because $K_{T/W}^A = 2$ and $K_{T/W}^B = 0.5$, A has greater affinity for toluene than B, and travels with the moving toluene layer faster than B; B has comparatively larger affinity for water and stays behind during the transfer process.

In general, the fraction of solute in each tube (in both upper and lower layers combined) after n transfers, $T_{n,r}$, can be calculated with the following expression, (Eq. 16):

$$T_{n,r} = \frac{n!}{(n-r)! \times r!} \times \left(\frac{K_{2/1}}{1 + K_{2/1}}\right)^r \times \left(\frac{1}{1 + K_{2/1}}\right)^{n-r} \qquad (16)$$

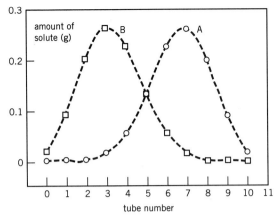

Figure 5.21 Distribution of A ($K_{T/W} = 2$) and B ($K_{T/W} = 0.5$) after 10 transfers.

where n is the number of transfers, r the tube number, and $K_{2/1}$ is the partition coefficient between the layers.

In general, the separation between two solutes A and B can be accomplished by countercurrent distribution only if they have *different* partition coefficients. The ratio of the partition coefficients is called the **separation factor**, β:

$$\beta = K^A/K^B \text{ (where } K^A > K^B\text{)}$$

The larger the separation factor, the easier the separation, or in other words, less transfers are necessary to achieve complete separation.

Carrying out a countercurrent distribution by transferring the moving phase by hand is a very tedious and time-consuming process. Automatic systems have been devised for this purpose. They were successfully used some 50 years ago to separate mixtures from natural and industrial sources. The explosive development of chromatography in the last three decades made countercurrent distribution an antiquated method. Chromatography has proven to be a more versatile and powerful technique for the separation of a wide spectrum of compounds, from small organic molecules to macromolecules such as nucleic acids and proteins. However, *the physical principles that rule chromatography and countercurrent distribution are very similar*. An uderstanding of how compounds with only slightly different partition coefficients can be separated by countercurrent distribution is the foundation of the understanding of chromatography.

Countercurrent distribution can be pictured as a process in which a mixture of solutes is partitioned between a **mobile phase** and a **stationary phase** (Fig. 5.22). In our previous example, the stationary phase is the water layer present in all the tubes from the beginning, and the mobile phase is the toluene layers that are transferred from tube to tube. We will come back to these ideas in Unit 7.

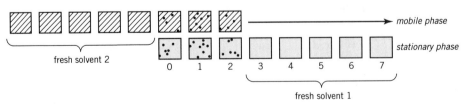

Figure 5.22 The countercurrent distribution process as a model for stationary and mobile phases. At this stage (after 2 transfers), the sample is in tubes 0–2 and distributed between both phases. In the next step, all the "squares" of mobile phase will move one position to the right.

BIBLIOGRAPHY

1. Partition Coefficients and Their Uses. A. Leo, C. Hansch, and D. Elkins, *Chem. Reviews*, **71**, 525 (1971).
2. Vogel's Textbook of Practical Organic Chemistry. A.I. Vogel, B.S. Furniss, A.J. Hannaford, P.W.G. Smith, and A.R. Tatchell. 5th ed. Longman, Harlow, UK. 1989.
3. Liquid-liquid Extraction: Are n Extractions with V/n mL of Solvent Really More Effective than One Extraction with V mL? D. Palleros, *J. Chem. Ed.*, **72**, 319–321 (1995).

EXERCISES

Problem 1

Given the following pairs of immiscible solvents, which solvent will be the upper layer and which one the lower layer? See Table 2.2 for the densities.

(a) water/diethyl ether

(b) petroleum ether/water

(c) cyclohexane/water

(d) toluene/water

(e) water/methylene chloride

(f) chloroform/water

Problem 2

The solubility of lorazepam (a tranquilizer found in Emotival, Lorax, Wypax, and other prescription medicines) is 0.08 mg/mL and 3 mg/mL, in water and chloroform, respectively. Estimate the partition coefficient of lorazepam between chloroform and water.

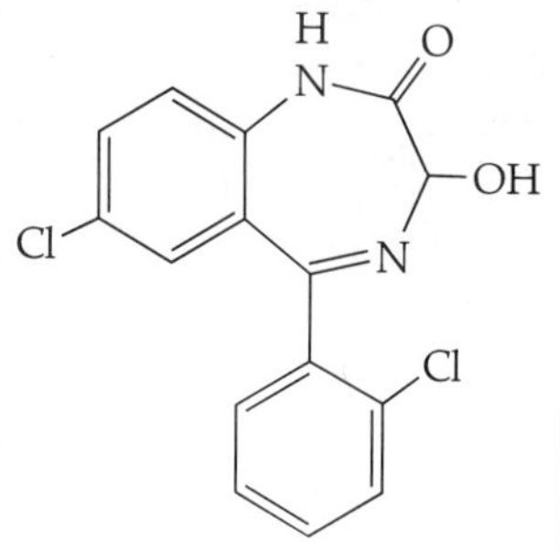

lorazepam

Problem 3

One gram of aspirin dissolves in 300 mL of water and in 17 mL of chloroform at 25°C.

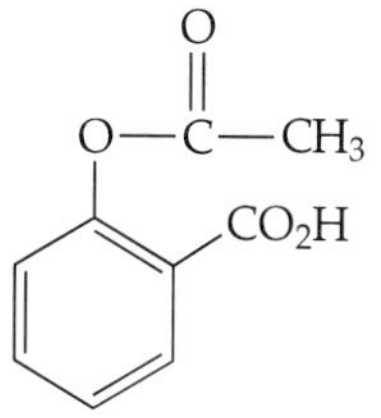

(a) Calculate the solubility of aspirin (g/mL) in both solvents at 25°C.

(b) Estimate the partition coefficient of aspirin between chloroform and water, $K_{C/W}$.

(c) An aqueous solution of aspirin (100 mL) was extracted with 100 mL of chloroform. The organic layer was separated and dried over magnesium sulfate. After removal of the drying agent and evaporation of the solvent 180 mg of aspirin were obtained. Estimate how much aspirin was present in the original aqueous phase.

aspirin

Problem 4

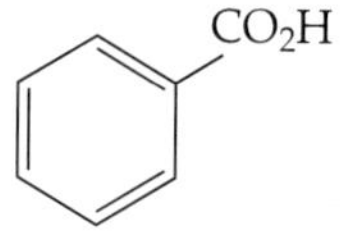

The solubility of benzoic acid at 25°C in water and diethyl ether is 3.4 mg/mL and 0.333 g/mL, respectively. Estimate the % Recovery of benzoic acid in the organic phase, if a solution of benzoic acid in water is extracted with an equivalent volume of ether.

benzoic acid

Problem 5

The partition coefficients between 1-octanol (O) and water (W) at room temperature for different fluoro-alcohols are given below:

	$K_{O/W}$
CF_3CH_2OH	2.57
$CF_3CF_2CH_2OH$	17.0
$CF_3CF_2CF_2CH_2OH$	64.6

Explain why the $K_{O/W}$ values increase from trifluorethanol to heptafluorobutanol.

Problem 6

The solubilities of cholesterol in water and chloroform are 0.2 mg/100 mL and 1 g/4.5 mL, respectively. Estimate the partition coefficient of cholesterol between chloroform and water. What conclusion can be drawn about the polarity of cholesterol?

Problem 7

Cholic acid is a steroid present in the bile of most vertebrates. Its solubility in water and chloroform is 0.28 g/L and 5.08 g/L, respectively. Estimate the partition coefficient of cholic acid between both solvents. What can be said about the polarity of cholic acid as compared to cholesterol? (See Problem 6)

Problem 8

How would you separate the following mixtures by liquid-liquid extraction?

(a) sodium chloride and butter (fats)

(b) acetanilide and glucose (a sugar)

acetanilide   glucose

Problem 9

The partition coefficients for acetanilide between chloroform (C), ethyl ether (E), benzene (B), and water (W) are: $K_{C/W} = 50$; $K_{W/E} = 0.1$; $K_{B/W} = 4$.

What solvent would you choose to extract acetanilide from an aqueous solution? Justify your answer.

Problem 10

(a) Two compounds A and B are partitioned between equal volumes of toluene (T) and water (W). If $K_{T/W}^A = 2$ and $K_{T/W}^B = 0.5$, calculate the relative amounts of A and B in each layer. Is this a good method to separate A from B? Why?

(b) Carry out the same calculations if $K_{T/W}^A = 0.001$ and $K_{T/W}^B = 500$. Is this a good method to separate A from B? Why?

EXPERIMENT 5

Isolation of Caffeine from Tea

E5.1   TEA, COFFEE, AND CAFFEINE

Caffeine is one of the most widespread stimulants in use today. Civilizations as diverse as the Chinese, the Arabs, the Aztecs, and the Guaraní People (in the heart of South America) were well acquainted with the effects of caffeine present in beverages extracted from indigenous plants such as tea, coffee, cocoa beans, and maté. Caffeine is an **alkaloid** of the **methylxanthine** family (alkaloids are naturally occurring *organic bases* containing nitrogen [alkaloid from *alkaline*]), structurally related to the **purine bases** adenine and guanine, which form part of the nucleic acids RNA and DNA (see structures below). Coffee beans are the seeds of a fruit (a drupe) that is usually red when ripe. The amount of caffeine in coffee beans is typically between 0.8 and 2.0 weight percentage, depending on the species of coffee. Two species are commercially important, *Coffea arabica*, which is cultivated in Latin America, and *Coffea robusta*, grown primarily in Africa. The robusta species has a larger caffeine content but it is usually regarded as inferior in flavor. The amount of caffeine found in tea leaves (*Camellia sinensis*) is typically between 2 and 3.5%. The most important types of tea are green, black, and oolong, which differ only in their manufacture. To obtain black tea, the tea leaves are subjected to an enzymatic oxidation process, erroneously called fermentation, while green tea is obtained by drying the leaves without oxidation. Oolong tea is only partially oxidized. **Theobromine**, which literally means food for the gods (from the Greek: *theos*, god; *broma*, food), is the most abundant alkaloid in cocoa beans; the concentration of caffeine in cocoa is only approximately one-eighth that of theobromine. Caffeine is also present in cola nuts, from which an extract is obtained to produce soft drinks, and in cassina and maté, aboriginal plants of North and South America, respectively.

A cup of typical American coffee contains between 70 and 100 mg of caffeine. A cup of black tea has approximately 20–35 mg of caffeine. A chocolate candy bar (30 g) contains about 2 mg of caffeine and 20 mg of theobromine, and a can of a cola-drink typically has about 40 mg of caffeine. Caffeine is absorbed in the gastrointestinal tract; maximum concentration in the bloodstream is reached within one hour of ingestion. Drinking one cup of coffee gives a peak plasma concentration of about 5–10 μM in a 70-kg (155-pound) person, which results in sleep latency and increase of alertness. At a plasma concentration between 15 and 30 μM, caffeine has more severe effects such as diuresis, and respiratory, gastric, and cardiovascular stimulation. At a concentration of 500–1000 μM, which is equivalent to 100 cups of coffee drunk in about 30 minutes, caffeine is lethal. The half-life of caffeine (the time required to decrease the concentration in plasma to 50%) is about 5 hours and is influenced by several factors such as gender, age, and smoking. The mechanism of action of caffeine seems to involve the blocking of adenosine receptors in the brain. Adenosine is a widespread compound that regulates the cardiovascular and nervous systems. Caffeine can be considered an addictive drug because individuals subjected to continuous exposure may develop tolerance and suffer withdrawal symptoms after cessation of caffeine intake.

Caffeine interferes with the repair of DNA, which is not surprising given the similarities between caffeine and the purine bases adenine and guanine. However, caffeine is not a mutagenic compound (at least in bacteria) and does not seem to be carcinogenic.

Roasted Coffee

The delicious flavor of coffee develops during the roasting process. The green coffee beans are subjected to high temperatures (200–240°C) for short periods of time, usually 3–12 minutes, to give them the characteristic aroma and brown color. Pyrolysis (from Greek; *pyros*, fire and *lysis*, loosening) of sugars and amino acids takes place during roasting; carbon dioxide and carbon monoxide form; the volume of the bean almost doubles and there is a weight loss of about 13–18%. More than 800 volatile compounds have been identified in roasted coffee but only about 70 of them contribute to its typical smell; among them, furfurylthiol, furfuryl methyl sulfide, and furfuryl alcohol seem to play a very important role in the aroma of coffee.

Decaffeination

The first method for the removal of caffeine from coffee was patented in Germany at the beginning of the twentieth century and it was used to produce commercial decaffeinated coffee before the onset of World War I. However, it was only after World War II that the production of decaffeinated coffee took off. Today, about 10% of all green coffee produced is later decaffeinated. There are essentially three different processes for decaffeinating coffee. They involve extraction of the beans using organic solvents, water, or liquefied carbon dioxide. To minimize the loss of flavor, decaffeination is carried out on the green bean, which is then roasted by conventional methods. Decaffeination of tea is performed in a similar way.

In the water-decaffeination process, the coffee beans are treated with a water solution that is saturated in all the water-soluble compounds present in coffee beans except caffeine. The lack of caffeine in the water solution is the driving force that allows the extraction of caffeine from the beans. At the end of the extraction, caffeine is removed from the aqueous solution by selective adsorption on activated carbon or liquid-liquid extraction with methylene chloride. Once the caffeine has been removed from the aqueous phase, this solution is recycled to decaffeinate a new batch of coffee beans.

E5.2 CAFFEINE FROM TEA: OVERVIEW OF THE EXPERIMENT

In this experiment you will isolate caffeine from tea leaves by performing a **solid-liquid extraction** first (boiling tea leaves in water) and then a **liquid-liquid extraction** of the "tea" with methylene chloride. The liquid–liquid extraction separates caffeine from the other components in the tea. You will then dry the methylene chloride solution with anhydrous $MgSO_4$ and evaporate the solvent. Finally, you will purify caffeine by sublimation or recrystallization.

Tea leaves contain **carbohydrates** such as **cellulose**, which is the structural component of plants, and small amounts of **glucose** and **starch**. Other components of tea leaves are: **proteins** and **amino acids** (3–5%), **caffeine** (2–3.5%), **polyphenols** (also known as **tannins**), **pigments**, and small amounts of **saponins**. Tannins are very prominent in tea leaves to which they confer their characteristic astringency and bitter taste. Most of the chlorophylls and other pigments of green tea leaves are decomposed by the oxidation process during manufacture. Saponins are compounds widespread in plants where they are used as a defense weapon against predators (saponins have a bitter taste and are poisonous to some animals). Saponins are molecules with a polar end and a long nonpolar body that behave similarly to soap from which they borrow their name (from Latin; *sapo*, soap). Saponins foam profusely in water and induce the formation of emulsions between water and organic solvents. Due to this emulsifying effect, saponins, although present in small amounts in tea leaves, pose a problem in the isolation of caffeine by liquid–liquid extraction, as discussed in the Procedure.

The compounds extracted along with caffeine into the tea are tannins, residual pigments, and very small amounts of glucose, free amino acids, some proteins, and saponins. Cellulose (see structure below), due to the presence of the HO— groups, is a very polar compound; however, it is insoluble in water by virtue of its extremely large molecular mass (500,000 g/mol).

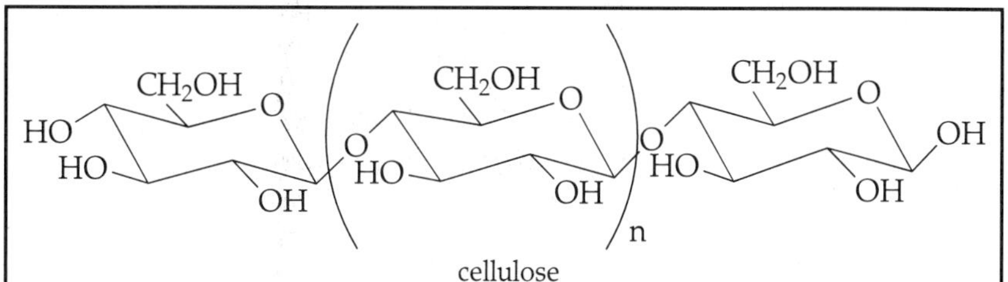

cellulose

Caffeine can be separated from the other components in the "tea" by liquid-liquid extraction with an organic solvent such as methylene chloride because of the difference in polarity among the compounds in the "tea." Caffeine is a compound of medium polarity, soluble in water and in organic solvents. Its partition coefficient between methylene chloride and water is approximately 10 ($K_{meth.chl./water} \approx 10$), indicating that caffeine is preferentially extracted into the organic solvent. Glucose, amino acids, and proteins are polar compounds soluble in water and insoluble in most organic solvents. They stay in the aqueous layer after extraction with the organic solvent. Pigments, such as chlorophylls and their oxidation products, are soluble in water and organic solvents and partition between both layers; they are also extracted along with the caffeine into the organic layer. However, they are present in very small amounts and can be eliminated at a later stage by recrystallization or sublimation of caffeine. Finally, tannins are compounds of medium polarity that are somewhat soluble in methylene chloride. To minimize their solubility in the organic solvent they must be chemically transformed into very polar (ionic) compounds, insoluble in most organic solvents.

Tannins are a family of organic compounds with molecular mass in the range 600 to 3000 Da. The type of tannins most prominent in tea leaves is called **hydrolyzable tannins**. They consist of a glucose molecule where one or more of the −OH groups is condensed with **gallic acid**; the linkage between glucose and gallic acid (galloyl and digalloyl groups, see structures below) is an **ester group** (−O−CO−). Ester bonds are easily broken by water in the presence of a base such as calcium carbonate; this type of reaction is known as **hydrolysis** (from the Greek; *hydor*, water; *lysis*, loosening). The hydrolysis of tannins in the presence of calcium carbonate affords glucose, which is a very polar compound because of the −OH groups, and also calcium salts of gallic acid, which are ionic. Due to their high polarity, glucose and the salts of gallic acid are *not soluble in methylene chloride and they stay in the aqueous layer after extraction with the organic solvent.*

The chemical transformation involved in the hydrolysis of tannins is given below to illustrate the point that hydrolyzed tannins are *very polar* compounds. It is not important at this stage to have a detailed understanding of the chemistry behind this reaction. The reactions of esters is the subject of a later unit. A flowchart for the isolation of caffeine from tea leaves is also shown (Fig. 5.23).

Components of hydrolyzable tannins

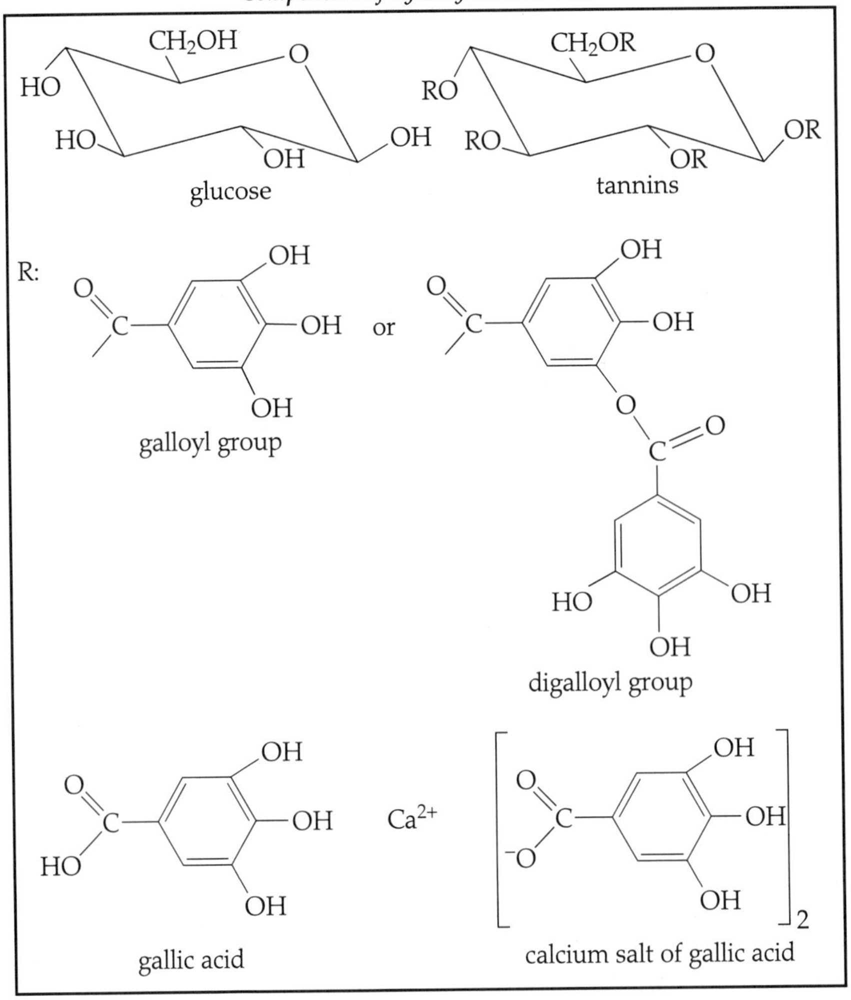

Hydrolysis of tannins

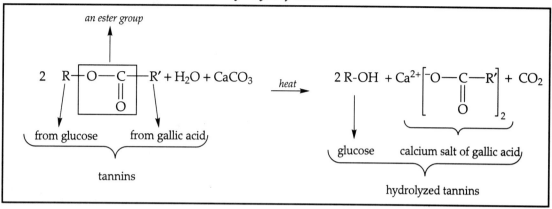

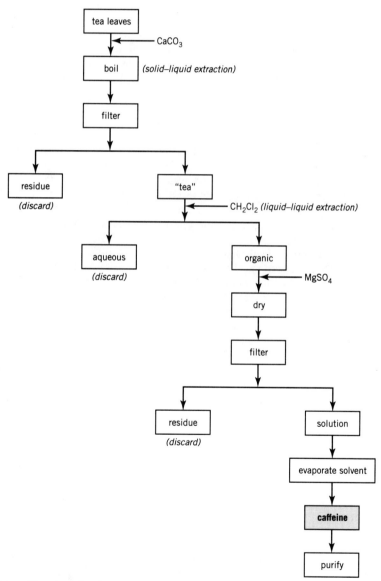

Figure 5.23 Flowchart for the extraction of caffeine from tea leaves.

BIBLIOGRAPHY

1. Coffee and Health; Banbury Report 17; B. MacMahon and T. Sugimura, ed. Cold Spring Harbor Laboratory; 1984.

2. Coffee, Vol. 1: Chemistry; R.J. Clarke and R. Macrae, ed. Elsevier, London; 1985.

3. Extraction of Natural Products Using Near-Critical Solvents; M.B. King and T.R. Bott, ed. Blackie Academic & Professional, Glasgow; 1993.

4. Tea; T. Eden; 3rd edition; Longman, London; 1976.

having many excellent vertues, closes the Orifice of the Stomack, fortifies the heart within, helpeth Dijestion, quickeneth the Spirits, maketh the heart lightsome, is good against Eyesores, coughs and cold, Rhumes, Consumptions, Head-ache, Dropsie, Gout, Scurvey, King's Evil and many other

Coffee advertisement in an English newspaper (1657)

The caffeine cartel After crude oil, coffee is the most widely traded commodity in the world; curiously, coffee is also a black symphony of chemicals rooted in the sandy banks of the Red Sea. The coffee plant is indigenous to Ethiopia where it was cultivated in the Arabian colony of Harar as early as in the ninth century. Legend has it that the stimulant effects of coffee were discovered in Yemen by a goatherd named Kaldi while tending his herd. He observed that his goats were agitated after chewing on berries of a wild bush. He tried these fruits himself and found that they helped him stay awake. Soon after, the monks from a nearby monastery were roasting and brewing these berries to make a beverage that kept them alert during their long prayers.

The word *coffee* derives from the Turkish word *kahweh* which is the equivalent of the Arabic *quahweh*, a poetic word for wine. Large-scale cultivation of coffee began in Arabia in the fifteenth century and for 200 years the Arabs had the monopoly of coffee. Seeds had to be rendered infertile by soaking them in boiling water before they could leave the country where coffee production was a very well-guarded secret. From the banks of the Red Sea coffee consumption spread to the rest of the Islamic world. By the beginning of the seventeenth century coffee had entered Europe through the door of Constantinople. In that age of European subjugation of the world, it did not take long for coffee to reach the most remote corners.

The fight to dominate the coffee market had inadvertently started, but not without intrigue and passion. In 1616 a Dutch merchant stole one coffee plant from the Arabs, which was then planted in the Amsterdam Botanical Gardens from where it was propagated by the Dutch to Sri Lanka and later Java. By the end of the seventeenth century, the Arabic monopoly had dwindled down, overshadowed by the Dutch coffee empire, but not for long. In 1713, the mayor of Amsterdam presented King Louis XIV of France with a coffee plant as a token of friendship after signing the peace treaty of Utrecht. This plant was kept in the *Jardins des Plantes* in Paris and can be considered the great-grandmother of the millions of coffee trees that grow in South America today. In 1723, a French captain named Desclieux took a seedling of this plant to the French colony of Martinique in the Antilles, from where it propagated to other French possessions in the West Indies. In 1727 a young Brazilian officer named Francisco de Mello Palheta was visiting Cayenne in the French Guiana to settle a border dispute; at his departure, as a sign of affection from the Governor's wife, he received a small coffee plant concealed in a bouquet of flowers. After returning to Brazil, Palheta started his own plantation near the mouth of the Amazon; from there coffee expanded south to São Paulo, which can be considered the coffee capital of the world today.

In the meantime, coffee houses flourished in Europe. By 1625 coffee was sold in Rome and from there the habit of coffee drinking spread to England. In 1650 the first coffee house opened in Oxford and by the beginning of the eighteenth century more than 2000 coffee houses were in operation in London. These coffee houses, the gathering place for intellectuals, merchants, and politicians, became an English institution that lasted for almost two centuries. Among them was Edward Lloyd's Coffee House in Tower Street, which opened between 1680 and 1689 and where seamen congregated to discuss the trade of their business and lament their losses to piracy and the inclement weather. To protect their interests they started the insurance of ships and their cargoes; this coffee house later became Lloyd's of London, one of the most powerful financial institutions in the world today. The proliferation of coffee houses did not go unnoticed within the moralist circle of seventeenth-century society, which saw in coffee a "vile and worthless foreign novelty." License to roast coffee had to be obtained in Germany. In England, King Charles II in 1675 issued a "Proclamation for the Suppression of Coffee Houses" on the grounds that they were the resort of seditious persons. The edict was extremely unpopular and had to be withdrawn only eleven days after being promulgated. After the Boston Tea Party, the United States has become the major consumer of coffee in the world. Since 1962 the price of coffee and its trade is regulated by the International Coffee Organization with headquarters in London.

BIBLIOGRAPHY

1. A History of Coffee. R.F. Smith, in "Coffee, Botany, Biochemistry and Production of Beans and Beverage." M.N. Clifford and K.C. Willson, ed. Avi Pub. Co., Westport, CT, 1985.

2. Caffeine and Health. J.E. James. Academic Press, San Diego, CA, 1991.

3. Introduction. A.W. Smith, in "Coffee" Vol. I: Chemistry. R.J. Clarke and R. Macrae, ed. Elsevier, London, 1985.

4. Sugar, Gold and Coffee. F. Reichmann. Cornell University Library, Ithaca, New York, 1959.

5. Brazil, the Infinite Country. W.L. Schurz. E.P. Dutton and Co. New York, 1961.

PROCEDURE

E5.3   ISOLATION OF CAFFEINE

Background reading: 5.1–3; 5.5–6;
3.3; 3.5
Estimated time: 3 hours

Open a tea bag and weigh its contents. This information will be used to calculate the percentage yield of the extraction. Dispose of the tea leaves in the waste container once you are done with them (do not discard them in the sink; they will clog the drain). Place 6 tea bags, 5 g of calcium carbonate powder, and 180 mL of water in a 500-mL Erlenmeyer flask. Bring the mixture to a gentle boil on a heating plate. Use a glass rod to stir the mixture and prevent bumping (be careful not to puncture the bags). Boil the mixture for 20 minutes. Remove the tea bags, gently squeezing them against the flask with the glass rod. Let the solids settle and vacuum-filter the system while still hot using coarse filter paper (Whatman 4) and a 250-mL filter flask; see Fig. 3.12.

Allow the filtrate to cool down by placing the flask in an ice-water bath. When the filtrate has reached ambient temperature, add 5 g of sodium chloride and extract the solution with a 25-mL portion of methylene chloride (dichloromethane) using a 250-mL separatory funnel; the addition of sodium chloride decreases the miscibility of the aqueous and organic layers and helps in avoiding the formation of unwanted emulsions. In this experiment, the presence of saponins in the extract favors the formation of stubborn emulsions that make the separation of layers difficult. To minimize this problem *do not shake the separatory funnel vigorously* (as is customary in liquid-liquid extraction), instead, just invert and rotate the separatory funnel gently several times for a period of about five minutes. During this process, vent the funnel often by slowly opening the stopcock to release any pressure buildup (see Figs. 5.7 and 5.8). *When releasing the pressure, point the stem of the separatory funnel away from your face and others.*

Safety First

- Methylene chloride is a possible carcinogen. Handle it in a well-ventilated place.

- When releasing the pressure, point the separatory funnel stem away from your face and others.

- Acetone and petroleum ether are flammable.

Separate both layers as follows: support the separatory funnel on a ring, remove the stopper and carefully open the stopcock, collecting the lower layer (the organic phase) in a 125-mL Erlenmeyer flask (Fig. 5.9). Some emulsion may be present in the organic phase. The organic layer is then treated with a drying agent with the dual purpose of drying it and breaking the emulsion. Add about 4 g of anhydrous magnesium sulfate to the organic phase. Cover the flask with a cork stopper; set it aside for about 10 minutes with occasional swirling. Meanwhile, repeat the extraction on the remaining aqueous phase with a fresh 25-mL portion of methylene chloride. Collect the organic layer in the flask containing the first methylene chloride extract. Add one more gram of anhydrous magnesium sulfate, stopper the flask, and let the solution dry for 5 minutes with frequent swirling. Filter the organic phase using a small piece of cotton in the neck of a glass funnel. With a glass rod apply gentle pressure to secure the cotton plug to the neck of the funnel while doing the filtration. Collect the filtrate in a pre-tared, dry, 100-mL round-bottom flask. Rinse the magnesium sulfate with 2 mL of fresh methylene chloride. The filtrate should be clear without traces of water or magnesium sulfate. If water is still present, transfer the liquid back to the Erlenmeyer flask and repeat the drying with about 1–2 g of anhydrous magnesium sulfate. If magnesium sulfate passes through the cotton plug, refilter the solution using a larger piece of cotton. Remove the solvent using a rotatory evaporator (rota-vap) equipped with a lukewarm water bath and connected to a cold trap (Fig. 5.17). Weigh the product. Calculate the percent yield based on the amount of tea originally used.

E5.4   PURIFICATION OF CAFFEINE BY SUBLIMATION

> **Background reading:** 4.7
> **Estimated time:** 1/2 hour

Scrape as much of the crude caffeine as possible out of the round-bottom flask into the bottom of a clean, dry Petri dish. Place the Petri dish on a hot plate and cover the dish with a stack of three disks of filter paper (such as Whatman 1). Cover the disks with a 250-mL beaker or a 250-mL Erlenmeyer flask containing 30–50 mL of water (see Fig. 5.24). The beaker or Erlenmeyer flask helps keep the filter paper in place and provides cooling. Turn on the heat to a low-medium setting. After a few minutes you will observe, when looking from the side, the white vapors of caffeine inside the Petri dish. Let the sublimation continue for some 5 minutes more and then turn off the heat. Let the system cool down at room temperature. In the meantime start cleaning glassware and putting things away.

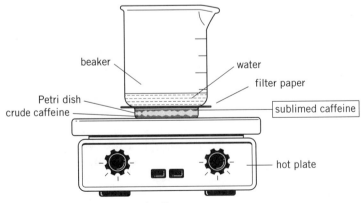

Figure 5.24   Microscale sublimation of caffeine.

Carefully pour out the water from the beaker or Erlenmeyer flask and scrape the purified caffeine from the filter paper onto a piece of preweighed weighing paper. Obtain the melting point of your product. Weigh the purified caffeine and determine the yield of the sublimation. Note the color and shape of crude and sublimed caffeine crystals.

E5.5   PURIFICATION OF CAFFEINE BY RECRYSTALLIZATION (ALTERNATIVE PROCEDURE)

> **Background reading:** 4.4
> **Estimated time:** 1/2 hour

The crude caffeine obtained upon evaporation of the methylene chloride can be purified by recrystallization from a mixture of solvents (section 4.4). Scrape as much of the crude caffeine as possible out of the round-bottom flask into a piece of preweighed weighing paper; determine the mass of caffeine to be recrystallized. Transfer the crude caffeine to a large test tube; dissolve the solid by adding hot acetone while heating in a water bath on a hot plate (less than 5–6 mL of acetone should be necessary to dissolve the crude caffeine). Stir constantly with a glass rod. Add hot petroleum ether drop-wise until the cloudiness that forms is persistent. Add a few drops of hot acetone just enough to make the cloudiness go away. Cool the solution and collect the crystals by vacuum filtration using a Hirsch funnel. Allow the solid to dry on the filter with the vacuum on. Weigh the purified caffeine and determine the yield of the recrystallization. Obtain the melting point of your product. Note the color of the product.

Cleaning Up

- Dispose of the tea bags and tea leaves in the trash can.
- Dispose of the aqueous layer after the extraction in the container labeled "Caffeine–Aqueous Waste."
- Dispose of the calcium carbonate, filter paper, magnesium sulfate, and cotton plug in the container labeled "Caffeine–Solid Waste."
- Discard the mother liquor from the recrystallization into the container labeled "Caffeine–Organic Solvents Waste."
- Turn in the caffeine in a labeled vial to your instructor.
- At the end of the lab period the instructor will empty the contents of the rota-vap traps (methylene chloride) into a "Reclaimed Methylene Chloride" bottle.

EXPERIMENT 5 REPORT

Pre-lab

1. What other compounds are extracted along with caffeine into the "tea"? Make a flowchart for the isolation of caffeine (similar to that shown in Fig. 5.23) and indicate the final destination of the following compounds: cellulose, amino acids, glucose, hydrolyzed tannins, chlorophylls, calcium carbonate.

2. Why do you use $CaCO_3$ in the extraction of caffeine? Explain why $CaCO_3$ is a base.

3. Why do you add NaCl before extracting the aqueous layer with methylene chloride? How does magnesium sulfate break the emulsion?

4. Could you use methanol instead of methylene chloride to perform the liquid-liquid extraction? (Hint: check Miscibility Chart, (Fig. 5.1).

5. Make a list of the glassware needed in the isolation and purification of caffeine.

6. Make a table with the physical properties (molecular mass, b.p., m.p., density, solubility, flammability [for solvents only], and toxicity/hazards) of caffeine, methylene chloride, calcium carbonate, acetone, and petroleum ether.

7. One gram of caffeine dissolves in 55 mL of water, 5.5 mL of chloroform, 530 mL of diethyl ether, and 100 mL of benzene. Calculate the solubility of caffeine in these four solvents in mg/mL. Estimate the partition coefficient of caffeine between chloroform and water, diethyl ether and water, and benzene and water. Which solvent would you choose to extract caffeine from an aqueous solution? Why?

8. Chloroform and benzene were widely used in the past as extracting solvents for liquid-liquid extractions. They have been replaced by other solvents such as methylene chloride and toluene. Why? (Hint: check P.E.L., Table 2.2.)

9. Foam usually develops when tea is brewed but not in coffee. Can you give a chemical explanation for this observation?

In-lab

1. Summarize the learning objectives of this experiment. Mention the new techniques introduced.

2. Calculate the percentage yield of the extraction:

$$\% \text{ yield} = \frac{\text{mass crude caffeine}}{\text{mass tea leaves}} \times 100$$

Compare % yield with the typical amount of caffeine found in tea leaves. Discuss your results.

3. Calculate the % Recovery of the sublimation or recrystallization process and discuss sources of error.

$$\% \text{ Recovery} = \frac{\text{mass purified caffeine}}{\text{mass crude caffeine}} \times 100$$

4. Report the melting point of caffeine after recrystallization or sublimation. Compare it with literature values.

If you purified caffeine by sublimation, answer the following questions:

5. Why is it possible to sublime caffeine?

6. Explain in a few words the process of sublimation.

If you purified caffeine by recrystallization, answer the following questions:

7. Which solvent is more polar acetone or petroleum ether? Base your answer on their structures (check section 2.3 and Table 2.2).

8. Why is petroleum ether added?

Distillation
Separation and Purification of Organic Liquids

Distillation is one of the oldest methods ever developed for the purification of liquids. It has been used for centuries to concentrate dilute alcoholic solutions such as wine and beer and to obtain perfumes from fruits and flowers. In a few words, distillation consists of heating a liquid to its boiling point and separating the condensed vapors. Different types of distillation are used in the organic chemistry laboratory such as **simple, fractional, vacuum**, and **steam distillation**. Each type has different applications as you will see in the next sections. To understand distillation we should first take a look at the boiling point of liquids.

6.1 BOILING POINT

At any given temperature a liquid is in equilibrium with its vapor. As the temperature increases, the *vapor pressure* of the liquid also increases. When the vapor pressure equals the *applied pressure* (usually atmospheric pressure) the liquid boils. The *normal boiling point* of a liquid is the temperature at which the vapor pressure equals 760 mm Hg (760 mm Hg = 760 Torr = 1 atm). Boiling points are quite sensitive to the external pressure. As the external pressure decreases the boiling point also decreases (Fig. 6.1). For example, it is a very well-known phenomenon that at high altitudes it takes longer to cook food by boiling. The reason for this is the decrease in the boiling point of water caused by the low atmospheric pressure; the most prevalent chemical reaction in cooking is the denaturation of proteins, a process dependent on temperature and time. To compensate for a lower temperature, longer cooking times are needed.

Like the melting point, the boiling point is an important physical property used to characterize organic materials. Pure compounds normally boil within a range of 1°C. Contrary to melting points, which are largely unaffected by changes in the external pressure, the marked effect of the external pressure on the boiling points makes them less reliable to assess purity and identify compounds. Boiling points should always be reported with the external pressure at which they were determined. For example, the boiling point of toluene at 760 Torr is 110.6°C and at 28 Torr is 25°C. The pressure is usually indicated by a subscript:

$$bp_{760} \ 110.6°C \qquad \text{and} \qquad bp_{28} \ 25°C$$

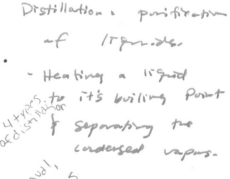

Figure 6.1 Boiling point as a function of the external pressure.

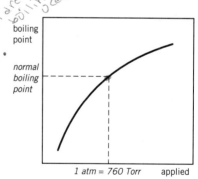

113

The boiling point of a liquid is a measure of its **volatility** or *its propensity to vaporize*. For example, in a mixture of methanol (b.p. 64.7°C) and ethanol (b.p. 78.5°C) methanol is the most volatile component.

6.2   BOILING POINT AND MOLECULAR STRUCTURE

The boiling point of a liquid depends on its polarity, its molecular mass, and the overall size and shape of the molecule. For a liquid to boil, the **intermolecular interactions** that keep the molecules together should be disrupted. These interactions include H-bonds and dipole-dipole interactions (Section 2.1). Note that *no covalent* bonds are broken during boiling!

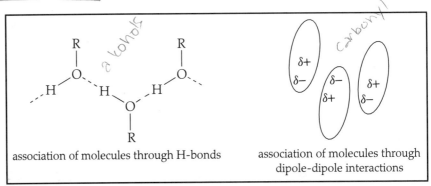

association of molecules through H-bonds   association of molecules through dipole-dipole interactions

Molecules with a hydrogen atom attached to oxygen or nitrogen are **H-bond donors**. These molecules associate through a network of H-bonds, which should be broken before the molecule can pass from the liquid to the vapor phase. As a result, these compounds have higher boiling points than similar compounds unable to form H-bonds. Examples of compounds associated through H-bonds include water and alcohols (methanol, ethanol, etc.). For example, the boiling point of *n*-butanol is 117.7°C, and the boiling point of *n*-pentane, an alkane with similar molecular mass, is 36.1°C.

n-butanol MM 74.12	*n*-butanal MM 72.11	*n*-pentane MM 72.15	neopentane MM 72.15
$CH_3CH_2CH_2CH_2OH$	$CH_3\ CH_2\ CH_2\ \overset{\overset{O}{\|\|}}{C}{-}H$	$CH_3CH_2CH_2CH_2CH_3$	$CH_3\overset{CH_3}{\underset{CH_3}{\overset{\|}{\underset{\|}{C}}}}CH_3$
bp 117.7°C	bp 74.8°C	bp 36.1°C	bp 9.5°C

Dipole-dipole interactions take place between molecules with permanent dipoles such as the carbonyl group. The dipole of one molecule interacts with the dipole from another as shown below. The stronger the interactions, the more energy has to be given to disrupt them, and the higher the boiling point. For example, the boiling point of *n*-butanal (a carbonyl compound) is 74.8°C, and the boiling of *n*-pentane, an alkane with similar molecular mass is, as already mentioned, 36.1°C.

$$\delta{-}\ O\cdots C\ \delta{+}$$

The carbonyl is polarized with a δ− on the oxygen and a δ+ on the carbon, because oxygen is more electronegative than carbon. The δ− of one molecule interacts with the δ+ of another molecule.

Usually, dipole-dipole interactions are weaker than H-bonds, and the boiling points for compounds associated exclusively through dipole-dipole interactions are lower than for those of similar mass but with H-bond association. For example, compare the boiling point of *n*-butanal (74.8°C), a compound incapable of self-association by H-bonds, with that of *n*-butanol (117.7°C).

The shape of the molecule is also important in determining the boiling point. Branched molecules usually have lower boiling points than linear molecules. The branching makes the molecules more spherical, reducing the area of contact between them. This results in weaker interactions between molecules and lower boiling points. For example, the boiling point of neopentane (2,2-dimethylpropane), a branched isomer of *n*-pentane, is 9.5°C.

The molecular mass also affects the boiling point. In general, the larger the molecule, the higher the boiling point. For example, the boiling point of linear alcohols increases as the molecular mass increases. A similar tendency is observed in other families of organic compounds.

alcohols stronger than carbonyls

branched easier to break than linear.

methanol	ethanol	*n*-propanol	*n*-butanol
MM 32.04	MM 46.07	MM 60.09	MM 74.12
CH_3OH	CH_3CH_2OH	$CH_3CH_2CH_2OH$	$CH_3CH_2CH_2CH_2OH$
bp 64.7°C	bp 78.5°C	bp 97.1°C	bp 117.7°C

6.3   SIMPLE DISTILLATION

The Distillation Apparatus

Distillation is one of the major tools used to separate and purify liquids (see "A Brief History of Distillation," presented in the box). The two most common types of distillation are **simple and fractional**. Simple distillation is a practical operation to separate:

- a solid from a liquid
- two liquids with very different boiling points ✳ *essential that the b.ps of the two b/very diff.*

The apparatus for simple distillation is shown in Figure 6.2. It consists of a round-bottom flask, called a **distillation flask**, where the liquid is placed; a **distillation head** with an attached **thermometer** connects the flask to a **condenser**; and a distillate **take-off adaptor** connects the condenser to the **receiving flask**. The condenser consists of a glass tube with a jacket through which tap water circulates against gravity; water should enter the condenser at the lower end and exit at the upper end as indicated in the figure; circulation of water in the opposite direction results in an incomplete filling of the water jacket. When liquids with boiling points higher than 150°C are distilled, breakage of the water condenser may occur due to thermal shock caused by the difference in temperature between the vapors and the cooling water. For such distillations an empty condenser without circulating water, called an **air condenser**, is recommended. *don't boil over 150°C*

Before the distillation is started, a few pieces of a porous material the size of a rice grain, called **boiling chips**, are added to the liquid. They produce a smooth bubbling and prevent bumping of the liquid during distillation. Fragments of porcelain and ✳ broken dinnerware make good boiling chips. To avoid boil-over, the liquid should fill no more than 1/2 to 2/3 of the total capacity of the distillation flask. Depending on the boiling point to be reached, the system is heated with the aid of a heating mantle, a sand bath, or water bath on a hot plate. Small bubbles form constantly on the surface of the boiling chips. Vapors ascend through the distillation head and condense

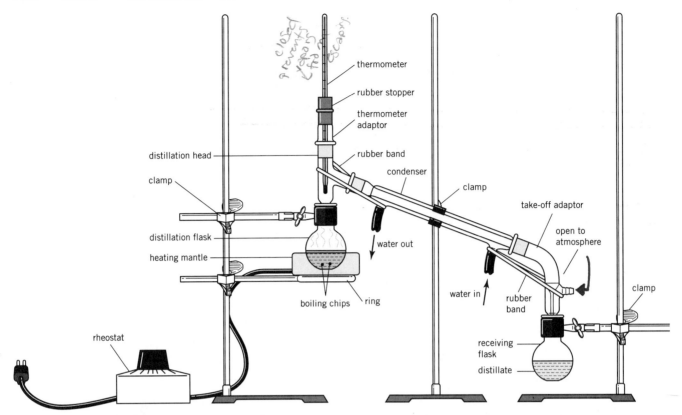

closed prevents vapors from escaping.

Figure 6.2   Simple distillation apparatus.

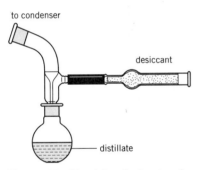

Figure 6.3   Receiving flask setup for hygroscopic liquids.

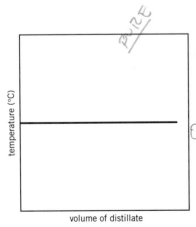

Figure 6.4   Ideal distillation curve for a pure liquid.

around the thermometer and the condenser, which prevents vapors from escaping. The distillate is collected in the receiving flask. The distillation apparatus should always have an opening to the atmosphere at the end, where the distillate is collected. **Never heat a closed distillation system because an explosion will occur.** When hygroscopic liquids are distilled, the receiving flask should be tightly connected to the take-off adaptor through a ground-glass joint, as shown in Figure 6.2; a tube with desiccant (silica, calcium chloride, etc.) should be placed at the end of the side arm open to the atmosphere (Fig. 6.3). This prevents atmospheric moisture from reaching the distillate.

temp should be constant for pure liquid

Liquid-Vapor Equilibrium

During the distillation of a pure liquid the temperature read on top of the distillation head remains constant as long as liquid and vapor are present in the system. If we plot the temperature readings as a function of the volume of distillate we would obtain a horizontal line as depicted in Figure 6.4. This kind of representation where the temperature is shown as a function of the volume of distillate is called a ***distillation curve.***

What happens when a mixture of two liquids with different boiling points is distilled? Does the temperature also remain constant during the distillation or does it increase? What is the composition of the distillate? To answer these questions, let's consider a mixture of two liquids A and B. Let's assume that A has a boiling point of 70°C and B a boiling point of 120°C. In principle, any mixture of A and B is expected to boil between 70°C and 120°C (this is not always the case, as we will see later in Section 6.6). As the proportion of B (the less volatile component) increases, the boiling point of the mixture also increases. This is shown in Figure 6.5a by the curve marked "liquid," which gives the boiling point of the mixture as a function of its composition (in the left corner of the *x* axis there is 100% A, and in the right corner 100% B). A mixture of 50% A and 50% B, for example, boils at 90°C.

What is the composition of the vapor in equilibrium with the boiling liquid? If the liquid has a composition of 50% A and 50% B, is the vapor also 50% A and 50% B? The answer is no. Because A is more volatile than B, the vapor is richer in A than the liquid. The vapor may have, for example, 70% A and 30% B (point *b*, Fig. 6.5*a*). *The vapor is always richer than the liquid in the more volatile compound regardless of the composition of the liquid*. The composition of the vapor in equilibrium with the liquid is given by the "vapor" curve in Figure 6.5*b*. You can see that at any temperature the vapor has proportionally more A than the liquid. This graph is called a **liquid-vapor phase diagram** and is very useful in explaining how distillation works.

[handwritten margin note: so even if liquid has less B compos. it will have more A comp in vapor]

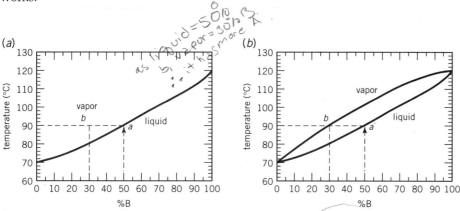

[handwritten note above graph: Liquid = 50 B. vapor = 30 B. as it has more A]

Figure 6.5 *a*) Boiling point of a liquid mixture as a function of the composition; *b*) vapor-liquid phase diagram for a typical mixture.

A Brief History of Distillation The development of distillation, more than any other physico-chemical process, is strongly tied to the early evolution of chemistry. Before the concept of microorganisms as cause of infections was well established toward the middle of the nineteenth century, the general belief was that infectious diseases were caused by the miasma or unpleasant smells present in the air. The stench of rotten flesh, the offensive emanations from sewers and slaughterhouses, were thought to cause serious health problems that could be somehow prevented by the inhalation of pleasant aromas. The use of saffron, musk, thyme, lavender, sandalwood, among many other natural extracts, was recommended to treat the sick and fumigate dwellings. Through the centuries, the profession of perfumer was highly regarded since it was the equivalent of our modern pharmacist and pest-control specialist. Not surprisingly, the oldest written record of a chemical process describes the preparation of perfumes by maceration of myrrh, calamous, and cypress in "fresh, good water of the well," followed by slow distillation. These early perfumery operations are described in cuneiform Akkadian characters inscribed in tablets found in Mesopotamia (dated around 1250 BC); mentioned also in these tablets are the names of the earliest chemists on record: Tapputi-Belatekallim, the Perfumeress and [...] ninu, the Perfumeress. These women can be considered true predecessors of modern chemists since they worked out their own preparation methods involving specialized operations such as extraction, sublimation, and distillation. One of the most commonly used containers for distillation in ancient Mesopotamia was the *diqaru* (see the figure), which is the oldest still known. Diqaru vessels dating back to ca. 3500 BC. were found in northern Mesopotamia; one of them can be admired at the University Museum in Philadelphia. The mixture to be distilled was placed in the inner cavity (with a capacity for more than 30 liters), the diqaru was covered with a lid and heated in holes specially dug in the ground; the distilled liquid was collected in the trough around the rim (with a capacity of 2 liters) and retrieved by soaking it up with a bandcloth. The diqaru can be considered the oversized ancestor of the current Hickman still used in microscale distillations.

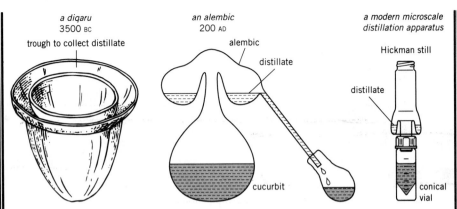

Evolution of the distillation apparatus through the centuries (not on the same scale).

Alexandrian alchemists of the first century AD devised the *alembic*, where the condensed vapors were collected and retrieved as shown in the figure; the alembic was directly attached to the *cucurbit* (flask); by extension the whole setup was known as alembic. For centuries alembics were used by Arabic alchemists to prepare elixirs and aromatic oils to treat the sick. It is believed that the water-cooled condenser is a European invention of the twelfth century related to the distillation of alcohol. The term "alcohol" was introduced in Europe in the sixteenth century; it is derived from the Arabic words 'al-khol," which can be translated as "finely divided spirit." Although the distillation of fermented beverages is probably a very old procedure (the Chinese may have distilled alcohol from wine as early as the fourth century AD), it did not become widespread until the eighteenth century, probably due to the lack of suitable distillation apparatus for massive production. In the mid-nineteenth century, distilleries producing brandy, whisky, rum, gin, and vodka mushroomed in Europe and the Americas. Around that time, Charles Mansfield, a student in August W. von Hofmann's laboratory in London, developed fractional distillation for the separation of benzene and toluene from tar oil. This pioneer work can be considered the birth of petroleum chemistry and the beginning of a new era in the history of our civilization. Curiously, Mansfield himself was the first casualty of an oil spill; in 1855 he accidentally died in a benzene fire during the operation of his process; he was 36 years old.

Bibliography

1. For a delightful exploration of the hidden connections between smell and culture the reader is referred to "Scent, the Mysterious and Essential Powers of Smell," A. Le Guérer, Turtle Bay Books, New York, 1992.

2. Perfumery in Ancient Babylonia. M. Levey, *J. Chem. Ed.* **31**, 373–375, 1954.

3. Babylonian Chemists, M. Levey, in "Great Chemists". E. Farber ed. Interscience, New York, 1961.

4. The Norton History of Chemistry. W.H. Brock. Norton, New York, 1992.

5. Arabic Chemists, M. Levey, in "Great Chemists," E. Farber ed. Interscience, New York, 1961.

The Distillation Process

How can we make use of Figure 6.5*b* to explain what happens during a simple distillation? Let's consider again a 50–50 mixture of A and B. As the liquid is heated, the temperature increases; at 90°C the liquid begins to boil (Fig. 6.6). The first vapor in equilibrium with the original mixture, as already shown, is richer in A than the liquid (point b, 70% A and 30% B). The vapor condenses on the distillation head and is collected in the receiving flask. This removal of vapor (which is richer in A than the

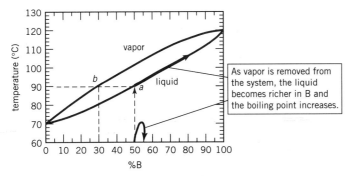

Figure 6.6 Vapor-liquid phase diagram showing the composition of the first drop of distillate, b, for a liquid of composition a.

original liquid) reduces the proportion of A in the liquid left behind in the distillation flask. After collecting the first drop of condensed vapor, the composition of the liquid has changed from 50% A–50% B to one slightly richer in B; for instance, 49% A–51% B. This mixture enriched in B now boils at a temperature slightly higher than 90°C and produces a vapor with a lower percentage of A than the "first" vapor (although still richer in A than the liquid). As this vapor is removed from the system by condensation, the liquid becomes even richer in B. Again the boiling point increases.

You should bear in mind that distillation is a continuous process; this illustration with "first" and "second" vapors is an oversimplification to facilitate the visualization of the process. If we plot the boiling point as the distillation progresses we would obtain a graph similar to that depicted in Figure 6.7, where the temperature increases steadily during the distillation.

If we collect the distillate in small fractions, changing the receiving flask every five-degree increase in the temperature, and analyze each fraction to determine the percentage of A and B (this type of analysis can be done by gas chromatography, as you will see in the next experiment), we may find results similar to those shown in Table 6.1.

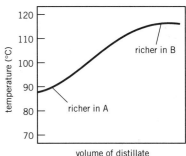

Figure 6.7 Distillation curve obtained by simple distillation of an A–B mixture.

Table 6.1 **Composition of Different Fractions After Simple Distillation**

Fraction	Boiling range	%A	%B
1	85–90	70	30
2	90–95	62	38
3	95–100	54	46
4	100–105	42	58
5	105–110	30	70
6	110–115	18	82
7	115–120	5	95

As can be observed from the data gathered in the table, none of the fractions is pure. The first fraction is richer in A (the more volatile compound) and the last fraction is richer in B, but the separation obtained is very poor. The separation of both components can be improved by redistillation of each fraction. For example, if fraction 1 (70% A–30% B) is subjected to simple distillation again, the first fraction would be richer in A (for example, 90% A–10% B). Fractions with the same boiling range from the first and second distillation can be combined and redistilled. This process can be repeated over and over until one eventually obtains pure A and pure B. As you can guess, this is a highly time-consuming operation that we hardly ever perform. The same results can be achieved by **fractional distillation** in one single step.

Let's close this section by emphasizing that what makes separation of two liquids by distillation possible is that at any given temperature and fixed applied pressure *the vapor in equilibrium with liquid is richer than the liquid in the more volatile compound*. If vapor and liquid in equilibrium have the same composition, separation by distillation is not possible (at least at that applied pressure). We will come back to this in Section 6.6 (**Azeotropic mixtures**).

6.4 FRACTIONAL DISTILLATION

In fractional distillation, a **fractionating column** is inserted between the distillation flask and the distillation head (Fig. 6.8). There are several kinds of fractionating

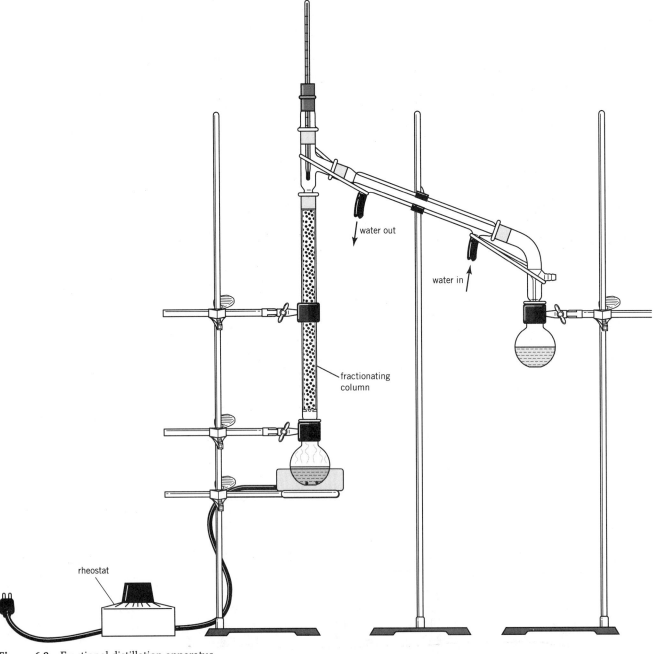

Figure 6.8 Fractional distillation apparatus.

columns as we will discuss in Section 6.7. The feature common to all of them is that they provide a large surface area in which the mixture to be separated can be continuously vaporized and condensed.

To understand how fractional distillation works, let us consider a mixture of A (10%) and B (90%), L_1, (Fig. 6.9), which is subjected to distillation using a *very efficient column*. The mixture is heated and boils at approximately 112°C. The vapor in equilibrium with the liquid mixture has composition V_1, which is richer in A than the liquid. The vapors ascend through the column and condense upon heat exchange with the cooler surface of the column. The liquid formed by condensation of the vapors has identical composition to the vapor (L_2); notice that a vertical line (which means no change in composition) connects V_1 and L_2. As long as heat is supplied to the system, more vapors form and ascend through the column. These vapors transfer heat to the liquid of composition L_2, which then vaporizes to give vapor of composition V_2; the temperature at this stage in the column is approximately 103°C. These vapors move up in the column to a cooler region and condense again to give liquid L_3. L_3 is in equilibrium with V_3 at 93°C. As can be observed, the composition of the vapors V_1, V_2, and V_3 is increasingly richer in A, the more volatile compound. The process of vaporization-condensation continues along the fractionating column as long as heat is provided. At the top of the column, liquid L_6 is in equilibrium with V_6, which is nearly 100% A; the temperature at this point is 70°C, the boiling point of A. V_6 emerges from the column and condenses. This fractionating process goes on along the column as long as A is present in the distillation flask. While A is being removed from the system the temperature at the top of the column remains constant at its boiling point of 70°C. When A has been completely removed, component B distills out and the temperature suddenly rises to 120°C.

(a) (b)

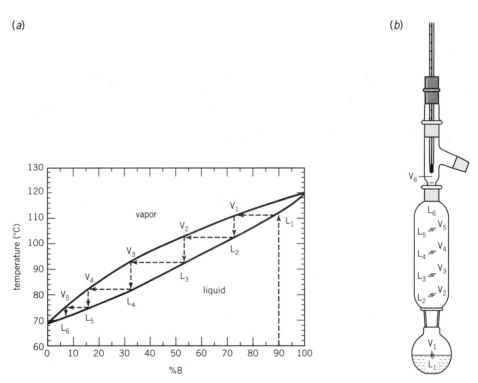

Figure 6.9 *a*) Vapor-liquid phase diagram showing vaporization-condensation cycles; *b*) vaporization-condensation cycles inside the column.

A distillation curve that represents this process is shown in Figure 6.10 along with a typical curve obtained by simple distillation. In an ideal fractional distillation, like the one described here, two distinct fractions are obtained. The first corresponds to the component with lower boiling point and the second to the high-boiling point

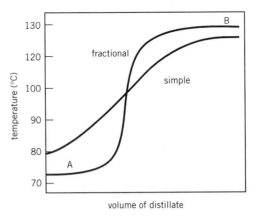

Figure 6.10   Simple and fractional distillation curves.

component. What characterizes a good fractional distillation is the sudden increase in temperature between both fractions, or in other words, a very small volume distilled at temperatures other than the boiling points of the pure liquids. In simple distillation, on the other hand, a much more gradual increase in temperature is observed, reflecting the impure nature of the distillate.

Theoretical Plates

The success of the separation of two liquids depends largely on the difference in their volatilities and the efficiency of the fractionating column. In general, the larger the difference in boiling points, the easier the separation. The phase diagram for a typical binary mixture is shown in Figure 6.11. A total of five condensation-vaporization cycles are needed to separate A from B; therefore, a fractionating column used to distill this mixture should operate under conditions that guarantee that at least five condensation-vaporization cycles take place along the column. Each one of these cycles is called a **theoretical plate**. In a fractionating column a theoretical plate is defined as the hypothetical region of the column in which the vapors emerging from the top of the region have the composition of the vapors that would be in stationary equilibrium with the liquid entering the bottom of the region (Fig. 6.12), as determined by the liquid-vapor phase diagram.

Good fractionating columns have a large number of theoretical plates. This number is determined by the geometry, packing, and operating conditions of the column, and also by the mixture to be separated. To separate liquids with a difference in boiling point between 20 and 40°C, 10–20 theoretical plates are usually required. If the difference in boiling points is only of a few degrees, a column with as many as 100 theoretical plates may be needed.

The separation of liquids by distillation is based on the physical separation of the vapor from the liquid. The vapor is richer than the liquid in the more volatile component and thus when the vapor condenses the resulting liquid is enriched in this component. If the vapor had the same composition as the liquid, separation by distillation wouldn't be possible. The boiling point difference is only a crude indicator of the ease of a separation. The true indicator is the difference between the "liquid" and "vapor" curves in the phase diagram. If both curves run very close to each other, then vapor and liquid in equilibrium have similar compositions and thus the separation is difficult. If, on the contrary, the vapor curve is well separated from the liquid curve, the vapor has a very different composition from the liquid and the separation is easier. This is illustrated in Figure 6.13*a* and *b*, with two different mixtures with the same boiling points for the pure components but different phase diagrams. Mixture C-D is easier to separate than mixture E-F because C-D mixtures have very different "liquid" and "vapor" curves. It takes about four cycles to separate C from D, while it takes more than seven to separate E from F.

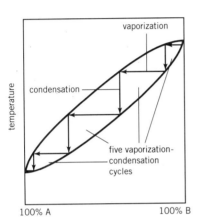

Figure 6.11   Temperature versus composition phase diagram for a hypothetical mixture with 5 vaporization-condensation cycles.

Figure 6.12   The same column as in Figure 6.9, showing six theoretical plates.

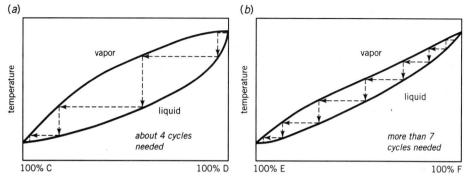

Figure 6.13   Two mixtures with the same boiling points for the pure components but different phase diagrams. Mixture C–D is easier to separate than mixture E–F.

6.5   IDEAL AND NONIDEAL SOLUTIONS

The vapor-liquid phase diagrams shown so far correspond to **ideal solutions** of two liquids. An ideal solution is one in which Raoult's law is obeyed by each component over the full range of concentration. It was found experimentally by Raoult that for a mixture of A and B, the vapor pressure of A, p_A, when A is in very large excess, is:

$$p_A = x_A p_A^o \tag{1}$$

where x_A is the mole fraction of A in the liquid, and p_A^o is the vapor pressure of pure A at a given temperature. In other words, when A is the solvent and B is the solute, the vapor pressure of A is directly proportional to its mole fraction. Likewise, for dilute solution of A in B (B is the solvent) the vapor pressure of B is given by Raoult's law:

$$p_B = x_B p_B^o \tag{2}$$

The total pressure for the system is:

$$p_T = p_A + p_B \tag{3}$$

When a binary mixture of liquids A and B obeys Raoult's law in the entire range of concentrations the mixture is called ideal. A vapor pressure-versus-composition diagram for an ideal solution is shown in Figure 6.14a. Only a limited number of pairs

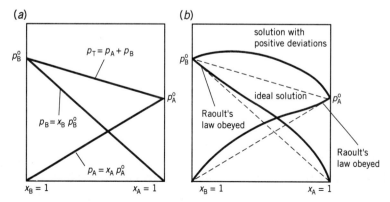

Figure 6.14   Vapor pressure versus composition diagrams: *a*) ideal solution; *b*) binary mixture with positive deviations from Raoult's law.

of liquids behave ideally. Benzene-toluene, *n*-hexane-*n*-heptane, and *ortho*-xylene-*meta*-xylene are a few examples of ideal solutions. Ideal solutions can be separated into their components by fractional distillation.

In general, two liquids form an ideal solution when they are very similar in size, shape, and intermolecular forces. More often than not, the molecular interactions between two liquids A and B (A...B) will be different from the molecular interactions between the molecules of pure A (A...A) and pure B (B...B); when this happens, **nonideal solutions** are obtained. A typical vapor-pressure-versus-composition diagram for a system of two liquids that deviate from Raoult's law is shown in Figure 6.14*b*. The vapor pressure for each component is represented by the curves. Raoult's law is obeyed by A only when x_A approaches 1 and by B when x_B is close to 1. At intermediate concentrations neither compound obeys Raoult's law. In the case illustrated in the figure, the actual vapor pressures are higher than the expected values for an ideal solution (dotted lines). Solutions for which the vapor pressure is higher than expected are said to have positive deviations from Raoult's law. These mixtures boil at temperatures lower than the boiling points of the individual components. It is also possible to find systems in which the vapor pressure of each component is lowered by the presence of the other component; such systems have negative deviations from Raoult's law and boiling points higher than the boiling points of the components.

6.6 AZEOTROPIC MIXTURES

Mixtures such as ethanol-water, ethanol-*n*-heptane, chloroform-acetone, and dioxane-formic acid form nonideal solutions. In these mixtures, *the volatility of each component is strongly affected by the presence of the other component.*

When the attraction between the molecules of the two components in the mixture is weaker than the attraction between the molecules of each component, the volatility of the liquids is enhanced when they are mixed. As a result, there is an increase in the vapor pressure and, consequently, a decrease in the boiling point; these systems present strong positive deviations from Raoult's law. Examples of such mixtures are ethanol-water, ethanol-*n*-hexane, and acetone-*n*-pentane.

A typical phase diagram for a mixture of A and B with strong positive deviations from Raoult's law is shown in Figure 6.15. The boiling point of pure A is 78°C; the boiling point of B is 69°C. There is a particular mixture represented by point a in Figure 6.15 (19% A and 81% B) for which the compositions of vapor and liquid in equilibrium are identical. This mixture has a sharp boiling point of 59°C, lower than the boiling points of A and B; at this temperature the liquid boils without change in composition. This mixture is called an **azeotrope** (from the Greek: to boil unchanged). Because the boiling point of this azeotrope is lower than the boiling points of A and B, this is called a **minimum-boiling azeotrope**. When the azeotropic mixture is heated (point a, Fig. 6.15), the vapor has the same composition as the liquid, and consequently, *the distillate has the same composition as the original liquid and no separation is achieved by distillation.*

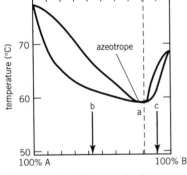

Figure 6.15 Minimum-boiling aze-otrope.

Azeotropic mixtures behave like pure liquids. *Pairs of compounds that form azeotropic mixtures cannot be completely separated by fractional distillation.* For instance, mixtures of A and B with composition different from the azeotrope can be only partially separated into their components by fractional distillation. Any mixture richer in A than the azeotrope (for example, point b, Fig. 6.15) gives azeotrope and pure A by distillation. Any mixture richer in B than the azeotrope (for example, point c, Fig. 6.15) produces the azeotrope and then B. The distillation curves for these three cases are shown in Figure 6.16. Azeotropic mixtures can be separated by distillation by changing the applied pressure because the shape of the phase diagram, including the composition and boiling point of the azeotrope, depends strongly on the applied pressure.

Mixtures where the boiling point of the azeotrope is higher than the boiling point of each component, called **maximum-boiling azeotropes**, are also observed although

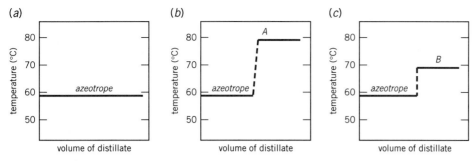

Figure 6.16 Distillation curves for A-B mixtures of: *a*) azeotropic composition, point a, Figure 6.15; *b*) composition b, Figure 6.15; *c*) composition c, Figure 6.15.

they are less frequent than minimum-boiling point azeotropes. These mixtures occur when the interactions between the two components in the mixture are stronger than the interactions between the individual components. Pairs of compounds with strong interactions and maximum-boiling azeotropes include acetic acid-pyridine and dioxane-formic acid. The components of these mixtures are associated by strong H-bonds.

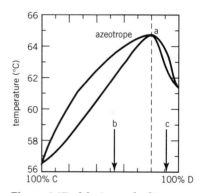

Figure 6.17 Maximum-boiling azeotrope.

A typical phase diagram for a maximum-boiling azeotrope is shown in Figure 6.17. The boiling points of C and D are 56.6°C and 61.5°C, respectively. The azeotropic mixture (point a) boils unchanged with a composition of 20% C and 80% D at 64.7°C. Any mixture with more C than the azeotrope (for example, point b) produces pure C and the azeotrope upon fractional distillation, while any mixture with more D than the azeotrope (for example, point c) gives pure D and the azeotrope.

6.7 FRACTIONATING COLUMNS

The success of a fractional distillation depends largely on the efficiency of the fractionating column. A good fractionating column should provide: a) a large surface area where the ascending vapors come in contact with the descending liquid; b) a good mixing between vapor and liquid; c) minimum heat losses, especially when liquids with high boiling points are distilled; d) maximum number of theoretical plates. The types of columns most commonly used in the organic chemistry laboratory are: **Vigreux, Hempel**, and **spinning-band** (Fig. 6.18).

The Vigreux column consists of a glass tube (10–60 cm long and 1–2 cm diameter) with indentations pointing inward. The indentations direct the liquid toward the center of the column and provide a larger surface area where cycles of condensation-vaporization occur. Vigreux columns are very popular ones; they are relatively inexpensive but provide only a moderate separating power. Hempel columns consist of a glass tube (10–100 cm long and 1–3 cm diameter) filled with packing material such as stainless steel sponge or small beads, rings, or helices made of glass; indentations at the lower end of the tube prevent the packing material from falling into the distillation flask. Hempel columns provide larger surface area than

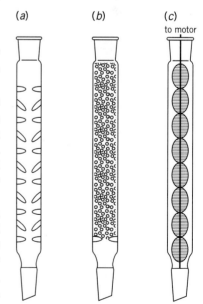

Figure 6.18 Different types of fractionating columns; *a*) Vigreux; *b*) Hempel; *c*) spinning-band.

Vigreux columns of similar dimensions and are more efficient because they offer a larger number of theoretical plates. Proper column packing is critical to achieve good separations. *Underpacking* reduces the surface area and therefore less condensation-vaporization cycles are possible along the column. *Overpacking* results in flooding of the column as the condensed vapors are not able to flow down freely and liquid fills the entire length of the column; flooding should be avoided because it greatly reduces the efficiency of the separation. When using stainless steel sponge, about 3 grams of this packing material is enough to pack a column 16 cm long and 1.3 cm in diameter. Spinning-band columns consist of a glass tube in which a spiral made of Teflon or stainless steel gauze fits tightly without touching the walls. The spiral rotates at 600–3000 rpm driven by a motor usually located on the top of the column. The turbulence caused by the rotation of the spiral favors a good interaction between the ascending vapors and the descending liquid. Spinning-band columns are very efficient and recommended for difficult separations; their only drawback is their high price.

To increase the efficiency of the fractionating column, heat losses along the column should be minimized; to this end, in the laboratory, the glass tube can be wrapped in aluminum foil.

6.8 MICROSCALE DISTILLATION

Purification of small volumes of liquids is not as easy as the purification of a similar amount of solid. Simple and fractional distillation of a volume of at least 5 mL can be easily achieved with the apparatus shown in Figures 6.2 and 6.8 with minor variations depending on the total volume to be distilled. Microscale glassware is used when the volume is between 0.5 and 5 mL. Distillation of less than 0.5 mL of liquid is usually not recommended due to mechanical losses. Preparative gas chromatography and high performance liquid chromatography are better suited to purify very small liquid volumes. A typical microscale simple distillation setup is shown in Figure 6.19.

The sample to be distilled is placed in a small vial or round-bottom flask, a boiling chip is added, and a **Hickman still** head is placed on top of the vial. The Hickman still head has a trough where the condensed vapors collect. Depending on the boiling point of the liquid, a condenser is added on top of the Hickman still. If the boiling point is higher than 150°C, no condenser is necessary; the cool surface provided by the upper neck of the Hickman still is enough to condense the vapors. If the boiling point is between 90°C and 150°C, an air condenser may be placed on top of the Hickman still. If the liquid boils below 90°C a water-jacketed condenser should be used. A mini-thermometer is placed inside the still so that the bulb is located below the trough. The thermometer is supported by a clamp or by a notched stopper that loosely sits on top of the still and allows venting. The distillate collected in the trough can be periodically retrieved with a Pasteur pipet. Hickman still heads with side ports are also commercially available; they are particularly useful to retrieve the distillate when a condenser is used. The side port has a cap and a septum for syringe retrieval of the liquid.

Microscale fractional distillations can be carried out using a small fractionating column provided the **hold-up** volume of the column is small. The hold-up volume is the minimum volume of liquid necessary to wet the column; in other words, it is the amount of liquid that stays in the column at the end of the distillation and therefore is lost. A microscale air condenser filled with stainless steel sponge makes a good column when the volume to be distilled is between 4 and 5 mL; its hold-up volume is 1–2 mL depending on the tightness of the packing. To distill volumes between 0.5 and 5 mL, microscale spinning-band columns have been devised. Although different types are commercially available, they all consist of a piece of threaded Teflon with a small magnet at the bottom; the Teflon shaft is placed in a vacuum-jacketed glass tube or in a modified Hickman still, called a Hickman–Hinkle column (Fig. 6.20). The

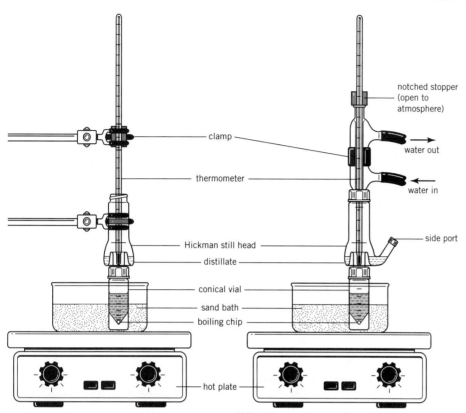

Figure 6.19 Microscale simple distillation apparatus: left, without condenser; right, with water-jacketed condenser.

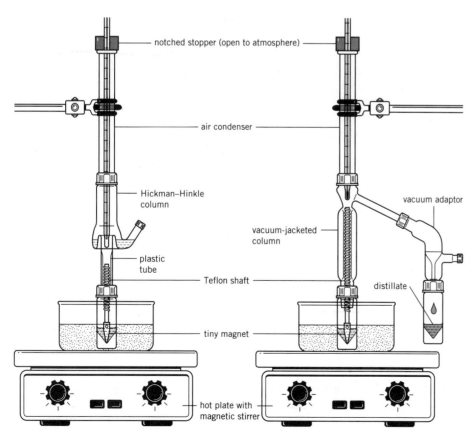

Figure 6.20 Microscale fractional distillation setups: left, with Hickman–Hinkle still; right, with spinning-band column.

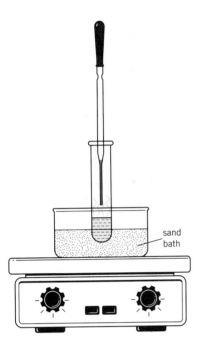

Figure 6.21 Distillation setup for very small volumes.

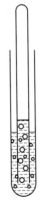

Figure 6.22 Siwoloboff micro-boiling point determination.

spinning of the Teflon shaft is controlled by a conventional magnetic stirrer. In the Hickman–Hinkle setup, the Teflon shaft is surrounded by an insulating plastic tube.

If only a few drops of purified liquid are required, for example, to obtain a spectrum or to use the compound as a standard for chromatography, the following method can be applied. The liquid (approximately 0.5 mL) is placed in a small test tube (12 × 75 mm) along with a small boiling chip; the test tube is *carefully* heated in a sand bath so that the vapors condense one or two centimeters above the liquid level (Fig. 6.21). This operation should be carried out in the fume hood. The air is expelled from a Pasteur pipet with the aid of a rubber bulb; with the rubber bulb still squeezed, the pipet is placed inside the test tube with the tip in the space where the vapors condense. Liquid accumulates in the tip of the pipet and is collected by releasing the bulb. The distillate is transferred to a clean test tube and the process is repeated if more sample needs to be collected. This method is particularly useful to separate liquids with relatively high boiling points from nonvolatile contaminants. Low-boiling point liquids can also be purified in this fashion, but a larger test tube should be used to provide a larger cold surface to condense the vapors.

6.9 BOILING POINT DETERMINATION

Boiling points can be determined using any of the distillation setups previously shown (for example, Figs. 6.2 and 6.19). In the microscale apparatus, the thermometer should be positioned so that the bulb is located in the neck of the still (Fig. 6.19). In the macroscale distillation apparatus (Fig. 6.2), the bulb should be just below the sidearm of the distillation head. To avoid faulty temperature readings the bulb should not touch the glass walls. The heat supply should be such that a gentle boiling is observed. The vapors condense on the thermometer and the temperature increases; a reading is taken when the temperature has finally stabilized. The temperature so determined corresponds to the condensation point of the vapors rather than the boiling point of the liquid; for pure compounds both points are identical. Boiling points determined in this fashion should not be used to assess the purity of the liquid because nonvolatile impurities are left behind during the distillation and therefore they do not affect the condensation point of the vapor. With this limitation in mind, the boiling point determined during distillation is good enough for most applications in the organic chemistry lab. Boiling points should never be determined by immersing the thermometer into the boiling liquid because its temperature could be higher than the true boiling point. When a liquid boils, the temperature is not homogeneous across the entire volume of the liquid; a temperature gradient is established with lower values near the surface and higher values in the proximity of the heating source. This phenomenon is called **superheating**. Thorough stirring of the liquid can substantially reduce the problem but it cannot eliminate it altogether. Due to differences in pressure inside the liquid, the boiling point on the surface of the liquid is always lower than the boiling point in its interior where the pressure is higher. For liquids with density close to 1 g/mL, an increase of about 0.1°C for every 3–4 cm increase in depth is normally observed.

When only a very small volume of liquid is available to determine the boiling point, the micromethod of Siwoloboff is recommended (Fig. 6.22). The liquid (100–200 μL) is introduced in a capillary tube with one sealed end. A second capillary tube of smaller diameter, with one of its ends also sealed, is placed sealed end up inside the first capillary tube. The ensemble is placed in a melting point apparatus (such as those discussed in Unit 4). Heat is slowly supplied to the system. Before the boiling point is reached a few bubbles escape from the small capillary tube; at the boiling point of the liquid a constant stream of bubbles is produced. The presence of volatile as well as nonvolatile impurities affects the boiling point determined by the Siwoloboff's method. Nonvolatile impurities decrease the vapor pressure of the liquid and result in an increase in the boiling point of the liquid. The presence of volatile impurities usually gives a reading below the boiling point of the pure liquid.

6.10   VACUUM DISTILLATION

The distillation of high-boiling-point liquids presents some problems. First of all, very high temperatures are difficult to reach and hold constant in the laboratory and, most importantly, some liquids are oxidized by atmospheric oxygen at high temperatures. To avoid decomposition of high-boiling-point liquids, the distillation should be conducted under reduced pressure, which results in a lower boiling point. This technique is known as "vacuum distillation." This operation is very similar to a regular distillation with the difference that the apparatus is connected to a vacuum line provided by an oil pump or a water aspirator. The receiving flask has three or four receptacles to collect different fractions during the distillation. Changing from one receptacle to the next is easily done by rotating this piece of glassware often called "cow" or "pig" (Fig. 6.23).

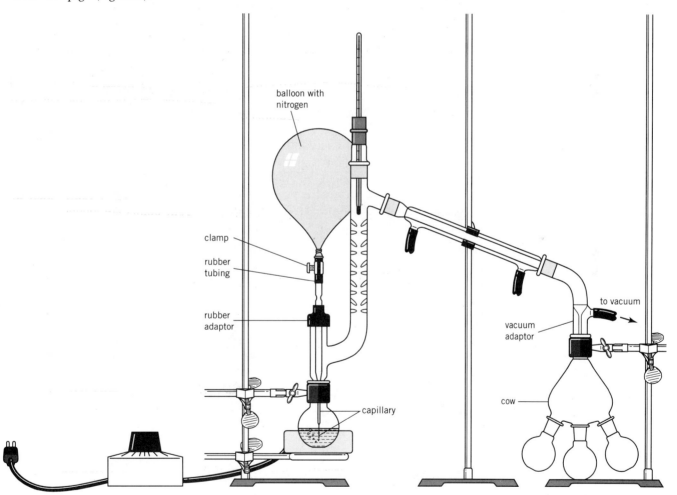

balloon with nitrogen

clamp

rubber tubing

rubber adaptor

capillary

to vacuum

vacuum adaptor

cow

Figure 6.23   Setup for macroscale vacuum distillation.

When the system is connected to the vacuum line, degassing of the liquid occurs as it loses the air dissolved in it; this lack of dissolved gases makes boiling chips rather ineffective in preventing bumping during boiling. Bumping can be avoided by insertion of a capillary tube with very small bore (less than 0.5 mm) to provide a continuous stream of air bubbles into the liquid. To prevent the oxidation of the liquid by air, the capillary tube is normally connected to a reservoir of nitrogen or other inert gases. A party balloon filled with nitrogen provides the necessary amount of gas to perform a macroscale distillation. The balloon is connected to the capillary by means of a piece of rubber tubing equipped with a clamp to control the gas flow. A good capillary tube with very small bore can be obtained by drawing the tip of a Pasteur pipet using a micro-burner. Alternatively, if a very good magnetic stirrer is available, the capillary can be omitted and bumping of the liquid prevented by *very fast* spinning of a stir bar. When performing a vacuum distillation, the system should be connected to the vacuum line first, and once the pressure has stabilized heat is applied. Performing these operations in reverse order, first heat and then vacuum, may result in a violent boil-over. To end a vacuum distillation, the heat source is removed and the vacuum slowly released; if a capillary tube is used, the tip of the capillary is broken by moving it sideways and pressing it against the walls of the flask; this fills the system with nitrogen and prevents liquid from entering the capillary.

6.11 STEAM DISTILLATION

Volatile compounds immiscible with water can be purified by **codistillation** with water, an operation also known as **steam distillation**. The combination of two immiscible liquids always results in a lower boiling point for the mixture than for the separate liquids, as we shall shortly see. Steam distillation is used in the purification of compounds with very high boiling points and in the isolation of volatile compounds from plants.

Let's assume that two immiscible liquids A and B are mixed together and the system is very well stirred. Since A and B are immiscible, the presence of one does not affect the properties of the other. Their partial pressures in the mixture of vapors are the vapor pressures of the pure liquids. In a way, two immiscible liquids behave as if they were separated by an impermeable wall as shown in Figure 6.24. The total pressure inside the container, p_T, is the sum of the vapor pressures of pure A and B at that temperature:

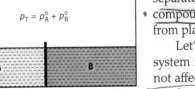

Figure 6.24 Vapor pressure of two immiscible liquids.

$$p_T = p_A^o + p_B^o \tag{4}$$

The vapor pressures of pure A and B and the total vapor pressure, p_T, as functions of the temperature are shown in Figure 6.25.

In this particular example, pure A has a lower boiling point than pure B. It can also be observed from the figure that when A and B are mixed, the boiling point of the mixture (the temperature at which the total vapor pressure equals 760 Torr) is lower than the boiling points of both A and B. In steam distillation one of the components is always water, and thus the boiling point is always lower than 100°C regardless of the boiling point of the other component. Steam distillation is particularly useful to distill compounds with very high boiling points, including volatile solids, which may decompose during normal distillation. A drawback of the technique is that at the end of the steam distillation, the compound must be separated from the water. This can be done by liquid-liquid extraction or by filtration (if the compound is a solid).

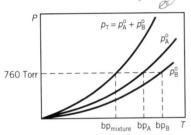

Figure 6.25 Vapor pressures of two immiscible compounds and their mixture as functions of the temperature.

The distillate of a steam distillation consists of water and the volatile compound in question. The amount of each compound can be calculated with the help of Dalton's law of partial pressures. According to this law, the mole fraction of each component (y_A and y_B) in the vapor phase is given by the ratio of its partial pressure over the total pressure (Eq. 5):

$$y_A = \frac{p_A^o}{p_T} \tag{5a}$$

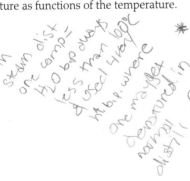

and

$$y_B = \frac{p_B^o}{p_T} \qquad \text{moles } B = \frac{\text{Pres. } B}{\text{Total } p} \qquad (5b)$$

If the vapor is condensed, the resulting distillate has the same composition. The ratio of the mole fractions for A and B in the distillate is then given by Equation 6, which results from dividing Equation 5a by Equation 5b:

$$\frac{y_A}{y_B} = \frac{p_A^o}{p_B^o} \qquad (6)$$

Since $y_A = n_A/n_T$ and $y_B = n_B/n_T$, where n_A, n_B are the number of moles of A, B and n_T is the total number of moles in the distillate, Equation 6 can be rewritten as:

$$\frac{n_A/n_T}{n_B/n_T} = \frac{p_A^o}{p_B^o} \qquad \Longrightarrow \qquad \frac{n_A}{n_B} = \frac{p_A^o}{p_B^o} \qquad (7)$$

The number of moles for each compound is given by its mass, m, divided by its molecular mass, MM: $n_A = m_A/MM_A$ and $n_B = m_B/MM_B$. Thus, substituting into Equation 7 and rearranging it, Equation 8 is obtained:

$$\boxed{\frac{m_A}{m_B} = \frac{p_A^o \times MM_A}{p_B^o \times MM_B}} \qquad (8)$$

This equation shows that the mass of each component in the distillate is directly proportional to its molecular mass and its vapor pressure at the temperature of the distillation. The higher the vapor pressure and the larger the mass of a compound, the larger its yield in the distillate.

To illustrate the principle of steam distillation let us consider Figure 6.26. The vapor pressure for water and limonene is shown as a function of the temperature. Limonene, a liquid of pleasant smell found in spearmint oil and lemon and orange peels, is immiscible with water, has a normal boiling point of 175°C, and a molecular formula: $C_{10}H_{16}$, MM = 136 g/mol. It can be isolated (although not in a pure form) by steam distillation of citrus peels. What is the maximum percentage of limonene that can be found in the distillate? Figure 6.26*b* shows the total vapor pressure of mixtures of limonene and water at temperatures near 100°C; the boiling point of the mixture (the temperature at which the total vapor pressure equals 760 Torr) is 97.4°C. From the same graph, the vapor pressure of water at this temperature can be estimated as

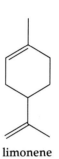

limonene

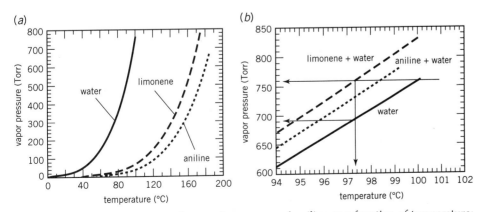

Figure 6.26 *a*) Vapor pressure of water, limonene, and aniline as a function of temperature; *b*) vapor pressure of limonene + water, aniline + water, and water in the vicinity of 100°C; (aniline is used in Problem 8).

690 Torr; thus the vapor pressure of limonene is: $(760 - 690)$ Torr $= 70$ Torr. Applying Equation 8, the ratio of limonene to water in the distillate can be calculated as:

$$\frac{m_{\text{limonene}}}{m_{\text{water}}} = \frac{p^{\text{o}}_{\text{lim}} \times \text{MM}_{\text{lim}}}{p^{\text{o}}_{\text{water}} \times \text{MM}_{\text{water}}} = \frac{70 \times 136}{690 \times 18} = 0.767$$

This indicates that the distillate has 0.767 g of limonene per 1 g of water. This is equivalent to 43% limonene.

Steam distillation is a very important technique in the isolation of **essential oils** from plants. Essential oils are volatile mixtures rich in compounds with applications in medicine, perfumery, and flavoring. To isolate them, the specimen from the plant (leaves, bark, flowers, seeds, etc.) is finely ground, mixed with water and subjected to distillation using a simple distillation apparatus; a milky emulsion of the essential oil in water usually distills at temperatures between 95 and 99°C. The distillation is continued until the distillate consists of clear drops of water. The essential oil can then be separated from the water by liquid-liquid extraction with an organic solvent.

If a long distillation is planned, water should be periodically added to the distillation flask to compensate for the amount of water already distilled. Sometimes, water vapor generated in a separate flask is passed through the solution to be steam-distilled. This assembly has the advantage that it can operate longer without interruption of the distillation process. An apparatus with steam generator is shown in Figure 6.27. Instead of the regular distillation head, a **splash head** is used to prevent boil-over from reaching the condenser. Sometimes, it is necessary to heat the distillation flask (where the sample is located) to prevent excessive condensation of the water vapor. When the distillation is finished, heat is removed from the vapor generator, the safety valve is then opened, and the valve connecting the distillation

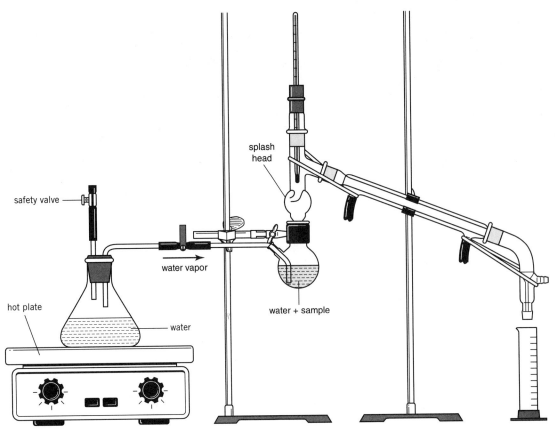

Figure 6.27 Steam distillation apparatus.

flask and the steam generator is finally closed to avoid suction of the liquid into the generator.

As mentioned before, steam distillation is not limited to liquids. Volatile solids immiscible with water and with a vapor pressure of at least 5 Torr around 100°C can also be isolated by this technique. When carrying out a steam distillation of a solid, precautions should be taken to avoid its crystallization in the condenser, which may cause clogging and lead to a pressure buildup. Stopping the flow of water in the condenser or using an empty condenser may prevent premature solidification.

BIBLIOGRAPHY

1. Elements of Fractional Distillation. C.S. Robinson and E.R. Gilliland. 4th ed. Series in Chemical Engineering. McGraw-Hill, New York, 1950.
2. Vogel's Textbook of Practical Organic Chemistry. A.I. Vogel, B.S. Furniss, A.J. Hannaford, P.W.G. Smith, and A.R. Tatchell. 5th ed. Longman, Harlow, UK. 1989.
3. Micro and Semimicro Methods, N.D. Cheronis, in Technique of Organic Chemistry, Vol. VI. A. Weissberger, ed. Interscience, New York, 1954.
4. Determination of Boiling and Condensation Temperatures, W. Swietoslawski and J.R. Anderson, in Technique of Organic Chemistry, Vol. I, Part I. A. Weissberger, ed. Interscience, New York, 1960.
5. Azeotropic Data. Dow Chemical Co. Advances in Chemistry Series VI. American Chemical Society. Washington, D.C., 1952.

EXERCISES

Problem 1

Using the conversion factors listed below complete the following table of pressures:

atm	mm Hg	Torr	psi	pascal
1	760	760	14.6960	1.01325 10⁵
0.592				
		10⁻⁴		

Problem 2

Francisco José de Caldas (1768–1816), a Colombian scientist who fought and was killed in the South American Independence Wars, lived in the Andean village of Popayan at 1760 m (5774 ft) above sea level. Around 1802 he discovered that the boiling point of liquids decreases with increasing altitude. His thermometer broke in one of his scientific expeditions and after he repaired it he observed that the boiling point of water in his home town was 94.6°C. The following equation (for a derivation see R. Ferreira, *J. Chem. Ed.*, **70**, 493 [1993]) relates the altitude of a place (h, in meters) with the boiling point of water (t, in °C)

$$h = 1.410^5 \frac{100 - t}{273 + t}$$

How far off was Caldas' thermometer?

Problem 3

The vapor pressure (P) of a pure liquid A as a function of the temperature (T) is given in the table below.

(a) Plot P versus T.
(b) What is the normal boiling point of A?

(c) If you want to distill 100 g of A under reduced pressure, what would be the boiling point if the applied pressure is: i) 600 mm Hg; ii) 0.592 atm; iii) 200 Torr?

T (°C)	P (mm Hg)	T (°C)	P (mm Hg)
70	15	140	490
80	45	150	615
90	90	160	740
100	140	170	860
110	205	180	1000
120	285	190	1200
130	395		

Problem 4

The table below shows the compositions (% mass) of the vapor and liquid phases at equilibrium for solutions of A and B at the indicated temperatures:

Temperature (°C)	% A in vapor	% A in liquid
70	100	100
75	97	79
80	93	65
85	85	55
90	77	45
95	69	37
100	60	28
105	50	20
110	40	12
115	25	6
120	0	0

(a) Construct a vapor-liquid phase diagram for this system.

(b) A mixture of 30 g of A and 70 g of B is distilled using a simple distillation apparatus. At what temperature would the first drop distill?

(c) A mixture of 50 g of A and 50 g of B is distilled using a very efficient fractionating column. Construct a distillation curve (temperature versus mL of distillate) for this system knowing that the density for A and B is 1.05 g/mL and 0.96 g/mL, respectively.

Problem 5

An ideal solution of two liquids A (bp 110°C) and B (bp 150°C) is separated by distillation using an ideal fractionating column.

(a) Draw a schematic[1] phase diagram (T versus % composition) for this mixture.

(b) Draw the distillation curve for a mixture that initially contained 70 mL of A and 50 mL of B.

Problem 6 (advanced level)

Mixtures of formic acid-dioxane show a maximum boiling point azeotrope at 113.4°C with the following composition (% mass): formic acid: 43%; dioxane: 57%. The boiling points of formic acid and dioxane are 100.8°C and 101.4°C, respectively.

(a) Construct a schematic[1] phase diagram for the system.

(b) Describe the behavior upon distillation of 1000 g of a 20% formic acid in dioxane mixture using a very efficient fractionating column. Construct the distillation curve (T versus mass of distillate) and calculate how many grams of pure formic acid, dioxane, and azeotrope (if any) will be obtained.

---

[1] The actual shape of the liquid and vapor curves (which must be determined experimentally) is unimportant for this problem.

Problem 7 *(advanced level)*

When a mixture of pyridine (50 g) and water (50 g) is distilled using a very efficient fractionating column, the distillation curve shown in part *a* of the figure below is obtained. When another mixture of water (40 g) and pyridine (10 g) is distilled using the same column, the curve shown in part *b* is obtained.

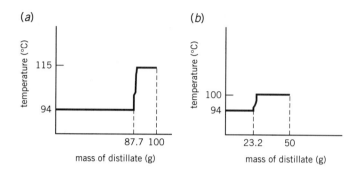

Construct a schematic[1] phase diagram for this system. Indicate relevant boiling points and compositions.

Problem 8

Using the graph shown in Figure 6.26*b*:

(a) Estimate the boiling point of a mixture of aniline (C_6H_7N) and water.
(b) Calculate the percentage (in mass) of aniline that codistills with water.

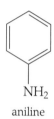

NH$_2$
aniline

Problem 9

(Previous Synton: Problem 10, Unit 4)   It is Ester's birthday and Synton wants to surprise her with some homemade lavender perfume that would remind her of Provence, the land of her grandparents and a place that she would like to visit someday. He collects 500 g of lavender from his garden and performs a codistillation with water. One of the main components in the lavender oil is linalyl acetate (MM 196 g/mol), which has a vapor pressure of approximately 20 mm Hg at the boiling point of the mixture with water. What would be the composition of the distillate?

EXPERIMENT 6A

Distillation of Alcohols

This experiment is divided into several parts. Check with your instructor to find out which ones you will perform. In the first part you will distill methanol from a mixture with ferric chloride. You will observe whether methanol distills at a constant temperature. You will determine the boiling-point range of methanol. In the second part you will separate a mixture of methanol and water by simple distillation and will construct the distillation curve. In the third part, you will try the same separation by fractional distillation. You will collect 8–10 separate fractions and construct the distillation curve. By comparing the distillation curves obtained by simple and fractional distillation you will determine which process was more efficient in separating methanol from water. To confirm or disprove your conclusions you will analyze the fractions from the fractional distillation by density (part E6A.4) or save them until the following lab period when you will analyze them by gas chromatography (Exp. 7A). Consult with your instructor to find out if you will do the analysis by density or gas chromatography (or both).

PROCEDURE

E6A.1   DISTILLATION OF METHANOL FROM A MIXTURE WITH FERRIC CHLORIDE

> **Background reading:** 6.1–3; 6.5; 6.9
> **Estimated time:** 70 minutes

You will distill methanol from a mixture with ferric chloride (0.1% w/v) using a simple distillation apparatus. Assemble the apparatus as illustrated in Figure 6.28 using a 50 mL round-bottom flask as a distillation flask, and a graduated cylinder as a receiving flask.

Assembling the Distillation Apparatus   Secure the distillation flask to a ring stand with the aid of a clamp; the flask should be resting on a heating mantle supported by a ring attached to the ring stand. Connect the heating mantle to a rheostat but do not plug it into the outlet yet. The rheostat allows heat control. Fasten the condenser to a second ring stand; connect the water hoses to the condenser so that water flows upward (see Fig. 6.28). The condenser's nipples should be pointing up when the condenser is first filled with water, otherwise it will not fill up. Make sure that the hoses are not crimped at the nipples and the water flows smoothly. Once the condenser is filled, rotate it about 90° so that the nipples are horizontal or slightly pointing down. This will avoid crimping of the hoses.

Place 15 mL of the sample in the flask, add a boiling chip, and attach the distillation head with the thermometer. Position the thermometer bulb just below the side arm of the distillation head so it reads the temperature of the vapors as they go into the condenser; the bulb should not touch the walls of the distillation head. Attach the condenser to the distillation head; use a rubber band or a plastic clip (Fig. 3.22) to keep it in place. The rubber band should never rest on the thermometer. Finally, attach the distillate take-off adaptor to the end of the condenser; use a rubber band or plastic clip to secure it. Place a 25-mL graduated cylinder to collect the distillate. Make sure that all the ground-glass joints are sealed. Ask your instructor to check your apparatus at this point. If you are by a door or window, wrap the

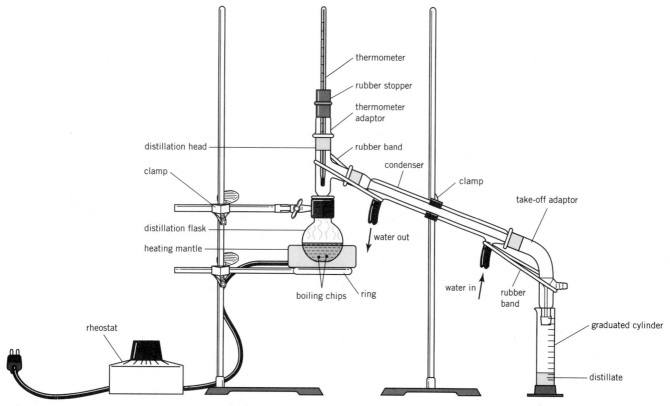

Figure 6.28   Simple distillation apparatus with graduated cylinder as receiving flask.

distillation head with aluminum foil to prevent sudden drops in temperature due to drafts. Plug the rheostat to the wall outlet, and adjust the heat control to a medium setting.

Performing the Distillation   Bring the methanol to a boil over low–medium heat. When the first drop of distillate is collected in the graduated cylinder, read the temperature and, if necessary, adjust the heat supply so that the distillate drips at a rate of about one drop per second. During the distillation the thermometer's bulb should be wet with dripping liquid. Record the temperature after the first drop has been collected in the receiving flask, and after 2.5, 5, 7.5, 10, and 12.5 mL. Stop the distillation by turning off the heat and removing the heating mantle. Ferric chloride should not be in the distillate, which should look colorless.

> **Cleaning Up**
>
> - With the aid of a Pasteur pipet transfer the liquid residue in the round-bottom flask to the container labeled "Methanol–Ferric Chloride Waste."

E6A.2   SEPARATION OF METHANOL-WATER BY SIMPLE DISTILLATION

> **Background reading:** 6.1–3; 6.5
> **Estimated time:** 70 minutes

Use the same apparatus as in E6A.1 to distill a mixture of 15 mL of methanol and 15 mL of water. If you haven't done E6A.1, read "Assembling the Distillation Apparatus" from the previous section, on page 136. Use recycled methanol from the previous distillation, if available. Wash the round-bottom flask from E6A.1 thoroughly with water before pouring in the sample. Bring the mixture to a boil over a low–medium heat. The distillate should drip at a rate of one drop per second. Record the temperature after the first drop has been collected and after every 2.5 mL of distillate until 25 mL of distillate have been collected. Save the distillate and the residue for the fractional distillation.

E6A.3   SEPARATION OF METHANOL-WATER BY FRACTIONAL DISTILLATION

Background reading: 6.1–4; 6.5; 6.7
Estimated time: 70 minutes

Before you start, label 10 scintillation vials with your name, the name of the chemical to be collected (methanol + water) and the fraction number (1–10). Pour 2.5 mL of water from a 5-mL graduated pipet into one of the vials and mark the level of the liquid with a marker; use this vial as a guide to indicate the 2.5-mL mark on the other vials. Pour out the water and dry the vial. All vials should be clean and dry before you start the fractional distillation.

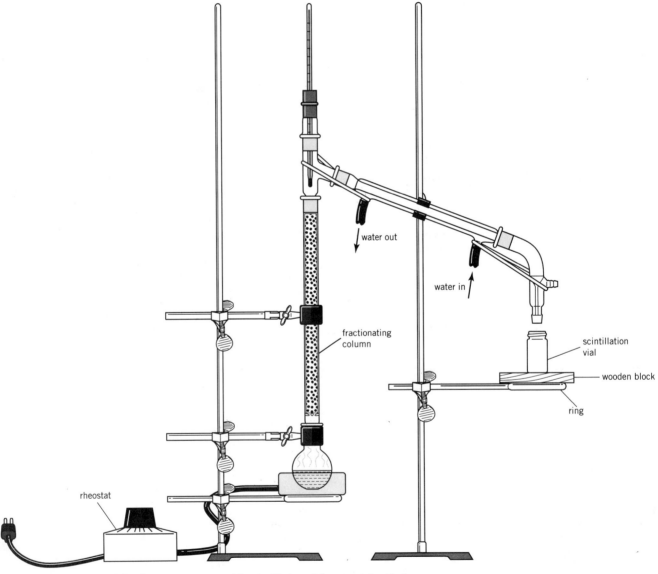

Figure 6.29   Fractional distillation apparatus with scintillation vial as receiving flask.

Rinse the condenser, distillation head, and take-off adaptor with methanol from a squeeze bottle; collect the washes in a beaker and dispose of them in an appropriate waste container; with a paper towel carefully blot away as much methanol as possible from the condenser, the distillation head, and adaptor. Combine the distillate and residue from the simple distillation in the distillation flask; add 5 mL of a 50–50 methanol-water mixture to make up for losses during the previous distillation. To the simple distillation apparatus shown in Figure 6.28, add a fractionating column as indicated in Figure 6.29.

The packing of the fractionating column with stainless steel sponge should be loose rather than tight; very tight packing prevents good flow along the column and reduces its efficiency. About 3 g of stainless steel sponge are enough to pack a 20-mL capacity column (approximately 16 cm long, 1.3 cm diameter). Handle the stainless steel sponge only with tweezers. Cut it in small pieces with scissors before packing. To prevent drafts, insulate the column with aluminum foil if you are close to a window or door. Record the temperature when the first drop is collected and every 2.5 mL of distillate until 25 mL of distillate have been collected. To keep the distillation at a rate of 1 drop per second you may need to increase the heat during the distillation. Plot your results in the same graph generated for the simple distillation. Collect each 2.5 mL fraction in a different vial. Tightly cap the vials and use them in the next part (E6A.4) or save them until the following period for use in the experiment on gas-liquid chromatography (Exp. 7A).

> **Cleaning Up**
>
> - Rinse the fractionating column with a little methanol. Collect the rinses in a beaker, and pour the liquid in the container labeled "Methanol Waste."
> - Condenser, distillation head, and take-off adaptor do not need to be rinsed.

E6A.4   ANALYSIS OF THE DISTILLATION FRACTIONS BY DENSITY

> **Estimated time:** 45 min.

The composition of the methanol-water fractions from the distillation can be determined by density measurements. The densities of pure methanol and water at 20°C are 0.791 and 0.998 g/mL, respectively. It was found experimentally that the density values of mixtures of methanol-water increase linearly between 0.791 and 0.998 g/mL as the mole fraction of water increases. In other words, a plot of density values versus the water mole fraction gives a straight line as shown in Figure 6.30.

The equation that describes the linear correlation between **water mole fraction,** X_{water}, and the **density of the mixture**, δ, is given below:

$$\text{Water mole fraction} = X_{water} = \left(\frac{1}{\delta_{water} - \delta_{methanol}}\right)\delta - \frac{\delta_{methanol}}{\delta_{water} - \delta_{methanol}} \quad (9)$$

Replacing $\delta_{methanol}$ and δ_{water} by 0.791 and 0.998 g/mL, respectively, the following equation is obtained:

$$X_{water} = 4.831\,\delta - 3.821 \quad (10)$$

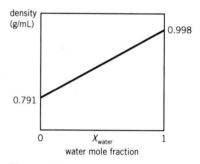

Figure 6.30 Linear correlation between density and water mole fraction for methanol mixtures at 20°C.

This equation can be used to determine the composition (X_{water}) of any mixture of methanol-water once its density, δ, is measured at 20°C. However, density values are strongly affected by temperature. Because the temperature of the room can fluctuate between 15 and 27°C, taking the density values for methanol and water at 20°C will introduce error in the calculations. This error can be eliminated by measuring the densities of these two compounds directly in the laboratory.

You will measure the density of pure methanol and water ($\delta_{methanol}$ and δ_{water}) at the temperature of the room. These figures will be plugged in Equation 9 to derive an equation similar to Equation 10 but with slightly different numbers. Then you will measure the density of the distillation fractions, and with the help of the equation,

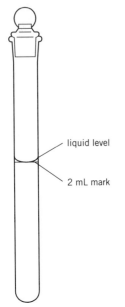

liquid level

2 mL mark

Figure 6.31 A 2-mL volumetric tube.

estimate their mole fractions. Finally, you will convert mole fraction into % mass composition using the equations given in the next section.

Measuring the Densities

Clean and dry a 2-mL volumetric flask. The flask can be dried with a wick of paper towel. Weigh the flask with the stopper in place. Use an analytical balance with a precision of 0.0001 g. With the help of a Pasteur pipet fill the flask to the mark with methanol (Fig. 6.31). With a wick of paper towel, remove any drop of liquid adhered to the walls of the tube above the mark. Make sure the cap is dry as well. Cap the tube and weigh it. Determine the mass of methanol by difference.

If time permits (consult with your instructor) determine the mass of methanol again by emptying the flask and drying it with a wick of paper towel until the paper comes out dry. Also dry the stopper. Refill the tube with methanol and determine the mass of the liquid by difference. The masses of methanol from both determinations should not differ by more than 1%. Take the average value and calculate the density of methanol by dividing by 2 mL:

$$\text{density} = \text{mass (g)}/2 \text{ mL}$$

Use the same procedure to determine the density of water.

Plug into Equation 9 the density values for pure methanol and water just determined, and derive an equation similar to Equation 10.

Dry the volumetric tubes as already described, and determine the density of the first, a middle, and the last fraction of the fractional distillation. If time permits, you can determine the density of the remaining fractions.

With the density values for the fractions, δ, and the derived equation, calculate their water mole fraction, X_{water}. Then convert the water mole fraction, X_{water}, into % mass composition (which is a more common unit) using the following equation where MM stands for molecular mass.

$$\text{water\%(mass)} = 100\frac{\text{MM}_{water}\, X_{water}}{\text{MM}_{methanol} - (\text{MM}_{methanol} - \text{MM}_{water})X_{water}}$$

$$\text{water\%(mass)} = 100\frac{18.02\, X_{water}}{32.04 - 14.02\, X_{water}} \tag{11}$$

Cleaning Up

- Dispose of the methanol-water fractions in the container labeled "Methanol Waste."

EXPERIMENT 6A   REPORT

Pre-lab

1. Make a table showing the physical properties of methanol, water, and ferric chloride. (Molecular mass, chemical structure, b.p., density, solubility in water and organic solvents, flammability [for solvents only], and toxicity/hazards).

2 Explain briefly why:

   (a) boiling chips are added to the distillation flask;

   (b) cooling water runs through the condenser against gravity;

   (c) the distillation apparatus should have a vent to the atmosphere;

   (d) boiling chips should never be added to a hot liquid;

   (e) the thermometer bulb should not touch the walls of the distillation head.

3. Discuss briefly the validity of the following statements:

   (a) Ideal mixtures of two liquids cannot be separated by fractional distillation.

   (b) Along a fractionating column the temperature increases from top to bottom.

   (c) The boiling point of a mixture of two liquids is a function of the composition of the mixture.

4. Why is it not recommended to immerse the thermometer in a boiling liquid to determine its boiling point?

5. Rearrange the equation given in Problem 2 and calculate the boiling point of water in the following places: Denver (5280 ft), Los Angeles (340 ft), Santa Fe, NM (6950 ft), and Mexico City (7347 ft) (1 foot = 0.3048 m).

6. Draw a simple and a fractional distillation apparatus in your lab book. Indicate the name of the different parts.

In-lab

1. Summarize the learning objectives of these experiments by mentioning the new techniques introduced.

Distillation of Methanol from a Mixture with Ferric Chloride

2. Make a table showing mL of distillate versus temperature. Plot the results.

3. Determine the boiling point of methanol as the average of your readings. Discuss your results.

Separation of a Methanol-Water Mixture

4. Make a table showing mL of distillate versus temperature (without and with fractionating column).

5. Plot temperature versus volume of distillate (distillation curve) for the simple and fractional distillation of the methanol-water mixture. Do both in the same graph and label each curve.

6. Which procedure was more efficient at separating the mixture into its components? Discuss your results.

7. Why is it important to rinse the condenser and distillation head with methanol before starting the fractional distillation? (Hint: think about the next part, E6A.4, or E7A.)

Analysis of the Fractions by Density

8. Make a table showing the fraction number, the distillation temperature range, the density, the mole fraction, and the % mass composition for the fractions analyzed. Discuss your results.

EXPERIMENT 6B

Isolation of Anise Oil

E6B.1   OVERVIEW

Anise is one of the oldest flavoring ingredients of which we have record. The anise plant (*Pimpinella anisum* L.) probably originated in the Middle East, from where it expanded to the entire Mediterranean region. A sweet-smelling oil is obtained by steam distillation of the dried ripe fruits (sometimes simply referred to as "anise seeds"). The main components of anise oil are *trans*-anethole, estragole, and *p*-anisaldehyde. Estragole, an isomer of anethole, is the main component of tarragon oil. Anise seeds are used as an aromatic ingredient in French and Italian pastries. The oil, with its distinctive licorice flavor, is employed in the confection of candies and beverages.

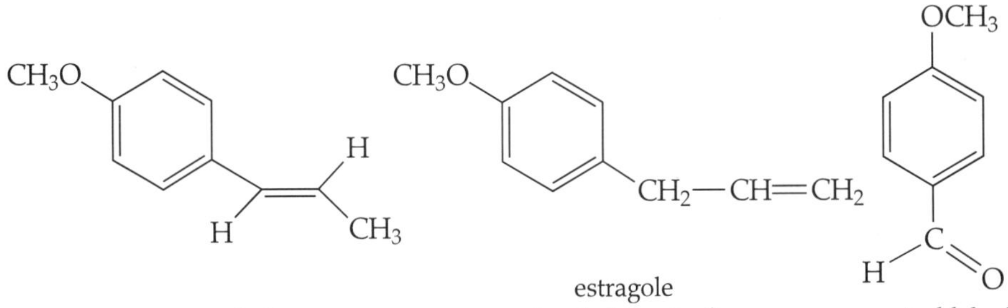

trans-anethole	estragole (methyl chavicol)	*p*-anisaldehyde

In this experiment you will isolate anise oil by steam distillation of anise seeds. You will also analyze and characterize the oil by physical and chromatographic methods in Exp. 7B. You will identify the main component in the oil as either estragole, *p*-anisaldehyde, or *trans*-anethole by gas chromatography (Exp. 7B).

PROCEDURE

E6B.2   ISOLATION OF ANISE OIL

> **Background reading:** 6.1; 6.2; 6.11; 5.3; 5.5
> **Estimated time:** 1.5 hr

Safety First

- *t*-Butyl methyl ether is flammable.

Crush 50 g of anise seeds in a blender for 10 seconds. Using a funnel for solids transfer the crushed material into a 500-mL round-bottom flask. Add 200 mL of distilled water to the flask and assemble a simple distillation apparatus as shown in Figure 6.2. Read "Assembling the distillation apparatus" on page 136. In a 100-mL round-bottom flask, pour 60 mL of water and mark the level of the liquid with a marker. Pour the water out and use this flask as the receiving flask for the distillate. Heat the distillation flask using a heating mantle. Adjust the heat supply so the liquid distills at an approximate rate of one drop per second. Collect approximately 60 mL of distillate. Read the temperature in the distillation head during the distillation.

Add 1 g of sodium chloride to the distillate, stir with a glass rod to mix, and transfer the liquid to a 125-mL separatory funnel. Extract the aqueous layer with 25 mL of *t*-butyl methyl ether (see Figs. 5.6–5.9). Separate the layers collecting the organic layer in a 125-mL Erlenmeyer flask. Extract the remaining aqueous layer again with another 25-mL portion of *t*-butyl methyl ether. Collect the organic layer in the same 125-mL Erlenmeyer flask. To the combined organic layers, add enough magnesium sulfate until the drying agent runs freely in solution (a spatulaful is usually enough). Cap the Erlenmeyer flask with a cork and let the suspension stand for about 5 minutes on the bench top with occasional swirling. Filter the suspension by gravity filtration using a piece of fluted paper (Fig. 3.8). Collect the liquid in a **dry** and pre-tared 100-mL round-bottom flask. Evaporate the solvent in a rota-vap (Fig. 3.17). Weigh the flask with the oil and calculate the mass of oil by difference.

EXPERIMENT 6B   REPORT

Pre-lab

1. Make a table showing the physical properties (molecular mass, b.p., density, solubility, refractive index, flammability [for solvents only], and toxicity/hazards) of *trans*-anethole, estragole, *p*-anisaldehyde, and *t*-butyl methyl ether.

2. Make a flowchart for the isolation of anise oil.

3. Why do you add sodium chloride before performing the extraction with *t*-butyl methyl ether?

4. In the liquid-liquid extraction with *t*-butyl methyl ether, is the organic layer the top or the bottom layer?

5. Why do you extract the aqueous layer with two 25-mL portions of *t*-butyl methyl ether instead of with one 50-mL portion?

6. How would you prepare *p*-anisaldehyde (4-methoxybenzaldehyde) from *trans*-anethole? (optional)

7. The vapor pressures of water and the main component of anise oil between 99 and 100°C are given in the table below. Estimate the normal boiling point of an anise oil-water mixture. Calculate the % of anise oil in the distillate assuming a molecular mass for the main component of 140 g/mol.

Vapor Pressure (Torr) of Water and the Main Component of Anise Oil as a Function of Temperature

Temperature (°C)	Water	Main comp. anise oil	Temperature (°C)	Water	Main comp. anise oil
99.0	733.24	7.46	99.6	749.20	7.70
99.1	735.88	7.50	99.7	751.89	7.74
99.2	738.53	7.54	99.8	754.58	7.78
99.3	741.19	7.58	99.9	757.29	7.82
99.4	743.85	7.62	100.0	760.00	7.86
99.5	746.53	7.66			

In-lab

1. Calculate the % yield of anise oil from anise seeds.

$$\% \text{ yield} = \frac{\text{mass oil}}{\text{mass seeds}} \times 100$$

2. Report and discuss the boiling point of the mixture.

3. Why does the distillate look "milky"?

4. Discuss any source of error during the isolation that may have resulted in a lower yield.

UNIT 7

Gas Chromatography

7.1 INTRODUCTION

The term *chromatography* was coined by the Russian botanist Mikhail Tswett (1872–1919), whose main interest was the study of plant pigments. He separated chlorophylls and carotenes from green leaf extracts using a narrow glass tube filled with calcium carbonate Figure 7.1. In Tswett's own words, "Like light rays in the spectrum, the different components of a pigment mixture, obeying a law, separate on the calcium carbonate column." Tswett called this separation process the **chromatographic method** (from Greek: *chromo*; color). Despite Tswett's first report in 1903, chromatography remained almost unexploited for more than 25 years with only occasional reference in the chemical literature. Since the publication of several papers by Kuhn and Lederer in the decade of the 1930s; chromatography has grown at an explosive rate, particularly after World War II. Today chromatography encompasses a wide variety of separation techniques with applications in all areas of chemistry and biochemistry. Chromatography is without doubt the most powerful and versatile tool available to separate chemicals.

What are the laws that Tswett referred to? What physical principles govern chromatographic separations? The answers to these questions form the subject of this unit.

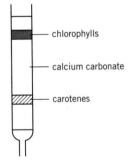

Figure 7.1 Separation of plant pigments on a calcium carbonate column.

- chlorophylls
- calcium carbonate
- carotenes

7.2 THE CHROMATOGRAPHIC METHODS

In the most general sense, chromatography can be defined as a separation technique based on the selective distribution of chemicals between a **stationary phase** and a **mobile phase**. The mixture to be separated is placed as a well-defined band or spot on the stationary phase (a solid or a liquid) and the mobile phase (a gas or a liquid) is allowed to pass through the system (Fig. 7.2*a*). Due to the selective interaction of chemicals with the stationary and mobile phases, separation is achieved after a certain period of time (Fig. 7.2*b*).

Before we discuss the mechanisms of chromatographic separation, we need to introduce some important names and concepts.

Chromatographic methods can be divided into different categories depending on the *physical process* ruling the separation. Separation can be achieved due to molecular differences in adsorptivity, solubility, charge, size, and specific binding.

144

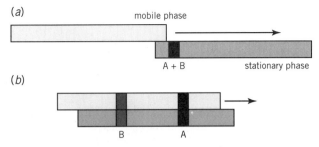

Figure 7.2   Schematic representation of the chromatographic process. Mixture of A + B: *a*) before separation; *b*) after interaction with mobile phase.

Each of these properties gives rise to a different type of chromatography as indicated in Table 7.1.

***Table 7.1* Classification of Chromatographic Methods Based on the Separation Process**

Property	Type of Chromatography
adsorptivity	adsorption
solubility	partition
charge	ion exchange
size	size exclusion
specific binding	affinity

Adsorption and **partition chromatography** are widely used in the organic chemistry laboratory and will be discussed in some detail in this book. Ion-exchange, size-exclusion, and affinity chromatography are chiefly employed in biochemistry and will be only briefly considered (Unit 10).

The main difference between adsorption and partition chromatography lies in the nature of the stationary phase. Whereas in *partition chromatography the stationary phase is a liquid*, in *adsorption chromatography it is a solid*. In either case the mobile phase can be a liquid or a gas. In this unit we will concentrate on partition chromatography, with special emphasis on gas chromatography, and in Units 8 and 9 we will deal with adsorption chromatography.

We can also classify chromatographic methods based on the technique or equipment used to carry out the separation regardless of the physical process that rules the separation. The most important types of chromatography based on this criterion are: **gas chromatography** or **GC** (the mobile phase is a gas), **thin-layer chromatography** or **TLC** (the stationary phase is spread on a sheet of metal, glass, or plastic), **column chromatography** (the method used by Tswett), **paper chromatography** (the stationary phase is a sheet of paper), and **high-performance liquid chromatography** or **HPLC** (a liquid mobile phase is forced to pass under pressure through the stationary phase packed inside a column).

Table 7.2 shows the most important types of chromatography used in the organic chemistry laboratory.

We must regard Tables 7.1 and 7.2 as simplified summaries of the diversity of chromatographic methods. For instance, the boundaries between adsorption and partition chromatography are not always well defined, as we shall discuss later. Also, due to the growing interest in the improvement of chromatographic methods, new techniques combining different types of chromatographies are constantly being developed.

Table 7.2   **Different Types of Adsorption and Partition Chromatography**

Type of chromatography based on separation process	Stationary phase	Mobile phase	Type of chromatography based on technique
adsorption	solid	gas	GC
		liquid	TLC, column, HPLC
partition	liquid	gas	GC
		liquid	TLC, column, paper, HPLC

7.3   PARTITION CHROMATOGRAPHY

As anticipated in Unit 5, partition chromatography is closely related to countercurrent distribution. Both methods separate chemicals on the basis of their **partition coefficients** by using a stationary and a mobile phase. While countercurrent distribution is a stepwise operation carried out in test tubes, chromatography is a continuous process that takes place inside small particles. In partition chromatography, the stationary phase is a **liquid** coating small particles of an inert **solid support**, and the mobile phase is a **gas** or a **liquid** immiscible with the stationary phase.

Let's see how the process works. For clarity, let's suppose that both the mobile and the stationary phases are liquids. Let's also suppose that the liquid stationary phase is coating fine particles of a solid support contained in a narrow column (the nature of the stationary phase and the solid support are unimportant for now). The liquid mobile phase, immiscible with the stationary phase, flows at a constant rate through the column. A small amount of sample containing a compound A is deposited (it does not matter how, for now) on the stationary phase near the inlet end of the column (Fig. 7.3).

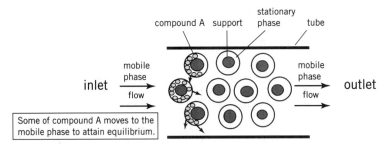

Figure 7.3   A longitudinal cross-section of a chromatographic column showing the particles of solid support coated with stationary phase. The small arrows indicate that compound A moves from the stationary to the mobile phase.

As the mobile phase flows through the column, some of compound A moves from the stationary phase to the mobile phase in order to reach equilibrium. At first approximation, we can assume that true partitioning equilibrium is attained between the stationary phase and the surrounding mobile phase. The concentration of A in the mobile phase, $[A]_m$, and its concentration in the stationary phase, $[A]_s$, are related by the partition coefficient $K_{m/s}^A$.

$$K_{m/s}^A = \frac{[A]_m}{[A]_s} \tag{1}$$

As you may recall, the partition coefficient is an equilibrium constant that measures the relative affinity of a chemical for both phases (Unit 5). If the partition coefficient is, for example $K_{m/s}^A = 10$, it follows from Equation 1 that at equilibrium $[A]_m = 10 \times [A]_s$, or in other words, that compound A prefers the mobile phase ten times over the stationary phase. The partition coefficient can be estimated by the ratio of the solubilities in both phases S_m^A and S_s^A:

$$K_{m/s}^A \approx \frac{S_m^A}{S_s^A} \qquad (2)$$

According to this equation, the larger the affinity for the mobile phase (as measured by the solubility), the larger the partition coefficient.

Equation 1 can be rearranged into Equation 3:

$$[A]_s = \frac{[A]_m}{K_{m/s}^A} \qquad (3)$$

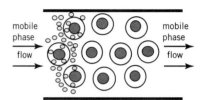

This equation indicates that there is a linear relationship between the concentrations in the stationary and mobile phases at equilibrium. If one increases, the other should increase proportionally.

Let's go back to our compound in the column. The compound is now partitioned between the stationary and the mobile phase according to its partition coefficient (Eq. 1), Figure 7.4.

Figure 7.4 Compound A partitioned between the mobile and the stationary phases.

Now, as the mobile phase flows, the compound dissolved in the mobile phase is carried toward the outlet end of the column (Fig. 7.5). When this solution of compound A in the mobile phase comes in contact with new particles of support plus stationary phase, some of A moves to the stationary phase to attain equilibrium. At the same time, as fresh solvent comes in contact with the stationary phase where A was first deposited, A redistributes between both phases to attain equilibrium.

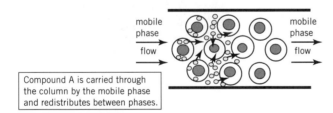

Compound A is carried through the column by the mobile phase and redistributes between phases.

Figure 7.5 Compound A is carried by the mobile phase. The small arrows indicate the redistribution of A.

Now the compound is partitioned between the two phases at a new location in the column (Fig. 7.6). This process keeps repeating itself and, as a result, compound A travels along the column carried by the mobile phase.

The speed at which the compound moves along the column depends on the flow rate of the mobile phase and, most importantly, on the relative affinity of the compound for the mobile and stationary phases, or in other words, on its partition coefficient. In general, *the larger the affinity for the mobile phase (or the larger the partition coefficient as defined in Equation 1), the faster the compound travels.*

Let us now assume that we load two compounds A and B with different partition coefficients, and that $K_{m/s}^A > K_{m/s}^B$. This means that A has higher affinity for the mobile phase than B. *Because of this, A will travel along the column faster than B.* If we plot the concentration of A in the stationary phase, $[A]_s$, as a function of its concentration in the mobile phase, $[A]_m$, according to Equation 3 we should obtain a straight line with slope $(1/K_{m/s}^A)$; this is shown in Figure 7.7. A similar plot is obtained for B, but with a

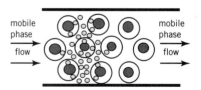

Figure 7.6 Compound A at a new location in the column.

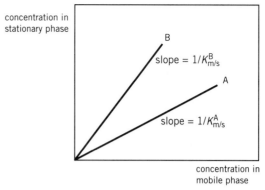

Figure 7.7 Partition isotherms for compounds A and B. Notice that $K_{m/s}^A > K_{m/s}^B$, thus A travels faster than B.

larger slope ($1/K_{m/s}^B$) because $K_{m/s}^B < K_{m/s}^A$. These lines are called **partition isotherms** because they describe the partitioning process at constant temperature (from Greek; *iso*; equal; *therme*; heat). We will come back to this in Unit 8. Figure 7.8 shows how the mixture of A and B separates inside the chromatographic column. Because A has a higher partition coefficient, it travels faster than B.

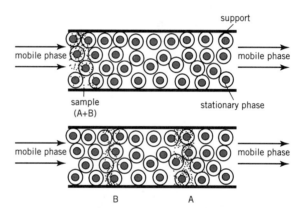

Figure 7.8 Partitioning of compounds A and B between the stationary and the mobile phases for the case $K_{m/s}^A > K_{m/s}^B$. A has higher affinity for the mobile phase than B and comes out of the column first (compare to Fig. 5.21).

The results shown here are very similar to those found in countercurrent distribution (Fig. 5.21). In countercurrent distribution the number of transfers (or steps) is limited by the number of tubes; in partition chromatography each step is performed on the microparticles with the stationary phase. The chromatographic column can be visualized as being divided into a large number of regions, called **theoretical plates**, where equilibrium between the stationary and mobile phase is attained. A theoretical plate in partition chromatography corresponds to a tube in countercurrent distribution. Generally, the larger the number of theoretical plates, the better the separatory power of a column.

A mixture of two or more compounds can be separated by partition chromatography as long as their partition coefficients are different. Because the partition coefficient is an equilibrium constant that depends on the temperature, the quality of the separation is strongly affected by the temperature, as we will see later.

7.4 GAS CHROMATOGRAPHY

In gas chromatography, **GC**, the mobile phase is a gas and the stationary phase can be a solid or a liquid. GC can be described as follows: a solution of the sample to be analyzed is introduced into the **gas chromatograph** where it is **vaporized** and carried by the mobile phase, or **carrier gas**, through a narrow column filled with the stationary phase. Separation of the mixture into its components is achieved in the column by partition or adsorption mechanisms. Single components come out of the column at different times and pass through a detection device, which sends a signal to a recorder. The result is a **gas chromatogram** or **GC trace**. Figure 7.9 shows a typical chromatogram where a mixture of A and B has been successfully separated. When the compounds are clearly separated like those in the figure, we say that they are **resolved**.

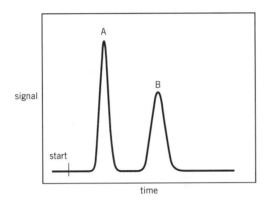

Figure 7.9   A gas chromatogram of a mixture of A and B.

The partition process described in the preceding section for a *liquid* mobile phase and a *liquid* stationary phase is also true if the mobile phase is a *gas* and the stationary phase is a *liquid*, as is the case of most GC applications. The partition coefficient is then defined as:

$$K_{m/s}^A = \frac{[A]_m}{[A]_s} = \frac{[A]_{gas}}{[A]_{liq}} \tag{4}$$

The equilibrium concentration of compound A in the liquid stationary phase, $[A]_{liq}$, is related to its solubility in the stationary phase, as we already discussed. The concentration of A in the gas mobile phase, $[A]_{gas}$, on the other hand, depends on the *volatility* of the compound, or in other words, its tendency to be in the vapor state. This tendency, in turn, depends on the temperature at which the operation is carried out. The higher the temperature, the larger the concentration of the chemical in the gas phase, the larger the partition coefficient, and the faster the compound comes out of the column.

Since most applications have been developed using liquids as the stationary phase, this technique is often called gas-liquid-chromatography (g.l.c.) or vapor-liquid-chromatography. Occasionally, however, solid stationary phases are used for GC analysis. In these cases the separation mechanism is adsorption instead of partition (adsorption chromatography is discussed in Units 8 and 9). The operational technique for both types of gas chromatography is the same and will be discussed in the following sections.

A simplified diagram of a typical gas chromatograph is shown in Figure 7.10. A gas chromatograph consists of the following parts:

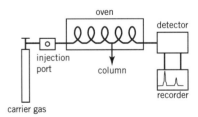

Figure 7.10   A diagram of a gas chromatograph.

- *A cylinder of gas (the carrier gas) such as nitrogen or helium.*
- *The injection port where the sample is injected and vaporized.*

- *The chromatographic column. The column is the heart of the GC instrument. The separation takes place along the column.*

- *An oven in which the column is located.* The temperature of the oven can be varied from ambient temperature to 250°C or higher.

- *A detector.* As the compounds come out of the column they are "seen" by the detector which sends a signal to the recorder.

- *A recorder.* The signal from the detector is transformed into vertical pen motion. The chart paper moves horizontally at constant speed. An increase in the output signal from the detector results in a peak on the chart paper. This trace is called a chromatogram. Today, computers are replacing chart recorders.

The Chromatographic Run

To carry out a GC separation we introduce the sample, a gas, a pure liquid, or a solution, into the injection port using a microsyringe. The solvent and the sample vaporize due to the high temperature at which the injection port is kept. A stream of carrier gas flows through the injection port and carries the sample into the column where the separation takes place. In general, the higher the volatility of the compound, the faster it travels; but also, the higher the affinity of the compound for the stationary phase, the slower the motion along the column.

Before the components of the sample reach the detector, only the carrier gas flows through it; this gives rise to a constant signal that is recorded on the chart paper as a horizontal line, the **baseline**. As each component comes out of the column and reaches the detector, a new signal (proportional to the amount of material passing through the detector) is sent to the recorder and a peak is drawn on the chart paper. When the component has completely emerged from the column and passed through the detector, the original signal for the carrier gas is sent to the recorder, which responds by producing the baseline again until a new component of the sample reaches the detector. The time elapsed between the injection of the sample and the detection of the peak is called the **retention time**, t_R (Fig. 7.11).

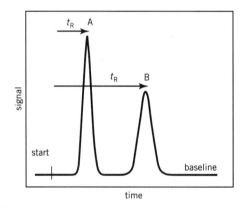

Figure 7.11   A typical gas chromatogram.

In general, different compounds have different retention times under identical experimental conditions. The retention time depends on physical properties of the compound such as its **boiling point**, its **molecular mass**, and its **polarity**. In general, for chemicals of similar structure, *the retention time increases as the boiling point increases.* Also, other things being equal, *the larger the molecular mass, the larger the retention time;* and *the larger the polarity, the larger the retention time.* For example, if we were to separate carvone and limonene (two compounds found in spearmint and caraway oils) by gas chromatography we would expect the limonene peak to come out before the carvone peak because limonene has a lower boiling point than carvone.

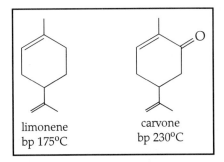

limonene
bp 175°C

carvone
bp 230°C

In analyzing chemicals with very different structures, generalizations may fail to predict the order in which they come out of the column. We should remember that the retention time is the result of the balance between two opposite forces: the tendency of the chemical to be in the vapor phase (volatility) and its interaction with the stationary phase (solubility).

The retention time depends strongly on the experimental conditions chosen for the GC run, such as **type of stationary phase**, **column temperature**, **flow rate** of carrier gas, and **length** and **diameter of the column**. When comparing retention times it is crucial that the experimental conditions be the same.

7.5 THE GAS CHROMATOGRAPH

A simple gas chromatograph is illustrated in Figure 7.12. This instrument has two columns (and thus two injection ports) but only one is used at any given time.

Injection Port

The sample (a gas, a pure liquid, or a solution in acetone, toluene, methanol, or another appropriate solvent) is injected using a microsyringe. The most common type of microsyringe is called the **Hamilton syringe** (Fig. 7.13). Hamilton syringes allow the handling of very small volumes in the microliter range (1 μL = 10^{-3} mL). Typical injection volumes for GC are 0.1–10 μL. To give you an idea of how small these volumes are, remember that *a drop of water is approximately 30 μL*. Hamilton syringes should be handled with extreme care since they are made of glass and are very fragile. The needle should not be bent during the injection process; it should be inserted in the septum (a rubber disk) with a firm and gentle stroke. The injection port is kept at a high temperature (70–250°C) to vaporize the sample.

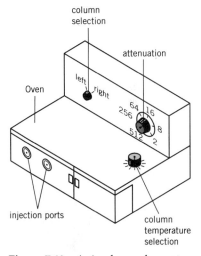

Figure 7.12 A simple gas chromatograph with two columns.

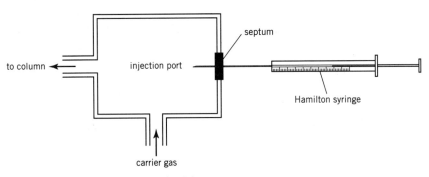

Figure 7.13 A simplified diagram of a GC injection port.

The Carrier Gas

The carrier gas, which is also the mobile phase, should constantly flow through the system while the column is heated. The flow rate of the carrier gas can be varied according to the manufacturer's directions for the GC instrument and the column under operation. In general, as the flow rate increases the quality of the separation, called **resolution**, improves very rapidly until the **optimum flow rate** is reached. By increasing the flow rate beyond this point the resolution deteriorates. The optimum flow rate varies from sample to sample and should be experimentally determined. Gases such as helium, argon, and nitrogen are all suitable mobile phases. Due to their low cost and inertness, helium and nitrogen are often preferred.

The Column

The success of a GC separation depends largely on the selection of the column. The nature of the stationary phase, the characteristics of the solid support, and the length and diameter of the column are important factors to consider when planning a GC analysis. To obtain optimum separation the column should provide a large surface area for the partition (or adsorption) process to take place. This can be achieved by using a finely divided solid support in which the stationary phase is embedded. The resulting columns are called **packed columns**. Supports often used for packed columns include calcined diatomaceous earth, crushed firebrick, glass beads, and powdered Teflon. Solid supports are commercially available under the trademark **Chromosorb** (Johns Mansville Corp.).

The tubes for packed columns are normally made of stainless steel, aluminum, or glass. Typical packed columns are 1–4 m long, have a 2–5-mm internal diameter, and are coiled up in a circular shape to make them compact. A porous material placed at both ends of the column allows the passage of the gas but prevents the stationary phase from escaping.

In contrast to packed columns, **capillary columns** do not use a solid support. A thin and uniform film of the liquid stationary phase directly coats the walls of a long and narrow tube. The tubes for capillary columns are usually made of glass or fused silica and have diameters between 1 mm and 0.25 mm, and length between 10 and 60 m. Capillary columns have better separatory power than packed columns. Depending on the stationary phase and the procedure used to pack the column, the number of theoretical plates for a typical packed column can be as high as 10,000. In contrast, capillary columns operating under optimal conditions can have up to 1,000,000 theoretical plates.

The Stationary Phase

There are hundreds of different stationary phases available on the market. There is an ideal stationary phase for almost any separation imaginable. Analyses of hydrocarbons, solvents, pesticides, spirits, steroids, and alkaloids, just to mention a few, can be performed with maximum efficiency if the appropriate stationary phase is used. Before selecting a stationary phase you should always check the catalogs of several manufacturers to see what products are available. Table 7.3 lists a few commonly used stationary phases and some of their applications. As a rule of thumb, the polarity of the stationary phase should be similar to the polarity of the mixture to be separated. When a sample contains compounds of very different polarity, a phase of medium polarity is recommended. In Table 7.3, T_{max} is the maximum temperature at which the phase can be used without deterioration.

The most common stationary phases for GC are high-boiling-point liquids. Examples include nonpolar hydrocarbons of high molecular mass such as squalane, low and medium-polarity polymeric siloxanes (for example SF96, OV17, etc.), and polar polyethers. Solid stationary phases made of polymeric materials such as

Table 7.3 **Common GC Stationary Phases**

Name	Composition	Polarity	T_{max} (°C)	Applications
Squalane	2,6,10,15,19,23-hexamethyltetracosane (alkane)	nonpolar	140	saturated hydrocarbons
Apiezon (Type K,L,M,N)	high boiling point petroleum fraction (alkanes)	nonpolar	300	general use; high boiling point compounds
SF96 OV1 OV101 SE30	polydimethylsiloxane	low	330–350	steroids; amines; alcohols; ketones; alkaloids; hydrocarbons
OV17 SE52	polyphenylmethylsiloxane	medium	330	phenols; ethers; nitrocompounds; esters; aromatic hydrocarbons
Porapak	porous polystyrene beads	medium	190–250	alcohols; glycols; esters; ketones; aldehydes
Carbowax	polyethylene glycol M.M. 1500-2000	high	225	aldehydes; esters; ethers; essential oils

polystyrene are also used. As was already mentioned, the separation mechanism with solid stationary phases is adsorption instead of partition.

The Oven

A precise control of the temperature at which the column operates is necessary to obtain reproducible GC runs. This is achieved by placing the column in an oven equipped with an electrical heating system. By controlling the column temperature we can drastically affect the retention times. *In general, the higher the temperature of the column, the faster the compounds come out.*

When the temperature is kept constant throughout the GC run, the run is called **isothermic.** Isothermic runs are adequate to separate compounds of similar boiling points but they may lead to poor and lengthy separations when the mixture contains compounds with very different volatilities. To separate these mixtures, modern gas chromatographs are equipped with **temperature programmers**, which make it possible to increase the temperature during the run; this improves the efficiency of the separation and shortens its duration. The run starts at a low temperature, allowing only the most volatile compounds to come out while the

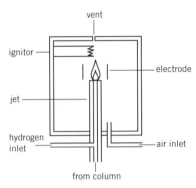

Figure 7.14 Diagram of a flame ionization detector.

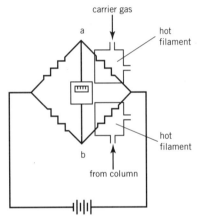

Figure 7.15 Diagram of a thermal conductivity detector.

others are retained in the column. As these very volatile compounds emerge from the column, the temperature is increased, forcing the less volatile ones out.

The Detector

No matter how good a GC separation is, it is rendered useless without a way of "seeing" the compounds as they come out of the column. Most modern GC instruments are equipped with one of two main types of detectors: a **flame ionization detector** (**FID**), or a **thermal conductivity detector** (**TCD**, also called **catharometer**). These detectors differ in sensitivity, operation, and applications. The TCD, less sensitive than the FID, is used in simple instruments and for very specific applications.

Flame Ionization Detector   A simplified picture of an FID is shown in Figure 7.14. The carrier gas passes through a flame of burning hydrogen in air. In the absence of other chemicals, the flame produces a low number of ions but when an organic compound reaches the flame the number of ions considerably increases.

The flame is kept in an electric field and the ions are attracted to the electrode, which produces a current that is amplified and measured. The larger the flow of ions through the flame, the larger the electrical current and, therefore, the larger the signal sent to the recorder.

Flame ionization detectors are not sensitive to inorganic compounds such as rare gases, nitrogen, oxygen, H_2O, NH_3, CO, CO_2, SO_2, and SH_2; they respond to all organic chemicals except formic acid. For optimum performance, FIDs have to be cleaned from time to time to remove carbon particles that accumulate around the jet.

Thermal Conductivity Detector   In contrast to FIDs, thermal conductivity detectors do not destroy the sample, which can be recovered intact at the end of the GC run. Thermal conductivity detectors are less sensitive than FID but are easier to maintain and operate. A diagram of a thermal conductivity detector is shown in Figure 7.15.

The detector consists of a Wheatstone bridge circuit, where four resistors are connected to a power supply as shown in the figure. Two of the resistors are hot filaments kept in heated chambers. Only the carrier gas passes through one of the chambers, and the effluent from the GC column passes through the other. When only carrier gas comes out of the column, both filaments are at the same temperature, the bridge is balanced (no electric current flows between points a and b), and no signal is detected. Chemicals coming out of the column change the thermal conductivity of the gas and produce a change in the temperature of the filament. Since the resistance of the filament depends on its temperature, this change in temperature causes the bridge to be imbalanced; now current flows between points a and b, and is sent as a signal to the recorder.

The difference in thermal conductivity between the carrier gas and the sample determines the sensitivity of the detector. Helium is an excellent carrier gas with TCD because its thermal conductivity is very different than that of most organic compounds.

FID and TCD do not provide information about the structure of the compounds passing through the column. To obtain structural information, mass spectrometers (Unit 34) and infrared spectrophotometers (Unit 31) are used as detectors. They not only detect the compounds as they come out of the column, but also give molecular information. When a mass spectrometer is used as a GC detector, one can find the molecular mass and the functional groups for each of the components in the sample, and ultimately identify all of them. This technique, called **GC-MS**, is widely used, for example, in the detection of illegal drugs in the urine of athletes.

The Recorder

The signal from the detector is sent to the recorder, where it is transformed into vertical pen motion. Two parameters have to be set beforehand for proper recorder operation: **chart speed** (which controls the x axis) and **attenuation** (which controls the y axis). The chart speed "shapes" the chromatogram: very fast chart speeds result in broad and ill-defined peaks; very slow ones cluster the peaks together. In both cases the interpretation of the chromatogram is difficult. Intermediate chart speeds (1–4 cm/min) are recommended for most purposes.

A peak can be scaled to fit the chart height (y axis) by changing the attenuation that controls the size of the peak. *As the attenuation increases, the size of the peak decreases.* Attenuation values are usually given in powers of 2 (2, 4, 8, 16, etc.). The choice of the right attenuation depends on the sample concentration and the injected volume. High attenuations are used for concentrated samples, and low attenuations for dilute ones. For samples in which the concentrations of chemicals vary widely, we may need to change the attenuation during the run if all peaks are to be on scale.

In modern gas chromatographs, chart recorders have been replaced by computers and printers. The computer collects the data and displays the GC traces on the screen. Hard copies of the chromatograms can be printed out.

7.6 MEASURING THE RETENTION TIME

In general, GC peaks are symmetrical and bell shaped (Fig. 7.16*a*). The retention time for these peaks is measured from the beginning of the injection to the crest of the peak. With simple recorders we measure the distance in cm and calculate the retention time using the chart speed.

$$\text{retention time (min)} = \text{distance from start (cm)/chart speed (cm/min)} \qquad (5)$$

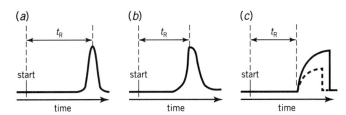

Figure 7.16 GC peaks. *a*) symmetrical; *b*) early maximum; *c*) late maximum; dotted peak obtained with less sample.

The retention time is, in principle, independent of the size of the peak. This is particularly true for symmetrical peaks. Very often, however, GC peaks deviate from the ideal bell shape. Two extreme cases of asymmetrical peaks are shown in Figure 7.16*b* and *c*. In **case b** the maximum is near the front of the peak and is called an early maximum; this phenomenon, also called **tailing**, is observed when adsorption rather than partition rules the separation (discussed in Unit 8). **Case c** shows an asymmetrical peak with a late maximum. This type of distortion occurs mainly with strongly polar samples and nonpolar stationary phases.

The retention time for a peak with an early maximum (Fig. 7.16*b*) is also measured by the distance from the beginning of the injection to the crest of the peak, and is fairly independent of sample size. For peaks with a late maximum (Fig. 7.16*c*), however, the distance from the start to the maximum increases as the sample size increases. The beginning of the peak, on the other hand, stays approximately constant and should be taken as the retention time.

In some applications the retention time should be corrected for the **dead time** of the run. The dead time is the time it takes for a compound without affinity for the stationary phase (very large $K_{m/s}$) to come out of the column. Air is a gas with little affinity for most stationary phases and is normally used to measure the dead time when thermal conductivity detectors are used (FIDs are insensitive to air). To do this correction we must inject some air along with the sample. The corrected retention time, $t_{R\,cor}$, is calculated by subtracting the retention time of the air peak from the retention time of the peak in question (Fig. 7.17):

$$t_{R\,cor} = t_R - t_{R\,air} \qquad (6)$$

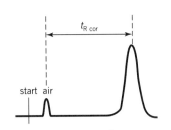

Figure 7.17 Corrected retention time.

7.7 INTEGRATION

The area under a GC peak is proportional to the amount of material present in the sample. Some recorders are equipped with integrators that directly calculate the areas under the peaks. Simple recorders, however, do not possess this capability and the integration has to be done manually.

In general, the area under the peak can be estimated by **triangulation**. GC peaks are usually bellshaped and closely resemble a triangle (Fig. 7.18). Their areas can be approximated as:

$$\boxed{A = h \times w_{1/2}} \qquad (7)$$

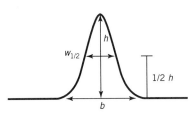

Figure 7.18 Triangulation of a bell-shaped peak.

where h is the **height of the peak**, and $w_{1/2}$ is the **width of the peak at half its height**. Equation 7 is equivalent to the more familiar formula for the area of a triangle: $A = (h \times b)/2$, where b is the base of the triangle. The use of Equation 7, however, is preferred because for bell-shaped peaks, the width at half the height is much easier to measure than the base. We can still apply Equation 7 when the peak is slightly asymmetric.

Triangulation of strongly asymmetric peaks gives unsatisfactory results. In such cases we can divide the peak into several regions (Fig. 7.19), compute their individual areas as triangles or rectangles, and add them up. Alternatively, we can estimate the peak area by weight. To accomplish this, we make a photocopy of the chromatogram, cut out the peaks with scissors or a razor blade, and weigh them using an analytical balance. Assuming the paper is of homogeneous thickness, the areas should be directly proportional to the weights.

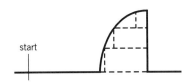

Figure 7.19 Integration of an asymmetrical GC peak.

A problematic situation arises when the peaks are not well resolved, for example, when they overlap as shown in Figures 7.20 and 7.21. Such cases should be individually considered. If it is possible to measure $w_{1/2}$ for each peak (Fig. 7.20), the triangulation method using Equation 7 should still give a reasonably good estimate of the area of each peak.

If one of the peaks is only a shoulder of a bigger peak (Fig. 7.21), the triangulation method still can be applied as follows: 1) draw a vertical line that passes through the center of peak A and measure its height, h^A; 2) repeat the same operation with peak B; 3) measure one-half of the width at half the height for peaks A and B, (ab) and (cd) respectively. The area for peak A can be estimated as $h^A \times 2 \times$ (ab), and the area under peak B as $h^B \times 2 \times$ (cd).

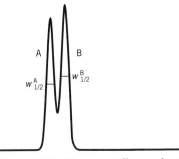

Figure 7.20 Two partially resolved peaks.

The integration of ill-resolved peaks, like those shown in Figure 7.21, can also be solved with the aid of computers. The GC data (as intensity versus time) is collected on a computer equipped with the proper interface. Commercially available programs can do a **curve fitting** of complex GC traces. The program deconvolutes the overall trace and gives the bell-shaped curves corresponding to each peak, as shown in Figure 7.21 by the dotted lines. The area under each peak is also part of the output of the program.

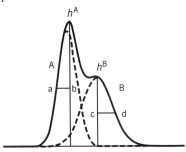

Figure 7.21 Triangulation of two overlapping peaks.

7.8 QUANTITATIVE ANALYSIS

The area, A, under a GC peak is directly proportional to the amount of material, m, passing through the detector (Eq. 8). The proportionality constant, f, is called the **response factor**:

$$A = f \times m \qquad (8)$$

The amount of material that reaches the detector is, in general, identical to the amount injected. This amount is given by the product of the volume injected, V, and the concentration, C, of the chemical in the sample, $m = V \times C$. Therefore, we can rewrite Equation 8 as follows:

$$A = f \times V \times C \qquad (9)$$

If the concentration is expressed in units of mass/volume, f represents a **mass response factor**; on the other hand, if C is given in moles/volume, f is called the **molar response factor**. Different compounds display distinctive response factors under identical conditions because their interactions with the detector are different. However, compounds that are closely related, such as members of the same chemical family, usually have similar response factors. When only a preliminary determination of the quantitative composition of a mixture is needed, we can assume that the components have identical response factors. *Under this assumption, the peak area is a direct measure of the amount of the compound in the sample.* This assumption enables a rapid and easy estimation of the **percentage composition** of a mixture, as discussed in the following example.

Example 1 A mixture of carvone and limonene from spearmint oil is analyzed by GC. The chromatogram is shown in Figure 7.22. Using a triangulation method calculate the % composition of carvone and limonene in the mixture assuming they have identical mass response factors.

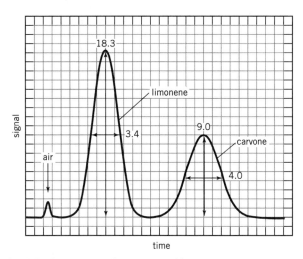

Figure 7.22 The GC of a mixture of carvone and limonene.

The heights (h) and widths at half the height ($w_{1/2}$) for the limonene and carvone peaks are indicated in the figure. The areas (A) can be calculated as follows:

$$A_{\text{lim}} = h_{\text{lim}} \times w_{1/2\ \text{lim}} = 18.3 \times 3.4 = 62.22$$

$$A_{\text{car}} = h_{\text{car}} \times w_{1/2\ \text{car}} = 9.0 \times 4.0 = 36.00$$

Since we are assuming that the mass response factors for carvone and limonene are identical, the areas under the peaks measure the mass of carvone and limonene present in the sample. Therefore, the % composition is:

$$\% \text{ Limonene} = 100 \times \frac{A_{\text{lim}}}{A_{\text{lim}} + A_{\text{car}}} = 100 \times \frac{62.22}{98.22}$$

$$\% \text{ Limonene} = 63.3\% \text{ and } \% \text{ Carvone} = 36.7\%$$

In general, the percentage of a component X in the sample can be estimated by its area using Equation 10:

$$\%X = \frac{\text{area of X}}{\text{total area}} \times 100 \tag{10}$$

The total area is the sum of the areas of the individual peaks in the gas chromatograph excluding the solvent and the air peaks.

7.9   QUANTITATIVE ANALYSIS: A CLOSER LOOK (ADVANCED LEVEL)

The response factor as defined in Equation 9 represents an **absolute response factor**. Absolute response factors can be experimentally determined, but this requires the injection of exactly measured volumes, which is very difficult to do manually. To circumvent this problem, **relative response factors** are usually determined. Figure 7.23 shows the GC of a mixture of known concentrations in two compounds A and B.

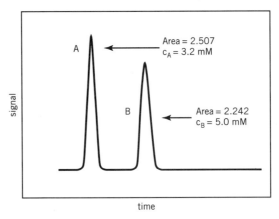

Figure 7.23   GC of a mixture of A and B with known concentrations.

The area under each peak has been determined by integration and is given in arbitrary units. Applying Equation 9 for each peak, Equations 11 and 12 are obtained:

$$A_A = f_A \times V \times C_A \tag{11}$$

$$A_B = f_B \times V \times C_B \tag{12}$$

Since A and B are injected together, V is the same in both equations. Dividing Equation 11 by Equation 12, the following relationship is obtained:

$$\frac{A_A}{A_B} = \frac{f_A}{f_B} \times \frac{C_A}{C_B} \tag{13}$$

The ratio of the two constants f_A and f_B, f_A/f_B, is a new constant that we call the **response factor of A relative to B**, $R_{A/B}$; thus Equation 13 is rewritten as:

$$\frac{A_A}{A_B} = R_{A/B} \times \frac{C_A}{C_B} \tag{14}$$

If the concentrations C_A and C_B in the sample are known, $R_{A/B}$ can be calculated from a single chromatographic run by measuring A_A and A_B. Rearranging Equation 14 we obtain:

$$R_{A/B} = \frac{A_A \times C_B}{A_B \times C_A} \tag{15}$$

In the example presented in Figure 7.23, the relative response factor $R_{A/B}$ is:

$$R_{A/B} = \frac{A_A \times C_B}{A_B \times C_A} = \frac{2.507 \times 5.0}{2.242 \times 3.2} = 1.75$$

Since the concentrations C_A and C_B used in this example are in molarity units, this response factor $R_{A/B}$, is the relative *molar* response factor.

Once the relative response factor has been measured for a given pair of compounds, it can be used to determine the percentage composition of any other mixture of the same two chemicals, provided the experimental conditions (GC instrument, column, detector, etc.) are the same (see Example 2).

Example 2   Recalculate the **percentage composition** of the mixture of carvone and limonene analyzed in Example 1, knowing that the mass response factor of limonene relative to carvone is $R_{\text{lim/car}} = 1.2$.

Applying Equation 15 to limonene and carvone and rearranging it, we obtain:

$$R_{\text{lim/car}} \times \frac{A_{\text{car}}}{A_{\text{lim}}} = \frac{C_{\text{car}}}{C_{\text{lim}}}$$

where $R_{\text{lim/car}}$ is the mass response factor of limonene relative to carvone, A_{car} and A_{lim} are the areas of the carvone and limonene peaks, respectively, and C_{car} and C_{lim} are the concentrations (in mass/volume) of carvone and limonene that we want to calculate.

Since $R_{\text{lim/car}} = 1.2$ and $\dfrac{A_{\text{car}}}{A_{\text{lim}}} = \dfrac{36.00}{62.22}$ (from Example 1), it follows that:

$$\frac{C_{\text{car}}}{C_{\text{lim}}} = 1.2 \times \frac{36.00}{62.22} = 0.694$$

The ratio of the concentrations $C_{\text{car}}/C_{\text{lim}}$ is identical to the ratio of the masses because the volume is the same (remember that both compounds are injected together):

$$\frac{C_{\text{car}}}{C_{\text{lim}}} = \frac{V \times m_{\text{car}}}{V \times m_{\text{lim}}} = \frac{m_{\text{car}}}{m_{\text{lim}}}$$

thus:

$$\frac{m_{\text{car}}}{m_{\text{lim}}} = 0.694$$

which implies:

$$m_{\text{car}} = 0.694 \times m_{\text{lim}}$$

Dividing both sides of the previous equation by the total mass ($m_{total} = m_{car} + m_{lim}$) and multiplying by 100 we obtain:

$$\frac{m_{car}}{m_{total}} \times 100 = 0.694 \times \frac{m_{lim}}{m_{total}} \times 100$$

or

$$\% \text{ Carvone} = 0.694 \times (\% \text{ Limonene})$$

Knowing that % Carvone + % Limonene = 100, we can solve the two-equation system for % Carvone and % Limonene and obtain:

$$\% \text{ Carvone} = 41.0\%, \text{ and } \% \text{ Limonene} = 59.0\%$$

Compare these numbers with the percentages obtained in Example 1.

Internal Standard Method

The methods described above are very useful to determine the percentage composition of mixtures but are difficult to apply for obtaining absolute concentrations. The **internal standard method** allows us to determine absolute concentrations (either in g/mL or molarity) in a simple and reproducible manner. The internal standard method works as follows: a known amount of a well characterized chemical (**the internal standard**) is added to the sample to be analyzed. The mixture is injected and the areas measured by routine procedures. Since we know the concentration of the standard in the mixture, we can calculate the absolute concentration of the components in the sample if we know their response factors relative to the internal standard. To apply the internal standard method, we must have previously measured the relative response factors of the compounds in the sample. The method is illustrated in Example 3.

Example 3   The composition of a mixture of ethanol and *n*-octanol in acetone as a solvent is determined by GC. *n*-Hexanol was chosen as internal standard. The molar response factors of ethanol and *n*-octanol relative to *n*-hexanol were previously determined in an independent chromatographic run and they are: $R_{ethanol/hexanol} = 0.62$ and $R_{octanol/hexanol} = 1.15$. To determine the concentrations of ethanol and *n*-octanol in the sample, 1 mL of a 4-mM *n*-hexanol solution in acetone is mixed with 1 mL of sample, and an aliquot of the mixture injected. By mixing equal volumes both solutions are diluted by a factor of two. Thus the concentration of *n*-hexanol is 2 mM.

The areas are measured by standard methods and they are (in arbitrary units): $A_{ethanol} = 4500$; $A_{octanol} = 740$ and $A_{hexanol} = 1100$. By applying Equation 15 the concentration of ethanol and *n*-octanol can be calculated as follows:

$$R_{ethanol/hexanol} = 0.62 = \frac{A_{ethanol} \times C_{hexanol}}{A_{hexanol} \times C_{ethanol}} = \frac{4500 \times 2 \text{ mM}}{1100 \times C_{ethanol}}$$

$$\implies \quad C_{ethanol} = 13.2 \text{ mM.}$$

$$R_{octanol/hexanol} = 1.15 = \frac{A_{octanol} \times C_{hexanol}}{A_{hexanol} \times C_{octanol}} = \frac{740 \times 2 \text{ mM}}{1100 \times C_{octanol}}$$

$$\implies \quad C_{octanol} = 1.17 \text{ mM}$$

You should notice that these concentrations are the concentrations of the alcohols *after mixing the sample with the internal standard solution*, which resulted in a dilution of the sample by half. Thus, the actual concentrations of ethanol and *n*-octanol in the original sample are 26.4 mM and 2.34 mM, respectively.

When planning a quantitative determination by the internal standard method we should first choose a compound as the internal standard. To be a good internal standard the compound should fulfill several criteria:

(a) The internal standard should give a well-resolved peak under the GC conditions used for the analysis.

(b) Ideally, the retention time of the standard should be in between those of the components to be analyzed.

(c) The internal standard should not interact with the chemicals in the sample.

(d) The relative response factors should be close to unity.

(e) The internal standard should be stable under the GC conditions.

(f) The internal standard should be a pure and accessible chemical.

There is an important limitation to the use of the internal standard method. The method can be applied only to chemicals that have previously been isolated in a reasonably pure state. This arises because we need to make solutions of known concentrations to determine the relative response factors.

7.10 QUALITATIVE ANALYSIS

The applications of gas chromatography stretch far beyond the organic chemistry laboratory. GC is an invaluable tool in all areas of chemical and biochemical research. Samples from diverse sources can be analyzed in a matter of minutes and the composition of complex mixtures can be determined in routine operations. In the organic chemistry lab, GC is used to follow the course of reactions, to investigate the presence of side products, to determine the concentration of chemicals (as discussed in sections 7.8 and 7.9), and to help elucidate the structure of unknown chemicals with the aid of a mass spectrometer as a detector (GC-MS). In this section we will consider how GC is applied in qualitative analysis.

One of the most common uses of GC is to determine how many chemicals are present in a mixture. The identity of the chemicals can also be asserted if pure samples, called **standards**, are available to be run under the same conditions. Identity based solely on retention times is of little use if the mixture under investigation is complex and no other information about its chemical composition is available. However, in analyzing a family of compounds, such as a homologous series, the retention time can be used to determine the identity of the chemical. There exists a correlation between the retention time and the molecular size in a family of compounds. A plot of **log $t_{R\,cor}$** versus the molecular mass or the number of carbons usually gives a straight line (Fig. 7.24), called the **standard curve**. Once the standard curve is known for a family of compounds, we can determine the number of carbons of an unknown chemical of the family by measuring its retention time.

We can use gas chromatography to analyze most organic chemicals provided they are volatile enough to vaporize in the injection port and do not decompose during the chromatographic run. Compounds such as amino acids, carboxylic acids, and sugars have to be transformed into more volatile **derivatives** before injection. Biopolymers, such as proteins, polysaccharides, and nucleic acids cannot be analyzed by GC due to their extremely low volatility.

In most GC analyses we are just interested in determining the composition of the sample and do not recover the compounds at the end of the run. This mode of operation is called **analytical**. Sometimes, however, we want to *collect the products* at the end of the column for further study and characterization. Under such circumstances the operational mode is called **preparative.** Preparative GC requires the use of larger columns that can be loaded with larger samples without losing resolution. Since the compounds must be recovered intact, thermal conductivity detectors are usually

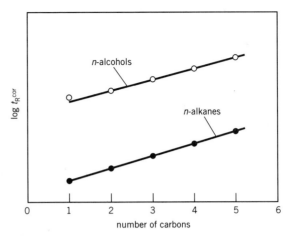

Figure 7.24 Correlation between the retention time and the number of carbons for two homologous series: *n*-alcohols (open circles) and *n*-alkanes (closed circles).

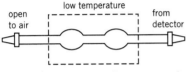

Figure 7.25 A GC collection vessel.

employed. After passing through the detector, the compounds are collected in a receiving vessel (Fig. 7.25), which is cooled at 0°C or lower to allow the condensation of the vapors emerging from the chromatograph. In most preparative columns only a few milligrams of sample can be injected at once; thus, preparative GC is useful only for the separation of small samples (0.1–20 mg).

7.11   RUNNING GC: STEP BY STEP

Getting Started

1. Get familiar with the gas chromatograph and make sure it is ready for use. Consult with your instructor to find out which controls you can set and which ones should not be touched.

2. Turn the chart recorder on and select an intermediate chart speed (1–4 cm/min). If the system is run by a computer, turn it on.

Cleaning the Syringe

3. Before the first injection, and always after an injection, clean the syringe by slowly drawing solvent several times (3–5 times). If you are going to inject a solution, use the same solvent as the sample's. If you are injecting an undiluted liquid sample, use methanol or acetone to clean the syringe.

4. Get familiar with the volume graduation of the barrel.

Loading the Sample

5. Rinse the syringe with your sample 3–5 times. Finally draw your sample slowly and make sure that no air bubbles are entrapped in the barrel. Draw more liquid than the actual volume that you are going to inject. Push the plunger to remove the excess liquid until you reach the desired volume to be injected. Now, if you are using a thermal conductivity detector, draw air into the syringe (1–10 μL). You are ready to inject (Fig. 7.26).

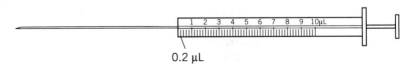

Figure 7.26 A typical 10-μL GC syringe.

Injecting the Sample

6. Insert the needle into the septum with a gentle and firm stroke. Push the plunger quickly all the way in and *immediately* remove the needle from the injection port. Be careful not to bend the needle or the plunger.

7. Mark the beginning of the injection on the chart paper, or start the data collection on the computer.

8. Immediately after use, clean the syringe with solvent as indicated in step 3.

9. Label the chromatogram and write down the experimental conditions (chart speed, column temperature, flow rate, stationary phase, solid support, etc.).

Analyzing the Chromatogram

10. Allow for all the expected peaks to come out before doing another injection.

11. For dilute samples, the solvent peak may be off scale. This will not interfere with your analysis. The solvent usually comes out before the sample.

12. The sample peaks should be on scale. If they are not, change the injected volume *or* the attenuation accordingly (lower attenuation to get bigger peaks, and vice versa) and reinject your sample.

13. Measure the retention time for each peak.

14. Integrate the peaks and calculate the % area. Unless instructed to do otherwise, do not integrate the solvent peak.

7.12   GAS CHROMATOGRAPHY DO'S AND DON'TS

Follow these recommendations to obtain good GC results.

Do's

- Clean the syringe with solvent immediately after each injection.
- Rinse the syringe with sample (3-5 times) before each injection.
- Inject quickly and remove the needle from the septum as soon as possible.
- Make sure the peaks are on scale. If not, reinject the sample.
- To obtain peaks on scale, change the injected volume or the attenuation.
- Label each chromatogram.
- Write down GC conditions (chart speed, column temperature, flow rate, etc.).
- Turn off the chart recorder when you are finished.

Don'ts

- Do not bend the needle.
- Do not overload the column.
- Do not change the GC settings without consulting with your instructor.
- Never heat a column without carrier gas flowing through.
- Do not heat the column beyond its maximum recommended temperature.

GC in the ER   At 2 in the morning a 2-year old girl is rushed to the emergency room. The toddler is drooling and trembling, and has glossy eyes with pinpoint pupils. Earlier that evening the mother had found an open bottle of a flea and tick dog dip in the child's play area. She called the local poison control center and was told that the girl would have had to drink 2 to 3 oz. of the liquid to show any ill effects. At that point the toddler had no symptoms of poisoning and the mother put her to bed and kept a watchful eye. The girl woke up in the middle of the night with tremors and was taken to the ER.

In the ER the girl is moaning and almost paralyzed. She is immediately put under IV therapy for poisoning and a urine sample is taken for a toxicology screen. The sample is treated with activated charcoal to adsorb organic chemicals present in the urine. The charcoal is filtered off and treated with methylene chloride to remove the adsorbed compounds. The solution is separated from the charcoal and the methylene chloride evaporated. The solid residue is dissolved in methanol and injected into a gas chromatograph using a mass spectrometer as a detector. The GC-MS analysis confirms the presence of chlorpyrifos, the insecticide's active ingredient, in the urine. The treatment is continued and the child completely recovers in a few days.

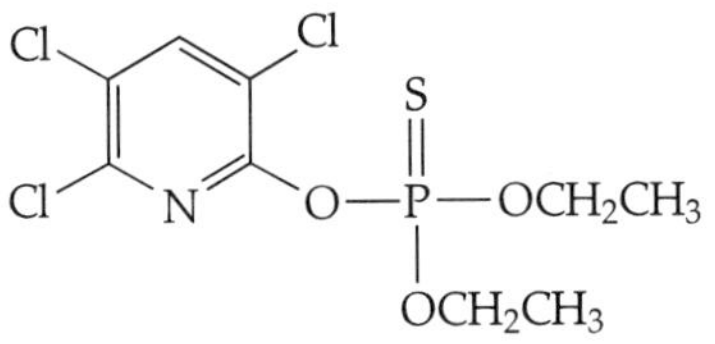

chlorpyrifos: an insecticide

Bibliography

Based on a true clinical case reported by Dr. Nancy Standler, M.D., Ph.D., and Dr. Mohamed A. Virji, M.D., Ph.D. http://path.upmc.edu/cases/case83.html

BIBLIOGRAPHY

1. Gas Chromatography. A Practical Approach. P.J. Baugh, ed. IRL Press. Oxford, 1993.
2. Gas Chromatography. Principles, Techniques and Applications. A.B. Littlewood. Academic Press, New York, 1970.
3. Chromatographic Methods. A. Braithwaite and F.J. Smith; 4th ed. Chapman and Hall, London, 1990.
4. Chromatography. E. Heftmann, ed. Reinhold, New York, 1967.
5. Capillary Gas Chromatography. D.W. Grant. Wiley, New York, 1996.
6. Gas Chromatography and Mass Spectrometry: A Practical Guide. F.G. Kitson, B.S. Larsen, and C.N. McEwen. Academic Press, San Diego, 1996.
7. Modern Practice of Gas Chromatography. R. L. Grob, ed. 3rd ed. Wiley, New York, 1995.

EXERCISES

Problem 1

Complete the following paragraph using the words shown below. Each word or group of words should be used once.

In gas chromatography the mobile phase is a _____ and the stationary phase can be a _____ or a _____ . The separation process can be _____ or _____ depending on the nature of the stationary phase. Two of the most commonly used carrier gases are _____ and _____ . We inject the sample using a _____ syringe. Separation takes place inside the column and the compounds reach the detector at different times. _____ detectors and _____ detectors are available for gas chromatographs. In general, FID are more _____ than TCD. The attenuation controls _____ . A decrease in the attenuation results in a peak size _____ . For a given compound, the retention time depends on the _____ and _____ of the column, the _____ of the carrier gas, and the _____ of the oven. An increase in the temperature normally results in a _____ in the retention time.

adsorption; decrease; diameter; flame ionization; flow rate; gas; Hamilton; helium; increase; length; liquid; nitrogen; partition; peak size; sensitive; solid; temperature; thermal conductivity.

Problem 2

The retention times of *n*-pentane, *n*-hexane, *n*-heptane, *n*-octane, *n*-nonane, *n*-decane, benzene and toluene are 3.61; 4.46; 6.37; 8.92; 11.5; 15.5; 18.0, and 22.3 min, respectively (column packing: OV-101 on Chromosorb; temperature program: 70°C to 110°C increased at a rate of 2°C/min).

(a) Check the boiling points for the hydrocarbons and determine whether they come out in the same order of their boiling points.

(b) Plot $\log(t_R)$ versus number of carbons. Why do benzene and toluene behave anomalously?

(c) A sample is suspected to be contaminated with an aliphatic hydrocarbon; its GC analysis shows a peak at 28.4 min (under the same experimental conditions). Based on your results from part b, find the most likely size for the hydrocarbon.

Problem 3

A sample of gasoline is tested for aromatic hydrocarbons by GC using a column of medium polarity. In what order do you expect the following hydrocarbons to come out of the column?: toluene, benzene, ethylbenzene, *o*-, *m*-, and *p*-xylene (use the *Aldrich Catalog* or *Merck Index* to answer this question).

Problem 4

Volatile aliphatic amines were analyzed by GC using a Carbowax column and the following retention times were obtained (min): CH_3NH_2: 1.2; $(CH_3)_2NH$: 1.3; $CH_3CH_2NH_2$: 1.4; $(CH_3)_3N$: 1.5; $CH_3CH_2CH_2NH_2$: 2.0. The boiling points of the amines are (in that order): −6.3°C; 7°C; 16.6°C; 2.9°C; and 49°C. Justify the retention time order using boiling point and molecular mass arguments.

Problem 5

(*Previous Synton: Problem 9, Unit 6.*)   Before going to grad school Synton used to work in a forensic lab as an intern. In one of his assignments he had to analyze a piece of evidence that looked like cookie crumbs, and determine if they could have actually belonged to a peppermint flavored cookie found at the crime scene. The GC-MS analysis of the volatile components of the crumbs showed the presence of menthol and menthone, the two main components of peppermint oil.
Which compound do you think came out of the GC column first? Why?

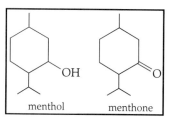

menthol   menthone

Problem 6

A mixture of cyclopentanol, 2-pentanol, and 4-methyl-2-pentanol was analyzed using a Carbowax column. The GC is shown in the figure below.

(a) Identify the peaks with the help of a boiling point table of alcohols.

(b) Estimate the % of each compound in the mixture; assume identical response factors for all.

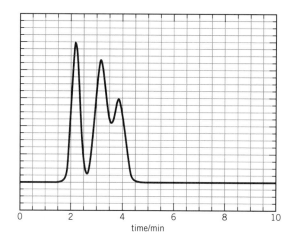

time/min

Problem 7 (*advanced level*)

The concentration of compounds 1 and 2 is determined by GC using the internal standard (IS) method. A standard solution (100 mL) containing 1 (0.98 g); 2 (0.63 g); and IS (1.50 g) is analyzed; the GC is shown in part *a* of the figure below. The integration for the peaks of 1, 2, and IS gives the following areas: $A_1 = 3.92$; $A_2 = 2.75$; $A_{IS} = 5.60$ (in arbitrary units).

(a) Calculate the response factor of 1 and 2 relative to IS ($R_{1/IS}$ and $R_{2/IS}$).

(b) 0.75 g of solid IS is added to 100 mL of a solution of 1 and 2 (concentration unknown) and the mixture analyzed by GC (part *b* of the figure below). The areas for 1, 2, and IS are: 4.03; 8.23; and 3.36, respectively. Calculate the concentrations of 1 and 2 in solution.

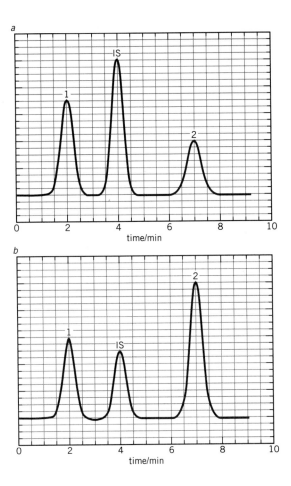

Gas Chromatography of Alcohols

In this experiment you will analyze a series of aliphatic alcohols and determine their retention times. You will correlate the retention times with the number of carbon atoms. You will also analyze an unknown mixture of alcohols and identify its components. You will integrate the peaks and determine the percentage composition of the mixture. Finally, you will analyze the mixtures of methanol and water that you obtained by fractional distillation in Experiment 6A. You will determine the percentage of the two substances in different distillation fractions.

The GC conditions will be set for you and the instruments will be ready for use. If a simple chart recorder is used, you will be allowed to change the attenuation and the chart speed. Your instructor will give you precise instructions on the use of the chromatograph and the syringes; follow them. For the analysis of alcohols a column of low polarity, such as SF-96, is recommended; for the analysis of the methanol-water fractions a more polar column gives better results (such as Carbowax 20M, registered mark of Union Carbide Corp.). Record the conditions used for your chromatographic separation in your lab book.

Important Note
• Use the same GC instrument throughout your analysis.

- Carrier gas:
- Stationary phase:
- Column temperature:
- Injection port temperature:
- Length of column:
- Diameter of column:
- Flow rate of carrier gas:
- Chart speed: you set
- Attenuation: you set

PROCEDURE

Background reading: 7.1–7.8; 7.10–7.12
Estimated time: 5–10 minutes per injection

E7A.1   ANALYSIS OF ALCOHOLS

A standard mixture of five alcohols (methanol, ethanol, *n*-propanol, *n*-butanol and *n*-pentanol) will be analyzed using a column of low polarity (for example, SF-96 column, at 75°C). Inject 0.2–3 µL (your instructor will tell you the right volume for your instrument) of the standard mixture of alcohols making sure that there is an air bubble in the syringe. As you are injecting, make a mark with pencil on the chart paper to indicate the beginning of the run (time zero) if you are using a chart recorder, or start the data collection on the computer. Write down the sample name, the chart speed (if applicable), and the attenuation (without this information you cannot complete your analysis). When the last peak has come out of the column, inject a sample of one of the pure alcohols and note its retention time. This will help you "anchor" the chromatogram of the mixture and identify all of its peaks. To avoid cross-contamination, always rinse the syringe with acetone between injections. Rinse several times with the

Safety First
• Acetone and the alcohols are flammable.

sample to be analyzed before injection. Before introducing the needle into a new liquid remove any traces of liquid from the needle by wiping it with a piece of tissue paper.

Determine the retention time for each alcohol. Remember that the retention time should be given in units of time (minutes) rather than distance (Eq. 5). Plot *log* t_R versus number of carbons. Correct the retention time ($t_{R\ cor}$) for each component by subtracting the t_R of the air peak (Eq. 6). Calculate the log of the corrected retention times and plot these values versus number of carbons.

Inject your **unknown** and determine the retention time of each component and measure the area under each peak (use a triangulation or weight method). Identify the alcohols and calculate the percentage composition of the unknown mixture assuming that all the alcohols have similar response factors, (Eq. 10). In case of doubt about the identity of your unknown, mix a few drops of your unknown with one or two drops of the suspected compounds (one at time) and inject the mixtures. If the suspected compound does not match any of the components of your mixture you will observe a new GC peak. This technique is called "**spiking**" and it is very useful for identifying unknowns.

<table>
<tr><td>

Cleaning Up

- Make sure that at the end of the experiment the syringes have been rinsed with acetone.
- Dispose of the methanol-water fractions and the alcohol mixtures in the container labeled "Alcohols Waste."

</td></tr>
</table>

E7A.2 SEPARATION OF METHANOL AND WATER

Inject a sample of pure methanol into the GC column (for example, Carbowax at 75°C). Using the same procedure, analyze fraction 1, the last fraction, and a middle fraction from the fractional distillation of methanol-water (Exp. 6A). If time permits analyze the other fractions as well. Measure the retention time of the methanol and water peaks. Measure the area under each peak and calculate the percentage composition of each fraction assuming similar response factors for water and methanol.

EXPERIMENT 7A REPORT

Pre-lab

1. Define briefly the following concepts: partition chromatography, mobile phase, stationary phase, solid support, integration, and retention time.

2. Four samples contain a mixture of two or more of the hydrocarbons listed below. They are analyzed by GC using an SF-96 column. The four chromatograms (run under identical conditions) are shown in the figure below.

   (a) Identify the components of each sample.

   (b) Estimate the % of each compound in sample 1. Hydrocarbon (bp): 2-methyl-heptane (117°C), cyclohexane (81°C), *n*-pentane (36°C), *n*-octane (126°C), *n*-heptane (98°C), and 2-methylhexane (90°C).

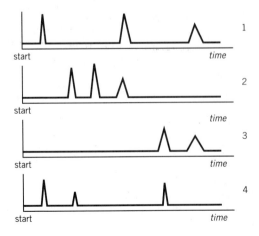

3. Make a table showing the physical properties (b.p., molecular mass, solubility, density, toxicity/hazards) of methanol, ethanol, *n*-propanol, *n*-butanol, *n*-pentanol, and water.

In-lab

1. Summarize the learning objectives of this experiment by mentioning the new techniques introduced.

2. Record the GC conditions: type of solid support; liquid stationary phase; length and diameter of the column; temperature of the column; temperature of the injection port; type of carrier gas; and flow rate.

3. Use a table format to report the retention times (corrected and uncorrected) for air, methanol, ethanol, *n*-propanol, *n*-butanol, *n*-pentanol, and the components of the unknown sample. Justify the order in which they come out.

4. Plot *log* t_R and *log* $t_{R\,cor}$ versus number of carbons for the standard mixture of alcohols. Which plot gives a better linear correlation? Discuss your results briefly.

5. Identify the alcohols in your unknown mixture.

6. Report in a table format the percentage composition of the unknown mixture. Show calculations and discuss your data.

7. Use a table format to report the retention times of methanol and water and the percentage composition of the methanol-water fractions. Discuss your GC data in relation to the fractional distillation results. Do your GC results support your conclusions about fractional distillation?

8. Attach the chromatograms to your report and label each peak.

EXPERIMENT 7B

Analysis of Anise Oil

In this experiment you will analyze a sample of anise oil obtained by steam distillation in Experiment 6B. If this is not available, a sample of commercial anise oil can be used instead. The main components of anise oil are *trans*-anethole, its isomer estragole (also known as methyl chavicol), and *p*-anisaldehyde. One of them accounts for more than 80% of the oil. In this experiment you will determine which one.

These three aromatic compounds can be separated by GC using a low-polarity column (for example, SF-96). The boiling points at reduced pressure are indicated below.

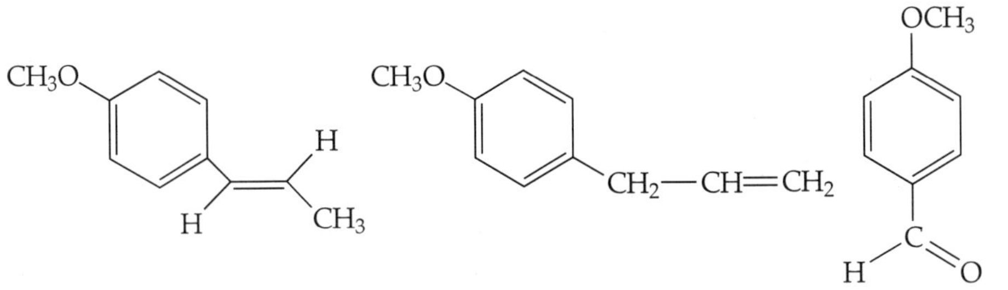

trans-anethole
$bp_{2.3}$ 82°C

estragole
(methyl chavicol)
bp_4 75°C

p-anisaldehyde
$bp_{1.5}$ 90°C

The GC conditions will be set for you and the instrument will be ready for use. If a simple chart recorder is used, you will be allowed to change the attenuation and the chart speed. Your instructor will give you precise instructions on the use of the chromatograph and the syringes; follow them. Record the conditions used for your chromatographic separation in your lab book. See the list on page 167. (Exp. 7A).

Important Note

- Use the same GC instrument throughout your analysis.
- To avoid cross-contamination, rinse the syringe with acetone after each injection, and rinse it 3–5 times with the sample to be injected before each injection. Before you introduce the needle into a new liquid sample, clean it with tissue paper to remove traces of liquid from previous injections.

PROCEDURE

Background reading: 7.1–7.8; 7.10–7.12
Estimated time: 5–10 minutes per injection

E7B ANALYSIS OF ANISE OIL

Safety First

- Acetone is flammable.
- Estragole and *p*-anisaldehyde are irritants.

GC Analysis Inject 0.2–3 μL (your instructor will tell you the right volume for your instrument) of anise oil into the GC (for example, SF-96 column at 150°C). Mark the beginning of the run on the chart recorder if you are using one, or start the data collection on the computer. Write down the sample name, the chart speed (if applicable), and the attenuation. If the peaks are not on scale, repeat the injection using a different attenuation. Calculate the retention times of the peaks and integrate them.

Inject pure samples of *trans*-anethole, estragole, and *p*-anisaldehyde. Record their retention times and compare them with those from anise oil. Identify the peaks in anise oil. Determine which compound is the main component of the oil.

If time permits, you can further analyze the anise oil by refractive index and freezing. To obtain good results in these tests, the oil should be solvent free. If you obtained the oil by steam distillation and liquid extraction (Exp. 6A), blow a gentle stream of nitrogen on the oil contained in a round-bottom flask to eliminate traces of solvent.

Refractive Index   Determine the refractive indexes of the oil, and also of pure samples of its components. (Refractive index is discussed in Unit 11.)

Freezing   Transfer a few drops each of anise oil, *trans*-anethole, estragole, and *p*-anisaldehyde to four labeled small test tubes. Place the test tubes in an ice-water bath and observe what happens.

Cleaning Up

- Make sure that at the end of the experiment the syringes have been rinsed with acetone.

- Dispose of the standard samples in the container labeled "Anise oil–GC–waste."

- Turn in the anise oil in a capped, labeled vial to your instructor.

EXPERIMENT 7B   REPORT

Pre-lab

1. Briefly define the following concepts: partition chromatography, mobile phase, stationary phase, solid support, integration, and retention time.

2. Make a table showing the physical properties (molecular mass, bp, mp, density, solubility, refractive index, and toxicity/hazards) for *trans*-anethole, estragole, and *p*-anisaldehyde.

3. The boiling points at reduced pressure for *trans*-anethole, estragole, and *p*-anisaldehyde are given on page 170. The pressures at which these boiling points were recorded are different (2.3, 4, and 1.5 Torr, respectively) and, thus, in principle, we cannot compare the b.p. However, in this case we can predict the order of the boiling points at one of the applied pressures by applying the simple rule that the boiling point increases (decreases) with an increase (decrease) of the applied pressure. Determine the order of the boiling points at 2.3 Torr, and predict the order in which you expect these compounds to come out of the GC column.

In-lab

1. Summarize the learning objectives of this experiment by mentioning the new techniques introduced.

2. In a table format, report the retention times of the oil components, and those of the pure samples of *trans*-anethole, estragole, and *p*-anisaldehyde. Do the retention times follow the prediction made based on their boiling points? If not, can you speculate why?

3. Compare the retention times of the oil components with those of the standards and identify the main component of the oil.

4. Integrate the peaks in the GC trace for the oil and calculate the % composition of the oil. Present your data in a table format. Discuss your results briefly.

If applicable:

5. Compare the refractive indices with literature values. Explain any source of error.

6. Report and discuss your freezing experiment.

UNIT 8

Thin-Layer Chromatography

8.1 OVERVIEW

Thin-layer chromatography (TLC) is one of the most versatile and rapid separation techniques available to the organic chemist. Analyses of complex mixtures can be successfully carried out in a matter of minutes with relatively inexpensive equipment. In TLC the separation takes place on a thin layer of **solid stationary phase** spread on a solid support (a glass, aluminum, or plastic plate) as a **liquid mobile phase**, also simply called the **solvent**, moves along the plate. Separation can be achieved by partition, adsorption, ion-exchange, or size-exclusion processes. In the organic chemistry laboratory, however, most applications of TLC are based on adsorption. Thus, in our discussion we will primarily concentrate on TLC as **adsorption chromatography**. A brief description of other types of TLC is also offered (section 8.7). An introduction to chromatography and the discussion of partition chromatography are presented in Unit 7.

In TLC the sample, a nonvolatile liquid or a solution in a volatile solvent, is spotted on the adsorbent with the aid of a micropipet at a distance of about 1.5 cm from the lower edge of the plate (Fig. 8.1). After the solvent has evaporated from the spot, the plate is placed in a chamber that contains the mobile phase of choice to a height of about 0.5 cm. The chamber is capped to avoid evaporation of the solvent and to ensure liquid-vapor equilibrium inside. The solvent rises on the TLC plate through capillary action, pulling along the components present in the sample. In general, different compounds travel at different speeds because they have specific interactions with the stationary phase. When the solvent has traveled 80–95% of the length of the plate, the plate is removed from the chamber, and the level reached by the solvent,

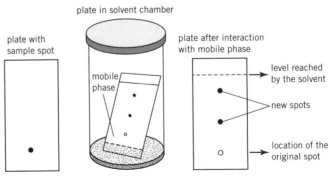

Figure 8.1 TLC plates before, during, and after interaction with the mobile phase.

called the **solvent front**, is marked. The solvent is allowed to evaporate, and the plate is analyzed.

In Figure 8.1, the original sample spot has separated into two spots after interaction with the mobile phase, indicating that the sample contains at least two different compounds.

8.2   THE ADSORPTION PROCESS

Adsorption is a process by which molecules of a gas, liquid, or solid in solution, called the **solute**, interact with the molecules *on the surface of a solid*, called the **adsorbent**. Adsorption is strictly a surface process that depends on electrostatic forces between adsorbent and sample. These forces arise from **dipole-dipole** and **ion-dipole interactions**, and **H-bonds** (polarity and H-bonds are discussed in section 2.1). The surface of the adsorbent is far from being perfectly smooth; it has discontinuities, crevices, and crests with centers of positive and negative charge density. The solute binds to the adsorbent through electrostatic attraction between its own centers of charge density and those on the surface of the adsorbent. The places on the surface of the adsorbent where the sample binds are called **binding sites** (Fig. 8.2). Ions and molecules with permanent dipole moments readily bind to the adsorbent.

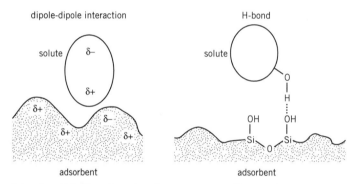

Figure 8.2   Adsorbent binding sites showing two types of interactions: dipole-dipole and H-bond.

The most commonly used adsorbents in TLC are **silica gel** and **alumina**. Silica gel (SiO_2 with an undefined number of water molecules) is obtained by hydrolysis of silicates. It has polarized $Si-O$ and $O-H$ bonds that interact with dipoles in the solute. It can also form H-bonds, especially with H-bond donors such as alcohols ($R-OH$), phenols ($Ar-OH$), amines ($R-NH_2$), amides ($R-CO-NHR'$), and carboxylic acids ($R-COOH$). Alumina is aluminum oxide (Al_2O_3 with an undefined number of water molecules) obtained by dehydration of aluminum hydroxide. Both alumina and silica gel are available with different particle sizes. The types normally used for TLC have an average diameter of about 0.025 mm with a wide distribution; some particles are as small as 0.005 mm and others as large as 0.045 mm.

To understand how separation occurs on the surface of the adsorbent, we should look closer at the interactions between the **adsorbent** and **solutes** and between the **adsorbent** and **mobile phase**. In the following discussion we will assume that the mobile phase dissolves equally well all the constituents of the sample for which it has similar affinities. Therefore, the interactions between **mobile phase** and **different solutes** can be considered quite similar and do not play a crucial role in the separation (Fig. 8.3).

Figure 8.4*a* displays a close-up view of a TLC plate cross-section showing the interactions on the surface of the adsorbent. The sample has two different components,

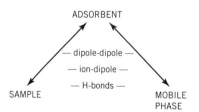

Figure 8.3   Interactions playing important roles in adsorption.

represented by ovals and triangles. The solvent molecules are shown as circles. Because there is a limited number of binding sites on the surface of the adsorbent, *solute and mobile phase molecules must compete* with each other to bind to the adsorbent; the stronger the interactions, the tighter the binding. In Figure 8.4a, the thickness of the lines indicates the strength of the interaction.

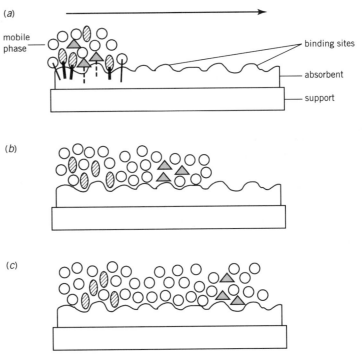

Figure 8.4 Separation of two compounds on a TLC plate (ovals and triangles). The arrow indicates the direction of the mobile phase flow (see text).

The oval molecules have a very strong interaction with the adsorbent as indicated by a very thick line, while the triangles have a weaker interaction as shown by the dashed line. The mobile phase molecules, on the other hand, have a stronger interaction than the triangles, but a weaker one than the ovals. As the mobile phase ascends along the plate, *those mobile phase molecules close to the adsorbent compete with the solutes for the binding sites.* If the interaction between mobile phase and adsorbent is stronger than the interaction between adsorbent and solutes, such as in the case of triangle molecules, the mobile phase displaces the solute molecules from their binding sites, moving them farther away from the surface of the adsorbent. As the mobile phase keeps flowing, these solute molecules are carried to a new location on the plate (Fig. 8.4b). Because the mobile phase has stronger interaction with the adsorbent, the triangle molecules do not bind efficiently and keep traveling along the plate at high speed. At the end of the chromatographic run, the triangles are found near the solvent front (Fig. 8.4c).

If the interaction between solute and adsorbent is stronger than the interaction between mobile phase and adsorbent, the mobile phase still moves the solute in the direction of flow. This is a result of mass action; the mobile phase molecules, being much more numerous than the solute molecules, eventually displace the solute molecules from their binding sites, even if the solute has stronger affinity for the adsorbent. However, the stronger the interaction between adsorbent and solute, the more difficult it is for the solvent to move the solute. As shown in Figure 8.4c, the oval molecules, which display the strongest affinity for the adsorbent, stay very close to the place where they were originally spotted.

8.3 SELECTION OF TLC CONDITIONS

Planning the Separation

The success of a TLC separation depends on the careful selection of the stationary and mobile phases. Alumina and silica gel are by far the most commonly used adsorbents; however, other stationary phases occasionally make their way into the organic chemistry lab, especially for very specific separations. For example, **kieselguhr** and **Celite** are diatomaceous earths with very limited adsorption strength, and are frequently used in the separation of very polar compounds. **Starch** and **powdered sugar** can be used to separate chlorophylls.

The adsorption strength or **adsorptivity** of the stationary phase is determined not only by the chemical nature of the adsorbent, but also by its particle size and the amount of water it contains. Water is bound to the adsorbent covalently and noncovalently through H-bonds. When aluminum and silicon oxides react with water, the water molecules are incorporated into their chemical structures as −OH groups.

Noncovalently bound water molecules interact with the adsorbent by H-bonding and occupy some of its binding sites. This interaction decreases the adsorbent's adsorptivity. The amount of water adsorbed in silica gel and alumina, which can be as high as 15%, can be decreased by heating the adsorbent at high temperatures. The removal of water at high temperatures is called **activation**. The **activity** of the adsorbent is usually denoted by a Roman numeral varying from I to V; I is for adsorbents with high activity (low water contents) and V for low activity (high water contents).

A list of different adsorbents in order of decreasing adsorptivity is given in Table 8.1. This table should be regarded solely as a rough guide, since ultimately, the power of the adsorbent depends on its activity and particle size.

The other parameter that plays a crucial role in the separations is the *polarity of the mobile phase*. The dielectric constant is usually taken as an indicator for polarity; however, other indices (based on chromatographic separations) have been devised (Table 2.2). Some useful solvents for TLC are given in Table 8.2 in order of increasing polarity.

Table 8.1 **Adsorbents in Order of Decreasing Strength**

activated charcoal
alumina
silica gel
calcium phosphate (hydroxyapatite)
diatomaceous earths
starch

Table 8.2 **Selected Solvents in Order of Increasing Polarity**

Solvent	Dielectric constant
hexanes	1.89
cyclohexane	2.02
toluene	2.38
diethyl ether	4.34
ethyl acetate[a]	6.02
methylene chloride[a]	8.93
acetone	20.7
methanol	32.7
water	80.1

[a]On the basis of chromatographic data, ethyl acetate is considered more polar than methylene chloride (Table 2.2).

The use of solvent mixtures is very common in TLC analysis. The polarity of the mobile phase can be changed within a wide range by mixing solvents of different polarities. For example, mixtures of hexane and methylene chloride with increasing proportion of the latter provide a series of solvents of increasing polarity.

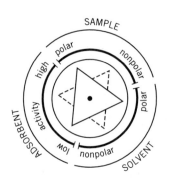

Figure 8.5 Diagram relating the polarity of the sample to the polarity of the solvent and the activity of the adsorbent for a good TLC separation. The triangle in the middle is free to rotate (after Ref. 2 and references therein).

In selecting the solvent and the adsorbent for a given TLC separation, there are some rules of thumb that we can apply. One says that *the polarity of the solvent should be comparable to the polarity of the solutes,* and the other that *the activity of the adsorbent should be low for highly polar solutes, and high for nonpolar solutes.* These rules are summarized in the diagram shown in Figure 8.5. In this figure, the central triangle is free to rotate around its center. The triangle should be rotated so that one of its vertices indicates the polarity of the sample; the other two vertices then indicate the activity of the adsorbent, and polarity of the solvent recommended for the separation.

According to these rules, the separation of **very polar compounds** is better carried out using **deactivated adsorbents** (low adsorption strength) and **polar solvents**. The use of high-strength adsorbents is not recommended for separating very polar compounds because the interaction between them would be so exceptionally strong that even very polar solvents would not be able to break them. Polar solvents are recommended with polar substances because they must be strong competitors for the binding sites on the adsorbent in order to displace the solute molecules and move them up.

On the other hand, the separation of **nonpolar compounds** is better carried out with the use of **active adsorbents** and **nonpolar solvents**. Adsorbents of low activity are not recommended, because their binding to nonpolar solutes would be too weak and easily disrupted by the solvents (even nonpolar ones). The use of nonpolar solvents is recommended because polar solvents would bind to the adsorbent more tightly than the solutes, and would displace *all the solutes equally well* from their binding sites without achieving any separation.

Some Examples

As we have seen, the separation in adsorption chromatography is largely based on differences in polarity and the ability to form H-bonds. Compounds capable of donating an H-bond adsorb to silica and alumina more strongly than similar compounds with no H-donor capabilities. For example, carboxylic acids (R—COOH) are adsorbed more tightly than esters (R—COOR′). Alcohols (R—OH) have a stronger interaction with the adsorbent than ethers (R—O—R′). A list of different types of compounds in order of increasing polarity is given in Table 8.3. This list should be regarded only as a rough guide.

Table 8.3 Families of Organic Compounds in Order of Increasing Polarity

Family of compounds	Structure
aliphatic hydrocarbons	R—H
alkyl halides	R—X
unsaturated hydrocarbons	R—CH=CH—R
aromatic hydrocarbons	Ar—H
aryl halides	Ar—X
ethers	R—O—R
esters	R—COOR
ketones	R—CO—R
aldehydes	R—CO—H
amides	R—CO—NH$_2$
amines	R—NH$_2$
alcohols	R—OH
phenols	Ar—OH
carboxylic acids	R—COOH
amino acids	H$_3$N$^+$—CHR—COO$^-$

In general, compounds that are only H-bond acceptors (such as ethers, esters, tertiary amines, nitrocompounds, etc.) are absorbed less strongly than H-bond donors (alcohols, phenols, carboxylic acids, amines, etc.). An interesting case is provided by molecules capable of forming *intramolecular H-bonds* as compared with similar compounds that can only form *intermolecular H-bonds*, for example, *o*-nitrophenol and *p*-nitrophenol.

In the case of *o*-nitrophenol, there is a strong intramolecular H-bond between the hydroxyl and the nitro group; therefore, *o*-nitrophenol is not able to interact as an H-bond donor with the adsorbent. On the other hand, *p*-nitrophenol can act as donor or acceptor of H-bonds and, thus, it is adsorbed strongly on silica gel. As a result, a mixture of *p*- and *o*-nitrophenol can be easily separated by adsorption chromatography on silica gel with the *ortho* isomer moving faster than the *para* (Fig. 8.6).

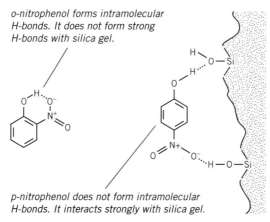

Figure 8.6   Interaction of nitrophenols with silica.

To illustrate the relation between polarity and the success of the separation, we will now consider three hypothetical cases (Fig. 8.7). In all cases, the mixture contains

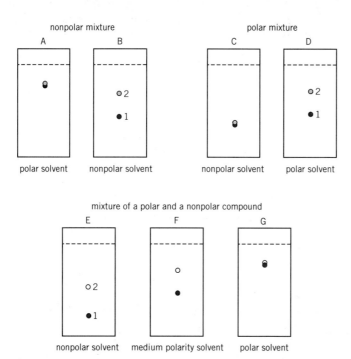

Figure 8.7   Separation of hypothetical mixtures. In all cases compound 1 is more polar than compound 2.

two compounds of different polarity. In the first case, both compounds are relatively nonpolar, for example, an alkene and an aromatic hydrocarbon. In this case a polar solvent does not separate the mixture (A); instead it moves both compounds similar distances. A nonpolar solvent, on the other hand, separates the two components (B). The alkene (less polar) travels farther than the aromatic hydrocarbon (more polar).

In the second case the mixture contains two polar compounds, for example, an ester and an alcohol. Neither compound moves with a nonpolar solvent (C), but they are moved and separated by a polar solvent (D). The less-polar ester travels farther than the more-polar alcohol.

Finally, the third case is a mixture of two compounds of very different polarities. This type of mixture is relatively easy to separate using nonpolar or medium polarity solvents. For example, a mixture of an aromatic hydrocarbon and a phenol can be separated using a nonpolar solvent; the aromatic hydrocarbon moves farther, and separates from the more polar phenol (E). Using a solvent of medium polarity increases the distances traveled by both compounds but they are still separated from each other (F). Finally, if a polar solvent is used, both compounds move with similar speeds and no separation is achieved (G).

8.4   RUNNING TLC PLATES

Selecting the Plates

Commercial TLC plates are available with a large variety of adsorbents and with glass, aluminum, or plastic backing. If money is not an issue, commercial plates are usually preferred over home-made ones because they give more reproducible results. For experimental procedures on TLC plate preparation, see, for example, Reference 5.

There are different types of silica gel and alumina available on the market. "Silica Gel G" (Merck) has calcium sulfate as a **binder** (G stands for "Gypsum," a synonym for calcium sulfate), which facilitates the adhesion of the adsorbent to the glass plate. "Silica Gel H" (Merck) contains no binders. Alumina is available in three different types: neutral (pH ≈ 7.5), basic (pH ≈ 9.5), and acidic (pH ≈ 4). "Aluminum Oxide G" (Merck) is alumina that contains calcium sulfate as a binder and has basic pH. For most applications, silica gel or neutral alumina plates give reasonably good separations.

Plates can be activated in an oven (70–105°C) for different periods of time depending on the application. For routine use, however, activation of commercial plates is not necessary.

Spotting the Plates

A solution of the sample to be analyzed is made in a suitable solvent. The solvent must dissolve the sample completely to ensure an accurate representation of the sample's composition. Volatile solvents such as acetone and methanol are particularly recommended because they evaporate rather quickly once the sample is spotted on the plate.

A small aliquot of the solution is taken with a capillary tube or micropipet (with both ends open) and applied on the TLC plate by touching the adsorbent with the tip of the micropipet lying flat (Fig. 8.8). Gentle pressure is applied to deliver the liquid, which penetrates the adsorbent. When the size of the spot is about 3–4 mm in diameter, the micropipet is raised to stop the flow of liquid. The solvent is allowed to evaporate and another application is made, if necessary.

Ordinary capillary tubes used for melting point determinations and commercial Pasteur pipets are too wide to be used as spotting devices for TLC. Narrow capillary tubes (0.3–0.7 mm internal diameter) should be used instead.

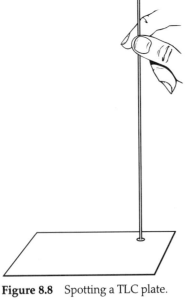

Figure 8.8   Spotting a TLC plate.

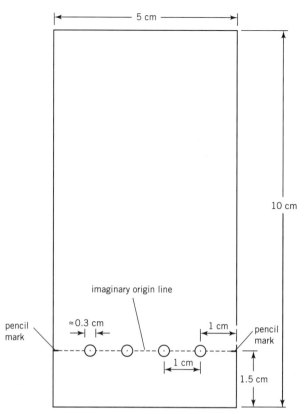

Figure 8.9 Positioning spots on a TLC plate.

For regular TLC plates (5 × 10 cm), spots should be applied at a distance of about 1.5 cm from the bottom edge of the plate, and about 1 from the sides of the plate (Fig. 8.9). If more than one sample is analyzed per plate, a distance of at least 1 cm should separate the spots. A mark made with pencil on both sides of the plate should indicate the **imaginary origin line** where all the spots are applied. For microplates (2.5 × 5 cm), the spots should be applied at a distance of about 1 cm from the bottom edge and 0.5 cm from the sides; the separation between spots should be at least 0.7 cm. All marks on the TLC plates should be made with pencil, never with ink because the components in ink may separate during the run and interfere with the TLC analysis.

A question frequently asked by the students is, "How many applications should we do per spot?" The number of applications per spot should be less than 10, if possible. The actual number depends on the concentration of the solution to be analyzed and the nature of the components. Beyond the overly general principle that the lower the concentration of the sample, the larger the number of applications, no other prediction can be made, because the optimum amount of material required for a good separation varies from sample to sample. It also depends on the mobile and stationary phases used. The recommended concentration range for the solutions is 0.1–5%. One should perform trials with different numbers of applications (1–10) until the optimum number is found. In an ideal situation, at the end of the run the spots should be clearly delineated, circular in shape, and without streaks. If more than 10 applications are needed to see the spots, the sample should be concentrated to avoid repeated applications that may damage the surface of the adsorbent.

Developing the Chromatogram

The process of allowing the mobile phase to run along the plate is called **development**. The solvent in which the solutes are originally dissolved should have completely

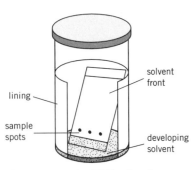

Figure 8.10 A TLC development chamber.

evaporated from the spots before the plate is placed inside the developing chamber. The chamber should be saturated with the mobile phase. For chambers used to analyze long plates (at least 10 cm long), a piece of filter paper with the bottom edge dipped into the solvent should cover at least 2/3 of the surface of the chamber's walls. With the aid of a pipet or by just swirling the chamber, the paper should be soaked with solvent. This lining ensures that solvent vapors are saturating the entire volume of the chamber. A lid should always cover the mouth of the chamber (Fig. 8.10).

The level of the mobile phase in the chamber should be below the spots on the TLC plate; otherwise the sample might leach into the solvent, which would lead to faulty results. Once the plate has been placed inside the chamber, the whole assembly should not be moved or relocated because the solvent may splash over the plate. When the solvent has traveled about 80–95% of the length of the plate, the plate is carefully removed from the chamber by holding it by its edges. The level reached by the solvent, or **solvent front**, is marked with pencil, and the solvent allowed to evaporate in a well-ventilated place (fume hood). If the plate is left in the chamber after the solvent front has reached the top edge, diffusion of the sample in all directions occurs, which causes the spots to appear broad and ill defined.

Visualizing the Chromatogram

With colored solutes, visualization on the plate is trivial. However, more often than not, samples are colorless and therefore we need to resort to other means for visualization. Silica gel and alumina are available with **fluorescent indicators**. These indicators adsorb UV radiation at either 254 (short-wave) or 366 nm (long-wave) and emit green or blue fluorescence, respectively. They are homogeneously distributed on the plate and are not disturbed by most solvents during development. If the sample absorbs UV radiation, the spots on the plate appear dark, because no light reaches the indicator and fluorescence does not occur (Fig. 8.11). Adsorbents with fluorescent indicators are usually labeled as F254, F366, or F254 + 366; the numbers allude to the absorption wavelength of the indicator present. The use of indicators is one of the most widespread visualization techniques.

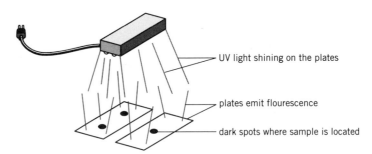

UV light shining on the plates

plates emit flourescence

dark spots where sample is located

Figure 8.11 TLC plates with fluorescent indicators visualized by UV light.

Another universal way of visualizing TLC plates is by exposure to iodine vapors. I_2 forms brown complexes with many organic compounds, especially if they have aromatic rings and double bonds, giving brown spots on the TLC plates. The plate is placed for several minutes in a chamber that contains a few crystals of iodine. It is removed from the chamber and examined immediately because the iodine evaporates from the spots rather quickly.

The spots can also be rendered visible by spraying the plate with specific reagents for different functional groups. For example, a mixture of *p*-anisidine and phthalic acid is used to visualize carbohydrates (Unit 24).

Analyzing the Chromatogram

Once the spots have been visualized and circled with pencil, the distance traveled by the spot from the origin is measured along with the distance traveled by the solvent front. The ratio between these two distances is called **ratio to the front (R_f)**:

$$\text{ratio to the front} = R_f = \frac{\text{distance traveled by spot}}{\text{distance traveled by solvent}} = \frac{X}{Y}$$

The distance traveled by the sample is measured from the origin to the middle of the spot (Fig. 8.12). With very large and ill-defined spots, as for example, spots with "tails," R_f values are meaningless because the middle of the spot varies with the amount of sample applied to the plate. If the spot streaks or runs with a tail, it is an indication that **too much** sample was applied. A decrease in the volume of sample spotted should be tried first; if this does not correct the problem, then a different adsorbent/solvent system for the separation should be tried.

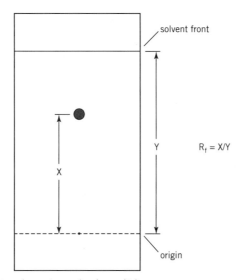

Figure 8.12   Calculating the ratio to the front (R_f).

The ratio to the front depends on several variables such as:

- the thickness of the adsorbent;
- the nature of the stationary phase and its degree of activation;
- the mobile phase;
- the amount of material applied.

R_f values are difficult to reproduce, especially between different laboratories. For this reason, the R_f value should never be used as a criterion to identify an unknown. If possible, standards should be run side by side with the unknown compound. If unknown and standard run with similar R_f values, this can be taken as an *indication* that they *may be* the same compound. It should be borne in mind, however, that different compounds may run with virtually identical R_f values under certain experimental conditions. Further analysis should be conducted to determine the identity of the unknown. For example, different solvents and adsorbents should be tried out to see if the spots still run with similar R_f values; if they do, the identity of the unknown should be finally confirmed by other methods such as isolation and spectroscopic studies.

Oftentimes the solvent does not run with an even front but rather slants or curves toward the edges. This phenomenon leads to nonuniform traveling distances, which may be problematic especially when comparing unknowns with standards run side by side. To circumvent this problem a **co-chromatography** can be run. The unknown and the standard are applied on the same spot. If after development they give one single spot, it means that standard and unknown have the same R_f value; if on the other hand, they give rise to two separate spots, it implies that they are different compounds. This is perhaps one of the most powerful applications of TLC. If two spots have different R_f values under identical conditions, they *must* correspond to at least two different compounds. Therefore, in analyzing TLC, *the number of spots indicates the minimum number of components in a sample.*

8.5 ADSORPTION ISOTHERMS *(advanced level)*

The quality of a chromatographic separation depends, among other factors, on the sample concentration. If too much sample is applied on the TLC plate, one usually observes after development that the spots are far from circular and show streaks and tails, which make the analysis difficult. Why is this? To understand this phenomenon we should consider the **adsorption isotherms**, explained in the next paragraph.

There is a limited number of binding sites on the surface of the adsorbent, and thus, there is a limit to the amount of material that can be adsorbed per unit area. A typical plot showing the relationship between the concentrations in the stationary ($[A]_s$) and mobile ($[A]_m$) phases for a hypothetical compound A is presented in Figure 8.13. Such representation is called an **adsorption isotherm**. As we can see, at very low concentrations there is a linear correlation between $[A]_s$ and $[A]_m$, but as the concentration in the mobile phase, $[A]_m$, increases, the concentration in the stationary phase levels off. This leveling off is due to a saturation of the stationary phase as all of its binding sites become occupied with sample.

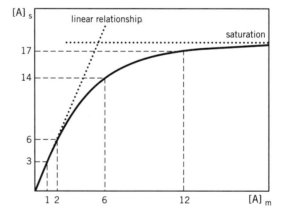

Figure 8.13 Diagram showing a typical adsorption isotherm.

At low concentrations, where there is a linear correlation between $[A]_s$ and $[A]_m$, the solute travels along the plate at a speed that is independent of the concentration. In this part of the isotherm, if the concentration in the mobile phase doubles, so does the concentration in the stationary phase. At higher concentrations, however, as the curve approaches saturation, a doubling of the concentration in the mobile phase results in less than a doubling of the concentration in the stationary phase. This is to say that at higher concentrations a larger proportion of the compound remains in the mobile phase, resulting in a faster motion of the compound along the plate. In this part of the isotherm, the compound travels at a speed that is dependent on its concentration. The higher the concentration, the faster it travels. Only at very low concentrations, where the adsorption isotherm is linear, can the speed be considered constant and independent of the concentration. It should be pointed out that this behavior is different from that described in partition chromatography (Unit 7). Contrary to adsorption isotherms, partition isotherms (Fig. 7.7) are linear in a very wide range of concentrations, making the speed at which the solute travels independent of its concentration.

The consequence of having different speeds for different concentrations is a distortion in the shape of the spots after development. Let us assume that a compound is spotted on a TLC plate. Because of diffusion, its concentration on the spot is not homogeneous but follows a bell-shaped distribution with a maximum at the center of the spot. This is represented in Figure 8.14, where the peaks on top of the spots represent the concentration distribution along the axis of the spot that coincides with the direction of the mobile phase flow. When low concentrations are used (linear portion of the isotherm), all parts of the spot travel at the same speed because the

At low total concentrations (linear portion of isotherm): speed is independent of the concentration and spot maintains shape as it travels

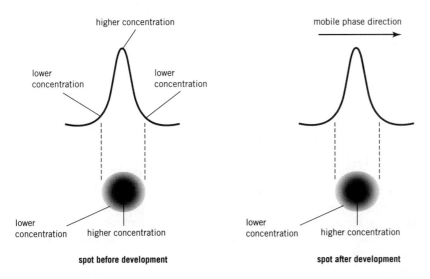

At high total concentrations (nonlinear portion of isotherm): speed depends on concentration and spot is distorted as it travels.

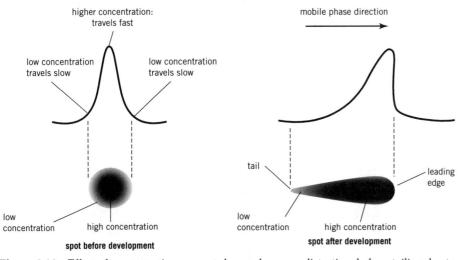

Figure 8.14   Effect of concentration on spot shape: above, no distortion; below; tailing due to high concentration.

speed is independent of the concentration. As a result, the spot keeps its shape as it moves up. On the other hand, if the concentration on the spot is high and beyond the linear portion of the isotherm, a distortion in the shape of the spot occurs. This happens because in this portion of the isotherm higher concentrations travel faster than lower concentrations. This causes an accumulation of solute in the **leading edge** of the spot (the one nearest to the solvent front) and a tail in the other. To minimize spot distortion it is very important not to overload the plates, so the solute concentrations are in the linear part of the isotherm. It is worth mentioning that sample overloading is also responsible for the peak distortion often observed in GC traces.

8.6 APPLICATIONS OF TLC

Uses and Limitations of TLC

The big advantage of TLC as compared to other chromatographic methods such as GC and HPLC (Unit 10) is its affordability. With a modest investment of a few hundred dollars a lab can be equipped to perform multiple types of TLC analysis. Another advantage of TLC is its speed. Complex mixtures can be separated in a matter of minutes. If necessary, different mobile phases can be simultaneously tested until the optimum conditions are found. Another advantage, especially compared to HPLC, is the small amount of sample needed to perform the analysis (in the microgram range).

TLC analysis finds many applications in the organic and biochemical lab. TLC is routinely used to follow the course of reactions, to determine the number of components in a mixture, to identify compounds, to follow chromatographic columns (Unit 9), and to check the purity of samples.

There are also limitations to TLC analysis. For example, volatile compounds cannot be analyzed directly, because they evaporate from the TLC plate before or during separation. They can be analyzed provided they are first derivatized, or in other words, transformed into nonvolatile compounds by means of simple and well-known chemical reactions. Another disadvantage is that quantification of samples is much harder to perform by TLC than by GC or HPLC. The determination of the amount of material present in a given spot can be achieved by excising or scraping off the stationary phase, extracting the compound with a suitable solvent, and determining its concentration by an independent method (UV, IR spectroscopy, liquid scintillation counting for radioactive samples, etc.). Alternatively, if the spots are colored, the concentrations can be determined by **densitometry**. Densitometry is based on the absorption of light by the spots, as the plate is scanned by an intense light beam.

If the sample is radioactive, the TLC plate can be placed in contact with a photographic plate for a given period of time ranging from minutes to months depending on the level of radioactivity. In those sites where the photographic plate is exposed to radiation, black spots of silver metal are observed. The intensity of the black spots can then be measured by densitometry. This technique is called **autoradiography**, and is commonly employed in biochemistry and molecular biology.

Preparative TLC

Most of the routine applications of TLC are at the **analytical** level, where the mixture is analyzed *to determine the number and nature of its components*, without isolating them. However, TLC can also be used to *separate and isolate* compounds from a mixture. This process is called **preparative TLC**. The principles of preparative and analytical TLC are very similar; the main difference between these two modes of running TLC plates is the amount of sample and adsorbent used. Whereas in analytical TLC the thickness of the layer is about 0.25 mm, in preparative TLC (also abbreviated **prep-TLC**) it is at least 3 mm. The amount of material loaded on a single preparative plate (20 × 20 cm) could be as large as 200 mg, whereas in analytical TLC only a few µg are spotted. The sample is loaded as a streak (instead of spots) using a Pasteur pipet. The chromatogram is developed in the usual manner and then visualized. The best way to visualize prep-TLC is with the help of fluorescent indicators. If iodine vapors or other reagents are used to locate the spots on the plate, only a lateral portion of the plate is subjected to the visualization treatment to avoid contamination of the compounds on the rest of the plate. In the case of iodine, this can be achieved with the help of a little vessel shaped like a narrow boat. A small volume of a solution of iodine in acetone (0.1–0.5%) is placed inside the boat and the solvent allowed to evaporate in a well-ventilated fume hood. The boat with the iodine deposited on the walls is placed, bottom-up, on a side of the TLC plate (Fig. 8.15). Brown spots develop on the plate under the boat where the compounds are.

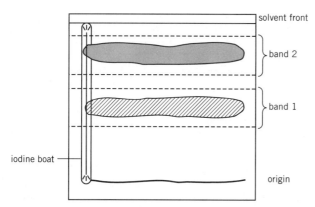

Figure 8.15 A preparative TLC plate.

After the components have been located on the plate, the bands can be scraped off with a spatula. The mixture of adsorbent and sample is placed in a beaker or Erlenmeyer flask and treated with an appropriate solvent to remove the compound from the adsorbent. For most organic compounds acetone is a good solvent; it readily extracts most organic compounds but it does not remove the fluorescent indicator from the adsorbent. The mixture of adsorbent and solvent is filtered and the extraction of the compound from the adsorbent is repeated, if necessary. The solvent is finally evaporated. The process of removing a compound from the stationary phase with the help of a solvent is called **elution**.

Bidimensional TLC

In analyzing complex mixtures it is sometimes difficult to find a single solvent, or a mixture of solvents, that gives an adequate separation. In such cases, bidimensional TLC (2D-TLC) could be of great help. In 2D-TLC, the sample is spotted in a corner of a square plate and the plate is developed using a solvent of choice. It is removed from the chamber, allowed to dry, and then developed with a different solvent in a direction perpendicular to the direction of the first development (Fig. 8.16). This results in the separation of the components in two dimensions. As can be observed in the figure, spots that are ill-resolved after the first development get separated by the second solvent.

Bidimensional TLC analysis is very useful in the separation of structurally similar compounds such as the mixtures of amino acids obtained after hydrolysis of proteins.

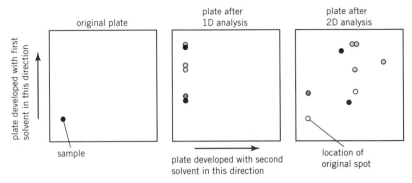

Figure 8.16   Bidimensional TLC.

8.7   OTHER STATIONARY PHASES

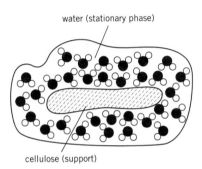

Figure 8.17   A cellulose particle with bound water.

TLC is not restricted to adsorbents as stationary phases. Separation can also be achieved by partition, ion-exchange, and size-exclusion processes. Cellulose is a commonly used stationary phase for partition chromatography. It can be used directly in the form of paper (**paper chromatography**), or it can be deposited on a solid support (cellulose-TLC) or packed inside a column (cellulose-column chromatography). In cellulose-based chromatography, the true stationary phase is **water** bound to cellulose fibers that act as the solid support (Fig. 8.17); the amount of water bound to cellulose is typically 15–20%. Traditionally, strips of filter paper (such as Whatman 1) were used, but the technique required long development times (2–24 hours). Nowadays, TLC plates covered with microcrystalline cellulose particles can be purchased; they have the advantage over filter paper that the cellulose is present as small particles of uniform diameter ($\approx 5\ \mu m$), instead of long and intermingled fibers. This allows an easy flow of solvent and considerably reduces the development time.

Chromatography on cellulose is useful in separating very polar compounds such as amino acids, some plant pigments, and sugars. The mobile phase for chromatography on cellulose is normally a polar organic solvent only partially miscible with water (*n*-butanol, ethyl acetate, etc.) that has been saturated with water. The mobile phase should be saturated with water to avoid stripping water molecules from the cellulose as the mobile phase interacts with the cellulose particles. Sometimes bases (ammonia, pyridine, etc.) and acids (formic acid, acetic acid, etc.) are added to the solvent. These compounds increase the solubility of water in the organic solvent, making the solvent more polar.

Cellulose can be chemically modified to afford products with acid or base properties. Groups such as $-CH_2COOH$ (carboxymethyl, CM), $-SO_3H$ (sulfonic), and $-CH_2CH_2N(CH_2CH_3)_2$ (diethylaminoethyl, DEAE) can be incorporated into cellulose to give materials with ion-exchange properties. For a brief discussion on ion-exchange chromatography, the student is referred to Unit 10.

8.8   TLC DO'S AND DON'TS

The following list summarizes some of the things you should do and the mistakes you should avoid when running TLC plates.

Do's

- Clean and **dry** the solvent chambers before use.
- Keep the solvent chambers capped.
- Handle TLC plates only by the edges.

- Use narrow micropipets to spot samples.
- Allow the solvent to evaporate between applications and before placing the plate in the solvent chamber.
- The spots should be above the solvent level in the chamber.
- Let the solvent evaporate before visualizing.
- All marks should be made with pencil.
- Handle iodine chambers in the hood.

Don'ts

- Do not use ink to mark the plates.
- Do not touch the plates with your fingers.
- Do not scratch the plates.
- Do not leave the plates in the chamber after the solvent has reached the top.
- Do not move the chamber during developing.
- Do not look directly into the UV-light.

BIBLIOGRAPHY

1. Chromatographic Methods. A. Braithwaite and F.J. Smith. 4th ed. Chapman and Hall, London, 1990.
2. Thin Layer Chromatography. K. Randerath. Verlag Chemie, Weinheim, 1965.
3. Thin-layer Chromatography. A Laboratory Handbook. E. Stahl. Translated by M.R.F. Ashworth. 2nd ed. Springer, Berlin, 1969.
4. Chromatography. E. Heftmann, ed. Reinhold, New York, 1967.
5. Vogel's Textbook of Practical Organic Chemistry. A.I. Vogel, B.S. Furniss, A.J. Hannaford, P.W.G. Smith, and A.R. Tatchell, 5th ed. Longman, Harlow, UK. 1989.
6. Hydrogen Bonding and Retention on Silica. A. Feigenbaum. *J. Chem. Ed.*, **63**, 815–817 (1986).
7. CRC Handbook of Chromatography. Vol. II. G. Zweig and J. Sherma, ed. CRC Press, Cleveland, 1972.

EXERCISES

Problem 1

Complete the following paragraph. Each blank corresponds to one of the words shown at the end. Each word should be used only once.

In _____ chromatography the separation is based on _____ -dipole, _____ -dipole interactions and _____ . The most commonly used _____ phases are _____ and _____ . The latter has higher _____ than the former. Very polar compounds can be effectively separated using a solvent of _____ polarity and an adsorbent of _____ activity. The process of allowing the _____ phase to run along the plate is called _____ . The distances traveled by the solvent and the sample are used to calculate the _____ values. In general, the higher the polarity of the sample, the _____ the R_f. _____ and _____ are used as visualization reagents. While _____ TLC is used to determine the _____ of components in a sample, _____ TLC is employed to _____ the components of a mixture. When microcrystalline _____ is used, the separation is based on _____ .

adsorption, adsorptivity, alumina, analytical, cellulose, development, dipole, fluorescent indicators, H-bonds, high, iodine, ion, isolate, low, lower, mobile, number, partition, preparative, ratio-to-the-front, silica gel, stationary.

Problem 2

Alkaloids were analyzed using TLC on silica gel G with three different solvents: solvent 1: chloroform/acetone/diethylamine (5 : 4 : 1); solvent 2: chloroform/diethylamine (9 : 1), and solvent 3: cyclohexane/chloroform/diethylamine (5 : 4 : 1). The R_f for different alkaloids are given in the table below (adapted from Ref. 2 and D. Waldi, K. Schnackerz, and F. Munter, *J. Chromatogr.* **6**, 61 [1961]).

You are running a forensic laboratory. Which solvent would you choose to analyze:

(a) a sample seized by the police in a drug smuggling case and suspected to contain morphine, cocaine, and aconitine?

(b) a mixture of cocaine and reserpine?

(c) a poison containing strychnine and reserpine?

Alkaloid	Solvent 1	Solvent 2	Solvent 3
morphine	0.10	0.08	0.00
quinine	0.19	0.26	0.07
codeine	0.38	0.53	0.16
cocaine	0.73	0.90	0.65
strychnine	0.53	0.76	0.28
aconitine	0.68	0.90	0.35
reserpine	0.72	0.80	0.20

Problem 3

The course of a chemical reaction (A + B $\longrightarrow$ C) was followed by TLC analysis at different times after the reaction was initiated. The TLC plates of the reaction mixture at time zero, and after 15 and 30 minutes of reaction, run under identical conditions, are shown in the figure below.

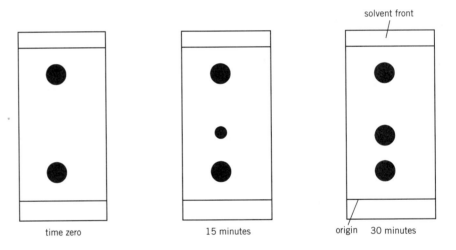

The R_f values of pure sample of A and B (under identical conditions) are 0.2 and 0.8, respectively.

(a) Calculate the R_f for all the spots.

(b) Identify the spots and explain the TLC results.

(c) What compound was the limiting reagent in the preparation of C? Briefly justify your answer.

Problem 4

The following mixtures are analyzed by TLC on silica gel using a solvent of medium polarity. Arrange each mixture in order of decreasing R_f values. Briefly justify your answer.

(a)

(b)

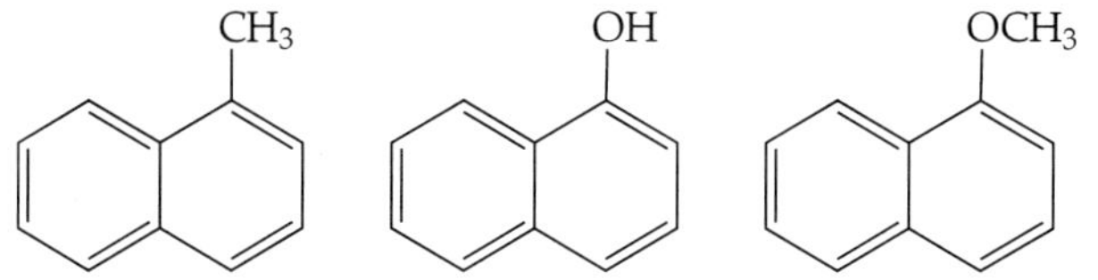

A   B   C

(c)

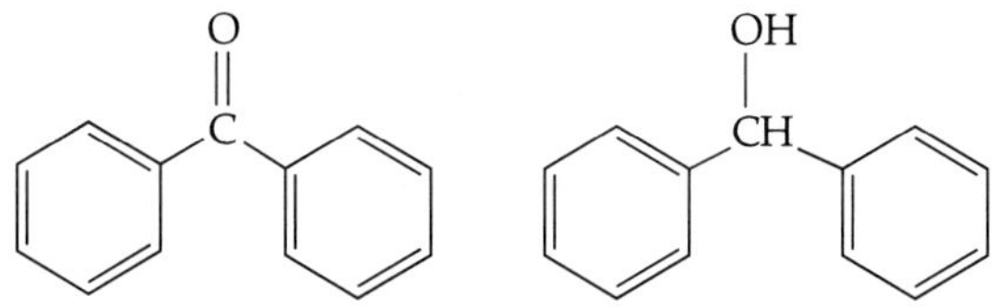

D   E

Problem 5

The following phenols were analyzed by TLC on silica gel G using a mixture of toluene, methanol, and acetic acid (10:2:1) as a mobile phase. The measured R_f values were: 0.20; 0.41; 0.63, and 0.84. Match the R_f values with the structures.

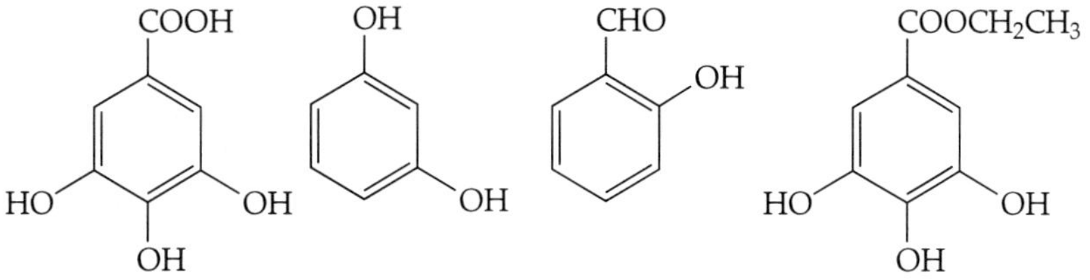

gallic acid   resorcinol   salicylaldehyde   ethyl gallate

EXPERIMENT 8

TLC Analysis of Vegetable Extracts

E8.1 PLANT PIGMENTS

The most important types of plant pigments are **chlorophylls**, **carotenoids**, **flavonoids**, and **tannins**. Chlorophylls contain a ring system formed by four pyrroles linked by four methine bridges, a Mg^{2+} ion in its center, and a long nonpolar hydrocarbon chain ($C_{20}H_{39}$). Chlorophylls are cousins of other biologically important molecules such as vitamin B_{12} and the heme found in hemoglobin; they all have a tetrapyrrole ring system. There are two main types of chlorophylls present in higher plants, **a** and **b**. Chlorophyll **a** is more abundant than chlorophyll **b** by a ratio of 3 to 1. The only structural difference between them is that a methyl group in **a** has been replaced by a formyl group, CHO, in **b**.

X = CH₃ : chlorophyll a

X = C=O : chlorophyll b
         |
         H

X = CH₃ : pheophytin a

X = C=O : pheophytin b
         |
         H

Chlorophylls are found in the chloroplast in association with small proteins. These protein-chlorophyll complexes are crucial in photosynthesis. The chlorophyll molecule acts as an antenna for visible radiation. The light absorbed is used to make carbohydrates ($C_6H_{12}O_6$) and oxygen using carbon dioxide and water as raw materials.

$$6CO_2 + 6H_2O \xrightarrow[chlorophyll]{h\nu} C_6H_{12}O_6 + 6O_2$$

Chlorophylls are labile compounds. In the presence of acids the central Mg^{2+} ion is easily replaced by protons and **pheophytins** are produced. In the summer months the leaves produce and degrade chlorophylls at a fast rate. Every fall, as the daylight dwindles, less chlorophyll is produced but its degradation continues, giving way to the colors of autumn: yellows, reds, purples, and browns. These colors are largely due to the other leaf pigments, such as carotenoids, flavonoids, and tannins, which are masked by an abundance of chlorophyll during the summer months.

Carotenes are polyunsaturated hydrocarbons that belong to the family of terpenes (Unit 11). Carotenes have 40 carbon atoms per molecule. They are found in the chloroplasts in association with chlorophylls and proteins where they, too, play an important role in photosynthesis. They are auxiliary pigments in the light-harvesting process and protect the chloroplasts against photooxidation.

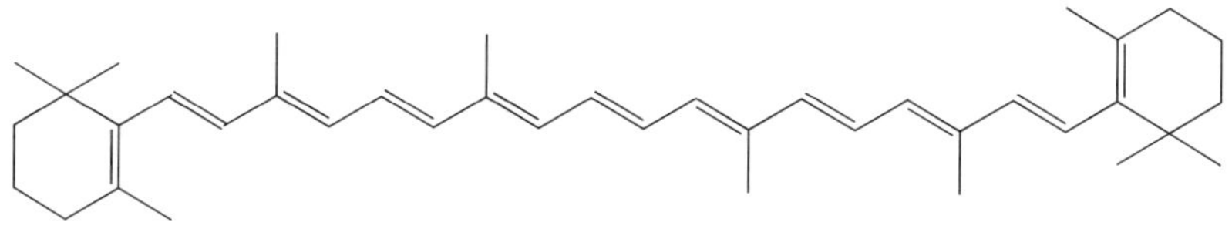

β-carotene

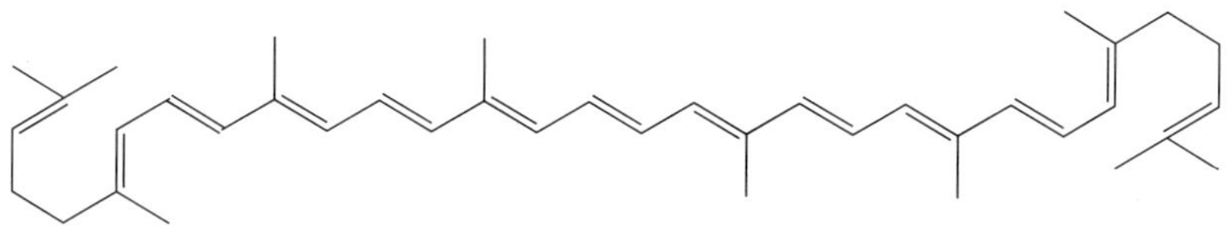

lycopene

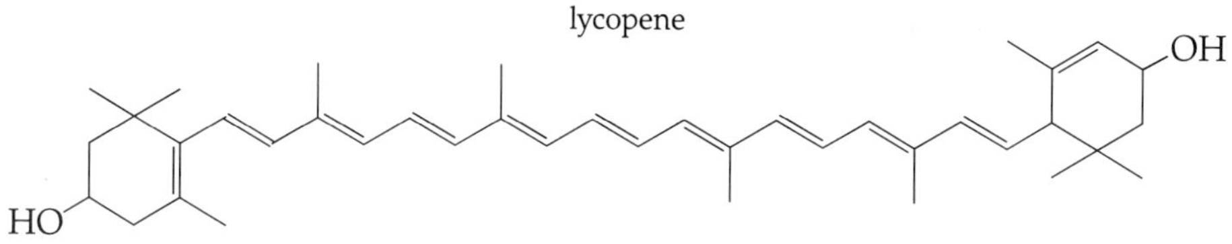

lutein

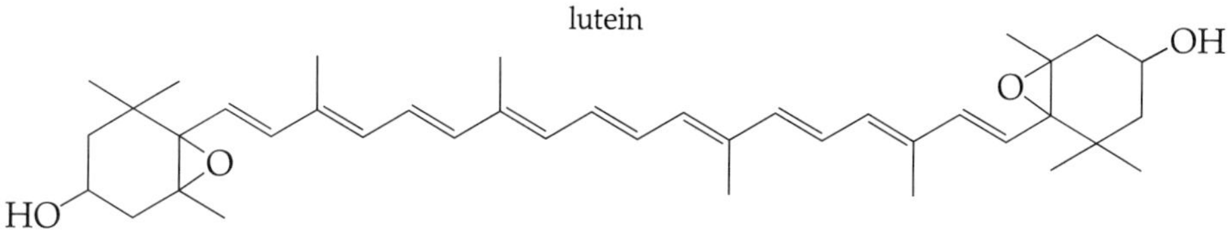

violaxanthin

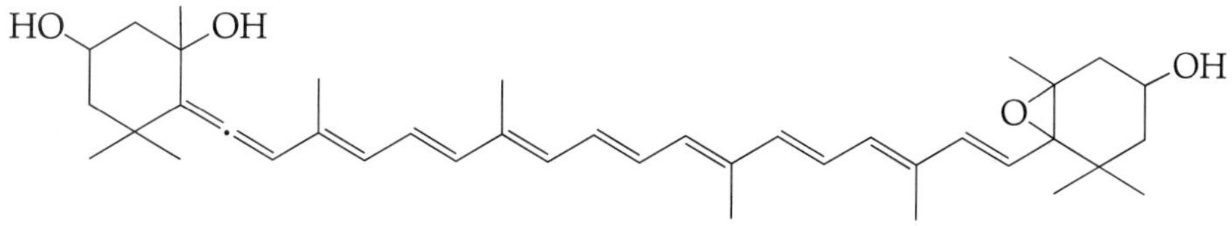

neoxanthin

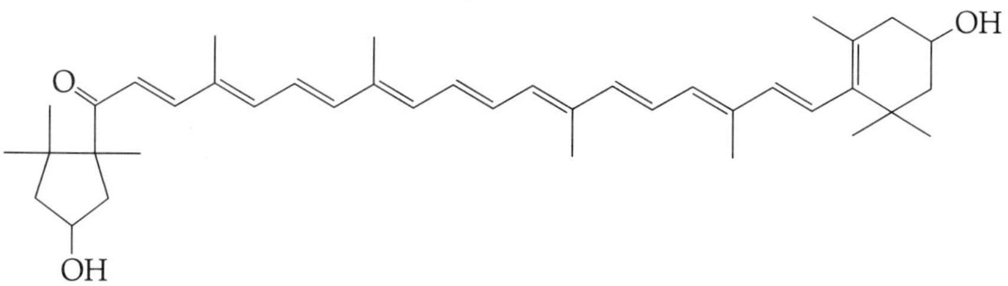

capsanthin

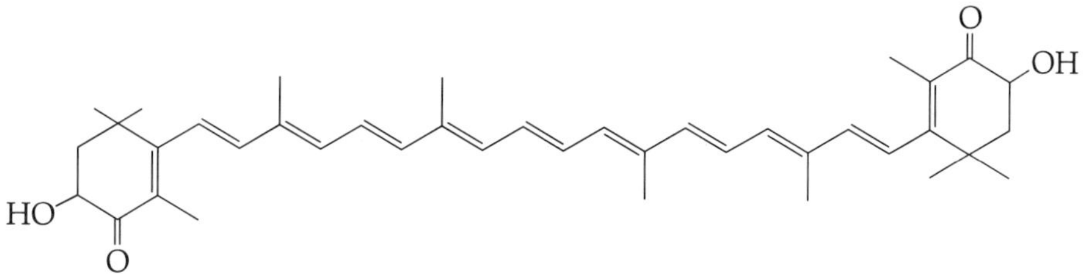

astaxanthin

β-Carotene, found in carrots, sweet potatoes, and green leaves, is one of the most abundant members of the family, and the precursor of vitamin A. Other members include α-carotene, an isomer of β-carotene with only half of its vitamin A activity, and lycopene, a red pigment found in tomatoes and watermelons that has no vitamin A activity. Carotenes are easily oxidized, especially in processed food, and must be protected from light and air to maintain their dietary value. The oxygen-containing products derived from carotenes are called **xanthophylls**. Xanthophylls and carotenes form the **carotenoid** family. The most abundant xanthophylls found in green leaves are lutein, violaxanthin, and neoxanthin. Capsanthin is the red pigment found in red peppers, and astaxanthin is the pigment responsible for the pink color of lobsters, shrimp, and salmon muscle. In shrimp and lobsters, astaxanthin exists in association with proteins that stabilize the pigment and change its color to gray-blue. When lobsters and shrimp are boiled, the proteins are denatured and the carotenoids are released from their complexes, giving these crustaceans their characteristic pink color. (I hope that this bit of information does not lead you to believe that the green color of alligators is due to chlorophylls!)

Flavonoids are pigments present in leaves, fruits, and flowers. All flavonoids contain the three-ring system shown in the structures below. They have a variable number of —OH groups attached to various positions of the rings. Flavones and flavonols are yellow (from Latin: *flavus*, yellow). **Anthocyanins** (from Greek: *anthos*, flower; *kyanos*, dark blue) are **ionic flavonoids** that are found coupled to sugars such as glucose. They change color with pH; they are normally red in acidic conditions, purple at neutral pH, and blue in the presence of bases. Unlike chlorophylls and carotenoids, which are soluble in fat and have limited solubility in water, flavonoids and anthocyanins are very polar compounds found primarily in the cytoplasm. Flavonoids are the most prevalent pigments of flowers but they also occur in leaves where their colors are generally masked by the chlorophylls until the onset of autumn.

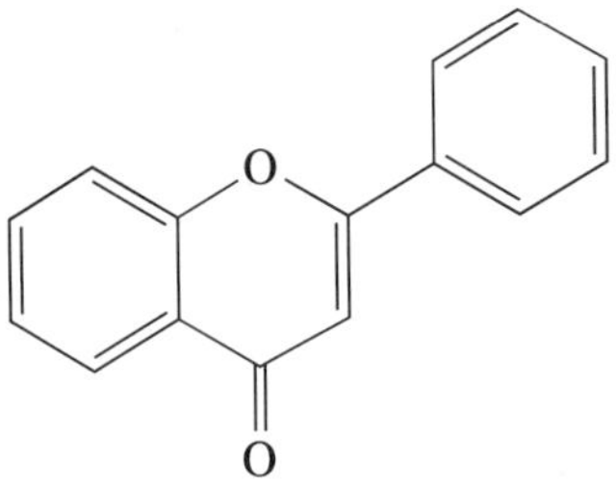

flavone skeleton 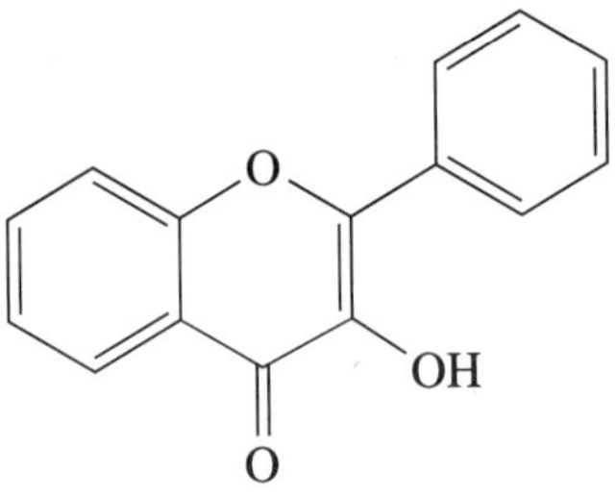   flavonol skeleton

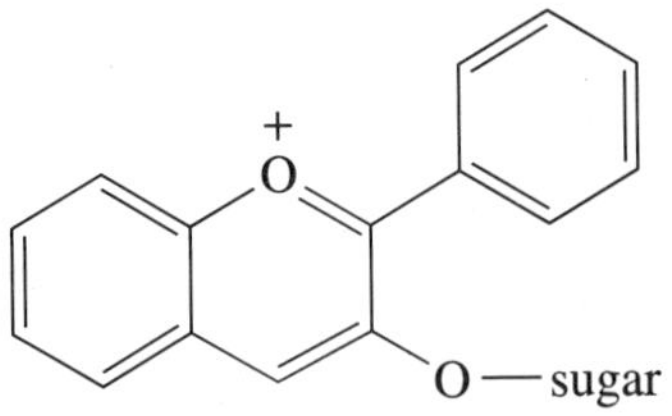

anthocyanin skeleton

Tannins are polyphenolic compounds. They confer brown and black hues to leaves and trunks. We have already seen tannins in Unit 5.

E8.2   ISOLATION AND ANALYSIS OF PLANT PIGMENTS

Background reading: 5.3; 5.5 and
8.1–6; 8.8
Estimated time: extraction 1.5 hour;
analysis 1.5 hour

In this experiment you will isolate pigments from spinach leaves or other vegetables. The procedure given below can be applied with minor modifications to the extraction of chloroplast pigments (chlorophylls and carotenes) from almost all types of leaves. It can also be used to isolate the pigments from bell peppers. Students can work with either red, yellow, or green bell peppers, and compare their results at the end. As already mentioned, the pigments from green leaves normally include carotenoids (β-carotene, lutein, violaxanthin, and neoxanthin), chlorophylls (a and b), and pheophytins. Pheophytins may be present in the leaves, but they may also be generated during the extraction process. As the cell membranes are broken, naturally occurring acids from the cytoplasm and other cell organelles come in contact with chlorophylls (from the chloroplasts), producing pheophytins.

The extraction is performed by crushing the leaves with methanol first, and then with a mixture of methanol and hexanes (two immiscible solvents). The first extraction with methanol breaks the cell membranes and removes water. This extract is discarded because it is poor in pigments. The second extraction with methanol-hexanes removes the pigments that go preferentially to the hexane layer. Methanol helps in detaching the pigments from their cellular complexes with proteins.

The extracts will be analyzed using TLC plates with silica gel as the stationary phase. You will investigate five different solvents as mobile phases and determine their power in separating the chlorophylls (green), the xanthophylls (yellow), the pheophytins (green-gray), and the carotenes (yellow-orange). You will observe that not all the solvents are able to move the pigments from the origin.

PROCEDURE

Vegetable Extract

Weigh 10 g of fresh spinach leaves and chop them using scissors. If you are using frozen chopped spinach, let the spinach thaw (at room temperature for about 2 hours or in a microwave oven for a few seconds — do not cook it!) and *remove as much water as possible by squeezing it before weighing.*

> ### Safety First
> - Hexane, toluene and acetone are flammable.

Place the spinach in a large mortar, add 12 mL of methanol, and crush the leaves with the pestle for about 3 minutes. With the aid of the pestle or a spatula, squeeze the spinach against the side wall of the mortar to remove as much methanol as possible. Transfer the liquid into a 250-mL Erlenmeyer flask labeled "methanol-water" and set aside. The contents of this flask will eventually be discarded.

Extract the remains of the leaves with a mixture of 15 mL hexanes and 5 mL methanol, crushing the tissue with the pestle for about 5 minutes. The extract should be deep green. Leaving behind as much solid as possible, transfer the liquid (a mixture of hexanes and methanol) to a 100-mL beaker. Filter it using a glass funnel (4–6 cm diameter) with a cotton plug. Collect the filtrate directly in a 125-mL separatory funnel supported on a ring (make sure the stopcock is closed).

Add 5 mL of water to the separatory funnel, shake, vent, and allow the layers to separate. Which one is the aqueous layer? Collect the aqueous layer in the Erlenmeyer flask labeled "methanol-water." Extract the remaining hexane layer with 5 mL of water. Collect the aqueous layer along with any emulsion present in the "methanol-water" flask.

Transfer the organic layer to a clean and dry 50-mL Erlenmeyer flask; add a small spatulaful of granular sodium sulfate to dry the organic layer. Cap the flask with a stopper and swirl occasionally. After about 5 minutes, filter the suspension by using a clean and dry micro-funnel (about 2.5 cm in diameter) and a cotton plug. Collect the filtrate in a dry 50-mL round-bottom flask. Using a rota-vap (Fig. 3.17), evaporate the solvent until the volume is approximately 2–3 mL. The color of the final extract should be deep green. With the aid of a Pasteur pipet, transfer the liquid to a labeled scintillation vial or test tube.

TLC Analysis

Clean and thoroughly dry five wide-mouth screw-cap jars. Label them with the names of the solvents to be used:

- hexanes
- toluene
- toluene-acetone (9 : 1)
- toluene-acetone (7 : 3)
- acetone

To each chamber, add solvent to a height of about 5 mm. Cover the jars with their lids.

Obtain five silica gel TLC plates (2.5 × 6.6 cm). Label them on the back with the name of the solvent on a piece of paper tape; the tape should be positioned far away from the lower edge of the plate. On the adsorbent side of the plate and at a distance of about 1 cm from the lower edge, make two pencil marks about 2-mm long each to indicate the origin (Fig. 8.9).

Determining the Optimum Number of Applications Take the plate labeled "toluene-acetone" (9 : 1). This plate will be used to determine the optimum number of applications needed to visualize the separation of pigments. On this plate, the sample will be applied on three different spots, each one with a different number of applications (1, 4 and 7). Immerse a micropipet (a narrow capillary tube with both ends open) in the extract; the liquid will rise through capillary action. Apply the liquid to the plate by touching the plate with the tip of the capillary tube and applying slight pressure. If necessary, gently wiggle the capillary to make the liquid flow. The spot should be about 0.5 cm from the side edge of the plate. Raise the capillary tube to stop the flow of liquid when the diameter of the spot is about 3 mm. On the same plate, apply a second and a third spot at a distance of about 0.5 cm from each other. The number of applications for the second spot should be 4, and for the third spot 7. Allow the solvent to evaporate between applications. Failure to do so will result in very large spots.

With the aid of tweezers, place the TLC plate in the chamber with **toluene-acetone (9 : 1)** and allow it to run until the solvent has risen to a height of about 6 cm (0.5–1 cm from the top of the plate). Remove the plate from the chamber (use tweezers); immediately mark the solvent front with pencil and allow the solvent to evaporate in a well-ventilated place (fume hood). Once the solvent has evaporated circle the spots with pencil. Draw the results in your lab notebook, paying special attention to the colors of the spots. Are they yellow, pale green, deep green, gray, orange? Notice any color spot at the origin. This analysis should be done without delay because the colors of the spots fade very quickly (within hours). Determine how many applications gave you the best results, or in other words, clear and well-delineated spots without streaking. If the spots are too faint to be seen, the extract should be concentrated by further evaporation of the solvent. If the spots run with tails, too much sample has been applied. Either reduce the number of applications or dilute your extract with hexanes.

Analyzing your Extract Spot the extract on the remaining plates, doing as many applications as you have already determined to be optimum. ***Develop all the plates simultaneously***. Calculate the R_f values for each pigment in each solvent. Decide which solvent gives the best separation. Be sure to write down in your notebook all the data that are necessary to reproduce your R_f values (stationary phase, solvent, number of applications). Draw the plates in your notebook and indicate colors. Pay attention to the differences between the plates. Where are the carotenes? Where are the pheophytins? Where are the xanthophylls? Where are the chlorophylls? Which is the best solvent to separate all the pigments? Answer these and the other in-lab questions (especially 3–12) before leaving the lab. To guide your interpretation,

Cleaning Up

- Let the remains of the spinach leaves dry in the hood and dispose of them in the trash. Do not put them in the sink. They will clog the drain.

- Dispose of the contents of the "methanol-water" flask in the container labeled "Organic Solvents–TLC Waste."

- Dispose of the sodium sulfate and cotton plug in the container labeled "TLC–Solid Waste."

- Dispose of the organic solvents used as mobile phases in the container labeled "Organic Solvents–TLC Waste."

- Dispose of used and unusable TLC plates, micropipets, and Pasteur pipets in the "TLC–Solid Waste" container.

- At the end of the lab period the instructor will empty the contents of the rota-vap traps (hexanes) into a "Reclaimed Hexanes" bottle.

remember that under optimum separation conditions these pigments usually show in the following order (in order of increasing R_f): chlorophylls (green), followed by xanthophylls (yellow), pheophytins (green-gray), and then carotenes (yellow-orange) closer to the solvent front.

BIBLIOGRAPHY

1. Plant Pigments. T.W. Goodwin, ed. Academic Press, London, 1988.
2. Chemistry and Biochemistry of Plant Pigments. T.W. Goodwin, ed. Academic Press, London, 1976.
3. The Chemical Pigments of Plants. J. Alkema and S.L. Seager, *J. Chem. Ed.*, **59**, 183–186 (1982).
4. Modifications of Solution Chromatography Illustrated with Chloroplast Pigments. H.H. Strain and J. Sherma. *J. Chem. Ed.*, **46**, 476–483 (1969).

EXPERIMENT 8   REPORT

Pre-lab

1. Make a table showing the physical properties for hexanes, methanol, toluene, and acetone (include dielectric constant, polarity-index [see Table 2.2], boiling point, density, flammability, miscibility in water, toxicity/hazards; for hexanes, use the values for *n*-hexane).

2. Mention the factors that influence the R_f of a given compound.

3. Using the table from 1, arrange the following solvents in order of increasing polarity: acetone, hexanes, toluene, toluene-acetone (9 : 1), toluene-acetone (7 : 3).

4. Why is TLC analysis limited to nonvolatile samples?

5. Why must ink be avoided in marking TLC plates?

6. Two compounds have the same R_f (0.74) under identical conditions. Does it mean that they have identical structures? Explain.

7. The R_f values for compounds I, II, and III in three different solvents are:

Solvent	Compound		
	I	**II**	**III**
A	0.10	0.80	0.80
B	0.20	0.30	0.35
C	0.70	0.70	0.20

A mixture of I, II, and III has to be separated by *preparative TLC*.

(a) Would you choose solvent B to carry out the separation? Why?

(b) Outline a brief procedure to separate the mixture using the solvents mentioned above (avoid mixture of solvents).

8. On the basis of their chemical structures, briefly explain why xanthophylls are more polar than carotenes, and chlorophylls more polar than pheophytins.

9. Which compounds do you expect would have a larger R_f value, carotenes or xanthophylls?

10. Arrange the following compounds in order of increasing polarity: lutein, neoxanthin, β-carotene, violaxanthin.

11. Which chlorophyll, a or b, is more polar? Which one would display larger R_f in a solvent of medium polarity?

12. Do you think that the hexanes extract would contain flavonoids? Briefly explain.

In-lab

1. What role does methanol play in the extraction? Could it be replaced by ethanol? (Hint: check "Solvent Miscibility Chart," Fig. 5.1).

2. Would you carry out the extraction of the pigments at high temperatures? Justify your answer briefly.

3. Draw the TLC plates in your lab notebook. Indicate colors. Identify as many pigments as possible.

4. Calculate the R_f values. Show your work. Using a table format, report the R_f values for the identified pigments.

5. Discuss the separation power of the solvents. Consider the polarity of the solvents in your discussion. Do you observe any correlation between the separation power and the polarity?

6. Why do carotenes have larger R_f values than xanthophylls?

7. Why do pheophytins have larger R_f values than chlorophylls?

8. How can you explain the differences observed in the plates run with toluene and with toluene-acetone (9 : 1)?

9. Which solvent separates the xanthophylls into different spots?

10. Is acetone a good solvent to separate the pigments?

11. Which solvent would you choose if you have to separate carotenes from the other pigments?

12. Is toluene a good solvent to separate the different chlorophyll types? Which solvent would you choose to carry out such separation?

Column Chromatography

9.1 OVERVIEW

In this unit we will study the separation of chemicals by **column chromatography**. In column chromatography the stationary phase is packed inside a glass tube, the sample is loaded into the top of the column, and the mobile phase is allowed to flow down. The sample components separate by their differential interactions with the stationary phase producing well-defined bands (Fig. 9.1). These bands are collected in separate containers as they come out of the column. Traditionally, column chromatography was performed in open tubes at atmospheric pressure. In the last 20 years, high-performance liquid chromatography (HPLC, Unit 10), where the mobile phase is pumped through the column at high pressure, has displaced open-column chromatography in many of its applications. However, the technique is still commonly used in the organic chemistry laboratory because it is rather inexpensive.

Separation by column chromatography depends on the type of stationary phase used and can take place by different mechanisms, such as adsorption, partition, reversed phase, ion-exchange, affinity, and size exclusion. Except for affinity chromatography, all the other types are commonly used in the organic chemistry lab, with adsorption being the most widely employed. In this unit the emphasis will be on adsorption column chromatography. In Unit 10, we will see reversed-phase, ion-exchange, and size-exclusion chromatographies when we discuss HPLC.

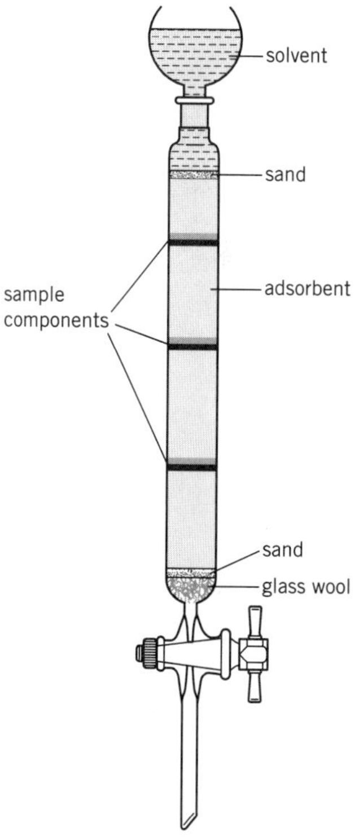

Figure 9.1 A chromatographic column.

9.2 PRACTICAL ASPECTS

Packing the Column

Before the separation is attempted, the sample should be analyzed by TLC (Unit 8) to determine the mobility of the components with different solvents. The selection of the adsorbent and mobile phase is made following the guidelines outlined in section 8.3. For column chromatography, as for TLC, the most common adsorbents are silica gel and alumina. **Calcium phosphate** (hydroxyapatite), a stationary phase of limited adsorptivity, is used in the purification of proteins and other biological macromolecules; **activated charcoal** and **starch** are used for special applications.

There are two main methods for packing a regular chromatographic column: **dry** and **wet**. The wet method is the most commonly used, and we will discuss it here. Dry columns afford separation in less time than conventional columns, and are used to separate complex mixtures; a disadvantage of dry columns, however, is that larger amounts of adsorbent are required to accomplish the separation. For more information about dry columns see Reference 3.

The quality of the separation obtained by column chromatography depends on the adsorption equilibrium in the column. Although equilibrium is actually never reached because the solvent is constantly flowing, experimental conditions can be set so that the column operates under near-equilibrium conditions. The **amount of adsorbent**, its **particle size**, the **dimensions of the column**, and the **flow rate** are all important parameters that determine the success of the separation.

To carry out a separation by the wet method, the amount of stationary phase needed is approximately 20–50 times the weight of the sample. For difficult separations involving compounds of similar polarity, this ratio can be increased to 100–200. Normal particle size for column chromatography is 0.15–0.5 mm. Particle size smaller than 0.1 mm results in very low flow rates. Larger particle size implies less surface area per mass unit of adsorbent, and therefore, less adsorptivity. The dimension of the column should be such that the ratio between its length and its diameter is in the range 10–20. In general, the longer the column, the better the separation. However, there is a practical limit to this imposed by the slow flow rate obtained with very long columns.

The flow rate plays an important role in the separation. Fast flow rates usually do not give good separations because the column operates under conditions far from equilibrium. On the other hand, very slow flow rates not only lead to lengthy analyses but also give poor separations as the solutes tend to remix by diffusion in the column. The optimal flow rate depends on the specific separation, the type of adsorbent used, and the geometry of the column. Normally, flow rates in the range 1–60 mL per hour are employed.

To pack a chromatographic column, the adsorbent of choice is mixed with about five volumes of the mobile phase and the slurry is poured into a dry glass column that is fitted with a sintered glass plate or a plug of glass wool and a layer of sand to support the stationary phase. The column should be clamped in a vertical position. Usually the glass tube is equipped with a stopcock to control the solvent flow through the column. As the solid stationary phase settles down, the column is gently tapped with a wooden rod or a spatula. This avoids the formation of channels and air bubbles inside the adsorbent. The stopcock is opened and the solvent allowed to run before more slurry is added to the column. To avoid the formation of boundaries in the stationary phase, the new portion of adsorbent should be added before the previous portion has settled down completely. At no point should the solvent level be allowed to run below the level of the adsorbent. If this happens, the column should be emptied and repacked. Once the column has been filled to the desired height, a circle of filter paper or a 0.5-cm layer of sand is placed on top of the stationary phase; this prevents disturbances in the adsorbent as the sample and solvent are added to the column. The solvent is drained so its meniscus is just above the surface of the sand. The column is now ready to be loaded (Fig. 9.2a).

Loading the Sample

A solution of the sample, typically in the same solvent as the one used to pack the column, is loaded at the top of the column with the aid of a Pasteur pipet (Fig. 9.2b). The volume of the sample solution should be kept to a minimum; for a column about 40 cm long and 2 cm in diameter, a sample volume of 1–2 mL is ideal. Very large sample volumes result in poor separations. The stopcock at the outlet end of the column is opened and the sample allowed to penetrate the adsorbent (Fig. 9.2c). Once the sample has been completely loaded and the meniscus of the sample solution is just above the surface of the sand, a small aliquot of fresh solvent is added to wash the walls of the column (Fig. 9.2d); this volume is allowed to penetrate the column (Fig. 9.2e) and then more solvent is added to fill the column (Fig. 9.2f). Allowing the sample to penetrate the column before the solvent is added prevents unwanted dilution of the sample and leads to better results. A solvent reservoir is attached to the top of the column. Solvent flow should not be stopped once the column is running because it results in broadening of the bands due to diffusion.

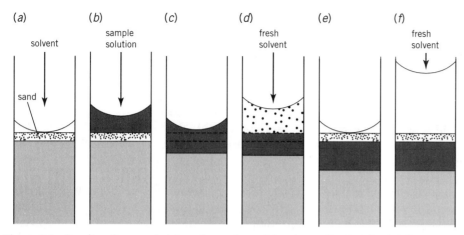

Figure 9.2 Loading the sample into a chromatographic column. See text for details.

Sometimes the sample is not totally soluble in the solvent used to run the column but it dissolves in more polar solvents. Loading a partially soluble sample directly on top of the column is unacceptable, because it leads to very poor separations. Loading the sample in a solvent more polar than the solvent used to run the column leads to poor separations as well, as we will discuss in the next section. The problem can be circumvented by dissolving the sample in the polar solvent, adding a small amount of adsorbent to the solution (2–10 times the weight of the sample), and evaporating the solvent in a rota-vap. As the solvent evaporates the sample gets adsorbed to the stationary phase. The solid mixture of sample and adsorbent, called a **pellet**, is then placed on top of the column and the solvent allowed to run as usual.

Development and Elution

The rules that govern the separation in adsorption column chromatography are the same as in adsorption TLC (Unit 8). Nonpolar compounds have weaker interactions with the adsorbent, move faster than polar compounds and come out first. Similar to TLC, the process of running solvent through the stationary phase is called *development*. However, there is a big difference between TLC and column chromatography. While TLC development is always done with a pure solvent or a single mixture of solvents, column chromatography development can be done in three different ways: **isocratic**, **stepwise**, or **gradient development**. In **isocratic development** (from the Greek; *isokratos*, equal strength) only one solvent is used throughout the development. In a way, this is similar to TLC. Isocratic development is useful in separating components of similar polarity; for example, a mixture of chlorophyll a and b can be separated by using a single solvent of medium polarity such as ethyl acetate or a mixture of toluene-methanol. Isocratic development is also employed when we want to isolate only the least polar compound of a mixture. The complex mixture is loaded into the column filled with a solvent of similar polarity to that of the desired compound. The solvent is run through the column, and only the compound (or compounds) with polarity similar to that of the solvent moves; the more polar compounds remain adsorbed at the top of the column.

After removal of the nonpolar compounds, if we want to remove compounds that are strongly adsorbed to the stationary phase, solvents of higher polarity are needed. We can do this in a **stepwise** manner by adding successive aliquots of solvents of increasing polarity. A series of *solvents of increasing polarity* is called an **eluotropic series**. For example: pentane, toluene, diethyl ether, acetone, and methanol constitute an eluotropic series. For a list of solvent polarities see Table 8.2 and Table 2.2. In stepwise development each solvent is added sequentially. Sudden changes in the composition of the mobile phase should be avoided because they are usually accompanied by exo or endothermic mixing processes, which result in abrupt temperature variations that may lead to the cracking of the stationary phase. In using the eluotropic series mentioned above, for example, mixtures of pentane with an increasing concentration of toluene (such as 10, 50, and 70%) should be run sequentially after pure pentane and before pure toluene. When changing from aprotic to protic solvents, for example, from acetone to methanol, the concentration of the protic solvent should be increased even more gradually than in the case of aprotic solvents (such as pentane and toluene). For example, a gradual increase of methanol in methanol-acetone mixtures would be 1, 2, 5, 10, 20, 30, 40, 70, and 100% methanol.

Sometimes it is convenient to increase the polarity of the mobile phase more gradually than in a stepwise manner. We can achieve this with a **gradient development** generated by mixing two solvents of different polarity. The solvent of high polarity is allowed to flow, at a flow rate R_1, into a mixing chamber that contains the less-polar solvent connected to the column with a flow rate R_2 (Fig. 9.3). Different gradients can be generated depending on the relative flow rates of both solvents. When $R_2 = 2R_1$ the concentration of the more-polar solvent in the mixing chamber increases linearly with time; this is called a **linear gradient**. Gradient development in column chromatography is the equivalent of temperature programming in GC. It makes it possible to separate in a single chromatographic run mixtures of compounds with a wide variety of polarities. We will come back to gradient development when discussing HPLC (Unit 10).

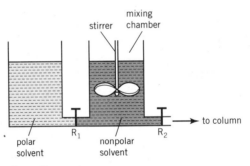

Figure 9.3   A gradient formation system.

Summarizing the types of column development we can say that in **adsorption chromatography, the polarity of the mobile phase is either kept constant, as in isocratic development, or increased during the development**, *but never decreased*.

The removal of the compounds from the column is normally done by allowing the solvent to flow through and collecting different fractions as the components come out of the column. This process is called **elution** and the liquid coming out of the column is called the **eluent**.

Colorless compounds must be detected by indirect methods. Unlike TLC, it is not customary to run chromatographic columns using adsorbents with fluorescent indicators because these indicators may leach into the elution solvent and contaminate the sample. TLC, GC, and UV-visible spectroscopy are all very useful ways to follow the development of a chromatographic column. Fractions of equal volume can be collected with the aid of an automatic fraction collector (Fig. 9.4) and analyzed by GC, TLC, or UV; those fractions that show similar compositions are pooled together and the solvent is evaporated. Automatic UV-visible detectors are also available. The detector is connected at the end of the column and monitors the absorbance of the eluent, at a given wavelength, as it comes out of the column. If the sample absorbs in the UV-visible region of the spectrum, peaks are observed as the compounds elute from the column. The use of this method is limited to mobile phases with little or no absorption in the UV-visible region. A more detailed description of detectors for column chromatography is offered in Unit 10, where HPLC is treated.

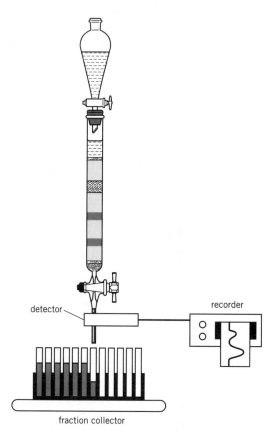

Figure 9.4   A chromatographic column with fraction collector, detector, and recorder.

A less frequently used elution method consists of extruding the stationary phase from the column and cutting the bands. The column is first allowed to run dry, then the stationary phase is removed by extrusion by forcing compressed air through the column. The bands of interest are cut with a spatula and the compounds are eluted from the adsorbent with acetone or other suitable solvents. This operation is similar to the elution of compounds from preparative TLC plates (section 8.6).

Microcolumns

When the total amount of sample is less than 200 mg, the separation can be performed by **microscale column chromatography** (and also by preparative TLC, section 8.6). Pasteur pipets are perfectly suited for this purpose. Pasteur pipets can accommodate about 1 g of adsorbent and can be assembled in a matter of minutes.

The tip of a short-tipped disposable Pasteur pipet is inserted in a piece of Tygon tubing. A screw clamp is used to control the flow of solvent. With the aid of a notched cork or rubber stopper, the Pasteur pipet is clamped in a vertical position. A small plug of cotton is placed at the bottom, followed by a small amount of sand (0.3 cm high). The column is filled with solvent, and the screw clamp opened to allow solvent flow. This removes air bubbles trapped in the cotton and sand; the removal of air can be sped up by gently tapping the column with a microspatula. The column is replenished with solvent, so its level is only 1 mm below the rim (Fig. 9.5a). A slurry of adsorbent and solvent is prepared and taken up into a second Pasteur pipet with the tip broken off. The second Pasteur pipet is placed on top of the column filled with solvent so the broken tip comes in contact with the column of solvent. If there are air bubbles in the tip of the top pipet, they should be expelled by applying gentle pressure to the rubber bulb. Once the solvent in the column has come in contact with the slurry in

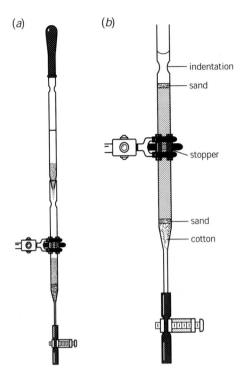

Figure 9.5 Microcolumns: *a)* preparing a microcolumn; *b)* assembled microcolumn.

the top Pasteur pipet, the adsorbent settles down, displacing the solvent and forcing it to move up. It is not necessary to apply pressure by squeezing the rubber bulb; the adsorbent spontaneously falls down.

This process is repeated as many times as necessary until the adsorbent reaches the indentation level of the Pasteur pipet. A small portion of sand is added on top of the column (about 0.3 cm high) and the column is ready to be loaded. Loading and development is performed as with regular columns (Fig. 9.2). A big difference between regular and microcolumns is the time required to separate mixtures. While regular columns can take several hours to run, microcolumns can be run in a matter of minutes. The big limitation is, of course, that only small samples can be separated at a time.

If solvents such as methylene chloride are used, they may produce swelling of the Tygon tubing, causing it to come loose. This can be prevented, to some extent, by minimizing the elapsed time from the moment the column is made until it is used. The problem can be avoided altogether by using tubing made of materials resistant to organic solvents, such as polytetrafluoroethylene (PTFE, Teflon).

9.3 APPLICATIONS

Column chromatography is always a preparative technique; this is to say that the compounds are isolated or collected for further studies. Families of compounds such as terpenoids, lipids, alkaloids, steroids, vitamins, aromatic hydrocarbons, and phenols, just to mention a few, can be successfully separated by adsorption column chromatography. In deciding which adsorbent to use, we should consider the stability of the sample under the separation conditions. For example, basic alumina should be avoided with samples that contain base-sensitive compounds such as esters.

Very polar or charged compounds, such as amino acids, nucleic acids, and proteins are better separated by other types of chromatography, such as ion-exchange and partition instead of adsorption chromatography. Ion-exchange column chromatography is particularly suited for compounds with net charge or those that can be ionized by a change of pH (proteins, polypeptides, nucleic acids, etc.). Macromolecules such as synthetic polymers, polysaccharides, and proteins can be separated by size-exclusion chromatography based on differences in molecular size and shape. We will discuss these types of chromatography in Unit 10.

The separation of enantiomeric mixtures is not possible by chromatography using conventional stationary phases, such as silica gel or alumina, and typical solvents. They can be separated provided that *either the adsorbent or the mobile phase, or both, is chiral*. The separation using an optically active stationary phase takes place as *one enantiomer interacts with the chiral adsorbent more tightly than the other*. The use of chiral stationary phases makes possible the separation and isolation of stereoisomers with a high degree of purity. This is of paramount importance in biochemical research because usually only one enantiomer has biological activity. Most chiral stationary phases are prepared by binding optically pure compounds, such as amino acids, sugars, and proteins, to silica gel or cellulose.

9.4 COLUMN CHROMATOGRAPHY DO'S AND DON'TS

Follow these recommendations to obtain good results with chromatographic columns.

Do's
- Clamp the column in a vertical position.
- The column should be free of air bubbles.
- Avoid sudden changes in the composition of the mobile phase.
- Load small sample volumes.

Don'ts
- Do not let the column run dry.
- Do not use extreme flow rates.
- Do not stop solvent flow once the sample has been loaded.

BIBLIOGRAPHY

1. Chromatographic Methods. A. Braithwaite and F.J. Smith. 4th ed. Chapman and Hall, London, 1990.
2. Chromatography. E. Heftmann, ed. Reinhold, New York, 1967.
3. Vogel's Textbook of Practical Organic Chemistry. A.I. Vogel, B.S. Furniss, A.J. Hannaford, P.W.G. Smith, and A.R. Tatchell. 5th ed. Longman, Harlow, UK. 1989.
4. CRC Handbook of Chromatography. Vol. II. G. Zweig and J. Sherma, ed. CRC Press, Cleveland, 1972.
5. A Practical Approach to Chiral Separations by Liquid Chromatography. G. Subramanian. VCH. Weinheim, 1994.

EXERCISES

Problem 1

A mixture containing three organic chemicals (A, B, and C) has to be separated by column chromatography. Preliminary TLC analyses using different solvents gave the results:

(a) Discuss the use of methanol, toluene, and cyclohexane to isolate A, B, and C by column chromatography. Which solvent(s) would you choose to successfully develop the column and isolate A, B and C?

(b) Which solvent(s) would you choose to run the column if only compound A is to be isolated and B and C discarded?

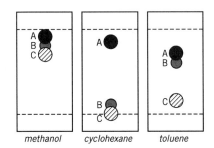

methanol   cyclohexane   toluene

Problem 2

The separation of a mixture of two compounds was attempted using two columns of different dimensions packed with the same stationary and mobile phases (see figure). In both cases band distortion due to uneven packing was observed. Column 2 has a larger length/diameter ratio than column 1. Which column gives a better separation? Explain.

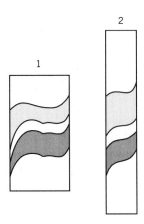

Problem 3

The separation of a mixture of alkaloids was attempted by column chromatography; 135 fractions (3 mL each) were collected using an elutropic series of toluene, diethyl ether, methylene chloride, and methanol. Fractions 1, 15, 30, 45, and so on, were analyzed by TLC (using ethyl acetate as solvent), and the following results were obtained (see figure).

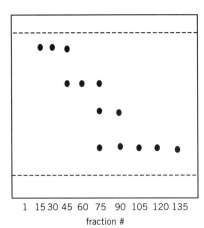

(a) How many alkaloids are present in the mixture?

(b) Which fractions can be pooled together without further analysis?

(c) Which other fractions would you analyze by TLC before you do the final pooling?

Problem 4

Arrange the following compounds in the order that you expect them to come out of a silica gel column using a gradient of petroleum ether and acetone.

(a)

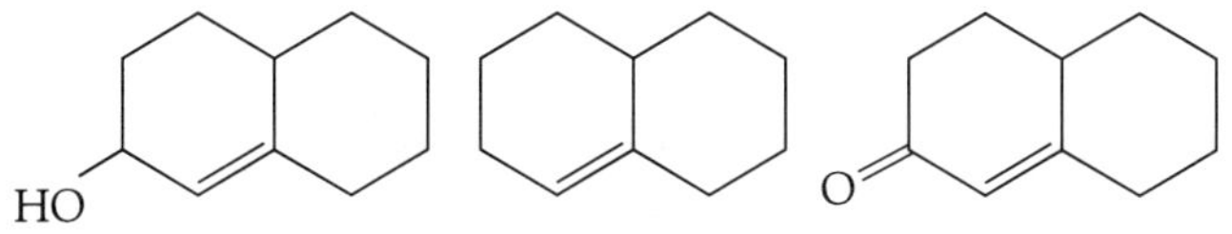

HO   A   B   C

(b)

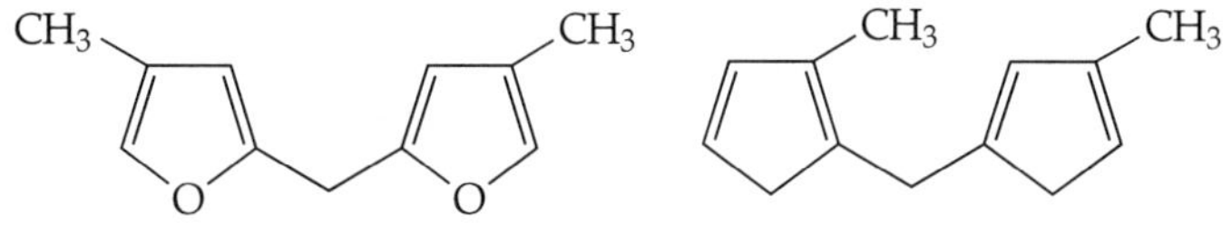

D   E

Problem 5

Which of the following solvent series are elutropic series? Which series has the widest polarity range?

(a) hexane, toluene, ethyl acetate

(b) chloroform, acetone, methanol, water

(c) hexane, cyclohexane, benzene, methylene chloride, butanone, methanol, water

(d) Methylene chloride, toluene, ethanol, acetone

Problem 6

How would you separate the following mixtures? Name the technique that you consider the best (only one per mixture). Possible techniques to use: recrystallization, filtration, sublimation, liquid-liquid extraction, solid-liquid extraction, simple distillation, fractional distillation, vacuum distillation, GC, preparative GC, TLC, preparative TLC, and column chromatography.

(a) 1 mg of methanol and water

(b) 150 mg of ß-carotene and chlorophyll a

(c) 10 g of acetanilide dissolved in methanol

(d) 5 g of cellulose and tannins

(e) 7 g of sodium chloride and caffeine

(f) 200 mg of caffeine and ß-carotene

(g) 2 g of methanol and water

(h) 1.5 g of silica gel and hexanes

(i) 3 g of ß-carotene and chlorophyll a

(j) 200 mg of caffeine and hydrolyzed tannins

EXPERIMENT 9

Isolation of C₆₀ from Fullerene Soot

E9.1   FULLERENES

Since the discovery in the early nineteenth century that diamond was just another form of carbon, it was believed that carbon existed only in two allotropic forms: diamond and graphite. This belief crumbled in 1985 when Harold Kroto, Richard Smalley, Robert Curl, and colleagues reported the production of a new stable form of carbon consisting of a cluster of sixty atoms, C_{60}. They proposed that the high stability of this molecule was due to its unusual shape, similar to a soccer ball. Because of its resemblance to the geodesic domes designed by the American architect Buckminster Fuller, they called this new molecule **buckminsterfullerene** (Fig. 9.6). The nickname **buckyball** was shortly coined.

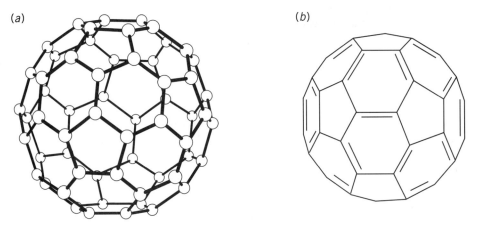

Figure 9.6   Buckminsterfullerene: *a*) stick-and-ball model; *b*) showing single and double bonds.

The study of this new form of carbon did not take off until 1990, when Donald Huffman and Wolfgang Kratschmer reported a method for the production of large quantities of buckminsterfullerene. Vaporization of graphite rods in a helium atmosphere afforded a soot from which C_{60} could be extracted and isolated in bulk amounts.

The carbon atoms in the C_{60} molecule are sp^2 hybridized and each one is bound to three others. They occupy the vertices of a truncated icosahedron, forming a closed structure with 32 faces (20 hexagons and 12 pentagons). Each pentagon is surrounded by hexagons, and no two pentagons share a carbon-carbon bond. All sixty carbon atoms in the buckminsterfullerene molecule are equivalent. This was demonstrated by its ^{13}C-NMR spectra, which shows one single peak in the chemical range typical of sp^2-hybridized carbons (143 ppm).

When it was first discovered, it was believed that C_{60} was some sort of super-aromatic molecule. Single bonds alternate with double bonds and 12,500 resonance forms can be written! However, it was later shown by crystallographic studies that the double and single bonds are rather localized. There are no double bonds on the sides of the pentagons; double bonds are confined only to the sides shared by two hexagons. This is reflected by two different carbon-carbon bond lengths. The bond length between two hexagonal rings is 1.38 Å while the bond length between a hexagon and a pentagon is greater, 1.45 Å. Far from being a superaromatic, nonreactive molecule, C_{60} reacts readily in solution as an electron-deficient polyene. It undergoes cycloadditions, nucleophilic additions, radical additions, hydrogenation and halogenations, just to mention a few transformations.

The excitement created by the discovery of the "roundest of all possible round molecules," as Smalley called the C_{60} molecule, permeated almost all areas of chemistry and physics. Applications to use these remarkable molecules as cages to trap atoms and ions, as molecular containers, as drug-delivery agents, and even as superconductors were rapidly sought. While many applications have been realized, others still remain unfulfilled. In 1991 buckminsterfullerene was named the Molecule of the Year by the *Science* magazine. In the editorial article, Daniel E. Koshland, Jr. wrote: "Part of the exhilaration of the fullerenes is the shock that an old reliable friend, the carbon atom, has for all these years been hiding a secret life-style. We are all familiar with the charming versatility of carbon, the backbone of organic chemistry, and its infinite variation in aromatic and aliphatic chemistry, but when you got it naked, we believed it existed in two well-known forms, diamond and graphite. The finding that it could exist in a shockingly new structure unleashes tantalizing new experimental and theoretical ideas." Kroto, Smalley, and Curl were awarded the 1996 Nobel Prize in Chemistry.

Buckminsterfullerene is not the only round molecule made exclusively of carbon atoms. Other closed molecules with a variable number of carbon atoms have been synthesized and characterized. C_{70} can be isolated along with C_{60} in the soot resulting from the vaporization of graphite (Fig. 9.7). Similar to C_{60}, C_{70} has pentagons surrounded by hexagons; it is an elongated molecule that resembles a football. Other molecules isolated from the same soot include C_{76}, C_{84}, C_{90} and C_{94}. This new family of compounds are called **fullerenes**.

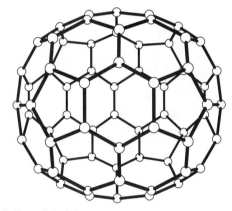

Figure 9.7   Stick-and-ball model of C_{70}.

E9.2 ISOLATION OF FULLERENES

C_{60} and C_{70} can be isolated from fullerene soot by extraction with toluene. Fullerenes are surprisingly soluble in aromatic hydrocarbon solvents such as toluene, benzene, and *o*-dichlorobenzene, and only slightly soluble in alkanes. Solubility values for selected solvents are shown in Table 9.1.

Solutions of C_{60} in toluene have a beautiful magenta color. When the solvent is evaporated and the solid deposits as a thin film it takes on a mustard-yellow color. In contrast, solutions of C_{70} in the same solvent are red-orange. These two fullerenes can be separated by column chromatography on neutral alumina by using mixtures of hexane and toluene as elution solvents. C_{60} has weaker interactions

Table 9.1 Solubility (mg/mL) of C_{60} and C_{70} at 30°C in Selected Solvents (Refs. 2 and 3)

Solvent	C_{60}	C_{70}
acetone	0.001	0.019
n-hexane	0.040	0.013
methylene chloride	0.254	0.080
toluene	2.75	1.40
carbon disulfide	5.16	9.87
o-dichlorobenzene	24.6[a]	36.2

[a] at 25°C.

Table 9.2 Molar Absorptivities (L mol^{-1} cm^{-1}) for C$_{60}$ and C$_{70}$ at Selected Wavelengths (Hexanes) (Refs. 7, 8, 17)

Wavelength (nm)	C$_{60}$	C$_{70}$
470	250	14,500
540	710	6,050

with the stationary phase and elutes first. This can be visualized by a magenta band followed by a red-orange one. C$_{60}$ can be obtained reasonably pure by this method, with a contamination of C$_{70}$ of less than 2%. On the other hand, this method does not afford pure samples of C$_{70}$, which contain variable amounts of C$_{60}$.

The fractions eluting from the column can be analyzed by UV-visible spectroscopy to determine their purity. C$_{60}$ and C$_{70}$ have sufficiently different absorptions in the visible portion of the spectrum (400–700 nm) to allow their differentiation. The UV-visible spectra of C$_{60}$ and C$_{70}$ are shown in Figure 9.8. Their molar absorptivities at selected wavelengths are gathered in Table 9.2.

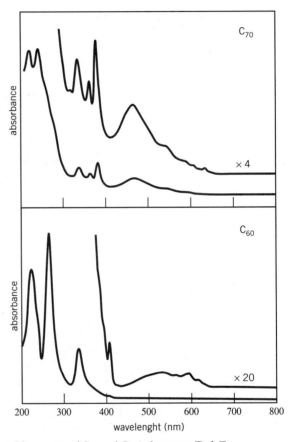

Figure 9.8   UV-visible spectra of C$_{70}$ and C$_{60}$ in hexanes (Ref. 7).

E9.3   A CHEMICAL TEST FOR FULLERENES

Fullerenes behave like electron-deficient alkenes and react easily with nucleophiles, especially amines. The mechanism seems to involve the transient formation of radicals that display a green color. The green color finally disappears and a brown product forms. A variable number of amine molecules, normally between 1 and 12, are added to the skeleton of the fullerene. Additions on C$_{70}$ lead directly to the observation of a brown color.

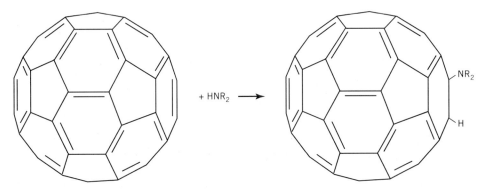

E9.4   OVERVIEW OF THE EXPERIMENT

In this experiment you will extract fullerenes from fullerene soot by using toluene. The soot contains more than 7% fullerenes, of which buckminsterfullerene is the main component; C_{70} is present in smaller amounts. The extraction of the fullerenes will be performed by stirring with the solvent at room temperature. Extraction at higher temperatures is not recommended because the solubility of fullerenes decreases as the temperature increases in the range 10–50°C.

The mixture of fullerenes will be separated on a microcolumn by using neutral alumina as stationary phase and hexanes-toluene as elution solvent. Because the fullerenes have very limited solubility in hexanes (Table 9.1), the mixture must be loaded into the column as a pellet (see section 9.2). To make the pellet you will mix the fullerene toluene extract with a small amount of alumina and evaporate the solvent in a rota-vap. The powder obtained after evaporation (the pellet) will be loaded on top of the column, the solvent will be added, and the column eluted as usual. For a successful separation, the alumina must be activated at 100°C for at least 12 hours before use.

You will start the separation by using 5% toluene in hexanes as the mobile phase and then you will increase the polarity by using 20% toluene in hexanes. If the separation works you will see a light magenta band coming out of the column, followed by a red-orange band. You will collect several fractions and analyze them by UV-visible spectroscopy (Unit 32). You will also obtain the UV-visible spectrum of the original toluene extract. Using the extinction coefficients listed in Table 9.2, you will calculate the concentration of C_{60} and C_{70} in the fractions, and assess the success of the separation.

You will finally perform a simple chemical test with ethanolamine. Both C_{60} and C_{70} undergo a nucleophilic addition with ethanolamine. The reaction with C_{60} gives an aqua color while C_{70} produces a brown color.

PROCEDURE

> **Estimated time:** 3–4 hr.
> **Background reading:** 9.1–4; 32.1–5

E9.5   ISOLATION OF C$_{60}$ FROM FULLERENE SOOT

Extraction of fullerenes

In a perfectly dry scintillation vial weigh 100 mg of fullerene soot.[1] Do this operation carefully to avoid spattering the solid. Add 3 mL of toluene, a small stir bar, and

---

[1] Fullerene soot can be obtained at a reasonable price from the MER Corporation, Tucson, Arizona, (520) 574–1980. According to the label, the soot contains more than 7% fullerenes.

Safety First

- Toxicological studies on C_{60} and C_{70} are incomplete. Treat them with caution. Avoid spattering the solid fullerene soot.

- Toluene and hexanes are flammable solvents.

stir on a magnetic stirrer for about 10 minutes. In the meantime, prepare a vacuum filtration apparatus by using a small filter flask and a Hirsch funnel fitted with a circle of Whatman 1 paper. Make sure that the paper covers the holes but does not touch the conical wall of the funnel. Filter the suspension. Wash the scintillation vial with 1 mL of toluene and transfer the rinse to the funnel.

To characterize the extract by UV-visible spectroscopy, take a 200-μL aliquot of the extract with an automatic pipet and dilute it with 2.80 mL of toluene (automatic pipet). Keep this solution for UV-visible analysis (next section).

Separation of Fullerenes

Activating the Adsorbent The alumina (neutral, Brockmann I, 250 mesh) must be activated in an oven at 100°C for at least 12 hours. Place about 2 g of alumina in a beaker; cover it with a piece of aluminum foil with a few holes punched. Place the beaker in the oven at least the day before the experiment. When ready to assemble the column, remove the beaker from the oven and place it in a desiccator (this operation may have been already performed by the instructor). Once the alumina has cooled off, rapidly weigh 2 g in a small beaker and immediately add about 4 mL of 5% toluene in hexanes. Swirl the slurry to mix.

Preparing the Pellet In a 25-mL round-bottom flask weigh approximately 100 mg of activated alumina and add the fullerene extract. Evaporate the solvent in a rota-vap with a *boiling water* bath (Fig. 3.17). The evaporation of solvents in the presence of fine powders usually results in spattering inside the rota-vap, especially toward the end of the process. To avoid this, when about one-half of the solvent has evaporated, remove the flask from the rota-vap, and place a piece of tissue paper around the neck of the rota-vap. Reattach the flask and continue the evaporation as usual. The paper precludes a perfect fit but the solvent still evaporates quickly. Now if spattering occurs, the fine powder will not be lost inside the rota-vap's neck. Stop the evaporation once the alumina becomes runny. The pellet is ready to be loaded.

Packing and Running the Column Using a clamp and a notched stopper, secure a small Pasteur pipet in the vertical position. Insert the tip of the column into a length of Tygon tubing (2 inches long) and position a screw clamp on the tubing (see Fig. 9.5). The clamp will control the flow of liquid. Place a small cotton plug at the end of the column and add a thin layer of sand no more than 2 mm thick. Add 2 mL of 5% toluene in hexanes to the column and open the screw clamp to drain most of the liquid, leaving some above the sand. This operation will get rid of any air bubbles in the cotton. Fill the column completely with solvent.

Using several layers of paper towel to protect your hands, break off the tip of a small Pasteur pipet. To facilitate this operation, *first etch the Pasteur pipet with a file*. Carefully dispose of the broken glass in the broken-glass container. Mix the alumina slurry in the beaker and draw some of it with the wide-mouth Pasteur pipet. Position this pipet on top of the column and watch the alumina settle down while the liquid ascends (Fig. 9.5). You do not need to squeeze the rubber bulb, since the alumina goes down just by gravity. Remove the wide-mouth pipet and add more solvent to refill the column to the rim. Reposition on top of the column the wide-mouth Pasteur pipet with more slurry. Let the alumina flow down. Repeat this operation until the alumina level is about one inch from the top of the column. This method of packing the column avoids the formation of air bubbles inside the column and results in better separations. For this method to work, the solvent should fill the column to the rim before you position the wide-mouth pipet on top. Gently tap the sides of the column with a microspatula. This will compact the alumina inside the column. Add a thin layer of sand on top of the column.

Before you load the column with the sample make sure that you have everything ready. Number eight screw-cap vials Fraction 1 to 8. Get at least 15 mL of 5% toluene in hexanes in an Erlenmeyer flask and 15 mL of 20% toluene in hexanes in another flask. Have these two flasks perfectly labeled and covered.

Open the screw clamp and drain the solvent. Leave a solvent head of about 3–4 mm above the sand and close the clamp. Make a small funnel with a piece of weighing paper and use it to load the alumina pellet on top of the column. Once the pellet is loaded, add solvent to completely wet the pellet and start collecting fractions. Keep adding solvent 0.5 mL at a time, making sure that the column never runs dry and there is always solvent on top of the pellet; keep track of the volume of solvent added. Fraction 1 should be colorless. Once the magenta band of C$_{60}$ starts coming out of the column, change the vial and start collecting Fraction 2. Keep collecting the magenta fraction until about 2.5–3 mL have eluted. Then change to Fraction 3. Collect fractions of about 3 mL each until the red-orange band of C$_{70}$ starts coming out of the column. When the red-orange band is about to come out of the column, change the vial to the next fraction, and start adding 20% toluene in hexanes on top of the column in 0.5 mL portions. Collect fractions of about 3 mL each until most of the fullerenes have eluted from the column as indicated by a pale red-orange color. With a 10-mL graduated pipet or cylinder, measure the volume of the magenta fraction.

Analysis of Fullerenes

Obtain the UV-visible spectrum of the dilute original extract in the range 700–400 nm; use toluene as a blank. Measure the absorbance values at 470, 540, and 700 nm and calculate the concentrations of C$_{60}$ and C$_{70}$ as indicated below. Some light scattering due to the presence of big particles in solution may occur. Light scattering increases the absorbance values but the effect can be corrected in part by subtracting the absorbance value at 700 nm (where absorption is almost exclusively due to light scattering) from the values at 470 and 540 nm. This correction is important in the analysis of the original extract, but it is negligible in the case of the column fractions and will be omitted.

Analyze the fraction with the deepest magenta color by UV-visible spectroscopy in the range 400–700 nm. Measure the absorbance at 470, 540, and 700 nm. Also analyze the fraction with the deepest red-orange color, measuring the absorbance at the same wavelengths. If time permits, also analyze some of the intermediate fractions.

Use the following system of two equations with two unknowns (see section 32.4) to calculate the concentration of C$_{60}$ and C$_{70}$ in each sample. Use the absorbance values (A), corrected for light scattering if necessary, and the molar absorptivity values (ε) at the specified wavelengths (Table 9.2):

$$A^{470} = \varepsilon^{470}_{C_{70}} \times [C_{70}] \times \ell + \varepsilon^{470}_{C_{60}} \times [C_{60}] \times \ell \qquad (1)$$

$$A^{540} = \varepsilon^{540}_{C_{70}} \times [C_{70}] \times \ell + \varepsilon^{540}_{C_{60}} \times [C_{60}] \times \ell \qquad (2)$$

In Equations 1 and 2, ℓ is the length of the cuvette (1 cm).

Substituting the molar absorptivities with the values given in Table 9.2 and solving for [C$_{60}$] and [C$_{70}$], we obtain the following equations:

$$[C_{60}] = \frac{A^{470} \times 6050 - A^{540} \times 14500}{(6050 \times 250 - 14500 \times 710)} \qquad (3)$$

$$[C_{70}] = \frac{A^{540} - 710 \times [C_{60}]}{6050} \qquad (4)$$

In Equations 1–4, A^{470} and A^{540} are the absorbances at 470 and 540 nm (in the case of the original extract these values are corrected for light scattering by subtracting the absorbance at 700 nm). [C$_{60}$] and [C$_{70}$] indicate molar concentrations.

Chemical Test Place about 0.5 mL of the fractions analyzed by UV in small test tubes. Add two drops of ethanolamine to each tube, shake, and observe the results. Ethanolamine does not dissolve in the fraction's solvent and any color change will be observed at the interface.

Cleaning Up

- Dispose of the Pasteur pipets in the container labeled "Fullerenes–Solid Waste."
- Dispose of the unused alumina slurry in the container labeled "Recycled Alumina."
- Dispose of the fullerene soot with filter paper in the container labeled "Fullerenes–solid waste."
- Dispose of the liquid fractions in the container labeled "Fullerenes–Liquid Waste."
- Dispose of the ethanolamine test solutions in the container labeled "Fullerenes–Liquid Waste."
- At the end of the section, the instructor will empty the rota-vap traps into the container "Fullerenes–liquid Waste."

BIBLIOGRAPHY

1. C$_{60}$: Buckminsterfullerene. H.W. Kroto, J.R. Heath, S.C. O'Brien, R.F. Curl, and R.E. Smalley. *Nature*, **318**, 162–163 (1985).

2. Recent Advances in the Chemistry and Physics of Fullerenes and Related Materials, K.M. Kadish and R.S. Ruoff, ed. Electrochemical Society, Pennington, New Jersey, 1994.

3. The Chemistry of Fullerenes. R. Taylor, ed. World Scientific, Singapore, 1995.

4. Buckminsterfullerenes. W.E. Billups and M.A. Ciufolini, ed. VCH Publishers, New York, 1993.

5. Fullerene C$_{60}$. History, Physics, Nanobiology, Nanotechnology. D. Koruga, S. Hameroff, J. Withres, R. Loutfy, M. Sundareshan. North-Holland, Amsterdam, 1993.

6. Fullerenes. Synthesis, Properties, and Chemistry of Large Carbon Clusters. G.S. Hammond, and V.J. Kuck. ACS, Washington, DC, 1992.

7. Characterization of the Soluble All-Carbon Molecules C$_{60}$ and C$_{70}$. H. Ajie, M.M. Alvarez, S.J. Anz, R.D. Beck, F. Diederich, K. Fostiropoulos, D.R. Huffman, W. Krätschmer, Y. Rubin, K.E. Schriver, D. Sensharma, and R. Whetten. *J. Phys. Chem.*, **94**, 8630–8633 (1990).

8. Two Different Fullerenes Have the Same Cyclic Voltammetry. P.-M. Allemand, A. Koch, F. Wudl, Y. Rubin, F. Diederich, M.M. Alvarez, S.J. Anz, and R.L. Whetten. *J. Am. Chem. Soc.*, **113**, 1050–1051 (1991).

9. Preparation and UV/Visible Spectra of Fullerenes C$_{60}$ and C$_{70}$. J.P. Hare, H.W. Kroto, and R. Taylor. *Chem. Phys. Lett.*, **177**, 394–398 (1991).

10. Molecule of the Year. D.E. Koshland. *Science*, **254**, 1705 (1991).

11. Fullerenes. R.F. Curl, and R.E. Smalley. *Scientific American*, **265**, 54–63 (1991).

12. Preparation of Fullerenes with a Simple Benchtop Reactor. A.S. Koch, K.C. Khemani, and F. Wudl. *J. Org. Chem.*, **56**, 4543–4545 (1991).

13. Improved Chromatographic Separation and Purification of C$_{60}$ and C$_{70}$ Fullerenes. P. Bhyrappa, A. Penicaud, M. Kawamoto, and C.A. Reed. *J. Chem. Soc., Chem. Commun.*, 936–937 (1992).

14. The Discovery of Bucky Balls. K. Poon. *UCLA Und. Sci. J.*, 56–62 (1993).

15. Improved Chromatographic Separation of C$_{60}$ and C$_{70}$. A.D. Darwish, H.W. Kroto, R. Taylor, and D.M. Walton. *J. Chem. Soc., Chem. Commun.*, 15–16 (1994).

16. Using High Performance Liquid Chromatography to Determine the C$_{60}$:C$_{70}$ Ratio in Fullerene Soot. M.C. Zumwalt and M.B. Denton. *J. Chem. Ed.*, **72**, 939–940 (1995).

17. Extraction, Isolation, and Characterization of Fullerene C$_{60}$. S.P. West, T. Poon, J.L. Anderson, M.A. West, and C.S. Foote. *J. Chem. Ed.*, **74**, 311–312 (1997).

18. Fullerene Nanotubes: C$_{1,000,000}$ and Beyond. B.I. Yakobson, R.E. Smalley. *American Scientist*, **85**, 324–337 (1997).

EXPERIMENT 9 REPORT

Pre-lab

1. Make a table with the physical properties, including toxicity/hazards, of C$_{60}$, C$_{70}$, toluene, hexanes, and ethanolamine. Include the molar absorptivities and the solubility values in toluene and *n*-hexane (Tables 9.1 and 9.2). Include the flammability and dielectric constants of the solvents. For hexanes, use the values listed for *n*-hexane.

2. How many types of carbon atoms are in C$_{60}$? How many types of carbon-carbon bonds?

3. The C$_{70}$ molecule (Fig. 9.7) has five different types of carbon atoms. Study this molecule and identify them (hint: look for symmetry elements: planes, axes).

4. What happens during the activation of the alumina? What are the consequences of activation?

5. Why is the mixture of fullerenes loaded as a pellet instead of a solution?

6. Would the separation work if the column is eluted with 20% toluene in hexanes followed by 5% toluene in hexanes?

7. If the fullerene soot does not contain more than 7% fullerenes, what is the maximum amount of C_{60} and C_{70} in mg in the original toluene extract?

In-lab

1. Using Equations 3 and 4, and the absorbance values at 470 and 540 nm (corrected for light scattering), calculate the molar concentration of C_{60} and C_{70} in the UV-visible solution and in the original extract. To determine the concentration in the original extract, multiply the concentrations obtained for the UV-visible solution by the dilution factor of 15 (the original extract was diluted 0.200 mL to 3.00 mL; $15 = 3/0.2$).

2. Using the molar masses of the fullerenes and the total volume of the original extract (4 mL), calculate the total mass of C_{60} and C_{70} in mg present in the extract. How does this figure compare with the value calculated in Pre-lab 7?

3. Using Equations 3 and 4, calculate $[C_{60}]$ and $[C_{70}]$ in the fractions analyzed by UV-visible spectroscopy.

4. Calculate the purity of C_{60} in the magenta fraction. Was the column effective in separating C_{60}?

5. Using the molar concentration of C_{60}, the total volume of the magenta fractions, and the molar mass of C_{60}, calculate the mass of C_{60} obtained.

6. What percentage of the total C_{60} was recovered in the magenta fractions? Where is the rest? Discuss your results.

7. Did you obtain any fraction with C_{70} of high purity? Explain based on the UV analysis of the red-orange fractions.

8. Report and discuss your observations with ethanolamine.

9. Make a table with the fraction number, the volume of the fraction, the color observed, $[C_{60}]$, $[C_{70}]$, and the ethanolamine results (see In-lab 3 and 8).

High-Performance Liquid Chromatography

10.1 OVERVIEW

Since the late 1970s, the use of **high-performance liquid chromatography** (HPLC) in the organic chemistry laboratory has been growing steadily. In HPLC the sample is separated along a column filled with finely divided particles of the stationary phase as the liquid mobile phase is forced to flow through at high pressure. Compounds are detected as they elute from the column by using some physical property that distinguishes them from the solvent background (UV-visible absorption, fluorescence emission, refractive index, etc.).

Depending on the type of stationary phase used, HPLC separation can be based on adsorption, partition, size-exclusion, ion-exchange or reversed-phase processes. In this unit we will discuss **reversed-phase chromatography**, which is the type of HPLC most commonly used in the organic chemistry lab. We will also briefly consider size-exclusion and ion-exchange.

The principles of open-tube column chromatography presented in Unit 9 also apply to HPLC. However, HPLC's high separation efficiency coupled with its reproducibility and ease of analysis puts this technique at a great distance from regular column chromatography. What makes HPLC a superior chromatographic technique is the small particle size of the stationary phase. In standard column chromatography, particle size is in the range 0.15–0.5 mm, while in HPLC it is customary to use stationary phases with a particle size as small as 0.003 mm. As particle size decreases, chromatographic columns operate closer to equilibrium conditions and this leads to better separations (we will come back to this subject in section 10.9). This decrease in particle size is accompanied by a large decrease in flow rate, which is overcome by pumping solvent through the column at high pressure.

There are a few other differences between HPLC and standard (open-tube) column chromatography. One is that open-tube column chromatography is always preparative (the compounds are isolated at the end), while HPLC is mostly used for analytical purposes. Another difference is that most open-tube columns are used once or twice and then discarded, while HPLC columns are repeatedly used and, with proper care, have a very long lifetime.

10.2 HPLC SYSTEMS

A basic HPLC system consists of the following parts: **solvent reservoirs**, one or more **pumps**, a **mixing chamber** for the solvents, an **injection port**, a **column**, a **detector**, and a **recorder**. The eluent from the column is collected in a waste container or in the test tubes of a fraction collector, depending on whether we are carrying out analytical or preparative work. In modern HPLC instruments, the detector, the recorder, and the pumps are controlled by a computer.

The mixing of the solvents can take place either *before* or *after* they pass through the pump; both types of systems are diagrammatically shown in Figures 10.1 and 10.2. Solvent mixing *before* the pump is called **low-pressure mixing**, and requires only one pump (Fig. 10.1). When the solvents are pumped independently and *then* mixed, the process is called **high-pressure mixing** (Fig. 10.2). Because two pumps are required for high-pressure mixing, these HPLC systems are considerably more expensive than those using just one pump. In return, they offer much better control of the proportions of solvents mixed.

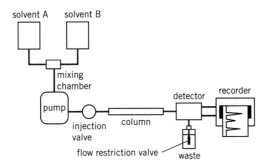

Figure 10.1 Low-pressure mixing HPLC instrument.

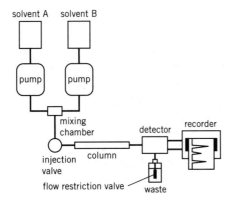

Figure 10.2 High-pressure mixing HPLC instrument.

10.3 HPLC VERSUS GC

HPLC makes possible the analysis and separation of macromolecules such as proteins, nucleic acids, polysaccharides, and synthetic polymers, which are not suitable for gas chromatography (GC) studies because of their low volatility. HPLC also offers a further advantage over GC in the fact that relatively large amounts of material, on the order of a tenth of a gram, can be easily separated and recovered in the preparative mode. A disadvantage of HPLC in relation to GC is the lack of a universal detector of high sensitivity. In GC, most organic compounds can be detected, even if present in trace amounts, by using flame ionization detectors. In HPLC, the detector should be chosen according to the type of mixtures to be analyzed. Most standard detectors have a sensitivity of about one-hundredth of that of GC. HPLC and GC should be regarded as complimentary techniques, and very often it is the combination of both that makes the study and characterization of unknown samples possible.

10.4 SOLVENTS

Similar to open-column chromatography, HPLC runs can be done isocratically or using a solvent gradient (Unit 9). The most common solvents for HPLC are methanol, acetonitrile, water, and aqueous buffer solutions. Other organic solvents, such as methylene chloride and hexane, are also used for specific separations.

The solvents used in HPLC must be dust-free. Dust particles cause pump troubles and may clog and damage the column. One can obtain dust-free solvents by filtration using special polymeric membranes with small pores. Filter paper should never be used to filter HPLC solvents because it sheds cellulose particles. For most applications, membranes with pores of 0.45 μm are adequate. There are different types of membranes on the market made from different polymeric products; some are not compatible with organic solvents. Information about the chemical resistance of membranes is provided by the manufacturers and should be checked before use. Once the solvent has been filtered it must be stored in a dust-free container. In addition to filtering the solvent with a polymer membrane, it is recommended that an **inlet filter** made of sintered metal (commonly called "**the stone**") be used at the end of the tubing that pumps the solvent from the reservoir (Fig. 10.3). This inlet filter prevents unnoticed particles from passing to the pump.

An insidious problem often encountered in HPLC analyses is the formation of air bubbles in the detector. These air bubbles generate noisy baselines and interfere in the analysis by causing faulty readings. The formation of air bubbles is especially noticeable when mixtures of solvents are used. Oxygen and nitrogen, the two main components of air, are *more soluble in single solvents* (such as methanol, acetonitrile, or water) *than in mixtures of solvents*. Therefore, just by mixing two solvents saturated with air, bubble formation occurs. This bubbling is called **outgassing**. If the solvents are mixed before the pump (low-pressure mixing, Figure 10.1) outgassing occurs in the mixing chamber and undesirable air bubbles reach the pump and column. If the solvents are mixed under pressure (Fig. 10.2), because the pressure increases the solubility of the gases in the liquid, outgassing does not occur until the solvent reaches the detector, where there is a substantial drop in pressure. Outgassing can be avoided by **degassing** the solvents before use. **Sparging** the solvent with helium is a good way of doing it (Fig. 10.3). Helium has very limited solubility in HPLC solvents; thus, as helium is flushed through the solvent it carries away dissolved air, leaving the solvent almost gas-free. Sparging the solvents for 10–15 minutes before use, and keeping them under a positive pressure of helium afterward, minimizes outgassing in the HPLC system. Even with exhaustive sparging, some air bubbles may still form in the detector when the pressure drops. This can be avoided by the use of a **flow restriction valve** positioned after the detector at the end of the waste tubing. The function of this valve is to avoid a sudden pressure drop in the detector. Working under isocratic conditions (one solvent), degassing of the solvent prior to use is not so important because there is no mixing of solvents, and no outgassing in the pump or column occurs; outgassing may still occur in the detector where the pressure drops, but this can be avoided with the use of the flow restriction valve.

In selecting mixtures of solvents for HPLC, we should pay attention to their miscibility. The solvents should be completely miscible in the range of concentration used for analysis. A chart of solvent miscibility is given in Unit 5. If aqueous buffers are mixed with organic solvents, care should be taken to avoid salt precipitation by carefully choosing the salt concentration and the amount of organic solvent used. The larger the proportion of organic solvent, the smaller the salt concentration that can be tolerated without precipitation.

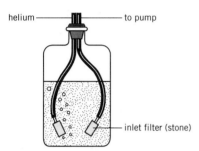

helium ——— to pump

——— inlet filter (stone)

Figure 10.3 Sparging the mobile phase with helium.

UV-visible spectroscopy is one of the most common ways to detect compounds as they elute from the column. To maximize the sensitivity of the analysis it is important that the absorbance of the solvent be small at the wavelength used to detect the sample. Each solvent has a **cutoff value**, which is the lowest recommended wavelength for the solvent. At wavelengths below the cutoff value the solvent absorbance is too large (larger than 1). A list of cutoff values for common solvents is given in Table 10.1. Sometimes the solvents contain impurities that absorb in the wavelength under consideration. This is undesirable and can be avoided by purchasing HPLC-grade solvents. HPLC-grade solvents need not be filtered if they haven't been mixed with other chemicals. Buffers made by mixing HPLC-grade solvents with salts must be filtered before use to remove dust particles that may have been present in the salts.

Table 10.1 **UV Cutoff of Selected Solvents**

Solvent	UV cutoff[a] (nm)
acetic acid	255
acetone	330
acetonitrile	190
butanone	329
tert-butyl methyl ether	210
chloroform	245
cyclohexane	200
ethyl acetate	256
n-hexane	190
isopropanol	205
methanol	205
methylene chloride	230
tetrahydrofuran	220
toluene	284
water	<190

[a]Cutoff: wavelength at which the absorbance is 1, using a 1-cm path cell.

10.5 PUMPS

Different types of pumps are available in commercial instruments. **Reciprocating pumps** (Fig. 10.4) are among the most commonly used ones. The piston is driven by a cam-shaft mechanism that generates a practically constant flow rate. HPLC pumps should be able to deliver solvents at pressures as high as 7000 psi (470 atm) and the pump head should have a small volume (<1 mL) to allow fast change in solvent composition. The materials used to build the inside parts of the pump (those in contact with solvents) should be inert. Solvents such as hydrochloric and hydrobromic acids should be avoided because they corrode even resistant stainless steel. There are also metal-free pumps available that are less susceptible to corrosion and are especially recommended when metals such as Fe, Cr, and Ni, which unavoidably dissolve from the stainless steel, may pose a problem in the analysis. Pumps should be handled with care; they should never run dry. Buffers, especially those containing chloride, *should never be left stagnant because they cause corrosion.*

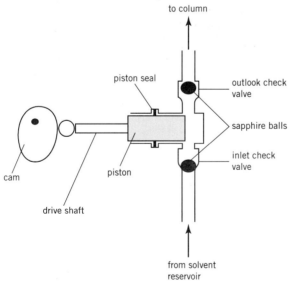

Figure 10.4 A reciprocating pump head.

10.6 INJECTION PORT

The most widely used injection port for HPLC is the six-port valve (Fig. 10.5). The valve has two positions: *load* and *inject*. With the valve in the load position, the sample

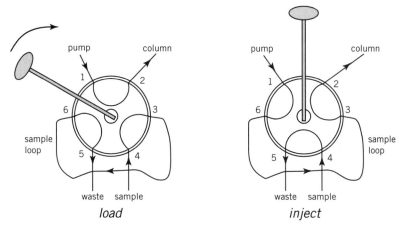

Figure 10.5   A six-port HPLC injection valve.

is introduced via a syringe (port 4, Fig. 10.5) and goes directly to the **sample loop**, a piece of tubing that holds specific volumes. Any excess of sample drains into the waste (port 5, Fig. 10.5). Most valves are equipped with removable sample loops that can hold from 5 μL up to 5 mL. When the valve is in the load position, the solvent is pumped directly into the column, bypassing the loop. Once the sample has been loaded, the handle is moved to the inject position, where the loop is now connected to the pump and to the column. To ensure that the sample is completely injected, one should leave the valve in the inject position until the next run. Before a new sample is loaded the loop (in the load position) must be washed a few times with solvent to avoid cross-contamination. No buffer solutions should be left on the loop overnight. If aqueous buffer solutions are used, the loop should be flushed with water at the end of the day.

For analytical work the sample volume injected into the column is normally 20 μL. The sample solution should be dust-free. This can be achieved by filtration using a 0.45 μm membrane, or by centrifugation (at 10,000–14,000 rpm) for a few seconds. To avoid detection problems it is recommended that the sample be dissolved in the same solvent used as the mobile phase.

10.7   COLUMNS

The column is the heart of the HPLC instrument. The quality of the separation depends on the type of stationary phase used. The number of stationary phases available for HPLC is reduced compared to GC; however, this is not a real drawback because we can also control the efficiency of the separation by the choice of mobile phase, thus making HPLC columns very versatile. We will discuss several types of stationary phases and different separation mechanisms in sections 10.9–11.

For most analytical applications, HPLC columns are normally made of stainless steel and have diameters in the range 2–7 mm and lengths of 10–60 cm. Preparative columns, with larger dimensions, are also available. For metal-free analysis, columns made of glass and glass-lined steel columns can be purchased. Columns must be packed with materials that can withstand very high pressures. Different stationary phases have different pressure limits, and these should be checked in the literature supplied by the manufacturer before the column is used. Subjecting a column to pressures higher than the maximum recommended will permanently ruin it.

Compatibility between the mobile and the stationary phase should also be checked before the column is used because some solvents may cause irreversible damage to the stationary phase. For example, silica columns used for size-exclusion chromatography (section 10.11) are resistant to most organic solvents and can be used with water solutions in the pH range 2–7. Outside this pH range they suffer significant damage due to partial dissolution of the silica as silicic acid (H_2SiO_3). This can be avoided by saturating the mobile phase with silicic acid. To this end a **pre-column** packed with silica is inserted before the injection valve. The silica from the pre-column is slowly dissolved by the mobile phase, which gets saturated with silicic acid and cannot dissolve more silica.

To ensure a long lifetime for the column, a **guard column** made with exactly the same stationary phase is usually inserted between the injection valve and the column. The guard column is a shorter version of the main column and prevents particles and other contaminants from reaching the long column. Guard columns should be replaced frequently and are especially recommended when biological fluids, which contain certain compounds that may damage the stationary phase, are analyzed.

Most HPLC analyses are carried out at room temperature. It is a little bit cumbersome to maintain the temperature of the solvent and the column at values other than ambient. However, there are commercially available thermostatic mantles that make it possible to do just that. Usually, there is no advantage in performing the analysis at higher temperatures because any possible gain in performance that may result from changing the temperature can be more easily obtained by changing the mobile or stationary phases.

For optimum performance, the column must be equilibrated with the mobile phase before use. This is done by running *at least five column volumes of solvent* through the column at a moderate flow rate (0.3–1.0 mL/min). For example, if the column has a volume of 10 mL we must run at least 50 mL of solvent prior to use.

10.8 DETECTORS

There are three main types of detectors: **UV-visible**, **fluorescence**, and **refractive index**. In principle, any property that differentiates solute from solvent can be used to detect the sample components as they elute from the column.

In the detector, the sample goes to a small **flow cell** where the physical property is measured. A schematic diagram of a **multi-wavelength UV-visible detector** is shown in Figure 10.6. This type of detector uses both deuterium and tungsten lamps as light sources. Deuterium lamps emit continuous UV radiation up to 340 nm, while tungsten lamps emit in the near UV and visible range (340–850 nm). The wavelength chosen to follow the separation is selected by the rotation of a grating (a grating is a dispersive instrument that separates radiation into individual frequencies). The light is split into two beams; one goes to the sample flow cell, and the other to the reference cell, which normally contains just air. The amount of radiation absorbed by the sample is determined by comparing the intensities coming out of the reference and the sample cells. The absorbance is proportional to the molar absorptivity, the concentration, and also the length of the cell (Beer's law, section 32.4). To increase the sensitivity of the method, long cells are preferred. Normal pathlengths for flow cells are in the range 0.6–1 cm. Multi-wavelength detectors allow us to work at the wavelength where the sample has maximum absorbance, or in other words, where the sensitivity is maximum.

For routine analysis and simple applications, single-wavelength detectors are an affordable option. **Mercury lamps** emit at 254 nm and are useful to detect compounds that absorb at this particular wavelength. Aromatic compounds, carbonyl containing compounds, and conjugated alkenes all have strong absorption around 254 nm and therefore can be detected with mercury lamps. **Zinc lamps** emit at a shorter wavelength (214 nm) where, with the exception of alkanes and cycloalkanes which absorb below 190 nm, almost all organic compounds can be detected. Under favorable conditions, UV-visible detectors can detect concentrations as low as 0.01 μM. Besides the UV-visible detectors already mentioned, **diode-array** models (Unit 32)

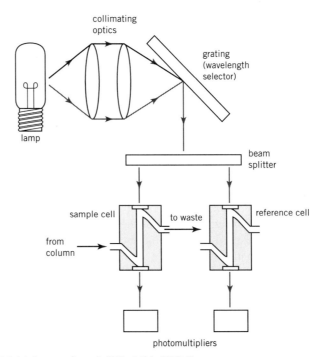

Figure 10.6 Multiple-wavelength UV-visible HPLC detector.

are becoming more and more popular; these detectors are very fast and allow the collection of the *whole UV-visible spectrum of the sample* as it comes out of the column.

Historically, **refractive index detectors** were the first to be used. They detect differences in refractive index (Unit 11) between solvent and sample; their sensitivity is about a thousandth of that of UV-visible detectors and can be used only for isocratic runs. Today, this type of detector is used mainly for preparative work and very specific applications.

Fluorescence detectors are about 10 times more sensitive than UV-visible ones. They work by irradiating the sample at a predetermined wavelength (where the sample has strong absorption) and detecting the light emitted by the sample at a 90° angle with respect to the direction of the incident light. Not all compounds are fluorescent, but most aromatic compounds, nucleic acids, and proteins are, and therefore, this type of detector offers a very sensitive way for studying these substances.

10.9 WHY IS HPLC HIGH PERFORMANCE?

The secret of HPLC's high performance is in the stationary phase *particle size*. Silica gel used as the stationary phase in column chromatography consists of small particles that are penetrated by the mobile phase and solute molecules through small pores. The type used for open-tube column chromatography has an average particle size of about 200 μm with a pore size in the range of 60–100 Å (0.006–0.010 μm). Solvent and solute molecules diffuse inside these particles (remember that typical organic molecules have diameters of less than 50 Å) where they can spend a long time (they get trapped) before they leave the particle (Fig. 10.7). Under such circumstances some of the mobile phase and solute molecules are stagnant inside the silica particles, reducing the efficiency of the mass transfer process, which is essential for separation. This retardation of molecules inside the stationary phase particles ultimately leads to broadening of the peaks and decreases resolution. In HPLC, **microparticles** of silica gel are used instead. These microparticles have a diameter in the range 3–10 μm,

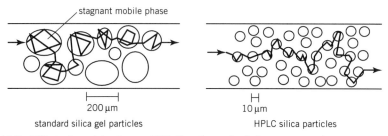

stagnant mobile phase

200 µm
standard silica gel particles

10 µm
HPLC silica particles

Figure 10.7   Effect of particle size on HPLC and standard chromatography.

and therefore are several thousand times smaller in volume than silica particles used in standard column chromatography. The smaller particle size allows a more rapid mass transfer and a better separation. The microparticles are also very uniform in size, which leads to a better flow of the solvent through the column and results in better column performance.

An additional advantage offered by the small particles is an increase in the specific surface area (m^2/g) of the silica, which results in higher loading capacity. HPLC columns have, on average, a tenfold better performance than standard ones. One might wonder why smaller particles are not used in standard column chromatography. There is a simple reason. As the particle size decreases, the flow rate also decreases. To achieve reasonable flow rates, pressure has to be applied to the column, which leads to HPLC.

10.10   REVERSED-PHASE CHROMATOGRAPHY

Reversed-phase chromatography (RPC) is one of the most common applications of HPLC. In reversed-phase chromatography the stationary phase is relatively nonpolar and the polarity of the mobile phase is either *kept constant* or *decreased during the run*. Because this is exactly the opposite of adsorption chromatography, where the stationary phase is relatively polar and the polarity of the solvent is *increased during the run*, this type of chromatography is called reversed-phase. Contrary to normal chromatography, in RPC polar compounds elute from the column first, and nonpolar compounds last.

The most commonly used stationary phase in RPC is **chemically modified silica gel**. Hydrocarbon chains are chemically bonded to the silica particles by reacting the free silanol groups (Si—OH) of silica with specific reagents. Octadecyl trichlorosilane is a reagent used for this purpose ($C_{18}H_{37}SiCl_3$). It has a hydrocarbon tail of 18 carbon atoms and the resulting stationary phase is called C_{18}. The long hydrocarbon chain of the octadecyl groups makes this stationary phase nonpolar. Other common stationary phases for RPC are C_1 and C_8, in which the number of carbon atoms attached to the silicon atom are 1 and 8, respectively.

Usually, an unspecified number of —OH groups remains attached to the silicon atoms in the reversed-phase column. Some are generated by hydrolysis of the Si—Cl bonds from the trichlorosilane reagent after coupling to the silica; others are unreacted silanol groups on the silica surface. The presence of silanol groups may cause tailing of the peaks due to adsorption processes. This problem can be minimized by reacting the free silanol groups with trimethylchlorosilane (($CH_3)_3SiCl$); this process is called **end-capping** (Fig. 10.8).

Separation by reversed-phase columns is based on a partition mechanism between the mobile phase and the bonded stationary phase. The stationary phase behaves like a nonpolar layer, permanently coating the silica solid support. When a polar mobile phase runs through the column, nonpolar solutes are preferentially partitioned **inside** the nonpolar stationary phase because they are **hydrophobic** (repel water). More polar solutes, on the other hand, are found preferentially in the mobile phase (Fig. 10.9). As

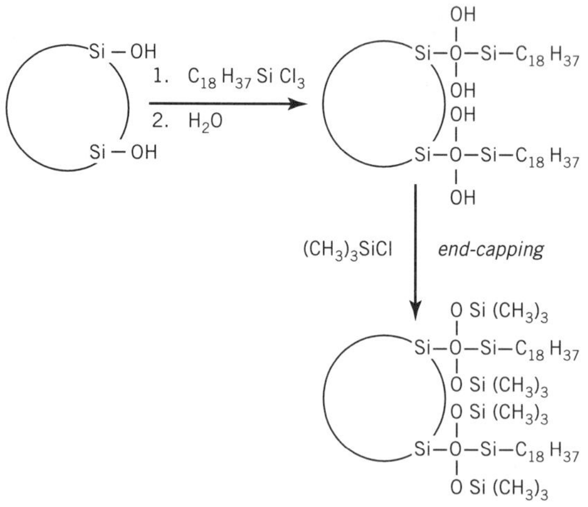

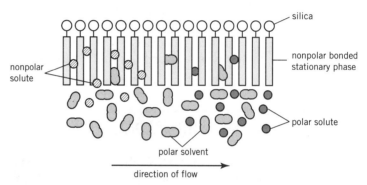

Figure 10.8   Modification and end-capping of silica particles.

Figure 10.9   Partition of solutes between a bonded stationary phase and the solvent.

a result, polar compounds move along the column faster and elute before nonpolar ones.

Depending on the mixture to be separated, reversed-phase HPLC can be run under isocratic conditions (one solvent) or with a gradient of *decreasing* polarity. In gradient development the analysis begins with polar solvents, which elute only polar compounds. The nonpolar compounds are preferentially retained inside the stationary phase and are removed by decreasing the polarity of the solvent.

To separate polar compounds, C_1 and C_3 stationary phases are recommended, whereas the separation of nonpolar compounds is better performed with long chain bonded phases (C_{18}). As a rule of thumb, the longer the hydrocarbon chain of the stationary phase, the longer the retention time.

10.11   OTHER CHEMICALLY BONDED STATIONARY PHASES

Bonded stationary phases with different degrees of polarity and with varied chemical and physical properties are commercially available. This wide selection allows the most difficult of separations. For example, stationary phases containing phenyl rings can be chemically modified (after binding to the silica support) to incorporate sulfonic acid or ammonium groups, which can be used for **ion-exchange chromatography**.

There are two types of ion-exchange chromatography: **anion-exchange** and **cation-exchange**. In cation-exchange chromatography, the H^+ from the stationary phase (normally in the form of sulfonic acid $R-SO_3H$) is exchanged for cationic solutes (M^+), which are retained by the column. Anion-exchange stationary phases, on the other hand, retain anionic solutes (X^-) by electrostatic interaction with positively charged centers, usually quaternary ammonium groups, $-NR_3^+$, attached to the solid support (Fig. 10.10).

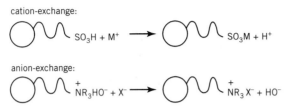

cation-exchange:

anion-exchange:

Figure 10.10 Ion-exchange processes. M^+ and X^- are ionic samples retained in the column.

The compounds are eluted from the column by reversing the process. Adding acid to cation-exchange columns regenerates the $-SO_3H$ groups and liberates M^+. Addition of base to anion-exchange columns regenerates the $-NR_3^+HO^-$ groups and elutes X^-. Ion-exchange chromatography is very useful in the separation of proteins and nucleic acids, and other charged molecules.

10.12   SIZE-EXCLUSION CHROMATOGRAPHY

Organic compounds can be separated by a principle totally different from that of adsorption, partition, and reversed phase discussed so far. This principle is *molecular size* and *shape*. The name of the technique is **size-exclusion chromatography**, also known as **gel-permeation** or **gel-filtration chromatography**.

In size-exclusion chromatography, the stationary phase is a porous solid with relatively well-defined pore diameter. Small molecules (smaller than the pore size) are able to penetrate the stationary phase particles, where they spend some time before they return to the bulk of the solvent. Big molecules, on the other hand, unable to go inside the stationary phase particles, move faster and leave the small molecules behind (Fig. 10.11).

There are different types of stationary phases available for size-exclusion chromatography. **Sephadex**, a modified dextran (dextran is a polymer of glucose), is especially suited for the separation of proteins. This stationary phase is very popular for open-tube column chromatography, but it cannot be used for HPLC columns because the dextran particles are not rigid enough to withstand very high pressures. Silica columns, which are unaffected by high pressures, have been developed for the separation of proteins by size-exclusion HPLC.

In HPLC in general and in size-exclusion chromatography in particular, it is customary to refer to the elution volume rather than the retention time. The elution volume, V_e, is calculated by multiplying the retention time, t_R, by the flow rate (Eq. 1):

$$V_e = t_R(\text{min}) \times \text{flow rate (mL/min)} \qquad (1)$$

As stated before, compounds that cannot penetrate the stationary phase particles are not retained in the column and are the first to come out. The volume of solvent required to elute such big particles is called the **void volume of the column**. On the other hand, very small molecules that penetrate the stationary phase particles follow a tortuous path and come out of the column later. Solvent molecules fall in this category. The volume required to elute such small molecules is called the **total**

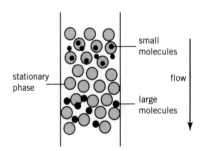

Figure 10.11 Separation of molecules based on size.

solvent-accessible volume of the column. For a given column, the elution volume of all other compounds falls between these two extreme volumes.

Size-exclusion chromatography is very useful for the estimation of molecular mass of macromolecules such as proteins. There is a linear correlation between the elution volume (V_e) of globular proteins and the logarithm of their molecular masses (*MM*) as shown in Equation 2, where *A* and *B* are empirical constants that depend on the column and conditions used.

$$V_e = A \log MM + B \qquad (2)$$

To estimate the molecular mass of an unknown protein, the elution volumes of a set of standard proteins of known molecular masses are first determined. The correlation between V_e and log *MM* is plotted to construct a **calibration curve** (a straight line in this case), which is then used to estimate the molecular mass of unknown proteins by measuring their elution volumes.

10.13 QUANTITATIVE DETERMINATIONS: STANDARD CURVE METHOD

To carry out quantitative determinations by HPLC, one can apply the same methods as those described for GC analyses, such as the internal standard method (section 7.9). But there is a method, called the **standard curve method**, which is primarily used in HPLC.

The area under an HPLC peak, *A*, is directly proportional to the concentration of the sample, *C*, and the volume injected, *V*, (Eq. 3):

$$A = f \times V \times C \qquad (3)$$

where *f* is the **absolute response factor**.

If the volume of injection is constant, a plot of *A* versus *C* for a given compound should give a straight line with slope $f \times V$. In HPLC, the injection loop delivers exactly the same volume in each injection. Thus, it is possible to obtain a good linear correlation when *A* is plotted against *C* because *V* is a constant. This differs from GC analyses, where it is very difficult to inject precisely the same volume in each run.

In HPLC we can determine the absolute response factor for a given chemical by measuring the areas obtained for different sample concentrations and by plotting *C* versus *A* (or *A* versus *C*). One can then use this plot, called the **standard curve**, to analyze samples of unknown concentrations. The area of the peak is measured, and with the help of the standard curve the sample concentration is determined. This method will be used in Experiment 10.

10.14 HPLC DO'S AND DON'TS

The following list is a summary of the things that you should do and those you should avoid at all cost when running HPLC.

Do's
- Use only HPLC-grade solvents.
- The mobile phase must be dust-free.
- Equilibrate columns with at least five column volumes of solvent before use.
- Use solvents compatible with the stationary phase.

- Inject only dust-free samples.
- Leave the injection valve in the inject position between runs.
- If you have used aqueous buffers, at the end of the day run at least 3 column volumes of HPLC-grade water through the system. Flush the sample loop with HPLC-grade water.
- When not in use for a long period, store the column with the proper solvent following manufacturer's instructions.

Don'ts
- Do not allow the pump to run on empty.
- Do not subject the column to sudden changes in pressure or to a pressure higher than its recommended maximum limit.
- Never leave aqueous buffers stagnant in the system.

BIBLIOGRAPHY

1. Chromatographic Methods. A. Braithwaite and F.J. Smith. 4th ed. Chapman and Hall, London, 1990.
2. HPLC. A Practical User's Guide. M.C. McMaster. VCH, New York, 1994.
3. Practical High-Performance Liquid Chromatography. V.R. Meyer. Wiley, Chichester, 1994.
4. Chromatography. 5th ed. E. Heftmann, ed. Elsevier, Amsterdam, 1992.
5. Maintaining and Troubleshooting HPLC Systems. D.J. Runser. Wiley, New York, 1981.

EXERCISES

Problem 1

Each word or group of words in Set A is related to one (and only one) word or group of words in Set B and vice versa. Find those matching pairs.

Set A
C_{18}, degassing, deuterium lamp, fast mass transfer, guard column, injection, large particle size, low-pressure mixing, mercury lamp, outgassing, pore size, solvent mixture, UV-cutoff.

Set B
Better resolution, column protection, fixed-wavelength detector, gel permeation, helium, miscibility, multi-wavelength detector, one pump, pressure drop, reversed-phase chromatography, six-port valve, solvent absorbance, stagnant solvent.

Problem 2

Which solvents would you not use with a fixed-wavelength detector at 254 nm: acetone, acetonitrile, butanone, cyclohexane, n-hexane, methanol, toluene?

Problem 3

The following phenols were separated by using a C_{14} reversed-phase column with methanol-water (6 : 4) as a mobile phase. Arrange the compounds in order of increasing elution volumes and briefly justify your answer.

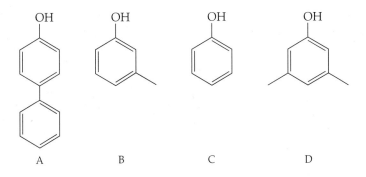

A   B   C   D

Problem 4

The following aromatic hydrocarbons were separated using a C_{18} column with a water-acetonitrile gradient (50–100%). Predict the elution order of these hydrocarbons and briefly justify your answer.

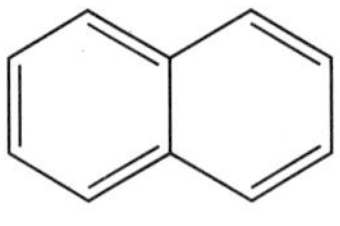

I 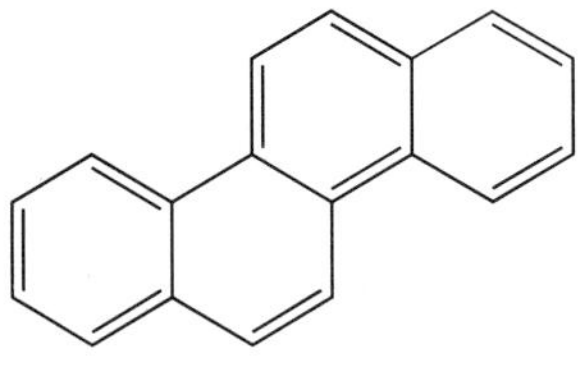   II

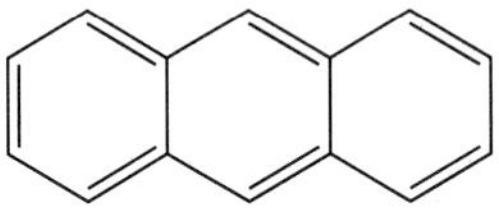

III    IV

Problem 5

A silica gel permeation column (60 cm × 7.8 mm) with particle size 5 µm and pore size 250 Å is used to determine the molecular mass of proteins. The mobile phase is sodium phosphate (0.02 M) with 0.2 M KCl, pH 6.5 at a flow rate of 1 mL/min. The following proteins are used as standards: bovine serum albumin dimer (MM = 134,000 Da, V_e = 14.9 mL), bovine serum albumin (MM = 67,000 Da; V_e = 17.3 mL), ovalbumin (MM = 43,000 Da; V_e = 18.8 mL), ribonuclease A (MM = 13,700 Da; V_e = 22.5 mL). Estimate the molecular mass for proteins with elution volume 16.5 mL and 20.3 mL.

EXPERIMENT 10

Vitamin Analysis — A Quantitative Study

E10.1 WATER-SOLUBLE AND FAT-SOLUBLE VITAMINS

Vitamins are a group of small organic compounds essential to sustain human life. Our bodies cannot produce vitamins and we must obtain them from our diet. There are at least thirteen different vitamins that fulfill diverse roles, from antioxidants and detoxifying agents to substances involved in the process of vision. Based on their solubility in water and oil, they can be classified into two main groups: **water-soluble** and **fat-soluble**.

Water-soluble vitamins do not accumulate in our bodies; they are transported in the bloodstream and are easily excreted in the urine. We need them in small and frequent doses. Vitamins B and C belong to this category. Vitamin C is a potent antioxidant that we obtain only from fruits and vegetables. Lack of vitamin C causes scurvy, an uncommon disease today but one that killed thousands of sailors in centuries past. Vitamin B is not a single compound but a group of eight unrelated compounds (B_1, B_2, B_3, B_6, B_{12}, biotin, folic acid, and pantothenic acid), each with a specific function. Vitamins of the B group are essential players in the metabolism of carbohydrates, fats, and proteins, and in the building of DNA molecules. We obtain these vitamins from varied sources but especially from meats and legumes.

Vitamins A, D, E, and K form the fat-soluble group. Vitamin A, found in carrots, colored vegetables, liver, and dairy products, plays a key role in the chemistry of vision. It is also necessary to maintain healthy skin. β-Carotene, the pigment of carrots, is one of the most important sources of vitamin A. β-Carotene breaks down in the liver to produce two molecules of vitamin A.

In contrast to water-soluble vitamins, of which we need a constant supply, vitamin A is stored in our bodies and released as needed. Our liver can store up to a year's supply of vitamin A. This vitamin is present in two main active forms: **retinol** (an alcohol), and **retinal** (an aldehyde). With four double bonds in the side chain, several cis-trans isomers of these molecules are possible. The all-*trans*-isomers are the ones with the greatest biological activity.

Vitamin E, or α-**tocopherol**, is present in seeds and grain oils and is a potent antioxidant. It protects cell membranes from oxidative damage caused by oxygen and free radicals. It reacts rapidly with oxidizing agents, removing them from the

Forms of Vitamin A

Because of its long hydrocarbon chain vitamin A is nonpolar. It does not dissolve in water but dissolves well in oils and fats.

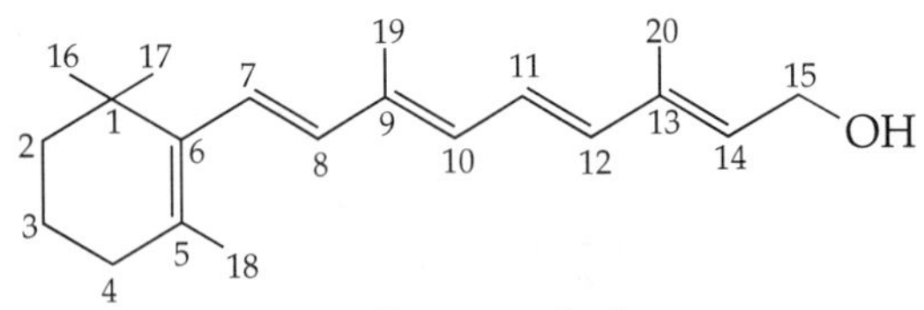

all-*trans*-retinol

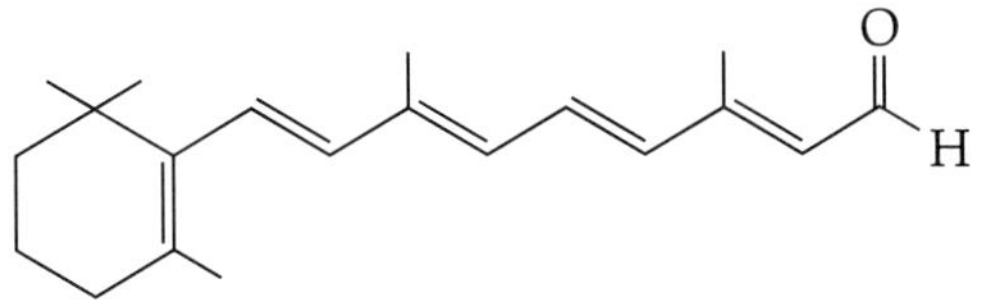

all-*trans*-retinal

splits into two retinol molecules in the liver

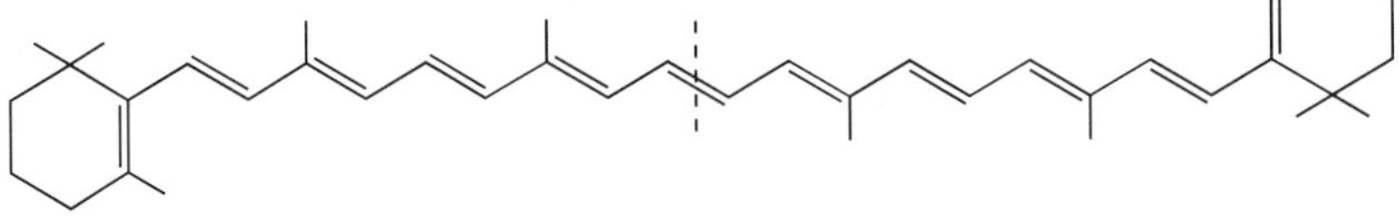

β-carotene

medium and preventing their reaction with other vital molecules. α-Tocopherol has three stereogenic centers at positions 2, 4′, and 8′; hence it can exist as eight different diastereomers, or four pairs of enantiomers. Naturally occurring vitamin E has R configuration in each of the three stereogenic centers.

Vitamin E

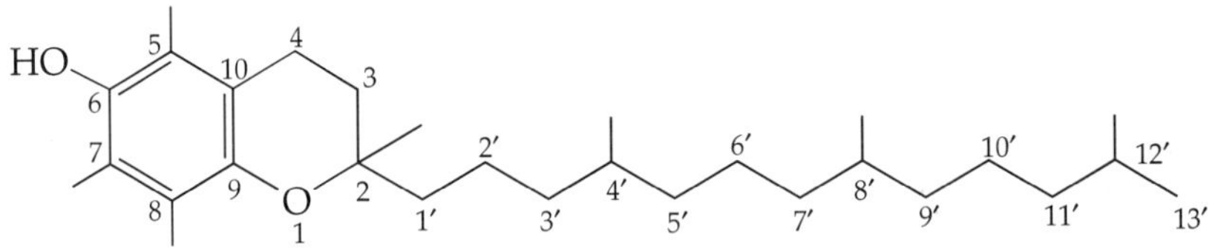

α-tocopherol

Vitamin D, which has two active forms, D_2 and D_3, is the only vitamin that our bodies can synthesize if exposed to sunlight. People living in cold regions with long

Vitamin D

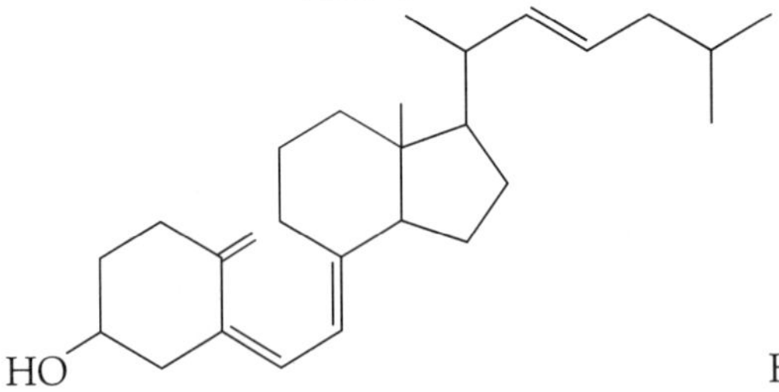

Ergocalciferol (D_2)

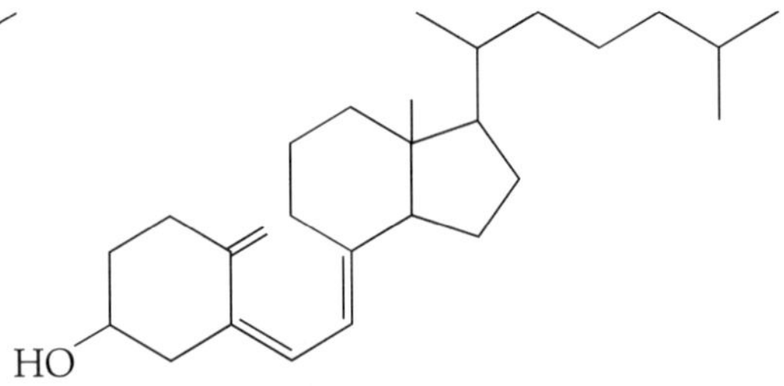

Cholecalciferol (D_3)

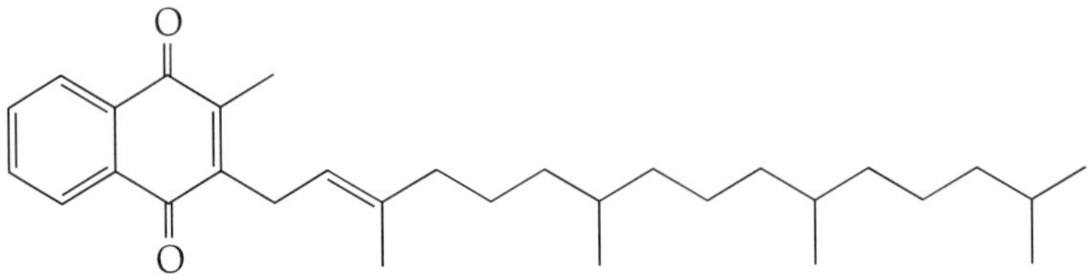

Vitamin K_1

winter seasons do not get enough sunlight to produce adequate levels of vitamin D, and they must obtain this vitamin from sources such as liver, eggs, and fortified milk. Deficiency of vitamin D is particularly damaging to children, who need it to sustain normal growth.

Vitamin K is essential for blood clotting. We obtain this vitamin from vegetables and from our own intestinal bacteria by absorption through the intestinal wall. There are several types of vitamin K; vitamin K_1 is shown in the figure above.

E10.2   VITAMIN STABILITY

Like most biologically active molecules, vitamins are affected by heat, light, and oxygen. Unlike proteins and carbohydrates, which are still good nutrients after prolonged processing, vitamins easily lose their nutritional value by cooking and storage.

Vitamin A becomes unstable when exposed to heat, light, air, and acids. In the presence of oxygen it undergoes oxidation reactions that lower its biological activity. Light promotes cis-trans isomerizations of the retinol side chain, yielding less-potent isomers. In vitamin supplements, vitamin A is present in more stable forms, such as **retinyl acetate** or retinyl palmitate. These esters are easily hydrolyzed to retinol in our bodies.

Vitamin E is more susceptible to oxidation than vitamin A. The phenol group is readily oxidized to a quinone, a process that leads to loss of activity. The acetate form of vitamin E is more stable than the free phenol and thus it is preferred for commercial preparations.

An oxidation product of Vitamin A

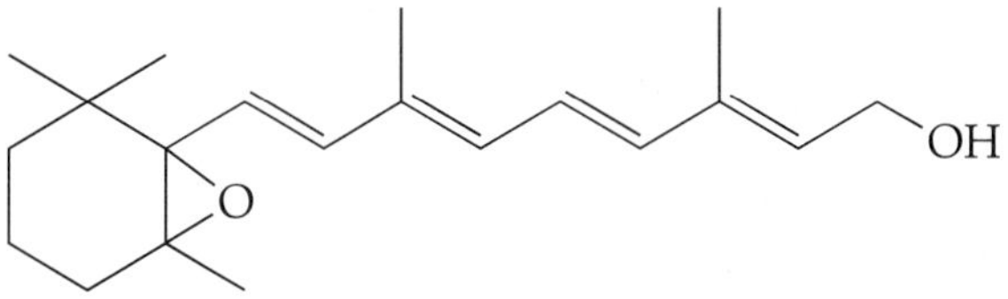

An isomerization product of Vitamin A

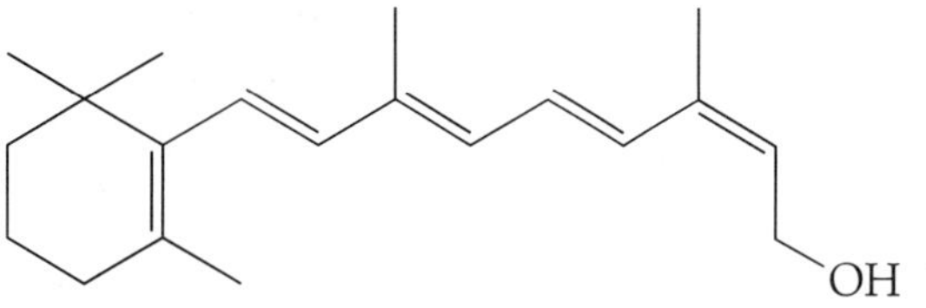

An oxidation product of Vitamin E

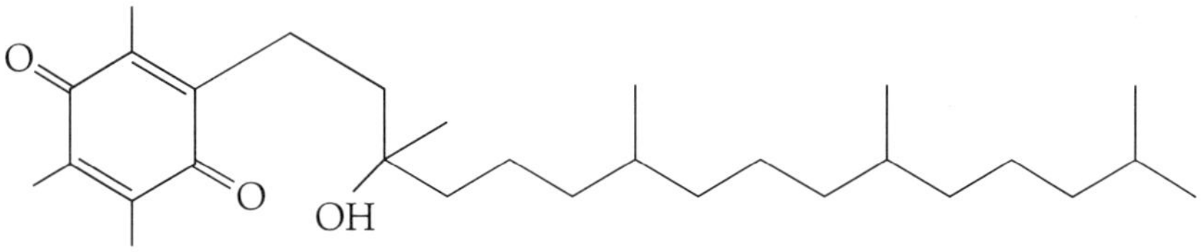

α-tocopherolquinone

Stable forms of vitamins A and E

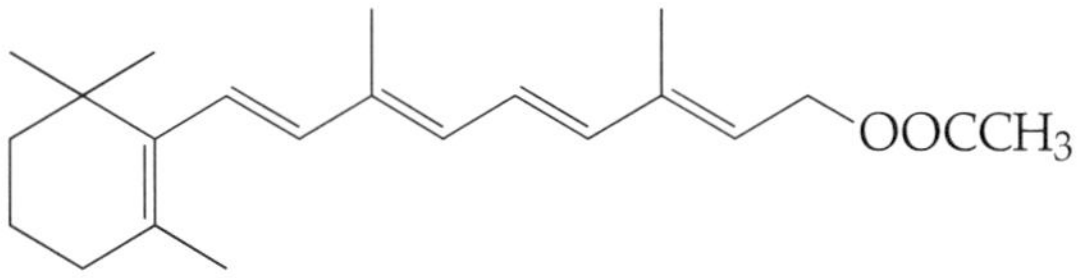

retinyl acetate

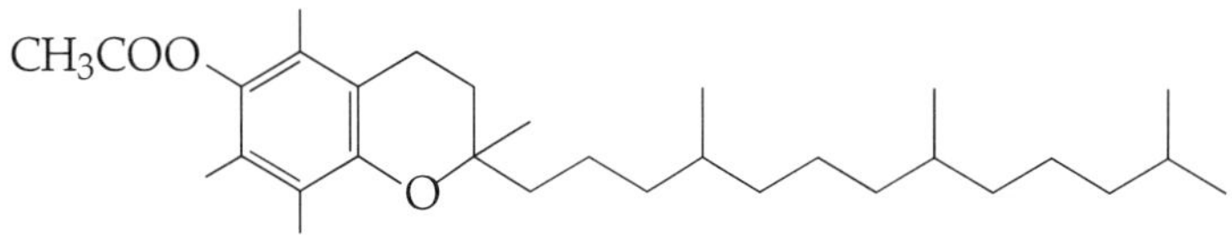

α-tocopheryl acetate

Vitamins can undergo decomposition reactions even in the solid state. After prolonged storage, vitamin tablets will inevitably contain some degradation products in amounts that depend, among other factors, on the concentration of vitamin in the tablet, the type of filler, moisture content, and storage conditions. Vitamin formulations have expected shelf lives. The manufacturer calculates the expiration date as the date when the vitamin activity has decreased to 90% of the label claim. To slow down their decomposition, vitamin supplements should be stored in well-capped bottles and protected from moisture, air, and light.

E10.3 INTERNATIONAL UNITS

The amounts of vitamins in foods and supplements are generally expressed in milligrams, micrograms ($1\,\mu g = 10^{-6}$ g), or **international units** (IU), depending on the type of vitamin. Vitamin A is usually given in international units; one IU of vitamin A is equivalent to 0.3 μg of all-*trans*-retinol, 0.344 μg of all-*trans*-retinyl acetate, or 0.6 μg of β-carotenes. The World Health Organization has proposed the use of a different unit for vitamin A, the **retinol equivalent** (RE). One RE equals 1 μg of all-*trans*-retinol.

Vitamins D and E are also expressed in IU. One IU of vitamin D equals 0.025 μg of vitamin D_2. One IU of vitamin E is equivalent to 1 mg of *d,l*-α-tocopheryl acetate. Contrary to what the name suggests, *d,l*-α-tocopheryl acetate is not a racemic mixture of two enantiomers, but rather an unspecified mixture of the four possible racemic mixtures of α-tocopheryl acetate (remember that α-tocopheryl acetate has three stereogenic centers; therefore there are eight stereoisomers grouped in four pairs of enantiomers). A more appropriate name for this species is all-*rac*-α-tocopheryl acetate.

E10.4 CHARACTERIZATION OF VITAMIN A

Polyenes containing at least three conjugated double bonds react with acids to give a transient blue color due to the formation of a carbocation. Carbocations are reactive intermediates with a fleeting existence. You may have seen them participating in reaction mechanisms as high energy intermediates that cannot be isolated. Here you will see that some carbocations can be *visualized* if the conditions are right. The reaction of vitamin A with acids has important practical applications. Under the name

rentinyl acetate

two of the resonance forms of the retinyl carbocation (blue)

products

of **Carr-Price method**, it is used to quantify the amounts of vitamin A present in foodstuff. In this experiment you will use this reaction to *identify* vitamin A in the HPLC eluent. The scheme above shows the reactions involved in the formation of the carbocation. The carbocation then reacts with nucleophiles or bases present in the reaction medium to give substitution or elimination products.

E10.5   OVERVIEW OF THE EXPERIMENT

> **Background reading**: 10.1–10; 10.14; 7.9 or 10.13; E10.1–E10.4; 32.4
> **Estimated time**: 2 periods of 3 hours

In this experiment you will extract the fat-soluble vitamins A and E from a multivitamin tablet. You will use HPLC to analyze the mixture of vitamins and quantify their contents in the tablets. The amount of vitamin D present in commercial tablets is generally very small and special techniques are required for its determination. This fat-soluble vitamin will not be analyzed in this experiment.

You will quantify vitamins A and E by one of two methods: the **internal standard method** or the **standard curve method**. Ask your instructor which one you will use. The internal standard method is treated in section 7.9. You will use 9-anthraldehyde as internal standard. The standard curve method is explained in section 10.13. Regardless of the method of choice, you will first make standard solutions of vitamins A and E. If you use the internal standard method, you will also prepare a stock solution of 9-anthraldehyde, and will determine the HPLC response factors for vitamins A and E relative to the internal standard. The detector used in this experiment measures the absorbance of the effluent at 254 nm, and thus the response factor is directly proportional to the chemical's molar absorptivity at 254 nm. If a different wavelength or other type of detector is used, the concentrations should be accordingly changed to match the detector's sensitivity. Ask your instructor about this issue. Before you go on, review Beer's law in section 32.4.

In this experiment you will also carry out a characterization test for vitamin A. You will treat samples of vitamins A and E with a drop of phosphoric acid and observe any change in color. You will collect the eluent from the HPLC corresponding to the vitamin peaks. You will treat these fractions with phosphoric acid and observe any color change. You will use this test to further identify each vitamin peak.

If time permits, you will also analyze vitamin tablets that have been stored at room temperature beyond the expiration date. You will compare their HPLCs with those obtained from fresh tablets, and will estimate the extent of deterioration by looking at the appearance of new peaks. Ask your instructor about working with a teammate. Each student can analyze a different vitamin tablet. You will work independently and compare notes at the end.

PROCEDURE

E10.5.1 Standard Solutions

Retinyl Acetate Solution   In a 50-mL volumetric flask, weigh approximately 25 mg of pure retinyl acetate using an analytical balance. Commercial retinyl acetate is generally available as a very concentrated solution (a **concentrate**) in vegetable oil. The product content is lower than 100% and is specified on the label by the manufacturer. In some commercial preparations it can be as low as 30%. In such a case, you should weigh about 83 mg (25/0.3) of the oily concentrate to obtain a mass of approximately 25 mg of retinyl acetate. Taking into account the label concentration, calculate how much of the oil you should weigh to obtain 25 mg worth of retinyl

---

Notes

- The solutions you are about to prepare are sensitive to air, light, and temperature. Keep the containers well capped and wrapped in aluminum foil. Store them in the refrigerator. They will keep for about a week if properly stored.

- Quartz UV cuvettes are expensive and fragile. Do not leave them unattended on the bench. Keep them in the proper cuvette holder when not in use.

- Check that your solution is particle free before injecting into the HPLC. Centrifuge if necessary to separate solid particles.

- Before injecting each sample, rinse the loop with 100 μL of the sample.

HPLC Conditions:

Column: Econosphere C18;
particle size: 10 µm

Dimensions: 250 × 4.6 mm.

Solvent: 100% methanol

Flow rate: 1 mL/min

Detection at 254 nm

Chart speed 2.5 cm/min

Injection loop: 20 µL

acetate. Add isopropanol to the mark, cap the flask, and invert it several times to dissolve the product completely. Label this solution A1.

The concentration calculated from the label values is only approximate. You will carry out a precise concentration determination by UV-visible spectroscopy. To this end, you must dilute solution A1 because the absorbance of the undiluted solution is too high (beyond the recommended maximum limit of 2). Using an automatic pipet, transfer 200 µL of solution A1 into a 10.0-mL volumetric flask and add isopropanol to the mark. Label this solution A2. Shake the flask several times and obtain the UV-visible spectrum of the solution in the range 200–500 nm using a 1-cm pathlength quartz cuvette and isopropanol as blank. Read the absorbance at 325 nm (A^{325}). The absorbance value should be below 2. (If the absorbance is higher than 2, take 100 µL of solution A1, dilute it to 10.0 mL in a volumetric flask using isopropanol and read its absorbance.)

Using Beer's law, the pathlength of 1 cm, the molar absorptivity of retinyl acetate of 50,800 L mol^{-1} cm^{-1} (at 325 nm), and its molecular mass (MM), calculate the concentration of solution A2 in g/L (equal to mg/mL; Eq. 4).

$$[A2](g/L) = \frac{A^{325}}{50,800 \text{ L mol}^{-1}\text{cm}^{-1} \times 1 \text{ cm}} \times MM \text{ (g/mol)} \qquad (4)$$

Using the dilution factor 0.2/10 (or 0.1/10 if the absorbance was higher than 2 and you had to make a new solution), calculate the concentration of A1 (Eq. 5).

$$[A1] = [A2] \times 10/0.2 \qquad (5)$$

α-Tocopheryl Acetate Solution   In a 25-mL volumetric flask, weigh approximately 100 mg of α-tocopheryl acetate using an analytical balance. Add isopropanol to the mark. Invert the flask several times to dissolve the sample. Label this solution E1. Calculate its concentration in mg/mL.

For a very precise determination of the concentration you can use UV-visible spectroscopy instead of weights. Use the UV-visible method described below only if indicated by your instructor or if you have reason to believe that your weights are wrong. As in the case of vitamin A, you must dilute solution E1 before measuring its absorbance. To this end, using an automatic pipet, transfer 500 µL of solution E1 to a 10.0-mL volumetric flask. Add isopropanol to the mark. Label this solution E2 and obtain its UV-visible spectrum in the 200–500 nm region using a 1-cm pathlength quartz cuvette and isopropanol as blank. Measure the absorbance at 285 nm (A^{285}). (If the absorbance is higher than 2, take 250 µL of solution E1, dilute it to 10.0 mL in a volumetric flask using isopropanol, and read its absorbance.)

Using Beer's law, the pathlength of 1 cm, the molar absorptivity for α-tocopheryl acetate of 2060 L mol^{-1} cm^{-1} (at 285 nm), and its molecular mass (MM), calculate the concentration of solution E2 in g/L (equal to mg/mL; Eq. 6).

$$[E2](g/L) = \frac{A^{285}}{2060 \text{ L mol}^{-1}\text{cm}^{-1} \times 1 \text{ cm}} \times MM \text{ (g/mol)} \qquad (6)$$

Calculate the concentration of E1 using the dilution factor 0.5/10 (or 0.25/10 if the absorbance was higher than 2 and you had to make a new solution), Equation 7.

$$[E1] = [E2] \times 10/0.5 \qquad (7)$$

The concentrations calculated from the mass of product and from the UV-visible spectrum should not differ by more than 10%.

E10.5.2 Internal Standard Method

9-Anthraldehyde Solution   Using an analytical balance weigh approximately 5 mg of 9-anthraldehyde in a 50-mL volumetric flask. Add isopropanol to the mark and shake to mix. Label this solution AN1. Calculate the concentration in mg/mL.

$$CHO$$

9-anthraldehyde

You can also determine the concentration of this solution by UV-visible spectroscopy. Do this if indicated by your instructor or if you want to double-check the concentration by an independent method. You must dilute this solution before measuring the absorbance. Transfer 1.00 mL of solution AN1 to a 10.0-mL volumetric flask using an automatic pipet and add isopropanol to the mark. Label this solution AN2. Obtain its UV-visible spectrum in the range 250–500 nm using a 1-cm pathlength quartz cuvette and isopropanol as blank. Measure the absorbance at 400 nm (A^{400}). (If the absorbance is higher than 2, take 500 μL of solution AN1, dilute it to 10.0 mL in a volumetric flask using isopropanol and read its absorbance.)

Using Beer's law, the absorbance value at 400 nm, the pathlength of 1 cm, the molar absorptivity of 9-anthraldehyde of 7000 L mol^{-1} cm^{-1}, and its molecular mass (MM), calculate the concentration of AN2 in g/L (equal to mg/mL; Eq. 8).

$$[AN2](g/L) = \frac{A^{400}}{7000 \text{ L mol}^{-1}\text{cm}^{-1} \times 1 \text{ cm}} \times MM \text{ (g/mol)} \qquad (8)$$

Calculate the concentration of AN1 using the dilution factor (1/10) (or 0.5/10 if the absorbance was higher than 2 and you had to make a new solution), Equation 9.

$$[AN1] = [AN2] \times 10/1 \qquad (9)$$

The concentration values calculated from the UV-visible spectrum and from the mass of product should not differ by more than 10%.

Relative Response Factors   In a scintillation vial or a test tube, mix 1.00 mL of solution A1, 3.00 mL of solution E1, and 1.00 mL of solution AN1, using an automatic pipet. Label this solution SM1 (standard mixture 1). Calculate the concentrations of the three components.

Set the detector absorbance range (sensitivity) at 0.5. An absorbance setting of 0.5 means that absorbance values in the range 0–0.5 will be on scale, and absorbance values larger than 0.5 will be off scale.

Using a Hamilton syringe, inject 20 μL of solution SM1 into the injection valve in the load position (Fig. 10.5). Move the handle to the inject position. Leave the valve in the inject position during the run.

If the peaks are off-scale or they are too small, change the absorbance range accordingly and reinject the sample. (If the peaks are off-scale, increase the range; if they are too small, decrease the range). Ideally, the absorbance setting should be in the range 0.1–1.

The elution order of the compounds is 9-anthraldehyde, then retinyl acetate, and finally α-tocopheryl acetate (in case of doubt in the assignment of peaks, inject solutions A1, E1, and AN1, separately). Calculate the area under the peaks for these three compounds by the triangulation method, the cut-and-weigh method (section 7.7), or using an integrator, if available. Calculate the **response factor** R of retinyl acetate and α-tocopheryl acetate relative to 9-anthraldehyde using the following equations:

$$R_{RA} = \frac{A_{RA}}{A_{AN}} \times \frac{C_{AN}}{C_{RA}} \tag{10}$$

$$R_{TA} = \frac{A_{TA}}{A_{AN}} \times \frac{C_{AN}}{C_{TA}} \tag{11}$$

where the subscripts RA, TA, and AN stand for retinyl acetate, α-tocopheryl acetate, and anthraldehyde, respectively; C is concentration in mg/mL in solution SM1, and A is area, in arbitrary units, of the HPLC peaks. Repeat the injection of SM1 two more times and calculate the average response factors.

E10.5.3 Standard Curve Method

You will prepare several diluted solutions (standard mixtures, SM) from stock solutions A1 and E1; you will inject them into the HPLC and determine the areas of the peaks at different concentrations. You will plot concentration versus area and use these plots as your standard curves. To minimize the number of injections, you will mix known volumes of solutions A1 and E1 to produce both standard curves simultaneously. The table below gives the volumes to use. *Prepare all these solutions in 10.0-mL volumetric flasks; use isopropanol to dilute to the mark.*

Standard mixture	Volume of A1 (mL)	Volume of E1 (mL)	C_{RA} (mg/mL)	C_{TA} (mg/mL)
SM2	2.00	2.00		
SM3	3.00	3.00		
SM4	4.00	4.00		
SM5	1.00	6.00		

Knowing the concentration of retinyl acetate (C_{RA}) in solution A1 and that of α-tocopheryl acetate (C_{TA}) in solution E1, calculate C_{RA} and C_{TA} in solutions SM2, SM3, SM4 and SM5.

Set the detector absorbance range (sensitivity) at 0.5. An absorbance setting of 0.5 means that absorbance values in the range 0–0.5 will be on scale, and absorbance values larger than 0.5 will be off scale.

Using a Hamilton syringe, inject 20 μL of the following solutions (one at a time) into the injection valve in the load position (see Fig. 10.5): E1, SM2, SM3, SM4, and SM5. Move the handle to the inject position. Leave the valve in the inject position during the run. If the peaks are off-scale or too small, change the absorbance range accordingly (increase it if the peaks are off-scale; decrease it if they are too small), and reinject the sample. Ideally, the absorbance setting should be in the range 0.1–1. All the samples should be analyzed using the same sensitivity setting.

Using the elution volume of vitamin E from the injection of E1, identify the peaks for vitamin E and A in all the runs. Calculate the peak areas using the triangulation method, the cut-and-weigh method (section 7.7), or an integrator, if available. Plot C_{RA} and C_{TA} in mg/mL versus peak area. Draw (or calculate by the least squares method) the best linear fit for each plot.

E10.5.4 Extraction of Vitamins A and E

Read the original vitamin container label and take as many tablets as needed to obtain approximately 9000 IU of vitamin A and between 30 and 60 IU of vitamin E. Generally, you will need one or two tablets. Do not fraction the tablets, take the whole number of tablets that gives the closest approximation to the IU mentioned above.

Crush the tablets using a mortar and pestle. With the help of a piece of weighing paper, transfer the powder to a 12-mL screw-cap test tube. Transfer the solid carefully,

---

Safety First

- Dimethylsulfoxide (DMSO) is an irritant. It enhances absorption of chemicals through the skin. Handle it with care.

minimizing the losses. Make sure that no solid is left in the mortar. With the help of a pluringe and a Pasteur pipet, add 5.0 mL of dimethylsulfoxide (DMSO) to the test tube. Cap it and shake it for 5 minutes at room temperature. Uncap the tube and place it in a warm water bath (about 40°C). Leave the test tube in the bath for about 5 minutes and swirl it occasionally. DMSO is an excellent solvent for polar and nonpolar compounds alike and extracts all the vitamins from the powder. Add 2.0 mL of water. Let the mixture reach room temperature and then add 4.0 mL of hexanes. Cap the test tube and shake it. Vent the tube immediately after mixing and occasionally afterward by carefully removing the screw cap. Shake it for about 5 minutes. Hexanes remove vitamins A and E from the DMSO-water layer.

Centrifuge the mixture for 3 minutes at 1500 rpm (or higher). Make sure that the centrifuge is properly balanced with another test tube of the same weight located opposite to your tube.

In the meantime, assemble a gravity filtration apparatus with a 15-mL round-bottom flask to collect the liquid, a small funnel (25 mm diameter), a small cotton plug and a spatulaful of anhydrous magnesium sulfate (Fig. 10.12).

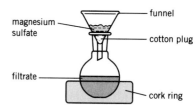

Figure 10.12   Gravity filtration setup.

With the aid of a Pasteur pipet draw the upper layer (hexanes) from the screw-cap test tube. *Keep the tip of the Pasteur pipet at least 5 mm away from the interface to avoid taking any of the lower layer.* Transfer the upper layer to the funnel. If you have accidentally taken some of the lower layer with the Pasteur pipet, squeeze the bulb to deliver the liquid back to the test tube and then rinse the Pasteur pipet several times with the upper layer by squeezing and releasing the rubber bulb until no traces of the lower layer are left in the pipet.

To remove the last traces of the upper layer without drawing any of the lower layer, add 1 mL of hexanes to the screw-cap test tube without disturbing the interface between layers. Transfer the upper layer to the funnel. Repeat this operation with one more mL of hexanes.

Evaporate the solvent in the round-bottom flask using a rota-vap (Fig. 3.17) and a warm water bath (about 50°C). After the liquid has evaporated and an oily residue is left in the round-bottom flask, remove the flask from the rota-vap and blow a gentle stream of nitrogen or helium through it to remove any traces of solvent.

If you are using the **internal standard method**, add with an automatic pipet 3.00 mL of the 9-anthraldehyde solution (AN1) to the flask. If you are using the **standard curve method**, add about 3 mL of isopropanol instead of solution AN1.

Swirl to ensure dissolution of the oily residue. Transfer the liquid to a 5-mL volumetric flask. Rinse the round-bottom flask by adding about 1 mL of isopropanol in two portions of approximately 0.5 mL. Swirl the liquid to remove traces of solution adhered to the round-bottom flask's walls, and transfer the rinses to the 5-mL volumetric flask. Add isopropanol to the mark. Label this solution S (for Sample).

E10.5.5 HPLC Analysis of the Vitamin Extract

Inject 20 μL of the sample solution S into the HPLC. If you are using the internal standard method, go to the next paragraph. If you are using the standard curve method, skip the next four paragraphs.

If you are using the **internal standard method**, identify the peaks for 9-anthraldehyde, retinyl acetate, and α-tocopheryl acetate by comparing the elution volumes for the peaks observed in this run with those obtained with SM1. The peaks should be on scale. If they are not, change the absorbance range and reinject the sample. (The absorbance range setting should be between 0.1 and 1.)

Determine the areas for each peak by the same method as before. Calculate the concentration of retinyl acetate (C_{RA}) and α-tocopheryl acetate (C_{TA}) in mg/mL using the following equations:

$$C_{RA} = \frac{1}{R_{RA}} \times C_{AN} \times \frac{A_{RA}}{A_{AN}} \tag{12}$$

$$C_{TA} = \frac{1}{R_{TA}} \times C_{AN} \times \frac{A_{TA}}{A_{AN}} \tag{13}$$

You can derive these equations from Equations 10 and 11. The symbols have the same meaning as before, but in Equations 12 and 13, A_{RA}, A_{TA}, and A_{AN} refer to the areas of retinyl acetate, α-tocopheryl acetate, and 9-anthraldehyde in the *HPLC run of the sample solution S, instead of the standard mixture. R_{RA} and R_{TA} are the response factors for both vitamins determined in section E10.5.2. C_{AN} in Equations 12 and 13 is the concentration (mg/mL) of 9-anthraldehyde in the sample solution S, which is equal to the concentration in solution AN1 multiplied by the dilution factor 3/5 (remember that 3.0 mL of solution AN1 were added to the 5.0-mL volumetric flask and diluted to the mark with isopropanol).

Calculate the total number of mg for both vitamins in the tablet(s) taking into account that the total volume of sample solution S was 5.0 mL. Repeat the injection of sample solution S at least one more time. Repeat the calculations and find the average values for each vitamin.

If you are using the **standard curve method**, identify the peaks for retinyl acetate and α-tocopheryl acetate in the sample, and determine the area for each peak. Measure the areas at the same absorbance setting as in E10.5.3. With the plots generated in section E10.5.3, calculate the concentrations of both vitamins in the sample solution S in mg/mL. Knowing that the total volume of sample solution S was 5 mL, calculate the total number of mg for retinyl acetate and α-tocopheryl acetate in the tablet(s). Repeat the injection of sample solution S at least one more time. Repeat the calculations and find the average values for each vitamin.

E10.5.6 Chemical Characterization

Transfer approximately 1 mL of solution A1 and E1 to two labeled test tubes, add a drop of concentrated phosphoric acid, and observe any color change.

Using labeled test tubes, collect the eluent from the HPLC corresponding to the vitamin peaks in the sample solution run. Add a drop of concentrated phosphoric acid to each test tube and note any change in color.

EXPERIMENT 10 REPORT

Pre-lab

1. Make a table showing the physical properties (MM, density, solubility, melting point, boiling point, toxicity/hazards, and flammability [for solvents only]) of retinyl acetate, α-tocopheryl acetate, 9-anthraldehyde (only if you are using the internal standard method), methanol, isopropanol, hexanes, DMSO, and phosphoric acid (synonyms: vitamin A acetate, vitamin E acetate). For hexanes, use the values listed for *n*-hexane.

2. Make a flowchart for the isolation and analysis of vitamins A and E.

3. Explain why the hexanes extract does not contain vitamins C or B.

4. Which of the following solvents are not recommended for HPLC analyses using a 254 nm UV-detector? Acetone, isopropanol, methanol, toluene, water, hexane.

5. The peak of retinyl acetate in the HPLC for a mixture of vitamins is almost full scale when an absorbance setting of 1.0 is used and the detection is at 325 nm. What

---

Safety First

- Phosphoric acid is corrosive. Handle it with care.

Cleaning Up

- Dispose of the waste solvent from the HPLC, the standard and sample solutions, and the solutions from the phosphoric acid test in the containers labeled "Vitamins — Liquid Waste."

- Dispose of the DMSO-water layer from the vitamin extraction (E10.5.4) in the container labeled "Vitamins — Liquid Waste."

- Dispose of the sodium sulfate residue in the container labeled "Vitamins — Solid Waste."

- At the end of the experiment, the instructor will empty the rota-vap traps.

absorbance range would you select to see a peak of similar size if the detection is done at 254 nm? The molar absorptivities of retinyl acetate at 325 and 254 nm are 50,800 and 5800 L mol^{-1}cm^{-1}, respectively. Possible absorbance settings: 2; 1; 0.5; 0.2; 0.1; 0.05; 0.02; and 0.01. (Hint: Remember that an absorbance setting of 2, for example, means that absorbance values in the range 0–2 will be on scale; an absorbance setting of 1 will have absorbance values in the range 0–1 on scale, and so on.)

6. Write all the possible resonance forms for the retinyl carbocation.

7. *Internal standard method*: The internal standard method was used to determine the concentration of retinyl acetate in baby formula. The sample solution was injected and the areas for vitamin A (A_{RA}) and 9-anthraldehyde (A_{AN}) (internal standard) peaks measured: $A_{RA} = 550$, $A_{AN} = 840$ in arbitrary units. If the concentration of the internal standard in the sample solution was 0.06 mg/mL and the response factor of retinyl acetate with respect to the internal standard (R_{RA}) was 0.08, calculate the concentration of retinyl acetate in mg/mL in the sample solution.

8. *Standard curve method*: Why is it possible to use the standard curve method to determine concentrations by HPLC while this method is not reliable for GC analyses?

In-lab

1. Calculate the concentrations of all solutions used. Show your work and present the results in a table.

2. *Internal standard method*: Calculate the response factors for retinyl acetate and α-tocopheryl acetate relative to 9-anthraldehyde. Present the data in a table. Include individual determinations as well as the average values.

3. *Standard curve method*: For each vitamin, plot concentration (mg/mL) in solutions SM2–SM5 versus peak areas (in arbitrary units). Draw the best linear fit for each plot and calculate the slopes of the straight lines. (Use a least squares algorithm if available.)

4. Calculate the concentration of retinyl acetate and α-tocopheryl acetate in mg/mL in your sample solution. Calculate the amount of each vitamin in mg and IU per tablet and compare with the label values.

5. If you have analyzed an old vitamin tablet, compare its HPLC with that for a fresh tablet. Indicate how many new peaks are seen and estimate their percentage in the mixture (disregard the internal standard peak for this calculation). Calculate the amounts of vitamin A and vitamin E left in the old tablet. Compare these figures with those from a fresh tablet.

6. Briefly discuss the most important sources of error in this experiment.

Refractometry and Polarimetry

In this unit we will discuss two optical methods commonly used in the laboratory for the characterization of organic chemicals: **refractometry** and **polarimetry**. The former is employed to measure the refractive index, which is an important tool in the analysis of liquid samples, and the latter is used to study chiral compounds with polarized light. Both techniques make use of light as a probe, and thus a brief discussion of the nature of light is included in this unit. The broad subject of light and its interaction with matter is treated in the units devoted to spectroscopy (Units 30–33).

11.1   REFRACTIVE INDEX

Light changes speed and direction as it passes from a medium to another. This phenomenon, called **refraction**, is illustrated in Figure 11.1, where the two different media are represented by m and M. Let's assume that m represents air or a vacuum, and M is a dense medium such as water or glass. When the light ray impinges at an angle α_m on the surface separating both media, it changes direction as it penetrates the denser medium. The new angle is α_M, smaller than α_m (Fig. 11.1a). When the light ray is perpendicular to the surface, it passes through both media without changing direction and no refraction is observed (Fig. 11.1b). The angles α_m and α_M are related to each other by a simple equation known as the **Snell's law** of refraction, which states that the ratio between the sine function of the angles is a constant called the

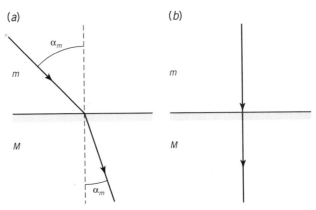

Figure 11.1   Light passing from one medium into another: *a*) refraction when light impinges at an angle; *b*) no refraction when light is perpendicular to the surface.

refractive index of medium M, n (Eq. 1).

$$n = \frac{\sin \alpha_m}{\sin \alpha_M} \qquad (1)$$

The refractive index depends on the temperature and also on the frequency of the light. When a beam of white light, which is a combination of all colors, is refracted it decomposes into a rainbow because each color has a different refractive index. This phenomenon is called *dispersion*.

As with other physical properties such as the melting point or the boiling point, the refractive index is used to identify compounds and investigate their purity. It is also useful in the determination of concentrations. Because the refractive index strongly depends on the frequency of the light and the temperature, it is important to use standardized conditions for its determination. It is customary to measure the refractive index at 20°C using yellow light of wavelength 589 nm, which corresponds to the maximum emission of a sodium lamp (**D line of sodium**). For most organic compounds the refractive index measured under these conditions, and represented by n_D^{20}, is in the range 1.35–1.50. Refractive indices can be measured with an accuracy of ± 0.00001; for example, the refractive index of water at 20°C determined with a precision refractometer using the D line of sodium is 1.33300. The refractive index of most organic liquids decreases with a temperature increase. Typical rates of change are in the range -3.5×10^{-4} to -5.5×10^{-4}/°C. An average temperature coefficient of -4.5×10^{-4}/°C can be used to correct refractive indices of organic compounds due to temperature effects.

11.2 THE REFRACTOMETER

The instrument used to measure refractive indices is called the **refractometer**. The most common refractometer in the organic chemistry laboratory is the Abbé refractometer. Because of its simplicity and accuracy, the Abbé refractometer has withstood the test of time (it was introduced more than a 100 years ago; see book cover). This refractometer operates on the principle of **critical angle of refraction**, explained below.

According to Snell's law, as the angle of the incident light (α_m) increases, so does the angle of the refracted light (α_M), so that the ratio of their sine functions remains constant (Eq. 1). In the limiting case when the incident light angle is 90° the angle of the refracted light reaches its maximum value, the **critical angle** (α^c) (Fig. 11.2). No light coming from medium m can ever reach the region in medium M beyond the critical angle. This creates a boundary between a dark and a light region, as indicated in Figure 11.2. The measurement of the critical angle is the basis for the determination of the refractive index using the Abbé refractometer. Applying Equation 1 to the case when $\alpha_m = 90°$, we obtain:

$$n = \frac{\sin 90°}{\sin \alpha^c}$$

and since $\sin 90° = 1$, it follows that:

$$\boxed{n = \frac{1}{\sin \alpha^c}} \qquad (2)$$

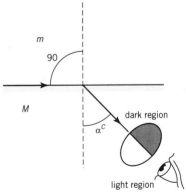

Figure 11.2 Refraction of light at the critical angle.

Therefore, the refractive index n can be easily determined if the critical angle α^c is measured.

A simplified diagram of the internal parts of an Abbé refractometer is shown in Figure 11.3. The sample is placed between two glass prisms, and white light is shone from a tungsten lamp in all directions. The rays of light are refracted by the sample at angles below a critical value, creating a light and a dark region. The rays of light

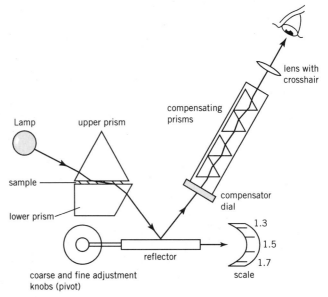

Figure 11.3 Simplified diagram of an Abbé refractometer.

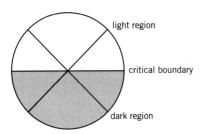

Figure 11.4 Field of view with cross-hairs.

reach a reflector mounted on a pivoted platform that can be moved with the help of adjustment knobs. The adjustment knobs are rotated until the critical boundary is observed at the intersection of the crosshairs on the eyepiece lens (Fig. 11.4). The refractive index is read directly on the scale.

Since the Abbé refractometer uses white light, dispersion into different colors occurs as the rays pass through the prisms. This produces a diffuse critical boundary, which shows the colors of the rainbow and leads to erroneous reading. To correct this problem and obtain a precise reading, the critical boundary is sharpened with the help of compensating prisms, also known as **Amici prisms**. These prisms converge the dispersed radiation back into white light without affecting the trajectory of the yellow light corresponding to the D line of sodium.

The accuracy of the Abbé refractometer is about ±0.0002. This level of accuracy requires a temperature control of ±0.2°C. The Abbé refractometer is simple to operate, can be used for a wide range of refractive indices (n_D = 1.3–1.7), and requires very small samples (less than 0.05 mL). It is not recommended for solutions that contain volatile components or for powders.

11.3 MEASURING THE REFRACTIVE INDEX

A picture of a commercial Abbé refractometer is shown in Figure 11.5. It is used to determine the refractive index of transparent liquids and solids.

Liquid Samples

Familiarize yourself with the refractometer that you will use. Several brands are available on the market and although they are all very similar, operational details may be different. Here you will find instructions for the most common type of Abbé refractometer. For determinations at temperatures other than ambient, connect the refractometer to a thermostatic bath using the nipples for hose attachment.

1. With the help of the lever, swing the upper prism sideways. Apply a drop of the liquid (approximately 0.05 mL) on the lower prism. If the sample is too viscous, spread it over the prism using a wooden applicator. Avoid contact between the

prism and glass or metal objects (such as Pasteur pipets and spatulas) because the surface of the prism is easily scratched.

2. Lower the upper prism and secure it in place.

3. Position the light source above the upper prism.

4. Momentarily depress the contact switch (not shown in Fig. 11.5) and look through the eyepiece. This will allow you to see the scale through the eyepiece. Rotate the coarse adjustment knob to set the scale near the expected refractive index value.

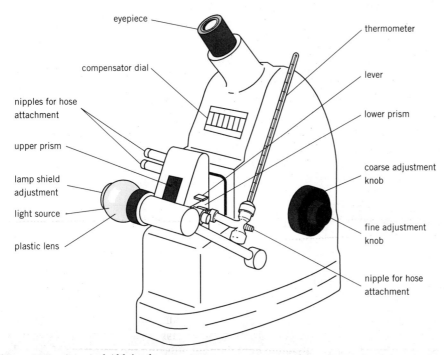

Figure 11.5 A typical Abbé refractometer.

5. Release the contact switch and bring the critical boundary into the field of view by rotating the coarse knob. Rotate the lamp shield to obtain optimum contrast between the dark and light regions. If the boundary appears colored, adjust the position of the compensator dial until the boundary appears almost achromatic.

6. With the help of the coarse and fine knobs, bring the critical boundary to the intersection of the crosshairs (Fig. 11.4).

7. Depress the contact switch and read the refractive index. Estimate the fourth decimal place.

8. Record the temperature.

9. Clean the prisms immediately after use with lens tissue dampened with water or ethanol. Soaps and detergents are not recommended because they fog the glass.

10. Accumulation of dust on the prism surface should be avoided. When not in use keep the prisms closed.

Solid Samples

1. The prism and sample surfaces must be scrupulously clean.

2. To ensure good contact between sample and prism, place a small drop (about 0.01 mL) of a contact liquid such as 1-bromonaphthalene on the lower prism. Other contact liquids can be used provided they have a refractive index higher than the

sample itself (for 1-bromonaphthalene, $n_D = 1.66$). Diiodomethane ($n_D = 1.74$) is recommended for refractive indices in the range 1.64–1.72.

3. Place the solid sample on the lower prism and move it around very gently to spread out the liquid.

4. Cut a small piece of white paper, roll it up in the shape of a one-layer cylinder, and place it between the lamp and the plastic lens in the light source. This paper screen is necessary to diffuse the light and obtain accurate readings (paper diffusers are not necessary for liquid samples).

5. Keep the upper prism swung open and bring the light source in line with the level of the lower prism.

6. Proceed as indicated in 4–10 for liquid samples.

11.4 POLARIMETRY

We are all familiar with polarized sunglasses. When we put them on, the surface of the ocean, lakes, and roads becomes clearer as the glare is eliminated; clouds that were barely visible with the naked eye become apparent. How is this possible? What is polarized light? What is the connection to organic chemistry? To answer these questions let's take a look at the nature of visible light.

Light can be described as waves of electromagnetic radiation with the *electric field* (E) and the *magnetic field* (H) vector components oscillating perpendicular to each other and also perpendicular to the direction of propagation (Fig. 11.6).

The electric field component of light is responsible for most of the effects that we observe when light interacts with matter. For this reason we will focus our discussion exclusively on the electric field, and will ignore, for now, the magnetic field.

The light emitted by the sun or by a fluorescent bulb has the electric field vector vibrating in all possible directions perpendicular to the direction of propagation. Such light is called *unpolarized*. When unpolarized light passes through special compounds, called **polarizers**, the intensity of the electric field vector vibrating *in a certain* direction is unchanged while the electric field vectors vibrating in all other directions are attenuated. The emerging light has the electric field vector vibrating in only one direction and is called **linearly polarized** or **plane-polarized light**. The privileged direction is determined by the molecular structure and orientation of the polarizer (Fig. 11.7).

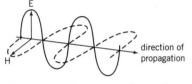

Figure 11.6 Representation of an electromagnetic wave.

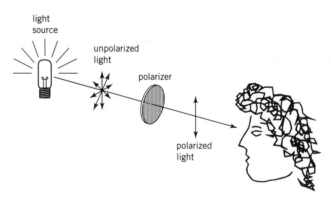

Figure 11.7 Polarization of light as it passes through a polarizer.

To understand how this process works, let us assume that unpolarized light shines on a polarizer, which has its privileged direction in the vertical position. The electric field vector (E in Fig. 11.8) can be decomposed into two components, one

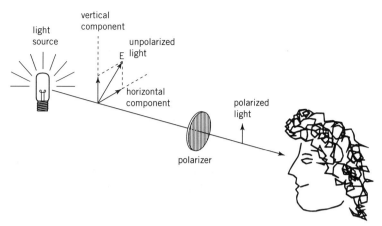

Figure 11.8 The electric vector decomposed into a vertical and a horizontal component. Only the vertical component, parallel to the polarizer's axis, passes through.

parallel to the privileged direction (vertical component), and the other perpendicular to it (horizontal component). The vertical component is not perturbed by the polarizer and passes through unchanged. The horizontal component, on the other hand, is absorbed by the polarizer and does not pass through. As a result, only light vibrating in the privileged vertical direction emerges from the polarizer. Because both vertical and horizontal components of unpolarized light are equally intense on average, the intensity of the emerging polarized light is half the intensity of the incident light.

If a second polarizer, called the **analyzer**, is placed in the path of linearly polarized light, the intensity of the light emerging from the analyzer will depend on the relative orientation of both polarizers. When the axis of the analyzer is parallel to the axis of the polarizer, the light impinging on the analyzer passes through unchanged (Fig. 11.9a). If the analyzer is rotated 90° with respect to the polarizer, the electric field vector is perpendicular to the axis of the analyzer and the light does not pass through. If the analyzer is at an intermediate angle, the light is partially absorbed by the analyzer.

You can test this at home with two pairs of polarized sunglasses, which are usually made of specially treated plastics known as Polaroid. Put one pair in front of the other facing any source of light. Rotate one with respect to the other and observe the light intensity. When the sunglasses are crossed almost no light will pass through. For an explanation on how polarized sunglasses reduce glare see the box "Cool shades!" at the end of this unit.

What Figure 11.9 shows is a primitive **polarimeter** that can be used to study the interaction of polarized light with chiral molecules. A chiral molecule is not superimposable with its mirror image. The two nonsuperimposable mirror images are called **enantiomers**. The most common cause of chirality, although not the only one, is the presence of a carbon atom attached to four different groups. Such carbon atoms are called **stereogenic**. For example, carbon 2 of 2-bromobutane is stereogenic (it is bonded to a methyl group, an ethyl group, a bromine, and a hydrogen) and the molecule is chiral. This compound can exist as two enantiomers. The configuration of stereogenic carbons is specified as R or S according to the Cahn-Ingold-Prelog rules. Consult your lecture textbook to review these rules.

CH$_2$CH$_3$ CH$_2$CH$_3$

C   C

H CH$_3$ CH$_3$ H

Br   Br

(R)-2-bromobutane   (S)-2-bromobutane

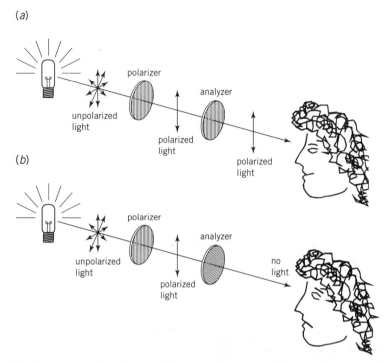

Figure 11.9 Polarizer and analyzer at different angles: *a*) parallel to each other, the light passes through; *b*) perpendicular to each other, no light emerges from the analyzer.

Enantiomers have the same physical properties (melting point, boiling point, density, etc.) but differ in their interactions with other chiral molecules and with polarized light. Chiral molecules change the plane of polarization of polarized light. When polarized light penetrates a chiral substance, the interaction of the electric field vector with the electron cloud of the chiral molecules is not symmetrical and, as a result, the electric field vector changes orientation. This effect is cumulative; the larger the number of chiral molecules in the path of the polarized light, the larger the change in the orientation of the plane of polarization.

This effect can be measured by placing the chiral substance between a polarizer and an analyzer. In the absence of a chiral substance, maximum intensity is observed when polarizer and analyzer are lined up (Fig. 11.9*a*), but when a chiral substance is placed in between the polarizer and the analyzer (Fig. 11.10), it changes the plane of polarization of light. To observe maximum intensity, now the analyzer must be rotated to make its axis parallel to the plane of the polarized light emerging from

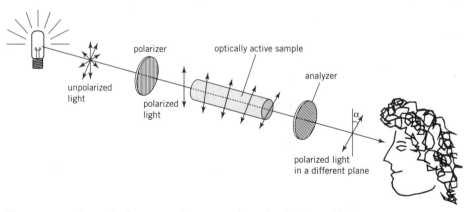

Figure 11.10 An optically active substance in the path of polarized light.

the sample. Compounds that rotate the plane of polarized light are called **optically active.** Those that rotate the plane of polarized light clockwise (to the right) are called **dextrorotatory,** represented by d, and those that cause the polarized light to rotate counterclockwise (to the left) are called **levorotatory**, represented by l. Enantiomers rotate the plane of polarized light in opposite directions. If one rotates the plane to the right, the other does it to the left.

Optical activity is measured by the angle, α, that the analyzer must be rotated to compensate for the rotation caused by the sample; this angle is called the **optical rotation**. The optical rotation increases with sample concentration, C, and the pathlength of the tube containing the sample, ℓ. The optical rotation per unit of concentration and length is called the **specific rotation** and is represented by $[\alpha]$ (Eq. 3).

$$[\alpha] = \frac{\alpha}{C \times \ell} \tag{3}$$

The unit for optical rotation is degree. By convention, the optical rotation is positive for dextrorotatory compounds and negative for levorotatory ones. The concentration, C, is in g/mL, and the length of the tube is in decimeters (1 dm = 10 cm = 0.1 m). The unit for specific rotation is given as degree by most handbooks. This is an unfortunate oversimplification. A dimensional analysis of Equation 3 indicates that the units of the specific rotation must be $degree \times (concentration\ unit)^{-1} \times (length\ unit)^{-1}$. Because the unit for concentration is g/mL, and for length is dm, the actual units of specific rotation are: $degree \times (g/mL)^{-1} \times dm^{-1}$.

The specific rotation depends on the structure of the chiral compound, the wavelength of the polarized light, the temperature, and the solvent. The light commonly used for polarimetric studies is the D line of sodium at 589 nm. The temperature, the solvent, and the concentration are generally specified next to the specific rotation. The effect of solvent and temperature can be rather large and should not be overlooked. On the other hand, the effect of the concentration on the specific rotation is small and, in principle, can be ignored. As an example, the specific rotation of cholesterol, a chiral natural product, is:

$$[\alpha]_D^{20} - 39.5° (c = 2\ in\ chloroform)$$

This means that the specific rotation was measured at 20°C using the D line of sodium from a 2 g/100 mL solution of cholesterol in chloroform. Notice that the concentration here is given in different units than those of Equation 3. While the concentration written next to the specific rotation is usually given in g/100 mL, C in Equation 3 is in g/mL.

To determine the specific rotation of **pure liquids**, the concentration C in Equation 3 is replaced by the density of the liquid, d, in g/mL (Eq. 4):

$$[\alpha] = \frac{\alpha}{d \times \ell} \tag{4}$$

11.5 THE POLARIMETER

A schematic diagram of a simple polarimeter is shown in Figure 11.11. To measure the optical rotation the sample is placed in a polarimeter tube of known length. Most tubes have two removable windows kept in place with the help of rubber gaskets and screw caps. One of the ends is usually wider than the other to accommodate any trapped air bubble away from the path of light (Figs. 11.11 and 11.12).

Most polarimeters use Nicol prisms as polarizers and analyzers, P and A in Figure 11.11a. A Nicol prism consists of two calcite (a form of calcium carbonate) crystals glued together by a layer of Canada balsam. The analyzer is attached to a scale

(a)

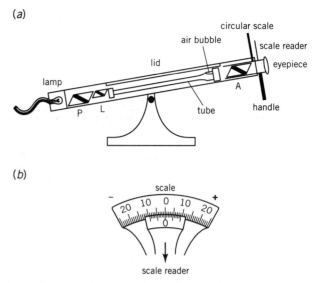

(b)

Figure 11.11 *a*) A schematic diagram of a polarimeter; *b*) partial view of scale and scale reader showing the sign of the rotation.

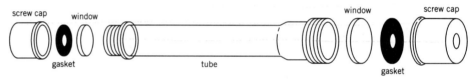

Figure 11.12 A typical polarimeter tube.

reader that is free to rotate over a circular scale where the optical rotation is measured from −90° to +90°.

Most polarimeters have an auxiliary Nicol prism, L in Figure 11.11*a*, which covers half the field of the polarizer P. This auxiliary prism has its plane of polarization slightly inclined to that of the polarizer P. The reason for this prism is to facilitate the reading. The human eye is not very adept at determining absolute brightness. If one rotates the analyzer with respect to the polarizer while searching for the point of maximum brightness, the measurement would have, on average, an error of ±5°, which is unacceptable for most applications. The auxiliary prism, L, divides the field into two halves and facilitates the location of the **end point**. *The end point is reached when both fields have equal brightness (Fig. 11.13b). If the analyzer is moved clockwise, one field becomes darker than the other (Fig. 11.13a), and if it is rotated in the other direction the darker field becomes brighter (Fig. 11.13c).*

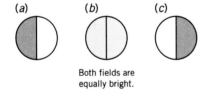

Both fields are equally bright.

Figure 11.13 The two fields of a polarimeter. The end point is reached when both fields are equally bright (b).

11.6 MEASURING THE OPTICAL ROTATION

Familiarize yourself with the polarimeter you will use. Although all polarimeters work on the same principle, their operational details are different. Turn the polarimeter on approximately 10 minutes before use because the lamp requires a few minutes to warm up. In the meantime prepare the solutions to be measured. First decide what tube you will use. They come in different lengths and thus they take different volumes. The most common ones are 1- and 2-dm long. Most 2-dm tubes take about 10 mL of solution. Weigh the appropriate amount of sample to make a 0.1–0.5 g/mL solution in a volumetric flask. Add solvent to the mark and shake well.

Determine the "zero" of the instrument with an empty polarimeter tube with both windows in position. When handling the windows, hold them by the edges. For a more precise determination of the "zero," the cell should be filled with the sample's solvent; however, this is not necessary for routine applications. As you look through the eyepiece, move the scale reader to the right and to the left around the zero mark, until both fields are equally bright (end point, Fig. 11.13b). *If you rotate the analyzer to the right or to the left of the end point you must see a sharp change in the relative brightness of the fields as already discussed* (Fig. 11.13a,c). Read the angle on the scale (it should be close to zero). **Note its sign** as shown in Figure 11.11b. Move the scale reader out of position, again locate the end point, and read the angle on the scale. Repeat this operation at least once more and take the average of your readings. This will be your "zero."

Unscrew the wider screw-cap and fill the tube with the help of a Pasteur pipet. The liquid should fill the tube to the rim. Position the other glass window by horizontally sliding it in place. This avoids trapping a big air bubble in the tube. Screw the cap with the rubber gasket in position. With a piece of lens paper, remove any trace of liquid from the windows.

Position the polarimeter tube inside the polarimeter. Tilt the body of the polarimeter to make sure that any trapped air bubble is in the wider section of the tube and out of the light path (Fig. 11.11a). Determine the optical rotation of the sample by rotating the scale reader right and left until you reach the end point where both fields have the same intensity, as already explained (Fig. 11.13). Read the optical rotation on the scale and note its sign as shown in Figure 11.11b. Move the scale reader from the end point position and relocate the end point. Repeat this operation at least once more, and take the average of your readings. Correct the optical rotation by subtracting the "zero" reading. After you are done with your measurements, clean the tube and the glass windows with ethanol.

When one reads the optical rotation on the polarimeter scale, one cannot distinguish, in principle, between a positive and a negative rotation. For example, a reading of $+15°$ may correspond to a positive rotation of $+15°$ or a negative rotation of $-345°$ ($-360° + 15°$). How do we know which one is the real rotation? In general, only very concentrated samples with unusually high specific rotations give optical rotations with absolute values larger than $90°$. A reading of $+15°$ in most cases is likely to correspond to an optical rotation of $+15°$ rather than $-345°$. However, if one wants to find out the optical rotation unequivocally, one must take measurements at two different concentrations. For example, the sample in the previous case can be diluted 10 times, and another measurement taken. According to Equation 3, by diluting the sample 10 times the optical rotation should also decrease 10 times. If the true optical rotation was $+15°$, the reading after a tenfold dilution would be $+1.5°$; on the other hand, if the optical rotation was $-345°$, the reading after dilution would be $-34.5°$.

11.7   OPTICAL ROTATION OF MIXTURES

The optical rotation of a mixture of optically active compounds is, in the first approximation, the sum of the individual optical rotations. For a mixture of two optically active compounds A and B, the total optical rotation, α_T, is:

$$\alpha_T = [\alpha]_A \times C_A \times \ell + [\alpha]_B \times C_B \times \ell \qquad (5)$$

where $[\alpha]_A$ and $[\alpha]_B$ are the specific rotations of A and B, and C_A and C_B are their concentrations. Dividing both sides of Equation 5 by the total concentration, C_T ($C_T = C_A + C_B$) and the pathlength of the tube, Equation 6 is obtained:

$$\frac{\alpha_T}{C_T \times \ell} = \frac{[\alpha]_A \times C_A \times \ell + [\alpha]_B \times C_B \times \ell}{C_T \times \ell} \qquad (6)$$

The left side of Equation 6 is the specific rotation of the mixture $[\alpha]_T$. The ratios C_A/C_T and C_B/C_T are the percentages (divided by 100) of A and B in the mixture. Thus, the specific rotation of the mixture is the sum of the specific rotations of the components multiplied by their proportions in the mixture (Eq. 7):

$$[\alpha]_T = [\alpha]_A \times \frac{\%A}{100} + [\alpha]_B \times \frac{\%B}{100} \qquad (7)$$

For example, according to Equation 7, the optical rotation of an equal-amount mixture of two enantiomers (a racemic mixture) is zero because the optical rotations of both enantiomers cancel out.

Cool Shades!   How can we explain the reduction of glare by polarized sunglasses? There are different mechanisms for light polarization; selective absorption by a polarizer, as already discussed, is one of them, but polarization also occurs when light reflects on the smooth surface of windows, wet pavement, and lakes. When unpolarized light shines on these surfaces, the electric field vector of the reflected light vibrates mainly in the direction perpendicular to the plane of incidence, and the reflected light is polarized. Maximum polarization of light reflected over bodies of water is observed when the sun is about 40° above the horizon. If we look directly at these surfaces through polarized glasses the glare is eliminated because the polarized sunglasses have their axis in the vertical direction and do not let the polarized reflection pass through.

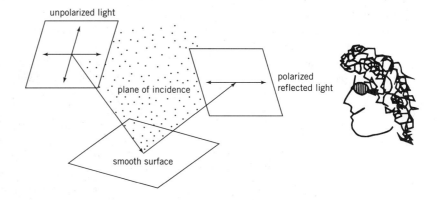

Sky light is also polarized due to scattering by air molecules. As sunlight enters the atmosphere, interaction with air molecules causes scattering in all directions. The scattered light is partially polarized with a maximum polarization found in a direction at 90 degrees from the sun (north and south). We can check this by looking at the sky in the north–south direction (preferentially when the sun is low) through a polarizer. If we rotate the polarizer we will observe that the light intensity alternates from high to low, indicating that the sky light is polarized. We are normally insensitive to light polarization, but bees and other insects use the natural polarization of sky light as a navigational tool.

BIBLIOGRAPHY

1. Techniques of Chemistry. Physical Methods of Chemistry. Vol. 1. Part IIIC. Polarimetry. A. Weissberger, ed. Wiley-Interscience. New York, 1972.

2. Introduction to Classical and Modern Optics. J.R. Meyer-Arendt. 2nd. ed. Prentice Hall. Englewood Cliffs, New Jersey, 1984.

3. Light and Color in the Outdoors. M. Minnaert. Springer-Verlag. New York, 1993.

Problem 1

The critical angles for crown glass, Plexiglas, and diamond (using the D line of sodium) are: 41° 2′ 28″; 42° 9′ 19″; and 24° 24′ 27″; respectively. Calculate the index of refraction for each substance.

Problem 2

A sample of oil of cedar leaf (3.5 g) is dissolved in acetone and the final volume is 25 mL. The optical rotation of the solution, using a 2-dm tube, is −4.2°. Calculate the specific rotation of the oil.

Problem 3

The oil of Siberian fir contains *l*-pinene, *l*-camphene, and *l*-bornyl acetate as the main components. The oil of cypress also has pinene, camphene, and bornyl acetate as the major ingredients. The specific rotation for the cypress oil is +12° $(g/mL)^{-1}$ dm^{-1}. What is the main difference between these two oils? If you mix the two oils, would you expect an increase, a decrease, or no change in the absolute value of the specific rotation?

Problem 4

A perfumer wants to make a mock caraway oil by mixing its main components, *d*-limonene and *d*-carvone. If the specific rotation of the mock oil should be +77° $(g/mL)^{-1}$ dm^{-1}, and the specific rotations of *d*-carvone and *d*-limonene are +62 and +100° (g/mL) dm^{-1}, respectively, in what proportions should the perfumer mix the two components?

Problem 5

By mistake a chemist mixes 25 mL of a solution of (−)-tartaric acid (2 g/100 mL) with 15 mL of a solution of (+)-tartaric acid (0.2 g/mL). The specific rotations of the pure enantiomers are −12 and +12° $(g/mL)^{-1}$ dm^{-1}, respectively. Calculate the optical rotation of the solution made by mistake if a 2-dm tube is used.

Problem 6

(*Previous Synton: Problem 5, Unit 7.*) When Synton was working as an intern in a forensic lab, one of his least pleasurable assignments was to characterize a sample of equilin, a steroidal hormone isolated from the urine of pregnant mares. He made a solution of equilin (0.45 g/mL), measured the optical rotation using a 2-dm tube, and read a value of −67.5° on the polarimeter scale. When he compared the specific rotation with the literature value (+325° $(g/mL)^{-1}$ dm^{-1}), he thought that there was something very wrong with his sample. Later in the day, while watching a rerun of "*Mr. Ed*," he suddenly realized that he had been careless in his calculation, and nothing was wrong with the horse sample. What was Synton's mistake?

equilin

EXPERIMENT 11A

Analysis of Essential Oils

E11A.1 TERPENOIDS

Terpenes encompass a large family of organic compounds widespread in nature and occurring in all organisms from bacteria to mammals. They are prevalent in higher plants, where they act as volatile chemical messengers to attract insects and also to defend the plant's territory. Mixtures of volatile and odoriferous compounds, very rich in terpenes, can be obtained by steam distillation of plant tissues. These mixtures are called **essential oils**.

From a structural standpoint all terpenes derive from the condensation of **isoprene** units. Isoprene (2-methyl-1,3-butadiene) is a five-carbon diene shown below. Carbons 1 and 4 of isoprene are called the "head" and the "tail" of the isoprene unit, respectively. Most terpenes are the product of "tail-to-head" condensations of isoprene units; this is the so-called **isoprene rule**. These condensations take place in nature with the help of biological catalysts (enzymes), and the final number of double bonds and their location vary from compound to compound and cannot be predicted by the isoprene rule. Oxidized terpenes in the form of ethers, alcohols, and ketones are often found in nature. The whole family of terpenes, including the oxygenated compounds, is called **terpenoids**. The isoprene rule for the cases of myrcene and limonene is illustrated below.

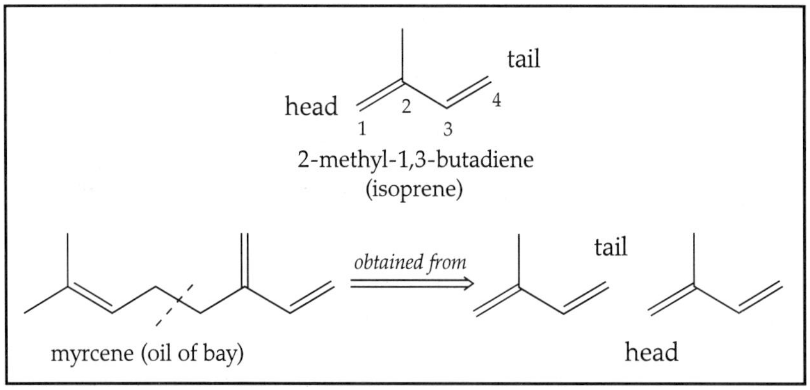

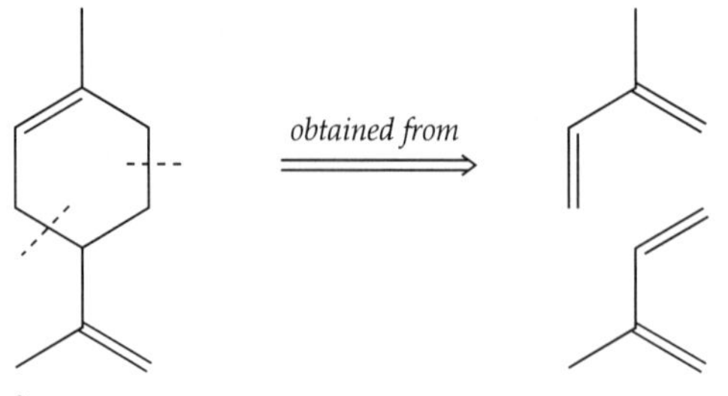

Terpenes are classified according to the number of carbon atoms. **Monoterpenes** have two isoprene units, and a total of 10 carbon atoms. Examples of monoterpenes include α- and β-pinene, cineole (also called eucalyptol), carvone, limonene, menthol, and citral.

Some terpenoids found in essential oils

(S)-(+)-carvone   (R)-(+)-limonene   *l*-menthol   *l*-menthone   (−)-α-pinene   (−)-β-pinene

cineole   myrcene   citral a (geranial)   citral b (neral)

Sesquiterpenes have 15 carbon atoms or three isoprene units. Farnesene and farnesol are examples of sesquiterpenes found in citronella oil; patchouli alcohol is the major component of patchouli oil, an oriental fragrance very popular in the 1960s.

farnesol (citronella oil)   patchouli alcohol

Terpenes with 20, 30, and 40 carbon atoms are called **diterpenes**, **triterpenes**, and **tetraterpenes**, respectively. Terpenes with 25 carbon atoms, **sesterterpenes**, are rarely found in nature. Unlike mono- and sesquiterpenes, which occur primarily in plants, di-, tri-, and tetraterpenes are found in animals. An example of a triterpene is squalene, an intermediate in the biosynthesis of cholesterol; it can be isolated in large amounts from the oil of shark liver. Squalene can be totally hydrogenated to give the corresponding alkane, squalane, which is widely used as a stationary phase for gas chromatography. Examples of tetraterpenes include carotenes and xanthophylls (Unit 8).

squalene (shark liver oil)

E11A.2 ESSENTIAL OILS

Essential oils obtained by steam distillation from bark, leaves, seeds, and other plant tissues are complex mixtures of volatile compounds (for the isolation of essential oils see Units 6 and 14). These compounds are synthesized by plants as part of their chemical weaponry and many of them have insecticide or herbicide properties. The composition of essential oils is unique to a given species; however, variations due to climate, terrain, and environment are often observed. Essential oils can be fully analyzed by gas chromatography or HPLC. Because of the high volatility of their components, GC is often the preferred method of analysis, especially in combination with mass spectrometry (GC-MS).

Table 11.1 gathers information about seven common essential oils. Figure 11.14 is a diagram representing specific rotation and refractive index ranges for the oils. Each oil shows in a different region of the $[\alpha]_D - n_D$ diagram. Only caraway and celery oil overlap in a small region. Thus, it is relatively easy to identify these oils by measuring their specific rotations and refractive indices. The combination of these two simple and inexpensive techniques is an invaluable tool in the study and characterization of essential oils. We can obtain further information about the oils, such as their chemical composition, by GC analysis.

Table 11.1 **Characteristics of Some Essential Oils**

Oil	Major components	Minor components	$[\alpha]_D$ (20°C) (°(g/mL)$^{-1}$dm^{-1})	n_D (20°C)	Source
caraway seed	d-carvone d-limonene	—	+70 to +80	1.484–1.497	Carum carvi
celery seed	d-limonene	β-pinene, myrcene	+48 to +78	1.480–1.490	Apium graveolens
eucalyptus	cineole	α- and β-pinene	−5 to +5	1.458–1.470	Eucalyptus globulus
lemongrass	citral a (geranial) citral b (neral)	limonene	−3 to +1	1.483–1.489	Cymbopogon citratus or Cymbopogon flexuosus
peppermint	l-menthol l-menthone	limonene, cineole, α- and β-pinene	−18 to −32	1.459–1.465	Mentha piperita
sweet orange	d-limonene	carvone α- and β-pinene	+94 to +99	1.472–1.474	Citrus sinensis
spearmint	l-carvone l-limonene	α- and β-pinene	−48 to −59	1.484–1.491	Mentha spicata or Mentha cardiaca

PROCEDURE

Safety First

- Acetone is a flammable solvent.
- d-Limonene is a possible carcinogen.

Background reading: 11.1–7
Estimated time: 3 hr.

In this experiment you will analyze two unknown essential oils. They will be among the ones listed in Table 11.1. You will identify the oils by measuring their optical rotations and refractive indices. You will also analyze the oils by GC to determine their chemical composition.

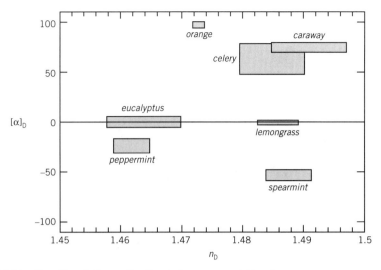

Figure 11.14   Specific rotation-refractive index representation for some essential oils.

E11A.3   SPECIFIC ROTATION

In 25.0 mL volumetric flasks weigh approximately 2.5 (±0.001) g of each oil and add acetone to the mark. Measure the optical rotation (α) by using a 2-dm pathlength tube. Before you load the tube with the solution, carefully rinse it with a small aliquot of the same solution. Calculate the specific rotation $[\alpha]_D$ with the help of Equation 3, section 11.4.

E11A.4   REFRACTIVE INDEX

Apply one drop of oil on the surface of the lower prism of the refractometer and secure the upper prism in place. Do not scratch the surface of the prism. Pasteur pipet tips, glass droppers, spatulas, and glass rods should not touch the prism. Determine the refractive index of each oil with the refractometer set at 20°C, following the directions outlined in section 11.3. Between determinations clean the prism surfaces with a few drops of ethanol and lens tissue.

Using the specific rotations and refractive indices and with the aid of Figure 11.14, identify the oils.

E11A.5   COMPOSITION OF ESSENTIAL OILS

Inject 0.2 µL (or an appropriate volume as indicated by your instructor) of each essential oil into a GC column of medium polarity (for example, SF-96 at 125–150°C). Identify the most prominent components in each oil by matching the most important peaks in each chromatogram with the major constituents listed in Table 11.1. When there is more than one major component, such as in the case of caraway, lemongrass, peppermint, and spearmint oil, base your peak identification on the boiling points of the components. Assume that the retention time increases as the boiling point increases. The normal boiling points for the compounds found in the essential oils are given in Table 11.2. If available, standards for the compounds can be injected to help in the identification process; however, for the oils listed in Table 11.1 the use of standards is not necessary.

Table 11.2 Normal Boiling Points of Some Terpenoids Found in Essential Oils

Compound	bp_{760} (°C)
α-pinene (d, l, or dl)	155–156
β-pinene (d, l, or dl)	164–165
myrcene	167
limonene (d, l, or dl)	175–176
cineole	176–177
l-menthone	207
l-menthol	212
citral (a or b)	≈229[a]
carvone (d, l, or dl)	≈230

[a]Citral b has a slightly lower bp than citral a.

Using boiling point considerations, identify the minor components in each oil. To see some of the minor components you may need to decrease the attenuation. Determine the composition of each oil by integrating the chromatograms.

BIBLIOGRAPHY

1. Essential Oils Analysis by Capillary Gas Chromatography and Carbon-13 NMR Spectroscopy. V. Formacek and K.-H. Kubeczka. Wiley, Chichester, 1982.
2. Analysis of Essential Oils by Gas Chromatography and Mass Spectrometry. Y. Masada. Wiley, New York, 1976.
3. The Encyclopedia of Essential Oils. J. Lawless. Barnes and Noble, New York, 1992.

EXPERIMENT 11A   REPORT

Pre-lab

1. Complete Table 11.1 by giving the densities of the oils. The oils can be found in the *Merck Index*. List uses and toxicity.
2. Explain why *lemongrass oil* and *eucalyptus oil* have specific rotations close to zero.
3. Consider the compounds shown in the box on page 251. How many stereogenic centers are in each one?
4. Speculate why carvone has a higher boiling point than menthone.

In-lab

1. Report the optical rotation and concentration of each oil. Calculate the specific rotation. Present your data in a table format.
2. In the same table, report the measured refractive indices.
3. Indicate where each oil falls in the specific rotation-refractive index diagram (Fig. 11.14). Identify the oils.
4. List the major and minor components you were able to identify in each oil. Give their retention times. Present the data in a table format.
5. Plot the retention times for the compounds identified versus their corresponding boiling points and draw the best curve that fits the data. Explain your results briefly.
6. Calculate the % composition of each oil. In calculating the total area, also consider those peaks that you were not able to identify. Do not consider solvent and air peaks in your calculations.
7. Discuss your data.

Separation of Carvone and Limonene

In this experiment, you will separate carvone and limonene, from either spearmint or caraway oil, by column chromatography. You will use a microcolumn with silica gel and an eluotropic series of hexanes and acetone. You will collect several fractions and analyze them by gas chromatography. You will also measure the optical rotation of your crude oil and, based on the sign of the optical rotation, confirm whether the oil is from spearmint or caraway. Before you go on, read the section on terpenoids (E11A.1) and essential oils (E11A.2).

E11B.1   CARVONE AND LIMONENE

Carvone is a naturally occurring ketone found in the essential oils of caraway, dill, and spearmint in association with other terpenoids such as **limonene**. Limonene is found in spearmint, caraway, lemon, and orange oils. Carvone and limonene, both monoterpenes, have only one stereogenic center and therefore they can exist in two enantiomeric forms, R and S. Enantiomers have the same physical and chemical properties, except that they interact differently with polarized light and other chiral molecules. The two enantiomers of carvone have different smells, one fresh and minty, and the other sweet and somehow unpleasant, especially at high concentrations. The olfactory receptors that line our nasal mucus are chiral ensembles with an exquisite sensitivity to the size, shape, and chirality of the odorant. Their interaction with individual enantiomers is often specific, resulting in the production of distinctive smells. The odor difference between the two enantiomeric forms of carvone is obvious to most people.

(S)-(+)-carvone
d-carvone
$[\alpha]_D = +62°$ (g/mL)$^{-1}$ dm^{-1}

(R)-(+)-limonene
d-limonene
$[\alpha]_D = +100°$ (g/mL)$^{-1}$ dm^{-1}

(S)-(+)-carvone, also known as d-carvone, is found in association with (R)-(+)-limonene (d-limonene) in caraway oil, which is **dextrorotatory**. The composition of the oil, obtained by steam distillation from dried, ripe fruits of *Carum carvi* L., varies according to the source of the fruits but normally contains 50–55% carvone and 40–45% limonene. (R)-(−)-carvone, also known as l-carvone, along with (S)-(−)-limonene (l-limonene) are the constituents of spearmint oil, which is **levorotatory**. Spearmint oil is obtained by steam distillation from the partially dried leaves of the flowering plants *Mentha spicata* L. Spearmint oil normally contains 13–21% limonene and 46–70% carvone and minor components such as α- and β-pinene (4–6%) and dihydrocarvone (3–12%).

The measurement of the optical rotation of the crude oils will help you identify the oil. A positive optical rotation implies caraway oil because this oil is made of (S)-(+)-carvone plus (R)-(+)-limonene. A negative optical rotation implies spearmint oil, made of (R)-(−)-carvone plus (S)-(−)-limonene.

> Of all the senses, none has been more mystified, albeit with unequal parts of glorification and vilification, than the sense of smell. Condemned as earthly and sensual by the early Christians, regarded as primitive by the philosophers of materialism, and praised as a catalyst for nostalgia by the romantics, the sense of smell has enjoyed a controversial status. Today, with an interest in the possible effects of odors on our moods and well-being, a new appreciation for the sense of smell has emerged. Perception of odors is one of the most interesting and active areas of research in neurobiology. Why do some molecules have pleasant smells, while others are offensive, and still others do not have any detectable odor at all? Odor is the result of the interaction between volatile molecules and the olfactory receptors. One of the earliest attempts to correlate odor with molecular structure suggested that, like the colors of the spectrum, there are some basic odors, such as floral, putrid, and ethereal, which produce all other odors when mixed. According to this theory, the olfactory receptors in the nose are geometrically designed to interact only with certain molecules that fit into them as keys in their locks. Although this theory has a kernel of truth, and receptors "recognize" the odorants by their molecular structures, modern research shows that there are no basic odor types, but rather many different types of receptors with varied sensitivity to numerous odorants. In general, many receptors are involved in the perception of a particular odor, and some receptors are more sensitive to certain odorants than to others. The olfactory receptors are specialized neurons that line the olfactory epithelium located in the rear of the nasal cavity. It is estimated that with a total of 6 to 12 million receptors divided into 500 to 1000 different types we are able to recognize 5000 different odors.

PROCEDURE

> **Background reading:** 11.4–7; 9.1–3.
> **Estimated time:** separation: 1 hr;
> analysis: 1 hr.

E11B.2   ANALYSIS OF THE CRUDE OILS

Safety First

- Acetone and hexanes are flammable.
- d-Limonene is a possible carcinogen.

Measurement of Optical Rotation   In order to economize, only one solution of each oil will be prepared per laboratory section. Consult with your instructor about this point. Obtain the caraway oil or the spearmint oil from your instructor and weigh, in a 25-mL volumetric flask, approximately 2.5 g (± 0.01 g) of the oil. Dilute it to the mark with acetone. Using a 2-dm polarimeter tube, measure the optical rotation of the solution following the instructions given in section 11.5–6. Determine which enantiomers are present in each oil.

GC Analysis   Analyze the crude oil by GC using a column of medium polarity (for example, SF-96 at 150°C). Determine the percentage of limonene and carvone in your oil. Use these percentages to estimate the optical rotation of the crude oil, assuming that only carvone and limonene contribute to the total optical rotation of the oil (see in-lab 6). Compare this estimate with the actual optical rotation.

E11B.3   SEPARATION OF CARVONE AND LIMONENE

A short-tipped disposable Pasteur pipet makes an appropriate column for microscale column chromatography. To prepare the column put a small piece of Tygon tubing over the tip of the pipet. Clamp the tubing with a screw clamp. With the help of a notched stopper, clamp the column, in a vertical position, to a ringstand. Place a small cotton plug at the bottom of the column and cover the cotton with a thin layer of sand (2 mm). Add hexanes to the column and let about 1 mL of the solvent run. This will get rid of air bubbles trapped in the cotton. Recycle this solvent for the next step. Obtain about 1 g of silica gel (silica gel 60 for column chromatography, 0.2–0.5 mm, 35–70 mesh). Mix the measured amount of silica gel with enough hexanes to make a pourable slurry.

Microcolumns run very quickly, so it is important that you have everything prepared (solutions, pipets, and receiving containers) before starting the column. In five labeled test tubes deliver the following solvent volumes (these solvents will be used in succession to run the column):

2 mL hexanes

2 mL hexanes

2 mL hexanes

2 mL 10% acetone in hexanes

2 mL 10% acetone in hexanes

Label five 20-mL scintillation vials 1–5. These vials will be used to collect the column fractions.

Follow the instructions given in section 9.2 to pack microcolumns. After the column has been packed, open the column outlet until the solvent level comes down to the top of the sand (not lower) and load the column with 150 microliters of the oil. Use a pluringe to measure this volume. Open the screw clamp and start collecting the eluent in the vial labeled "Fraction 1." Without allowing the column to run dry and after the sample has penetrated the column (the liquid level should be just above the sand), add little by little the first solvent from the list above (use a Pasteur pipet to do this). Keep collecting the eluent in the "Fraction 1" vial. After the first 2 mL of hexanes have penetrated the column (solvent level just above the sand), start adding the next solvent (2 mL of hexanes) and immediately change the receiving flask to the one labeled "Fraction 2." Keep adding the solvents listed above, in succession, and collecting fractions equal in volume to the solvent volumes. Allow the column to run dry when collecting the fifth fraction.

Place a boiling chip in fractions 1 and 5 and evaporate the solvent by placing the vials in a sand bath on a hot plate, in the fume hood, until about 0.2–0.3 mL remains (approximately 8–10 drops); **do not evaporate to dryness** because some of the limonene may also evaporate.

GC Analysis   Analyze fractions 1 and 5 (and the other fractions if time permits) by gas chromatography. To see the carvone and limonene peaks clearly, you will need a higher sensitivity (lower attenuation) than the one used for the crude oil because now the compounds are diluted with solvent. If after lowering the attenuation one or two steps the limonene and carvone peaks are still too small, evaporate the solvent further. In a good chromatograph the solvent peak is likely to be off-scale.

IR Analysis   Obtain the IR spectrum of the fraction rich in carvone using Teflon tape or NaCl plates (see section 31.5). Compare it with the spectrum of carvone shown below.

Cleaning Up

- Dispose of the Pasteur pipets and micro columns in the "Carvone–Glass Waste" container.

- Dispose of column chromatography fractions and the solution of the oil (used for optical rotation) in the container labeled "Spearmint–Caraway Oils Waste."

BIBLIOGRAPHY

1. Science of Olfaction. M.J. Serby and K.L. Chobor, ed. Springer-Verlag, New York, 1992.

2. Biological Psychology. M.R. Rosenzweig, A.L. Leiman, and S.M. Breedlove. Sinauer Associates. Saunderland, 1996.

3. Analysis of Essential Oils by Gas Chromatography and Mass Spectrometry. Y. Masada. Wiley, New York, 1976.

4. Bioactive Volatile Compounds from Plants. R. Teranishi, R.G. Buttery, and H. Sugisawa, ed. ACS Symposium Series 525. ACS. Washington, 1993.

5. The Essential Oils. Vol. II-IV E. Guenther. R.E. Krieger. Huntington, New York, 1975.

6. Essential Oils Analysis by Capillary Gas Chromatography and Carbon-13 NMR Spectroscopy. V. Formácek and K.-H. Kubeczka. Wiley, Chichester, 1982.

EXPERIMENT 11B REPORT

Pre-lab

1. Make a table with the physical properties of carvone, limonene, hexanes, and acetone (MM, b.p., solubility, toxicity/hazards, flammability [for solvents only], dielectric constant [for solvents only]). For hexanes, use the values listed for *n*-hexane.

2. In general, is there any correlation between R and S, and + and −? between *d*- and *l*-, and + and −?

3. Arrange the following compounds (A, B, and C) in the order you expect them to come out of a silica column with a solvent of medium polarity:

A

B

C

4. Discuss the validity of the following statements:

   (a) A mixture of (+)- and (−)-carvone can be separated by column chromatography on silica gel using hexanes-acetone as elution solvent.

   (b) In general, column chromatography on silica gel is run with a series of solvents of increasing polarity.

   (c) Carvone is more polar than limonene.

5. Which compound, limonene or carvone, do you expect would elute first from the silica gel column? Which one would have shorter retention time on the GC? Briefly justify your answer.

6. In the gas chromatograph of spearmint oil, where do you expect the peaks of α- and ß-pinene and dihydrocarvone in relation to the limonene and carvone peaks? The boiling points of α- and β-pinene, dihydrocarvone, limonene, and carvone are 157°C, 166°C, 221°C, 175°C, and 230°C, respectively.

7. Is it possible to separate carvone and limonene using the following series of solvents: acetone (100%), acetone-hexanes (50:50), and finally hexanes (100%)? Explain.

8. How could you distinguish carvone and dihydrocarvone by IR?

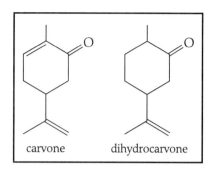

In-lab

1. Report the **measured optical rotation** of your oil. Calculate the **specific rotation** of the oil (Eq. 3). Compare the measured specific rotation with literature values (check the *Merck Index* under "Oil of Caraway" and "Oil of Spearmint").

2. Based on the sign of the optical rotation, identify the oil as spearmint or caraway. Tell which form of carvone and limonene is present in your essential oil.

3. Compare and discuss the gas chromatograms of the crude oil and fractions 1 and 5. Calculate the percentage of carvone and limonene in the crude oil and in the fractions analyzed by GC.

4. Briefly justify the elution order of limonene and carvone from the silica and GC columns.

5. Is it possible to isolate pure limonene from spearmint or caraway oil by using just hexanes as elution solvent? Discuss.

6. Using the equation given below, estimate the optical rotation of your crude oil. Compare the calculated optical rotation with the measured optical rotation for your oil.

$$\alpha_T = [\alpha]_c \times C_c \times \ell + [\alpha]_L \times C_L \times \ell$$

where

α_T = calculated optical rotation

$[\alpha]_c$ = specific rotation of carvone (-62 or $+62°$ $(g/mL)^{-1}dm^{-1}$)

$[\alpha]_L$ = specific rotation of limonene (-100 or $+100°$ $(g/mL)^{-1}dm^{-1}$)

C_c = concentration of carvone in g/mL

$$C_c = C_T \times \frac{\%\,\text{carvone}(GC)}{100}$$

where

C_T = total concentration of oil (weighed mass of oil / 25 mL)

C_L = concentration of limonene in g/mL

$$C_L = C_T \times \frac{\%\,\text{limonene}(GC)}{100}$$

where

C_T = total concentration of oil (weighed mass of oil/25 mL)

ℓ = length of the cell in decimeters (2 dm).

7. Interpret the IR of the carvone-rich fraction and compare it with the one shown below. Present your data as indicated in Tables 31.3 and 31.4. Along with the "observed," "expected (from the tables)," and "assignment" columns, add an extra column ("literature") with the values from the authentic sample shown in Figure 11.15.

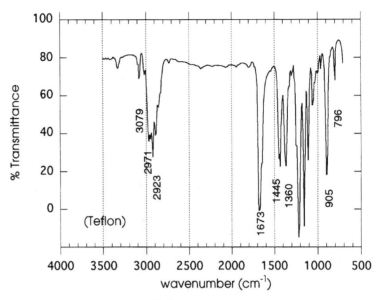

Figure 11.15 IR spectrum of carvone (Teflon).

UNIT 12

Alcohols and Alkenes

12.1 ALKENES FROM ALCOHOLS

Alcohols are compounds with an —OH group attached to an alkyl group; their general formula is R—OH. Some examples of simple alcohols are:

$$CH_3CH_2OH \qquad \underset{\underset{OH}{|}}{CH_3CHCH_3} \qquad \underset{\underset{OH}{|}}{\overset{\overset{CH_3}{|}}{CH_3CCH_3}}$$

ethanol · 2-propanol or isopropanol · 2-methyl-2-propanol or *tert*-butanol

Alcohols are easily converted into alkenes by elimination of a water molecule. This reaction, called a **dehydration**, is catalyzed by strong acids. The process is reversible; alkenes react with water in the presence of acids to give alcohols.

an alcohol · an alkene

In the laboratory, dehydrations are usually carried out by heating the alcohol in the presence of sulfuric or phosphoric acid. The equilibrium can be shifted to the right to increase the yield of the reaction by removing the alkene as it forms. This is usually done by a continuous distillation of the reaction mixture at a controlled temperature. The alkene distills off, leaving the alcohol behind because the alkene has a lower boiling point than the alcohol.

Sulfonic acids ($R-SO_3H$) are often employed as a source of protons instead of phosphoric and sulfuric acids. A convenient way of using sulfonic acids is to have the sulfonic acid group ($-SO_3H$) chemically attached to an inert polymeric resin. The result is a solid acid catalyst that donates protons to the medium but does not dissolve in the reaction mixture. The advantage of the polymeric resin over mineral acids, such as sulfuric and phosphoric acids, is that it is very easy to handle and can be recovered at the end of the reaction and reused.

An acidic polymeric resin

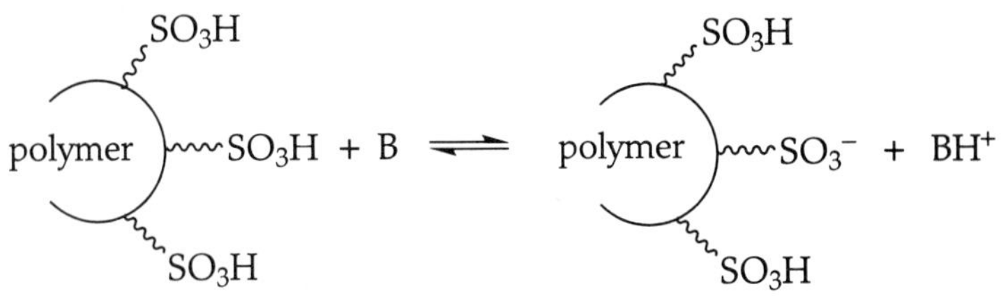

12.2 DEHYDRATION MECHANISMS

The dehydration of alcohols takes place by an *elimination mechanism*. For tertiary and most secondary alcohols, the reaction occurs in a three-step mechanism. Alcohols are weak Lewis bases. The lone electron pairs on the oxygen atom can take a proton from a strong acid to give a **protonated alcohol** in a fast reaction:

Fast reaction

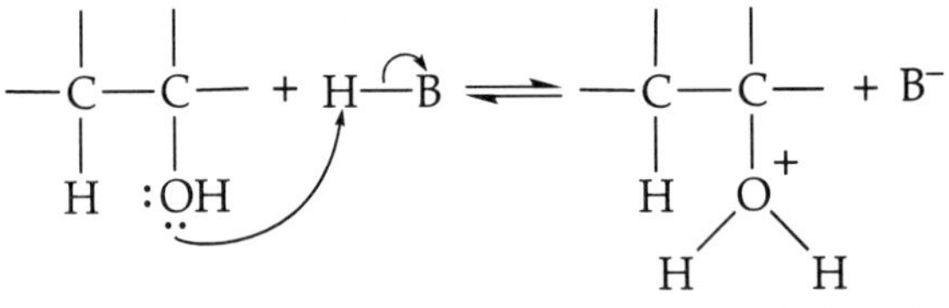

a protonated alcohol

The protonated alcohol is an unstable species; it loses water to give a **carbocation**. This step is the slowest one in the whole reaction, and thus it determines the reaction rate. It is called the **rate-limiting step** (the concept of rate-limiting step is further discussed in Unit 18).

Rate-limiting step

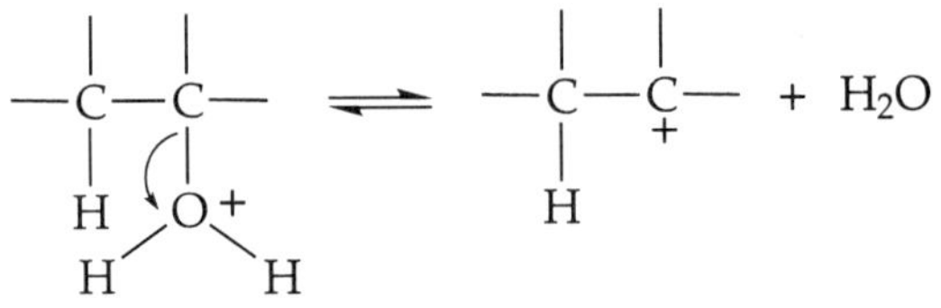

a carbocation

Carbocations are highly reactive intermediates. They react rapidly with electron-rich compounds to restore the octet of the positively charged carbon atom. The conjugated base of the acid, B$^-$, abstracts a proton from a carbon atom adjacent to the positive charge, a β-hydrogen, to give an alkene:

Fast reaction

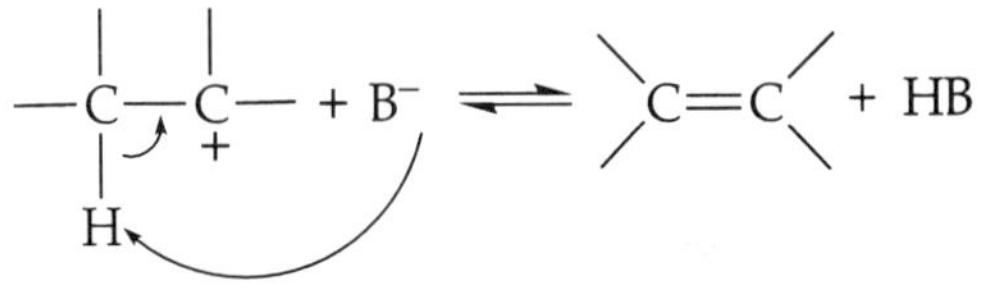

This mechanism is called **E1** or **unimolecular elimination** because only *one* molecule (the protonated alcohol) participates in the rate-limiting step (the elimination of the water molecule).

Primary alcohols do not react by an E1 mechanism because primary carbocations are very unstable. They dehydrate by an **E2** mechanism, also known as **bimolecular**

elimination. Similar to the E1 mechanism, in the E2 reaction the −OH group is protonated in the first step. This is followed by the simultaneous loss of a water molecule and a β-hydrogen. This step, which is rate-limiting, gives the name *bimolecular* to the mechanism because it involves *two* molecules: the protonated alcohol and a base.

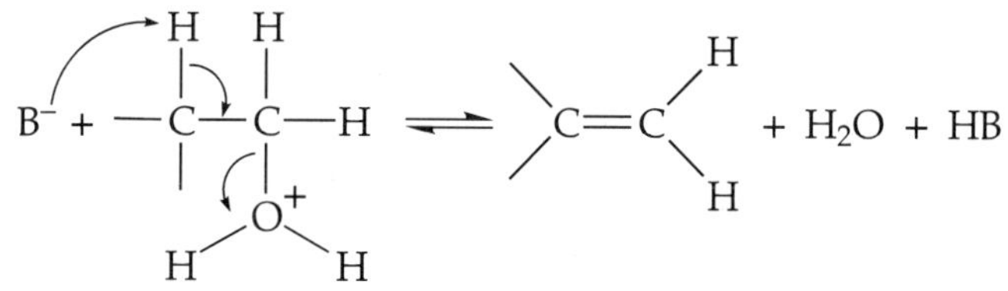

a protonated primary alcohol

Tertiary alcohols are easily dehydrated. The reaction takes place with only traces of acids and at relatively low temperatures (50–70°C). The dehydration of primary alcohols, on the other hand, is more difficult. It requires higher temperatures (150–170°C), higher concentration of acid catalyst, and does not constitute an important synthetic route in the laboratory. Secondary alcohols are intermediate between these two extremes. The different reactivity of alcohols is a result of the relative stability of their corresponding carbocations.

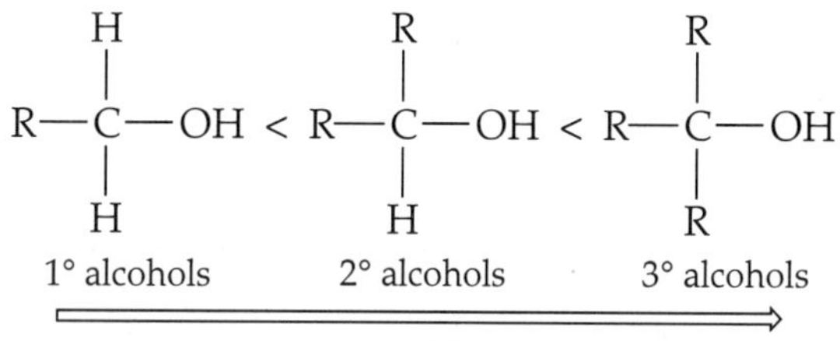

Some examples of dehydration of alcohols are shown below:

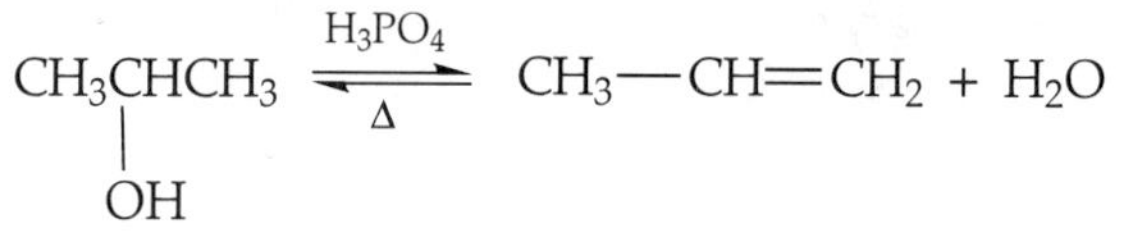

isopropanol   propene

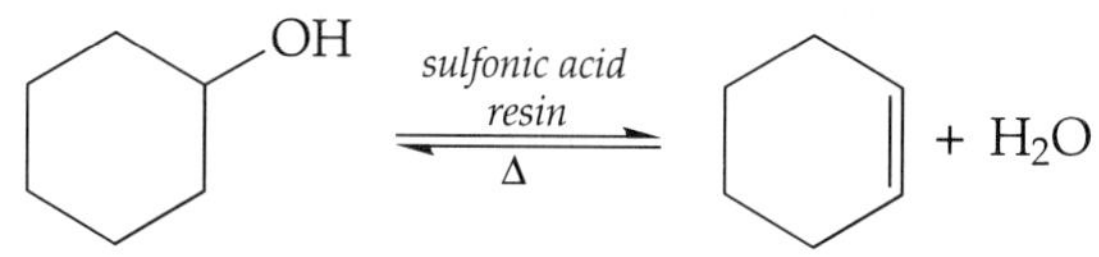

cyclohexanol   cyclohexene

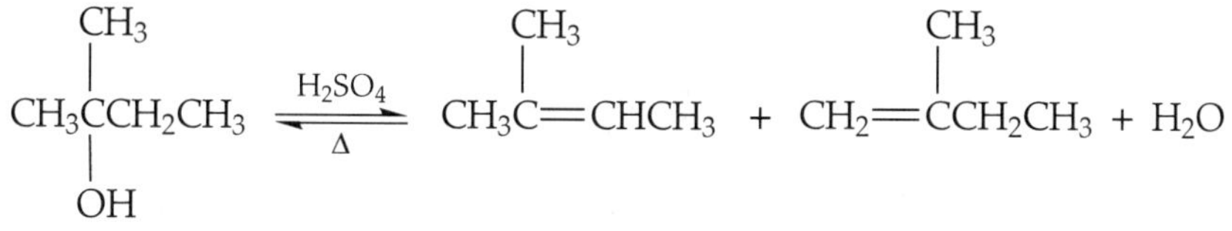

2-methyl-2-butanol   2-methyl-2-butene   2-methyl-1-butene
                            (a trisubstituted alkene)   (a disubstituted alkene)

When the alcohol has different types of β-hydrogens, as in the case of 2-methyl-2-butanol above, more than one product is possible. Which one predominates? In general, dehydration of alcohols follows **Zaitsev's rule**, which says that *the main product is the more substituted alkene*, or in other words, the more stable one. Thus, in the previous case of the dehydration of 2-methyl-2-butanol, 2-methyl-2-butene, an alkene with three substituents, predominates over 2-methyl-1-butene, a disubstituted alkene.

When both *cis* and *trans* isomers are possible, usually the *trans* isomer forms in larger amounts because it is more stable. The dehydration of 2-butanol with sulfuric acid gives a mixture of *trans*- and *cis*-2-butene as the main products; they are found in a 3:1 ratio. A small amount of 1-butene also forms.

2-butanol	*trans*-2-butene 74%	*cis*-2-butene 24%	1-butene 2%

The acid-catalyzed dehydration of alcohols is often accompanied by side-reactions. If good nucleophiles are present in the reaction mixture, for example, halide anions, substitution competes with elimination and alkyl halides are produced. Thus, mineral acids such as HCl or HBr should not be used as catalysts for the dehydration of alcohols. In general, elimination can be favored over substitution by working at high temperatures. Another undesirable side-reaction, especially in the dehydration of primary alcohols, is the formation of ethers in which two alcohol molecules react with loss of a water molecule. This side-reaction is unimportant in the dehydration of tertiary alcohols.

$$2\,R\,CH_2OH \xrightarrow[\Delta]{H_2SO_4} R\,CH_2O\,CH_2R + H_2O$$

Finally, another complication is the rearrangement of intermediate carbocations. Secondary carbocations can rearrange to more stable tertiary ones. This may involve hydride and methyl group shifts. For example, the dehydration of 3,3-dimethyl-2-butanol yields 2,3-dimethyl-2-butene as the major product. This compound results from the rearrangement of the secondary carbocation to a more stable tertiary one by the migration of a methyl group with its electron pair.

3,3-dimethyl-2-butanol	secondary carbocation	tertiary carbocation	2,3-dimethyl-2-butene (major product)

Even with all its drawbacks, the dehydration of alcohols, particularly of tertiary alcohols, is a very useful laboratory procedure to obtain alkenes.

12.3 CHARACTERIZATION OF ALKENES

A chemist can investigate whether or not a molecule contains a carbon-carbon double bond by simple chemical tests. The **addition of bromine** is one of them. Alkenes react rapidly with bromine to give dibromoalkanes. The reaction is carried out in an inert organic solvent such as carbon tetrachloride (CCl_4) or dichloromethane (CH_2Cl_2); the latter is usually preferred because of its lower toxicity.

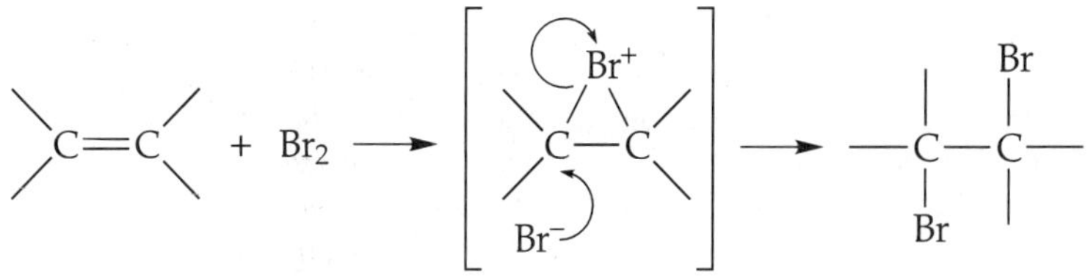

red-brown   colorless

When the alkene is added to a red-brown solution of **bromine in methylene chloride**, a decoloration is observed because the alkene reacts with bromine to give a colorless dibromocompound. This change in color is considered a positive test for alkenes. In contrast, saturated hydrocarbons do not change the color of bromine solutions. Alkynes also add bromine under the same conditions and produce a change in color. Other compounds such as phenols, amines, aromatic ethers, and some aldehydes and ketones also react with bromine, but they do so by *substitution instead of addition*. The substitution reaction produces HBr, which is a gas that doesn't dissolve in methylene chloride and can be noticed as bubbles escaping from solution. Therefore, if decoloration of a bromine solution is accompanied by the liberation of HBr, the reaction should be considered a negative test for alkenes. This is illustrated below with the substitution reaction of an aromatic ether.

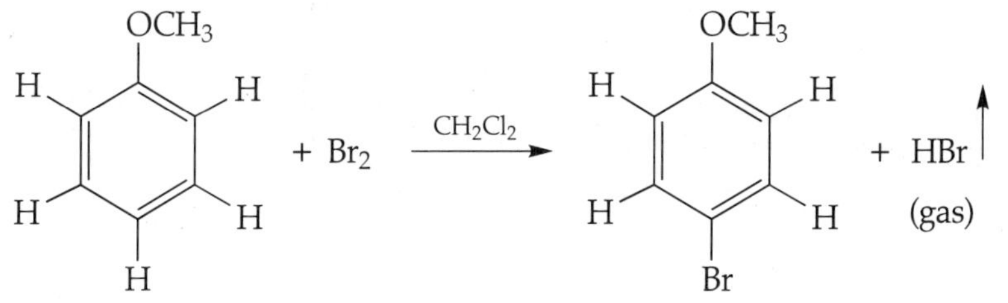

an aromatic ether   a substitution product

Alkenes also react with **potassium permanganate** at room temperature to give a diol and a brown-black precipitate of manganese dioxide. A positive test for alkenes is a change in color from purple to brown with the formation of a precipitate. This reaction is called the Baeyer test.

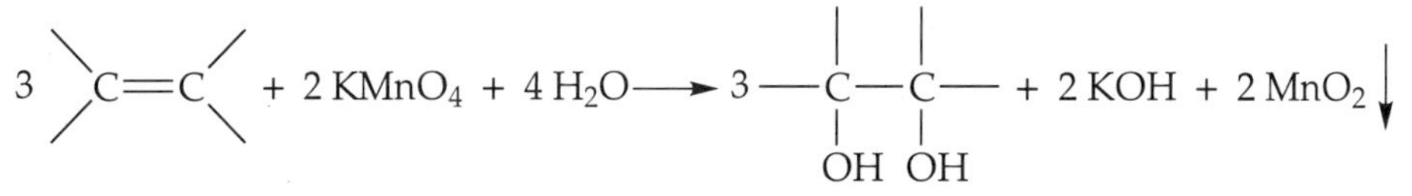

purple   brown-black

Other compounds such as alkynes, primary and secondary alcohols, aldehydes, and phenols also react with potassium permanganate under the same conditions.

The IR spectra of alkenes present characteristic bands in the $600-1000$ cm^{-1} region (C—H out-of-plane bending), and around 3000 cm^{-1} (C(sp^2)—H stretching). The C=C stretching band around 1650 cm^{-1} is usually of very low intensity.

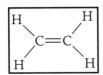

> **Ethylene** is a natural plant growth regulator, sometimes called the death hormone because it promotes maturation and aging. It is produced and released in large amounts by fruits during ripening, and is also synthesized by plants under conditions of stress. Apples, peaches, and pears are high ethylene producers while other fruits such as cherries and pineapples produce it in lower levels. The produce industry uses ethylene to artificially ripen avocados and bananas, which are harvested green for easy shipment. Ethylene affects many aspects of plant development and is particularly powerful on cut flowers, bulbs, and flower buds. It arrests the development of immature flowers, it damages bulbs, and hastens the aging of cut flowers. Carnations and orchids are especially sensitive to ethylene; roses and tulips seem to be less sensitive. To prevent damage, never store bulbs and flowers near ripening fruit.

EXERCISES

Problem 1

Predict the products of the following dehydration reactions. Indicate the major product if more than one is formed.

(a)

acid catalyst
Δ

(b)

acid catalyst
Δ

(c)

acid catalyst
Δ

Problem 2

(a) How would you use IR spectroscopy to distinguish between an alcohol and its dehydration product, an alkene?

(b) Dialkyl ethers are possible side-products in the dehydration of alcohols. How can you distinguish dialkyl ethers from alkenes by IR spectroscopy?

Problem 3

(*Previous Synton: Problem 6, Unit 11.*)   Synton was in a hurry. He had an appointment with his chiropractor that his stiff neck was urging him not to miss. He was in the laboratory preparing 1-methylcyclohexene by the dehydration of 1-methylcyclohexanol. He added some concentrated phosphoric acid to the alcohol, assembled a distillation apparatus, and distilled the product as it formed. To make sure that the reaction took place as planned, he took a drop of the distillate and without further treatment ran its IR spectrum. His face turned red with frustration when he saw a very broad band around 3500 cm^{-1}. He thought that the alcohol hadn't reacted and was recovered intact in the distillate. As he was rushing out of the lab to catch the bus, he was wondering what went wrong with his dehydration and was already making plans to repeat it the next day. At midnight his sore neck woke him up. As he was tossing in bed trying to go back to sleep, he suddenly realized that nothing had been wrong with his reaction. He smiled as he remembered that he had saved the product and fell asleep soundly. What was wrong with Synton's IR analysis? Why did he think that the dehydration didn't occur?

Problem 4

You carry out the dehydration of 2-methyl-3-pentanol with phosphoric acid. You place 625 μL (1 μL (microliter) = 1×10^{-3} mL) of the alcohol (density = 0.824 g/mL; molecular mass = 102 g/mole) in a microscale distillation apparatus and add 200 μL of concentrated phosphoric acid (85% w/w, density = 1.685 g/mL; molecular mass = 98 g/mole). You heat the reaction mixture on a sand bath and distill off the products. After drying the distillate over magnesium sulfate and filtering it, you obtain 300 mg of a product, which you analyze by GC. A schematic GC trace is shown below.

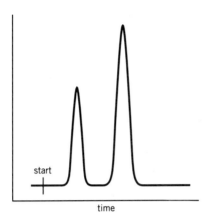

(a) Write a chemical reaction for the transformation and underline the major product.

(b) Calculate the number of mmoles of 2-methyl-3-pentanol and phosphoric acid used.

(c) What role does phosphoric acid play? Which compound is the limiting reagent?

(d) Calculate the theoretical yield of the reaction and the percentage yield.

(e) Find the boiling points of the products and use them to assign the GC peaks. Calculate the percentage of each product in the mixture.

Problem 5

A compound X gave the following results when characterization tests a–b were performed:

(a) Br_2/methylene chloride: +

(b) $KMnO_4$: +

Which one of the following compounds is most likely to be X? Justify your answer.

i) benzene; ii) cyclohexanol; iii) cyclohexane; iv) 1,2-dichlorocyclohexane; v) 1-methylcyclohexene; vi) bromobenzene.

EXPERIMENT 12

The Dehydration of Methylcyclohexanols

E12.1   OVERVIEW

In this microscale experiment you will perform the dehydration of 2-methylcyclo-hexanol and will characterize the products by gas chromatography. The reaction gives 1-methyl and 3-methylcyclohexene as major products. These two isomers can be separated by GC. To increase the yield of the reaction you will distill the products as they form.

2-methylcyclohexanol   1-methylcyclohexene 3-methylcyclohexene

According to Zaitsev's rule, which isomer do you expect to be the predominant product? You will investigate this issue by determining the relative amounts of the isomeric products in the distillate. To this end, you will analyze the mixture of products by GC and identify and integrate the GC peaks.

In order to identify the GC peaks, some students will be assigned the dehydration of 1-methylcyclohexanol instead of 2-methylcyclohexanol. The dehydration of the 1-methyl isomer gives 1-methylcyclohexene almost exclusively; only minute amounts of the much less stable isomer methylenecyclohexane are formed. In general, double bonds protruding outside a ring (*exocyclic* double bonds) are less stable than double bonds inside a ring (*endocyclic* double bonds). By analyzing the dehydration product of 1-methylcyclohexanol by GC you can unequivocally determine the retention time of 1-methylcyclohexene.

1-methylcyclohexanol   1-methylcyclohexene   methylenecyclohexane
                            (> 95%)   (formed in minute amounts)

In comparing the retention times of the products of both reactions, the same GC instrument should be used. Two students will team up (one doing 1-methylcyclo-hexanol and the other 2-methylcyclohexanol) to perform the analyses on the same instrument and share results.

As a source of protons you will use either concentrated phosphoric acid or a sulfonic acid resin; your instructor will decide on this. In this experiment, you will also perform some simple tests to confirm the presence of a carbon-carbon double bond in your product. You will obtain the IR spectra of products and the starting alcohols. This experiment has been adapted from References 1–3. You can obtain further mechanistic information in References 4 and 5.

PROCEDURE

Background reading: 6.8 & 12.1–12.3
Estimated time: 3 hours

E12.2 DEHYDRATION OF METHYLCYCLOHEXANOLS

Safety First

- Phosphoric acid is corrosive. Handle it with care.

1. With the help of a pluringe (Fig. 3.2), weigh about 750 μL (1 μL (microliter) $= 1 \times 10^{-3}$ mL) of 2-methylcyclohexanol (a mixture of *cis* and *trans*), *or* 1-methylcyclohexanol in a 3-mL conical vial (or in a 5-mL round-bottom flask). 1-Methylcyclohexanol melts around 24°C; to deliver the liquid in cold climates place the open container in a warm water bath for a few minutes.

2. If you are using phosphoric acid as a catalyst, add 225 μL of the concentrated acid to the vial with the help of a pluringe. Concentrated phosphoric acid is an 85% weight/weight (w/w) solution of H_3PO_4 in water (85 g of H_3PO_4 per 100 g of solution; density $= 1.685$ g/mL). Add a boiling chip and proceed to 4.

3. If you are using a sulfonic resin, add 100 mg (about 15 beads) of Nafion (trade name of Du Pont) NR50 resin (0.8×10^{-3} equivalent of sulfonic acid group/g of resin). A boiling chip is not necessary because the Nafion beads play the double role of acid catalyst and boiling chips.

4. Set up a microscale simple distillation apparatus with a Hickman still as shown in Figure 12.1a (see also Fig. 6.19). If you are using a conical vial, check that the screw-cap O-ring is in place (Fig. 12.1b). Carefully immerse the conical vial (or round-bottom flask) in a sand bath preheated at 170–200°C. Make sure that the conical vial (or round-bottom flask) is 2/3 immersed in the sand. Bring the liquid to a boil keeping the temperature of the sand bath around 170–200°C. Distill the reaction mixture until approximately one-half of the volume has been collected in the Hickman still. Turn off the heat.

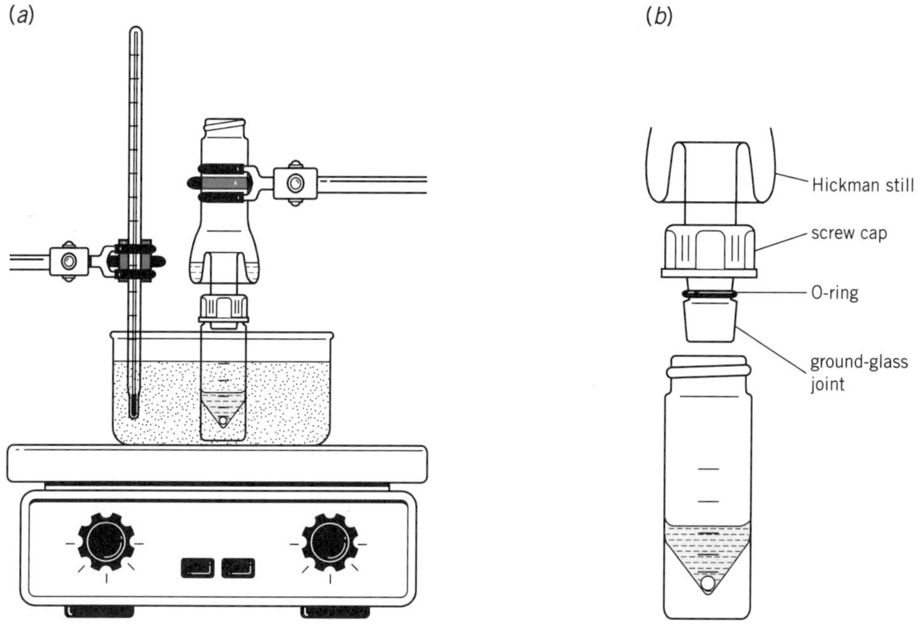

(a)

(b)

Figure 12.1 *a*) A microscale simple distillation apparatus; *b*) detail showing the correct position of the O-ring.

5. Using a Pasteur pipet, transfer the distillate to a small screw-cap vial or screw-cap test tube. Add *small* amounts of anhydrous $MgSO_4$ until the drying agent runs free in suspension. Cap the vial and let the system sit for about 5 minutes. Filter the suspension using a filter-Pasteur pipet (Fig. 3.9b); collect the liquid into a pre-tared, dry, labeled screw-cap vial. Weigh the product, calculate its percentage yield, and analyze it by GC, chemical tests, and IR spectroscopy.

E12.3 ANALYSIS OF THE PRODUCT MIXTURE

GC Analysis Inject $0.2\ \mu L$ (or the volume specified by your instructor) of your product mixture. Compare the retention times of the products obtained in the dehydration of 1- and 2-methylcyclohexanol (run on the same GC instrument). Inject the starting alcohol dissolved in acetone (approximately 50% v/v) and compare its retention time with those of the products. Suggested GC conditions: stationary phase: SF-96; column temperature: 80°C; flow rate: 30 mL/min (Helium); recorder speed: 2.5 cm/min; attenuation: you set.

IR Analysis Obtain the IR of the starting alcohol and the reaction products using NaCl plates (section 31.5). Compare the 3500–3000 and 1000–700 cm^{-1} regions of both spectra.

Chemical Tests

1. Bromine test Perform this test in the fume hood. In a 10×75 mm test tube add 0.5 mL of a 0.1 M solution of bromine in methylene chloride. To this solution add 2–3 drops of your mixture of products. Observe any change in color. The disappearance of the bromine color is a positive test for unsaturation. Repeat the test using cyclohexane, cyclohexene, and the starting alcohol.

2. Permanganate test (Baeyer test) In a 10×75 mm test tube add 2–3 drops of your product to 0.5 ml of a 0.5% potassium permanganate solution in water. Carefully shake the tube, and observe any change in color. For comparison, repeat the test using cyclohexane, cyclohexene, and the starting alcohol. The formation of a black-brown precipitate is considered a positive test.

BIBLIOGRAPHY

1. Dehydration of 2-Methylcyclohexanol. R.L. Taber and W.C. Champion. *J. Chem. Ed.*, **44**, 620 (1967).
2. Concerning Dehydration of 2-Methylcyclohexanol. A. Feigenbaum. *J. Chem. Ed.*, **64**, 273 (1987).
3. Replacing Mineral Acids in the Laboratory. Nafion-Catalyzed Dehydration and Esterification. M.P. Doyle and B.F. Plummer, *J. Chem. Ed.*, **70**, 493–495 (1993).
4. The Dehydration of 2-Methylcyclohexanol Revisited: The Evelyn Effect. D. Todd. *J. Chem. Ed.*, **71**, 440 (1994).
5. The Acid Catalyzed Dehydration of an Isomeric 2-Methylcyclohexanol Mixture. A Kinetic and Regiochemical Study of the Evelyn Effect. J.J. Cawley and P.E. Lindner. *J. Chem. Ed.*, **74**, 102–104 (1997).

EXPERIMENT 12 REPORT

Pre-lab

1. Make a table showing the physical properties (b.p., density, solubility, molecular mass, toxicity/hazards) of 2-methylcyclohexanol, 1-methylcyclohexanol, 1-methylcyclohexene, 3-methylcyclohexene, methylenecyclohexane, and

phosphoric acid. These compounds can be found in the *CRC Handbook of Chemistry and Physics*.

2. Make a flowchart for this experiment.

3. Draw a microscale distillation apparatus in your lab notebook and name its parts.

4. Write a balanced reaction for the dehydration of your alcohol. Show the mechanism.

5. Calculate the number of equivalents of sulfonic acid present in 100 mg of resin, knowing that the concentration of sulfonic groups in the Nafion resin is 0.8 milliequivalent/g of resin.

6. Calculate how many mmoles of 2-methylcyclohexanol, 1-methylcyclohexanol, and phosphoric acid are used. Calculate how many mmoles and milligrams of the mixture of isomeric methylcyclohexenes you expect (theoretical yield).

7. What role do the Nafion resin and H_3PO_4 play? Could you use concentrated HCl or H_2SO_4 instead? Briefly explain.

8. Write balanced reactions for the bromine and permanganate tests for 1-methylcyclohexene. Which one of the following compounds would give both tests positive: cyclopentane, 2-chloropentane, cyclopentene?

9. Would it be advantageous to reflux (see "Refluxing," section 3.4) the reaction mixture instead of performing a distillation? Briefly discuss.

10. Which compounds are present in the distillate?

11. Why do you add $MgSO_4$ to the distillate? What other chemicals could you use instead?

12. Do you think that you could distinguish between 3-methylcyclohexene and methylenecyclohexane by their GC retention times? Base your answer on their respective boiling points.

13. You heat 2,5-dimethyl-3-hexanol in the presence of catalytic amounts of concentrated sulfuric acid in a conical vial fitted with a Hickman still, and collect the distillate. Draw the structure of the products that you expect and underline the major product.

In-lab

1. Calculate the number of mmoles of product obtained (actual yield).

2. Calculate the percentage yield of the synthesis:

$$\% \text{ yield} = \frac{\text{actual yield}}{\text{theoretical yield}} \times 100$$

3. In a table format, report the GC retention times and peak areas for the products and starting alcohols. Write down the GC experimental conditions (type of column, carrier gas, flow rate, temperature).

4. Do 1-methyl and 3-methylcyclohexene elute from the GC column in the expected order based on their respective boiling points?

5. Compare the results for the dehydration of both alcohols. Using the GC areas, calculate the percentage of 1-methyl and 3-methylcyclohexene in the dehydration of 2-methylcyclohexanol.

6. Discuss the distribution of products in both dehydrations in terms of the dehydration mechanism and relative stability of intermediates and products. Is Zaitsev's rule obeyed?

7. Report and discuss the results of your unsaturation tests.

8. Interpret the IR spectra. Compare the IR spectra of the products with those of authentic samples of 1-methyl and 3-methylcyclohexene shown in Figure 12.2.

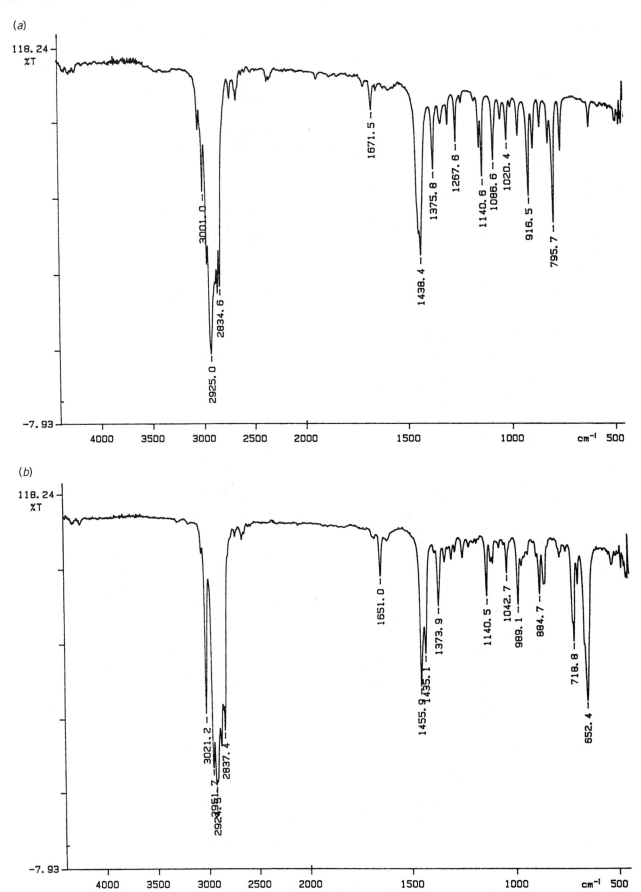

Figure 12.2 IR spectrum (neat) of: *a*) 1-methylcyclohexene; *b*) 3-methylcyclohexene.

UNIT 13

Alkyl Halides

Alkyl halides encompass a large family of organic compounds structurally derived from alkanes by replacing one or more hydrogen atoms by halogens. They are represented by the general formula: R–X, where R is an alkyl group and X is F, Cl, Br or I. For example:

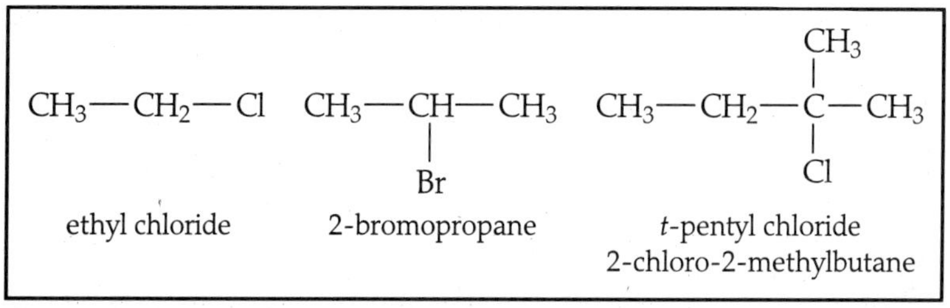

$CH_3—CH_2—Cl$
$CH_3—CH(Br)—CH_3$
$CH_3—CH_2—C(CH_3)(Cl)—CH_3$
ethyl chloride
2-bromopropane
t-pentyl chloride
2-chloro-2-methylbutane

Aryl halides, where R is a phenyl ring or other aromatic group, display chemical properties very different from those of alkyl halides.

13.1 PREPARATION

Alkyl halides can be prepared by the direct halogenation of alkanes, using a radical chain reaction (usually initiated by heat or light), or by the treatment of alcohols with hydrogen halides (HX), phosphorous halides (PX_3), or thionyl chloride ($SOCl_2$). Alkyl chlorides can also be prepared by the addition of hydrogen halides to alkenes. The different chemical routes to generate alkyl halides are shown in Figure 13.1.

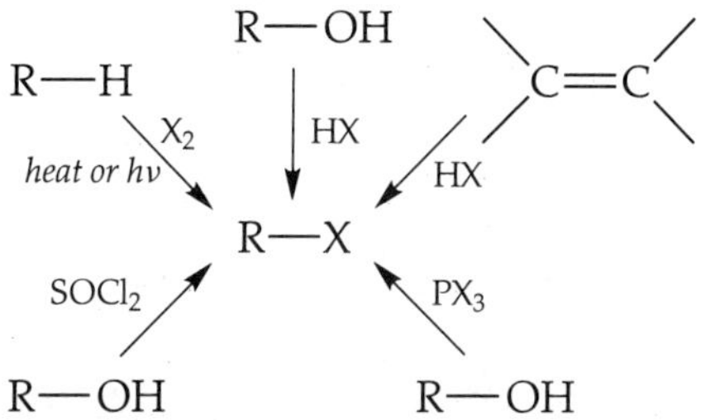

R—H
X_2
heat or hν
R—OH
HX
C=C
HX
R—X
$SOCl_2$
R—OH
PX_3
R—OH

Figure 13.1 Preparation of alkyl halides.

Alkyl halides have numerous technological applications. They are raw materials in the manufacture of pharmaceuticals, polymers, and pesticides. Polyhalogenated alkanes are, or have been, extensively used as fire retardants, aerosol propellants, refrigerants (Freons; see the box "The Ozone Hole"), anesthetics ($CHCl_3$, CH_3CH_2Cl, $CF_3CHBrCl$), and pesticides (CH_3Br). Chloroform ($CHCl_3$) and carbon tetrachloride (CCl_4) have been extensively used as solvents in the past; they are now known to cause liver damage upon prolonged exposure, and are also listed as possible carcinogens. Today their use is limited and strictly regulated in the United States; their use in the laboratory should be minimized. Halogenated compounds are also naturally produced. The oceans, with their high contents of Cl^-, Br^-, and I^-, are the most prolific sources of natural alkyl halides. Methyl chloride and methyl bromide (CH_3Cl and CH_3Br) are constantly being released into the atmosphere by the sea. More than a hundred different halogenated compounds (including $CHBr_3$, CH_2Br_2, and CCl_4) have been identified in edible varieties of red alga such as *Aspargopsis toxiformis*, which is highly regarded in Hawaii because of its appealing flavor.

13.2 REACTIONS

Alkyl halides are very versatile compounds with many applications in organic synthesis; they are precursors in the synthesis of a variety of compounds from amines to organometallics. Some important reactions of alkyl halides are summarized in Figure 13.2. These can be classified in three basic types: 1) substitutions (a–d); 2) eliminations (e); and 3) reactions with metals (f–h). Grignard reagents are very important intermediates in organic synthesis, as we will see in Unit 23.

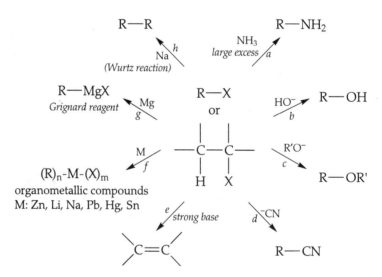

Figure 13.2   Transformations of alkyl halides.

13.3 ALKYL HALIDES FROM ALCOHOLS

The simplest method for obtaining alkyl halides is the treatment of alcohols with hydrogen halides. In alcohols, the C—OH bond is polar due to the difference in electronegativity between the carbon and the oxygen atoms. The carbon atom bears a positive charge density. This electron-deficient center is **electrophilic** (from Greek: electron-loving) and is attracted to electron-rich reagents or **nucleophiles**, such as chloride, Cl^-, and bromide, Br^-.

$$-\overset{|}{\underset{|}{C}}-OH + HX \longrightarrow -\overset{|}{\underset{|}{C}}-X + H_2O \quad (X = Cl, Br, I)$$

This type of reaction is known as **aliphatic nucleophilic substitution**. Aliphatic nucleophilic substitutions can take place by one of two different mechanisms. Primary and secondary alcohols react with hydrogen halides via an S_N2 mechanism, whereas tertiary alcohols react by an S_N1 mechanism, as discussed below.

Alcohols are very weak Lewis bases. They are protonated by strong acids to give a "protonated alcohol"; this reaction is always very fast. Depending on the nature of the R group, the solvent, the temperature, and other reactants present in the medium, the protonated alcohol will further react in different ways.

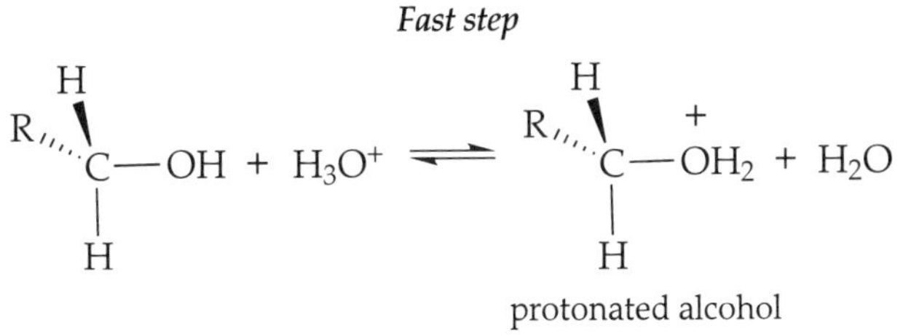

Fast step

protonated alcohol

In the presence of anions such as Cl^- and Br^-, which are *good nucleophiles*, primary protonated alcohols are attacked from the rear by the nucleophile and lose one molecule of water, the **leaving group**. This concerted attack by the nucleophile and departure of the leaving group is the step that *controls the reaction rate* or the **rate-limiting step** (the concept of rate-limiting step is discussed in Unit 18).

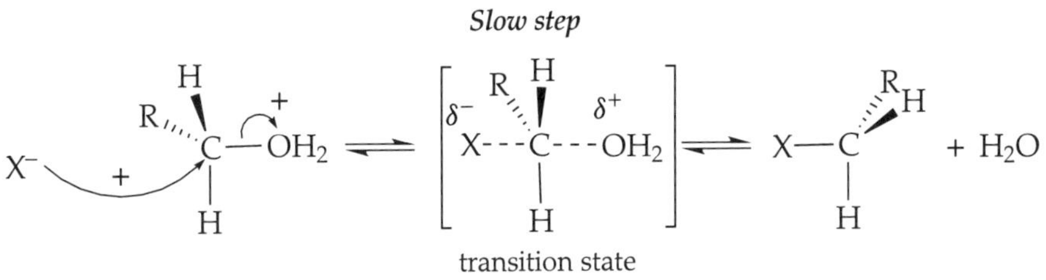

Slow step

transition state

This mechanism is called S_N2 or **bimolecular nucleophilic substitution**. The term *bimolecular* is employed because the *transition state results from the collision of two molecules*: the nucleophile and the protonated alcohol.

Protonated tertiary alcohols have a good deal of steric hindrance created by the three alkyl groups and cannot be attacked by the halide anion from the backside. Protonated tertiary alcohols decompose to give a **carbocation** and a molecule of water in a rate-limiting step:

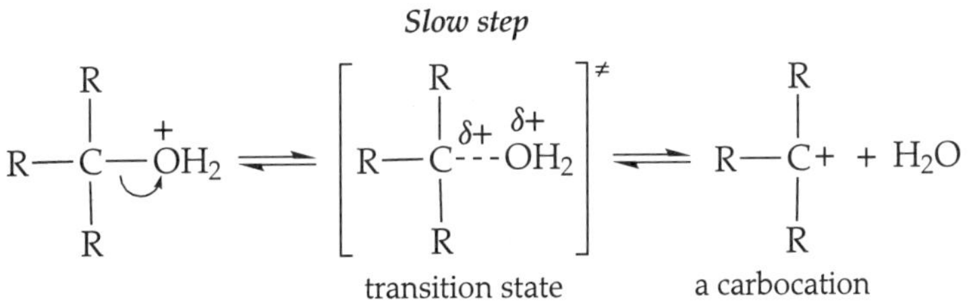

Slow step

transition state   a carbocation

The carbocation then reacts with the halide anion to give the product in a rapid reaction.

Fast step

$$
\underset{\underset{R}{\displaystyle |}}{\overset{\overset{R}{\displaystyle |}}{R-C+}} \; + \; X^- \;\longrightarrow\; \underset{\underset{R}{\displaystyle |}}{\overset{\overset{R}{\displaystyle |}}{R-C-X}}
$$

This mechanism is known as **S$_N$1** or **unimolecular nucleophilic substitution**. The term unimolecular implies that the *transition state involves only **one** molecule* (the protonated alcohol).

There are some important differences between the reactions of primary and tertiary alcohols, especially from an experimental point of view. The substitution of primary alcohols is generally slow and carried out at high temperatures. In contrast, tertiary alcohols react with hydrogen halides even at room temperature to give tertiary alkyl halides in high yields. In general, the order of reactivity of different types of alcohols is: $3° > 2° > 1°$. That tertiary alcohols react with hydrogen halides more rapidly than primary alcohols will be clearly demonstrated in the syntheses of *n*-butyl bromide and *t*-pentyl chloride in the experimental part.

During the preparation of alkyl halides by nucleophilic substitution, small amounts of alkenes may also be produced by elimination reactions (Unit 12). These side-products are removed from the crude material by extractions.

13.4   CHARACTERIZATION TESTS

Beilstein Test

When alkyl and aryl halides are burnt on a copper surface in an open flame, they give a characteristic green color to the flame. This is a very sensitive and easy test to determine the presence of chlorine, bromine, and iodine in an organic sample. The chemical transformations of organic compounds caused by an open flame are quite complex. Burning organic halides react with cuprous oxide on the copper surface to give *volatile cuprous halides*, which are responsible for the green color. Organic fluorides give negative tests because cuprous fluoride is nonvolatile. Some nitrites and urea also produce green flames when burnt on a copper wire.

Silver Nitrate in Ethanol

Alkyl halides react with silver nitrate in ethanol to give insoluble silver halides, which separate as a precipitate. The reaction is driven to completion by the precipitation of the silver halide:

$$
R-X \; + \; AgNO_3 \longrightarrow R-ONO_2 \; + \; AgX\downarrow
$$

The formation of a precipitate of AgX is a positive test. Other halogen-containing compounds such as acyl halides and organic ammonium halides also react with silver nitrate in ethanol. Alkyl iodides react faster than alkyl chlorides and bromides, and usually give a precipitate of silver iodide immediately after mixing with silver nitrate at room temperature. The reaction takes place by an S$_N$1 mechanism and the order of reactivity of alkyl halides is: $3° > 2° > 1°$. The reaction can be accelerated by heating at the boiling point of the solvent for a few seconds. Aryl halides give a negative test even upon heating.

Sodium Iodide in Acetone

Alkyl chlorides and bromide react with sodium iodide in acetone to give alkyl iodides and the corresponding sodium halide.

$$R\!-\!Cl + NaI \longrightarrow R\!-\!I + NaCl \downarrow$$

$$R\!-\!Br + NaI \longrightarrow R\!-\!I + NaBr \downarrow$$

This test is based on the limited solubility of sodium chloride and sodium bromide in acetone. These salts separate as precipitates, which is an indication of a positive test. The reaction takes place via an S_N2 mechanism and the order of reactivity is bromides > chlorides, and $1° > 2° > 3°$. The reaction can be accelerated by heating at 50°C for 5 minutes. Tertiary alkyl chlorides and aryl and vinyl halides do not react even upon heating. Acyl halides give a positive test at room temperature.

BIBLIOGRAPHY

1. Biological Halogenation: Roles in Nature, Potential in Industry. S.L. Neidleman and J. Geigert, *Endeavour*, New Series, **11**, 5–15 (1987).

2. Nucleophilic Reactivities of the Halide Anions. M.S. Puar, *J. Chem. Ed.*, **47**, 473–474 (1970).

3. Nucleophilic Substitution Reactions at Secondary Carbon Atoms. D.J. Raber, J.M. Harris, *J. Chem. Ed.*, **49**, 60–64 (1972).

EXERCISES

Problem 1

A compound Y gave an immediate reaction with silver nitrate in ethanol, and a negative reaction, after heating at 50°C, with sodium iodide in acetone. Which one of the following compounds is most likely to be Y? i) iodobenzene; ii) 1-chlorohexane; iii) *p*-bromochlorobenzene; iv) 3-chloro-3-methylpentane; v) 1-bromopentane.

Problem 2

In the synthesis of alkyl halides from alcohols, how would you distinguish the starting alcohol from the alkyl halide by IR spectroscopy? What precaution should be taken before running the IR of the product?

Problem 3

t-Butyl chloride was prepared by treating *t*-butanol with concentrated HCl. Concentrated HCl is a solution of HCl in water with an acid content of 37% (w/w).

(a) Write a chemical equation for the process.

(b) Given the information gathered below, fill in the spaces with "?". For HCl give the mass of the concentrated solution (density 1.2 g/mL) as well the mass of HCl present in that solution.

Compound	Volume used or obtained (mL)	Density g/mL	MM g/mole	Mass g	mmoles
t-butanol	1.56	0.786	74	?	?
t-butyl chloride	1.3	0.847	92.6	?	?
HCl (conc; 37% w/w)	3.0	1.2	36.5	?	?

(c) Which compound is the limiting reagent?

(d) Calculate the % yield of *t*-butyl chloride.

Problem 4

The boiling points of *t*-butyl alcohol, *t*-butyl chloride, *t*-butyl bromide, and *t*-butyl iodide are 82, 51, 73, and 98°C, respectively.

(a) Explain the order of boiling points.

(b) If you must prepare only one of these alkyl halides from *t*-butyl alcohol, which one would you synthesize if the last step involves purification of the product from unreacted alcohol?

Problem 5

Arrange the following compounds in order of decreasing reactivity toward a) silver nitrate in ethanol, and b) sodium iodide in acetone: i) 1-chloro-1-methylcyclohexane; ii) *n*-octyl bromide; iii) 2-chlorohexane; iv) bromobenzene.

The Ozone Hole   Life on Earth is possible thanks to a protective ozone layer found in the stratosphere (13–56 km above the surface of the earth). Ozone absorbs UV radiation in the wavelength range 210–320 nm where other natural light screens, such as oxygen, do not absorb. UV radiation has a tremendous impact on the biosphere; an increase in its intensity can lead to abnormal numbers of melanoma (skin cancer) and cataracts, and a decrease in crop yields and ocean productivity. **Chlorofluorocarbons** (CFCs, compounds with fluorine, chlorine, and carbon) have been widely used as solvents, refrigerants, and propellants for aerosol spray cans. These compounds are called **Freon.** Examples of freons are: CCl_3F, CCl_2F_2, $CHClF_2$, and CCl_2FCClF_2. CFCs are very inert compounds. They do not react in the troposphere (the lower part of the atmosphere), and due to their very low solubility in water they are not washed down by the rainfall. Once they are released to the atmosphere, they keep ascending to the stratosphere where they are eventually exposed to UV radiation that photodissociates them (below 240 nm):

$$CF_2Cl_2 + h\nu \ (< 240 \text{ nm}) \longrightarrow {}^{\bullet}CF_2Cl + Cl^{\bullet}$$

Chlorine atoms can start a catalytic chain reaction that decomposes ozone:

$$Cl^{\bullet} + O_3 \longrightarrow ClO^{\bullet} + O_2$$

$$ClO^{\bullet} + O^{\bullet} \longrightarrow Cl^{\bullet} + O_2$$

$$\overline{\text{neat: } O_3 + O^{\bullet} \longrightarrow 2O_2}$$

This is a self-sustained chain reaction because chlorine atoms are consumed and reformed with the overall effect of destroying ozone. Chlorine atoms are not the only species that can deplete the ozone layer; other radicals such as $HO^{\bullet}$ and NO can also react with ozone; however, the chlorine chain reaction is more efficient in the catalytic destruction of ozone. The first warning on the possible effects of CFCs on the environment was published by Molina and Rowland from the University of California, Irvine, in 1974. In their seminal paper, these authors estimated that the lifetime of CFC molecules in the atmosphere is 40–150 years; with such long lifetimes, the concentration of CFCs in the stratosphere is expected to increase over the next decades if their release is not halted. In the same year, there was a preliminary and independent study on the decrease in the ozone levels over Australia. However, it was not until 1985 that the first full report on the "ozone hole" over Antarctica was published. During the spring months of September to November in the Southern Hemisphere, a marked drop (about 50% below normal levels) in the amount of stratospheric ozone has been detected since the mid-1970s. During those months the "ozone hole" can extend to populated regions of Argentina, Australia, and Chile. The reason why the ozone hole is detected in the Southern but not in the Northern Hemisphere is still controversial, but it seems to be due to the lower stratospheric temperatures over the Antarctic region. In the Northern Hemisphere, the presence of larger land areas with high mountains (the Rockies and Himalayas) produces a more thorough mixing of the air over lower, warmer latitudes with the consequence of somewhat higher stratospheric temperatures. At the lower stratospheric temperatures of the Antarctic region, clouds of ice crystals are formed during the winter months (July and August), which absorb chlorine as HCl and ClNO$_3$. These molecules trapped in the stratospheric ice clouds combine to give molecular chlorine and nitric acid. In the spring, the sunshine easily dissociates the chlorine molecule into two chlorine atoms that react with ozone as discussed above.

$$HCl + ClNO_3 \longrightarrow Cl_2 + HNO_3$$

$$Cl_2 + h\upsilon \longrightarrow 2Cl^{\bullet}$$

Halogenated hydrocarbons, for example, methylene chloride (CH$_2$Cl$_2$), methyl chloride (CH$_3$Cl), and methyl bromide (CH$_3$Br), have reactive hydrogen atoms and are easily destroyed in the troposphere; their lifetime is less than a year, and therefore they do not seem to pose a serious threat to the ozone layer.

Bibliography

1. Atmospheric Chemistry: Fundamentals and Experimental Techniques. B.J. Finlayson-Pitts and J.N. Pitts, Jr. Wiley, New York, 1986.

2. Global and Regional Changes in Atmospheric Composition. E. Mészáros, Lewis Publishers, Boca Raton, 1993.

3. Stratospheric Sink for Chlorofluoromethanes: Chlorine Atom-catalysed Destruction of Ozone. M.J. Molina and F.S. Rowland. *Nature*, **249**, 810–812 (1974).

4. Large Losses of Total Ozone in Antarctica Reveal Seasonal ClO$_x$/NO$_x$ Interaction. J.C. Farman, B.G. Gardiner, and J.D. Shanklin. *Nature*, **315**, 207–210 (1985).

5. Trends in the Vertical Distribution of Ozone over Australia. A.B. Pittock. *Nature*, **249**, 641–643 (1974).

6. Mending the Ozone Hole: Science, Technology, and Policy. A. Makhijani and K.R. Gurney. MIT Press, Cambridge, 1995.

EXPERIMENT 13

Synthesis of n-Butyl Bromide and 2-Chloro-2-Methylbutane

In this experiment you will prepare *n*-butyl bromide from *n*-butyl alcohol by treatment with sodium bromide and sulfuric acid. You will also prepare 2-chloro-2-methylbutane (*t*-pentyl chloride) from 2-methyl-2-butanol (*t*-pentyl alcohol) and hydrochloric acid. Check with your instructor to find out if you will perform both parts. The products will be characterized by GC analysis, chemical tests, and infrared spectroscopy. This is a two-lab-period experiment. The formation of *n*-butyl bromide is an example of an S_N2 reaction, while the preparation of 2-chloro-2-methylbutane occurs via an S_N1 mechanism. A comparison of the very different experimental conditions that lead to the formation of each alkyl halide will be made.

E13.1 SYNTHESIS OF *n*-BUTYL BROMIDE

Primary alkyl bromides can be prepared by heating the corresponding alcohol in the presence of an excess of concentrated hydrobromic acid. The yield is improved and the reaction is faster if sulfuric acid is added to the reaction mixture. Sulfuric acid has several roles in this reaction. It acts as an acid catalyst by protonating the alcohol and making it more reactive in the nucleophilic substitution. Sulfuric acid also increases the yield of the reaction. Sulfuric acid is a highly hygroscopic material and absorbs the water produced in the reaction. This removal of water molecules by the acid increases the yield of the synthesis by shifting the equilibrium toward the products. The hydrobromic acid needed for the reaction can be added in the form of commercial concentrated acid, or generated *in situ* by the reaction of NaBr with H_2SO_4. In the case of *in situ* formation of HBr, the amount of sulfuric acid needed is larger.

In our experimental procedure, HBr is generated by mixing NaBr and H_2SO_4 as shown below:

$$NaBr + H_2SO_4 \rightleftharpoons HBr + NaHSO_4$$

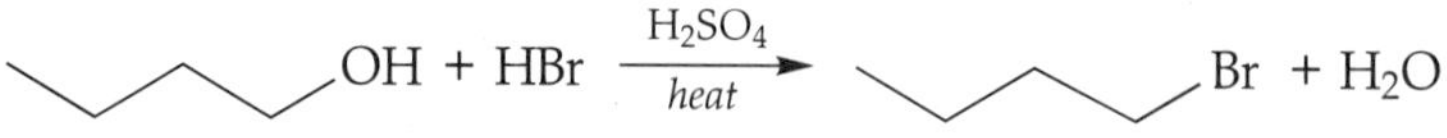

$$\diagdown\diagup\diagdown\diagup OH + HBr \xrightarrow[heat]{H_2SO_4} \diagdown\diagup\diagdown\diagup Br + H_2O$$

Primary alkyl chlorides can be obtained by a similar method; however, a Lewis acid such as anhydrous $ZnCl_2$, instead of H_2SO_4, is used as a catalyst. For the preparation of secondary and tertiary alkyl halides, the use of sulfuric acid is not recommended because it facilitates dehydration reactions that produce alkenes.

In this experiment, *n*-butyl alcohol is refluxed in the presence of sodium bromide, sulfuric acid, and water, and the reaction mixture is then distilled. The distillate is biphasic; one layer contains the alkyl halide (organic layer) and the other mainly water. In the distillate, besides *n*-butyl bromide and water, HBr, unreacted alcohol, and alkenes are also found. Alkenes (1-butene and 2-butene) form in small amounts by elimination reactions favored by high temperatures and the presence of sulfuric acid. The aqueous layer is separated from the organic alkyl halide layer, and the alkyl halide layer is then subjected to several consecutive extractions, called **washes**, to remove the contaminants. The unreacted alcohol and the alkenes can be eliminated by washing with cold, concentrated sulfuric acid. The reaction of alkenes and alcohols with sulfuric acid gives alkyl hydrogen sulfates:

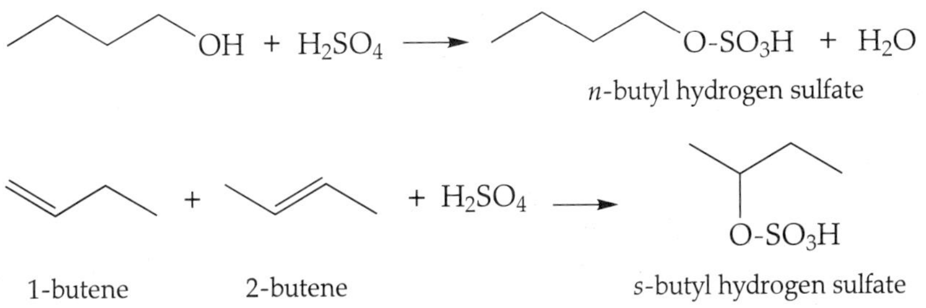

n-butyl hydrogen sulfate

| 1-butene | 2-butene | | *s*-butyl hydrogen sulfate |

The butyl hydrogen sulfates are polar compounds easily dissolved by sulfuric acid, and therefore they stay in the sulfuric acid layer. The acid layer is removed and the organic layer washed sequentially with water, a dilute solution of sodium hydroxide (to remove traces of acid), and finally water (to remove salts and sodium hydroxide). The pH of the final water wash layer should be neutral. This experiment involves several consecutive microscale extractions (washes); *label* all the layers and *keep them* until the end of the experiment. In case of doubt about which one is the organic and which one is the aqueous layer, use the drop method as described in section 5.3.

PROCEDURE

> **Background reading:** 5.4; 6.8; 13.1–4.
> **Estimated time:** synthesis: 3 hours;
> analysis: 1 hour

Synthesis

1. Place 2.15 g of sodium bromide and 2.7 mL of water in a 15-mL round-bottom flask equipped with a small magnetic bar. Stir with a magnetic stirrer until mixed.

2. With the aid of a Pasteur pipet and a pluringe (Fig. 3.2) **slowly** add while stirring 1.20 mL of cold concentrated sulfuric acid that has been sitting in a water-ice bath for a few minutes (concentrated sulfuric acid is 98% w/w, density = 1.84 g/mL). Have the cold water bath (cold water plus a small amount of ice) handy in case the flask gets too hot during the addition of the acid to the aqueous solution.

3. Add 1.8 mL of *n*-butyl alcohol (use a 5-mL graduated pipet or a 2-mL pluringe).

4. Slowly add an additional 1.20 mL of cold concentrated sulfuric acid using a Pasteur pipet.

5. Place the flask in a sand bath and set up a reflux apparatus (see "Refluxing", section 3.4) by attaching a microscale water-jacketed condenser on top (Fig. 13.3*a*). With the help of a septum or a rubber stopper connect the condenser to a gas trap made with a length of Tygon or latex tubing with one end placed in a test tube containing a moistened cotton wad. The cotton wad should completely surround the tubing. The tubing end should be far from the bottom of the test tube (Fig. 13.3*a*). The wet cotton wad absorbs HBr and prevents its release to the atmosphere.

6. Reflux the solution for 30 minutes. The sand bath temperature should be between 140 and 150°C.

7. Turn off the heat and allow the mixture to cool. Remove the condenser with the gas trap. Disassemble the gas trap but keep the condenser with the attached

> **Safety First**
>
> - Concentrated sulfuric acid is very corrosive; with the aid of a Pasteur pipet and a pluringe transfer just the amount of sulfuric acid needed for steps 2 and 4 to a labeled test tube. Do not take more sulfuric acid than the amount needed for the step you are about to perform. Do not remove the sulfuric acid bottle from the hood.
>
> - Sodium hydroxide is very caustic. Handle it with care.
>
> - Use lubricant (water) when inserting glass tubing into Tygon or latex tubing. To protect your hands, always wrap in paper towels or a cloth the piece of glass to be inserted into a hose.

(a)   (b)

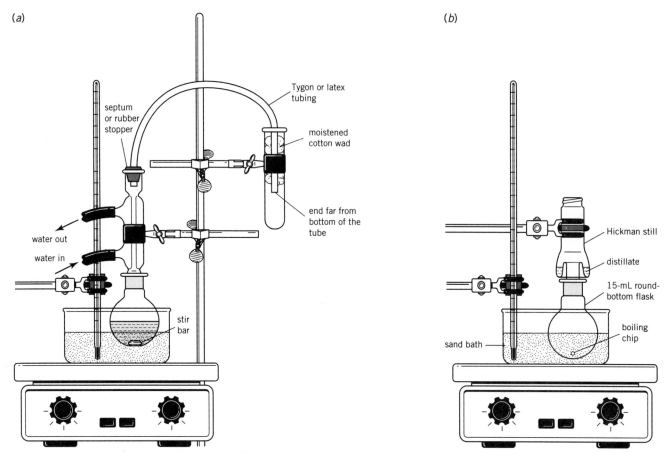

Figure 13.3 *a*) Reflux apparatus with gas trap; *b*) distillation apparatus with Hickman still.

hoses handy on the bench (you may need it later). Place a Hickman still head on top of the round-bottom flask (Fig. 13.3*b*). Apply heat again so the temperature of the bath is between 140 and 150°C; a mixture of the alkyl halide and water will codistill. Adjust the heat supply to make sure that the temperature of the bath is below 150°C; otherwise vapors will escape from the top of the Hickman still. If this happens, attach at once the condenser to the Hickman still and *lower* the heat supply.

8. **Periodically** remove the distillate with a Pasteur pipet and place it into a 5-mL conical vial or a 16 × 125 mm screw-cap test tube. (If a condenser is attached on top of the Hickman still, momentarily remove it to collect the distillate with the Pasteur pipet.) Distill the mixture in the round-bottom flask until the distillate is completely miscible in water. Check its miscibility in water by adding one drop of distillate to 0.5 mL of water in a test tube. When the distillate becomes completely miscible in water, no butyl bromide is present and the distillation is stopped.

9. Add 1 mL of water to the distillate. Cap the vial and shake. Vent the system. Allow the layers to separate (this process is called washing). Remove the organic layer using a Pasteur pipet and save it in a clean 3-mL conical vial. See "Microscale Liquid-Liquid Extraction", section 5.4. (Which layer contains the *n*-butyl bromide? What are the densities of *n*-butyl bromide and water?)

10. Wash the organic layer with 1 mL of cold concentrated sulfuric acid (see **Safety First** above; make sure that the cap of the vial does not leak). Let the mixture stand for a few minutes until the two layers completely separate. Save the organic layer. Wash again the organic layer with 0.5 mL of water, then with 10% NaOH

(0.5 mL), and finally with 0.5 mL of water. Check the pH of the water layer with pH paper. It should be close to 7; if not, repeat the water wash. Be careful to save the organic layer each time. See step 14 for cleaning-up procedures.

Important Tip *With five washes, it is important not to lose or spill anything, especially at the microscale level. Keep all the layers from different washes clearly labeled until the end of the experiment.*

11. Dry the organic layer with a small amount of MgSO$_4$ (about one-half the contents of a microspatula). Let it sit for 5 minutes and filter through a filter-Pasteur pipet (Fig. 3.9*b*). If you have more than 0.5 mL of liquid product, proceed to the final distillation (step 12); if not, go directly to step 13.

12. Place the *n*-butyl bromide in a 3-mL conical vial (or a 5-mL round-bottom flask); add a boiling chip and assemble a microscale distillation apparatus, adding an air condenser on top of the Hickman still to prevent loss of product (Fig. 13.4*a*). If you are using a conical vial, make sure that the O-ring is correctly positioned (above the inner ground-glass joint; Fig. 13.4*b*). Make sure that the vial or flask is 2/3 immersed in the sand bath. Collect the fraction that boils between 97–102°C. The sand bath temperature should be considerably higher (140–150°C).

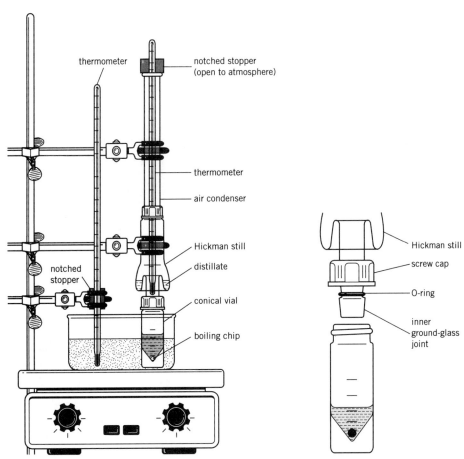

Figure 13.4 *a*) Microscale distillation apparatus with air condenser; *b*) detail of Hickman still joint.

13. *Analysis.* Transfer the liquid from the filter-Pasteur pipet (from step 11) or the Hickman still (from step 12) into a pre-tared scintillation vial. Weigh the product and calculate the % yield. Determine its purity by GC analysis using a column of medium polarity such as SF-96 (suggested temperature 80°C). Run the GC

of authentic samples of *n*-butyl bromide and *n*-butanol and compare with that of the reaction product. Perform the silver nitrate, sodium iodide, and Beilstein tests (E13.3). Obtain the IR spectra of product and starting material using sodium chloride windows (section 31.5).

14. In a 100 mL beaker, carefully add the sulfuric acid wash (use a Pasteur pipet) to all the combined water washes. Add 3 mL of water and then slowly add the **10% sodium hydroxide wash** with the aid of a Pasteur pipet. Then follow "Cleaning up" instructions.

E13.2   SYNTHESIS OF 2-CHLORO-2-METHYLBUTANE

Tertiary alkyl halides can be obtained by mixing the corresponding tertiary alcohol with a concentrated solution of the desired hydrogen halide in water at room temperature. The reaction takes place via an S_N1 mechanism, and does not require the presence of a catalyst. The same hydrogen halide protonates the alcohol leading to the formation of the carbocation. The overall reaction for the synthesis of 2-chloro-2-methylbutane is shown in the equation below.

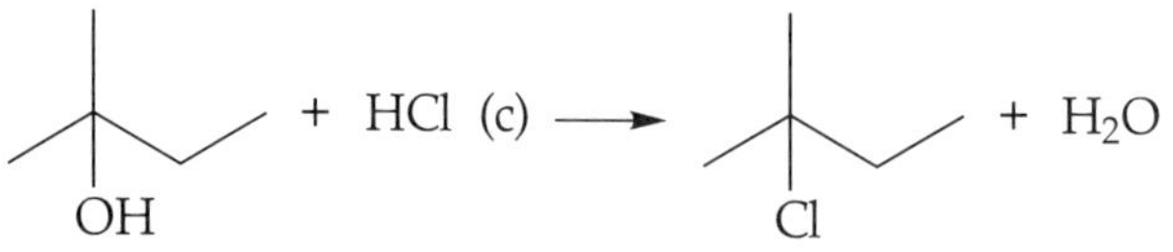

Because the reaction takes place at room temperature, side-products such as alkenes are not formed and, thus, the purification of the alkyl halide is straightforward. A few washes with sodium bicarbonate and water, to remove unreacted HCl, are enough to obtain a reasonably pure alkyl chloride. In the preparation of tertiary alkyl halides, contact with basic aqueous solutions should be kept to a minimum to avoid hydrolysis of the alkyl halide back to the alcohol.

PROCEDURE

> **Background reading:** 5.4; 6.8; 13.1–4.
> **Estimated time:** synthesis: 1.5 hours:
> analysis: 1 hour

Synthesis

1. Pipet 1.00 mL (Pasteur pipet and pluringe, Figure 3.2) of 2-methyl-2-butanol and 2.5 mL of concentrated HCl (37% w/w; d = 1.2 g/mL) into a 5-mL conical vial or 16 × 125 mm screw-cap test tube. Cap the vial and let the mixture stand for a minute. Then carefully shake the mixture with occasional venting. Be careful that none of the mixture leaks from the vial.

2. Allow a few minutes for the two phases to completely separate. Separate the layers with a Pasteur pipet and save the layer containing the alkyl halide. See "Microscale Liquid-Liquid Extraction," section 5.4. What are the densities of reagents and products?

Important Tips   *The following washes should be performed as quickly as possible because 2-chloro-2-methylbutane is hydrolyzed by basic aqueous solutions. Keep all the layers from different washes, clearly labeled, until the end of the experiment. With several washes, it is important not to lose or spill anything, especially at the microscale level.*

3. Wash the organic layer with 1 mL of water. Separate the phases and save the organic layer. Wash the organic layer with 1 mL of 5% aqueous sodium bicarbonate. Shake

the vial and vent carefully because carbon dioxide is produced in this step. Allow the layers to separate and save the organic layer. Wash again with 1 mL of water.

4. Dry the organic layer with a small amount of $MgSO_4$. Let it stand for at least 5 minutes.

5. Using a filter-Pasteur pipet (Fig. 3.9b), transfer the crude alkyl halide to a clean 3-mL conical vial or a 5-mL round-bottom flask. Add a boiling chip and assemble a microscale distillation apparatus similar to that depicted in Figure 13.4a adding an air condenser on top of the Hickman still to prevent loss of product. If you are using a conical vial, make sure that the O-ring is correctly positioned (Fig. 13.4b). Make sure that the distillation flask or conical vial is at least 2/3 immersed in the sand bath. Collect the fraction that boils between 80 and 86°C (the sand bath temperature should be 40–50°C higher).

6. *Analysis.* Place the product in a tared vial. Weigh the product and calculate the % yield. Analyze by GC using a column of medium polarity such as SF-96 (suggested temperature 80°C). Compare with the GC of an authentic sample of 2-chloro-2-methylbutane and 2-methyl-2-butanol. Perform the silver nitrate, sodium iodide and Beilstein tests (E13.3). Obtain the IR spectra of product and starting material using sodium chloride windows (section 31.5).

7. In a 100 mL beaker, carefully add the hydrochloric acid layer from the first separation to all the combined water washes. With the aid of a Pasteur pipet, slowly add the 5% sodium bicarbonate wash, if there is bubbling (CO_2), wait until it completely stops. Then follow "Cleaning Up" instructions.

> **Cleaning Up**
>
> - Pour the liquid from step 7 into the "Alkyl Halides Washes Waste" container.
> - Dispose of the magnesium sulfate and Pasteur pipets in the container labeled "Alkyl Halides Solid Waste."
> - Turn in the product in a labeled screw-capped vial to your instructor.

E13.3 CHARACTERIZATION TESTS

Beilstein Test

Burn the tip of a copper wire (about 10 cm long, attached to a cork as a handle) in the blue cone of a Bunsen burner flame until the flame is no longer green. Let the copper wire cool down and then dip the tip of the wire into a small portion of the sample to be tested. Burn the tip with the sample on it. A green coloration is a positive test. Try this test on your alkyl halide and the starting alcohol.

> **Safety First**
>
> - This test should be performed on a lab bench far away from flammable solvents and on-going experiments.
> - Do not start this test without your instructor's approval.

Silver Nitrate in Ethanol

To six clean, dry, and labeled test tubes add 0.5 mL of a 0.1 M solution of silver nitrate in ethanol and one drop of one of the following compounds: n-butanol, 2-methyl-2-butanol, n-butyl bromide (from E13.1 if available, otherwise use the commercial product), 2-chloro-2-methylbutane (from E13.2 if available, otherwise use the commercial product), t-butyl bromide, and bromobenzene. If no precipitate is observed within 5 minutes at room temperature, bring the solutions to a boil by heating the test tubes in a water bath at about 80°C. Record your observations and interpret your results. If you are doing the synthesis of both n-butyl bromide and 2-chloro-2-methylbutane, perform this test after both alkyl halides have been synthesized.

Sodium Iodide in Acetone

To six clean, dry, and labeled test tubes add 0.5 mL of a 15% (w/v) solution of sodium iodide in acetone and one drop of the same compounds mentioned in the silver nitrate test. If no precipitate is observed after 3 minutes at room temperature, heat in a water bath at 50°C for five minutes. Record your observations and interpret your results. If you are doing the synthesis of both n-butyl bromide and 2-chloro-2-methylbutane, perform this test after both alkyl halides have been synthesized.

> **Cleaning Up**
>
> - With the aid of a Pasteur pipet transfer the contents from the silver nitrate and sodium iodide test tubes to a container labeled "Alkyl Halides Tests Waste." Rinse the test tubes with a little ethanol and dispose of the rinses in the same container.

BIBLIOGRAPHY

1. Vogel's Textbook of Practical Organic Chemistry. A.I. Vogel, B.S. Furniss, A.J. Hannaford, P.W.G. Smith, and A.R. Tatchell. 5th ed. Longman, Harlow, UK, 1989.

2. The Systematic Identification of Organic Compounds. A Laboratory Manual. R.L. Shriner, R.C. Fuson, and D.Y. Curtin. 5th ed. Wiley, New York, 1964.

EXPERIMENT 13   REPORT

Synthesis of *n*-Butyl Bromide

Pre-lab

1. Make a table showing the physical properties (molecular mass, b.p., density, solubility in water and in organic solvents, toxicity/hazards) of *n*-butanol, sulfuric acid, *n*-butyl bromide, and sodium bromide.

2. Calculate how many moles of *n*-butanol, sulfuric acid, and sodium bromide were used. Calculate the theoretical yield of *n*-butyl bromide.

3. Write a balanced equation for the preparation of *n*-butyl bromide.

4. Discuss the mechanism of this reaction.

5. What side-products do you expect?

6. Make a flowchart showing the experimental steps of this preparation.

7. Draw the reflux apparatus with gas trap in your lab notebook. Name its parts.

8. Why is it important to use a gas trap?

9. What compounds are present in the first distillate?

10. What would happen if the reaction mixture is overheated?

11. Why do you wash the organic layer with cold concentrated sulfuric acid? why with sodium hydroxide solution?

12. Draw a microscale distillation apparatus in your lab notebook and name its parts.

In-lab

1. Calculate the percentage yield of the synthesis and discuss your results.

2. Report and discuss your GC results. Write down the GC experimental conditions.

3. Discuss your Beilstein, silver nitrate, and sodium iodide test results.

4. Interpret your IR spectra and compare them with that shown in Figure 13.5.

5. Which bands of the IR spectra can you use to determine if the conversion of the alcohol into the alkyl halide took place?

6. Interpret the ^{1}H-NMR and ^{13}C-NMR of *n*-butyl bromide shown in Figures 13.6 and 13.7 (advanced level).

7. Summarize the learning objectives of this experiment. Make a list of the new techniques and reactions introduced.

Synthesis of 2-Chloro-2-Methylbutane

Pre-lab

1. Make a table showing the physical properties (molecular mass, b.p., density, solubility, toxicity/hazards) of reactants and products (synonyms of 2-methyl-2-butanol: *tert*-pentyl alcohol, *tert*-amyl alcohol).

2. Calculate how many moles of 2-methyl-2-butanol and hydrogen chloride were used. Calculate the theoretical yield of the synthesis.

3. Write a balanced equation for the preparation of 2-chloro-2-methylbutane.
4. Discuss the mechanism of this reaction.
5. What side-products do you expect?
6. Make a flowchart showing the experimental steps of this preparation.
7. Why do you wash the product with sodium bicarbonate? Could you use sodium hydroxide instead?
8. Draw a microscale distillation apparatus in your lab notebook and name its parts.
9. An unknown sample may contain only one of the following compounds:

    i) 1-methylcyclohexene   ii) *n*-butyl chloride   iii) 1-chloropropene
    iv) cyclohexane

    (a) Write down the structure of each compound.

    (b) You have at your disposal a piece of copper wire and potassium permanganate (see Unit 12, Baeyer test). How would you determine the identity of the unknown? Present your answer in a table format.

10. Formulate a chemical reaction for a generic alkyl halide with silver nitrate in ethanol and with sodium iodide in acetone. In each case, which alkyl halides react faster?

In-lab

1. Calculate the percentage yield of the synthesis and discuss your results.
2. Report and discuss your GC results. Write down the GC experimental conditions.
3. Discuss your Beilstein, silver nitrate, and sodium iodide test results.
4. Interpret your IR spectra and compare them with that shown in Figure 13.8.
5. Which bands of the IR spectra can you use to determine if the conversion of the alcohol into the alkyl halide took place?
6. Interpret the ^{1}H-NMR and ^{13}C-NMR of 2-chloro-2-methylbutane shown in Figures 13.9 and 13.10 (advanced level).
7. Summarize the learning objectives of this experiment. Make a list of the new techniques and reactions introduced.

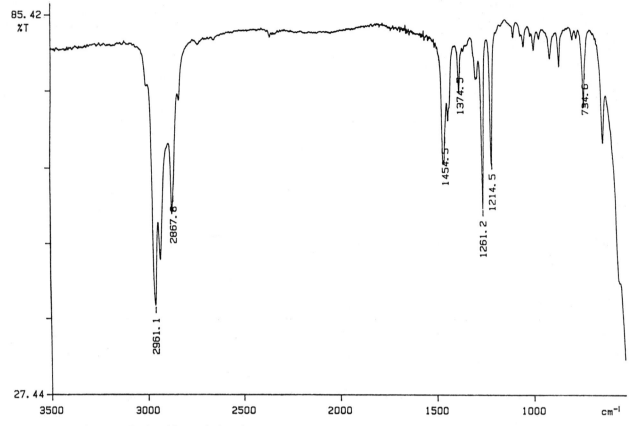

Figure 13.5 IR spectrum of *n*-butyl bromide (neat).

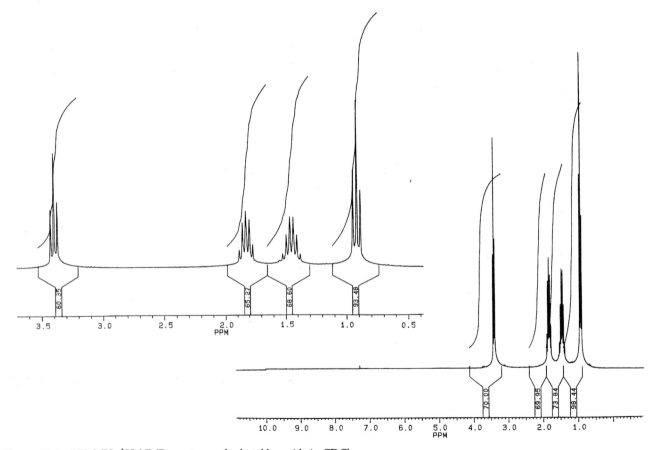

Figure 13.6 250-MHz ^{1}H-NMR spectrum of *n*-butyl bromide in CDCl$_3$.

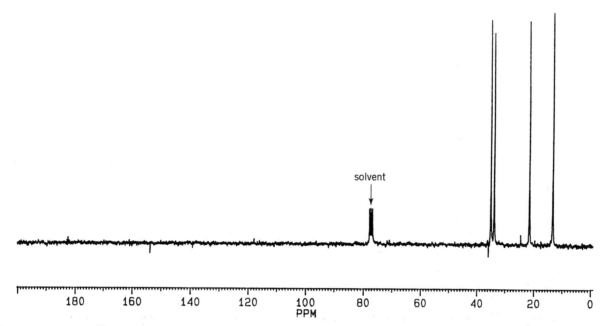

Figure 13.7   62.9-MHz ^{13}C-NMR spectrum of *n*-butyl bromide in CDCl$_3$.

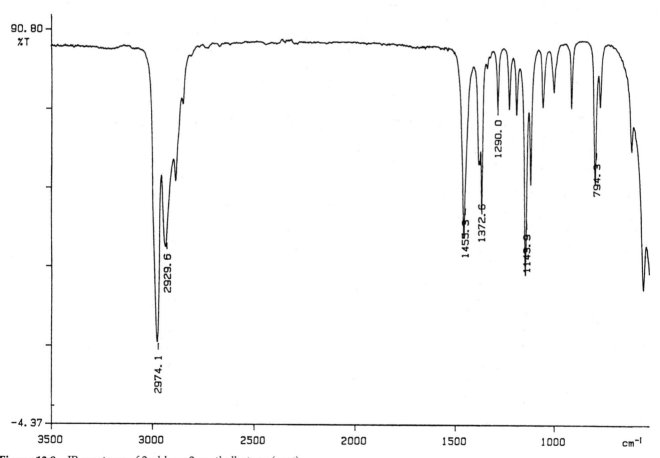

Figure 13.8   IR spectrum of 2-chloro-2-methylbutane (neat).

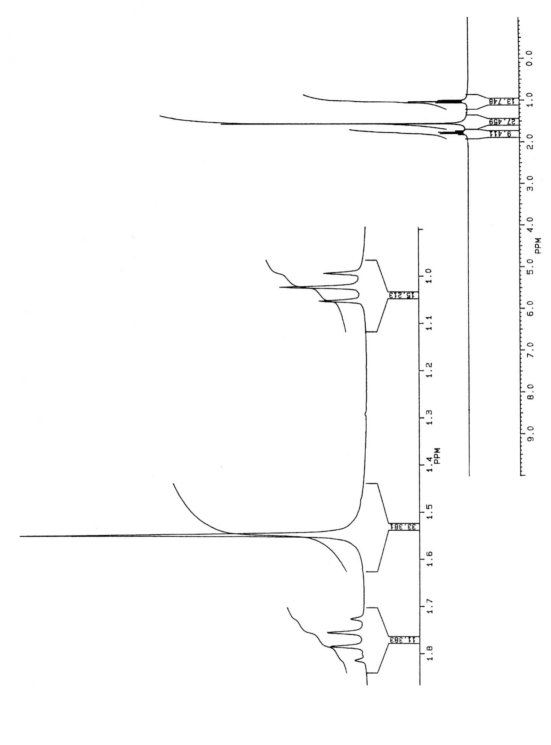

Figure 13.9 250-MHz ^{1}H-NMR spectrum of 2-chloro-2-methylbutane in CDCl$_3$.

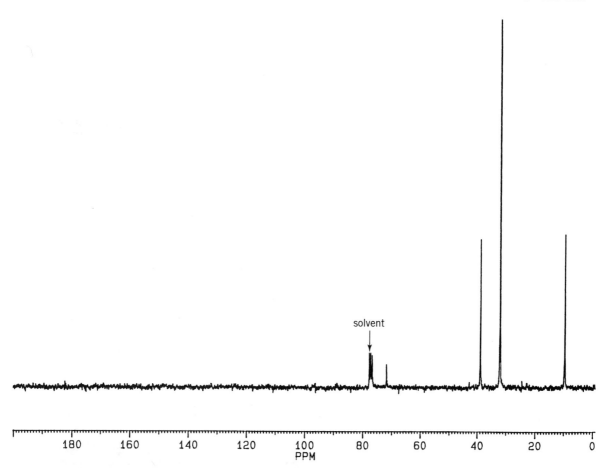

Figure 13.10   62.9-MHz ^{13}C-NMR spectrum of 2-chloro-2-methylbutane in CDCl$_3$.

Acid-Base Extraction

14.1 INTRODUCTION

As discussed in Unit 5, separation of chemicals by liquid-liquid extraction using an organic solvent and water is possible only if the compounds have very different polarities, and therefore, very different partition coefficients between the two solvents. If the compounds have similar affinities for both solvents, separation by liquid-liquid extraction is still possible if one of the components has acid-base properties.

Most organic compounds with six or more carbon atoms have limited solubility in water but are freely soluble in organic solvents such as methylene chloride or *tert*-butyl methyl ether. (Exceptions to this rule are ionic and very polar compounds, such as amino acids and sugars, which dissolve well in water but are insoluble in most organic solvents.) The solubility of water-insoluble organic acids and bases can be drastically changed by acid-base reactions. Organic acids and bases can be transformed into ionic species by reaction with inorganic bases and acids. The resulting *salts*, by virtue of their high polarity, are very soluble in water and have limited solubility in most organic solvents.

The use of acid-base reactions for the separation of organic chemicals by liquid-liquid extraction is the subject of the present unit. Before we attack the problem of acid-base extraction, let's review some important concepts about acids and bases.

14.2 ACIDS AND BASES

Almost all organic compounds display to some extent acid-base properties. Although there are several definitions of acids and bases, for the time being we will restrict the concept of **acids** to those **compounds which give protons to the medium**; a **base**, on the other hand, is **a compound capable of accommodating a proton in its structure** (Brønsted–Lowry theory). We will confine our discussion of acid-base properties to aqueous solutions.

Unlike inorganic compounds, most organic acids and bases are weak or moderately weak. In aqueous solution an organic acid (AH) exists in equilibrium with its conjugate base (A^-):

$$AH + H_2O \underset{}{\overset{K_a}{\rightleftharpoons}} A^- + H_3O^+ \qquad (1)$$

where

$$K_a = \frac{[A^-][H_3O^+]}{[AH]} \qquad (2)$$

The dissociation constant, K_a, is usually expressed by the pK_a value ($pK_a = -\log K_a$). Thus, if $K_a = 10^{-5}$, then $pK_a = 5$. Note that the stronger the acid, the larger the K_a, and the lower the pK_a.

Similarly, organic bases (B) in water are in equilibrium with the corresponding conjugate acid (BH^+) and produce hydroxide ions:

$$B + H_2O \overset{K_b}{\rightleftharpoons} BH^+ + HO^- \tag{3}$$

where

$$K_b = \frac{[BH^+][HO^-]}{[B]} \tag{4}$$

There exists a relationship between K_a and K_b for any conjugate acid and base pair. If the base is represented by A^-, Equation 3 can be rewritten as follows:

$$A^- + H_2O \overset{K_b}{\rightleftharpoons} AH + HO^- \tag{5}$$

where

$$K_b = \frac{[AH][HO^-]}{[A^-]} \tag{6}$$

The product of K_a by K_b is a constant, the **water constant**, K_w:

$$K_a \times K_b = \frac{[A^-][H_3O^+]}{[AH]} \times \frac{[AH][HO^-]}{[A^-]} = [H_3O^+] \times [HO^-] = K_w \tag{7}$$

Applying the function $-\mathbf{log}$ to both sides of Equation 7, and recalling that $K_w = 10^{-14}$, Equation 9 is obtained:

$$-\log K_a - \log K_b = -\log K_w \tag{8}$$

$$\Longrightarrow \boxed{pK_a + pK_b = 14} \tag{9}$$

Equation 9 indicates that a change in pK_a is compensated by a change in pK_b, so that their sum remains constant at 14. It follows that strong acids (low pK_a values) have weak conjugate bases (high pK_b), and vice versa.

Let's now consider the relationship between K_a and the pH. By applying $-\mathbf{log}$ to both sides of Equation 2, the following relationship is obtained:

$$-\log K_a = -\log[H_3O^+] - \log \frac{[A^-]}{[AH]} \tag{10}$$

therefore:

$$\boxed{pK_a = pH - \log \frac{[A^-]}{[AH]}} \tag{11}$$

When $pH = pK_a$, then $-\log \frac{[A^-]}{[AH]} = 0$, and $[A^-] = [AH]$. In other words, when the pH of a solution is equal to the pK_a value of the dissolved acid, the concentrations of the acid, $[AH]$, and its conjugate base, $[A^-]$, are identical. For example, the pK_a of acetic acid is 4.74. If acetic acid is dissolved in water and the pH adjusted by addition of a sodium hydroxide solution, when the pH reaches 4.74, the concentrations of acetic acid (AH) and sodium acetate (Na^+A^-) are the same. It can also be derived from Equation 11 that if $pH < pK_a$, then $[AH] > [A^-]$. This is equivalent to saying that at pHs lower than pK_a, the concentration of the acid (AH) is larger than the concentration of its conjugate base (A^-).

Some typical families of organic acids and bases are shown along with their pK_a values in Table 14.1.

Table 14.1 **pK_a Range of Some Typical Organic Compounds**

Acid	Conjugate base	Family	pK_a
R—COOH	R—COO$^-$	carboxylic acids	3–5
Ar—OH	Ar—O$^-$	phenols	7–10
R—SO$_3$H	R—SO$_3$$^-$	sulfonic acids	−2–3
Ar—SH	Ar—S$^-$	thiophenols	7–9
R—SH	R—S$^-$	thiols	11–13
R—CO—CH$_2$—CO—R	R—CO—C̄H—CO—R	β-dicarbonyl compounds	9–13
R—OH	R—O$^-$	alcohols	15–20
R—CH$_2$—COOR	R—C̄H—COOR	esters	19–25
R—NH$_3$$^+$	R—NH$_2$	aliphatic amines	9–11
Ar—NH$_3$$^+$	R—NH$_2$	aromatic amines	2–4

14.3 STRUCTURAL EFFECTS ON ACID-BASE PROPERTIES

Aliphatic Carboxylic Acids

The dissociation constants of organic acids and bases are strongly affected by the electronegativity of substituents near the acidic (or basic) center of the molecule. For example, substitution of one H atom by chlorine on the methyl group of acetic acid results in a 77-fold increase in its K_a. Two chlorine atoms increase the K_a value by a factor of 3000, and trichloroacetic acid is 13,000 times stronger than acetic acid. The pKa values for substituted acetic acids are shown in Table 14.2.

$$R\text{-COOH} + H_2O \underset{}{\overset{pK_a \approx 3\text{-}5}{\rightleftharpoons}} R\text{-COO}^- + H_3O^+$$

a carboxylic acid   a carboxylate anion

Table 14.2 **Dissociation of Acetic Acids**

Compound	K_a	pK_a
CH$_3$COOH	1.8 10^{-5}	4.74
ClCH$_2$COOH	1.3 10^{-3}	2.87
Cl$_2$CHCOOH	5.5 10^{-2}	1.26
Cl$_3$CCOOH	0.30	0.52

Electron-withdrawing substituents such as chlorine stabilize the anionic form of carboxylic acids, called the **carboxylate anion**, and therefore increase their acidity. The C—X bond is polarized, and the favorable orientation of this dipole, with the positive end near the negative carboxylate anion, is responsible for the acid-strengthening effect.

The more electronegative the substituent, the more pronounced the effect. The increase in acidity fades away as the substituent is farther away from the carboxyl group. In Table 14.3, the K_a values of different butanoic acids are gathered. Notice that 2-chlorobutanoic acid is 100 times more acidic than butanoic acid but 4-chlorobutanoic acid is only slightly more acidic.

Table 14.3 Dissociation Constants of Butanoic Acids

Compound	K_a	pK_a
$CH_3CH_2CH_2CO_2H$	1.5×10^{-5}	4.82
$CH_3CH_2CHCO_2H$ $\quad$ $\mid$ $\quad$ Cl	1.4×10^{-3}	2.86
$CH_3CHCH_2CO_2H$ $\quad$ $\mid$ $\quad$ Cl	8.9×10^{-5}	4.05
$ClCH_2CH_2CH_2CO_2H$	3.2×10^{-5}	4.50

By comparing the K_a values for acetic acid ($pK_a = 4.74$) and butanoic acid ($pK_a = 4.82$), it can be deduced that alkyl groups, because of their inductive effect, decrease K_a values slightly.

Benzoic Acids

Benzoic acid is only slightly more acidic than acetic acid but substituents on the aromatic ring have an important influence on the K_a values (Table 14.4).

S represents a substituent anywhere on the ring.

Table 14.4 pK_a Values of Substituted Benzoic Acids

Substituent	ortho	meta	para
$-H$	4.20	4.20	4.20
$-CH_3$	3.90	4.27	4.37
$-F$	3.27	3.87	4.14
$-Cl$	2.88	3.83	3.99
$-Br$	2.85	3.81	3.98
$-OCH_3$	4.09	4.08	4.48
$-OH$	2.98	4.08	4.59
$-NH_2$	4.79	4.72	4.85
$-CN$	3.14	3.60	3.55
$-NO_2$	2.18	3.46	3.44

Substituents affect the dissociation constant through inductive, resonance, and steric effects. Electron-withdrawing substituents, such as halogens, cyano, and nitro groups, increase the acidity regardless of their position. Electron-donating substituents, on the other hand (for example, methyl, hydroxyl, and amino groups), decrease the acidity of benzoic acids, especially from the para position. With the exception of $-NH_2$, all substituents on the ortho position have an acid-strengthening effect. Steric hindrance on the undissociated acid accounts for this phenomenon. With substituents on the ortho position, the carboxyl group can no longer be coplanar with the aromatic ring and therefore it is less stabilized by resonance. The coplanarity, and therefore the stabilization, is somewhat regained when the carboxylate anion is formed. The net effect is an increase in acidity.

Phenols

Phenols are four to five orders of magnitude less acidic than carboxylic acids, but they are much stronger acids than alcohols. The dissociated form of a phenol is called a **phenolate anion**.

S represents a substituent
anywhere on the ring

a phenolate
anion

Substituents on the aromatic ring have a pronounced effect on the acidity of phenols. K_a values for several substituted phenols are gathered in Table 14.5. The same effects that operate on benzoic acids are also present in phenols.

Table 14.5 pK_a Values of Substituted Phenols

Substituent	ortho	meta	para
$-H$	10.00	10.00	10.00
$-CH_3$	10.29	10.00	10.26
$-F$	8.81	9.28	9.81
$-Cl$	8.48	9.02	9.38
$-Br$	8.42	8.87	9.26
$-OCH_3$	9.98	9.65	10.20
$-CN$	—	8.61	7.95
$-NO_2$	7.22	8.36	7.15

The interaction between the electron pairs on the oxygen atom of phenols and the aromatic ring is very strong. Substituents such as the nitro group, with a powerful electron-withdrawing resonance effect, have a stronger effect on phenols than on benzoic acids, especially from the ortho and para positions. The stabilization of the phenolate anion by resonance accounts for this effect.

p-Nitrophenol is three orders of magnitude more acidic than phenol. 2,4-Dinitrophenol has a pK_a of 4.09 (note that it is even more acidic than acetic acid) and 2,4,6-trinitrophenol (picric acid) has p$K_a = 0.25$, and therefore is almost as acidic as trichloroacetic acid.

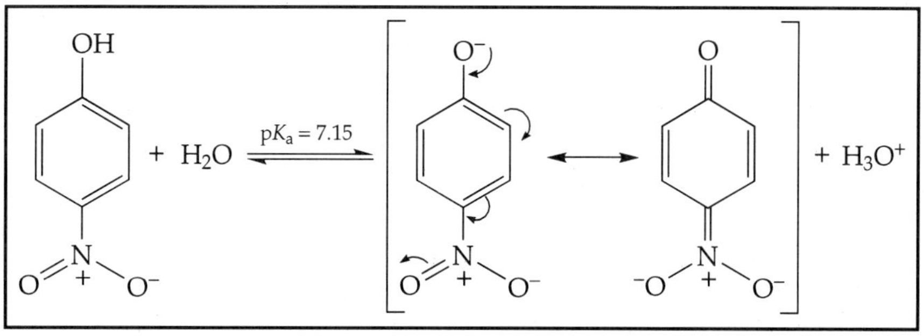

Amines

Because of the presence of the lone electron pair on the nitrogen atom, amines are Lewis bases. They can accommodate a proton, producing ammonium ions. Their reaction with water, in which HO^- is also produced, is shown below:

$$R_3N: + H_2O \xrightleftharpoons{K_b} R_3\overset{+}{N}H + HO^-$$

an ammonium ion

The extent of the protonation depends on the pK_b value of the amine and the pH of the solution. Aliphatic amines display pK_b values in the range 3–5. Instead of using pK_b values, it is customary to use the pK_a value of the conjugate acid, the ammonium ion. The more basic the amine, the higher the pK_a of the ammonium ion (the less acidic the ammonium ion).

$$R_3\overset{+}{N}H + H_2O \xrightleftharpoons{K_a} R_3N: + H_3O^+$$

Because of the inductive effect of the alkyl groups, amines are more basic than ammonia. For the same reason, secondary amines are, in general, more basic than primary amines. If only inductive effects were operating one would expect tertiary amines to be more basic than secondary amines. However, this is not the case; because of the steric hindrance around the nitrogen atom, tertiary amines are less basic than secondary amines. The pK_a values for the ammonium salts of some typical aliphatic amines are shown in Table 14.6.

Aromatic amines, **anilines**, are very weak bases. The aromatic ring attached to the nitrogen atom strongly decreases the basicity by delocalizing the lone electron pair.

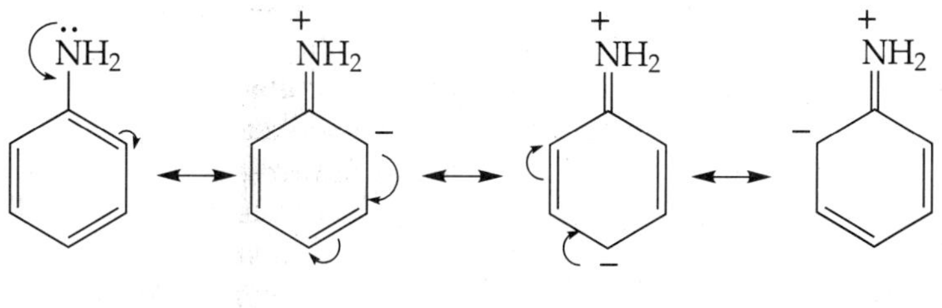

The pK_a values for the ammonium ions of selected anilines are shown in Table 14.7. Aromatic amines are about five to seven orders of magnitude less basic than aliphatic amines.

The same factors that increase the acidity of phenols, decrease the basicity of anilines. Electron-withdrawing groups such as halogens, cyano, and nitro groups decrease the basicity of the nitrogen atom.

Pyridines are an important class of aromatic amines. Unlike anilines, where the lone electron pair is delocalized in the aromatic ring, the lone electron pair on the nitrogen atom of a pyridine occupies an sp^2-hybrid orbital, which lies on the plane of

Table 14.6 pK_a Values for the Ammonium Ions of Some Typical Aliphatic Amines

Compound	pK$_a$
NH$_4$$^+$	9.74
CH$_3$NH$_3$$^+$	10.62
(CH$_3$)$_2$NH$_2$$^+$	10.77
(CH$_3$)$_3$NH$^+$	9.80
CH$_3$CH$_2$NH$_3$$^+$	10.63
(CH$_3$CH$_2$)$_2$NH$_2$$^+$	10.94
(CH$_3$CH$_2$)$_3$NH$^+$	10.72

Table 14.7 pK_a Values for the Ammonium Ions of Substituted Anilines

Substituent	ortho	meta	para
–H	4.63	4.63	4.63
–CH$_3$	4.41	4.70	5.08
–F	3.20	3.58	4.65
–Cl	2.64	3.52	3.99
–Br	2.53	3.53	2.88
–OCH$_3$	4.48	4.30	5.30
–CN	1.00	2.70	1.70
–NO$_2$	–0.20	2.50	1.00

the aromatic ring and does not overlap with the π electron cloud of the ring. Because of this, one would expect pyridines to be more basic than anilines, perhaps similar in basicity to aliphatic amines. However, the sp^2-orbital, where the pyridine lone electron pair resides, has more s-character than the sp^3 orbitals of aliphatic amines. As a consequence of this increased s-character, the lone electron pair of a pyridine is closer to the nitrogen nucleus, which results in a greater electrostatic attraction between the nucleus and the electrons, and in a lower basicity. Overall, pyridines are similar in basicity to anilines.

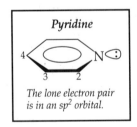

Pyridine

The lone electron pair is in an sp^2 orbital.

Substituents on the pyridine ring have effects similar to those in substituted anilines. The pK$_a$ values for some typical pyridines are shown in Table 14.8.

14.4 ACID-BASE EXTRACTION

Separation of Carboxylic Acids

The solubility of organic compounds in water depends strongly on their polarity and size. The number of polar and ionic centers, the charge distribution, and the molecular

Table 14.8 **pK_a Values for the Ammonium Ions of Substituted Pyridines**

Substituent	Position 2	Position 3	Position 4
$-H$	5.25	5.25	5.25
$-CH_3$	5.94	5.63	6.03
$-Cl$	0.49	2.84	3.84
$-Br$	0.90	2.84	3.78
$-OCH_3$	3.06	4.91	6.47
$-NO_2$	—	0.81	—

size and shape are all factors that affect the solubility behavior. In general, the larger the polarity, the larger the solubility in water. Most nonionic organic compounds with six or more carbon atoms are only sparingly soluble in water (with the exception of sugars and other very polar compounds). Ionic organic compounds, on the other hand, such as sodium and potassium salts of organic acids, are more soluble in water than the corresponding acids. For example, the solubility of sodium benzoate (the sodium salt of benzoic acid) in water at 20°C is 62.8 g/100 mL, whereas the solubility of benzoic acid is only 0.29 g/100 mL. At the same time, the undissociated free acids are more soluble in organic solvents, such as toluene or methylene chloride, than in water.

$$R\text{-COOH} + NaOH \longrightarrow R\text{-COO}^- Na^+ + H_2O$$

Carboxylic acids with six or more carbons are sparingly soluble in water but soluble in organic solvents.

Sodium salts of carboxylic acids are very soluble in water and insoluble in most organic solvents.

We can take advantage of the differential solubility of ionic and nonionic compounds in water and organic solvents to separate organic mixtures.

Let's suppose that we have a solution of benzoic acid and naphthalene in methylene chloride (other organic solvents immiscible with water, such as toluene or *tert*-butyl methyl ether, can be used instead). We add a few milliliters of water and shake the system. After equilibrium is attained, benzoic acid (B) and naphthalene (N) will be present in both phases.

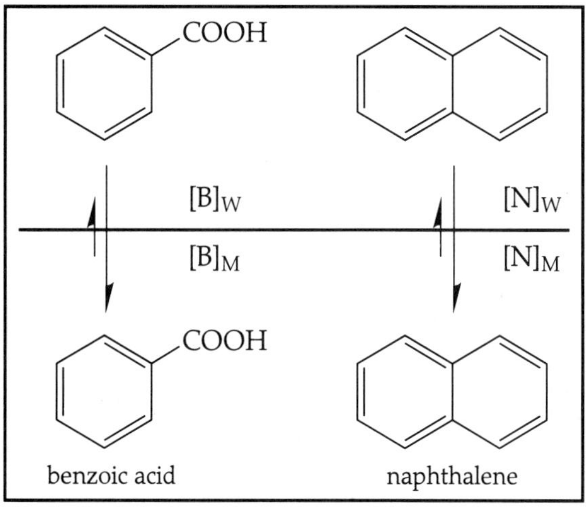

Water layer (W)

$[B]_W$ $[N]_W$

$[B]_M$ $[N]_M$

benzoic acid   naphthalene

Methylene chloride layer (M)

The concentration of each compound in each phase is determined by the partition coefficients of benzoic acid ($K^B_{W/M}$) and naphthalene ($K^N_{W/M}$) between water (W) and methylene chloride (M):

$$K^N_{W/M} = \frac{[N]_W}{[N]_M} \quad \text{and} \quad K^B_{W/M} = \frac{[B]_W}{[B]_M}$$

Since both compounds have six or more carbon atoms and are relatively nonpolar, they have more affinity for the organic solvent than for water, and therefore they remain mainly in the methylene chloride phase ($K^B_{W/M} \ll 1$ and $K^N_{W/M} \ll 1$).

Let's suppose now that the same mixture of benzoic acid and naphthalene in methylene chloride is shaken with a volume of a 10% NaOH aqueous solution. At the interface between the layers a neutralization reaction takes place. Benzoic acid reacts with sodium hydroxide to give sodium benzoate:

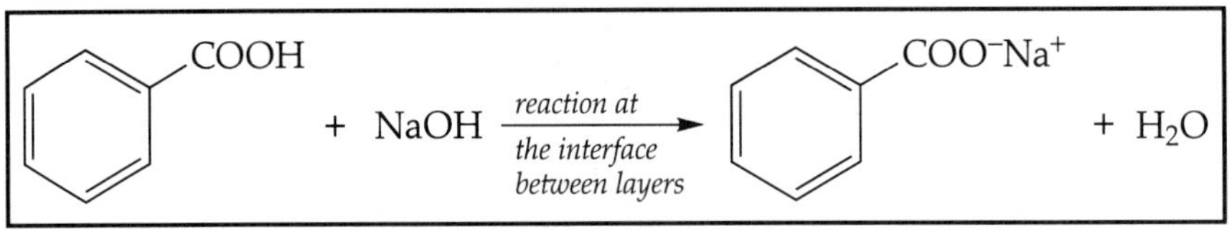

Sodium benzoate is an *ionic* compound and its solubility in water is larger than in methylene chloride. As a consequence of this reaction, benzoic acid is driven from the organic phase into the aqueous phase. Naphthalene, which is a neutral compound, remains mainly in the organic phase.

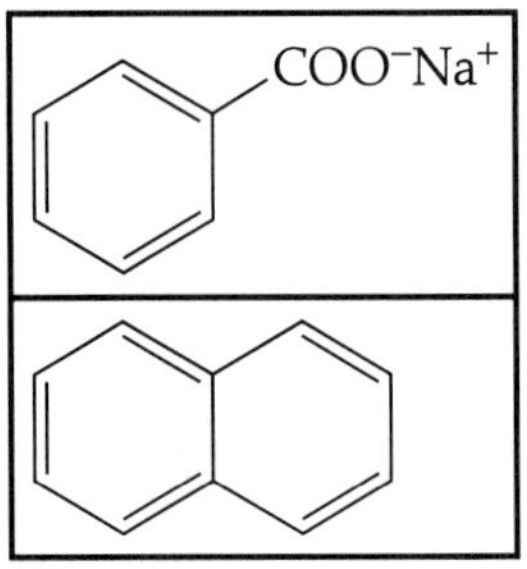

Basic water layer (W)

Methylene chloride layer (M)

In order to ensure that the separation is complete, several extractions (two or three) with aqueous sodium hydroxide are usually performed. After separation of the phases, addition of a strong mineral acid, such as HCl, to the aqueous phase regenerates benzoic acid.

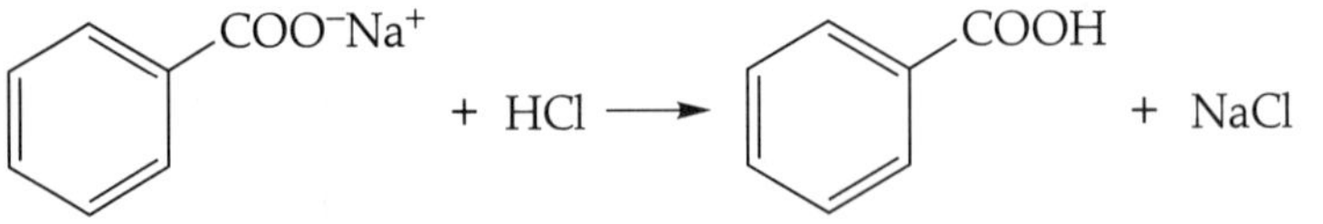

As mentioned before, carboxylic acids with six or more carbon atoms have a limited solubility in water. When HCl is added to the basic aqueous phase, the carboxylic acid precipitates from solution and can be recovered by filtration. The efficiency of the recovery is dictated by the solubility of the acid in water. A more general approach to recover the acid involves several extractions of the acid-containing aqueous phase with an organic solvent. In this way it can be guaranteed that the recovery of the acid is almost complete. The organic solvent is then removed by evaporation to give the pure acid. A flowchart showing the separation of naphthalene and benzoic acid is given below.

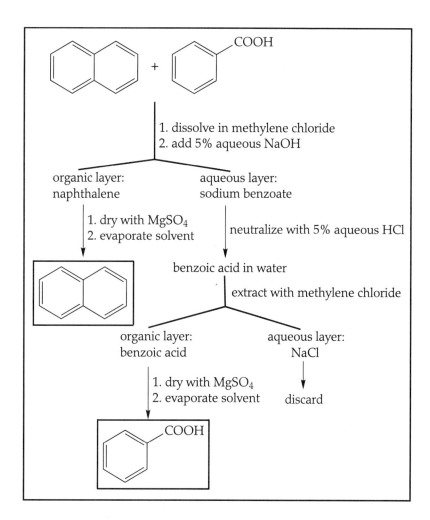

Separation of Phenols

Because the pK_a values of most phenols are in the range 7–10, they are fairly soluble in aqueous NaOH solutions provided the pH is higher than 10. Similar to carboxylic acids, phenols react with NaOH and can be separated from neutral compounds by extraction with aqueous NaOH solutions.

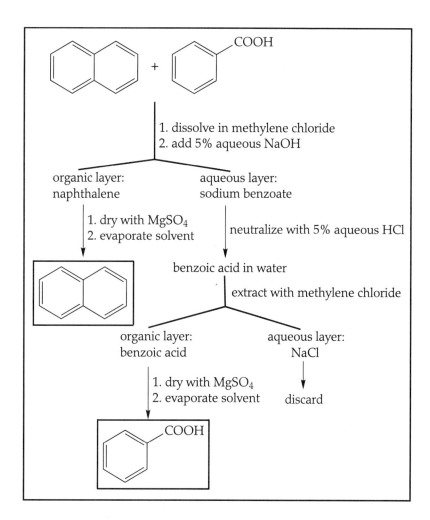

Most substituted phenols are sparingly soluble in water but soluble in organic solvents.

Sodium salts of substituted phenols are soluble in water but insoluble in most organic solvents.

Phenols can also be separated from carboxylic acids by taking advantage of their different pK_a values. Phenols do not dissolve in weakly basic solutions such as sodium bicarbonate in water. Sodium bicarbonate is a weak base and the maximum pH that can be achieved by dissolving it in water is around 8.5. This pH is not high enough for full dissociation of most phenols in water (only ortho and para nitrophenols are sufficiently acidic to be dissociated at pH 8). On the other hand, most carboxylic acids have a pK_a value in the range 3–5, making them completely dissociated and soluble in a bicarbonate solution of pH 7–8.

The differential solubility of carboxylic acids and phenols in sodium bicarbonate solutions can be successfully used to separate them. For example, a mixture of *p*-bromobenzoic acid (pK_a = 3.97) and *p*-bromophenol (pK_a = 9.26) can be separated by acid-base extraction using a 5% sodium bicarbonate solution (pH 8). At this pH, *p*-bromobenzoic acid is completely dissociated and thus it is extracted into the aqueous phase. The phenol, on the other hand, is not acidic enough to be dissociated by the sodium bicarbonate solution and remains in the organic phase. The flowchart for the separation is shown below.

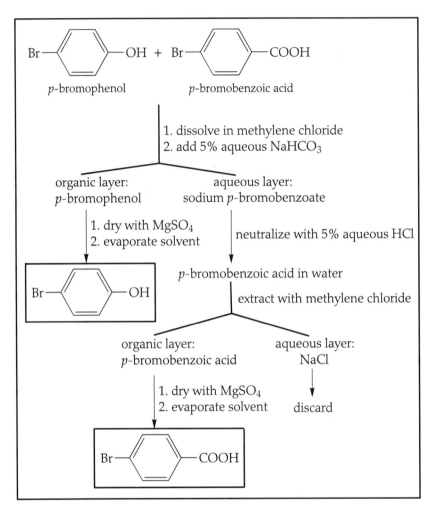

As usual, after the separation of the layers, the phenol can be recovered by evaporation of the organic solvent. The bromobenzoic acid can be collected by acidification of the aqueous solution with 5% HCl, followed by filtration of the precipitated free acid or, most efficiently, by extraction into an organic solvent followed by evaporation.

Separation of Amines

In dilute HCl aqueous solutions (pH 1–3) most amines, including anilines and pyridines, are protonated. The resulting ammonium salts are ionic and readily soluble in water.

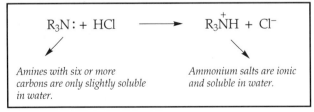

Acid-base extraction can be used to separate amines from neutral and acidic compounds. For example, a mixture of *n*-decane (b.p. 174°C) and *n*-octylamine (b.p. 180°C) is difficult to separate by distillation. Both compounds can be separated by acid-base extraction. The separation is outlined in the flowchart.

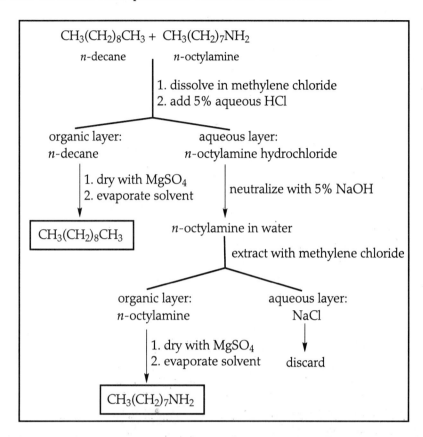

After the mixture is dissolved in methylene chloride (or some other appropriate organic solvent), a 5% aqueous solution of HCl is added. The alkane does not react with the acid and stays in the organic phase. The amine, on the other hand, forms the ammonium chloride, which by virtue of its ionic character is readily soluble in water and is extracted into the aqueous phase. After the layers have been separated, the amine is recovered by neutralization with a dilute NaOH solution and extraction with an immiscible organic solvent.

$$R_3\overset{+}{N}H + Cl^- + NaOH \longrightarrow R_3N + Cl^- + Na^+ + H_2O$$

14.5 OVERVIEW

Separation of a mixture of organic compounds by acid-base extraction is possible if the compounds are insoluble in water and at least one of them is a base or an acid. In either case the separation is based on the *ionization* of the compound with acid-base properties and its *transfer from the organic phase into the aqueous phase*. Organic acids are ionized by reaction with NaOH or KOH, and organic bases by HCl or other mineral acids.

Organic compounds can be classified into three categories depending on their acid-base properties: acidic, basic, and neutral, as depicted in Table 14.9. It should be borne in mind that this classification is based on acid-base reactions in aqueous environments where the pH range is usually 1–12. Compounds with very high pK_a, such as alcohols (pK_a 15–20), are not dissociated under the conditions of acid-base extractions and therefore are considered neutral for that purpose.[1]

Table 14.9 **Classification of Most Common Functional Group Families According to Their Acid-Base Properties in Aqueous Solutions**

Acids	Bases	Neutral
carboxylic acids: R—COOH	aliphatic and aromatic amines (1°, 2°, and 3°): R—NH$_2$; Ar—NH$_2$	hydrocarbons
		ethers
sulfonic acids: R—SO$_3$H	*pyridines and quinolines*	alcohols
		aldehydes
phenols: Ar—OH		ketones
		esters
thiophenols: Ar—SH		nitriles
		amides
		nitro compounds

Acid-base extraction can be successfully applied to separate binary, ternary, and even quaternary mixtures, provided the components have different acid-base properties. Possible types of binary mixtures that can be separated by acid-base extraction are summarized in Table 14.10.

Table 14.10 **Separation of Possible Binary Mixtures by Acid-Base Extraction**

Compound A	Compound B	Separation using
carboxylic acid[a]	neutral	NaOH
carboxylic acid[a]	amine	NaOH or HCl
carboxylic acid[a]	phenol[b]	NaHCO$_3$
phenol[b]	neutral	NaOH
phenol[b]	amine	NaOH or HCl
amine	neutral	HCl

[a]or sulfonic acid;
[b]or thiophenol

It should be pointed out that acid-base extraction is useful only when the compounds in their nonionized form are insoluble in water. If the nonionized compounds are already soluble in water (for instance, compounds with six carbon atoms or fewer, very polar compounds, and sugars) acid-base reactions will not change their solubility behavior, and therefore will not be useful to separate such mixtures.

BIBLIOGRAPHY

1. Vogel's Textbook of Practical Organic Chemistry. A.I. Vogel, B.S. Furniss, A.J. Hannaford, P.W.G. Smith, and A.R. Tatchell. 5th ed. Longman, Harlow, UK. 1989.

2. Introduction to Organic Chemistry. A. Streitwieser, Jr., and C. Heathcock. 2nd ed. Macmillan, New York, 1981.

3. Lange's Handbook of Chemistry. J.A. Dean, ed. 13th ed. McGraw-Hill, New York, 1985.

4. The Merck Index. M. Windholz, ed. 10th ed. Merck & Co., Rahway, N.J., 1983.

---

[1] Very strong bases and nonaqueous solutions are necessary to dissociate alcohols into R—O$^-$ to any significant degree.

Problem 1

Calculate the pH of the following aqueous solutions. Use the pK_a values given in Tables 14.2 and 14.8.

(a) 0.1 M acetic acid

(b) 0.1 M pyridine

Problem 2

Calculate the concentration of acetic acid (AH) and acetate anion (A^-) at the following pHs. The total concentration ($[AH] + [A^-]$) is 0.1 M.

(a) pH = 3

(b) pH = 5

(c) pH = 7

Problem 3 (advanced level)

The pK_a values of several substituted acetic acids are given in the table. Using the pK_a values of meta substituted benzoic acids given in Table 14.4, plot the pK_a values of substituted acetic acids against the pK_a's of meta substituted benzoic acids. If the pK_a of phenylacetic acid is 4.31, estimate the pKa of 3-carboxybiphenyl.

pK_a Values of $X-CH_2-COOH$

$-X$	pK_a
$-H$	4.74
$-CH_3$	4.87
$-Cl$	2.85
$-Br$	2.90
$-CN$	2.47
$-OH$	3.83
$-OCH_3$	3.57

Problem 4

Arrange the following compounds in order of decreasing acidity: hydrochloric acid; benzoic acid; *p*-nitrobenzoic acid; cyclohexylamine hydrochloride; trifluoroacetic acid.

Problem 5 (advanced level)

(a) Benzoquinuclidine is a tertiary aromatic amine. Explain why it is less basic than quinuclidine and more basic than *N,N*-diethylaniline. The pK_a values of the conjugate acids are given.

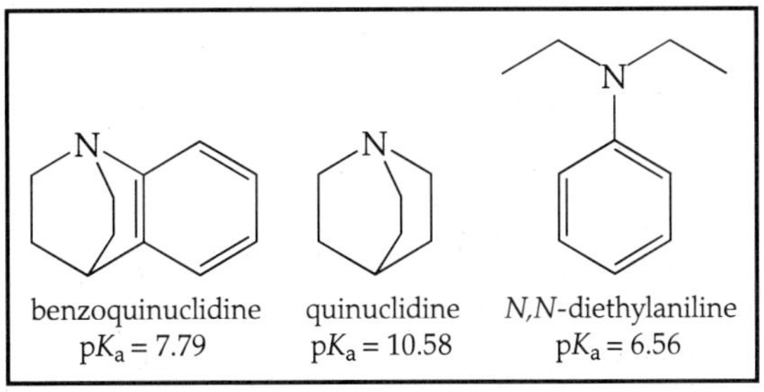

benzoquinuclidine quinuclidine *N,N*-diethylaniline
$pK_a = 7.79$ $pK_a = 10.58$ $pK_a = 6.56$

(b) The pK_a values of 2,5-dihydroxybenzoic acid and 3,5-dihydroxybenzoic acid are 2.97 and 4.04, respectively. Explain.

Problem 6

How would you separate 10 g of a mixture of *m*-chlorobenzoic acid and *m*-chlorotoluene by acid-base extraction?

Problem 7

How would you separate 5 g of a mixture of 2-hydroxybenzamide (salicylamide) and acetanilide (*N*-acetylaniline)?

salicylamide   acetanilide

Problem 8

How would you separate a mixture of 4-ethoxyaniline and 4-nitrophenol by acid-base extraction?

4-ethoxyaniline   4-nitrophenol

Problem 9

How would you separate a mixture of biphenyl, aniline, and 4-carboxybiphenyl by acid-base extraction?

biphenyl   aniline   4-carboxybiphenyl

Problem 10

How would you separate a mixture of eugenol and acetyleugenol by acid-base extraction?

eugenol   acetyleugenol

Isolation of Eugenol from Cloves

E14A.1 ANCIENT MEDICINE

Botanical medicine is a timeless art. The so-called "primitive cultures" have long been aware of the healing properties of plant extracts. The natives of Peru made an extract from the bark of the cinchona tree (*Cinchona officinalis*), that was used as an antipyretic and antimalarial agent. The beverage was introduced in Spain in 1639 and its use shortly spread to Italy and other European countries through a network of Jesuit priests. However, it was not until the end of the seventeenth century that the powers of this miraculous decoction were widely recognized, after saving the lives of Charles II, King of England, and the Dauphin of France. Two centuries later, it was found that the therapeutic properties of the extract were due to the alkaloid *quinine*. In 1856, in a failed attempt to prepare quinine from anilines, Henry Perkin discovered "aniline purple" or "mauve," a dyestuff that revolutionized Victorian fashion and launched the modern chemical industry (see Unit 20).

The Incas in South America used to chew coca leaves to withstand strenuous working conditions at 3,600 m (12,000 ft.) above sea level. This tradition still lives on among their descendants in Peru and Bolivia. The active ingredient in coca leaves, *cocaine*, is a potent local anesthetic that for quite some time in the nineteenth century was regarded as a wonder-drug. It was largely prescribed to ameliorate throat and mouth ailments. It was soon recognized, however, that cocaine had also devastating effects. Numerous reports on its addictive nature and acute toxicity precipitated its fall from grace. Today, its illegal trade involves more than 6 million Americans.

Mint, among many other herbs, was widely used by the Romans against heat and cold injuries, spasms, headaches, eye infections, dandruff, serpent bites, and even amorous dreams. Thyme was used as a disinfectant by the Greeks, who believed that the smoke of burning thyme would drive evil forces away.

Much of the early study in organic chemistry centered around the isolation, structure elucidation, and synthesis of natural compounds. Today, the search for natural products with useful medicinal applications is one of the major goals of pharmaceutical companies. As with other drugs, herbal medicines must be used with prudence; despite current public perception, the most potent venoms known to humans are not man-made but naturally occurring products.

Active ingredients from plants can be obtained by several methods; among them, steam distillation affords the so-called **essential oils**. As we have already discussed in Units 6 and 11, essential oils are complex mixtures of relatively volatile and odorous compounds immiscible with water. These compounds co-distill with water below 100°C, at temperatures substantially lower than their boiling points (section 6.11). Under these conditions only minimum decomposition and oxidation of the oils occur. Essential oils can be further separated into their components by chromatography, vacuum distillation, or acid-base extraction depending on the characteristics of the individual compounds.

E14A.2 EUGENOL FROM CLOVES: OVERVIEW OF THE EXPERIMENT

In this experiment you will isolate the essential oil from cloves by co-distillation with water. Cloves are the flower buds from a tree (*Eugenia caryophyllata*) native to the

Molucca Islands of Indonesia. Today, Madagascar and Tanzania (on the islands of Zanzibar and Pemba), off the east coast of Africa, are the major producers of cloves. The evergreen tree grows to about 50 ft high and yields up to 80 lb of the sun-dried buds. When ready to be picked, the flower buds are about 3/4 in. long and pink colored. The essential-oil content of cloves averages 17% by weight; usually only half of that is isolated by water co-distillation; the amount of oil in the aqueous distillate ranges between 0.6 and 0.8% (by weight).

In contrast with many other essential oils that are complex mixtures, the oil of cloves has only three major components: **eugenol, acetyleugenol** or **eugenol acetate**, and **β-caryophyllene**. Eugenol accounts for 70−90% of the oil, acetyleugenol ranges from 2−17%, and β-caryophyllene comprises 5−12%. Even though these three components together account for more than 99% of the clove oil, part of its fruity fragrance is due to a minor component: 2-heptanone.

Main components of clove oil

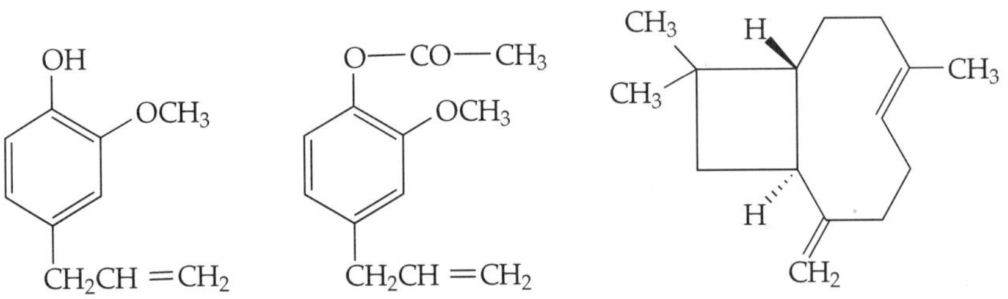

Eugenol
(≈ 85%)

Acetyleugenol
(≈ 10%)

β–Caryophyllene
(≈ 5%)

Eugenol is acidic because of the phenol group. Eugenol acetate is an ester, and caryophyllene is a hydrocarbon that belongs to the family of sesquiterpenes. Eugenol acetate and caryophyllene are neutral and can be successfully separated from eugenol by acid-base extraction. A flowchart of the separation is shown in Figure 14.1.

Clove oil is a food flavoring agent as well as a dental anesthetic. The Food and Drug Administration has granted eugenol and acetyleugenol GRAS (Generally Regarded As Safe) status for food and medicinal use.

PROCEDURE

E14A.3   ISOLATION OF EUGENOL

> **Background reading:** 5.3; 5.5; 6.11; 14.1−5
> **Estimated time:** isolation: 3 hr; analysis: 3 hr

Safety First

- *tert*-Butyl methyl ether is flammable.
- When releasing the pressure point the separatory funnel stem away from your face and others.
- NaOH and HCl are corrosive. HCl is highly toxic. Handle these solutions with care.

Isolation of the Essential Oil   Weigh about 10 g of fresh cloves (stored in a closed container) and grind them using a blender or a mortar and pestle. Transfer the powder to a 500-mL distillation flask an add 200 mL of distilled water. Assemble a simple distillation apparatus (Unit 6) as shown in Figure 14.2. If available, use plastic clips instead of rubber bands (Fig. 3.22). Read "Assembling the Distillation Apparatus" in Unit 6, page 136. In a 100-mL round-bottom flask, pour 60 mL of water and mark the level of the liquid with a marker or tape. Pour the water out and use this flask as the receiving flask for the distillate.

Heat the distillation flask using a heating mantle. Adjust the heat supply so the liquid distills at an approximate rate of one drop per second. Distill the mixture until about 60 mL of distillate has been collected. Read the temperature in the distillation head during the co-distillation.

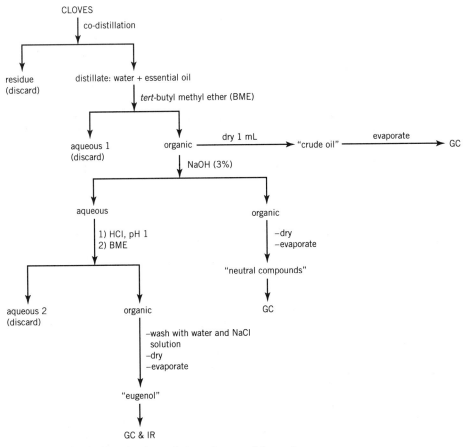

Figure 14.1   Flowchart for the isolation of eugenol from cloves.

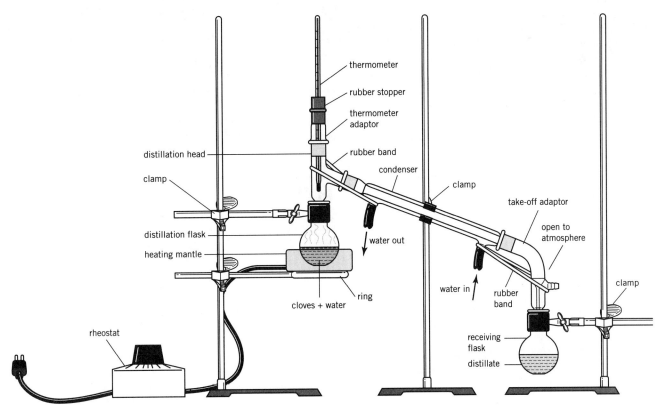

Figure 14.2   Distillation apparatus for the isolation of clove oil.

Separation of the Components Transfer the distillate to a 125-mL separatory funnel and extract it with a 25-mL portion of *tert*-butyl methyl ether (BME) by shaking well and venting occasionally (see section 5.3). Which layer is the organic layer? Separate the layers and extract the aqueous layer again with another 25-mL portion of BME. Collect the aqueous layer in a beaker labeled "aqueous 1." This layer will be kept until the end of the experiment and then it will be discarded. Collect the organic layers in a 125-mL Erlenmeyer flask and transfer a 1-mL aliquot to a screw-capped vial labeled "**crude oil**," add a little anhydrous magnesium sulfate to the vial, and set aside for GC analysis.

Transfer the combined organic layer back to the separatory funnel and extract it with a 25-mL portion of a 3% NaOH solution. Separate the layers and collect the aqueous layer in a 500-mL Erlenmeyer flask. Extract the remaining organic layer two more times with 25 mL (each time) of the 3% NaOH solution. Collect all the basic aqueous layers in the 500-mL Erlenmeyer flask and keep it aside for the next step. In the meantime, transfer the organic layer to a 125-mL Erlenmeyer flask labeled "**neutral compounds**." Add enough anhydrous magnesium sulfate until the solid runs freely in suspension; let it dry for 5 minutes. Filter the suspension using a glass funnel with a cotton plug, and collect the liquid in a dry and pre-tared 100-mL round-bottom flask. Evaporate the solvent using a rota-vap (Figure 3.17). Calculate the weight of the "neutral-compounds" fraction by difference. With the aid of 0.5–1 mL of BME transfer the residue from the round-bottom flask to a screw-capped vial labeled "**neutral compounds**," and set aside for GC analysis.

Acidify the combined aqueous layer, in the 500-mL Erlenmeyer flask, by slowly adding with a Pasteur pipet a 6 M HCl aqueous solution until the liquid turns permanently cloudy; the final pH[2] should be around 1. Transfer the acidified aqueous layer to a 250-mL separatory funnel and extract it with a 15-mL portion of BME. Separate the layers and repeat the extraction of the aqueous layer two more times with 15 mL of BME each time. Combine all the organic layers in a 125-mL Erlenmeyer flask labeled "**eugenol**." Transfer the aqueous layer to a beaker labeled "**aqueous 2**" and discard at the end of the experiment (see "Cleaning Up").

Transfer the combined organic layer labeled "eugenol" to a 125-mL separatory funnel and extract it with 10 mL of water. Separate the layers and extract the organic layer again with 25 mL of saturated sodium chloride solution in water. Transfer the organic layer to the 125-mL Erlenmeyer flask labeled "eugenol" (that has been rinsed and dried with paper towels). Dry the solution by adding anhydrous magnesium sulfate. After about 5 minutes filter the suspension using a glass funnel with a cotton plug. Collect the liquid in a pre-tared, 100-mL round-bottom flask and evaporate the solvent. Determine the weight of eugenol by difference. With the aid of 0.5–1 mL BME transfer the residue to a screw-capped vial and label it "**eugenol**."

Cleaning Up

- Separate the residue from the co-distillation into liquid and solid parts. Dispose of the solid in the trash can and the liquid in the "Eugenol Liquid Waste."

- Dispose of the "aqueous 1" and "aqueous 2" solutions and the water and sodium chloride solution used to wash the organic "eugenol" layer in the container labeled "Eugenol Liquid Waste."

- Dispose of the magnesium sulfate used to dry the organic layers, cotton plugs, and Pasteur pipets in the container labeled "Eugenol Solid Waste."

- At the end of the lab period the instructor will empty the contents of the rota-vap traps (BME) into a "BME Waste" bottle.

- At the end of E14A.4, turn in the "crude oil," "neutral," and "eugenol" vials, labeled, to your instructor.

E14A.4 GC AND IR ANALYSES

Remove the drying agent from the fraction labeled "crude oil" by filtration using a filter-Pasteur pipet (Figure 3.9*b*); if during storage the solvent has evaporated, add about 0.5 mL of BME before filtration. Evaporate the solvent from the fractions labeled "crude oil," "neutral," and "eugenol," on a sand bath in the hood until about 2–3 drops of the liquid remain (this operation may not be necessary as the solvent may have already evaporated during storage; in that case you may need to add a drop or two of BME). Analyze these concentrated fractions by GC using a

---

[2] To check the pH, cut small pieces of pH paper (about 0.5 cm long), arrange them on a watch glass, and touch them with a glass rod that was immersed in the solution to be tested. Never dip the pH paper into the solution to test because the dyes used as indicators may leach into the liquid.

column of medium polarity such as SF-96 at 165°C. Changing the attenuation and/or diluting/concentrating the samples may be necessary to obtain good GC peaks. The peaks should have a size that is easy to integrate. If time permits, inject standard samples of eugenol, eugenol acetate, and caryophyllene to unequivocally identify the GC peaks.

Obtain the IR spectrum of eugenol using Teflon tape (see IR unit, section 31.5) and compare it with the IR of an authentic sample shown in Figure 14.3. Load one drop of a 10–20% eugenol solution in BME on the Teflon tape; allow the solvent to evaporate before obtaining the IR spectrum. If necessary, apply more sample and scan the IR again.

BIBLIOGRAPHY

1. The Essential Oils: Vol. 4. E. Guenther. Krieger, Huntington, New York, 1972.
2. Green Pharmacy. A History of Herbal Medicine. B. Griggs. Viking Press, New York, 1981.
3. The Encyclopedia of Herbs and Herbalism. M. Stuart, ed. Orbis, London, 1979.
4. Monographs on Fragrance Raw Materials. D.L.J. Opdyke, ed. Pergamon Press, Oxford, 1979.
5. Ancient Herbs in the J. Paul Getty Museum Gardens. J. D'Andrea. J. Paul Getty Museum, Malibu, California, 1982.

EXPERIMENT 14A   REPORT

Pre-lab

1. Make a table showing the physical properties (molecular mass, b.p., density, solubility, flammability [for solvents only], and toxicity/hazards) of eugenol, acetyleugenol, β-caryophyllene, and *tert*-butyl methyl ether. Synonyms for eugenol: 2-methoxy-4-(2-propenyl)phenol and 4-allyl-2-methoxyphenol (usually listed under phenol); for acetyl eugenol: eugenol acetate; 2-methoxy-4-(2-propenyl)phenol acetate.

2. The boiling points of eugenol, acetyleugenol, and ß-caryophyllene at the pressures specified by the subscripts are: bp_{15}: 128–130°C; bp_{13}: 163–164°C, and bp_{14}: 128–130°C, respectively. Knowing that the boiling point increases with an increase in the applied pressure, and that the higher the boiling point, the longer the retention time, predict the order in which these three compounds may come out of the GC column.

3. Classify the following compounds according to their acid-base properties: eugenol, acetyleugenol, and β-caryophyllene.

4. The flowchart shown in Figure 14.1 summarizes the steps involved in the separation of the clove oil components. Write the acid-base reactions involved in the extraction with NaOH and the acidification with HCl.

5. Could you use sodium bicarbonate instead of NaOH to extract eugenol into the aqueous layer?

6. Why are acetyleugenol and β-caryophyllene not extracted into the basic aqueous solution? In which fraction are they recovered?

7. Could you use ethanol instead of BME to carry out the liquid-liquid extractions?

8. Is the BME phase the top or the bottom layer?

9. Why do you wash the final organic layer with distilled water? Why with saturated NaCl solution?

10. How would you separate a mixture of eugenol and cholesterol (see structure of cholesterol in Unit 5, page 90) by acid-base extraction?

11. (Optional) Knowing that the amount of clove oil in the co-distillate with water is about 0.7% by weight, and assuming that the oil is 100% eugenol, estimate the vapor pressure of water and eugenol during the co-distillation. Assume that the distillation is carried out under normal atmospheric pressure (760 Torr). (Hint: use Equation 8, Unit 6). Using Figure 6.26*b*, and the vapor pressure of water calculated above, estimate the boiling point during the co-distillation.

In-lab

1. Outline the learning objectives of this experiment. Mention the new techniques introduced.

2. Calculate the percentage yield of eugenol and neutral compounds from whole cloves.

3. Report the boiling point during the co-distillation.

4. (Optional) Compare the experimental boiling point with the expected one (pre-lab 11).

5. Why does the aqueous layer turn cloudy when 6 M HCl is added? What compound forms?

6. Report the retention time for eugenol, acetyleugenol, and caryophyllene. Do they come out in the order predicted (pre-lab 2)?

7. Assign as many bands as you can in the IR spectrum of eugenol. Compare your IR with that of an authentic sample shown in Figure 14.3. Present your data as indicated in Tables 31.3 and 31.4. Along with the "observed," "expected (from tables)," and "assignment" columns add an extra column ("literature") with the values from the authentic sample shown in Figure 14.3.

8. Using your GC peak areas calculate the percentage of eugenol, acetyleugenol, and caryophyllene in the crude oil, and in the eugenol and neutral fractions. Discuss your data. Was the separation of eugenol successful?

9. Interpret the ^{1}H-NMR and ^{13}C-NMR spectra shown in Figures 14.4 and 14.5 (advanced level).

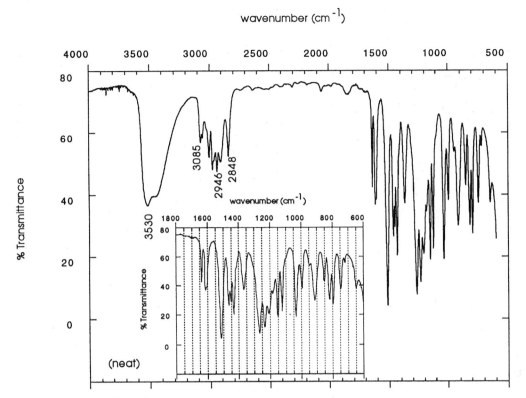

Figure 14.3 IR spectrum of eugenol (neat).

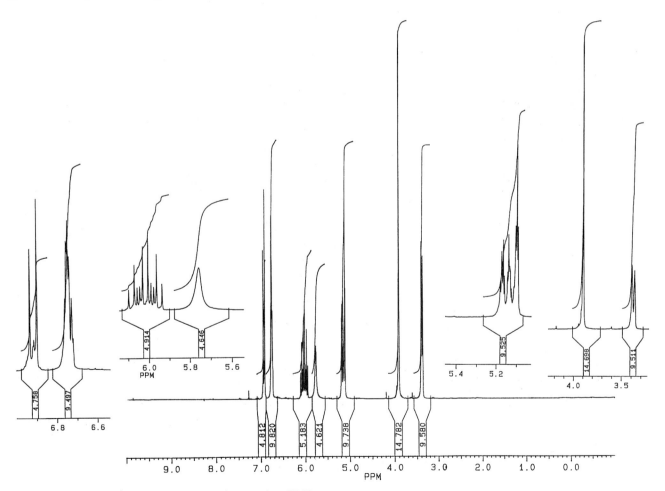

Figure 14.4 250-MHz ^{1}H-NMR spectrum of eugenol in CDCl$_3$.

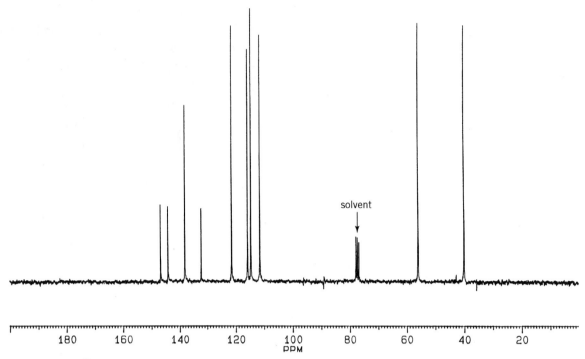

solvent

Figure 14.5 62.9-MHz ^{13}C-NMR spectrum of eugenol in CDCl$_3$.

Isolation of the Active Ingredients in an Analgesic Tablet

E14B.1 OVERVIEW

In this experiment you will isolate aspirin, acetaminophen, and caffeine from an Excedrin tablet. Excedrin is an over-the-counter analgesic with three active ingredients: aspirin (also called acetylsalicylic acid), acetaminophen (the main ingredient in Tylenol), and caffeine. A discussion on the mode of action of analgesics is postponed until Unit 15. A tablet of extra-strength Excedrin contains 250 mg of aspirin, 250 mg of acetaminophen, and 65 mg of caffeine. The tablet also contains a binder made of mostly insoluble materials such as modified cellulose.

aspirin   acetaminophen   caffeine

Aspirin has two functional groups attached to the aromatic ring: an ester group, which is neutral, and a carboxylic acid group. Overall, aspirin is an acidic compound with $pK_a = 3.49$. It is insoluble in neutral aqueous solutions but it dissolves in basic solutions.

Acetaminophen is a phenol ($pK_a \approx 10$) with a neutral acetamido group in *para* position. Acetaminophen is a very weak acid, about six orders of magnitude less acidic than aspirin. It dissolves in strong basic solutions (pH > 11) but it does not dissolve under mild basic conditions (pH = 8–9).

Contrary to the other two compounds, caffeine is polar and soluble in water. Because caffeine is a base, although a weak one, its solubility in water is increased in the presence of strong acids.

Because aspirin is more acidic than acetaminophen, in principle, these two chemicals could be separated by acid-base extraction using a weak base such as sodium bicarbonate. Sodium bicarbonate reacts with the carboxylic acid of aspirin, forming a sodium salt, but it does not react with the phenol group of acetaminophen. However, their separation can be accomplished in a much easier way because of the unusual solubility behavior of acetaminophen. This compound is only slightly soluble in organic solvents such as methylene chloride and *t*-butyl methyl ether, whereas aspirin is readily soluble in these solvents.

The first step in the separation of the three active ingredients of Excedrin is a solid-liquid extraction with acetone. Acetone dissolves the three compounds and leaves behind the tablet's binder. Evaporation of acetone renders a solid mixture of aspirin, acetaminophen, and caffeine, which can be separated by extraction.

You will treat the mixture of caffeine, aspirin, and acetaminophen with methylene chloride. This solvent dissolves caffeine and aspirin but does not dissolve acetaminophen, which is recovered on the filter. Then you will extract the solution of aspirin and caffeine in methylene chloride with 5% aqueous HCl. The acid removes the caffeine from the organic layer, leaving only aspirin in the methylene chloride. Evaporation of methylene chloride gives solid aspirin. Neutralization of the acid aqueous layer with sodium hydroxide and extraction with methylene chloride allows the recovering of caffeine after evaporation of the solvent. A flowchart with the outline of the separation is shown in Figure 14.6. You will assess the success of the separation by TLC analysis (Unit 8) of the different fractions and by melting point. If time permits, you will also obtain the IR spectra of the isolated products.

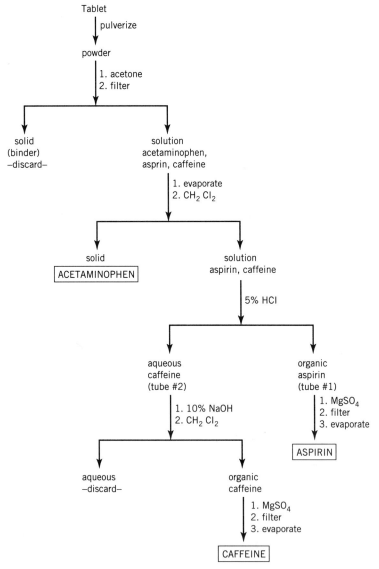

Figure 14.6 Flowchart for the separation of acetaminophen, aspirin, and caffeine from an Excedrin tablet.

In this experiment you will practice acid-base extraction at the microscale level. Screw-cap test tubes will be used instead of separatory funnels.

PROCEDURE

> **Background reading:** 5.4; 5.5; 14.1–5;
> 8.1–4; 8.6
> **Estimated time:** separation: 2.5 hr; analysis:
> 1 hr

E14B.2 SEPARATION OF THE ACTIVE INGREDIENTS IN EXCEDRIN

Weigh a tablet of Excedrin E.S. Pulverize it using a mortar and pestle. Weigh out 400 mg of the powder and transfer it to a **dry** 16×125 mm screw-cap test tube. Add 7 mL of acetone, cap the tube, and shake it well for about 2–3 minutes with occasional venting. Filter the suspension using a Pasteur pipet with a small cotton plug (Fig. 3.9a) and collect the filtrate in a pre-tared round-bottom flask. Transfer a few drops of this solution to a capped vial labeled "original mixture" and keep it for TLC analysis. Evaporate the solvent in a rota-vap and weigh the product.

Treat the solid residue in the flask with 5 mL of methylene chloride. It will not completely dissolve. Vacuum-filter the suspension using a Hirsch funnel. Use a Pasteur pipet to transfer the suspension. Wash the walls of the flask with 1–2 mL of methylene chloride and transfer the rinses to the filter. Save the solid in a capped vial with the name "Acetaminophen."

Transfer the filtrate to a 16×125 mm screw-cap test tube labeled "1." Extract the organic layer with 2 mL of a 5% aqueous HCl solution. Cap the tube and invert it about 50 times with frequent venting. Remove the organic lower layer with a Pasteur pipet and temporarily place it in a test tube. Remove the aqueous layer from the screw-cap test tube and place it in another screw-cap test tube labeled "2."

Transfer the methylene chloride layer from the temporary test tube to tube 1, and extract it two more times with 2 mL of a 5% aqueous HCl solution each time. Collect all the aqueous layers in tube 2.

Dry the methylene chloride layer with a little anhydrous magnesium sulfate. Remove the drying agent by gravity filtration using a Pasteur pipet and a cotton plug, and collect the filtrate in a dry and pre-tared round-bottom flask. Evaporate the solvent, and weigh and transfer the solid to a capped vial labeled "Aspirin."

In the meantime, treat the acid aqueous layer in tube 2 with 10% aqueous NaOH until the pH is basic. Extract the aqueous layer three times with 2 mL of methylene chloride each time. Collect all the organic layers in a dry screw-cap test tube. Dry it with anhydrous magnesium sulfate. Remove the drying agent by gravity filtration using a Pasteur pipet. Collect the liquid in a pre-tared round-bottom flask and evaporate the solvent using a rota-vap. Weigh the product and transfer it to a vial labeled "Caffeine."

> **Safety First**
>
> - HCl and NaOH are corrosive and toxic. HCl is highly toxic. Handle them with care.
> - Acetone is flammable. Methylene chloride is a possible carcinogen. Handle them in a well-ventilated place.

E14B.3 ANALYSIS

Take the melting points of the acetaminophen, aspirin, and caffeine fractions. Analyze the solids (dissolved in acetone) and the "original mixture" by TLC, using silica gel plates with fluorescence indicator and ethyl acetate as a solvent. Use standards of pure aspirin, acetaminophen, and caffeine to help in the identification of the spots.

Use the UV light to visualize the spots. Draw the plates in your lab notebook. Indicate colors observed under the UV light. Calculate the R_f values. Interpret your results.

If time permits, obtain the IR of each fraction using a Nujol mull (section 31.5). Compare them with the IR spectra shown in Figures 14.7–14.9.

> **Safety First**
>
> - Acetone and ethyl acetate are flammable.
> - Keep the solvent chamber perfectly closed.
> - Do not look directly into the UV light.

<table>
<tr><td>

Cleaning Up

- Dispose of all used organic solvents in the container labeled "Excedrin Solvents Waste."

- Dispose of all aqueous layers in the container labeled "Excedrin Liquid Waste."

- Dispose of the binder residue, any unused powdered tablet, and the Pasteur pipets in the container labeled "Excedrin Solid Waste."

- At the end of the experiment the instructor will empty the contents of the rota-vap traps in the container labeled "Excedrin Solvents Waste."

- Dispose of the TLC solution in the container labeled "Excedrin Solvents Waste."

- Dispose of the TLC plates in the "Excedrin Solid Waste" container.

- Turn in your products in a capped, labeled vial to your instructor.

</td></tr>
</table>

EXPERIMENT 14B   REPORT

Pre-lab

1. Make a table showing the physical properties (molecular mass, m.p., b.p., solubility, flammability [for solvents only], and toxicity/hazards) of aspirin, acetaminophen (synonym: N-acetyl-4-hydroxyaniline), caffeine, acetone, and methylene chloride.

2. Classify the following compounds as acidic, basic, or neutral: acetaminophen, aspirin, and caffeine. For aspirin and acetaminophen, identify the functional groups present in each molecule and determine which ones affect their acid-base properties.

3. Could you use NaOH to separate aspirin and acetaminophen by acid-base extraction? Briefly justify your answer.

4. Why do you use HCl in the extraction of caffeine?

5. Look at the structure of caffeine and decide which one of the four nitrogens is most basic. Why?

6. Could you separate a mixture of salicylic acid and aspirin by acid-base extraction? Briefly justify your answer.

In-lab

1. Calculate the amount in mg of aspirin, acetaminophen, and caffeine present in 400 mg of powder (use the weight of the whole tablet and the fact that each tablet contains 250 mg of aspirin, 250 mg of acetaminophen, and 65 mg of caffeine).

2. Report the mass of the solid mixture after evaporation of acetone. Compare this mass with the amount of acetaminophen, aspirin, and caffeine that should be present in 400 mg of tablet powder. Was the solid-liquid extraction effective?

3. Report the mass of acetaminophen, aspirin, and caffeine after the separation.

4. Calculate the % Recovery for each compound using the following equation:

$$\% \text{ Recovery} = \frac{\text{mass of isolated compound}}{\text{mass of compound in 400 mg powder}} \times 100$$

5. Report the melting points and compare them with literature values.

6. Report and discuss the TLC results. Make a table with the R_f values. Was the separation effective?

7. Interpret the IR spectra of aspirin, acetaminophen, and caffeine. Compare them with the ones shown in Figures 14.7–14.9.

8. (Advanced level) Interpret the ^{1}H- and ^{13}C-NMR shown in Figures 14.10–14.15.

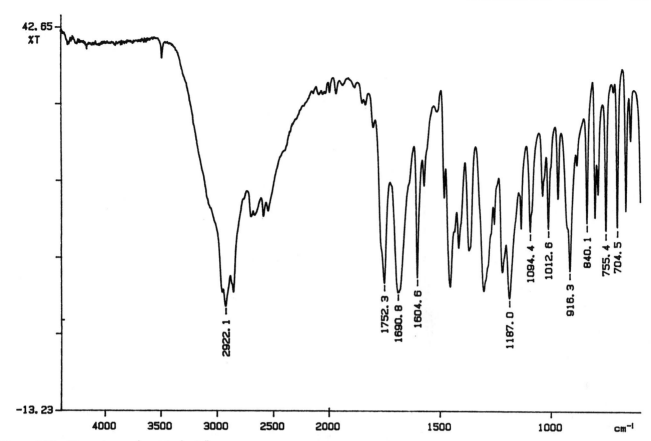

Figure 14.7   IR spectrum of aspirin (nujol).

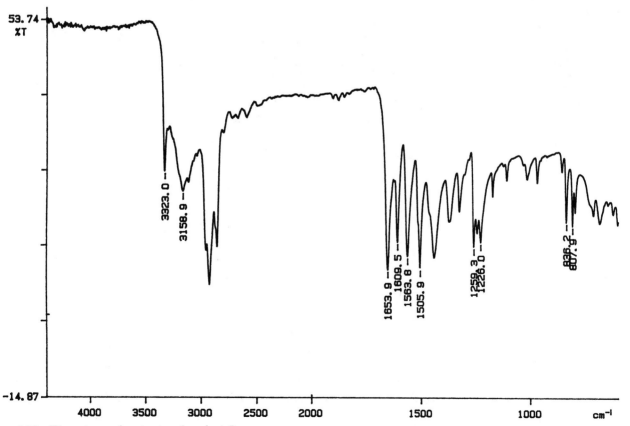

Figure 14.8   IR spectrum of acetaminophen (nujol).

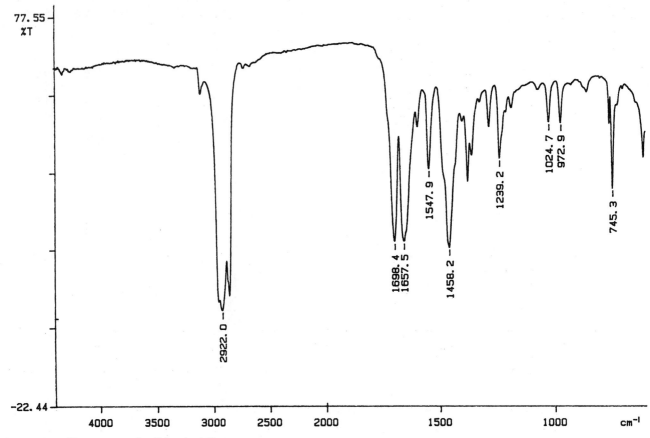

Figure 14.9   IR spectrum of caffeine (nujol).

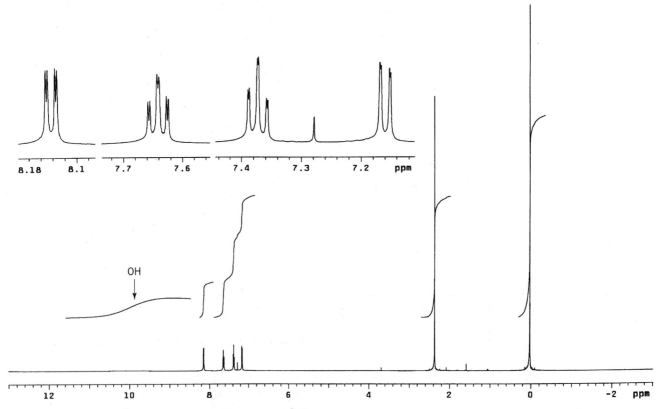

Figure 14.10   500-MHz ^{1}H-NMR spectrum of aspirin in CDCl$_3$.

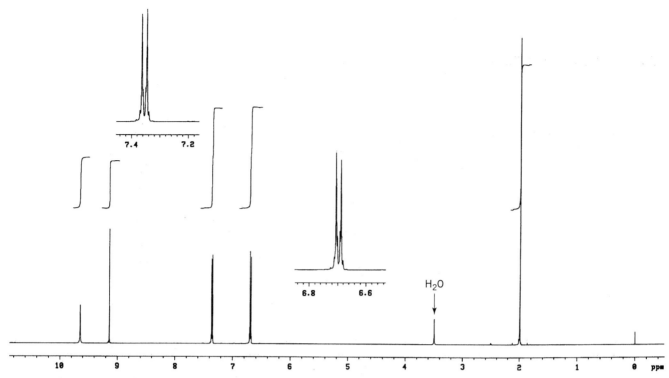

Figure 14.11   500-MHz ¹H-NMR spectrum of acetaminophen in DMSO-d₆.

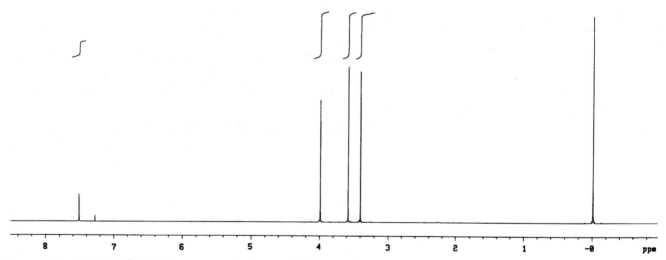

Figure 14.12   500-MHz ¹H-NMR spectrum of caffeine in CDCl₃.

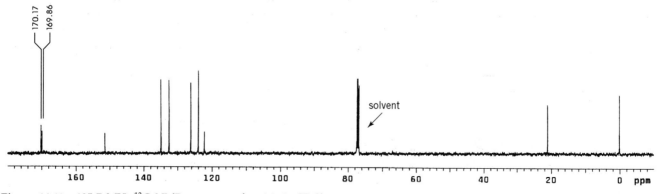

Figure 14.13   125.7-MHz ¹³C-NMR spectrum of aspirin in CDCl₃.

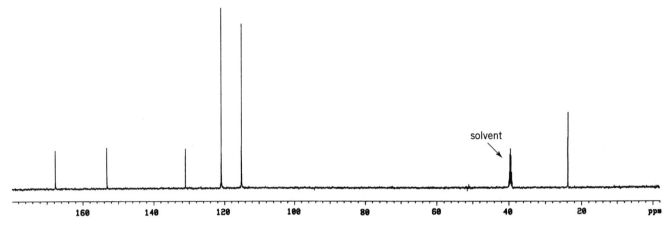

Figure 14.14 125.7-MHz ^{13}C-NMR spectrum of acetaminophen in DMSO-d$_6$.

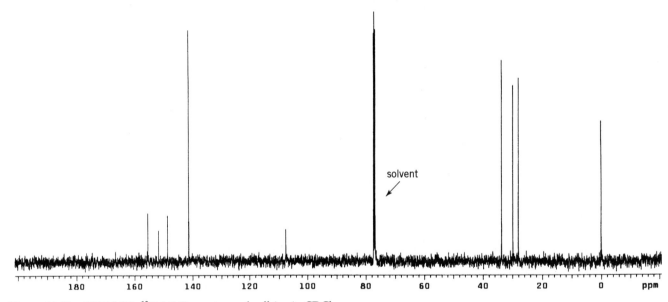

Figure 14.15 125.7-MHz ^{13}C-NMR spectrum of caffeine in CDCl$_3$.

Phenols and Ethers

15.1 PHENOLS AND ETHERS

Phenols are compounds with a hydroxyl group directly attached to an aromatic ring: ArOH. As was already discussed in Unit 14, phenols are acidic compounds with pK_a in the range 7–10. In the presence of strong bases such as NaOH most phenols are completely dissociated:

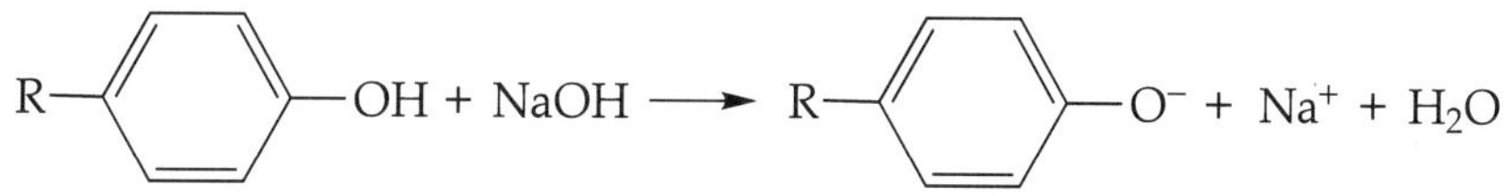

a phenol

a phenoxide
or phenolate anion

Ethers are compound with the general structure R—O—R′, where R and R′ are alkyl or aryl groups. Ethers are rather inert compounds. They do not react with most of the reagents used in organic chemistry except for strong acids. This lack of reactivity makes them excellent solvents. For instance, tetrahydrofuran (THF) and *tert*-butyl methyl ether (BME) are two common solvents.

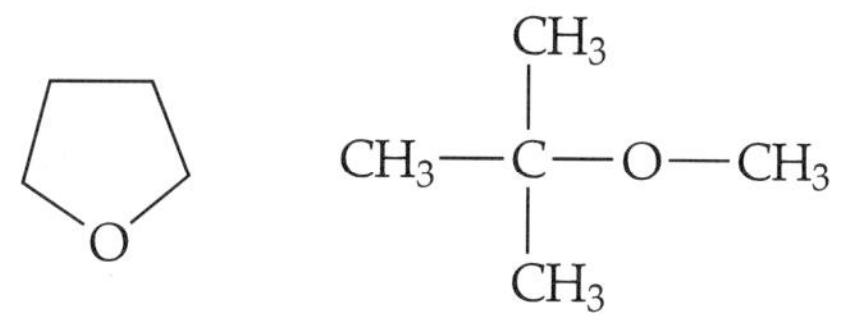

tetrahydrofuran
(THF)

tert-butyl methyl ether
(BME)

Ethers can be obtained by the reaction of alkoxides and phenoxides with alkyl halides. This reaction is called the **Williamson ether synthesis** and is the subject of the next section.

15.2 WILLIAMSON ETHER SYNTHESIS

In this reaction, primary alkyl halides react with alkoxides (conjugate bases of alcohols) in aprotic solvents to give ethers via an S_N2 mechanism. The alkoxide is formed by reacting the alcohol with a strong base, such as sodium hydride, in an aprotic medium.

$$\text{R---OH} + \text{NaH} \xrightarrow{\textit{aprotic solvent}} \text{R---O}^- + \text{H}_2$$

an alcohol   sodium hydride   an alkoxide

$$\text{R---O}^- + \text{R'CH}_2\text{---X} \xrightarrow[\text{S}_\text{N}2]{\textit{aprotic solvent}} \text{R---O---CH}_2\,\text{R'} + \text{X}^-$$

an alkoxide   a primary alkyl halide   an ether

Secondary and tertiary alkyl halides cannot be used in the Williamson ether synthesis because they undergo elimination reactions. The Williamson ether synthesis is particularly useful to prepare alkyl aryl ethers. Phenols are more acidic than alcohols (by a factor of 10^7) and can be converted into phenoxides under mild basic conditions (Unit 14). Potassium carbonate in acetone or butanone provides a sufficiently basic medium to establish an equilibrium between the phenol and the phenolate anion:

a phenol   a potassium
                        phenolate

By heating a phenolate anion with a primary alkyl halide in an aprotic solvent, an alkyl aryl ether is formed. In the Williamson ether synthesis, alkyl iodides are usually preferred over bromides and chlorides because iodide is a better leaving group than bromide and chloride.

an alkyl aryl ether

In general, the reaction does not go to completion and a mixture of ether and starting phenol is obtained. The mixture can be separated by acid-base extraction (Unit 14). Ethers are neutral compounds and are retained in the organic layer, while phenols are extracted by a basic aqueous solution.

15.3 CHARACTERIZATION OF PHENOLS: FERRIC CHLORIDE TEST

Phenols react with ferric chloride to form colored complexes; the color of these complexes ranges from green to red depending on the structure of the phenol. The development of a green, red, blue, or violet color is considered a positive test and is an indication of the presence of a phenol. If the yellow color of the ferric chloride solution does not change upon addition of the sample, the test is considered negative. Other compounds different from phenols, such as hydroxamic acids and hydroxy acids, also give a red-violet color in the presence of ferric chloride (section 22.5).

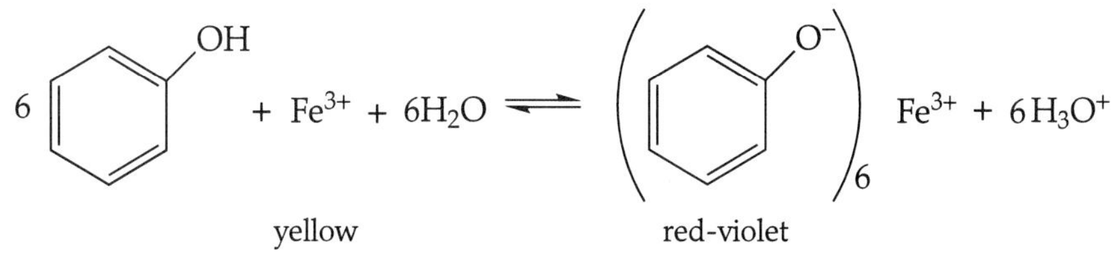

$$6 \text{ [phenol-OH]} + Fe^{3+} + 6H_2O \rightleftharpoons \left(\text{[phenol-O}^-] \right)_6 Fe^{3+} + 6H_3O^+$$

yellow   red-violet

15.4 IR AND NMR OF PHENOLS AND ETHERS

IR Spectra

Phenols (associated by H-bonds) show a broad O—H stretching band in the $3200-3500 \text{ cm}^{-1}$ region. The C—O stretching band (around $1180-1260 \text{ cm}^{-1}$) is intense but lies in the fingerprint region and has limited diagnostic value. Ethers present strong bands in the $1000-1300 \text{ cm}^{-1}$ region due to C—O—C stretching (asymmetric). Alkyl aryl ethers show this band around $1200-1280 \text{ cm}^{-1}$.

NMR Spectra

Phenolic hydrogens resonate around 5 ppm when they are not associated by H-bonds (at low concentrations), and between 5 and 11 ppm when the phenol is associated. In ethers, hydrogens from methyl, methylene, and methine groups attached to the oxygen atom are deshielded and resonate at high frequencies: between 3.2 and 3.5 ppm for aliphatic ethers and between 3.7 and 4.5 for phenyl ethers. Ether carbon atoms resonate at relatively high frequencies: between 50 and 80 ppm for aliphatic carbons and around 160 ppm for aromatic carbons.

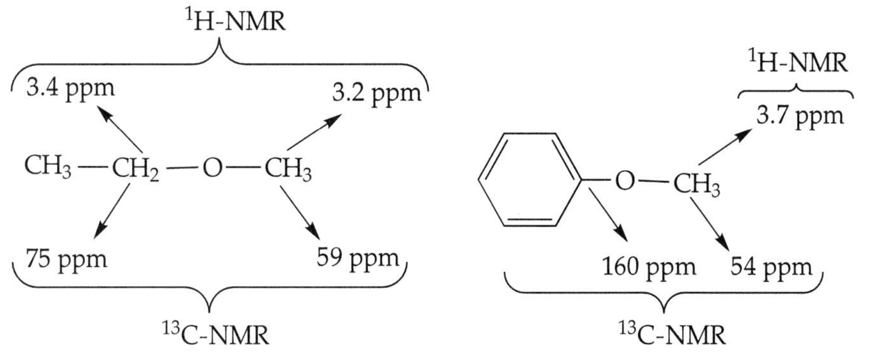

^{1}H-NMR

3.4 ppm   3.2 ppm

$CH_3 — CH_2 — O — CH_3$

75 ppm   59 ppm

^{13}C-NMR

^{1}H-NMR

3.7 ppm

[phenyl]—O—CH_3

160 ppm   54 ppm

^{13}C-NMR

EXERCISES

Problem 1

With the help of your textbook suggest two different pathways to prepare *N*-acetyl-4-ethoxyaniline (phenacetin) from 4-hydroxyaniline.

Problem 2

How would you separate a mixture of phenacetin and 4-ethoxyaniline by acid-base extraction?

Problem 3

Could you separate a mixture of salicylic acid and aspirin by acid-base extraction (see structures on page 328)? Briefly justify your answer. If the separation by acid-base extraction is not possible, propose a method to carry out the separation.

Problem 4

How can you tell apart salicylic acid and aspirin by IR spectroscopy? What differences do you expect in their ^{1}H-NMR spectra?

Problem 5

(*Previous Synton: Problem 3, Unit 12*) On a cold Friday morning after Thanksgiving, Synton was awakened by an early phone call. It was Vanadia, his father's sister, who in her sweet Russian accent reminded him of his promise made the night before to analyze a natural extract, called "*Canadian Forest*," which she had been using for her sciatica. She went on and on, arguing that the health-supplement store was making a bundle by overpricing cheap materials under the pretense of fancy names. Synton was annoyed. If he had promised anything it was to get her off his back. He tried to reason with her again. He didn't own the lab, analyses were costly, he didn't have time. . . . It was to no avail. A promise was a promise and he was a chemist after all, wasn't he? Synton gave up.

He took the mysterious, chewing-gum-smelling oil to his lab, and with his supervisor's permission, ran GC and NMR. The GC showed one main peak (accounting for more than 95%) and the ^{1}H-NMR (300 MHz, CDCl$_3$) gave the following signals: 3.9 (singlet, 3H); 6.85 (triplet, 1H); 6.96 (doublet, 1H); 7.45 (triplet, 1H); 7.82 (doublet, 1H); and 10.78 (singlet, 1H). The proton-decoupled ^{13}C-NMR showed the following peaks: 52.2; 112.3; 117.5; 119.1; 130.0; 135.6; 161.5; and 170.5. After analyzing the spectra, Synton had a very good idea about the structure of the chemical. Just to confirm his presumption he performed a chemical test with ferric chloride and observed a deep red color. He couldn't believe that the analysis had been so easy. He was going to score high with his aunt who was entirely right from the beginning. It was a very cheap material indeed. What is the structure of the component in the "*Canadian Forest*" oil? Check the *Merck Index* and find another name for the oil.

EXPERIMENT 15

Medicinal Chemistry: from Tylenol to a Banned Chemical

E15.1   ANALGESICS

The medicinal effects of willow bark extracts have been known for centuries. Hippocrates used a willow bark preparation to alleviate pains and reduce fever. Early in the nineteenth century it was recognized that the active principle in willow bark extracts, as well as in extracts from flowers of the meadowsweet plant, was *salicylic acid* (see structure in the first box on page 328). It was found that salicylic acid was a very effective **analgesic** (pain killer), **antipyretic** (fever reducer), **anti-inflammatory** (reduces swelling), and **uricosuric** (relieves symptoms of arthritis and gout) agent. It was also found that salicylic acid had undesirable side effects: gastrointestinal irritation.

Felix Hoffmann, a chemist at the Bayer Company in Germany, discovered in 1893 that a rather simple chemical modification of salicylic acid afforded a new compound that not only had superior pain-relief power than salicylic acid but also none of its side effects (he tested the new drug by administering it to his father, who suffered from chronic arthritis). The newly synthesized compound, acetylsalicylic acid, was clinically evaluated and its industrial production began in 1899 under the trademark of **Bayer Aspirin** (**a** from *acetyl* and **spirin** from the Latin name for the meadowsweet plant, *spirea*).

Aspirin is probably the most versatile drug ever devised. Besides the effects mentioned above, it has been shown that aspirin reduces the risk of heart attacks in patients with a history of cardiovascular disease. Consumption of aspirin in the United States is more than 15 million kilos a year, and the average American takes about 32 aspirin tablets a year.

Mode of Action

Although the mechanisms by which aspirin acts have not yet been completely elucidated, some light has been shed in recent years. The action of aspirin is closely related to the production of **prostaglandins**. Prostaglandins are twenty-carbon carboxylic acids that contain a five-carbon ring. They are natural constituents of most cells. They have been grouped into different classes, which are designated PGA, PGB, PGE, and PGF, followed by a subscript that indicates the number of double bonds outside the ring. Some examples of prostaglandins are shown in the second box on page 328. Although their biological roles are quite wide, they all modulate the action of hormones. Prostaglandins are involved in the activation of pain receptors (nerve endings), regulation of body heat (a function controlled by the hypothalamus region of the brain), production of edema and inflammation, clotting of blood platelets on the walls of arteries, and spontaneous abortion. Prostaglandins act like the "smoke alarm" of our bodies. They are responsible for the onset of symptomatic reactions (such as fever, pain, and inflammation) to let us know that something is malfunctioning. Studies indicate that aspirin inhibits the synthesis of prostaglandins and therefore helps alleviate the symptoms of different disorders.

327

Some drugs used as analgesics

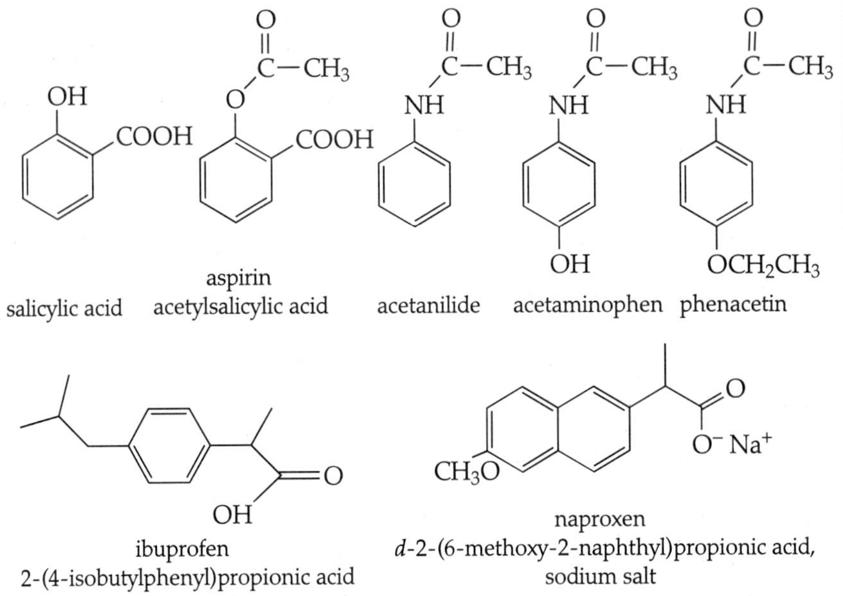

salicylic acid acetylsalicylic acid acetanilide acetaminophen phenacetin

aspirin

ibuprofen
2-(4-isobutylphenyl)propionic acid

naproxen
d-2-(6-methoxy-2-naphthyl)propionic acid,
sodium salt

Some examples of prostaglandins

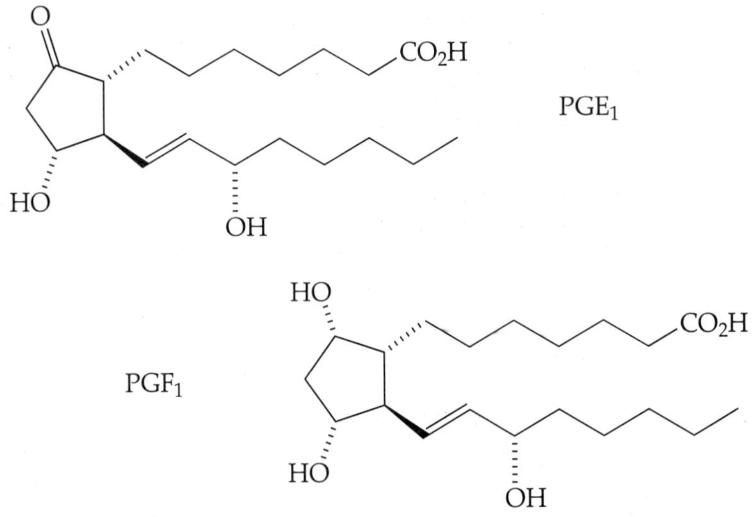

PGE$_1$

PGF$_1$

Although aspirin has low toxicity, approximately 0.2–0.9% of the population experience some allergic reactions to aspirin, such as skin rash or respiratory problems. Aspirin may be contraindicated for people with uncontrolled high blood pressure or liver or kidney disease. Other undesirable side effects of aspirin are gastrointestinal disturbances, especially when used in large doses.

Other analgesics have been developed in the last fifty years; among them **naproxen, ibuprofen**, and especially **acetaminophen** (the active ingredient in Tylenol) have gained wide acceptance. It was discovered by accident in 1886 that acetanilide was an effective antipyretic. Although it later proved to be toxic, its discovery led to the development of two new analgesics: **phenacetin** and **acetaminophen**.

Naproxen and ibuprofen (the active ingredients in Aleve and Advil, respectively) are propionic acids with an aromatic ring in position 2. Besides being potent analgesics they also have antipyretic and anti-inflammatory effects. A distinctive feature of naproxen is that it has a slow metabolic degradation in the organism, leading to a longer-lasting effect.

Acetaminophen is an effective analgesic although it acts more slowly than aspirin; it is also a potent antipyretic. Phenacetin is a pain reliever and a less potent antipyretic. Both phenacetin and acetaminophen have little value as anti-inflammatory agents. Excessive use of acetaminophen can cause **methemoglobinemia**, a blood disorder by which the transport of oxygen by hemoglobin in the bloodstream is seriously weakened. Phenacetin was found to cause liver and kidney damage when administered in large doses. Experiments on laboratory animals showed that phenacetin can induce cancer when taken orally. Its use in medicinal preparations was banned in the United States in 1981.

Acetaminophen and phenacetin are chemically related. Both have an acetamido group, but whereas there is a free —OH group in acetaminophen, phenacetin has an ethoxy group in its place. How can we convert acetaminophen, a phenol, into phenacetin, an ether? The answer is the **Williamson ether synthesis**.

E15.2   THE EXPERIMENT: OVERVIEW

In this experiment you will prepare, by the Williamson ether synthesis, phenacetin from acetaminophen contained in Tylenol tablets. You will heat a mixture of acetaminophen, ethyl iodide, and potassium carbonate in butanone for about one hour. At the end of this period some acetaminophen still remains in the reaction mixture. To obtain pure phenacetin you will separate the unreacted acetaminophen by acid-base extraction (section 14.4) taking advantage of the fact that acetaminophen is a phenol and, therefore, is acidic and easily removed by a NaOH solution, while phenacetin, a neutral compound, remains in the organic layer. You will purify phenacetin by recrystallization from a mixture of acetyl acetate and hexane (section 4.4). The experiment has been adapted from Reference 5.

In the second part of this experiment you will analyze several analgesics by TLC (Unit 8). Chromatography is an invaluable tool for the analysis of drugs. If a medicine is mislabeled, we can always devise an analytical procedure to identify the drug in question. If no information about the nature of the medicine is available, the process of identifying the chemicals can be painstaking. It usually involves chromatographic analysis, separation of the mixture into its components, preparation of standards, and spectroscopic analysis (IR, NMR, MS). It is always advisable to run a TLC analysis before engaging in more elaborate procedures. TLC is an inexpensive, fast, and reliable technique to find out how many components are present in a mixture. In this experiment you will analyze an unknown analgesic suspected of containing phenacetin. You will use your sample of purified phenacetin as a standard for the TLC analysis. Since your unknown may also contain some of the ingredients of common analgesics such as aspirin, acetaminophen, and caffeine, you will use Bayer

Table 15.1 **Composition of Some Over-the-Counter Analgesics (mg/tablet)**

| | Active ingredient | | |
Product	Aspirin	Acetaminophen	Caffeine
Alka-Seltzer	325	—	—
Anacin	400	—	32
Bayer aspirin	325	—	—
Bufferin	325	—	—
Excedrin	194	97	65
Excedrin (E.S.)	250	250	65
Aspirin-free Excedrin (E.S.)	—	500	65
Tylenol	—	325	—
Tylenol (E.S.)	—	500	—

aspirin, Tylenol, and Anacin (each one has a different composition; see Table 15.1) as standards. You will determine the possible composition of the unknown analgesic.

PROCEDURE

E15.3 CONVERSION OF ACETAMINOPHEN INTO PHENACETIN

> **Background reading:** 15.1–3; 14.4; 4.4; 4.5; 5.4.
> **Estimated time:** synthesis: 3–4 hours.

Safety First

- Ethyl iodide is a severe irritant. Wear gloves when dispensing this reagent and avoid skin contact. Dispense this chemical in the hood.
- Remove the gloves immediately.
- Phenacetin is a probable carcinogen and a mutagen. Harmful if swallowed. Avoid skin contact.
- *tert*-Butyl methyl ether is flammable.

---

1. Weigh an Extra-Strength Tylenol tablet. Pulverize the tablet with mortar and pestle. Weigh out 0.22 g and place it in a **dry** 15-mL round-bottom flask along with 0.28 g of finely pulverized K_2CO_3 (mortar and pestle) and 3.0 mL of butanone. Carefully add 0.28 mL of ethyl iodide with a pluringe.

2. Add a stir bar; attach a microscale water-cooled condenser to the flask. Heat the mixture under reflux directly on a hot plate at medium setting for 1 hour (see Fig. 15.1). In the meantime, obtain the IR of acetaminophen (see step 10).

3. Turn off the heat. Allow the mixture to cool down. Add 4 mL of water to the flask, and transfer its contents to a 16 × 125 mm test tube with a screw cap. Rinse the round-bottom flask 4 times with 1 mL of *tert*-butyl methyl ether (BME) and add the rinsings to the test tube.

4. Cap the test tube and shake the layers. Vent to release the pressure by unscrewing the cap momentarily. With a Pasteur pipet remove the lower (aqueous) layer. Transfer the aqueous layer to another 16 × 125 mm screw-cap test tube. Keep the organic layer in the original test tube.

5. Extract the aqueous layer with 2.5 mL of BME. Remove the aqueous layer with a Pasteur pipet (save it until the end in a labeled test tube, and then discard it). Transfer the organic layer to the screw-cap test tube with the original organic layer.

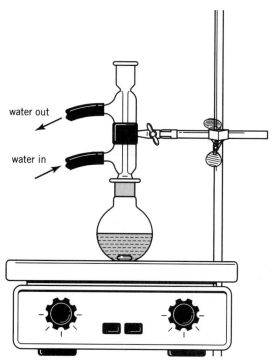

water out

water in

Figure 15.1 Reflux apparatus for the synthesis of phenacetin.

6. Add 2.5 mL of 5% aqueous NaOH to the combined organic layers and shake well. Let the layers settle and remove the lower aqueous layer with a Pasteur pipet. Repeat the extraction with another 2.5-mL portion of 5% NaOH. Save the aqueous layers until the end of the experiment.

7. Extract the organic layer with 2.5 mL of saturated sodium chloride solution. If emulsions form during these extractions, they can be broken by centrifugation (1000–2000 rpm for a few minutes). Dry the organic layer by adding magnesium sulfate little by little with a microspatula until the solid runs freely in the liquid. Let the system stand for 5 minutes with occasional swirling.

8. Filter the mixture using a Pasteur pipet with a cotton plug and receive the filtrate in a dry, pre-tared, 25-mL round-bottom flask. Evaporate off the solvent using a rota-vap (Fig. 3.17). Weigh the flask with the product. Calculate the crude % yield. Scrape out the crude phenacetin and determine its melting point.

9. Review recrystallization from mixed solvents (section 4.4) and microscale recrystallization (section 4.5). Using a Craig tube, recrystallize the product by dissolving it in the minimum amount of hot ethyl acetate. Perform this operation in the fume hood. Heat the system in a sand bath, and continuously stir it by rolling a microspatula between your fingers. To the hot solution add hot hexane dropwise until the solution becomes cloudy. Cool the solution to induce crystallization and collect the crystals by centrifugation. Determine the weight and melting point.

10. Obtain the IR spectrum of acetaminophen and phenacetin in a nujol mull using NaCl windows (section 31.5). Compare the IR of the product with the spectrum shown in Figure 15.2. If possible, obtain the ^{1}H-NMR of the product in CDCl$_3$, (section 33.20) (*Caution: CDCl$_3$ is a possible carcinogen; dispense this chemical in the hood*).

Cleaning Up

- Dispose of all the aqueous layers in the container labeled "Analgesics-Aqueous Waste."

- At the end of the lab period the instructor will empty the contents of the rota-vap traps (BME) into the "*tert*-Butyl Methyl Ether Waste" bottle.

- Dispose of the recrystallization solvent in the container labeled "Analgesics Organic Solvents Waste."

- Dispose of the Pasteur pipets and MgSO$_4$ in the container labeled "Analgesics Solid Waste."

- Dispose of the ferric chloride solutions in the container labeled "Analgesics Aqueous Waste."

- Dispose of the NMR solution in the "Chloroform Waste" container.

E15.4 FERRIC CHLORIDE TEST

Place 1 mL of a 0.1% aqueous solution of ferric chloride in each of four small, labeled test tubes. To each one, add a microspatulaful of one of the three following compounds: acetaminophen, phenacetin (from E15.3), and salicylic acid. Add a drop of water to the fourth tube, which is used as a blank. Observe the change in color. Colors ranging from green to red are considered a positive test; yellow is negative.

E15.5 ANALYSIS OF ANALGESICS BY TLC

Safety First

- Acetone and ethyl acetate are flammable.
- Keep solvent chambers closed.
- Do not look directly into the UV light.

Cleaning Up

- Dispose of all used organic solvents and analgesic solutions in the container labeled "Analgesics Organic Solvents Waste."
- Dispose of any unused powdered analgesic in the "Analgesics Solid Waste" container.
- Turn in your product in a capped, labeled vial to your instructor.
- Dispose of the TLC plates in the "Analgesics Solid Waste" container.

> **Background reading:** 8.4
> **Estimated time:** 2–3 hours.

1. Pulverize a tablet of each analgesic (Bayer aspirin, Tylenol, and Anacin) using a **clean** mortar and pestle (use a little acetone and tissue paper to clean mortar and pestle).
2. Transfer about one-eighth of each solid to a labeled test tube. Place about 20 mg of the unknown in a labeled test tube.
3. Add 2 mL acetone to each tube and stir with a clean microspatula. The insoluble solid is the binder of the tablet.
4. Dissolve about 1–2 mg of phenacetin in 0.5 mL of acetone.
5. Spot your samples on 6 × 10 cm silica gel plates with fluorescent indicator (section 8.4). Do not overload the spots. One or two applications will be enough. For a precise analysis, the unknown should be run side by side on the same plate with the other analgesics. Use ethyl acetate as the development solvent. Use the UV lamp to visualize the spots; notice any difference in color among the spots under the UV light. This may be useful in the identification of your unknown analgesic. Draw the plates in your lab notebook. Calculate the R_f for each spot. Knowing the composition of each analgesic, identify all the spots.
6. Determine the R_f values for each spot in your unknown and identify the components.

BIBLIOGRAPHY

1. A Hearty Endorsement for Aspirin. P. Aldhous. *Science*, **263**, 24 (1994).
2. Towards a Better Aspirin. J. Vane. *Nature*, **367**, 215–216 (1994).
3. Goodman & Gilman's the Pharmacological Basis of Therapeutics. Joel G. Hardman and Lee E. Limbird, ed. 9th ed. McGraw-Hill, New York, (1996).
4. Medicinal Chemistry of Aspirin and Related Drugs. *J. Chem. Ed.*, **56**, 331–333 (1979).
5. Drugs in the Chemistry Laboratory. The Conversion of Acetaminophen into Phenacetin. E.J. Volker, E. Pride, and C. Hough. *J. Chem. Ed.*, **56**, 831 (1979).
6. Analysis of APC Tablets. V.T. Lieu. *J. Chem. Ed.*, **48**, 478–479 (1971).

EXPERIMENT 15 REPORT

Pre-lab

1. Make a table showing the physical properties (MM, m.p., b.p., density, solubility, toxicity/hazards) of acetaminophen (*N*-acetyl-4-hydroxyaniline), ethyl iodide, potassium carbonate, butanone, phenacetin (*N*-acetyl-4-ethoxyaniline, also known as acetophenetidin or *p*-acetophenetide), and *tert*-butyl methyl ether.

2. Make a flowchart for the preparation and purification of phenacetin.

3. Why do you extract the reaction mixture with *tert*-butyl methyl ether?

4. Why do you extract the organic layer with a 5% NaOH aqueous solution? Why do you finally wash it with saturated NaCl solution?

5. Which bands do you expect to see in the IR of phenacetin? How can you tell phenacetin and acetaminophen apart by IR?

6. Drugs exert their effects by binding to specific proteins and receptors. To be recognized by the receptor, the molecule must contain certain atoms positioned at the right distances with respect to one another. Compare the structures of ibuprofen and naproxen and find all the structural features common to both molecules.

7. Based on polarity, which compound do you expect to have a larger R_f value in a solvent of medium polarity (such as ethyl acetate): acetaminophen or phenacetin?

8. Give two reasons why ethyl iodide is preferred over ethyl chloride in the Williamson ether synthesis. (Hint: check their boiling points).

In-Lab

1. Calculate how many mmoles of each reactant were used. Determine which one is the limiting reagent. Take into account the weight of the tablet and the amount of acetaminophen per tablet (see Table 15.1).

2. Calculate the theoretical yield of phenacetin in mg.

3. Report the actual yield of phenacetin in mg. Calculate the % yield of the synthesis. Discuss any source of product loss.

4. Report the m.p. of phenacetin. Compare with the literature value.

5. Report and discuss the results of the ferric chloride tests.

6. Assign as many bands as possible in the IR of phenacetin and the starting material. Compare the IR of the product with that from an authentic sample shown in Figure 15.2. Present your data in a table format.

7. Report the R_f values of phenacetin and the components of Bayer aspirin, Tylenol, and Anacin. Justify the R_f values using polarity arguments.

8. Report the possible composition of the unknown analgesic. Explain your reasoning.

9. Analyze the ^{1}H-NMR of phenacetin shown in Figure 15.3. Calculate the chemical shift of each type of hydrogen using the NMR correlation tables (Unit 33).

10. Interpret the ^{13}C-NMR of phenacetin shown in Figure 15.4.

11. (Advanced level) Interpret the mass spectrum of phenacetin shown in Figure 15.5. Propose a mechanism for the formation of the m/z 137 peak (loss of CH_2CO).

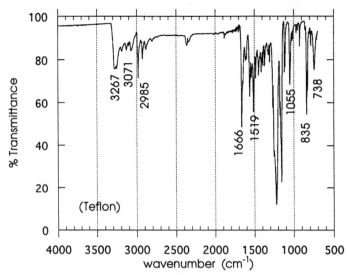

Figure 15.2 IR spectrum of phenacetin (Teflon).

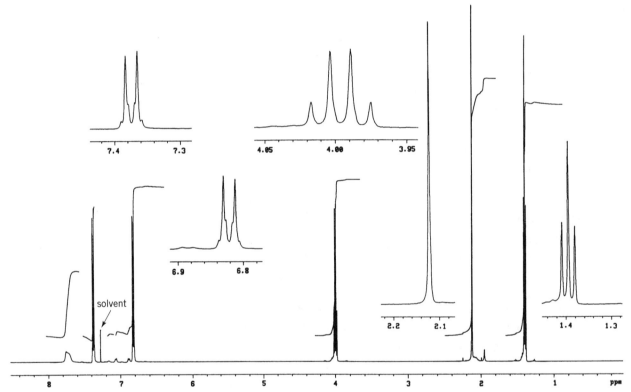

Figure 15.3 500-MHz ^{1}H-NMR spectrum of phenacetin in CDCl$_3$.

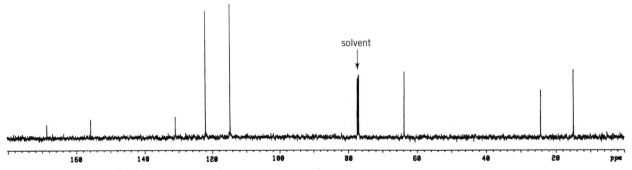

Figure 15.4 125.7-MHz ^{13}C-NMR spectrum of phenacetin in CDCl$_3$.

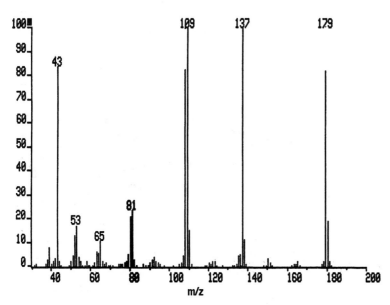

Figure 15.5   Mass spectrum of phenacetin (electron impact, 70 eV).

Electrophilic Aromatic Substitution

16.1 MECHANISM OF ELECTROPHILIC AROMATIC SUBSTITUTION

Understanding the reactivity of aromatic compounds is critical in the design of new chemicals and the planning of synthetic strategies, especially in the fields of pharmaceutical chemistry and the chemistry of dyes and polymers. Substituents on aromatic rings can be introduced by two basic reactions: **electrophilic** and **nucleophilic substitutions**. Modification of substituents already present on the ring is also possible through oxidations and reductions.

Substituents rich in electrons, such as $-NH_2$, $-OH$, $-OCH_3$, directly attached to aromatic rings increase the electron density at positions ortho and para with respect to the substituent. These electron-rich positions become more susceptible to react with **electrophiles** or electron-loving species. These reactions are called **electrophilic aromatic substitutions**, and take place via a common mechanism that proceeds in two steps:

1. The electrophile, E^+, is attacked by the π-electron cloud of the aromatic ring, preferentially from a position with high charge density, leading to the formation of an intermediate **carbocation**, called a σ-**complex** or a **Wheland complex**:

2. A proton from the position where the attack took place is lost and the aromatic character of the ring is restored. The driving force for this step is the stability of the aromatic ring.

Among the most important electrophilic aromatic substitutions are nitrations and halogenations.

16.2 NITRATION AND HALOGENATION

In nitration reactions the electrophile is NO_2^+ (nitronium ion), which is formed by the action of a strong acid on nitric acid:

$$O_2N-\ddot{O}H \underset{B}{\overset{BH^+}{\rightleftarrows}} O_2N-\overset{+}{\ddot{O}}H_2 \rightleftharpoons NO_2^+ + H_2O \qquad (1)$$

Once attached to an aromatic ring, the nitro group can be easily transformed into a number of different substituents via its reduction to an amino group. The versatility of nitro compounds as synthetic intermediates is illustrated below with a few reactions.

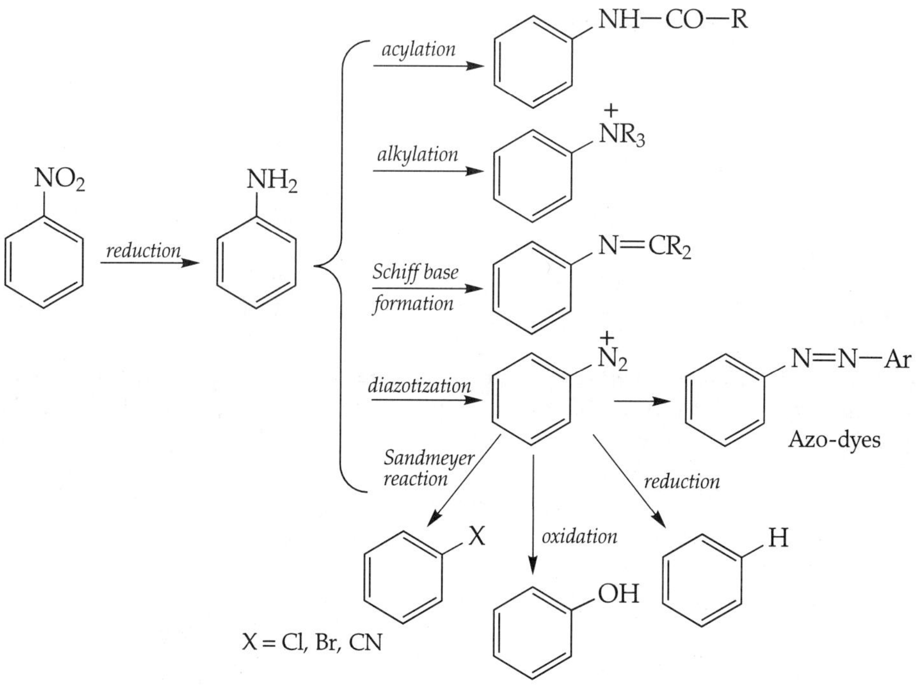

Halogenation of aromatic compounds involves treatment with Cl_2, Br_2, or I_2 with or without a catalyst. Fluorination of aromatic compounds by using F_2 is of little practical application because F_2 is too reactive and usually destroys the aromatic ring. Bromination of benzene requires the presence of a catalyst such as $FeBr_3$, which acts as a Lewis acid, forming a complex with Br_2. The Br—Br bond is polarized and finally gives $FeBr_4^-Br^+$, which reacts with the aromatic ring as if it were Br^+.

$$FeBr_3 + Br-\overset{\delta^-\ \ \delta^+}{Br} \longrightarrow [Br_3Fe\cdots Br\cdots Br] \longrightarrow Br_4Fe^-Br^+ \qquad (2)$$

To avoid the manipulation of liquid bromine, which is a powerful oxidizer that may cause severe skin burns, alternative brominating agents have been developed.

Particularly useful is a combination of potassium bromate, $KBrO_3$, and hydrobromic acid, which generates bromine when mixed according to the following reaction:

$$6\,HBr + KBrO_3 \longrightarrow 3\,Br_2 + KBr + 3\,H_2O \qquad (3)$$

Iodination of aromatic rings is a very important reaction with applications in biology, medicine, and pharmacology. In vertebrates, *in vivo* synthesis of thyroid hormones occurs via iodination reactions, which take place in the thyroid gland. These reactions are believed to proceed by an electrophilic aromatic substitution mechanism involving enzyme catalysis. In the following experiment, a procedure for the synthesis of a precursor of thyroid hormones, 3,5-diiodotyrosine, is presented.

Iodine (I_2) is the least reactive of the halogens in electrophilic aromatic substitution. The reaction between I_2 and aromatic rings (Eq. 4) is usually reversible and is rarely of any practical application. However, the reaction can be greatly accelerated, and the equilibrium shifted to products, in the presence of Ag^+ or oxidants such as HNO_3. These reagents react with I_2 and are believed to form I^+ (Eqs. 5 and 6), which is the electrophile that reacts with the aromatic ring.

$$ArH + I_2 \rightleftharpoons Ar{-}I + HI \qquad (4)$$

$$I_2 + Ag^+ \longrightarrow I^+ + AgI \qquad (5)$$

$$I_2 + 2\,H^+ + HNO_3 \longrightarrow 2\,I^+ + 1/2\,NO + 1/2\,NO_2 + 3/2\,H_2O \qquad (6)$$

The use of oxidizing agents is not always recommended because they may also produce undesirable oxidations in the substituents already present on the aromatic ring. Also, when HNO_3 is used as an oxidant, minor amounts of nitrated byproducts are formed, which must be separated at the end, adding an extra step to the preparation.

One of the cleanest methods to iodinate aromatic compounds is the use of iodine monochloride, ICl. By virtue of chlorine's greater electronegativity, the I—Cl bond is polarized as shown below:

$$\overset{\delta+}{I}{-}\overset{\delta-}{Cl}$$

This makes the iodine end of the molecule electrophilic. It is not clear whether the electrophile is free I^+ or the intact ICl molecule. Studies have suggested that depending on the solvent and the pH, the electrophilic species is different. For example, in aqueous solution at low pH, H_2OI^+ seems to be the electrophile (H_2OI^+ can be regarded as equivalent to H_3O^+, in which one of the H^+ has been replaced by I^+, or alternatively, as protonated HOI, hypoiodous acid), while at higher pH, ICl appears to be the electrophilic species.

16.3   EFFECTS OF SUBSTITUENTS

Electron-donating substituents such as $-NH_2$ and $-OH$ increase the electron density of the aromatic ring, making the ring more reactive than benzene. This effect is called **activation**. Such groups can donate an electron pair to the ring, increasing the electron density at the ortho and para positions and, therefore, directing the electrophilic attack to those positions. Because of these effects, these substituents are called ortho- and para-orienting. The following resonance structures for the amino group illustrate this point.

Because of the electronegativity of the nitrogen and oxygen atoms, the amino and hydroxy groups also have the effect of *pulling electrons away from the aromatic ring*, which would result in deactivation. However, the overall effect of these groups is strong activation, because the resonance effect predominates over the electronegativity effect (also called **inductive effect**). Alkyl groups are also activating (and also ortho- and para-orienting) but much less powerful than $-NH_2$ and $-OH$.

The activating effect of the amino group on the aromatic ring is such that, for example, the reaction of aniline with an excess of bromine gives, almost instantly, 2,4,6-tribromoaniline. Unlike the bromination of benzene, the bromination of aniline is fast and does not require a catalyst. It cannot be stopped at the mono- or dibromoaniline products. Because of its *electron withdrawing character*, bromine is a **deactivating** group; however, the electron-donating power of the amino group is such that it overrides the deactivating effect of the first bromine introduced, and the reaction continues by the incorporation of a second and a third bromine atom.

A list of activating and deactivating groups for electrophilic aromatic substitutions is given below. Many of the deactivating groups are unsaturated. Groups with double and triple bonds, such as C=O, N=O, and C≡N, pull electrons away from the ring due to a resonance effect, especially at ortho and para positions, and therefore orient the attack to position meta.

Halogens are also deactivating. In principle, they have both electron-withdrawing and electron-donating capabilities. On one hand, because of their electronegativity they pull electrons away from the ring. On the other hand, their lone electron pairs interact with the aromatic ring by resonance and create a greater electron density at the ortho and para positions. Contrary to $-NH_2$ and $-OH$, the *inductive effect of halogens predominates* over their resonance effect and therefore halogens are *deactivating substituents*. However, because of the resonance effect, they orient to ortho and para positions.

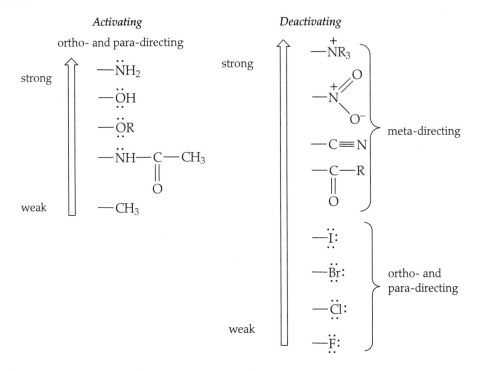

When two or more substituents are attached to the aromatic ring, the overall effect is the resultant of all the individual activating-deactivating effects. The more powerful activating group usually dominates the outcome of the substitution. For example, nitration of 4-methylphenol affords primarily 2-nitro-4-methylphenol, rather than 3-nitro-4-methylphenol.

BIBLIOGRAPHY

1. The Aromatic Substitution Game. M. Zanger, A.R. Gennaro, and J.R. McKee, *J. Chem. Ed.*, **70**, 985–987 (1993).

2. Electrophilic Aromatic Substitution. R. Taylor. Wiley, Chichester, 1990.

3. Aromatic Substitution Reactions. When You've Said Ortho, Meta, and Para, You Haven't Said it all. J.G. Traynham. *J. Chem. Ed.*, **60**, 937–941 (1983).

4. Bromination of Acetanilide. P.F. Schatz. *J. Chem. Ed.*, **73**, 267 (1996).

EXERCISES

Problem 1

The treatment of *o-*, *m-*, or *p*-toluidine (aminotoluene) with an excess of bromine gives three different compounds (one single compound in each reaction). Two of them are dibromo derivatives and the other is a tribromo derivative. What are the bromination products of these toluidines?

Problem 2

How could you use ^{1}H-NMR to determine the identity of the product of the reaction of aniline with an excess of bromine?

Problem 3

Predict the major product of mononitration of the following compounds:

(a) 4-methylbenzoic acid

(b) *m*-dibromobenzene

(c) *o*-chlorobenzoic acid

Problem 4

(a) Given the bond dissociation enthalpies, calculate the $\Delta H°$ for the chlorination, bromination, and iodination of benzene:

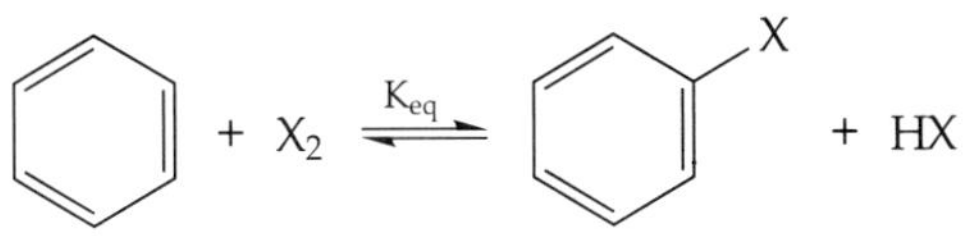

(b) Disregarding solvent and entropic effects, which halogenation is expected to have the smallest K_{eq}?

Bond Dissociation Enthalpies (kJ/mol)

H–Cl	431	Cl–Cl	243	C_6H_5–H	460
H–Br	368	Br–Br	193	C_6H_5–Cl	406
H–I	297	I–I	151	C_6H_5–Br	335
				C_6H_5–I	272

Problem 5

The nitration of 2-methylphenol with HNO_3 in acetic acid gives a mixture of isomeric products. By steam distillation of the mixture a solid compound, A, is isolated with 24% yield. The product is treated with K_2CO_3 and CH_3I and yields a liquid, B, that gives the following ^1H-NMR spectrum in CCl_4: 2.33 ppm (singlet, 3H); 3.85 ppm (singlet, 3H); 7.02 ppm (triplet, 1H); 7.35 ppm (double-doublet, 1H); 7.55 ppm (double-doublet, 1H). Propose a structure for A and B and justify the spectroscopic data. Speculate why it is possible to separate the product from the reaction mixture by steam distillation. What makes this product different from its isomers?

Problem 6

How would you separate 20 g of a mixture of *p*-nitroaniline and *o*-nitroaniline? How would you separate 10 mg? (Hint: solve Problem 5 first).

Problem 7

(*Previous Synton: Problem 5, Unit 15.*) In Synton's lab a container was labeled "bromonitro...ol...." Synton was in charge of keeping the inventory and did not like the idea of having a chemical so poorly labeled. He ran the ^1H-NMR of the unknown and found the following signals: 3.92 ppm (singlet, 3H); 6.9 ppm (doublet, 1H); 7.54 (double-doublet, 1H); 7.82 ppm (doublet, 1H). The mass spectrum showed the molecular ion peak at m/z 233/231 with an abundance of 70%/71%. What is the most likely identity for the compound?

EXPERIMENT 16A

Iodination of Tyrosine

E16A.1 SYNTHESIS OF 3,5-DIIODOTYROSINE

Iodoarenes are versatile intermediates in organic synthesis and important compounds from a biological point of view. Radioactive iodinated drugs are used in nuclear medical imaging, a powerful diagnostic tool. The active compounds secreted by the thyroid gland (**thyroid hormones**) are iodinated amino acids.

In this experiment you will carry out the iodination of L-tyrosine, a natural α-amino acid, using ICl in acetic acid. The product, 3,5-diiodotyrosine, will be characterized and identified by melting point and IR and NMR spectroscopy. The reaction is shown below.

$$\text{tyrosine} \quad + \ 2\,\text{ICl} \longrightarrow \quad \text{3,5-diiodotyrosine} \quad + \ 2\,\text{HCl}$$

The excess of ICl at the end of the reaction is eliminated by reduction with sodium thiosulfate:

$$\text{ICl} \ + \ 2\,S_2O_3{}^{2-} \longrightarrow \ I^- \ + \ Cl^- \ + \ S_4O_6{}^{2-}$$

thiosulfate   tetrathionate

L-tyrosine is one of about 20 natural α-amino acids that are the building blocks of proteins. In solid form as well as in solution (near-neutral pH), α-amino acids exist in the doubly charged form where the hydrogen from the carboxylic acid has reacted with the basic amino group, giving $-NH_3^+$ (we will come back to this in Unit 24). L-tyrosine undergoes iodination reactions *in vivo* to form 3,5-diiodotyrosine (DIT) and 3-iodotyrosine (monoiodotyrosine, MIT), two compounds that are the precursors of the thyroid hormones T_3 and T_4 (see the box, "Hormones of Change").

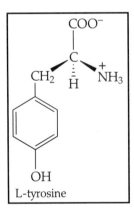
L-tyrosine

When Gregor Samsa woke up one morning from uneasy dreams, he found himself transformed in his bed into a gigantic insect. He was lying on his back as hard as armor plate, and when he lifted his head a little he saw his domelike brown belly, divided into stiff arched segments on top of which the bed cover could barely keep in position and was about to slide off completely. His numerous legs, which were pitifully thin compared with the rest of his body, were waving helplessly before his eyes.
"The Metamorphosis," Franz Kafka

Hormones of Change   The thyroid gland, located in the neck of all vertebrates, secretes two important hormones, **thyroxine**, also called T_4, and **3,5,3'-triiodothyronine**, T_3, which play a vital role in normal growth, development, and metabolism.

thyronine   3,5,3'-triiodothyronine (T$_3$)   thyroxine (T$_4$)

The whole process of thyroid hormone synthesis is complex and seemingly wasteful; it involves the iodination and coupling of two tyrosine residues of thyroglobulin, a constitutive protein of the thyroid, which must be hydrolyzed before T$_3$ and T$_4$ are released into the bloodstream. The iodination of tyrosine residues involves iodine in a +1 oxidation state, probably as hypoiodite (IO$^-$). Since the iodine circulating in the bloodstream is in the form of iodide (I$^-$), oxidation of the element must take place before it is able to react with tyrosine. The oxidation of iodide to I$^+$ is catalyzed by an enzyme called thyroid peroxidase, which uses H$_2$O$_2$ as the oxidant. The overall reaction is represented below:

$$I^- + H_2O_2 + 2\,H^+ \longrightarrow I^+ + 2\,H_2O$$

Normal thyroid function requires a sufficient iodine intake. A daily intake of 150–300 µg of iodine is usually recommended; iodine is abundant in seaweed (0.2% of dry material) and in sea fish (200–1000 µg/kg). It is also present, to a lesser extent, in eggs and milk. In some areas of the world far away from the oceans, seafood is a rarity and people are at risk of not getting enough iodine to sustain normal thyroid function. A low intake of iodine results in a decreased synthesis of T$_4$ and T$_3$. Severe iodine deficiency may lead to disorders such as *endemic goiter* and *cretinism*.

Endemic goiter is a condition in which the thyroid gland becomes enlarged; the disorder reverts by an increase of iodine ingestion. Cretinism is a condition present at birth, which if not treated may lead to serious and irreversible physical and mental abnormalities. Children affected by cretinism are dwarfed and mentally retarded with enlarged tongues and thick skin. Endemic cretinism is found only in areas where endemic goiter occurs and is always related to extreme iodine deficiency. Early detection of the condition by the screening of newborn infants usually leads to fully effective treatments. Endemic goiter and cretinism affect millions of people from poor as well as rich countries, from the highland plains of Bolivia to the Tyrolean Alps (see map below).

The occurrence of these disorders can be substantially reduced by simple, and relatively inexpensive, preventive measures. Correction of iodine deficiency has been implemented with success in certain regions of the world by addition of small amounts of iodine to table salt (at least 30 µg KI/g salt) or to drinking water, or by the injection of iodinated oil. However, the large-scale implementation of such programs, which would improve the quality of life of millions of people, has been hindered by inertia, ineffectiveness, and rampant bureaucracy. In 1978, in the village of Jixian in northeast China, there were about 1300 inhabitants of which more than 850 suffered from goiter and 150 from cretinism; among the people in nearby villages this place was known as the village of idiots. Iodinated salt was introduced in 1978 and after only four years, the rate of goiter dropped to 4% and no cases of cretinism were reported among newborns.

How the thyroid hormones regulate growth and development is still unclear, but it is recognized that they control gene expression and protein synthesis. The effect of thyroid hormones in development and differentiation is dramatically illustrated in the metamorphosis of tadpoles into frogs and toads. In normal development, a few days after

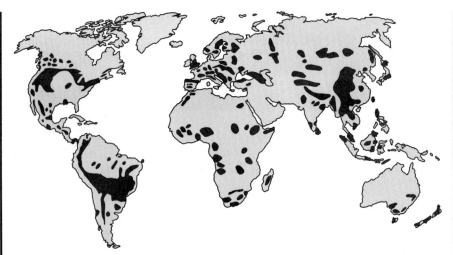

Regions of the world where endemic goiter has been detected (according to the World Health Organization).

the eggs have hatched the tadpoles have gills; after 10 weeks they have developed limbs; at three months the froglets still have a tail; finally the tail is completely reabsorbed in the adult frog (see below). If the thyroid gland is removed from the larva, the tadpole grows without undergoing metamorphosis; if thyroid hormones are injected into such an overgrown tadpole, it almost magically transforms into a frog.

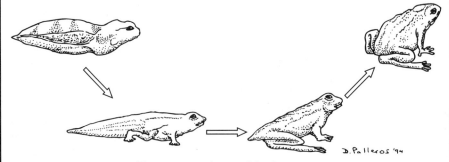

The metamorphosis of the frog.

Bibliography

1. Thyroid and Antithyroid Drugs by R.C. Haynes, Jr. in "Goodman and Gilman's The Pharmacological Basis of Therapeutics," A.G. Gilman, T.W. Rall, A.S. Nies and P. Taylor, ed. 8th ed., Pergamon Press, New York, 1990.

2. Goitre and Iodine Deficiency in Europe. P.C. Scriba et al. *The Lancet*, v. 1, 1289–1293 (1985).

3. Prevention and Control of Iodine Deficiency Disorders. *The Lancet*, v. 2, 433–434 (1986).

4. Amphibian Morphogenesis H. Fox, Humana Press, Clifton, N.J., 1984.

5. Biology of Animals. C.P. Hickman, Jr., L.S. Roberts, and F.M. Hickman, 3rd ed. C.V. Mosby, St. Louis, 1982.

6. Goiter and Endemic Cretinism: Before Iodine Prophylaxis. C.T. Sawin. *Endocrinologist*, **9**, 157–158 (1999).

PROCEDURE

> **Background reading:** 16.1–3
> **Estimated time:** synthesis: 1.5 hours; purification: 1/2 hour; analysis: 1 hour

Synthesis Warm about 200 mL of water in a 250-mL beaker on a hot plate; the temperature of the water should be between 60 and 70°C. Weigh about 0.2 g of L-tyrosine on a piece of weighing paper and transfer it to a 16 × 120-mm test tube equipped with a stir bar. With the aid of a 1-mL pluringe add 0.80 mL of acetic acid; tyrosine will not dissolve at this stage, it will form a suspension.

To the suspension of tyrosine, add 0.80 mL of a freshly made solution of iodine monochloride in glacial acetic acid (0.5 g/mL). This solution will already be made and ready to use.[1] Most of the tyrosine should dissolve at this point. Add 3 mL of water.

Cap the tube with a soft rubber stopper with a hole through which a length of Tygon tubing (30 cm long) has been inserted (a rubber thermometer adapter makes an excellent stopper). Place the other end of the Tygon tubing in a 13 mm × 100 mm test tube containing a cotton plug moistened with sodium thiosulfate solution (Fig. 16.1). This assembly works as a halogen trap, preventing iodine monochloride from escaping to the atmosphere. Place the test tube with the reaction mixture in the water bath kept at 60–70°C. Stir the reaction mixture for about 40 minutes, keeping the temperature of the bath between 60 and 70°C. At the end of this period remove the water bath and let the system cool down. Remove the Tygon tubing from the thiosulfate trap and place the reaction mixture in an ice-water bath. With the aid of a Pasteur pipet add

> ### Safety First
>
> - Acetic acid is corrosive. Iodine monochloride causes skin burns.
> - Handle the iodine monochloride solution with care. Wear gloves when handling it. Using a Pasteur pipet, a pluringe, and a small test tube, take only the volume needed.
> - Remove the gloves immediately.
> - Sodium hydroxide is corrosive and toxic. Handle it with care.

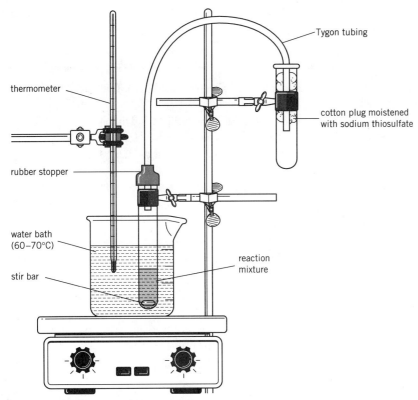

Figure 16.1 Setup for the iodination of tyrosine.

thermometer

Tygon tubing

cotton plug moistened with sodium thiosulfate

rubber stopper

water bath (60–70°C)

stir bar

reaction mixture

---

[1] To make the solution, do not weigh out solid iodine monochloride because it has a tendency to spatter. Instead melt it in a water bath at 30°C (m.p. 27°C) and dissolve 0.32 mL (1.0 g) of the liquid in 1.68 mL of glacial acetic acid.

9–15 drops of a 1.3 M sodium thiosulfate solution and mix well; a change in color from red-brown to yellow should be observed.

Add 6 M NaOH **dropwise** to precipitate the product; it takes about 20 drops or 0.6 mL of sodium hydroxide solution and the final pH should be 3–4. Stir well. If accidentally more sodium hydroxide than needed is added and the final pH is alkaline, the product will redissolve; it can be reprecipitated by lowering the pH (to a final value of 3–4) with the addition of 6 M HCl. Scratch the walls of the test tube to induce precipitation while cooling in an ice-water bath (do not scratch the bottom of the test tube, because the glass rod may puncture it). Collect the crystals by vacuum filtration on a Hirsch or Buchner funnel; wash the crystals with about 10–15 drops of cold water. Let the crystals dry for about 15 minutes on the filter with the vacuum on. Weigh the crude product on a piece of weighing paper.

Purification Transfer the solid to a 50-mL Erlenmeyer flask. Recrystallize the solid by adding 6 mL of water per each 0.10 g of crude solid. Bring the suspension to a **boil** on a hot plate while stirring continuously with a **small** glass rod. Let the system cool undisturbed on the bench top. This would favor the formation of long needles. Finally cool down the suspension in an ice-water bath. Filter the solid using a Buchner funnel. Let the crystals dry on the filter for about 10 min. A layer of crystals with a metallic texture and the appearance of aluminum foil can be peeled off from the filter paper if the solid is dry enough. Weigh the recrystallized product and calculate the % yield of the synthesis.

Analysis Determine the melting point of the purified material. Since the melting point is close to 200°C, to avoid unnecessary delays, heat the melting point apparatus at a rate of 10–15°C/minute up to a temperature of about 150°C and then decrease the heating rate to about 5°C/min. Note any decomposition and colored vapors produced during melting. Obtain the IR spectrum of the product using a nujol mull (section 31.5).

Cleaning Up

- Dispose of mother liquors from the synthesis and the recrystallization in the container labeled "Iodination Liquid Waste."
- Dispose of Pasteur pipets, cotton wad, filter papers, and capillary tubes in the container labeled "Iodination Solid Waste."
- Turn in the product in a labeled screw-capped vial to your instructor.

BIBLIOGRAPHY

1. The preparation of 3,5-diiodotyrosine has been adapted from: The Iodination of Tyrosine by Iodine Monochloride. P. Block, Jr. and G. Powell. *J. Amer. Chem. Soc.*, **65**, 1430–1431 (1943).

2. Aromatic Iodination: Evidence of Reaction Intermediate and of the σ-Complex Character of the Transition State. C. Galli. *J. Org. Chem.*, **56**, 3238–3245 (1991).

3. Advances in the Synthesis of Iodoaromatic Compounds. E.B. Merkushev. *Synthesis*, 923 (1988).

4. Radioiodination Techniques for Small Molecules R.H. Seevers and R.E. Counsell. *Chem. Rev.*, **82**, 575–590 (1982).

EXPERIMENT 16A REPORT ───────────────

Pre-lab

1. Write a chemical reaction for the iodination of tyrosine with iodine monochloride. Show the mechanism for a generic electrophilic aromatic substitution.

2. Write a balanced chemical equation for the reaction of iodine monochloride with sodium thiosulfate.

3. Following the directions given in the experimental procedure, write a flowchart for the preparation, purification, and characterization of 3,5-diiodotyrosine.

4. Why does the iodination of tyrosine afford 3,5-diiodo- instead of 2,6-diiodo-tyrosine?

5. Make a table showing the physical properties (molecular mass, m.p., b.p., density, solubility, flammability [for solvents only], and toxicity/hazards) of L-tyrosine, iodine monochloride, 3,5-diiodotyrosine, glacial acetic acid, and sodium thiosulfate.

In-Lab

1. Calculate the number of mmoles of reactants and products.
2. Calculate the theoretical yield of the reaction in mg.
3. Report the actual yield of crude and purified product and calculate the % yield before and after recrystallization. Discuss your results.
4. Report the melting point of the product and compare with a literature value. Did you observe any vapors during melting? of what color? What is the nature of the vapors? (Hint: in Greek, *iodes* means violetlike.)
5. Interpret your IR spectrum and compare with the one shown in Figure 16.2.
6. Interpret the ^{1}H-NMR of 3,5-diiodotyrosine shown in Figure 16.3. Account for spin-spin splitting and chemical shifts. Calculate the chemical shift of the aromatic hydrogens and the methylene and methine groups using ^{1}H-NMR tables (Unit 33).
7. (Advanced level) Figure 16.4 shows a detail (only for signals A and B) of the ^{1}H-NMR of 3,5-diiodotyrosine at 500 MHz. Read section 33.19 and interpret these signals. (Note: there are differences in the chemical shifts observed in Figures 16.3 and 16.4 because the spectra were obtained in different solvents.)
8. How would you distinguish 3,5-diiodotyrosine, 3-iodotyrosine, and tyrosine by ^{1}H-NMR?
9. Interpret the ^{13}C-NMR spectrum of 3,5-diiodotyrosine shown in Figure 16.5.

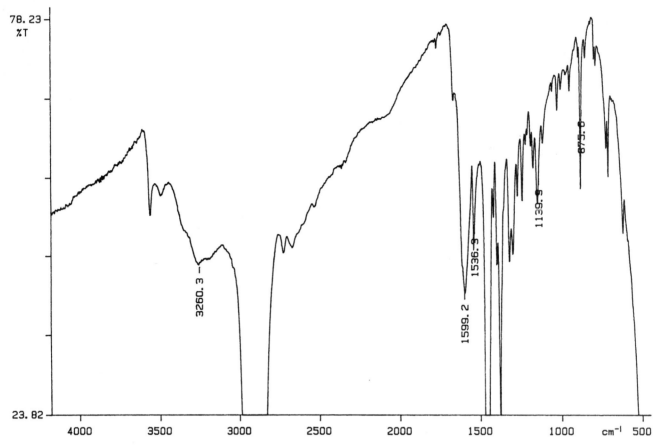

Figure 16.2 IR spectrum of 3,5-diiodotyrosine (nujol).

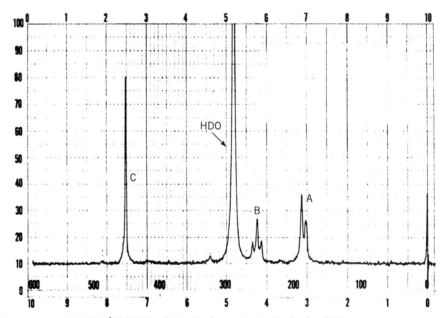

Figure 16.3 60-MHz ^{1}H-NMR of 3,5-diiodotyrosine in D_2O plus DCl.

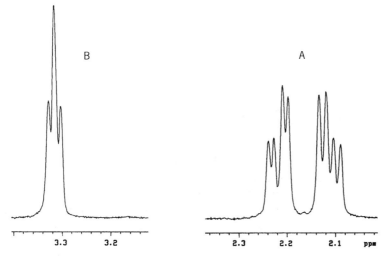

Figure 16.4   Detail of signals A and B in the ^{1}H-NMR of 3,5-diiodotyrosine at 500 MHz (in D_2O plus D_3PO_4).

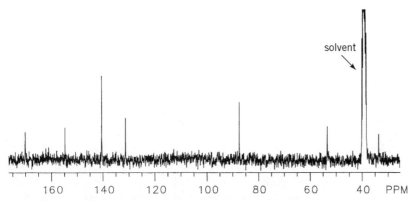

Figure 16.5   62.9-MHz ^{13}C-NMR of 3,5-diiodotyrosine in DMSO-d_6.

EXPERIMENT 16B

Two Substitution Puzzles

E16B.1 OVERVIEW

In this experiment you will carry out the nitration or the bromination of phenacetin (*N*-acetyl-4-ethoxyaniline). Phenacetin has two strong activators in para position. When this molecule is subjected to electrophilic aromatic substitutions, the new substituent can go, in principle, to any of the two vacant positions, ortho to the acetamido group, or ortho to the ethoxy group, because both are similarly activated.

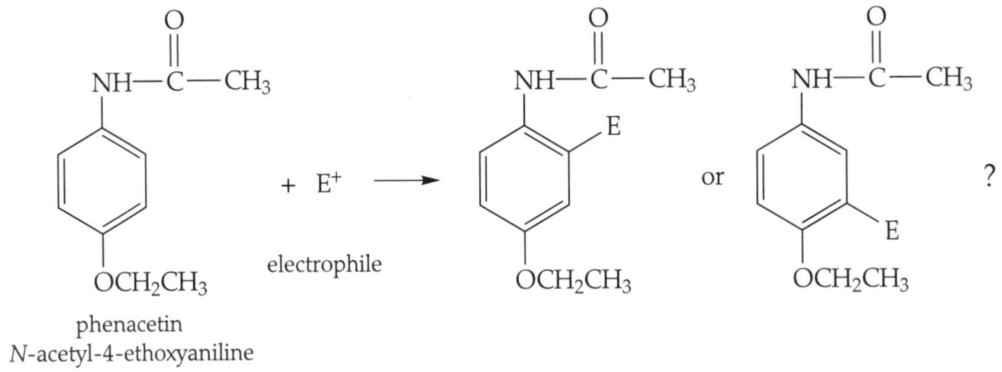

phenacetin
N-acetyl-4-ethoxyaniline

electrophile

Predicting which product will be predominant isn't easy. The final outcome of the substitution reaction depends on the size of the substituent, the stability of the product, and the experimental conditions such as solvent and temperature.

In the nitration of phenacetin described here only one substitution product forms. Is it the 2- or the 3-substituted product? The two possible products have different melting points and give rise to different ^{1}H-NMR spectra. The main difference in their spectra lies in the chemical shifts and coupling pattern of the aromatic hydrogens, as explained in the next section.

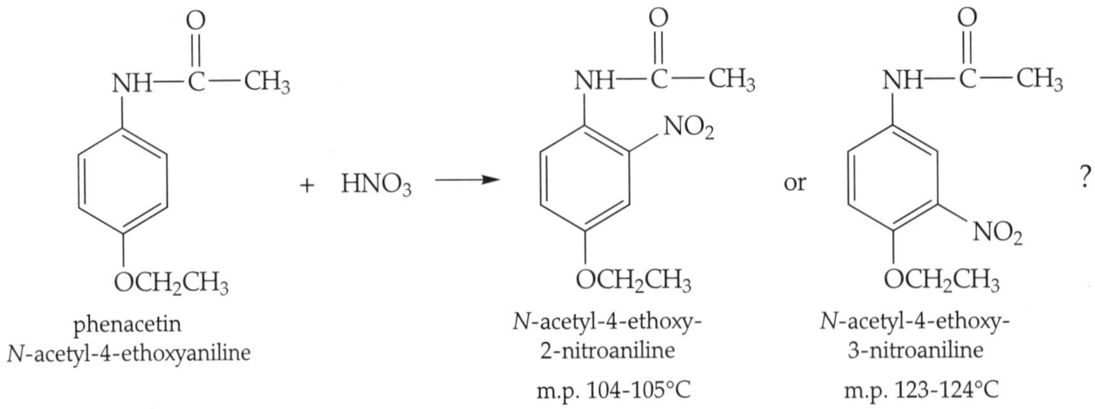

phenacetin
N-acetyl-4-ethoxyaniline

N-acetyl-4-ethoxy-
2-nitroaniline
m.p. 104-105°C

N-acetyl-4-ethoxy-
3-nitroaniline
m.p. 123-124°C

The bromination of phenacetin also affords a single compound that can be identified by its melting point and ^{1}H-NMR. Is it *N*-acetyl-2-bromo-4-ethoxyaniline or *N*-acetyl-3-bromo-4-ethoxyaniline?

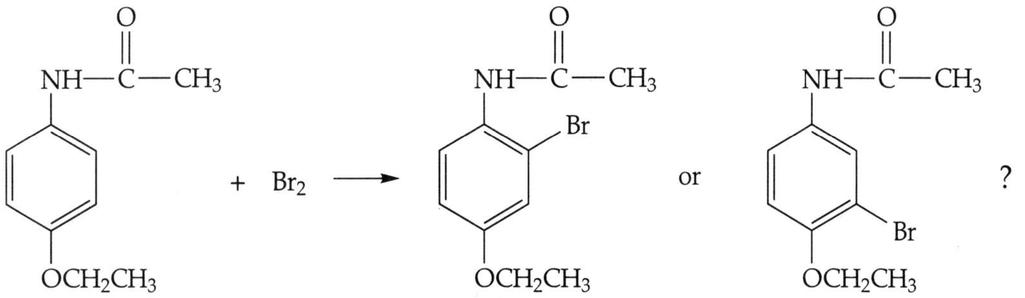

phenacetin
N-acetyl-4-ethoxyaniline

N-acetyl-2-bromo-
4-ethoxyaniline
m.p. 96-97°C

N-acetyl-3-bromo-
4-ethoxyaniline
m.p. 112-114°C

To take full advantage of this experiment you will work with a partner. Each student will perform one reaction, either the bromination or the nitration, and both will compare notes at the end.

E16B.2   NMR ANALYSIS

In each reaction, the two possible products have the same substitution pattern, namely two substituents in para position, and the third one in ortho position to one of them. These compounds, generally known as 1,2,4-trisubstituted benzenes, display a typical ^{1}H-NMR coupling pattern that is discussed in detail in section 33.18. The result is that H_A shows as a doublet with a coupling constant $J_{AM} \approx 7$–9 Hz. H_M shows as a double-doublet with two coupling constants $J_{AM} \approx 7$–9 Hz and $J_{MX} \approx 2$–3 Hz. H_X gives a doublet with $J_{MX} \approx 2$–3 Hz.

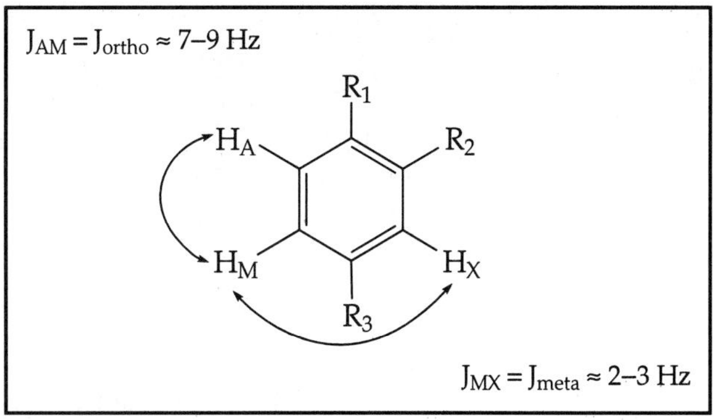

The chemical shifts of H_A, H_M, and H_X depend on the substituents R_1, R_2, and R_3. The nature of these substituents and their relative position with respect to the hydrogens determine the chemical shift for each hydrogen. Different 1,2,4-trisubstituted benzenes will show H_A, H_M, and H_X in different chemical shift order. Two such different orders are illustrated in Figure 16.6.

By analyzing the chemical shifts and the coupling pattern in the ^{1}H-NMR of your product, you should be able to identify the isomer that you obtained. In this experiment you will calculate the chemical shifts of the aromatic hydrogens for both possible isomeric products (Table 33.6, Unit 33) and will predict their coupling pattern as already explained. By comparing the predictions with the experimental values you will identify the product. In calculating the chemical shifts with Table 33.6, keep in mind that the values obtained are just approximate. It is not unusual that calculated values would differ from the experimental ones by ±0.3 ppm or more, especially if the aromatic ring has three or more substituents. Even with this limitation, the identification of isomers is possible.

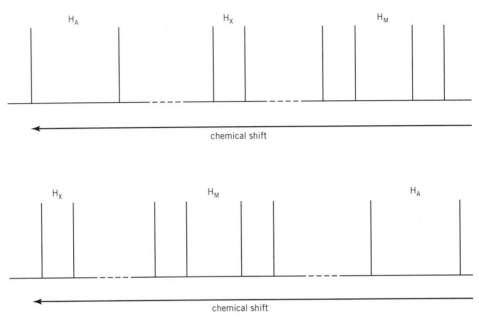

Figure 16.6 Two different orders of chemical shifts for H_A, H_M, and H_X.

PROCEDURE

E16B.3   NITRATION OF PHENACETIN

In a 15-mL round-bottom flask place 200 mg of phenacetin and 4 mL of glacial acetic acid. Add a magnetic stir bar, place the flask in a small water bath made with a small crystallizing dish (about 8 cm in diameter), and secure the flask to a ringstand. Stir the mixture to dissolve the solid. With the help of a Pasteur pipet and a pluringe add 0.25 mL of concentrated nitric acid (70% w/w) dropwise while stirring. Attach a microscale water condenser and heat the system until the water boils. Turn off the heat and let the system cool down.

With the help of a Pasteur pipet transfer the reaction mixture to a small beaker containing 15 mL of cold water. Cool the system in an ice-water bath, stirring well with a glass rod to crystallize the product.

Filter the solid under vacuum using a small Buchner funnel. Wash the solid on the filter with 10 mL of cold water. Let the solid dry on the filter with the vacuum on for about 10 minutes. Do not attempt to weigh the crude product at this stage because it is still wet.

Transfer the solid to a 50-mL Erlenmeyer flask and recrystallize it from water (section 4.2). Between 25 and 35 mL of water are necessary. Let the solid crystallize at room temperature undisturbed and then cool the system in an ice-water bath. Vacuum filter the suspension and let the shiny golden product dry on the filter with the vacuum on for about 10 minutes. The product is relatively dry at this point. Weigh the solid and transfer a portion to a porous plate to remove traces of water. Spread the solid on the plate to make a thin film. Scrape the solid off with a spatula. Do this operation gently to avoid etching the plate. Spread and scrape the crystals for about 5 minutes. Determine the melting point.

Dissolve the dry solid in $CDCl_3$ and obtain its ^{1}H-NMR spectrum (section 33.20). Obtain its IR spectrum using a nujol mull (section 31.5). Read "Cleaning Up" at the end of next section.

E16B.4   BROMINATION OF PHENACETIN

In a 25-mL Erlenmeyer flask weigh 200 mg of phenacetin. Weigh 0.063 g of potassium bromate on a piece of weighing paper and transfer the solid to the Erlenmeyer flask. Add 2.5 mL of glacial acetic acid and a stir bar. Stir the mixture to dissolve the phenacetin (the potassium bromate will not dissolve at this point).

With the help of a Pasteur pipet and a pluringe add dropwise 0.225 mL of a 48% (w/w) aqueous hydrogen bromide solution. Stir well during the addition and continue to stir for half an hour at room temperature. Upon addition of the hydrogen bromide, the solution will turn orange, indicating the formation of bromine.

At the end of the 30 minute period, transfer the liquid with a Pasteur pipet to a small beaker containing 20 mL of cold water. Cool the system in an ice-water bath and stir with a glass rod to crystallize the solid. If the solution is still orange add a few drops of a 1 M sodium thiosulfate aqueous solution to destroy the unreacted bromine. Vacuum-filter the suspension using a small Buchner funnel. Let the white solid dry on the filter with the vacuum on for about 10 minutes.

Transfer the solid to a 50-mL Erlenmeyer flask. Recrystallize it using about 25–30 mL of water (section 4.2). Let the system cool down and vacuum filter the solid. Let it dry on the filter with the vacuum on for about 10 minutes. Weigh the product.

Transfer a small portion to a porous plate and dry it by repeatedly spreading and scraping the solid. Determine its melting point. Obtain the ^{1}H-NMR spectrum using $CDCl_3$ as a solvent (section 33.20). Obtain the IR using a nujol mull (section 31.5).

Cleaning Up

- Dispose of the filtrate from the synthesis and the mother liquor from the recrystallization of nitrophenacetin (E16B.3) in the container labeled "Phenacetin Nitration Liquid Waste."

- Dispose of the filtrate from the synthesis and the mother liquor from the recrystallization of bromophenacetin (E16B.4) in the container labeled "Phenacetin Bromination Liquid Waste."

- Dispose of Pasteur pipets, filter papers, and capillary tubes in the "Phenacetin Solid Waste" container.

- Dispose of the $CDCl_3$ NMR solutions in the container labeled "Chloroform Waste."

- Turn in your product in a labeled vial to your instructor.

BIBLIOGRAPHY

1. The Nitration of Phenacetin. R.A. Russell, R.W. Switzer, R.W. Longmore, B.H. Dutton, and L. Harland. *J. Chem. Ed.*, **67**, 168–169 (1990).

2. Bromination of Acetanilide. P.F. Schatz. *J. Chem. Ed.*, **73**, 267 (1996).

3. Dictionary of Organic Compounds (CD-ROM, release 5 : 2). Chapman and Hall, 1997.

4. Deshielding Effects in the NMR Spectra of *ortho*-Substituted Anilides and Thioanilides. G.W. Gribble and F.P. Bousquet. *Tetrahedron*, **27**, 3785–3794 (1971).

EXPERIMENT 16B   REPORT

Pre-lab

1. Make a table with the physical properties, including toxicity/hazards, of the chemicals used in this experiment.

2. Write a balanced equation for the reaction that you will perform. Indicate possible isomeric products.

3. Write a mechanism for the reaction.

4. For E16B.3: What is the electrophilic species in the nitration? How is it formed?

5. For E16B.3: What are the hazards of working with nitric acid?

6. For E16B.4: Why do you use potassium bromate and hydrobromic acid in the bromination? What reagent are they replacing?

7. For E16B.4: What molar ratio of hydrobromic acid and potassium bromate should be used to generate Br_2? Consider Equation 3 on page 338 and answer this question assuming that HBr is the only source of H^+. Answer the same question considering the following equation and assuming that there are other sources of H^+ in the

medium such as acetic acid:

$$5\,Br^- + BrO_3^- + 6\,H_3O^+ \longrightarrow 3\,Br_2 + 9\,H_2O$$

8. Calculate the chemical shift of the aromatic hydrogens for both isomeric products. Predict the multiplicity for each hydrogen.

In-lab

1. Report and explain any observed change in color during the reaction.
2. Report and discuss the % yield of the product.
3. Report the melting point of the recrystallized product and compare it with those of the two possible isomers. Can you identify the product? Explain.
4. Compare the experimental ^{1}H-NMR with the predicted one (pre-lab 8). Describe the similarities and explain the differences. Identify your compound. To help you in the interpretation of the ^{1}H-NMR of the nitrophenacetin, check Reference 4 from the bibliography.
5. Compare the substitution patterns of the nitro and bromophenacetin obtained. Are they the same? Discuss your results.
6. Interpret the IR spectrum of the product.

Nucleophilic Aromatic Substitution

17.1 NUCLEOPHILIC AROMATIC SUBSTITUTION

There are three types of nucleophilic aromatic substitution reactions: 1) the displacement of a leaving group by a nucleophile on an aromatic ring bearing electron-withdrawing substituents; 2) reactions of aryl halides (without electron-withdrawing substituents) with very strong bases; and 3) decomposition of diazonium salts. In this unit only reactions of the first type will be studied. Reactions of the second type proceed via a different mechanism involving the formation of a benzyne intermediate; reactions of the third type, on the other hand, occur via an S_N1 mechanism.

An aromatic ring with electron-withdrawing groups, such as nitro and cyano, is susceptible to attack by nucleophiles at the ortho and para positions relative to the substituent; this attack is usually followed by the departure of a leaving group, X^-, leading to an aromatic substitution reaction (Eq. 1).

$$\text{(1)}$$

Although in Equation 1 the nucleophile is shown as having a negative charge, neutral nucleophiles, for example, amines, are also common in nucleophilic aromatic substitution reactions. The mechanism for this type of reaction involves two steps: 1) the attack of the nucleophile to give an intermediate species, called the **Meisenheimer complex**, and 2) the departure of the leaving group, X^- (Eq. 2):

$$\text{(2)}$$

Meisenheimer complex

Depending on the nature of the leaving group, the nucleophile, and the substituents on the aromatic ring, Meisenheimer complexes can be stable enough to be isolated or at least characterized by spectroscopic techniques. These complexes were isolated as potassium salts for the first time in 1902 by Meisenheimer, who prepared them by mixing potassium methoxide with 2,4,6-trinitroethoxybenzene (2,4,6-trinitrophenetole) or by adding potassium ethoxide to 2,4,6-trinitromethoxybenzene (2,4,6-trinitroanisole), see below. The isolation and characterization of Meisenheimer complexes give strong evidence in favor of the mechanism presented in Equation 2; this mechanism is sometimes referred to as *addition-elimination mechanism*.

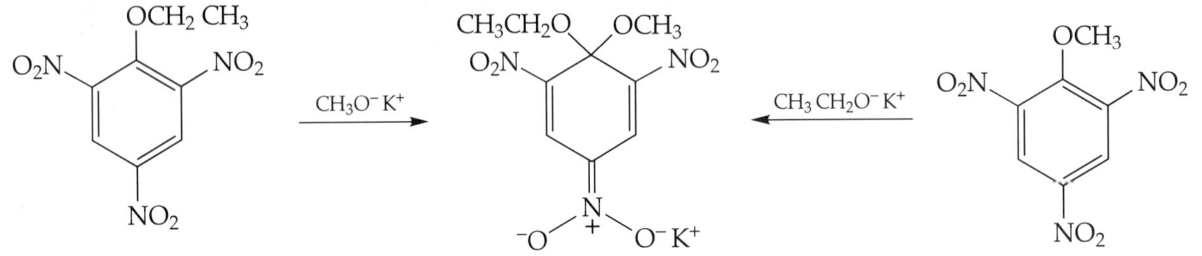

a Meisenheimer salt

Meisenheimer complexes are stabilized by resonance. The nitro groups can accommodate the negative charge brought by the nucleophile, as shown below in structure b, and stabilize these complexes. As the number of nitro groups at para and ortho positions increases so does the stability of the complex.

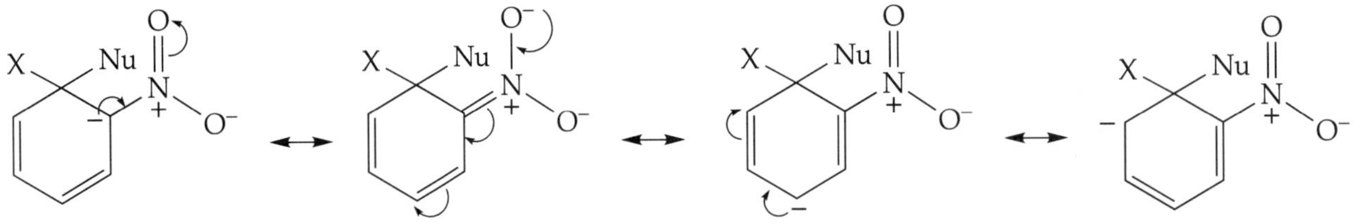

While electron-withdrawing substituents, such as $-NO_2$ and $-CN$, activate the ring for nucleophilic substitution, electron-donor groups ($-NH_2$, $-OH$, etc.) are deactivating. This is the opposite effect to that observed in electrophilic aromatic substitution.

With anionic nucleophiles, such as CH_3O^-, $CH_3CH_2O^-$, or HO^-, and halogens as leaving groups, the first step depicted in Equation 2 is rate-limiting and controls the reaction rate. The relative reactivity with halogens as leaving groups was found to be: F > Cl > Br > I. This order is the opposite to that found in S_N1 and S_N2 reactions. This reflects the fact that the rate is determined by the attack of the nucleophile (first step) rather than the departure of the halogen leaving group (second step). The more electronegative fluorine atom creates a more positive charge density around the aromatic carbon atom, which results in a faster attack by the nucleophile. On the other hand, with neutral nucleophiles, such as amines, either the first step or the second step can control the reaction rate. We will come back to this in a kinetic experiment (Exp. 18).

The leaving groups are not restricted to halogens. Alkoxide and phenoxide anions (for example, methoxide, ethoxide, and *p*-nitrophenoxide) can be displaced by nucleophiles.

17.2 DINITROANILINES

Amines are common nucleophiles in aromatic substitutions activated by nitro groups. For instance, 2,6-dinitroanilines encompass a family of potent herbicides (see box "Herbicides: They Are Organic, Are They?") that can be synthesized by treatment of the 2,6-dinitrochlorobenzenes with aliphatic amines (Eq. 3). Primary amines normally react faster than the more sterically hindered secondary amines. A byproduct of the reaction of halobenzenes with amines is the ammonium halide that separates as a precipitate when nonprotic solvents such as ethyl acetate and diethyl ether are used. At least a twofold excess of the amine over the halobenzene should be used to obtain a quantitative transformation.

X = F, Cl, Br   G = any substituent in position 3 or 4

Dinitroanilines are yellow-red compounds with absorbance maxima in the visible region (370–500 nm) of the spectrum.

Herbicides: They Are Organic, Are They?   There are three important reasons for the use of herbicides: 1) to increase crop yield by reducing competition from weeds; 2) to afford cleaner crops at the time of harvesting by reducing the volume of unwanted material; and 3) to reduce the risk of fungal and viral infections due to cross-contamination from other plant species. All these reasons ultimately result in cheaper crops that can be used to feed larger populations. There are also many reasons to be suspicious of the indiscriminate use of herbicides: hazards to the environment, to our health, and to all forms of life. There are advocates on both sides of the herbicide issue: those who would like to see a ban on the use of any pesticide and those who would like to ignore the health problems altogether. The rational approach lies somewhere in between. Solid and reliable scientific evidence should be used to resolve these issues. However, the research required to assess long-term effects of chemicals is lengthy, and in some cases it may take more than a generation to realize their detrimental effects.

According to their mode of action herbicides can be selective or nonselective; selective herbicides have deleterious effects on only a few species while nonselective herbicides kill all types of plants. Selectivity is a relative term. Herbicide action is not only dependent on plant species but it also depends on environmental factors and the method of application. For example, the type of soil and the amount of rainfall are critical in determining herbicidal action. Herbicides can also be classified according to their chemical structures; among them triazines, carbamates, dinitroanilines, and phenoxys constitute the most important families. Dinitroanilines have been extensively used since 1960, especially 2,6-dinitroanilines. Substitution at the 3- and 4-positions has a pronounced effect on their potency. For example, in 2,6-dinitroanilines, the degree of herbicidal power for substituents at the 4-position is $CF_3 > CH_3 > Cl > H$.

Trifluralin

α,α,α-Trifluoro-2,6-dinitro-
N,N-dipropyl-p-toluidine

The mode of action of dinitroanilines has not been completely elucidated. They are used as **preemergence** herbicides, which means that they are applied to the soil before weed seed germination. They are useful in the control of grasses and broad-leaved weeds in the cultivation of legumes, carrots, tomatoes, artichokes, cotton, and strawberries, among many other crops. Their primary effect is inhibition of root growth, apparently by suppression of mitosis; they kill the seeds as they germinate. Studies have shown that when these herbicides are absorbed, they cause changes in the sugar, amino acid, and nucleic acid content but little herbicide is translocated from the roots to the aerial parts of plants. These herbicides are relatively nontoxic to humans if handled as directed. The acute oral LD_{50} of trifluralin, one of the most commonly used dinitroanilines, for different animals is given in the table below:

Animal	LD_{50} (g/kg)
rats	>5
mice	0.5
dogs	2
rabbits	2
chickens	>2

In feeding trials it was found that trifluralin did not have ill effects on rats receiving diets with 2 g/kg for a 2-year period. However, it was found to be toxic to fish for which LC_{50}, the concentration necessary to kill 50% of a population, is in the range 0.01–0.09 mg/L. This herbicide should never be applied to water or wetlands. Trifluralin is absorbed by the soil and its low solubility in water (<1 mg/L) makes it very resistant to leaching. Its activity in the soil lasts only 6 to 8 months. Degradation reactions of trifluralin in the soil involve dealkylation (loss of propyl group), reduction of the nitro groups, and oxidation of $-CF_3$ to $-COOH$.

BIBLIOGRAPHY

1. Mechanisms and Reactivity in Aromatic Nucleophilic Substitution Reactions. C.F. Bernasconi. MTP *International Rev. Sci. Arom. Compds. Org. Chem. Series One*, **3**, 33 (1973). H. Zollinger ed., Butterworths, London.

2. Kinetic Evidence for a Methanolysis Intermediate in Aromatic Nucleophilic Substitution. N.S. Nudelman and D. Garrido, *J. Chem. Soc. Perkin Trans. II*, 1256–1260 (1976).

3. Mode of Action of Herbicides. F.M. Ashton and A.S. Crafts. 2nd ed., Wiley-Interscience, New York, 1981.

4. Herbicide Use and Invention. K.P. Parry, in "Herbicides and Plant Metabolism." A.D. Dodge ed. Cambridge University Press, Cambridge, 1989.

5. Dinitroanilines. G.W. Probst, T. Golab and W.L. Wright, in "Herbicides, Chemistry, Degradation and Mode of Action." P.C. Kearney and D.D. Kaufman, ed. Vol. 1, 2nd ed. Marcel Dekker, New York, 1975.

6. Agrochemical and Pesticide Safety Handbook. M.F. Waxman. Lewis Pub., Boca Raton, 1998.

7. The Pesticide Manual. A World Compendium. C.D.S. Tomlin ed. 11th ed., British Crop Protection Council. Farnham, Surrey, 1997.

EXERCISES

Problem 1

Classify the following substituents as activating or deactivating for the nucleophilic aromatic substitution: $-COOCH_3$, $-Br$, $-N(CH_3)_2$, $-NO_2$, $-SO_3H$, $-OCH_3$.

Problem 2

Complete the following reactions:

(a)

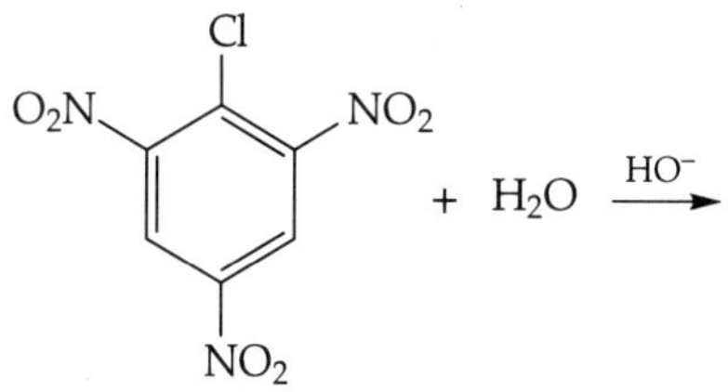

$+ \ H_2O \ \xrightarrow{HO^-}$

(b)

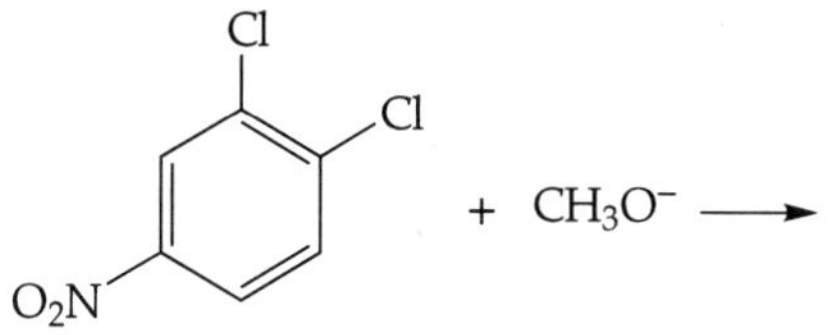

$+ \ CH_3O^- \ \longrightarrow$

(c)

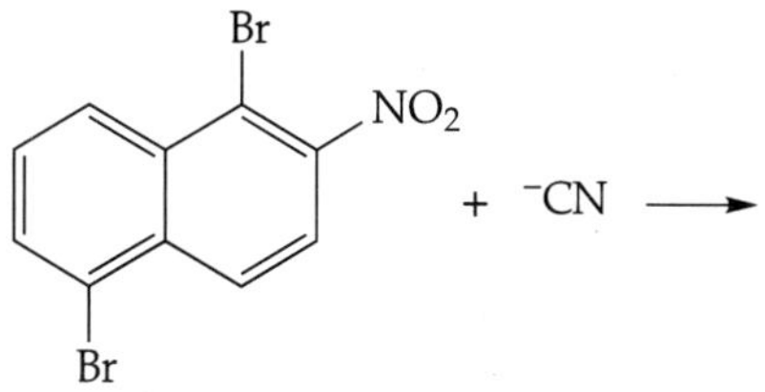

$+ \ ^-CN \ \longrightarrow$

(d)

$+ \ CH_3CH_2CH_2CH_2NH_2 \ \longrightarrow$

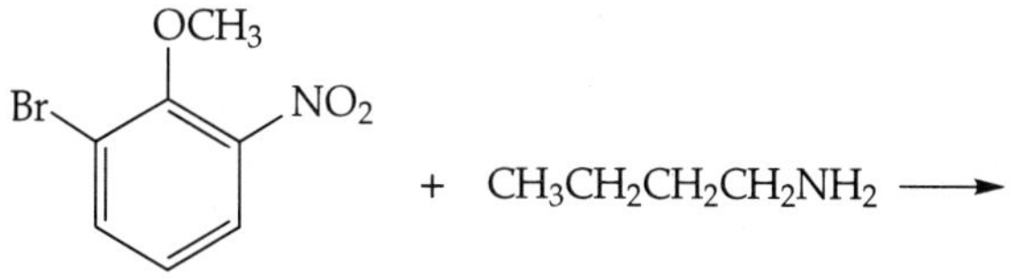

Problem 3

In the reaction of 1-chloro-2,4-dinitrobenzene with an excess of cyclohexylamine in methanol as a solvent, there was evidence for the transient formation of 2,4-dinitroanisole (Ref. 2). Give a reasonable explanation for this observation.

Problem 4

In the synthesis of N,N-dipropyl-2,4-dinitroaniline, 1-chloro-2,4-dinitrobenzene (0.40 g; MM: 202.5 g/mol) was dissolved in 10 mL of ethyl acetate and 0.50 mL of dipropylamine (MM: 101.2 g/mol; $\delta = 0.738$ g/mL) was added. After a few minutes at room temperature, white flakes separated out, which were collected by filtration and washed with a small amount of cold ethyl acetate to yield 200 mg of material.

(a) Write a chemical equation for the whole process.

(b) What is the flaky product?

(c) Calculate the theoretical yield in mg for the dinitroaniline. Which reagent is limiting?

(d) Assuming that the solubility of dipropylamine hydrochloride is negligible in ethyl acetate, calculate the % yield for the transformation of the dinitrochlorobenzene into dinitroaniline.

(e) How could the dinitrochlorobenzene and the dinitroaniline be separated?

(f) How would you improve the yield in dinitroaniline?

EXPERIMENT 17

Dinitrocompounds–Herbicides

In this experiment you will synthesize two dinitroanilines by nucleophilic aromatic substitutions and try their potency as preemergence herbicides. In parts E17.1 and E17.2 you will synthesize: *N,N*-dipropyl-2,6-dinitro-4-trifluoromethylaniline (trifluralin) and *N*-(*n*-butyl)-2,4-dinitroaniline, respectively.

In part E17.3 you will study their herbicide effect against two different weeds: crabgrass and velvetleaf. You will make a stock solution of each herbicide and prepare dilute solutions. You will investigate the effect of herbicide concentration on weed germination. Velvetleaf seeds need scarification in order to germinate. This can be done by adding to the seeds about six times their volume of hot water (at 70–80°C) and letting them soak for 12–24 hours at room temperature. The water is removed and the seeds are ready for use (for later use they can be stored in the refrigerator for about a week). The seeds will be already prepared for you to use.

You will work with a partner in this two-period experiment. In the first period, you will prepare one of the herbicides and your partner the other. You will make the stock and dilute solutions and set the seeds with different amounts of herbicides in Petri dishes. You will allow one week for the germination to take place and in the second week, you and your partner will analyze the results. You will compare the effect of each herbicide against the controls and against each other. You will also characterize the dinitroanilines by IR and ^{1}H-NMR.

Before doing any of the many parts of this experiment, read the "Safety First" box. Follow cleanup procedures given at the end of this experiment.

PROCEDURE

E17.1   SYNTHESIS OF TRIFLURALIN

> **Background reading:** 17.1–2
> **Estimated time:** 45 minutes

α,α,α-Trifluoro-2,6-dinitro-
N,N-dipropyl-*p*-toluidine

Trifluralin

Set a reflux apparatus using a 25-mL round-bottom flask, a water-jacketed condenser, and a sand bath on a hot plate with magnetic stirrer. Weigh 0.40 g of 4-trifluoromethyl-2,6-dinitrochlorobenzene (also known as 4-chloro-3,5-dinitro-benzotrifluoride) on a piece of weighing paper and transfer the solid to the round-bottom flask. Dissolve the solid with 10 mL of ethyl acetate. In the hood, add 1.0 mL of dipropylamine to the solution. Add a magnetic stir bar and heat the system under reflux for 15 minutes. At the end of this period turn the heat off and allow the system to cool to ambient temperature by removing the sand bath. Meanwhile, set up a gravity filtration apparatus using a short-stem funnel, fluted paper, and a pre-tared 50-mL round-bottom flask to collect the liquid. Filter the suspension using a Pasteur pipet to transfer the liquid. Carry out this operation in the hood. Rinse the flask with three 2-mL portions of ethyl acetate and transfer the rinses to the funnel. Let the solvent evaporate from the funnel by leaving it for a while in the hood. Meanwhile, evaporate the solvent in the 50-mL round-bottom flask using a rota-vap (Fig. 3.17). Cool down the mixture in an ice-water bath. Sometimes the product does not solidify but rather separates as a syrup (this is often observed with solids of low melting points). Leaving the syrup in an open flask for a couple of days usually leads to solidification. Weigh the product. Determine its melting point, if possible. Recrystallization of the product, which can be done from hexanes, is not necessary since the product is obtained with an adequate degree of purity.

After the ethyl acetate is completely evaporated from the funnel, transfer the dipropylamine hydrochloride to a pre-tared, labeled vial and determine its weight by difference.

Analysis   Obtain the IR of the product using a nujol mull (section 31.5). If possible, obtain its ^{1}H-NMR using CDCl$_3$ as a solvent (section 33.20) (*Caution: CDCl$_3$ is a possible carcinogen; dispense this chemical in the fume hood.*)

E17.2   SYNTHESIS OF *N*-(*n*-BUTYL)-2,4-DINITROANILINE ———

> **Background reading:** 17.1–2
> **Estimated time:** 45 minutes

Weigh about 200 mg of 1-chloro-2,4-dinitrobenzene on a piece of weighing paper. Transfer the solid to a 13 × 100 mm test tube and dissolve it in 3.0 mL of anhydrous methanol. Use a microspatula to stir the system until it dissolves. Add 1.2 mL of *n*-butylamine (in the hood) and stir well. Cover the test tube with a piece of Parafilm and leave it at room temperature on the bench top. Within 15 minutes a yellow product separates. Cool the system down in an ice-water bath. Filter it using a Hirsch funnel. Wash the solid with 0.5–1 mL of cold methanol. Carry out this operation in the hood. Let the crystals dry on the filter for about 15 minutes. Determine the melting point of the product. The product obtained is adequately pure, but it can be recrystallized from ethanol if necessary.

Analysis   Obtain the IR of the product using a nujol mull (section 31.5). If possible, obtain its ^{1}H-NMR using CDCl$_3$ as a solvent (section 33.20) (*Caution: CDCl$_3$ is a possible carcinogen; dispense this chemical in the fume hood.*)

E17.3 HERBICIDE EFFECTS OF TRIFLURALIN AND N-(n-BUTYL)-2,4-DINITROANILINE

> **Background reading:** 17.1–2;
> Herbicides (p. 357)
> **Estimated time:** 2 hours

You will study the effects of these two dinitroanilines on two different weeds: velvetleaf and crabgrass. One of them is resistant to this type of herbicide. You will also compare the relative potency of both dinitroanilines and determine the optimum concentration for weed control. Herbicide effect will be assessed by studying the degree of germination after a week and comparing with controls (no herbicides). A control with *dilute* acetone, the solvent used to dissolve the dinitroanilines, must also be carried out.

For each of the two dinitroanilines prepare the following solution. In a labeled 5-mL volumetric flask dissolve 25 mg of the dinitroaniline with 5.0 mL of acetone. Carry out the dilutions indicated in Table 17.1, solutions 1–4, using 25-mL volumetric flasks and automatic pipets.

Table 17.1 Herbicide solutions

Number	V (μL) of dinitroaniline solution	V (μL) of acetone	V (mL) of water	[Dinitroaniline] in mg/mL	Total amount of dinitroaniline in mg
1	500	0	24.5	0.1	2.5
2	250	250	24.5	0.05	1.25
3	50	450	24.5	0.01	0.25
4	5	495	24.5	0.001	0.025
Controls					
5	0	500	24.5	0	0
6	0	0	25.0	0	0

You will study the effects of the two dinitroanilines at four different concentrations on two different weeds. You will need 16 Petri dishes (bottom plus top) for this. Also, two controls, one with dilute acetone (number 5) and one with pure water (number 6), will be performed for each weed. Thus, a total of 20 Petri dishes, **adequately labeled**, are necessary; see (Fig. 17.1).

Cut circles of paper towel to fit the Petri dishes (9-cm diameter); you will need about 12 circles per Petri dish. Filter paper can also be used, but paper towels give better results because they retain more water. Place six circles in each of the 20 bottom parts of the Petri dishes. To ten of them, add 200 mg of crabgrass seeds, and 200 mg of scarred velvetleaf seeds to the other ten. Cover the seeds with six circles of paper towel. Water each dish with one of the solutions prepared according to the table. All the circles should be wet but should not drip water; a volume of about 12 mL is just right to wet 12 circles (about 3 g) of paper towels. This amount may vary with the type of paper towel used. Cover the dishes with the other half. Stack up the Petri dishes and leave them in a well lit and warm place (temperature 19–23°C) but not in direct sunlight.

After at least five days, **with the aid of tweezers** and **wearing gloves**, remove the circles covering the seeds, and place them in the proper waste container (see "Cleaning Up"). Observe germination and growth **by comparing with the controls**. Grade the growth from + + + + for maximum growth to 0 for no growth. Interpret

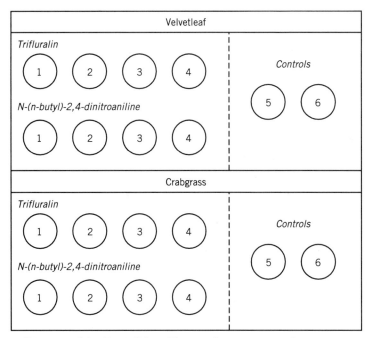

Figure 17.1 Diagram of the Petri dishes. The numbers correspond to the solutions indicated in the table.

your results before leaving the lab. Which dinitroaniline is a more potent herbicide? Which weed is more resistant? Is there any concentration effect (see in-lab questions)?

BIBLIOGRAPHY

1. Q.F. Soper, U.S. Patent 3,257,190. *Chem. Abstr.* **65**, 13606e (1966).
2. Kinetics of the Reaction of Fluoro- or Chloro-2,4-dinitrobenzene with *n*-Butyl, *s*-Butyl or *t*-Butylamine in Benzene. Further Evidence for an Addition Intermediate in Nucleophilic Aromatic Substitution. F. Pietra and D. Vitale, *J. Chem. Soc.*, (B), 1200–1203 (1968).

EXPERIMENT 17   REPORT

E17.1 Synthesis of Trifluralin

Pre-lab

1. Prepare a table with the physical properties (MM, m.p., b.p., density, solubility, flammability [for solvents only], and toxicity/hazards) of 4-trifluoromethyl-2,6-dinitro-chlorobenzene, dipropylamine, trifluralin, and ethyl acetate.
2. Write a chemical reaction for the transformation.
3. Calculate the number of mmoles of reactants used.
4. Calculate the theoretical yield of trifluralin and dipropylamine hydrochloride in mg.

In-lab

1. Calculate the % yield of the synthesis based on the mass of trifluralin obtained.
2. Calculate the % yield of the synthesis based on the mass of dipropylamine hydrochloride.

3. Compare the m.p. of your product with the literature m.p. 41–43°C (Ref. 1).

4. Interpret your IR and NMR spectra and compare them with those shown in Figures 17.2–17.4.

E17.2 Synthesis of *N*-(*n*-butyl)-2,4-dinitroaniline
Pre-lab

1. Prepare a table with the physical properties (MM, m.p., b.p., density, solubility, flammability [for solvents only], and toxicity/hazards) of 1-chloro-2,4-dinitrobenzene, *n*-butylamine, *N*-(*n*-butyl)-2,4-dinitroaniline, and methanol. The m.p. of *N*-(*n*-butyl)-2,4-dinitroaniline is 89.5–90°C (Ref. 2).

2. Write a chemical reaction for the transformation.

3. Calculate the number of mmoles of reactants used.

4. Calculate the theoretical yield of the substitution product in mg.

In-lab

1. Calculate the % yield of the synthesis based on the mass of product obtained.

2. Compare the m.p. of your product with the literature value.

3. Interpret your IR and NMR spectra and compare them with those shown in Figures 17.5–17.7.

E17.3 Herbicide Effects of Trifluralin and *N*-(*n*-butyl)-2,4-dinitroaniline
Pre-lab

1. What is the meaning of *preemergence herbicide*?

2. To which type of animals is trifluralin particularly toxic?

In-lab

1. Present your growth data in a table format.

2. Which dinitroaniline is more potent as a herbicide?

3. Are both weeds equally susceptible? Which one is more resistant?

4. Is there any effect due to concentration? Discuss your data.

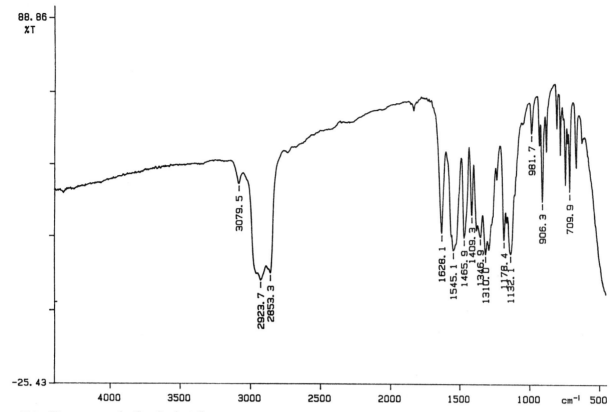

Figure 17.2 IR spectrum of trifluralin (nujol).

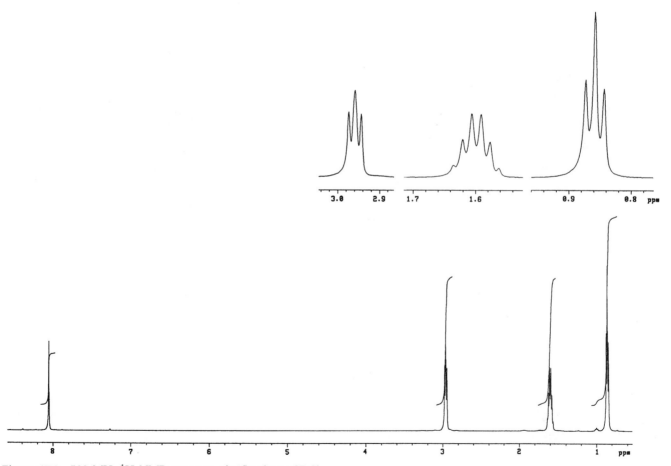

Figure 17.3 500-MHz ¹H-NMR spectrum of trifluralin in CDCl₃.

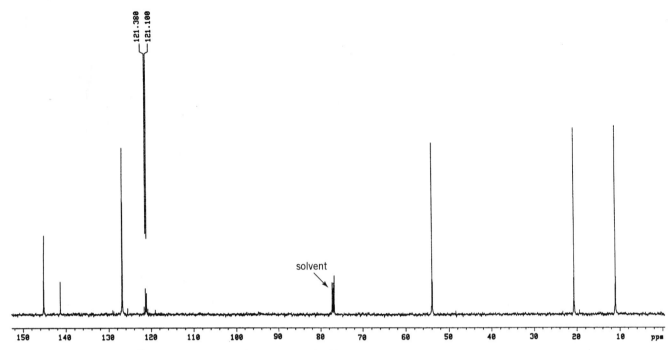

Figure 17.4 125.7-MHz ^{13}C-NMR spectrum of trifluralin in CDCl$_3$.

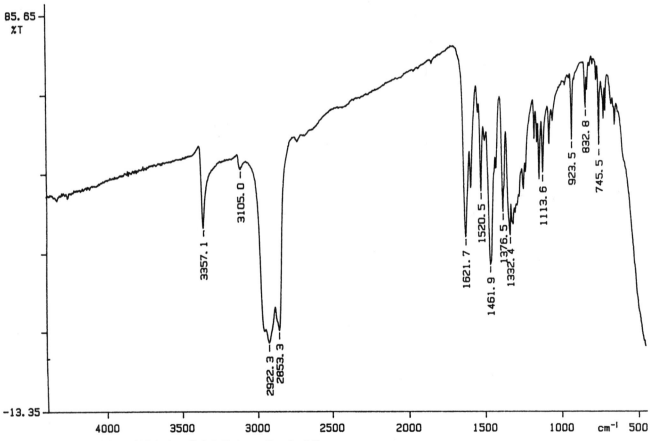

Figure 17.5 IR spectrum of N-(n-butyl)-2,4-dinitroaniline (nujol).

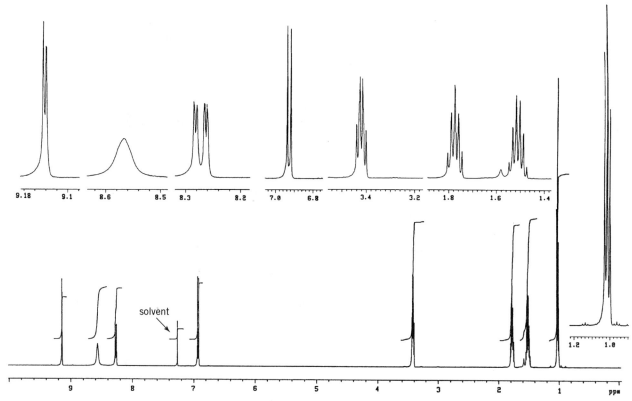

Figure 17.6 500-MHz ^{1}H-NMR spectrum of N-(n-butyl)-2,4-dinitroaniline in CDCl$_3$.

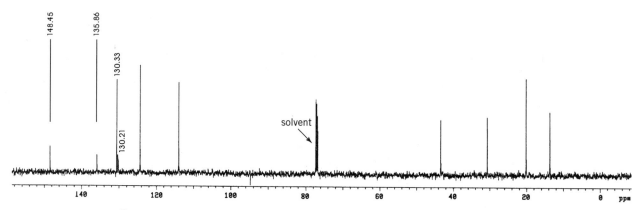

Figure 17.7 125.7-MHz ^{13}C-NMR spectrum of N-(n-butyl)-2,4-dinitroaniline in CDCl$_3$.

UNIT 18

Chemical Kinetics

18.1 MECHANISTIC STUDIES

Most organic reactions occur by complex mechanisms composed of several steps. Nucleophilic aromatic substitutions (Unit 17), for example, occur by a multistep mechanism in which a nucleophile, Nu^-, attacks the aromatic ring and forms an intermediate compound. The intermediate then decomposes to give the final products.

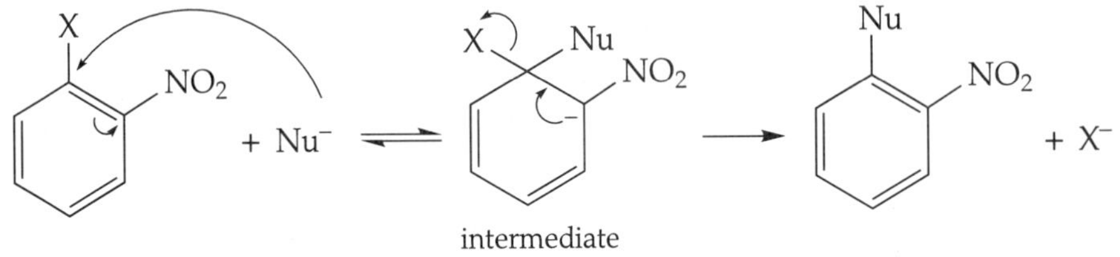

intermediate

How do we know that this reaction takes place in stages instead of in one single step? How fast is the reaction? Can we accelerate it? Does the reaction rate depend on the nature of the nucleophile? In this unit we will discuss the tools needed to answer these types of questions.

The study of reaction mechanisms is a challenging area of chemistry where different disciplines, from thermodynamics to quantum mechanics, make significant contributions. One of the most important tools in the elucidation of reaction mechanisms is **chemical kinetics**, or the study of reaction rates.

The study and understanding of reaction mechanisms has transformed organic chemistry from a purely descriptive into a highly predictive science. Yet, reaction mechanisms are seldom proven. In most cases all we can say is that a given mechanism is compatible with the experimental evidence, whereas others can be rejected because they do not agree with the data. As more sophisticated techniques for investigating reaction mechanisms become available, pathways that once seemed reasonable mechanistic routes may become unacceptable.

18.2 CHEMICAL KINETICS

Let's consider the following reaction where A collides with B giving product P without the formation of any intermediate:

$$A + B \longrightarrow P \tag{1}$$

Reactions like this are called **elementary reactions** because they take place in one single step. Elementary reactions are the building blocks of complex reaction mechanisms.

If we were to plot the concentration of reactants and products as a function of time for an elementary reaction, we would obtain graphs similar to those shown in Figure 18.1. These plots are called the **progress curves of the reaction** and describe how fast reactants are turned into products.

In elementary reactions the rate at which reactants disappear is equal to the rate at which products are formed. The rate is calculated by the change in concentration (ΔA, ΔB, or ΔP) for a given period of time, Δt (Eq. 2):

$$\text{rate} = -\frac{\Delta[A]}{\Delta t} = -\frac{\Delta[B]}{\Delta t} = \frac{\Delta[P]}{\Delta t} \qquad (2)$$

The minus sign in front of the rate of disappearance of reactants A and B makes the rate a positive figure since $\Delta[A]$ and $\Delta[B]$ are negative (notice that [A] and [B] *decrease* with time). Figure 18.2 illustrates how the reaction rate can be calculated from the product's progress curve.

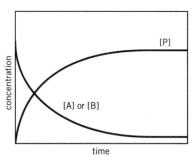

Figure 18.1 Progress curves for an elementary reaction.

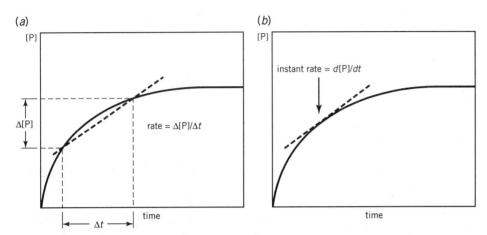

Figure 18.2 Calculating the reaction rate using the product's progress curve: (*a*) average rate for period Δt; (*b*) instant rate at time *t*.

If the period of time (Δt) used to calculate the rate is too long, average rate values are obtained. These average values give a poor description of the kinetics and are usually unacceptable. A much more accurate picture can be obtained by calculating the instant rate at different times. This is done by making Δt infinitesimally small. The instant rate, or simply the **reaction rate**, v, is given by the slope of the tangent line to the curve at each point. In mathematical terms, the instant rate is the *derivative of the concentration with respect to time* (Eq. 18.3):

$$v = -\frac{d[A]}{dt} = -\frac{d[B]}{dt} = \frac{d[P]}{dt} \qquad (3)$$

Most elementary reactions involve the collision of **two** molecules; these events are called **bimolecular**. The collision of three molecules is very unlikely but when it occurs it gives rise to **termolecular** reactions.

The rate at which A and B are consumed in reaction 1 is proportional to the concentrations of A and B because if the concentration of A or B increases, so does the chance of collision between them. The proportionality constant, k, is called the **rate constant** of the reaction (Eq. 18.4):

$$v = -\frac{d[A]}{dt} = -\frac{d[B]}{dt} = \frac{d[P]}{dt} = k \times [A] \times [B] \qquad (4)$$

An equation such as Equation 4, where the reaction rate, v, is given as a function of reactants concentrations, is known as **rate law** or **kinetic law**. The power to which each concentration is raised is called the **order** of the reaction in that reactant. In our example, the order in each A and B is one. The **overall order** is the sum of the individual orders, two in this case. For elementary reactions, the order in each reactant is equal to the stoichiometric coefficient. For reactions that take place via complex mechanisms the order cannot be predicted by the stoichiometric coefficients.

18.3 EXPERIMENTAL ASPECTS

The main objective in the study of chemical kinetics is the determination of the reaction orders and the calculation of the rate constants. The reaction can be followed by the rate of product formation or by the rate of reactant disappearance. If both rates can be measured, their comparison may shed light on whether the reaction involves several steps. If the rate of product formation is different from the rate of reactant disappearance, we can suspect a multistep mechanism. On the other hand, if both rates are the same, it does not necessarily follow that the reaction occurs in one step. Under certain conditions some intermediates may be present at undetectable levels and may not affect the measured reaction rates. Many important reactions in organic chemistry fall into this category.

Numerous experimental methods can be used to follow chemical reactions; any quantitative technique that allows the determination of concentration is potentially suitable for kinetics studies. UV-visible, IR, and fluorescence spectroscopies, nuclear magnetic resonance, high performance liquid chromatography (HPLC), and gas chromatography (GC) are just a few examples of the methods commonly used to investigate chemical kinetics.

Some of these methods, such as UV-visible and IR spectroscopy, allow virtually continuous monitoring of the reaction. The reaction mixture is placed in the path of light and the instrument reads the absorbance as a function of time. The number of readings is limited by the speed at which the instrument can take measurements. Other methods, such as HPLC and GC, on the other hand, require taking aliquots of the reaction mixture at different times and analyzing them one by one. When possible, a continuous monitoring of the reaction is preferred because extensive data can be easily collected and more reliable results are obtained. Continuous methods provide us with "a movie" of the reaction while methods such as HPLC and GC give us "snapshots." For very slow reactions, however, continuous methods are impractical. Slow kinetics are normally studied by taking isolated readings throughout the course of the reaction.

18.4 INTEGRATED RATE EQUATIONS

To obtain the rate constant from the progress curve, we could in principle draw the tangent lines and determine the slopes at different times, as shown in Figure 18.2b. However, this method is seldom used because the errors involved are too large. Instead, computer algorithms are used to find the mathematical function that best fits the data. This method, called **curve-fitting**, is very accurate and powerful, especially for the analysis of complex kinetic data. A brief discussion on curve-fitting is postponed until section 18.9.

When only a limited number of measurements of the reaction progress are made, as shown in Figure 18.3, we can also use a computer algorithm to perform a curve-fitting of the data, or we can simply apply an **integrated rate equation**, which does not necessitate a computer, as explained in the next paragraphs.

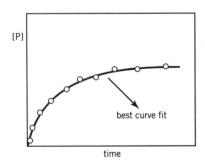

Figure 18.3 Kinetic data with the best curve fit.

Let us consider again the reaction:

$$A + B \longrightarrow P$$

and let us assume that we monitor the rate of disappearance of A. The rate law is:

$$v = -\frac{d[A]}{dt} = k \times [A] \times [B] \tag{5}$$

Equation 5 represents an overall second-order reaction because the rate depends on the first power of the concentrations of both A and B. Let's now assume that we use a *large excess* of B, so that its concentration does not change significantly during the course of the reaction and can be considered constant for all practical purposes (to this end, the initial concentration of B should be at least 10 times larger than the initial concentration of A). Under these conditions, the product $k \times [B]$ is a constant. Let's call this constant k_{obs} ("k observed"):

$$k \times [B] = k_{obs} \tag{6}$$

Substituting Equation 6 into Equation 5 we obtain:

$$v = -\frac{d[A]}{dt} = k_{obs} \times [A] \tag{7}$$

This rate law is now first order because it depends only on the first power of [A]. This type of reaction, where the *overall order is reduced to one* by using a large excess of one of the reactants, is called **pseudo-first-order kinetics**, and the constant k_{obs} is called the **pseudo-first-order rate constant**. In general, pseudo-first-order kinetics are easier to analyze than second-order ones, and are commonly employed in mechanistic studies.

Let's rename reactant A, S (for **Substrate**), and rearrange Equation 7:

$$-\frac{d[S]}{[S]} = k_{obs} \times dt \tag{8}$$

Equation 8 can be integrated using standard calculus methods. The result is shown in Equation 9, where $[S]_0$ is the initial substrate concentration and $[S]_t$ is its concentration at later times:

$$\ln \frac{[S]_t}{[S]_0} = -k_{obs} \times t \tag{9}$$

Equation 9 can be rearranged into Equation 10, which gives the substrate concentration as a function of time. This is called the integrated rate equation for first order, or pseudo-first-order, kinetics.

$$\boxed{\ln[S]_t = \ln[S]_0 - k_{obs} \times t} \tag{10}$$

To determine k_{obs} using Equation 10, ln $[S]_t$ is plotted as a function of time. A straight line with slope $-k_{obs}$ and intercept ln $[S]_0$, as indicated in Figure 18.4, is obtained. It should be noticed that k_{obs} is not a true rate constant because it depends on [B]. Once k_{obs} is known, we can determine the **second-order rate constant**, k, by dividing k_{obs} by [B].

$$k = \frac{k_{obs}}{[B]}$$

As already mentioned, integrated rate laws are very useful for the investigation of reaction rates when a limited number of data points are available. The minimum number of data points to obtain reliable results is about eight. A limitation of the

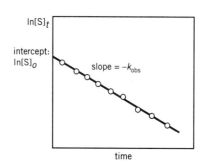

Figure 18.4 Plot of ln $[S]_t$ vs. time for the calculation of k_{obs}.

integrated rate equation method is that we need to know, or assume, the reaction order. In the previous example it was assumed that the order in S was one.

18.5 REACTION HALF-LIFE

An important concept frequently used in kinetic studies is that of **half-life**. The **half-life of a reaction**, τ, is the time it takes for the reagent to drop to half of its original concentration. For first-order or pseudo-first-order kinetics there is a simple relationship between the half-life and the rate constant. Substituting $[S]_t = [S]_0/2$ and $t = \tau$ in Equation 9 and rearranging it we obtain:

$$\tau = \frac{\ln 2}{k_{obs}} = \frac{0.693}{k_{obs}} \tag{11}$$

Equation 11 indicates that there is an inverse relationship between the half-life and the rate constant for a first-order reaction. One can be easily calculated from the other.

The time required for total conversion of reactants into products is, in principle, infinite. Yet, for practical purposes, the reaction can be considered complete when the time elapsed is at least 10 times the half-life of the reaction. For a first-order reaction this is equivalent to a 99.9% conversion (see Problem 2). Following a reaction to 99.9% conversion is sometimes impractical, especially for slow reactions, which are usually studied only to 2–70% conversion. When reactions are followed only to a small conversion percentage, the results must be taken with caution because the effects that products may have on the reaction rate (as catalysts, inhibitors, etc.) may not be noticeable at the beginning of the reaction when the product concentration is low. When possible, kinetics should be followed to at least 70% completion.

18.6 TRANSITION STATES

Elementary reactions have very simple free-energy diagrams. A plot of the free energy as a function of the reaction coordinate shows a single maximum that corresponds to the transition state (Fig. 18.5). In the transition state, represented by the symbol ‡, bonds are in the process of forming and breaking. As a result, the transition state is an unstable, short-lived species with a vanishingly small concentration. The transition state, which *cannot be isolated*, is always located at the top of the energy barrier.

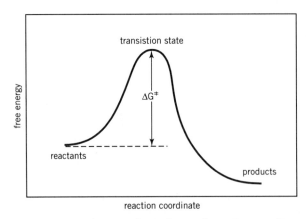

Figure 18.5 Free energy vs. reaction coordinate for an elementary reaction.

For an elementary reaction there is a relationship between the height of the energy barrier, called the **free energy of activation**, $\Delta G^{\ddagger}$, and the rate constant, k, given by Equation 12:

$$k = \frac{\mathbf{k}T}{h}e^{-\Delta G^{\ddagger}/RT} \tag{12}$$

where **k** and h are the Boltzmann and Planck's constants, respectively, T is the absolute temperature, and R is the gas constant. This equation indicates that the higher the energy barrier (the larger the $\Delta G^{\ddagger}$), the slower the reaction (the smaller the rate constant, k).

While elementary reactions show only one maximum in the free-energy diagrams, complex reactions display a number of maxima separated by *as many valleys as intermediates* formed in the reaction. The plot shown in Figure 18.6 represents a reaction that takes place by a two-step mechanism (Eq. 13), with the formation of one intermediate, I:

$$A \rightleftharpoons I \rightleftharpoons P \tag{13}$$

Intermediates are found in energy valleys and can be quite stable species (if the energy of the intermediate is lower than that of the reactants, the intermediate sometimes can be isolated). The student should notice the *intrinsic difference between a transition state and an intermediate*. Transition states are at the top of the energy barrier and *cannot be isolated*. Intermediates are found in valleys and *sometimes are stable enough to be isolated*. In the complex reaction represented in Figure 18.6 there are two transition states, one for each step, and one intermediate.

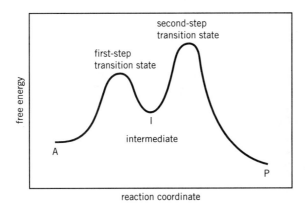

Figure 18.6 Free energy vs. reaction coordinate profile for a reaction that proceeds with the formation of an intermediate.

In a multistep reaction, one step usually controls the overall rate. This step is called the **rate-limiting step**. Here, we will consider as *rate-limiting step the step with the highest transition state in the energy diagram*. The rate-limiting step is the first step in Figure 18.7a and the second step in Figure 18.7b. This definition of the rate-limiting step applies to most organic reactions and is valid when the intermediate has higher free energy than the reactants. It may not apply, however, to cases where the intermediate is lower in energy than the reactants. For a detailed treatment of the subject see Reference 9.

In multistep mechanisms where the intermediate cannot be isolated, the experimentally measured rate constant, the **overall rate constant**, corresponds to the *total jump in free energy, $\Delta G^{\ddagger}$, from reactants to the highest transition state*, which is the transition state for the rate-limiting step. In Figure 18.7a this jump corresponds to the

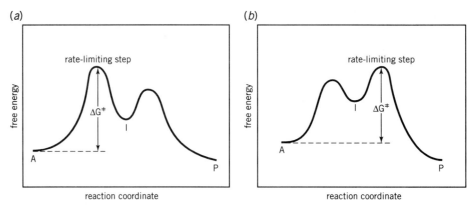

Figure 18.7   Free energy vs. reaction coordinate profile for a two-step reaction: (*a*) the first step is rate limiting; (*b*) the second step is rate limiting.

first transition state, and in Figure 18.7*b* to the second transition state. The overall rate constant is a function of the individual rate constants for each step.

Some books define the rate-limiting step as the slowest step of a multistep mechanism. Unfortunately this definition is incorrect. Although the rate-limiting step may well be the slowest step of a multistep mechanism, this is not always the case. Let's consider again the energy diagrams shown in Figure 18.7*a* and *b* (redrawn in Figure 18.8). In Figure 18.8*a*, the first step is rate limiting, because it has the transition state with the highest energy. It is also the slowest step because $\Delta G_1^{\ddagger} > \Delta G_2^{\ddagger}$. On the other hand, in Figure 18.8*b*, the second step is rate limiting because it has the transition state with the highest energy; however, the first step is the slowest one because $\Delta G_1^{\ddagger} > \Delta G_2^{\ddagger}$.

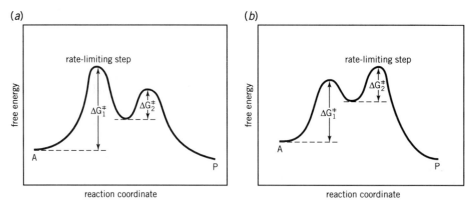

Figure 18.8   Free energy vs. reaction coordinate diagram for a two-step reaction: (*a*) first step is rate limiting and the slowest; (*b*) first step is slower than second step, but second step is rate limiting.

The rate-limiting step has important consequences in the kinetics of the overall reaction. The rate measured experimentally always depends on the rate of the rate-limiting step. Steps *previous* to the rate-limiting step *also have an effect* on the measured rate because they control the formation of the intermediates. On the other hand, steps *after* the rate-limiting step *have no impact* on the measured reaction rates and usually go unnoticed from a kinetic standpoint. Let us discuss this with the following two-step reaction. In the first step A reacts with B to give intermediate I, which can decompose back to A and B or collide with another molecule of B to give the final product P.

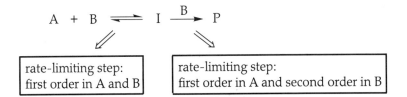

 shows:

A + B ⇌ I $\xrightarrow{B}$ P

| rate-limiting step: first order in A and B |
| rate-limiting step: first order in A and second order in B |

Let's also assume that the reaction is followed by the appearance of product, P, and that the intermediate cannot be isolated and is present in negligible amounts (this condition is true for many complex organic reactions). If the first step is rate limiting, we would observe second-order kinetics, first order in each A and B, because one molecule of A and one molecule of B participate in the transition state of the rate-limiting step. If, on the other hand, the second step is rate limiting, the overall order would be three: first order in A and second order in B, because two molecules of B participate in the transition state of the rate-limiting step.

18.7 DEPENDENCE OF THE RATE CONSTANT WITH TEMPERATURE: ARRHENIUS EQUATION

Toward the end of the nineteenth century, Arrhenius made the discovery that there is a logarithmic relationship between the rate constant and the inverse of the absolute temperature, T. He expressed this relationship in Equations 14 and 15.

$$\ln k = \ln A - \frac{E_a}{RT} \tag{14}$$

or

$$k = A \times e^{-E_a/RT} \tag{15}$$

A is called the **preexponential factor** and E_a the **Arrhenius activation energy**. The preexponential factor measures the probability of collision between molecules. Typical values are 10^{12}–10^{13} s^{-1} for first-order reactions and 10^{10}–10^{11} L mol^{-1} s^{-1} for second order reactions. The energy of activation is the minimum energy that the colliding species should have in order to proceed to products. Equations 14 and 15 indicate that the reaction rate increases with an increase in temperature. According to Equation 14 a plot of $\ln k$ versus $1/T$ should give a straight line with slope $-E_a/R$, as shown in Figure 18.9. The higher the activation energy, the faster the change in the rate constant with temperature. For example, a reaction with an activation energy of 10 kcal/mol (42 kJ/mol) doubles the reaction rate with a temperature increase of 10 K in the vicinity of 300 K. If the activation energy is 20 kcal/mol (84 kJ/mol), the rate increases 3 times for the same temperature increase.

The Arrhenius equation is particularly useful for elementary reactions, but the overall rate constant for complex mechanisms may not obey the Arrhenius equation.

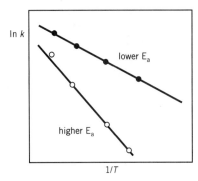

Figure 18.9 Arrhenius plots for two reactions. The reaction represented by solid dots has lower Arrhenius activation energy.

18.8 FOLLOWING THE KINETICS

One of the most important methods for following kinetics is UV-visible spectroscopy. The method is best applied when only the reactant *or* the product absorbs in the UV-visible region. We will first discuss the case when only the substrate absorbs, and then we will consider the case when the product absorbs. We will limit our discussion to first-order and pseudo-first-order kinetics.

The Substrate Absorbs

The best wavelength to follow the kinetics is the λ_{max} of the substrate. At the λ_{max}, the sensitivity of the method is optimum because any change in substrate concentration results in a maximum absorbance change.

At any point in time, substrate concentration [S] and absorbance are related by the Beer-Lambert's law (Eq. 16), where ε is the molar absorptivity of the substrate at the wavelength of choice, and ℓ is the length of the cuvette where the sample is placed.

$$A = \varepsilon \times [S] \times \ell \tag{16}$$

Thus, the substrate concentration can be calculated as:

$$[S] = \frac{A}{\varepsilon \times \ell} \tag{17}$$

Substituting [S] into Equation 9, Equation 18 is obtained where A_0 and A_t are the absorbances at time zero and at a later time. Notice that the factor $\varepsilon \times \ell$ cancels out and does not appear in the final equation.

$$\ln \frac{A_t}{A_0} = -k_{obs} \times t \tag{18}$$

Rearranging Equation 18, the following expression is obtained.

$$\boxed{\ln A_t = \ln A_0 - k_{obs} \times t} \tag{19}$$

A plot of $\ln A_t$ versus time gives a straight line with slope $-k_{obs}$ and intercept $\ln A_0$.

The Product Absorbs

The best wavelength to follow the kinetics is the λ_{max} of the product. The substrate and product concentrations at any time, [S] and [P], are related by the following equation:

$$[S]_t + [P]_t = [P]_\infty = [S]_0 \tag{20}$$

where $[P]_\infty$ is the product concentration at the end of the reaction (infinite time) and $[S]_0$ is the initial substrate concentration. Notice that both concentrations are identical. Thus, the substrate concentration at any time can be expressed as a function of the product concentration:

$$[S]_t = [P]_\infty - [P]_t \tag{21}$$

By substituting into Equation 9, Equation 22 is obtained:

$$\ln \frac{[P]_\infty - [P]_t}{[P]_\infty} = -k_{obs} \times t \tag{22}$$

The product concentration and the absorbance of the reaction mixture, A, are related by the Beer-Lambert's equation (Eq. 23):

$$A = \varepsilon \times [P] \times \ell \tag{23}$$

where ε is the molar absorptivity of the product and ℓ is the pathlength of the cuvette; [P] can then be calculated as:

$$[P] = \frac{A}{\varepsilon \times \ell} \tag{24}$$

Substituting [P] in Equation 22 and rearranging it, the following expression is obtained:

$$\ln(A_\infty - A_t) = \ln A_\infty - k_{obs} \times t \qquad (25)$$

where A_∞ is the absorbance of the solution at infinite time. When possible, A_∞ is determined experimentally by measuring the absorbance after at least 10 half-lives. This ensures a conversion of 99.9%. Sometimes it is not feasible to measure the absorbance after 10 half-lives. In such cases, A_∞ can be calculated with the help of Beer-Lambert's law (Eq. 23) if the initial substrate concentration is known. Taking into account that the final product concentration $[P]_\infty$ is equal to the initial substrate concentration, $[S]_0$, Equation 26 results:

$$A_\infty = \varepsilon \times [S]_0 \times \ell \qquad (26)$$

In Equation 26, ε is the molar absorptivity of the product at the chosen wavelength.

18.9 BEST FIT

In analyzing kinetics results, one must often find the best straight line that fits the data points. This is called the **best linear fit**. The best linear fit is always given by the straight line with minimum deviation from the actual points. If the deviation of the straight line from each point is D_1, D_2, D_3, and so on, as indicated in Figure 18.10, the best straight line is that for which the sum of the squares of the deviations is the smallest possible. In other words:

$$\sum (D_1)^2 + (D_2)^2 + \cdots + (D_n)^2$$

is minimum.

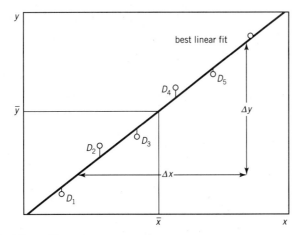

Figure 18.10 Best linear fit.

The squares of the deviations are used, instead of the deviations themselves, because the deviations can be positive or negative and their sum may be close to zero even when the individual deviations are large. The squares of the deviations, on the other hand, are always positive figures and cannot cancel each other out.

The best linear fit can be drawn by inspection of the data. A simple but approximate way of doing this is to make sure that the line connects as many points as possible, and an equal number of points are left below and above the line. The line should pass through the *center of gravity* of the data points. The center of gravity is defined by the $\bar{x}, \bar{y}$ pair, where $\bar{x}$ is the average of the x points, and $\bar{y}$ is the average of the y points, as shown in Equations 27 and 28, where n is the total number of points:

$$\bar{x} = \frac{\sum x}{n} \tag{27}$$

$$\bar{y} = \frac{\sum y}{n} \tag{28}$$

To calculate the slope, Δy and Δx are measured as shown in Figure 18.10. The slope is calculated with Equation 29:

$$\text{slope} = \frac{\Delta y}{\Delta x} \tag{29}$$

A precise way of obtaining the best linear fit is the use of the **least squares algorithm**. The slope, b, and the intercept, a, of the best straight line

$$y = a + bx \tag{30}$$

can be calculated with Equations 31 and 32. These equations are derived, with the help of calculus, from the premise that the sum of the squares of the deviations be minimum.

$$a = \frac{(\sum y)(\sum x^2) - (\sum x)(\sum xy)}{n \sum x^2 - (\sum x)^2} \tag{31}$$

$$b = \frac{n \sum xy - (\sum x)(\sum y)}{n \sum x^2 - (\sum x)^2} \tag{32}$$

For a derivation of these equations see Reference 17.

The least squares method is not limited to linear fits. Parabolas, polynomial, and exponential curves can all be fitted by the same principle of making the sum of the squares of the deviations minimum. Some pocket calculators and computer programs such as SigmaPlot, KaleidaGraph, and Microsoft Excel have the least squares algorithm.

BIBLIOGRAPHY

1. Chemical Kinetics: The Study of Reaction Rates in Solution. K.A. Connors. VCH, New York, 1990.
2. Chemical Kinetics and Reaction Mechanisms. J.H. Espenson. 2nd ed. McGraw-Hill, New York, 1995.
3. Basic Chemical Kinetics. H. Eyring, S.H. Lin, and S.M. Lin. Wiley, New York, 1980.
4. Kinetics and Mechanism: A Study of Homogeneous Chemical Reactions. A.A. Frost and R.G. Pearson. 2nd ed. Wiley, New York, 1961.
5. Chemical Kinetics. K.J. Laidler. 3rd ed. Harper & Row, New York, 1987.
6. Fundamentals of Chemical Kinetics. S.R. Logan. Longman, Essex, 1996.
7. Relaxation Kinetics. C.F. Bernasconi. Academic Press, New York, 1976.
8. Investigation of Rates and Mechanisms of Reactions. Techniques of Chemistry, Vol. VI. 3rd ed. A. Weissberger and E.S. Lewis, ed. Wiley, New York, 1974.
9. What is the Rate-Limiting Step of a Multistep Reaction? J.R. Murdoch. *J. Chem. Ed.*, **58**, 32–36 (1981).
10. An Intuitive Approach to Steady-State Kinetics. R.T. Raines and D.E. Hansen. *J. Chem. Ed.*, **65**, 757–759 (1988).
11. Rate-Controlling Step: A Necessary or Useful Concept? K.J. Laidler. *J. Chem. Ed.*, **65**, 250–254 (1988).

12. Just What is a Transition State? K.J. Laidler. *J. Chem. Ed.*, **65**, 540–542 (1988).

13. Transition States of Chemical Reactions. G.C. Schatz. *Science*, **262**, 1828–1829 (1993).

14. Real-Time Observation of the Vibration of a Single Adsorbed Molecule. F. Watanabe, G.M. McClelland, and H. Heinzelmann. *Science*, **262**, 1244–1247 (1993).

15. A Simplified Integration Technique for Reaction Rate Laws of Integral Order in Several Substances. J.G. Eberhart and E. Levin. *J. Chem. Ed.*, **72**, 193–195 (1995).

16. The Steady State and Equilibrium Assumptions in Chemical Kinetics. D.C. Tardy and E.D. Cater. *J. Chem. Ed.*, **60**, 109–111 (1983).

17. Application of Kinetic Approximations to the A $\rightleftharpoons$ B $\longrightarrow$ C Reaction System. G. Gellene. *J. Chem. Ed.*, **72**, 196–199 (1995).

18. Schaum's Outline of Theory and Problems of Statistics. M.R. Spiegel. 2nd ed. McGraw-Hill, New York, 1988.

EXERCISES

Problem 1

2,4-Dinitroanisole reacts with cyclohexylamine in cyclohexane to give *N*-cyclohexyl-2,4-dinitroaniline as shown below. The kinetics of the reaction was studied at 60°C. The concentration of 2,4-dinitroanisole as a function of time, $[S]_t$, is given in the table. The initial concentration of cyclohexylamine was 0.402 M.

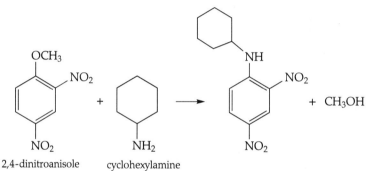

2,4-dinitroanisole   cyclohexylamine

(a) Is this a pseudo-first-order kinetic run? Briefly explain.

(b) Calculate k_{obs} and the second-order rate constant.

(c) Determine the half-life of the reaction.

(d) Calculate the half-life of the reaction if the concentration of cyclohexylamine is 0.1 M.

Time (min)	$10^5 \times [S]_t$ (M)	Time (min)	$10^5 \times [S]_t$ (M)
0	16.4	120	8.40
18	14.9	147	7.13
37	13.5	174	6.00
53	12.4	221	4.40
77	10.8	310	2.30
96	9.63	356	1.60

Problem 2

Calculate the % completion of a first-order reaction when the time elapsed from the beginning of the kinetic run is:

(a) 0.1 τ;   **(b)** τ;   **(c)** 10 τ;   **(d)** 100 τ

Problem 3

The half-life of a pseudo-first-order kinetic run is 70 minutes when the concentration of the reactant in excess is 0.1 M. Calculate the half-life of the same reaction when the concentration is 0.3 M.

Problem 4

Given the following reaction pathway:

(a) Draw schematic free-energy versus reaction coordinate profiles for each of these situations:

i) $k_{-1} \gg k_1 \gg k_2$;   ii) $k_2 \gg k_{-1} \gg k_1$

In both cases assume that the free energy of C is much lower than the free energy of B.

(b) Which is the rate-limiting step in each case?

Problem 5

The isomerization of [2.1.0]bicyclopentane to cyclopentene in the gas phase was studied (at relatively high pressures) at different temperatures. The first-order rate constant as a function of the temperature is given in the table below. Calculate the energy of activation and the preexponential factor for the reaction.

T (K)	$10^4 \times k$ (s^{-1})	T (K)	$10^4 \times k$ (s^{-1})
561.1	2.52	573.4	6.10
564.1	3.30	576.3	7.60
564.5	3.21	576.4	7.60
569.8	4.79	579.9	9.63
570.0	4.83	579.9	9.92
573.4	6.02	583.3	12.29

Data from M. Halberstadt and J. Chesick,
J. Amer. Chem. Soc., **84**, 2688–2691 (1962).

Nucleophilic Aromatic Substitution Kinetics

E18.1   THE PROBLEM

In this experiment you will study the kinetics of the reaction of 2,4-dinitroanisole (2,4-DNAn) with *n*-butylamine (*n*-BuNH$_2$) in methanol.

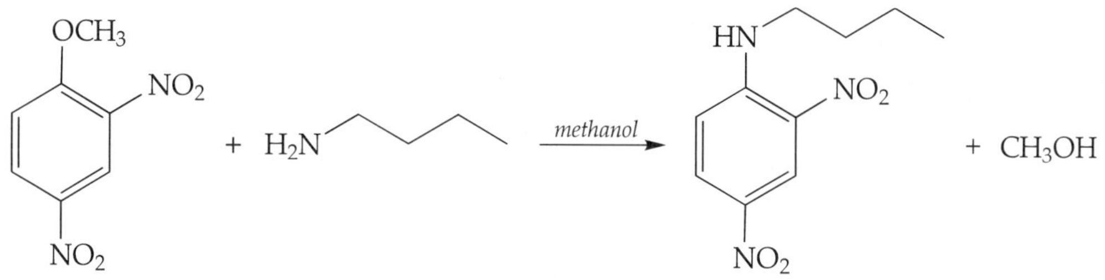

This is an example of a nucleophilic aromatic substitution reaction activated by electron-withdrawing groups. 2,4-Dinitroanisole is the activated substrate and the amine is the nucleophile that displaces the methoxy group. The final products are N-(*n*-butyl)-2,4-dinitroaniline and methanol. How does this reaction proceed?

If the reaction takes place in one single step we should observe overall second-order kinetics, first order in each reactant.

$$v = -\frac{d[2,4-\text{DNAn}]}{dt} = \frac{d[\text{product}]}{dt} = k \times [2,4-\text{DNAn}] \times [n-\text{BuNH}_2] \qquad (33)$$

For many nucleophilic aromatic substitutions it has been observed, however, that *the order in the amine is two*. Because a *one-step mechanism* involving a molecule of substrate and *two* molecules of amine implies a *termolecular* collision, which is a rare event, the second order in the amine has been interpreted as a strong piece of evidence for a *multistep mechanism*. The stoichiometry of the reaction indicates that only one amine molecule is required; thus, the second amine molecule should act as a catalyst that is recovered unchanged at the end. A scheme of the proposed mechanism is shown below:

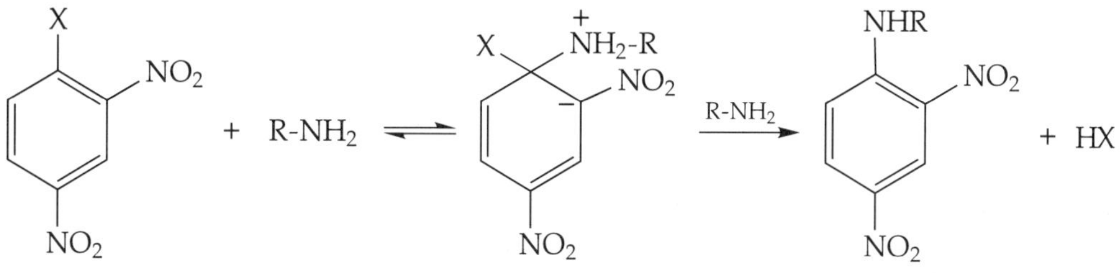

Meisenheimer complex

In the first step, the amine attacks the aromatic ring in position 1, activated by the nitro groups. The resulting intermediate, called a Meisenheimer complex, is unstable and decomposes to give the final products. The second amine molecule acts as a base and catalyzes the decomposition of the Meisenheimer complex by assisting in the removal of a H$^+$ from the nitrogen atom and giving it to the leaving group X$^-$.

Bases other than amines are also known to catalyze the second step. The occurrence of **base catalysis** depends on the nature of the leaving group X, the nucleophile, and the solvent.

What rate law would we observe if the first step is rate limiting? What would be the rate law if the second step is rate limiting? If the first step is rate limiting, we would observe that the reaction is first order in the substrate and in the amine (see section 18.6). This situation is kinetically indistinguishable from a one-step mechanism where the substrate and the amine collide to give products without the formation of the intermediate complex. Thus, the observation of an *overall second-order kinetics* does not say much about the mechanism of the reaction. It does not confirm a two-step mechanism but it does not rule it out either.

If, on the other hand, the second step is rate limiting, we would observe that the rate depends on the *square* of the amine concentration because *two* amine molecules participate in the rate-limiting step. One is already forming part of the Meisenheimer complex and the other acts as a catalyst. The overall order for the reaction would be three: first order in substrate and second in amine (Eq. 34):

$$v = k' \times [2,4-\text{DNAn}] \times [n-\text{BuNH}_2]^2 \tag{34}$$

In Equation 34, k' is called the third-order rate constant.

In this experiment you will determine whether the reaction is first or second order in the amine concentration. In other words, you will investigate whether Equation 33 or 34 is valid. To this end, you will study the reaction under pseudo-first-order conditions using a large excess of amine. Thus, for all practical purposes, the amine concentration can be considered constant during each kinetic run. Under these conditions Equation 33 becomes:

$$v = k_{obs} \times [2,4-\text{DNAn}] \tag{35}$$

where

$$\boxed{k_{obs} = k \times [n-\text{BuNH}_2]} \tag{36}$$

Under pseudo-first-order conditions, Equation 34 also becomes Equation 35, but where

$$\boxed{k_{obs} = k' \times [n-\text{BuNH}_2]^2} \tag{37}$$

In this experiment you will test whether your data fit Equation 36 or 37. To this end, you will obtain k_{obs} for each run at a given amine concentration, and then divide the k_{obs} values by the amine concentrations:

$$\frac{k_{obs}}{[n-\text{BuNH}_2]}$$

If Equation 36 is valid, this ratio will be constant and independent of the amine concentration and equal to k. If, on the other hand, Equation 37 is valid, the ratio will increase linearly with the amine concentration. Dividing both sides of Equation 37 by the amine concentration, Equation 38 is obtained:

$$\frac{k_{obs}}{[n-\text{BuNH}_2]} = k' \times [n-\text{BuNH}_2] \tag{38}$$

Thus, by studying the *behavior of k_{obs} as a function of the amine concentration*, you will determine whether the reaction is base catalyzed. In the experiment you will measure k_{obs} at four different amine concentrations.

You will also perform a temperature study of the kinetics. You will carry out kinetic runs at ambient temperature, 35°C and 45°C (at one amine concentration). You will calculate the Arrhenius energy of activation by plotting ln k_{obs} versus $1/T$.

E18.2   TO FOLLOW THE KINETICS

You will monitor the kinetics by measuring the absorbance at 400 nm where the product, N-butyl-2,4-dinitroaniline, absorbs and the starting dinitroanisole does not. To analyze the data you will use the integrated method discussed in sections 18.4. and 18.8 and apply Equation 25:

$$\ln(A_\infty - A_t) = \ln A_\infty - k_{obs} \times t \tag{25}$$

where A_∞ and A_t are the absorbances at "infinite time" and any other time, respectively.

To determine A_∞, you should, in principle, measure the absorbance of the reaction mixture after at least ten half-lives have elapsed from the beginning of the reaction. However, this is not possible within the time frame of the experiment; thus you will determine A_∞ with the help of Equation 26:

$$A_\infty = \varepsilon \times [S]_0 \times \ell \tag{26}$$

where $[S]_0$ is the initial substrate concentration and ε the product's molar absorptivity at 400 nm. The pathlength of the cuvette, ℓ, is 1 cm.

For each kinetic run, you will plot ln $(A_\infty - A_t)$ versus time, and determine the pseudo-first-order rate constant, k_{obs}, from the slope of the linear correlation (Eq. 25). You will then calculate the ratio $k_{obs}/[n\text{-BuNH}_2]$. If this ratio is constant and independent of the amine concentration, you can conclude that the reaction is first order in the amine and not base catalyzed. If the ratio depends linearly on the amine concentration, it would be an indication of base catalysis.

E18.3   OVERVIEW OF THE EXPERIMENT

Before any kinetic study is conducted, the starting material and product should be prepared and characterized. You will first prepare 2,4-dinitroanisole (E18.4) (the **substrate**) by the reaction of 1-chloro-2,4-dinitrobenzene with sodium methoxide in methanol. This reaction takes at least 24 hours. Alternatively, the substrate, which is commercially available, can be purchased. You will also obtain the final product of the kinetic study, N-(n-butyl)-2,4-dinitroaniline, by mixing 1-chloro-2,4-dinitrobenzene with n-butylamine as explained in Experiment 17.2. This compound could, of course, be obtained by the reaction of 2,4-dinitroanisole with n-butylamine, but the reaction is too slow to give the product in a reasonable time.

You will study the UV-visible properties of this substitution product, and determine its molar absorptivity (Beer's law) in methanol at 400 nm (E18.5 and E18.6). You will use this information to perform the kinetic analysis.

Your instructor may shorten this experiment by providing you with commercial 2,4-dinitroanisole and the UV-visible properties of the substitution product. In that case, parts E18.4, E18.5, and E18.6 will be omitted.

Before doing any of the many parts of this experiment, read the "Safety First" box. Follow the cleaning-up procedures given at the end of this experiment.

PROCEDURE

E18.4 SYNTHESIS OF 2,4-DINITROANISOLE

Background reading: 17.1
Estimated time: 45 minutes

Weigh 0.20 g of 1-chloro-2,4-dinitrobenzene (wear gloves) on a piece of weighing paper and transfer it to a perfectly dry test tube (13 × 100 mm). Add 4 mL of anhydrous methanol (absolute methanol) and 1.0 mL of a 1.0 M solution of sodium methoxide in anhydrous methanol. With a microspatula stir the suspension until the solid dissolves. Cover the test tube with a stopper and let the system react for at least 24 hours at room temperature (put it in a safe place in your drawer). Cool the system down in a water-ice bath for at least 15 minutes, occasionally scratching the walls of the test tube with a glass rod (do not scratch the bottom of the tube because you may puncture it). The solubility of the product is very temperature dependent, so it is very important that the solution be cold before filtering. Vacuum filter the solid using a Hirsch funnel. Wash the solid with 0.1 mL of cold methanol. Weigh the product and determine its melting point. In most cases, the product obtained is adequately pure for the subsequent applications. If needed, it can be recrystallized from methanol. Obtain the IR spectrum of the product using a nujol mull. Compare your spectrum with that shown in Figure 18.11.

E18.5 VISIBLE SPECTRUM OF
N-(*n*-BUTYL)-2,4-DINITROANILINE

Background reading: 17.1–2; 32.1–5
Estimated time: 1 hour

Obtain *N*-(*n*-butyl)-2,4-dinitroaniline as indicated in Experiment 17.2 and recrystallize it from ethanol. On a piece of weighing paper, weigh (analytical balance with a readability of at least 0.0001 g) about 10 mg of product. Transfer the solid *quantitatively* to a 10-mL volumetric flask. Use a little methanol to transfer any residue of the product from the paper to the flask. Add anhydrous methanol to the mark and shake to dissolve. Transfer the solution to a labeled scintillation vial. Rinse the volumetric flask with several small portions of anhydrous methanol and discard the washes.

Do a 1/100 dilution of the solution in the scintillation vial by transferring 100 µL (automatic pipet) of this solution to a 10-mL volumetric flask and adding anhydrous methanol to the mark. Calculate the molar concentration of the solution. Transfer the solution to the Spectronic 20 (or other spectrophotometer) cuvette and determine the visible spectrum of the compound by measuring % transmittance at 10-nm intervals, in the range 340–450 nm; blank against anhydrous methanol at each wavelength. Determine the λ_{max}. Compare your spectrum with that shown in Figure 18.12.

E18.6   BEER'S LAW OF *N*-(*n*-BUTYL)-2,4-DINITROANILINE

> **Background reading:** 32.4; 32.5
> **Estimated time:** 1 hour

Transfer the solution used to determine the visible spectrum (E18.5) to a labeled scintillation vial and use it to make the following dilutions with anhydrous methanol: 1/5; 2/5; 3/5; and 4/5. Prepare these solutions directly in the cuvettes or in clean scintillation vials by mixing the appropriate volumes (automatic pipet) of the *N*-(*n*-butyl)-2,4-dinitroaniline solution and methanol. The final volume in each case should be 5 mL. Measure the %*T* at 400 nm. Plot **absorbance** versus the molar concentration of the product. Calculate the slope, **molar absorptivity**, by the best linear fit. The path length of the cuvette is 1.0 cm.

E18.7   KINETICS OF THE REACTION OF 2,4-DINITROANISOLE WITH *n*-BUTYLAMINE IN METHANOL

> **Background reading:** 18.1–9; 32.4; 32.5
> **Estimated time:** 4 hr.

2,4-Dinitroanisole Stock Solution

Using an analytical balance with a readability of at least 0.0001 g, weigh about 23 mg of 2,4-dinitroanisole on a piece of weighing paper and transfer it quantitatively to a 10-mL volumetric flask. Add anhydrous methanol to the mark. Calculate the molar concentration of the 2,4-dinitroanisole stock solution. Transfer the solution to a labeled scintillation vial.

$$[2,4 - \text{DNAn}]_{\text{stock}} = \frac{\text{mass (g)}}{\text{MM (g/mol)} \times 0.010 \text{ L}} = \frac{\text{mass (g)}}{198.13 \text{ g/mol} \times 0.010 \text{ L}}$$

n-Butylamine Stock Solution

In a 10-mL volumetric flask, weigh (top-loading balance with a readability of at least 0.01 g) about 2.6 g (3.5 mL) of *n*-butylamine; carry out this operation in the hood. Add anhydrous methanol to the mark; cap the flask and shake it. Calculate the molar concentration of *n*-butylamine stock solution. Transfer the solution to a labeled scintillation vial.

$$[n - \text{BuNH}_2]_{\text{stock}} = \frac{\text{mass (g)}}{\text{MM (g/mol)} \times 0.010 \text{ L}} = \frac{\text{mass (g)}}{73.14 \text{ g/mol} \times 0.010 \text{ L}}$$

Kinetic Runs

Determine the temperature of the room using a thermometer with 1°C divisions. Label six colorimetric test tubes (cuvettes) 1–6. Cut several pieces of Parafilm (2 × 2 cm). Deliver a 0.50 mL aliquot (automatic pipet) of the 2,4-dinitroanisole stock solution into cuvette 1. Add 4.20 mL (automatic pipet) of anhydrous methanol to the cuvette, and 0.30 mL of the *n*-butylamine solution (automatic pipet) in that order, see Table 18.1. Cover the cuvette with a piece of Parafilm and invert it a few times. Start the stopwatch; read the %*T* at 400 nm and note the time of the reading. Remove the cuvette from the instrument and set it aside, protected from drafts and sudden changes in temperature. Read the %*T* every 5–10 minutes until about 10–15 points have been collected.

Table 18.1   Volumes for Kinetic Runs

Cuvette	Volume of 2,4-dinitroanisole stock solution (mL)	Volume of methanol (mL)	Volume of n-butylamine stock solution (mL)	Temperature (°C)
1	0.50	4.20	0.30	ambient ($\approx$22)
2	0.50	4.00	0.50	ambient ($\approx$22)
3	0.50	3.50	1.00	ambient ($\approx$22)
4	0.50	3.00	1.50	ambient ($\approx$22)
5	0.50	4.00	0.50	35
6	0.50	4.00	0.50	45

In cuvettes 2, 3, and 4 add the volumes of 2,4-dinitroanisole and n-butylamine stock solutions and methanol indicated in Table 18.1 and follow the kinetics by reading the %T of each cuvette every 3–5 minutes. Do not start kinetic runs 2, 3, and 4 at the same time. Allow at least 15 minutes between their beginnings. This will give you time to measure the first points of each run without rushing.

As the concentration of n-butylamine increases from cuvettes 1–4, readings should be taken more and more often because the kinetics are faster. One stopwatch is enough to follow all four kinetics; digital stopwatches are particularly suited for these kinetic measurements. The moment the cuvette is inverted should be taken as the time zero for that particular kinetic run.

Temperature Effect   When the reactions get slower and your readings become less frequent, prepare cuvettes 5 and 6 (one at a time) to study the temperature effect. Immerse them in water baths at 35 and 45°C, respectively. Allow one minute for the temperature to equilibrate and then read time zero on the stopwatch. Remove the cuvette from the bath, dry the walls with a piece of tissue, and immediately read the %T. Place the cuvette back in the water bath. The error introduced by removing the cuvettes from the bath is small if this operation is done quickly. Read the absorbance every 1–2 minutes at the beginning of the run and at longer intervals as the reaction slows down.

Calculations   Using the molar concentrations of the 2,4-dinitroanisole and n-butylamine stock solutions, and the dilutions introduced in preparing the solutions according to Table 18.1, calculate the initial concentrations of reactants for each kinetic run. For example, the initial concentration for the substrate in all cuvettes and for the amine in cuvette 1 are:

$$[2,4-\text{DNAn}]_0 = \frac{[2,4-\text{DNAn}]_{\text{stock}}}{5.0 \text{ mL}} \times 0.5 \text{ mL} = \frac{[2,4-\text{DNAn}]_{\text{stock}}}{10}$$

$$[n-\text{BuNH}_2]_0 = \frac{[n-\text{BuNH}_2]_{\text{stock}}}{5.0 \text{ mL}} \times 0.3 \text{ mL}$$

Using the molar absorptivity of the product at 400 nm and Equation 26, calculate the absorbance at infinite time (A_∞). For each kinetic run, plot $\ln(A_\infty - A_t)$ versus time (in seconds). Determine the pseudo-first-order rate constant, k_{obs}, from the slope. Calculate the ratio $k_{\text{obs}}/[n-\text{BuNH}_2]$ for each run. Plot $k_{\text{obs}}/[n-\text{BuNH}_2]$ versus $[n-\text{BuNH}_2]$ and determine if there is base catalysis, as explained in E18.1.

Plot $\ln k_{\text{obs}}$ versus $1/T$ (Eq. 14, where T is the absolute temperature in K) for runs 2, 5, and 6 and calculate the activation energy from the slope of the linear correlation, using Equation 39:

$$E_a = -\text{slope} \times R \tag{39}$$

In Equation 39, $R = 8.314$ J/(mol K) $= 1.987$ cal/(mol K).

Cleaning Up

- Dispose of the reaction mixture filtrates and unused stock and dilute solutions in the container labeled "Dinitrocompounds Liquid Waste."

- Before washing glassware with residual dinitrocompounds rinse it with a little ethanol and transfer the rinsing with a Pasteur pipet to the container labeled "Dinitrocompounds Liquid Waste."

- Turn in any unused portion of N-(n-butyl)-2,4-dinitroaniline and 2,4-dinitroanisole to your instructor in labeled, screw-capped vials.

- Dispose of Pasteur pipets, capillary tubes, filter papers, and paper towels contaminated with chemicals in the container labeled "Dinitrocompounds Solid Waste."

EXPERIMENT 18   REPORT ———————————————————

Pre-lab

Synthesis of 2,4-Dinitroanisole (E18.4)

1. Prepare a table with the physical properties (MM, m.p., b.p., density, solubility, flammability [for solvents only], and toxicity/hazards) of 1-chloro-2,4-dinitrobenzene, sodium methoxide, 2,4-dinitroanisole, and methanol.

2. Write a chemical reaction for the transformation.

3. Calculate the number of mmoles of reactants used.

4. Calculate the theoretical yield of substitution product in mg.

Visible Spectrum of N-(n-Butyl)-2,4-Dinitroaniline & Beer's Law (E18.5 and E18.6)

5. What is the relation between % transmittance and absorbance?

6. If the molar absorptivity of compound A is 15,000 M^{-1} cm^{-1} at 400 nm, what concentration range gives absorbance values in the range 0.05–1.5? Assume that a cuvette of 1-cm pathlength is used.

7. Calculate the molar absorptivity at 400 nm from the UV spectrum of N-(n-butyl)-2,4-dinitroaniline shown in Figure 18.12 (pathlength is 1 cm).

8. The following % T values were measured at 420 nm for different concentrations of compound B in toluene. Plot absorbance versus concentration and calculate the molar absorptivity for B in toluene at 420 nm.

10^5 × concentration (M)	% T
0.865	73.1
1.73	51.8
2.60	37.7
3.46	29.9
4.33	22.1

Kinetic Study (E18.7)

9. Make a table with the physical properties of 2,4-dinitroanisole, n-butylamine, N-(n-butyl)-2,4-dinitroaniline, and methanol. The melting point of N-(n-butyl)-2,4-dinitroaniline is 89.5–90°C.

10. Write a two-step mechanism for the reaction, including base catalysis in the second step.

11. Predict the order in substrate and amine if the first step is rate limiting and if the second step is rate limiting.

12. Outline the objectives of this experiment.

13. What does the Arrhenius activation energy measure? Briefly explain.

In-lab

Synthesis of 2,4-Dinitroanisole (E18.4)

1. Calculate the % yield of the synthesis based on the mass of product obtained.

2. Compare the m.p. of your product with the literature m.p.

3. Interpret your IR spectrum and compare it with that shown in Figure 18.11.

4. Interpret the ^{1}H-NMR and ^{13}C-NMR shown in Figures 18.13 and 18.14.

Visible Spectrum of N-(n-butyl)-2,4-Dinitroaniline & Beer's Law (E18.5 and E18.6)

5. Report your spectrophotometric data from E18.5 and E18.6 in a table format.

6. Plot (or show) the spectrum of N-(n-butyl)-2,4-dinitroaniline (absorbance versus wavelength).

7. Compare your spectrum with that shown in Figure 18.12.

8. Plot Absorbance at 400 nm versus molar concentration. Calculate the molar absorptivity. Compare it with that calculated in pre-lab 7.

Kinetic Study (E18.7)

9. Calculate A_∞.

10. For each kinetic run, calculate the initial molar concentration of 2,4-dinitroanisole and n-butylamine. Present the data as indicated below:

Kinetic run #	[2,4-DNAn]$_0$ =	[n-BuNH$_2$]$_0$ =	Temp. =		A_∞ =
Time (sec)	% Trans- mittance (%T_t)	$A_t =$ $-\log(\%T_t/100)$	$A_\infty - A_t$	$\ln(A_\infty - A_t)$

11. Plot $\ln(A_\infty - A_t)$ versus time. Calculate k_{obs} from the slope (Eq. 25).

12. Report k_{obs} and the ratio $k_{obs}/[n\text{-BuNH}_2]$ at ambient temperature, 35°C, and 45°C, and for each amine concentration investigated, as indicated below:

Temperature (°C)	[n-BuNH$_2$]	k_{obs}	$k_{obs}/[n\text{-BuNH}_2]$

13. Plot the ratio $k_{obs}/[n\text{-BuNH}_2]$ versus $[n\text{-BuNH}_2]$ (at ambient temperature). Is the reaction base catalyzed? Justify your answer.

14. Plot $\ln k_{obs}$ versus the inverse of the absolute temperature $(1/T)$ in Kelvin, and determine the energy of activation in kcal/mol and in kJ/mol, as indicated by Equations 14 and 39.

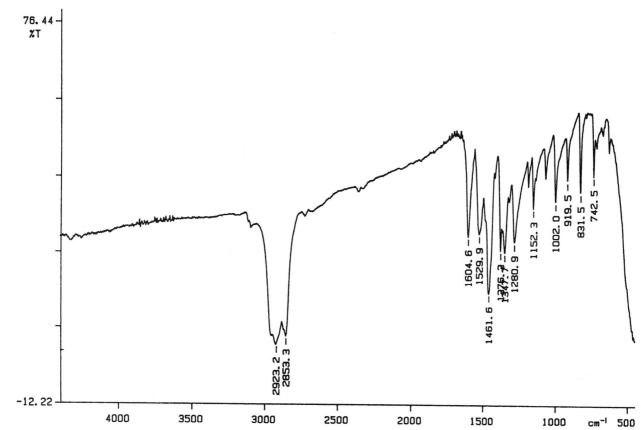

Figure 18.11   IR spectrum of 2,4-dinitroanisole (nujol).

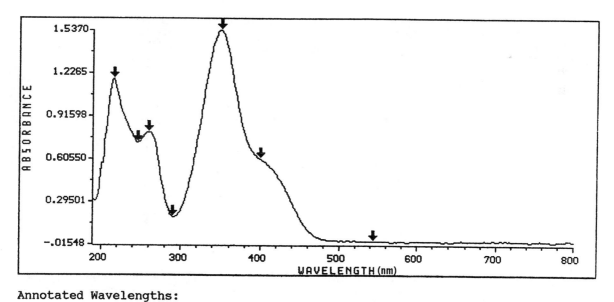

```
Annotated Wavelengths:
  1 : Wavelength = 544    Result =    0.006943
  2 : Wavelength = 400    Result =    0.608612
  3 : Wavelength = 350    Result =    1.536957
  4 : Wavelength = 290    Result =    0.186874
  5 : Wavelength = 260    Result =    0.800369
  6 : Wavelength = 246    Result =    0.725281
  7 : Wavelength = 216    Result =    1.182114
```

Figure 18.12   UV-visible spectrum of N-(n-butyl)-2,4-dinitroaniline in methanol (concentration: 8.6 10^{-5} M).

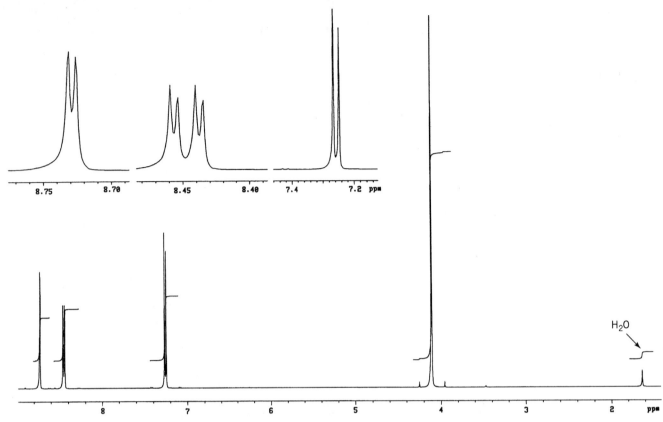

Figure 18.13 500-MHz ¹H-NMR spectrum of 2,4-dinitroanisole in CDCl₃.

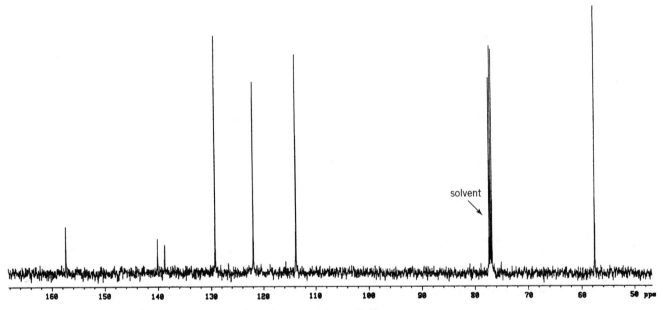

Figure 18.14 125.7-MHz ¹³C-NMR spectrum of 2,4-dinitroanisole in CDCl₃.

UNIT 19

Diels–Alder Reaction

19.1 DIELS–ALDER REACTION

Conjugated dienes react with alkenes to give six-membered rings:

a conjugated   an alkene   a six-membered
diene   ring

This reaction, discovered in 1928 by Otto Diels and Kurt Alder, is a cornerstone in organic synthesis because of its versatility and ease. The Diels–Alder reaction is very useful to build large molecules because it creates *two* C–C single bonds in one step. Diels and Alder were awarded the Nobel Prize in Chemistry in 1950.

Reactions like the Diels–Alder, where a ring is formed by the addition of two unsaturated molecules, are called **cycloadditions**. The conjugated diene is simply called the **diene**, the alkene is commonly referred to as the **dienophile** ("diene-loving"), and the product is called the **adduct**. In the Diels–Alder reaction the diene has a π system of 4 electrons and the dienophile has 2 π electrons; because of this, the reaction is known as a [4 + 2] cycloaddition.

In the Diels–Alder reaction, all bonds are formed and broken in one step. There are no intermediates involved and the reaction takes place through a unique transition state. This type of mechanism is called **concerted**.

diene   dienophile   transition state

The Diels–Alder reaction is not limited to ethylene and 1,3-butadiene. In fact, these two compounds react only under strong experimental conditions (200°C) and give a relatively poor yield in cyclohexene (20%). The most useful Diels–Alder reactions involve substituted dienes and dienophiles and take place under mild conditions. Dienophiles with electron-withdrawing groups attached to the double bond are particularly reactive in Diels–Alder reactions. For instance, maleic anhydride, methyl vinyl ketone, and methyl propenoate are all good dienophiles. In contrast, electron-donating substituents on the diene increase its reactivity.

Some typical dienes and dienophiles for the Diels–Alder reaction

Dienophiles

Dienes

maleic anhydride   methyl vinyl ketone   methyl propenoate   cyclopentadiene   1,3-cyclohexadiene

The Diels–Alder reaction is reversible. At high temperatures the adduct decomposes to give back the diene and the dienophile. This process is called **retro Diels–Alder reaction**. Thus, to maximize the yield in adduct, very high reaction temperatures should be avoided.

19.2   ENDO VERSUS EXO

Let's consider the reaction of 1,3-cyclopentadiene with maleic anhydride. The two molecules can approach each other in two different ways: the carbonyl groups of maleic anhydride can be directly below the diene, or away from the diene as shown in the diagrams below. These two approaches lead to two different products. When the diene and the substituents on the dienophile are directly below each other the approach is called *endo*, and the resulting product is the *endo* isomer. The other approach, with the dienophile substituents away from the diene, is called *exo*.

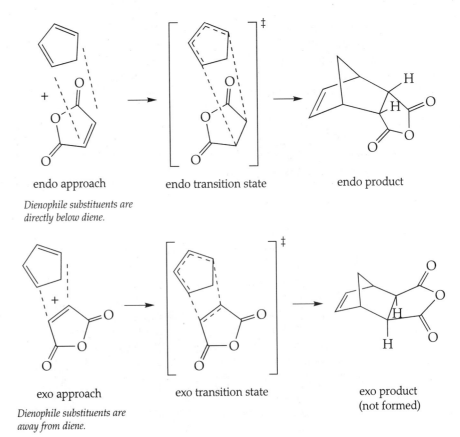

endo approach

endo transition state

endo product

Dienophile substituents are directly below diene.

exo approach

exo transition state

exo product
(not formed)

Dienophile substituents are away from diene.

The endo adduct is generally the main product of a Diels–Alder reaction because there is more orbital overlap between the diene and the dienophile pi systems in the transition state, which results in a more stable transition state and a lower activation energy. In the reaction shown above, maleic anhydride and 1,3-cyclopentadiene exclusively afford the endo adduct.

The endo-exo nomenclature applies to substituted bicyclic systems such as norbornane, shown below. Substituents *syn* to the larger bridge are called *endo*, while those *anti* to the larger bridge are *exo*.

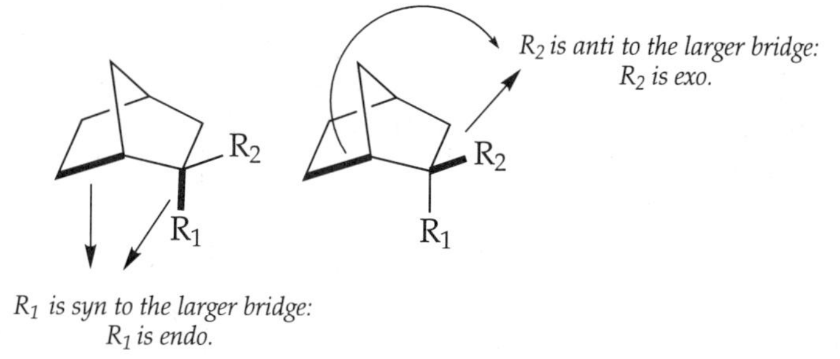

R_2 is anti to the larger bridge:
R_2 is exo.

R_1 is syn to the larger bridge:
R_1 is endo.

In the Diels–Alder reaction the endo-exo nomenclature has a broader meaning and refers to the products and also to the mode of approach between the diene and the dienophile.

19.3   SOLVENT EFFECTS

Because the Diels–Alder reaction takes place by a concerted mechanism, with no charge separation in the transition state, it is rather insensitive to the polarity of the solvent. Changing the solvent from nonpolar to polar usually has a negligible effect on the reaction rate. An exception is, however, the case of water. It was found that in many Diels–Alder reactions, water had an unexpected accelerating effect. For example, the reaction of cyclopentadiene with butenone is 740 times faster in water than in nonpolar isooctane. The proportion of endo-exo products is also affected by water, which favors the endo stereoisomer.

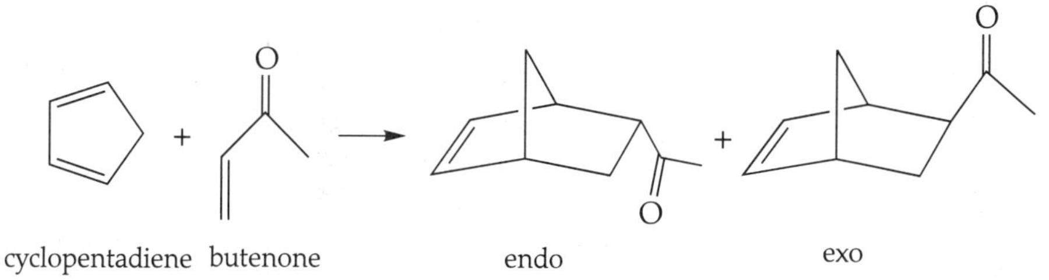

cyclopentadiene butenone   endo   exo

The accelerating effect of water in the Diels–Alder reaction cannot be simply explained by water's high polarity because it was found that other polar solvents often have a mild decelerating effect on the reaction rate. The rate increase in water is explained by the **hydrophobic effect**. When nonpolar molecules dissolve in water they disrupt the network of water molecules bound to one another through H-bonds, and create a cavity for themselves inside the bulk of the solvent. The creation of this cavity comes at the expense of a loss in entropy because now the water molecules are "frozen" in place, keeping the shape of the cavity. *To minimize this unfavorable process, several nonpolar molecules aggregate inside one cavity.* This is called the hydrophobic effect. It is believed that in the Diels–Alder reaction the hydrophobic effect forces the diene and dienophile to come into close proximity and thus speeds up the reaction.

A quantitative scale was devised to account for the hydrophobic effect in different solvents. Based on solubility data of noble gases and alkanes, **solvophobicity** values (*Sp*) were derived for a series of solvents. In the solvophobicity scale, water is assigned a value of 1, while *n*-hexadecane, a nonpolar solvent, has a solvophobicity of zero; all other solvents fall in between. The solvophobicity value measures the tendency of a solvent to behave like water and show hydrophobic effects. Polar solvents have higher solvophobicity values than nonpolar solvents because of their similarity to water. The solvophobicity values have been used to study solvent effects in the Diels–Alder reaction. A correlation between the logarithm of the rate constant and the solvophobicity values for fourteen different solvents is illustrated in Figure 19.1 for the reaction of cyclopentadiene and diethyl fumarate. As can be observed, a good linear relationship is obtained.

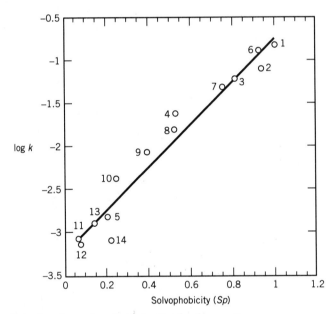

Figure 19.1   Plot of log *k* vs. solvophobicity (*Sp*) for the reaction of cyclopentadiene and diethyl fumarate. 1: H_2O; 2–4: 10%; 30%; 60% MeOH-water; 5: MeOH; 6–11: 10%; 30%; 50%; 60%; 70%; 90% dioxane-water; 12: dioxane; 13: dimethylformamide; 14: acetonitrile. (Adapted from Ref. 7, © 1986 by The Royal Society of Chemistry.)

Like the dielectric constant and other polarity indices (Unit 2), solvophobicity is a different way of measuring solvent polarity. Although there is a relationship between solvophobicity and dielectric constant, the solvophobicity scale has been especially devised to account for the hydrophobic effect, and is particularly useful in the study of reactions in water and in mixtures of water and organic solvents.

19.4   EXPERIMENTAL CONSIDERATIONS: FOLLOWING THE REACTION

In the laboratory, we can follow the course of a Diels–Alder reaction with the help of different spectroscopic techniques such as IR and NMR. They provide us with detailed information about the structure of the adduct and also give us stereo and regiochemistry information. UV-visible spectroscopy is also a valuable tool for the study of these reactions. At the heart of any Diels–Alder reaction lies the conversion of a conjugated diene into an isolated double bond. This process is always accompanied by a noticeable change in the UV-visible spectrum of the reaction mixture. Because of conjugation, the diene always absorbs at longer wavelengths than the adduct (the effect of conjugation on the UV-visible spectrum is discussed in section 32.1).

conjugated diene   isolated double bonds
$\lambda_{max} > 220$ nm $\lambda_{max} < 200$ nm

This difference between the starting diene and the final adduct is dramatically illustrated in the reaction of 9-anthraldehyde with maleic anhydride. Anthracene and anthracene derivatives react as dienes in Diels–Alder reactions. The central ring of the anthracene molecule is less stable than the lateral ones, making positions 9 and 10 more reactive than the rest.

9-anthraldehyde   maleic anhydride

An important difference between the starting anthracene and the adduct is the extent of conjugation; while the anthracene ring is totally conjugated, the adduct has two *separate* benzene rings (the central ring is no longer aromatic). This loss of conjugation can be directly observed by the difference in color between the starting material and the product. While 9-anthraldehyde is deep yellow and absorbs at 400 nm, the adduct is colorless and absorbs at shorter wavelengths (264 nm) in the UV region of the spectrum.

BIBLIOGRAPHY

1. Dienes in the Diels–Alder Reaction. F. Fringuelli and A. Taticchi. Wiley-Interscience, New York, 1990.

2. Cycloaddition Reactions in Organic Synthesis. W. Carruthers. Pergamon Press, Oxford, 1990.

3. Hydrophobic Effects on Simple Organic Reactions in Water. R. Breslow. *Accounts Chem. Res.,* **24**, 159–164 (1991).

4. Hydrophobic Acceleration of Diels–Alder Reactions. D.C. Rideout and R. Breslow. *J. Am. Chem. Soc.,* **102**, 7816–7817 (1980).

5. Chaotropic Salt Effects in a Hydrophobically Accelerated Diels–Alder Reaction. R. Breslow and C.J. Rizzo, *J. Am. Chem. Soc.*, **113**, 4340–4341 (1991).

6. Organic Reactions in Aqueous Media-With a Focus on Carbon-Carbon Bond Formation. C-J. Li. *Chem. Rev.*, **93**, 2023–2035 (1993).

7. Diels–Alder Reactions in Hydrophobic Cavities: a Quantitative Correlation with Solvophobicity and Rate Enhancements by Macrocycles. H-J. Schneider and N.K. Sangwan. *J. Chem. Soc., Chem. Commun.*, 1787–1789 (1986).

8. Changes of Stereoselectivity in Diels–Alder Reactions by Hydrophobic Solvent Effects and by ß-Cyclodextrin. H-J. Schneider and N.K. Sangwan. *Angew. Chem. Int. Ed. Engl.*, **26**, 896–897 (1987).

9. A Quantitative Measure of Solvent Solvophobic Effect. M.H. Abraham, P.L. Grellier, and R.A. McGill. *J. Chem. Soc. Perkin Trans. II*, 339–345 (1988).

EXERCISES

Problem 1

Complete the following reactions:

(a)

(b)

(c)

Problem 2

What product would you expect from the retro Diels–Alder reaction of the following compounds:

a)

b)

c)

d)

Problem 3

Classify the following compounds as *exo* or *endo*.

a)

b)

c)

Problem 4

Predict the adduct of the following Diels–Alder reaction. How would you distinguish it from the starting materials by IR?

Problem 5

Each term in group A is related to one in group B. Match them.

Group A: dienophile; hydrophobicity; solvophobicity; endo; 4 + 2 cycloaddition; exo; concerted.
Group B: *syn* to larger bridge; decrease in conjugation; no intermediates; diene-loving; water cavity; solvent polarity; *anti* to larger bridge.

EXPERIMENT 19A

Diels–Alder Reactions in Toluene

This experiment consists of two related but independent parts. Consult with your instructor to find out which part(s) you will perform. In part E19A.1 you will prepare the adduct of 9-anthraldehyde with maleic anhydride. You will isolate the solid product and characterize it by spectroscopic methods. UV-visible spectroscopy is especially useful in the study of this reaction because of its simplicity and valuable information. While the starting 9-anthraldehyde is deep yellow, the product of the reaction is colorless. This is reflected in the very different UV visible spectra exhibited by these compounds. You will also characterize the adduct by IR and ^{1}H-NMR. As an exercise in 2-D NMR, you will interpret the ^{1}H-^{1}H COSY spectrum of the adduct.

In part E19A.2, you will prepare and isolate the adduct of phencyclone (a diene), and norbornadiene (a dienophile), Figure 19.2. You will characterize the adduct by ^{1}H-NMR and confirm that it is the product of the endo approach between the diene and dienophile. The stereochemical study will be based on the abnormal chemical shifts of H_a and H_b, which resonate near or *below the TMS signal*. These hydrogens resonate at very low, even negative, chemical shifts because they are in the shielding cone of the phenanthrene ring system.

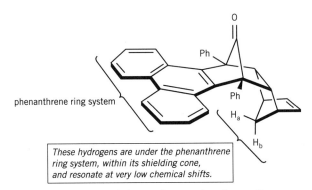

Figure 19.2 Adduct of the reaction of phencyclone and norbornadiene.

PROCEDURE

E19A.1   REACTION OF 9-ANTHRALDEHYDE WITH MALEIC ANHYDRIDE

Background reading: 19.1–4; 32.4; 4.4
Estimated time: 3 hours

9–anthraldehyde   maleic anhydride

Synthesis

Weigh 350 mg of maleic anhydride directly in a dry 15-mL round-bottom flask equipped with a microscale water-jacketed condenser and a magnetic stir bar. Using a piece of weighing paper, weigh 250 mg of 9-anthraldehyde (9-anthracenecarbox-aldehyde) and transfer it to the round-bottom flask. Add 2 mL of toluene. Make sure that the neck of the flask is clean; if not, clean it with a piece of tissue paper. Attach the condenser and reflux the mixture directly on a hot plate at a medium setting for 60 minutes (see "Refluxing," section 3.4 and Fig. 19.3). Cool down the system at ambient temperature and add 5 mL of hexanes. Cool in an ice-water bath. Filter the solid using a Hirsch funnel. If necessary, use some hexanes to help transfer the solid to the filter. Wash the solid in the funnel by turning the vacuum off and adding 2 mL of cold hexanes. Turn the vacuum on and let the solvent drain. Repeat this washing two more times with 2 mL of hexanes each time. Let the solid dry on the filter with the vacuum on for a few minutes. Weigh the product. Calculate the percentage yield.

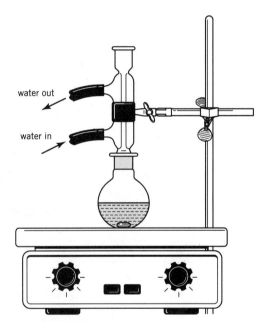

Figure 19.3 Microscale reflux apparatus.

Recrystallize about 150 mg of product from toluene-hexanes. Perform this operation in a **fume hood**. Transfer the solid to a 25 mL Erlenmeyer flask and add about 6 mL of toluene. While stirring with a glass rod, heat the system on a hot plate until the toluene boils and the solid dissolves. Add more toluene if necessary. Add about 8–10 mL of hexanes. Some product may already precipitate as this solvent is added. Cool the system down to ambient temperature first, and then place it in an ice-water bath. Filter the suspension. Wash the crystals on the filter with a few milliliters of hexanes. Let them dry on the filter. Weigh the product and determine its melting point.

Analysis

Determine the melting point of the recrystallized product and note any change in color during melting. The melting point of adduct is 238–240°C. Run the IR of the adduct and the starting aldehyde using a nujol mull (section 31.5). Note any difference in the frequency of the carbonyl stretching. Obtain the UV-visible spectrum of the starting aldehyde and the adduct in acetone as a solvent. Weigh about 5 mg of each compound in 10-mL volumetric flasks. Add acetone to the mark. Dilute each solution by taking 1.00 mL with a graduated or automatic pipet, and diluting it with the same solvent in a clean 10-mL volumetric flask. Obtain the spectrum in the range 350–500 nm, if possible. If not, determine the absorbance of each solution in the range 350–410 nm, at 10 nm intervals.

If possible, obtain the ^{1}H-NMR of the adduct using DMSO-d_6 as a solvent (section 33.20). See the UV-visible, IR and NMR spectra in Figures 19.4–19.8.

E19A.2   REACTION OF PHENCYCLONE WITH NORBORNADIENE

Background reading: 19.1–4
Estimated time: 3 hours

phencyclone   norbornadiene

Synthesis

In a 15-mL round-bottom flask, weigh 0.200 g of phencyclone, add 0.140 mL of norbornadiene (bicyclo[2.2.1]hepta-2,5-diene) with the help of a pluringe and a Pasteur pipet, and 2 mL of toluene. Add a stir bar and attach a reflux microscale water condenser. Reflux the system directly on a hot plate at a medium setting for one hour (read "Refluxing", section 3.4 and see Fig. 19.3). The solution will turn from blue-green to yellow as the adduct is formed. Let the system cool at room temperature and then in an ice-water bath. Separate the solid product by vacuum filtration. A second crop of product can be obtained by adding petroleum ether to the mother liquor and filtering the precipitate. Let the product dry on the filter with the vacuum on. Dry it on a piece of porous plate. Weigh the product. If desired, the product can be recrystallized from toluene-petroleum ether. Adapted from References 1–3.

Analysis

Determine the melting point of the adduct (m.p. 230–232°C) and observe any decomposition reaction with heat. The adduct undergoes a decarbonylation reaction followed by a retro Diels–Alder reaction promoted by heat. The loss of CO releases strain from the multiring system. The retro Diels–Alder reaction produces a fully aromatic compound, 1,4-diphenyltriphenylene and cyclopentadiene.

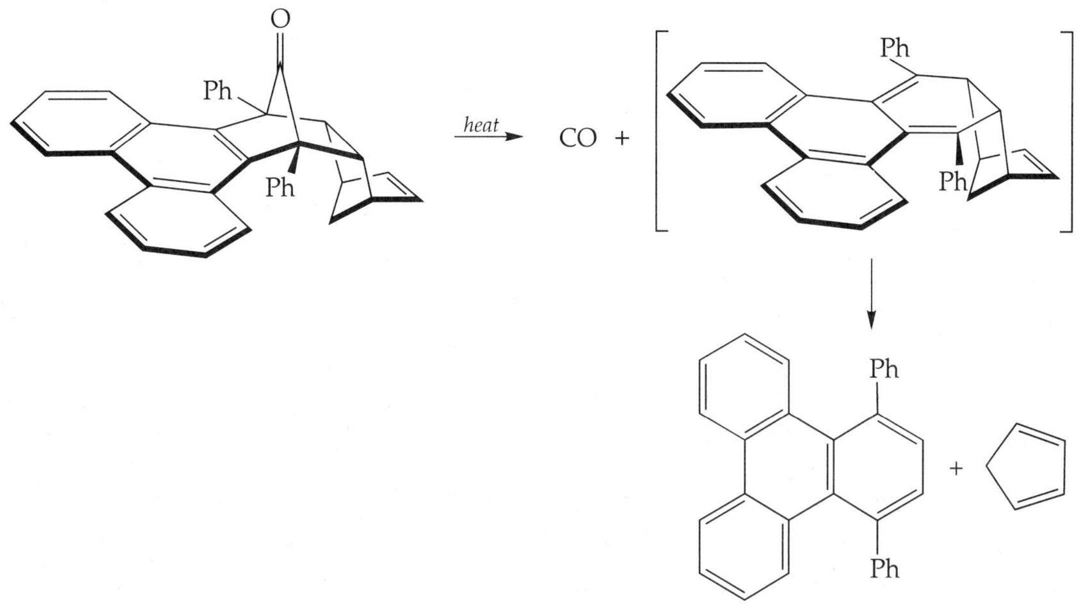

1,4-diphenyltriphenylene

Obtain the IR of the product using a nujol mull (section 31.5) and compare it with the IR shown in Figure 19.9. Determine the absorption frequency of the carbonyl group. Obtain the ^{1}H-NMR using DMSO-d$_6$ as solvent (section 33.20). *Pay special attention to the signals that show near or below the TMS signal at about +0.2 and −0.6 ppm*.

Cleaning Up

- Dispose of the reaction mixture filtrate, the toluene-petroleum ether filtrate, and the DMSO-d$_6$ adduct solution in the container labeled "Diels–Alder Liquid Waste."
- Dispose of Pasteur pipets, filter papers, and capillary tubes in the container labeled "Diels–Alder Solid Waste."
- Turn in your product in a capped and labeled vial to your instructor.

BIBLIOGRAPHY

1. Synthesis of a Bicyclo[2.2.1]heptene Diels–Alder Adduct. E.A. Harrison Jr. *J. Chem. Ed.*, **68**, 426–427 (1991).

2. Molecular Design by Cycloaddition Reactions. XXV. High Peri- and Regiospecificity of Phencyclone. T. Sasaki, K. Kanematsu, and K. Iizuka. *J. Org. Chem.*, **41**, 1105–1112 (1976).

3. Bicyclo[2.2.1]heptadiene in the Diels–Alder Reaction. K. Mackenzie. *J. Chem. Soc.*, 473–483 (1960).

4. Assignment of ^{13}C Resonances in the Phencyclone-Norbornadiene Adduct Via 2D NMR. Y. Xu, L.A. LaPlanche, and R. Rothchild. *Spectroscopy Letters*, **26**, 179–196 (1993).

EXPERIMENT 19A   REPORT

Pre-lab

E19A.1
1. Write a chemical equation for the reaction of 9-anthraldehyde with maleic anhydride. Are *endo* and *exo* isomers possible in this reaction?

2. Make a table showing the physical properties (molecular mass, b.p., density, solubility, toxicity/hazards, flammability [for solvents only]) of 9-anthraldehyde, maleic anhydride, and toluene. For the adduct, calculate its molecular mass and list its m.p. (238–240°C). Include in the table millimoles of reactants used.

3. What important differences in the IR spectrum of the adduct as compared to those of the starting materials do you expect to see?

E19A.2

4. Make a table showing the physical properties of phencyclone, norbornadiene, the adduct, and toluene. For the adduct, calculate its molecular mass and list its m.p. (230–232°C). Include in the table the millimoles of reactants used.

5. There are two possible products of the endo approach of norbornadiene and phencyclone: the product already discussed, and the one shown below (which does not form). Consider the exo approach and draw the two possible products that can result from it. Convince yourself that in none of these products are the methylene hydrogens (H_a and H_b) in the shielding cone of the phenanthrene ring system.

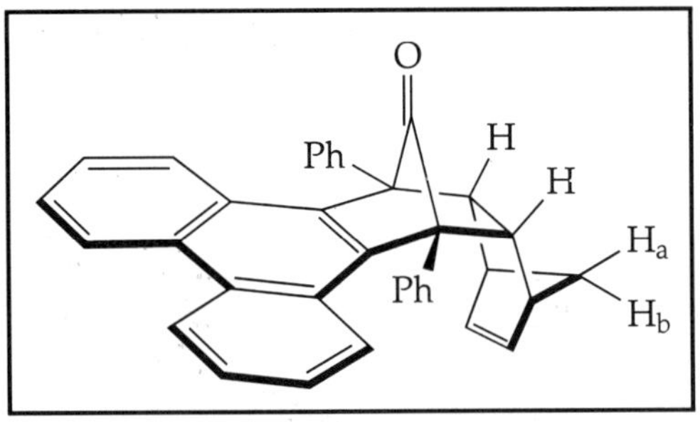

In-lab

For E19A.1 and E19.2

1. Calculate the % yield of the synthesis.

2. Report the m.p. of the product.

3. What happened during melting? Discuss.

For E19A.1

4. Why did you add hexanes in the last step of the synthesis? Why did you add hexanes in the recrystallization?

5. Briefly explain why 9-anthraldehyde is yellow and the adduct is colorless. Explain the differences in their UV-visible spectra. Compare your UV-visible spectra with that of 9-anthraldehyde shown in Figure 19.4.

6. Interpret your IR spectrum of the adduct and compare it with those of the starting materials and with that shown in Figure 19.5.

7. Interpret the ^{1}H-NMR of the product (Fig. 19.6). Make a molecular model of the adduct and estimate the dihedral angle between H_a and H_b and between H_a and H_c (see section 33.12). Using the Karplus equation that relates the coupling constant (J) to the dihedral angle (θ): $J = -0.28 + 8.5 \cos^2 \theta$, calculate the expected coupling constants and compare them with the measured values.

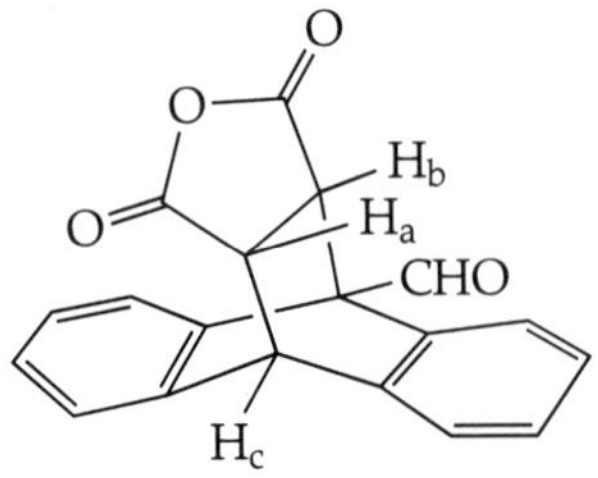

8. Interpret the ^{1}H-^{1}H COSY spectrum of the adduct shown in Figure 19.7 (advanced level).

9. Interpret the ^{13}C-NMR spectrum of the adduct shown in Figure 19.8.

For E19A.2

10. Interpret the IR of the adduct and compare it with the one shown in Figure 19.9.

11. The carbonyl absorbs at a particularly high frequency. Why?

12. Interpret the ^{1}H-NMR of the adduct (Fig. 19.10). Explain the chemical shifts of H_a and H_b. Calculate their coupling constant.

13. Interpret the ^{13}C-NMR shown in Figure 19.11 (see Ref. 4 for help).

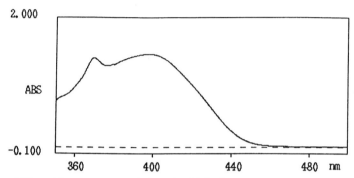

Figure 19.4 UV-visible spectrum of 9-anthraldehyde in acetone (concentration: 2.4 10^{-4} M).

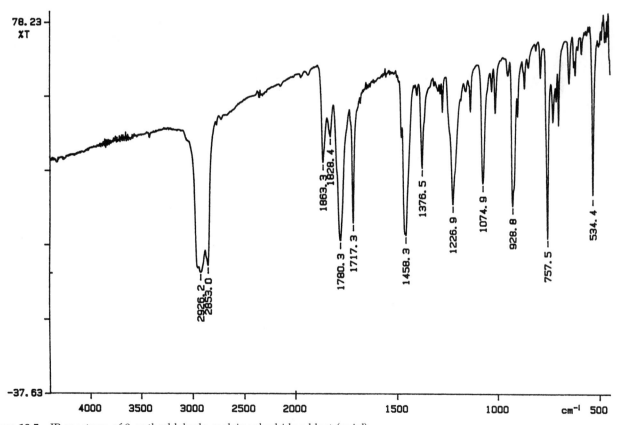

Figure 19.5 IR spectrum of 9-anthraldehyde-maleic anhydride adduct (nujol).

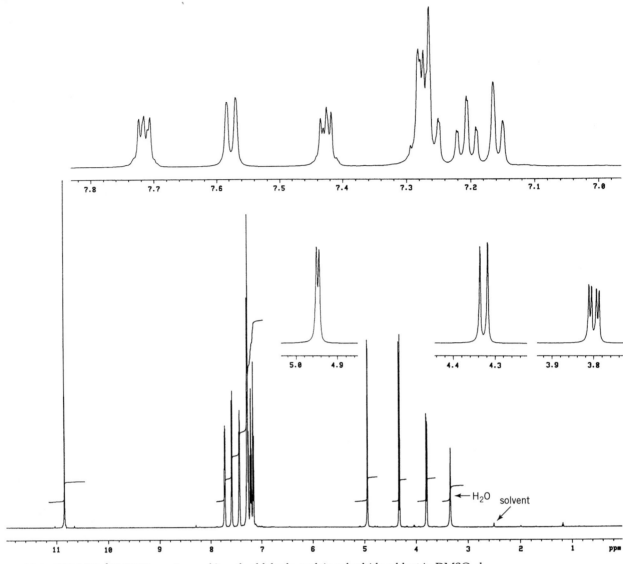

Figure 19.6 500-MHz ^{1}H-NMR spectrum of 9-anthraldehyde-maleic anhydride adduct in DMSO-d_6.

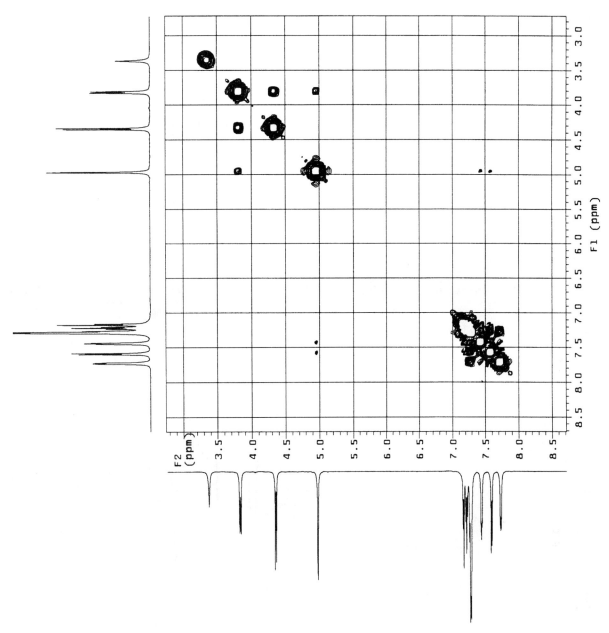

Figure 19.7   500-MHz ^{1}H-^{1}H COSY spectrum of 9-anthraldehyde-maleic anhydride adduct in DMSO-d_6.

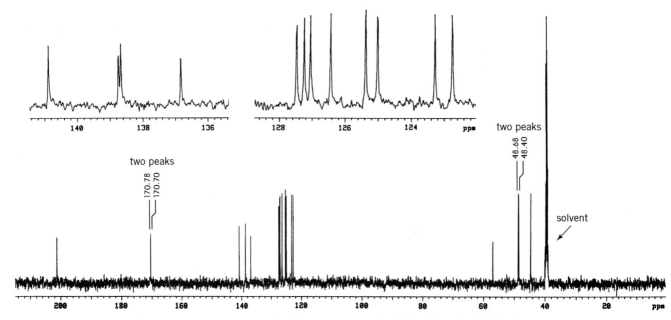

Figure 19.8 125.7-MHz ^{13}C-NMR spectrum of 9-anthraldehyde-maleic anhydride adduct in DMSO-d_6.

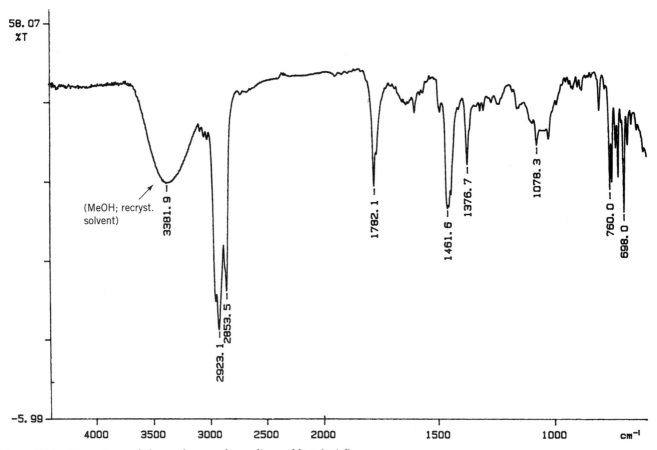

Figure 19.9 IR spectrum of phencyclone-norbornadiene adduct (nujol).

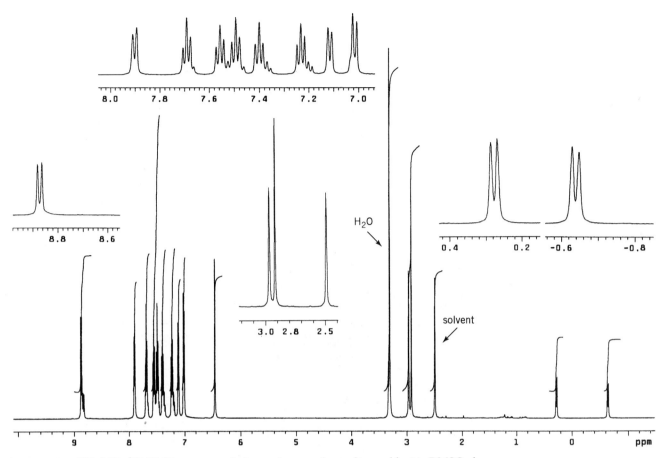

Figure 19.10   500-MHz ^{1}H-NMR spectrum of phencyclone-norbornadiene adduct in DMSO-d$_6$.

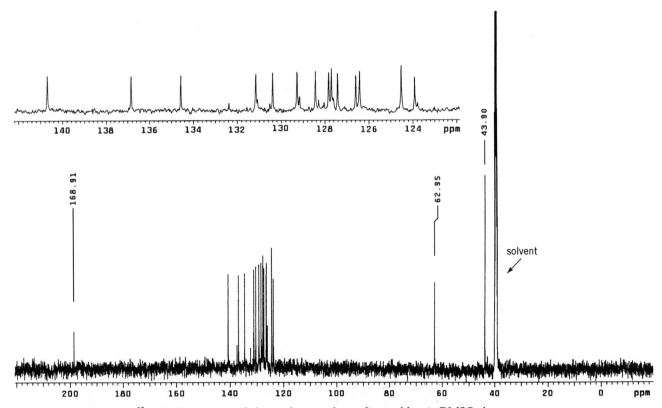

Figure 19.11   125.7-MHz ^{13}C-NMR spectrum of phencyclone-norbornadiene adduct in DMSO-d$_6$.

EXPERIMENT 19B

Diels–Alder Reaction in Water

In this experiment you will carry out the Diels–Alder reaction of 9-anthracenemethanol and *N*-ethylmaleimide in water and will investigate its kinetics. The experiment is divided into two parts: synthesis and kinetics. Each part will be performed in a different lab period. Check with your instructor to find out if you will perform both parts. You will first synthesize and characterize the adduct. By comparing the UV-visible spectra of the starting material and the product, you will determine the appropriate wavelength to follow the kinetics. In the kinetic study you will investigate the hydrophobic effect by comparing the reaction rates in different methanol-water mixtures. You will follow the reaction by UV-visible spectroscopy at 45, 50, and 55°C and will calculate the activation energy. Before you proceed to the kinetic study, read sections 18.1–8.

PROCEDURE

E19B.1 REACTION OF 9-ANTHRACENEMETHANOL WITH *N*-ETHYLMALEIMIDE

> **Estimated time:** 3 hours
> **Background reading:** 19.1–4

9-anthracenemethanol   N- ethylmaleimide

Synthesis

In a 15-mL round-bottom flask weigh 0.200 g of 9-anthracenemethanol (9-hydroxymethylanthracene) and 0.150 g of *N*-ethylmaleimide (**NEM**). Add 6 mL of water and a stir bar. Attach a microscale water condenser and reflux the system with stirring for one hour (read "Refluxing", section 3.4 and see Fig. 19.3, page 399). Heat directly on a hot plate at a medium setting. Occasionally dislodge any solid adhered to the walls of the flask above the liquid by loosening the clamp and swirling the whole assembly. Let the system cool down at room temperature and vacuum filter the reaction mixture using a Hirsch funnel. Wash the product with 2 mL of cold water and let it dry for about ten minutes on the filter with the vacuum on. Transfer the product to a porous plate and dry it with the help of a spatula. Weigh the crude product and determine its melting point.

Recrystallize the product from a mixture of toluene-petroleum ether. In a medium-size test tube dissolve your product in the minimum amount of hot toluene while heating on a sand bath. Stir constantly with a microspatula. After the solid has completely dissolved, add petroleum ether little by little while still heating and stirring. This produces a turbidity that disappears as the solvents mix. Add more petroleum ether until the turbidity persists; then add a few more drops of hot toluene to make the turbidity go away. Let the system crystallize at room temperature first, and then in an ice-water bath. Filter the solid under vacuum by using a Hirsch funnel. Let it dry on the filter. Weigh the product and determine its melting point.

Analysis

UV-visible Weigh about 4 mg of the adduct on a piece of weighing paper using an analytical balance. Transfer it quantitatively to a 25-mL volumetric flask. Add methanol to the mark and obtain the UV-visible spectrum of the product in the range 200–400 nm, if possible. If not, measure the absorbance in the range 340–400 nm at 10 nm intervals.

Also obtain the spectrum of 9-anthracenemethanol. To this end weigh 40 mg of the compound on a piece of weighing paper, using an analytical balance. Transfer the solid quantitatively to a 25-mL volumetric flask and add methanol to the mark. This will also be your **stock solution for the kinetic study** (E19B.2). Take an aliquot of the stock solution to make a dilute solution, as described in the next paragraph, and keep the rest in a well-capped scintillation vial sealed with Parafilm and protected from light. You will use this stock solution again in the next lab period to perform the kinetics.

Take a 0.100-mL aliquot of the stock solution and dilute it in a 10-mL volumetric flask with methanol. Use this dilute solution to obtain the UV-visible spectra in the same wavelength range. Compare the spectrum of the adduct with that of the starting material. Calculate the molar absorptivities of starting material and product at the λ_{max} of the starting material in the range 340–400 nm, and determine the best wavelength to follow the kinetics.

IR Obtain the IR specta of product and starting material using a nujol mull. Compare them with the spectrum shown in Figure 19.12.

NMR If possible, obtain the ^{1}H-NMR of the product using DMSO-d$_6$ as a solvent. Compare it with the spectrum shown in Figure 19.13.

Cleaning Up

- Dispose of the reaction mixture filtrate, the toluene-petroleum ether filtrate, the 9-anthracenemethanol dilute methanol solution, adduct methanol solution, and DMSO-d$_6$ adduct solution in the container labeled "NEM Adduct Liquid Waste."

- Dispose of Pasteur pipets, filter papers, and capillary tubes in the container labeled "NEM Adduct Solid Waste."

- Keep the 9-anthracenemethanol stock solution in a well-capped and labeled vial until the next period.

- Keep your solid adduct until the next period in a well-capped and labeled vial.

E19B.2 KINETICS OF THE REACTION OF 9-ANTHRACENEMETHANOL WITH NEM

Background reading: 19.1–4; 18.1–8; 32.4; 32.5
Estimated time: 3 hours

You will study the kinetics of the reaction of 9-anthracenemethanol with NEM in methanol-water mixtures. Because 9-anthracenemethanol is only slightly soluble in water, you will prepare a stock solution of this reagent in methanol and dilute it with water for the kinetic runs. The amount of methanol introduced in this fashion will be minimal, accounting for about 1% in volume. You will work under pseudo-first-order conditions using a large excess of NEM. You will follow the kinetics at the λ_{max} of 9-anthracenemethanol, where the adduct has negligible absorbance. You will monitor the decrease in absorbance due to the disappearance of 9-anthracenemethanol. NEM does not absorb significantly in the 340–400 nm region and does not interfere in this study.

Preparing the Stock Solutions

The stock solution of 9-anthracenemethanol was already prepared in E19B.1. To prepare the stock solution of NEM, weigh approximately 40 mg of the solid compound on a piece of weighing paper using an analytical balance. Transfer the solid quantitatively to a 25-mL volumetric flask and add water to the mark. Shake well to dissolve the crystals. Calculate the molar concentrations of 9-anthracenemethanol and NEM in the stock solutions.

Safety First

- *N*-ethylmaleimide is a corrosive solid.

- Methanol is flammable.

Experimental Conditions

You will determine the rate constants at four different methanol concentrations. You will work in a team of three people. Each team member will perform a complete kinetic study (four different methanol concentrations) at one temperature (45, 50, or 55°C), and then all the members will combine their results to calculate the activation energy. The instructions given here are to be followed when a Spectronic 20 or other similar spectrophotometers are used. The cuvettes hold about 6–7 mL of liquid and can be directly heated at the desired temperature in a water bath. Changes in the procedure may be necessary if a different spectrometer is used.

Cut five pieces of Parafilm (2 × 2 cm) and wash and dry five cuvettes. Wash five 10-mL volumetric flasks well. The flasks need not be dry. Label the flasks as follows: blank; 1%; 10%; 20%; and 40%. These numbers represent the amount of methanol present in each flask. The blank is used as a reference for the absorbance measurements; it contains the solvent and NEM in the same amounts as in the kinetic runs. Ideally, four different blanks should be used because there are four different methanol concentrations. However, the difference in absorbance introduced by variable amounts of methanol is negligible and only one blank (with 20% methanol) will be used for all kinetic runs.

Blank

With the help of an automatic pipet dispense 2.00 mL of the NEM stock solution in the 10-mL volumetric flask labeled "blank." Add about 5 mL of water and 2.00 mL of methanol with an automatic pipet (Table 19.1). Fill to the mark with water and shake well. Transfer about 5 mL of this solution to a cuvette. Cover it with a piece of Parafilm to avoid evaporation and introduce it in the water bath. After about five minutes adjust the 100% transmittance of the spectrometer with the blank in the cuvette compartment.

Table 19.1   **Volumes Mixed for Each Run**

Run	NEM (mL)	Initial water volume (mL)	Methanol (mL)	9-Anthracenemethanol (mL)
blank	2.00	5	2.00	0
1%	2.00	7	0	0.100
10%	2.00	6	0.90	0.100
20%	2.00	5	1.90	0.100
40%	2.00	3	3.90	0.100

Kinetic Runs

Add 2.00 mL of the **NEM stock solution** to the 10-mL volumetric flasks labeled 1%; 10%; 20%; and 40%. To the flask labeled 1%, add about 7 mL of water and with the help of an automatic pipet add 0.100 mL (100 μL) of the **9-anthracenemethanol stock solution** in methanol followed by **water** to the mark. Shake well. Transfer about 5 mL of this solution to one of the cuvettes and cover it tightly with Parafilm. Place a label on the mouth of the cuvette. After the 100%T has been adjusted with the blank, read the %T (or absorbance) at the chosen wavelength. This will be your reading at time zero. Immerse the cuvette in the water bath and start the stopwatch. The water level in the bath should be above the liquid level in the cuvette. Take a second reading after about 5 minutes to allow the reaction mixture to reach the desired temperature. Take readings every 4–6 minutes for about 1.5 hours.

To prepare the 10%, 20%, and 40% methanol reaction mixtures employ the same procedure. Mix the volumes indicated in Table 19.1 in the labeled volumetric flasks before adding water to the mark. *Do not add the 9-anthracenemethanol solution until you are ready to follow the kinetics!*

Follow each run as indicated for the 1% solution. You do not need to reset the 100%*T* every time you take a measurement, but do it occasionally during the experiment. For the 40% run, take readings every 10–15 minutes and do it for a period of about 2.5 hours.

Calculations

Second-order rate constant   For each run plot the natural logarithm of the absorbance at different times, $\ln A_t$, versus time (in minutes or seconds). In each case you should obtain a straight line (a linear correlation), which obeys the following equation:

$$\ln A_t = \ln A_0 - k_{obs} \times t$$

(This is Eq. 19 from Unit 18.) The slope of the straight line is $-k_{obs}$, and the intercept is $\ln A_0$; k_{obs} is the pseudo-first-order rate constant and A_0 is the absorbance at time zero.

The best straight line that fits the points can be drawn by simple inspection of the plot. The slope can be algebraically calculated as indicated in section 18.9. However, it is much easier and more accurate to perform the calculation using a least squares algorithm. The least squares method is available in some pocket calculators and computer programs such as KaleidaGraph, SigmaPlot, and Microsoft Excel. Once the k_{obs} is calculated for each methanol concentration, divide it by the NEM concentration and obtain the second order rate constant, k.

$$k = \frac{k_{obs}}{[NEM]}$$

Solvent Solvophobicity Correlation   For the four methanol concentrations investigated at each temperature, plot the common logarithm of the second-order rate constant, $\log k$, as a function of the solvent solvophobicity, Sp, given in Table 19.2. Calculate the best linear fit for the data. From the linear fit at 45°C extrapolate the values for the expected rate constants in 100% methanol, 100% water, 40% 1-propanol (60% water), and 40% 2-propanol (60% water) using the solvophobicities provided in Table 19.2.

Table 19.2   **Solvophobicity Values (Ref. 1)**

Solvent	Sp	$\log k$
Methanol-water		
1% MeOH	0.9942	your
10% MeOH	0.9417	experimental
20% MeOH	0.8806	values
40% MeOH	0.7270	
100% MeOH	0.1998	
100% H$_2$O	1.0000	you
40% 1-Propanol	0.6430	extrapolate
40% 2-Propanol	0.6398	

Cleaning Up

- Dispose of the reaction mixtures, the 9-anthracenemethanol methanol stock solution, and the NEM stock solution in the container labeled "NEM Adduct Liquid Waste."

- Turn in your adduct in a labeled vial to your instructor.

Activation Energy   Plot the natural logarithm of the second-order rate constant, ln k, versus the reciprocal of the **absolute** temperature ($1/T$). Do this at the four methanol concentrations investigated. For each plot calculate the slope of the best linear fit. Calculate the Arrhenius activation energy, E_a, at each methanol concentration with the help of the following equation:

$$E_a = -\text{slope} \times R$$

where R is the gas constant ($8.314\,\text{J K}^{-1}\,\text{mol}^{-1}$).

BIBLIOGRAPHY

1. A Quantitative Measure of Solvent Solvophobic Effect. M.H. Abraham, P.L. Grellier, and R.A. McGill. *J. Chem. Soc. Perkin Trans. II*, 339–345 (1988).
2. Hydrophobic Acceleration of Diels–Alder Reactions. D.C. Rideout and R. Breslow. *J. Am. Chem. Soc.*, **102**, 7816–7817 (1980).
3. Chaotropic Salt Effects in a Hydrophobically Accelerated Diels–Alder Reaction. R. Breslow and C.J. Rizzo, *J. Am. Chem. Soc.*, **113**, 4340–4341 (1991).

EXPERIMENT 19B   REPORT

Pre-lab

1. Write a chemical equation for the reaction of 9-anthracenemethanol with *N*-ethylmaleimide. Are *endo* and *exo* isomers possible in this reaction?
2. Make a table showing the physical properties (molecular mass, b.p., density, solubility, toxicity/hazards, flammability [for solvents only]) of 9-anthracenemethanol, *N*-ethylmaleimide, methanol, toluene, and petroleum ether. For the adduct (not described in the literature), calculate the molecular mass and list its m.p. (175–177°C) only.
3. Explain the meaning of the following terms: pseudo-first-order rate constant; second-order rate constant; Arrhenius activation energy. Be brief and concise.
4. Briefly explain the hydrophobic effect.

In-lab

1. Calculate the % yield of the synthesis.
2. Report the melting point of the product.
3. Explain the differences in the UV-visible spectra of starting material and product.
4. Interpret the IR of the adduct.
5. The ^{1}H-NMR of the adduct (Fig. 19.13) shows a triplet at 0.26 ppm, which is an abnormally low chemical shift. Why? A model of the adduct will help you find the answer.
6. Interpret the NMR spectra of the adduct (Figs. 19.13 and 19.14).
7. Calculate the pseudo-first-order and second-order rate constants for each methanol concentration at each temperature. Present your data in a table format.
8. From the log k versus Sp at 45°C, extrapolate the k values for 100% methanol, 100% water, 40% 1-propanol (60% water), and 40% 2-propanol (60% water). The Sp values are given in Table 19.2. Compare the extrapolated values with results given below: For 100% methanol, $k = 3.44\ 10^{-3}\ \text{L mol}^{-1}\text{s}^{-1}$ (Ref. 2 & 3); 100% water, $k = 230\ 10^{-3}\ \text{L mol}^{-1}\ \text{s}^{-1}$ (Refs. 2 & 3); 40% 1-propanol, $k = 33.2\ 10^{-3}\ \text{L mol}^{-1}\text{s}^{-1}$; and 40% 2-propanol, $k = 39.5\ 10^{-3}\ \text{L mol}^{-1}\text{s}^{-1}$.
9. Report the activation energy at each methanol concentration. Present the data in a table format. Discuss your results.

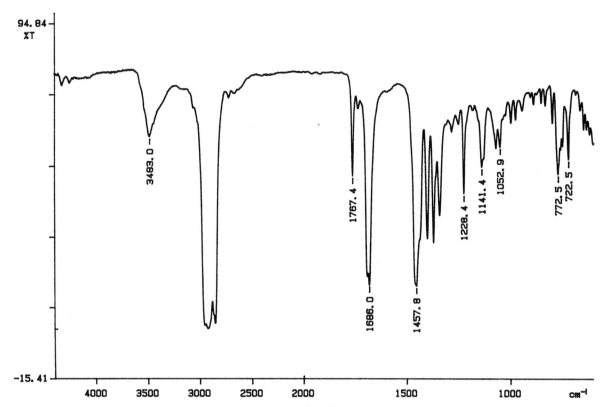

Figure 19.12   IR spectrum of 9-anthracenemethanol-NEM adduct (nujol).

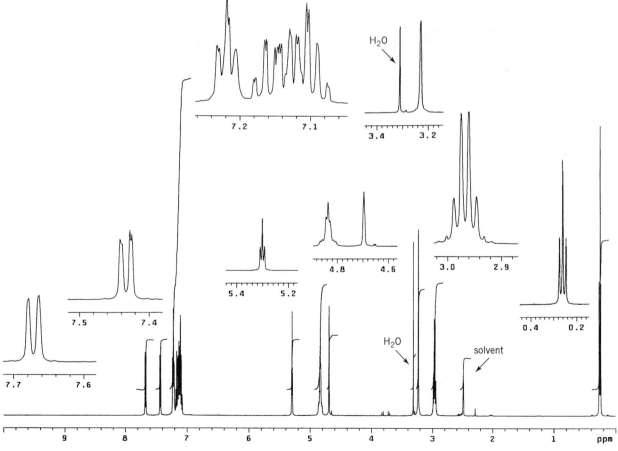

Figure 19.13   500-MHz ^{1}H-NMR spectrum of 9-anthracenemethanol-NEM adduct in DMSO-d$_6$.

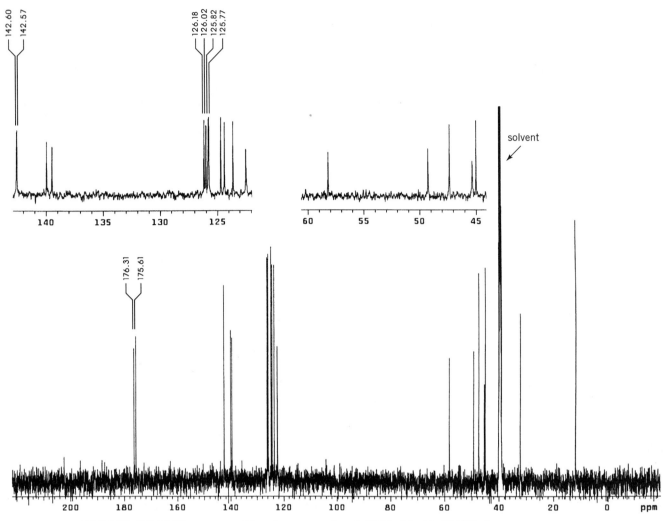

Figure 19.14 125.7-MHz ^{13}C-NMR spectrum of 9-anthracenemethanol-NEM adduct in DMSO-d$_6$.

Aldehydes and Ketones

20.1 INTRODUCTION

Aldehydes and ketones play a pivotal role in organic chemistry. Their functional group is the **carbonyl group**, *a carbon-oxygen double bond*, C=O. Aldehydes have at least one hydrogen atom attached to the carbonyl while ketones have two alkyl or aryl groups instead.

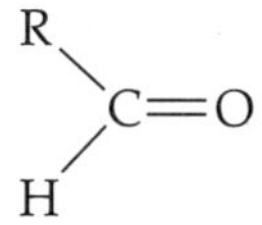

an aldehyde 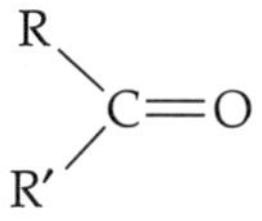   a ketone

Aldehydes and ketones are found widely distributed in the animal and plant kingdoms. Many carbonyl compounds are responsible for the sweet and penetrating aroma of fruit and flowers. Plants release volatile and odoriferous compounds to attract insects to aid in pollination, and also as a defense mechanism to repel parasites and intruders. Aldehydes and ketones are part of this chemical weaponry. Other aldehydes and ketones, for example, glucose and ribose, are components of larger biomolecules such as cellulose and nucleic acids. In the box below you will find a few examples of naturally occurring aldehydes and ketones.

Some examples of naturally occurring aldehydes and ketones.

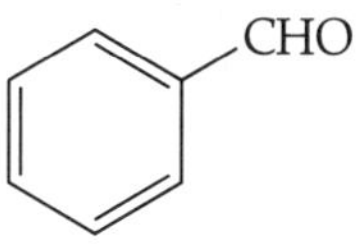

benzaldehyde (oil of almonds)   vanillin (vanilla)   menthone (peppermint oil)   carvone (caraway and spearmint oils)

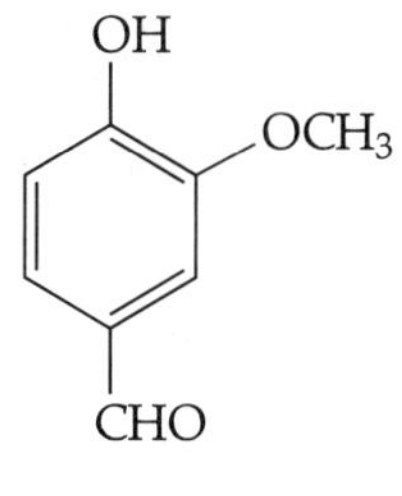

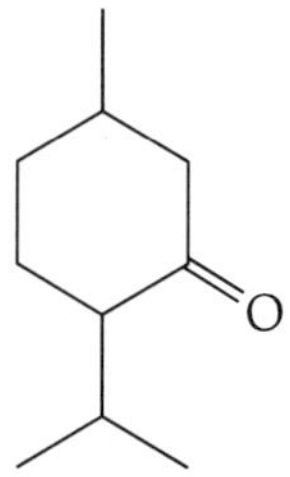

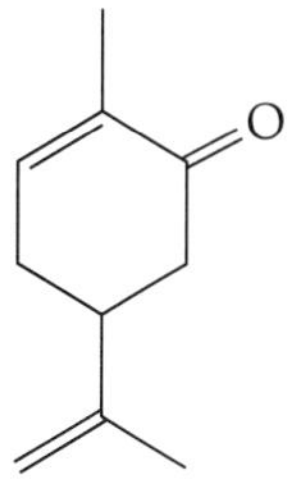

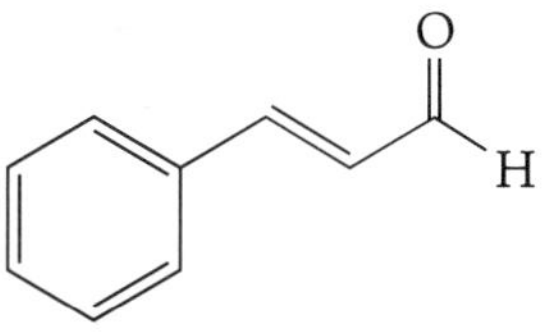

trans-cinnamaldehyde (cinnamon)   β-ionone (violets)   muscone (male musk deer)

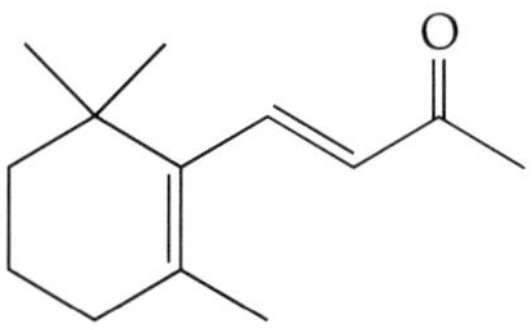

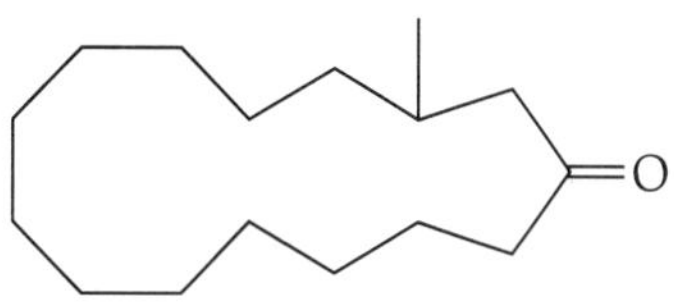

415

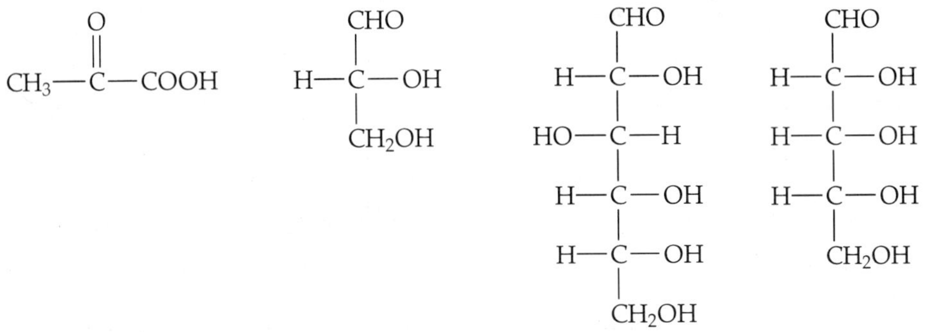

$$CH_3 - \underset{\underset{O}{\parallel}}{C} - COOH$$

pyruvic acid
(metabolite of sugars)

$$\begin{array}{c} CHO \\ | \\ H - C - OH \\ | \\ CH_2OH \end{array}$$

glyceraldehyde
(metabolite of sugars)

$$\begin{array}{c} CHO \\ | \\ H - C - OH \\ | \\ HO - C - H \\ | \\ H - C - OH \\ | \\ H - C - OH \\ | \\ CH_2OH \end{array}$$

glucose
(cellulose, starch, etc.)

$$\begin{array}{c} CHO \\ | \\ H - C - OH \\ | \\ H - C - OH \\ | \\ H - C - OH \\ | \\ CH_2OH \end{array}$$

ribose
(nucleic acids)

20.2   PREPARATION

Aldehydes and ketones can be prepared by the oxidation of primary and secondary alcohols, respectively. The oxidation of secondary alcohols to ketones can be performed with different reagents under a variety of conditions. For example, a mixture of CrO_3 in dilute sulfuric acid (Jones' reagent) or sodium dichromate ($Na_2Cr_2O_7$) in acetic or sulfuric acid are good oxidizing agents for this purpose.

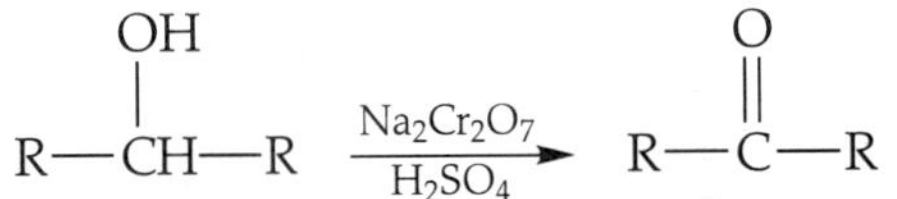

$$R - \underset{\underset{\underset{\text{alcohol}}{\text{a secondary}}}{\underset{OH}{|}}}{CH} - R \quad \xrightarrow[H_2SO_4]{Na_2Cr_2O_7} \quad R - \underset{\underset{\text{a ketone}}{\underset{O}{\parallel}}}{C} - R$$

The oxidation of primary alcohols under the same conditions gives carboxylic acids instead of aldehydes. Most aldehydes are rapidly oxidized by these powerful oxidizing agents and cannot be isolated. Special reagents are used to oxidize primary alcohols to aldehydes; among them pyridinium chlorochromate (PCC, $C_6H_6NClCrO_3$) is particularly useful.

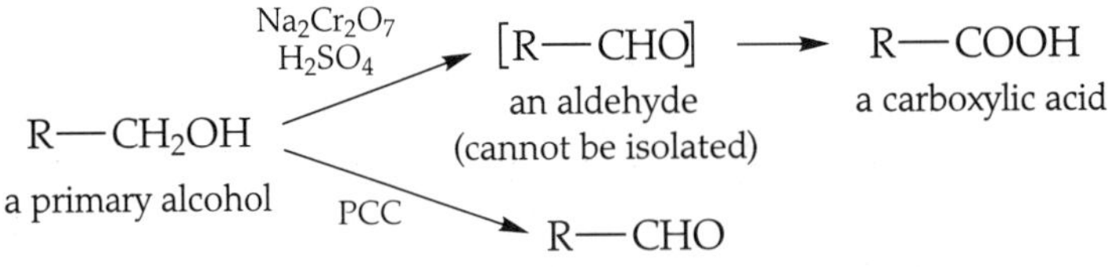

$$R - CH_2OH \quad \begin{array}{l} \xrightarrow[H_2SO_4]{Na_2Cr_2O_7} \quad [R - CHO] \longrightarrow R - COOH \\ \qquad\qquad\qquad \text{an aldehyde} \qquad \text{a carboxylic acid} \\ \qquad\qquad\qquad \text{(cannot be isolated)} \\ \xrightarrow{PCC} \quad R - CHO \\ \qquad\qquad \text{an aldehyde} \end{array}$$

a primary alcohol

We will see these and other reactions to prepare aldehydes and ketones from alcohols in Unit 21.

20.3   REACTIONS OF ALDEHYDES AND KETONES

The reactions of aldehydes and ketones can be divided into three categories:

1. Nucleophilic additions
2. Oxidations
3. Reactions on the α-carbon

Nucleophilic Addition

Due to the electronegativity of the oxygen atom the carbonyl group is polarized, with the carbon atom bearing a positive charge density. This makes the carbon atom the subject of nucleophilic attacks.

The reaction of aldehydes and ketones with nucleophiles is often acid-catalyzed. The protonation of the carbonyl leaves the carbon atom with a positive charge density, activating it for nucleophilic attacks.

Activation of the carbonyl

Nucleophilic attack

a tetrahedral intermediate

The attack of the nucleophile leads to a **tetrahedral intermediate**. Depending on the nature of the nucleophile and the structure of the carbonyl compound, the tetrahedral intermediate can be stable enough to be isolated or it can further decompose to yield stable products. Nucleophiles such as **bisulfite ion** and **cyanide ion** give stable intermediates; on the other hand, as we will see later, amines (NH_2R) and other nitrogen-containing nucleophiles form tetrahedral intermediates that spontaneously decompose, losing a water molecule.

In the laboratory we will make use of several nucleophilic addition reactions to **characterize** and **identify** aldehydes and ketones. **Characterization** is the study of the chemical and physical properties of unknown substances. In Experiment 20A you will characterize an unknown carbonyl compound by a series of chemical tests. By performing these characterization tests we can place an unknown sample in a specific family of compounds, such as aromatic aldehydes, aliphatic aldehydes, methyl ketones, and so forth. **Identification**, on the other hand, implies an exact knowledge of the structure of the compound. The identification is carried out after the sample has been fully characterized. Identification usually requires a combination of chemical and spectroscopic methods. A useful tool for this purpose is the formation of **derivatives**. Derivatives are crystalline compounds with well-defined melting points that we prepare from unknown samples. The unknown compound is identified by comparing the melting point of its derivative with those of known samples. Large tables with the melting points of derivatives for thousands of compounds have been compiled for this purpose. To be useful as a derivative, the solid product should be easily made and purified. Only unknowns that have been previously described in the literature can be identified in this fashion.

Let's now see the characterization and identification reactions for aldehydes and ketones that you will perform in the laboratory.

Bisulfite Addition Compounds *Aldehydes, unhindered methyl ketones and unhindered cyclic ketones react with saturated sodium bisulfite (sodium hydrogen sulfite) aqueous solutions to form white crystalline addition products. The nucleophile is the bisulfite ion.*

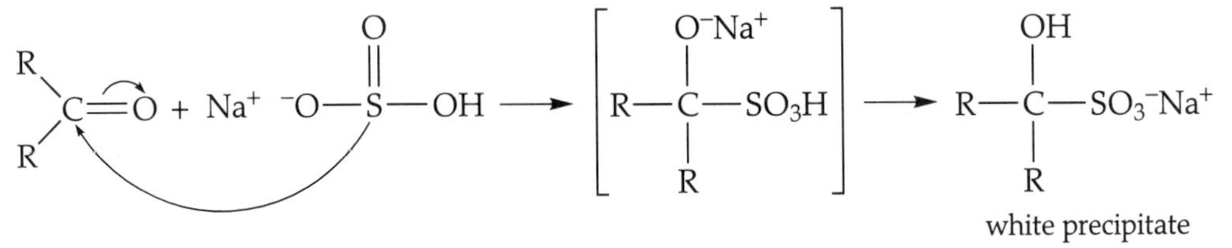

white precipitate

Bisulfite addition products crystallize from aqueous solutions and can be separated by filtration. The aldehyde or ketone can be regenerated by adding either a strong acid or base:

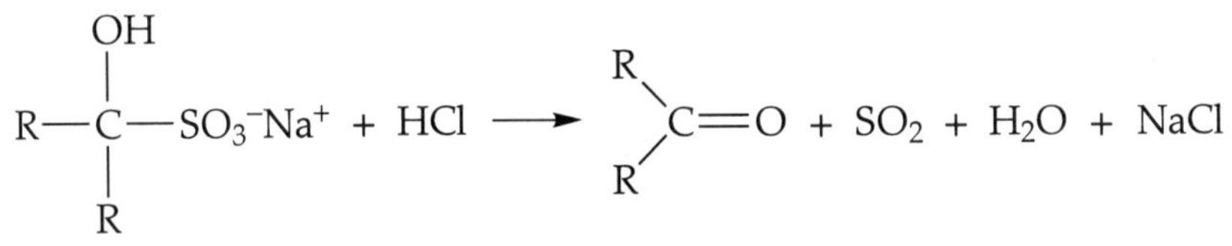

In the laboratory one can use the formation of a bisulfite addition product to separate aldehydes and ketones from mixtures containing other substances such as hydrocarbons, alcohols, and esters, which do not react with sodium bisulfite.

2,4-Dinitrophenylhydrazones *Aldehydes and ketones react with 2,4-dinitrophenylhydrazine, under acidic conditions, to give* **2,4-dinitrophenylhydrazones** *(2,4-DNPH).*

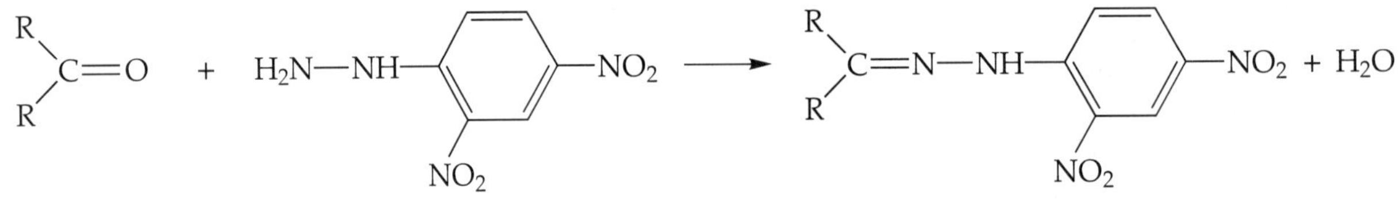

an aldehyde or a ketone   2,4-dinitrophenylhydrazine   a 2,4-dinitrophenylhydrazone
(yellow-red solid)

2,4-Dinitrophenylhydrazones are yellow-red solids that can be used to **characterize** aldehydes and ketones. The formation of 2,4-dinitrophenylhydrazones is a useful characterization test because most aldehydes and ketones react immediately, yielding a yellow-red precipitate. Other carbonyl-bearing functional groups, such as esters and amides, do not react with this reagent. 2,4-Dinitrophenylhydrazones have well-defined melting points, making them excellent **derivatives** for aldehydes and ketones. Tables of melting points of 2,4-dinitrophenylhydrazones are available to identify unknown aldehydes and ketones.

The mechanism for the formation of 2,4-DNPH is shown below:

Semicarbazones

Semicarbazones   *Aldehydes and ketones react with semicarbazide* (H_2N-$NHCONH_2$) *to produce* **semicarbazones**. Semicarbazide solutions are unstable and the reagent is better handled in the hydrochloride form. To carry out the reaction, the semicarbazide must be freed by the addition of a base such as sodium acetate.

| acetate anion | semicarbazide hydrochloride | acetic acid | semicarbazide |

| an aldehyde or a ketone | semicarbazide | a semicarbazone |

The mechanism for this reaction is similar to that for 2,4-dinitrophenylhydrazine. Semicarbazones are also crystalline compounds with well-defined melting points and, therefore, excellent derivatives for aldehydes and ketones. However, unlike the formation of 2,4-DNPH, the formation of semicarbazones is not widely used as a characterization test because most of them have no distinctive color.

Oxidation Reactions

Aldehydes can be converted into carboxylic acids by oxidation under mild conditions.

| an aldehyde | a carboxylic acid |

Ketones are not oxidized under the same conditions because they have no hydrogen attached to the carbonyl group. To oxidize ketones more drastic conditions (high temperatures and stronger oxidizing agents) are needed to break the C—C bond. Several laboratory tests take advantage of this difference to distinguish aldehydes from ketones. Among them are Tollens' and Fehling's tests, described below.

Tollens' Test   *Aldehydes react with a solution of silver ion in ammonia,* **Tollens' reagent**, *to give carboxylic acids and metallic silver. If the test tube in which the reaction is carried out is clean and grease free, the silver deposits on the walls as a silver mirror; otherwise it forms a black precipitate. Ketones are not oxidized by Tollens' reagent.*

$$\underset{\text{an aldehyde}}{R-\overset{\overset{\displaystyle |\!|}{O}}{C}-H} + \underset{\text{Tollens' reagent}}{2\,Ag(NH_3)_2^+} + 2\,HO^- \longrightarrow \underset{\text{a carboxylate}}{R-\overset{\overset{\displaystyle |\!|}{O}}{C}-O^-} + NH_4^+ + 3\,NH_3 + H_2O + \underset{\text{metallic silver}}{2\,Ag^0}$$

Tollens' reagent is made by adding sodium hydroxide to a silver nitrate solution. A black precipitate of Ag_2O forms first. This precipitate is redissolved by adding ammonium hydroxide, which forms a complex with the silver ion: $Ag(NH_3)_2^+$. Tollens' reagent should be prepared immediately before use and should never be stored because it forms explosive silver nitride (Ag_3N) on standing (see Experiment 20A).

$$2\,Ag^+NO_3^- + 2\,NaOH \longrightarrow \underset{\text{black precipitate}}{Ag_2O} + 2\,Na^+NO_3^- + H_2O$$

$$Ag_2O + 4\,NH_3 + H_2O \longrightarrow \underset{\underset{\text{(colorless)}}{\text{Tollens' reagent}}}{2\,Ag(NH_3)_2^+} + 2\,HO^-$$

Fehling's Test   The oxidizing agent in this test is Cu^{2+} in alkaline solution. Cu^{2+} is kept in solution by complexing it with tartrate (otherwise it precipitates as $Cu(OH)_2$ in the basic medium). *Aliphatic aldehydes reduce Fehling's reagent whereas aromatic aldehydes do not.* In a positive test, the solution turns from blue to green and finally a reddish precipitate of cuprous oxide (Cu_2O) develops. Sometimes, the reduction of copper proceeds one step further and a spectacular copper mirror (Cu^0) deposits on the walls of the tube. Simple ketones do not react but certain hydroxy ketones and carbohydrates do (Unit 24).

tartrate

$$\underset{\underset{\text{an aldehyde}}{}}{R-\overset{\overset{\displaystyle |\!|}{O}}{C}-H} + \underset{\text{blue}}{2\,Cu^{2+}} + 5\,HO^- \xrightarrow{\text{tartrate}} R-\overset{\overset{\displaystyle |\!|}{O}}{C}-O^- + 3\,H_2O + \underset{\text{reddish precipitate}}{Cu_2O\downarrow}$$

Reactions on the α-Carbon

Hydrogen atoms located on the α-carbons to a carbonyl group are acidic. They can be removed by bases to produce a carbanion, which is stabilized by resonance. Such an anion is called an **enolate anion**.

$$\underset{R}{\overset{H}{R-\overset{|}{\underset{|}{C}}-\overset{\overset{\displaystyle |\!|}{O}}{C}-R}} \underset{BH}{\overset{B^-}{\rightleftharpoons}} \left[R-\overset{|}{\underset{|}{C}}-\overset{\overset{\displaystyle |\!|}{O}}{C}-R \longleftrightarrow \overset{R}{\underset{R}{C}}=\overset{R}{\underset{O^-}{C}} \right]$$

an enolate anion

Enolate anions can act as nucleophiles in many reactions. The **iodoform test** and the **aldol** and **Perkin condensations**, explained below, illustrate this point.

The Iodoform Test Enolate anions rapidly react with iodine to give α-iodo carbonyl compounds.

$$R—CO—CH_3 \ + \ HO^- \ \rightleftharpoons \ R—CO—CH_2^- \ + \ H_2O$$

a methyl ketone   a ketone
enolate anion

$$R—CO—CH_2^- \ + \ I—I \ \longrightarrow \ R—CO—CH_2I \ + \ I^-$$

an α-iodo carbonyl
compound

Iodine is an electron-withdrawing atom and increases the acidity of the remaining hydrogens on the α-carbon. As a result, iodination of methyl ketones stops only when all three hydrogens have been replaced by iodine.

$$R–CO–CH_2I \xrightarrow[\text{}]{\text{HO}^- \quad I_2} R–CO–CHI_2 \xrightarrow[\text{}]{\text{HO}^- \quad I_2} R–CO–CI_3$$

The triiodo methyl group ($-CI_3$) is a strong electron-withdrawing group and also a relatively good leaving group. The HO^- group attacks the carbonyl and the relatively stable triiodo carbanion leaves; a subsequent proton transfer between the carboxylic acid and the $^-CI_3$ anion gives iodoform, a yellow crystalline solid.

$$R—COO^- \ + \ \boxed{HCI_3}$$

iodoform
(yellow precipitate)

The iodoform reaction is an excellent characterization test for *methyl ketones*, $R–CO–CH_3$, which give a yellow precipitate of iodoform. However, a positive test also results from alcohols with structure $R–CH(OH)–CH_3$ *(methyl carbinols)*. It happens that under the experimental conditions, methyl carbinols are oxidized to methyl ketones by iodine. The methyl ketone then reacts in the usual way to give iodoform.

$$R—\underset{\underset{OH}{|}}{C}H—CH_3 \xrightarrow{I_2 / HO^-} R—\underset{\underset{O}{||}}{C}—CH_3$$

a methyl carbinol   a methyl ketone

Aldol Condensation Enolate anions of aldehydes and ketones can act as nucleophiles and attack other carbonyl groups. This reaction is very useful in synthesis because *it forms a new carbon-carbon bond*. The reaction first gives a β-hydroxy carbonyl compound (an aldol), which easily dehydrates to produce an α,β-unsaturated aldehyde or ketone.

$$R-CO-CH_2R \;+\; B^- \;\rightleftharpoons\; R-CO-\overset{-}{C}HR \;+\; HB$$

a β-hydroxy carbonyl compound
(an aldol)

$-H_2O$

an α,β unsaturated carbonyl compound

You will apply this reaction for the synthesis of ionones (compounds responsible for the smell of violets) in Experiment 23B.

Perkin Condensation In this reaction, the enolate anion of an acid anhydride attacks an aromatic aldehyde (an aldehyde without α-H). The reaction is catalyzed by a carboxylate salt of the same acid that forms the anhydride. The most commonly used acid anhydride is acetic anhydride. The reaction between benzaldehyde and acetic anhydride gives trans-cinnamic acid. The Perkin condensation is named after William Perkin, an extraordinary British chemist (see the box "The Color of Success").

| benzaldehyde | acetic anhydride | trans-cinnamic acid | acetic acid |

The Perkin condensation is very useful in synthesis because it provides *a way of adding two carbon atoms* to the side chain of an aromatic compound. The mechanism for the Perkin condensation is complex. It involves the attack of the enolate anion of the acid anhydride on the aromatic aldehyde. A series of rearrangements follows, with the overall effect of forming a carbon-carbon double bond (loss of water) and hydrolyzing the anhydride group. Usually, the *trans* isomer forms preferentially because it is more stable. You will perform a Perkin condensation in Experiment 20B.

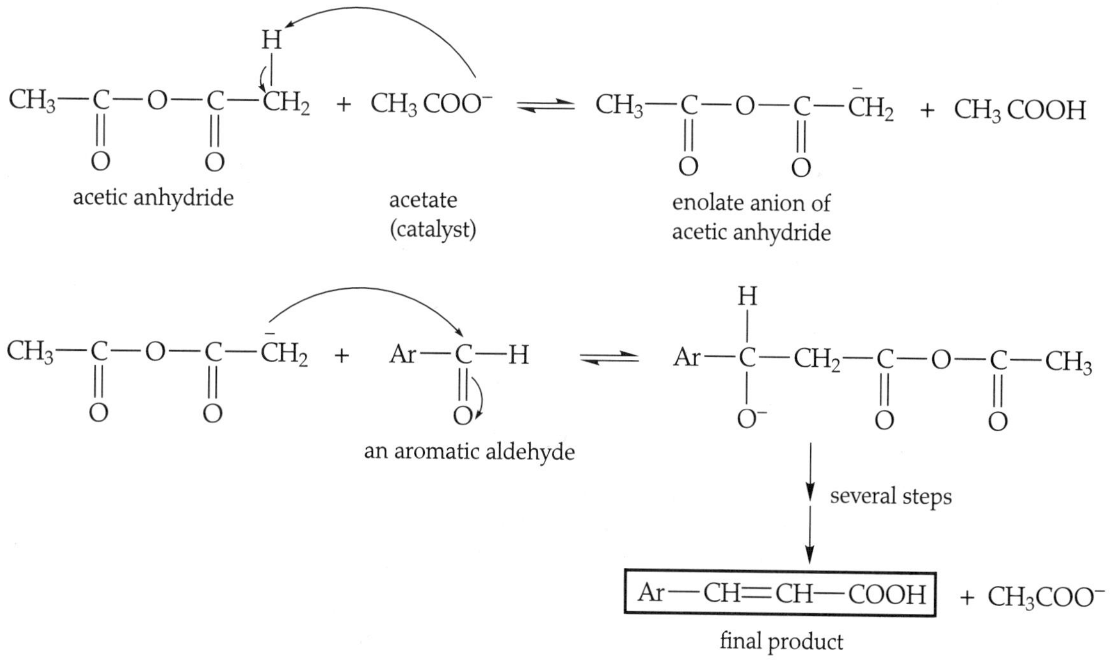

$$CH_3-\underset{O}{\overset{||}{C}}-O-\underset{O}{\overset{||}{C}}-\overset{H}{\underset{|}{C}}H_2 \;+\; CH_3COO^- \;\rightleftharpoons\; CH_3-\underset{O}{\overset{||}{C}}-O-\underset{O}{\overset{||}{C}}-\bar{C}H_2 \;+\; CH_3COOH$$

acetic anhydride acetate   enolate anion of
     (catalyst)   acetic anhydride

$$CH_3-\underset{O}{\overset{||}{C}}-O-\underset{O}{\overset{||}{C}}-\bar{C}H_2 \;+\; Ar-\underset{O}{\overset{||}{C}}-H \;\rightleftharpoons\; Ar-\underset{O^-}{\overset{H}{\underset{|}{C}}}-CH_2-\underset{O}{\overset{||}{C}}-O-\underset{O}{\overset{||}{C}}-CH_3$$

an aromatic aldehyde

several steps

$$\boxed{Ar-CH=CH-COOH} \;+\; CH_3COO^-$$

final product

20.4 IR AND NMR OF ALDEHYDES AND KETONES

IR Spectra

Aliphatic ketones absorb around $1715 \, cm^{-1}$; aliphatic aldehydes around $1720-1740 \, cm^{-1}$. These bands are due to C=O stretching and are usually very intense. Conjugation with a double bond or an aromatic ring lowers the frequency by $30-40 \, cm^{-1}$. Aldehydes also show two bands around 2720 and $2820 \, cm^{-1}$, due to C–H stretching (of the CHO group) and an overtone of C–H bending. These bands are absent in ketones.

NMR Spectra

The aldehyde hydrogen resonates around 9–10 ppm for aliphatic aldehydes and between 9.6 and 10.5 ppm for aldehydes conjugated with double bonds or aromatic rings. In ^{13}C-NMR, the carbonyl resonates around 200 ppm for aliphatic aldehydes and around 206–212 ppm for aliphatic ketones. Conjugation with an aromatic ring or a double bond lowers the chemical shift by about 10 ppm.

BIBLIOGRAPHY

1. Vogel's Textbook of Practical Organic Chemistry. A.I. Vogel, B.S. Furniss, A.J. Hannaford, P.W.G. Smith, and A.R. Tatchell. 5th ed. Longman, Harlow, UK, 1989.

2. The Systematic Identification of Organic Compounds. R.L. Shriner, C.K.F. Herman, T.C. Morrill, and D.Y. Curtin. 7th ed. Wiley, New York, 1998.

3. Semimicro Qualitative Organic Analysis, N.D. Cheronis, J.B. Entrikin, and E.M. Hodnett, 3rd. ed., Interscience Pub., New York, 1965.

4. CRC Handbook of Tables for Organic Compound Identification. Z. Rappoport, 3rd ed. CRC Press, Cleveland, 1977.

The Color of Success   Let's now travel to London. It is the Easter break of 1856 and in an improvised home laboratory William Perkin is trying to synthesize quinine, an antimalarial drug badly needed to bring back to health the rubber-tree workers who are dying in the remote confines of the empire. Perkin has the molecular formula of quinine ($C_{20}H_{24}N_2O_2$) as the only firm clue to lead his research. He thinks that by combining two molecules of allyltoluidine ($C_{10}H_{13}N$) in the presence of a strong oxidizer to supply the oxygen, the elusive quinine would be obtained. He is wrong; he obtains instead a brown mess. He tries again, this time with aniline instead of allyltoluidine, and a black precipitate is formed. He is about to toss away the unwanted solid. He pours some alcohol into the flask and, to his surprise, observes that the alcohol takes on a beautiful lilac color like nothing seen before. Perkin doesn't throw away the product; he sees people dressed in purple where others would have seen just a black and ugly solid. This moment of wonder and amazement sparks the beginning of the modern chemical industry. In less than a year Perkin sets up a factory in West London for the production of the first manmade aniline dye, which he calls *mauveine*. In the affluent Victorian society of 1850s, hungry for new fashions, mauveine is an instant success. William Perkin is only eighteen.

Perkin's factory can be considered the prototype of the modern chemical plant with research and development facilities. He didn't just make mauveine, he also prepared some of the reagents needed for the process and developed the synthesis of other dyes as well. Despite England's early lead in the chemical industry, Germany soon took over. The Germans cultivated a strong tie between industry and academia, had a better patent law, and also a better chemical education with emphasis in experimentation and research. All these factors contributed to the switch in leadership. By the 1880s Germany was in the forefront of chemical research and selling chemicals to the rest of the world. Some of the German companies created in the mid-nineteenth century are Bayer, BASF, and Agfa, all of them still in operation today.

In 1874, at the age of 36, Perkin sold his factory and retired to dedicate himself fully to chemical research. The reaction that carries his name (Perkin condensation) was developed in the 1870s, and it was used to make cinnamic acid and coumarin. Coumarin is found in tonka beans and has the odor of freshly cut grass. It is used in the manufacture of perfumes and deodorizers, but it has been banned as a food additive because of its toxicity.

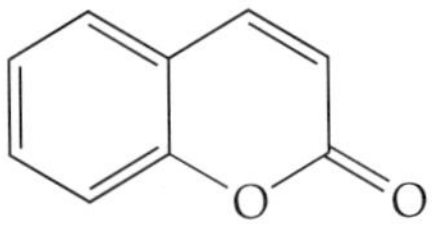

coumarin

In 1994, almost 150 years after it was first made, two scientists, also from England, analyzed an original sample of mauveine using modern chemical methods (TLC, NMR, MS). They found that the dye consisted of two major products with the structures shown below (Ref. 1). These molecules derive from the oxidation and condensation of aniline and toluidine. Had Perkin used pure aniline instead of the crude product available then (which was contaminated with toluidines), mauveine would have not been obtained. To end this note where it started, let's add that it took 88 years from that Easter break for chemists to figure out how to synthesize quinine. But that's a different story.

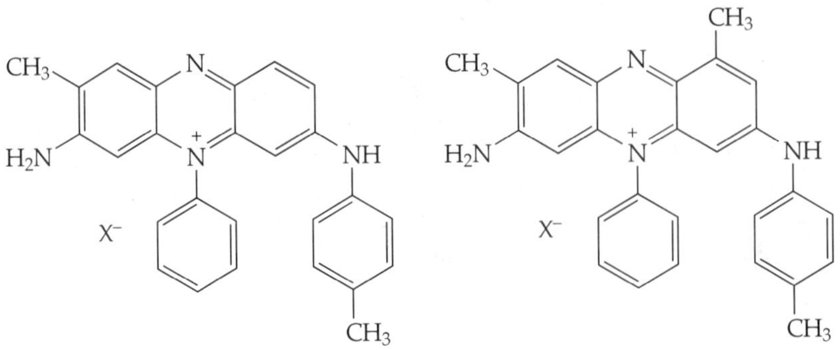

The two main components of mauveine.
In Perkin's original dye, X⁻ was acetate anion.

Bibliography

1. What did W.H. Perkin Actually Make When He Oxidised Aniline to Obtain Mauveine? O. Meth-Cohn and M. Smith. *J. Chem. Soc. Perkin Trans. 1*, 5 (1994).
2. The Story of Chemistry. F.L. Darrow. Bobbs-Merrill Co. Indianapolis, 1927.
3. The Norton History of Chemistry. W.H. Brock. W.W. Norton and Co. New York, 1993.

EXERCISES

Problem 1

Complete the following sentences:

(a) Acetaldehyde gives a _____ iodoform test.

(b) 3-Bromobenzaldehyde gives a _____ Fehling's test and a _____ Tollens' test.

(c) Semicarbazones are good _____ for aldehydes and ketones.

(d) Sodium bisulfite gives a white precipitate with _____ and _____ methyl and cyclic ketones.

(e) 2,4-Dinitrophenylhydrazine reacts with aldehydes and ketones to give _____ precipitates.

(f) Butanal gives a _____ iodoform test.

Problem 2

How would you separate 10 g of the following mixtures:

(a) Benzaldehyde (b.p. 179°C) and 2-octanol (b.p. 179°C)?

(b) *Trans*-cinnamic acid and *trans*-cinnamaldehyde?

(c) Acetone (b.p. 56°C) and benzaldehyde (b.p. 179°C)?

How would you separate mixtures a and c if you had only 1 mg?

Problem 3

Three isomers X, Y, and Z with molecular formula C_8H_8O gave the tests indicated below. The IR spectra of X, Y, and Z showed intense bands at the indicated wavenumbers (cm^{-1}). The proton-decoupled ^{13}C-NMR spectra showed 6 peaks for each compound.

Compound	2,4-DNPH	Tollens'	Fehling's	Iodoform	IR
X	+	+	+	−	1720
Y	+	−	−	+	1690
Z	+	+	−	−	1700

Propose structures for X, Y, and Z in agreement with the data.

Problem 4

How could you tell apart the following compounds by characterization tests (2,4-DNPH, bisulfite, Tollens', Fehling's, iodoform)? Make a table indicating + or −.

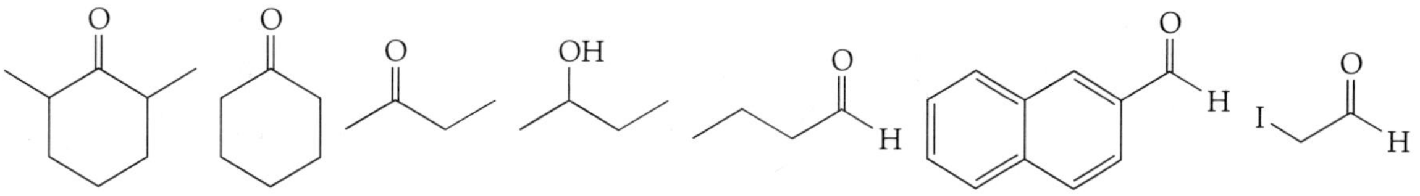

Problem 5

How would you prepare the following compounds by aldol condensation reactions?

Indicate the starting materials that you would use in each case.

Problem 6

Compound P ($C_8H_6O_2$) reacts with acetic anhydride in the presence of potassium acetate to give solid Q ($C_{12}H_{10}O_4$). The ^{1}H-NMR of Q (300 MHz, DMSO-d_6) shows two doublets at 6.65 and 7.65 ppm (J = 20 Hz) and two singlets at 7.75 and 12.5 ppm. The singlet at 12.5 ppm is broad. The IR spectrum of Q presents a broad band at 3200–2200 cm^{-1} and two strong bands at 1677 and 828 cm^{-1}. Give structures for Q and P in agreement with the data.

Problem 7

How would you obtain coumarin (see structure on p. 424) using a Perkin condensation reaction?

Identification of Aldehydes and Ketones

In this experiment you will **characterize** several aldehydes and ketones using the tests already discussed. You will also prepare the semicarbazone of *trans*-cinnamaldehyde and determine its melting point. This will give you some practice on how to prepare a derivative. You will also obtain an unknown from your instructor and will **identify** it. The unknown may be any of the aldehydes and ketones listed in Table 20.1 or a noncarbonyl compound. Hence, the first test should be with 2,4-dinitrophenylhydrazine; if this test is negative, report accordingly and obtain another unknown. You will characterize your unknown carbonyl compound by determining whether it is an aldehyde or a ketone. You will also investigate the type of aldehyde or ketone (methyl ketone, aliphatic vs. aromatic aldehyde, etc.) by a series of tests. To help interpret these tests you will simultaneously assay other compounds of known structure. You will run the IR spectrum of your unknown and measure its refractive index. You will also prepare a derivative (2,4-dinitrophenylhydrazone or semicarbazone). All this information should allow you to identify your unknown. To confirm your analysis, you will finally obtain from your instructor the ^{1}H- and ^{13}C-NMR spectra of your

Table 20.1 **List of Possible Unknowns**

Compound	Structure	b.p. (°C)	MM (g/mol)	Density (g/mL)	n_D^{20}	Semi-carbazone m.p. (°C)	2,4-DNPH m.p. (°C)
		Aldehydes					
butanal	$CH_3(CH_2)_2CHO$	74.7	72.11	0.802	1.3791	96	123
pentanal	$CH_3(CH_2)_3CHO$	103.4	86.13	0.810	1.3947	oil	98
2-ethylbutanal	$(CH_3CH_2)_2CHCHO$	116	100.16	0.814	1.4018	96	134
hexanal	$CH_3(CH_2)_4CHO$	131	100.16	0.834	1.4035	106	104
heptanal	$CH_3(CH_2)_5CHO$	155	114.19	0.822	1.4125	109	108
octanal	$CH_3(CH_2)_6CHO$	171	128.12	0.821	1.4183	98	106
t-cinnamaldehyde	Ph CH = CHCHO	252	132.16	1.050	1.6219	215	255
		Ketones					
acetone	CH_3COCH_3	56	58.08	0.791	1.3592	187	126
2-butanone	$CH_3COCH_2CH_3$	80	72.11	0.805	1.3791	135	116
3-pentanone	$CH_3CH_2COCH_2CH_3$	102	86.13	0.814	1.3922	138	156
2-pentanone	$CH_3CO(CH_2)_2CH_3$	102.3	86.13	0.810	1.3902	112	144
2-hexanone	$CH_3CO(CH_2)_3CH_3$	129	100.16	0.812	1.4000	122	106
cyclopentanone		130.7	84.12	0.951	1.4370	203	142
4-heptanone	$CH_3(CH_2)_2CO(CH_2)_2CH_3$	145	114.19	0.817	1.4070	133	75
3-heptanone	$CH_3CH_2CO(CH_2)_3CH_3$	150	114.19	0.818	1.4080	100	70
2-heptanone	$CH_3CO(CH_2)_4CH_3$	151	114.19	0.820	1.4080	127	89
cyclohexanone		156	98.15	0.948	1.4500	166	162
acetophenone	$PhCOCH_3$	205	120.15	1.024	1.5322	198	238
propiophenone	$PhCOCH_2CH_3$	218	134.18	1.011	1.5258	174	190

unknown and interpret them. This situation is similar to that encountered very often in research, where we perform a detailed spectroscopic study only after we have gathered chemical information about the sample.

PROCEDURE

Background reading: 20. 1–3   **Estimated time:** 3–4 hours

E20A.1   CHARACTERIZATION TESTS

For the following tests use 13 × 100 mm test tubes. The test tubes should have cork stoppers and should be labeled. In addition to the chemicals to be tested (indicated in each test), always run a "blank" reaction by adding a drop of water to the reagent instead of the sample. Do not contaminate the reagents. Always use clean Pasteur pipets! The reagents will be already prepared for you to use. For a better interpretation of each test, try all the samples (including your unknown) simultaneously.

Bisulfite Test

Place 0.5 mL of the sodium bisulfite reagent in a test tube and add 2 drops of the liquid sample to be tested. Stopper the test tube and shake well. Observe the results. A positive test is the formation of a white precipitate. Aldehydes, unhindered methyl ketones, and unhindered cyclic ketones give a positive test. Perform this test with heptanal, acetone, acetophenone, and the unknown.

Reagent   Add 60 mL of 95% ethanol to 240 mL of a saturated aqueous solution of sodium bisulfite (sodium hydrogensulfite). Filter off and discard any precipitate that forms. Use shortly after preparation. Carry out this preparation in a fume hood.

Cleaning Up
• With a Pasteur pipet transfer the contents of the tubes to the container labeled "Aldehydes Liquid-Waste."
• Dispose of the Pasteur pipets in the container labeled "Aldehydes Solid Waste."

2,4-Dinitrophenylhydrazine

Place 0.5 mL of the reagent in a test tube and add a drop of the liquid sample. Stopper the test tube and shake. If precipitation doesn't occur remove the stopper and warm the tube in a water bath (at 50°C) for a few minutes and gently scratch the inner walls of the test tube by moving a glass rod vertically. A positive test is the formation of a yellow-red precipitate. Most aldehydes and ketones give a positive test. Try heptanal, acetone, 2-propanol, and your unknown.

Reagent   Dissolve 5.0 g of 2,4-dinitrophenylhydrazine in a mixture of 35 mL of water, 25 mL conc. sulfuric acid, and 100 mL 95% ethanol.

Cleaning Up
• With a Pasteur pipet transfer the contents of the tubes to the container labeled "Aldehydes Hydrazines Waste." Do not mix with other waste.
• Dispose of the Pasteur pipets in the container labeled "Aldehydes Solid Waste."

Safety First

- Sodium bisulfite is an irritant to the eyes, nose, and throat. It may cause asthmatic attacks in sensitive individuals. Handle it with care in a well-ventilated place.
- Hydrochloric acid is highly toxic and corrosive. Handle it with care.
- Many hydrazines are suspected carcinogens. Avoid skin contact.
- Prepare Tollens' reagent immediately before use. Do not store it. Tollens' reagent forms explosive silver nitride on standing. Destroy the reagent immediately after use following the guidelines given below.
- Avoid skin contact with silver nitrate and Tollens' reagent. Unsightly brown stains form.
- Iodine is highly toxic and corrosive.
- Use fresh 1,2-dimethoxyethane only. It is a peroxide forming and flammable liquid.

Tollens' Test

Place 1 mL of a 5% aqueous solution of silver nitrate in a test tube cleaned very well with soap and water. Add two drops of a 10% sodium hydroxide solution. Stopper the tube and shake well. Add dropwise of a 2% ammonium hydroxide solution until the black silver oxide (Ag_2O) just redissolves. Add one drop of your liquid sample. Stopper the test tube and shake. A positive test is the formation of a silver mirror on the walls of the test tube or the formation of a black precipitate of reduced silver. If no positive test is observed in about 10 minutes, remove the stopper and warm the test tube *in a water bath* at approximately 40°C. Aldehydes give a positive test. Ketones give a negative reaction. Try heptanal, acetone, and your unknown.

Following the guidelines below, immediately destroy any excess of Tollens' reagent and the contents of the test tubes used for this test. Do not keep any silver mirror!

Destroying Tollens' Reagent

With a Pasteur pipet transfer the contents of all the test tubes and unused portions of Tollens' reagent to a beaker. Remove the silver mirror from the test tubes with a small amount of dilute nitric acid (*Caution: nitric acid is a strong oxidant; avoid skin contact!*) and transfer the nitric acid solutions to the beaker with the help of a Pasteur pipet. Acidify the mixture in the beaker with dilute nitric acid to destroy unreacted Tollens' reagent. Dispose of the liquid properly.

Cleaning Up

- Dispose of the destroyed Tollens' reagent in the container labeled "Silver Nitrate Waste." Do not mix with other waste.
- Dispose of the Pasteur pipets in the container labeled "Aldehydes Solid Waste."

Fehling's Test

Place in a test tube 0.5 mL of Fehling's reagent A and 0.5 mL of Fehling reagent B. Add two drops of the liquid sample and heat the mixture in a boiling water bath for about 10 minutes. A positive test is the formation of reddish precipitate (Cu_2O) or metallic copper deposited on the walls of the tube. Aliphatic aldehydes give a positive test; aromatic aldehydes and ketones do not react. Try benzaldehyde, acetone, heptanal, and your unknown.

Reagent   Fehling's A: 6.9 g of pentahydrate copper sulfate ($CuSO_4 \cdot 5H_2O$) in 100 mL of water; heat, if necessary, to dissolve. Fehling's B: 34.6 g of sodium potassium tartrate (Rochelle salt) in 100 mL of 10% sodium hydroxide solution.

Iodoform Test

Dissolve one drop of sample in 0.5 mL of water. If the sample doesn't dissolve in water use 1,2-dimethoxyethane as a solvent instead. Add 0.5 mL of a 10% aqueous sodium hydroxide solution followed by 1 mL of iodine reagent. A positive test is the formation of a yellow precipitate. Methyl ketones and methyl carbinols give positive reactions. Perform this test using acetone, 3-pentanone, 2-propanol, and your unknown.

Reagent   Dissolve potassium iodide (20 g) in 80 mL of water; add 10 g of iodine and stir well until all dissolves. Carry out this preparation in a fume hood.

Cleaning Up

- With a Pasteur pipet transfer the contents of the tubes from Fehling's test to the container labeled "Aldehydes Liquid Waste."
- With a Pasteur pipet transfer the contents from the iodoform test to the container labeled "Iodoform Waste." Do not mix with other waste.
- Dispose of the Pasteur pipets in the container labeled "Aldehydes Solid Waste."

E20A.2   PREPARATION OF DERIVATIVES

Semicarbazone of *trans*-Cinnamaldehyde

Place in a 13 × 100 mm test tube 1.0 mL of semicarbazide reagent, add 0.25 mL of *trans*-cinnamaldehyde (measured with a pluringe), and 1 mL of methanol. Mix well. Crystals begin to separate immediately. Cool the solution in an ice-water bath and collect the solid by vacuum filtration using a Hirsch funnel. Wash the crystals on the filter with cold water followed by a small amount of cold methanol. Let the solid dry for 10–15 minutes on the filter and then in an oven at 70°C for a few minutes. Determine the melting point of the derivative.

The reaction between *trans*-cinnamaldehyde and semicarbazide is so fast that there is no need to free the semicarbazide from its hydrochloride by adding sodium acetate. However, for most aldehydes and ketones the addition of the base is necessary. Use the technique outlined for *trans*-cinnamaldehyde to prepare the derivative of your unknown. If the product does not precipitate, add 200 mg of sodium acetate, stopper the test tube, and shake well. Remove the stopper and warm the solution in a hot water bath for a few minutes. If no precipitation occurs, scratch the inner walls of the tube with a glass rod. Collect the product as indicated above. Dry it and determine its melting point. The product can be recrystallized from ethanol (or ethanol-water mixtures if the solid is too soluble in ethanol).

Reagent   Dissolve 2.2 g of semicarbazide hydrochloride in 10 mL of water.

2,4-Dinitrophenylhydrazones

To 5 mL of the 2,4-dinitrophenylhydrazine reagent in a 13 × 100 mm test tube, add 0.1 mL (100 μL) of the unknown (measured with a pluringe) and mix well. Allow the solution to stand at room temperature until crystallization is complete. If precipitation doesn't occur, warm the solution in a water bath for a few minutes. Carry out this operation in the fume hood. Remove it from the heat; if no precipitate appears after 15 minutes, scratch the inner walls with a glass rod. If precipitation does not occur, add water dropwise to the warm solution until it is cloudy, heat to redissolve the precipitate, and cool. Collect the solid by vacuum-filtration using a Hirsch funnel. Recrystallize the product from ethanol (or ethanol-water if the solid is too soluble in ethanol) and determine its melting point.

Reagent   See p. 428.

Cleaning Up

- With a Pasteur pipet transfer the filtrates from the semicarbazone and 2,4-DNPH reactions to a container labeled "Aldehydes–Hydrazines Waste." Do not mix with other waste.
- Dispose of the Pasteur pipets, filter papers, and capillary tubes in the container labeled "Aldehydes Solid Waste."
- Turn in your derivatives in labeled containers to your instructor.

E20A.3   THE UNKNOWN

Obtain an unknown from your instructor. Write down in your lab notebook the unknown's number. Using the 2,4-dinitrophenylhydrazine test, determine whether it is an aldehyde or a ketone. If the unknown tests negative for aldehydes and ketones, report your result and obtain another sample. Characterize your sample by measuring its refractive index (Unit 11) and performing the following tests: Tollens', Fehling's, bisulfite, and iodoform as indicated above (E20A.1). Obtain the IR spectrum of your unknown and assign as many bands as possible. The unknown is selected from the compounds listed in Table 20.1. Based on your characterization tests make a preliminary list of those compounds that match your results. Choose a suitable derivative that would allow you to identify your unknown by its melting point. Prepare the derivative following the procedures outlined in E20A.2. After you have interpreted your chemical tests and the IR spectrum, obtain the ^{1}H- and ^{13}C-NMR spectra from your instructor.

EXPERIMENT 20A   REPORT

Pre-lab

1. Write an equation for the reaction of a generic aldehyde with the following reagents:
   (a) 2,4-Dinitrophenylhydrazine
   (b) Semicarbazide
   (c) Bisulfite
   (d) Tollens' reagent
   (e) Fehling's reagent
   (f) I_2/HO^- (Iodoform test)

2. Predict whether the following compounds give positive or negative tests. Write + or − according to your predictions:

Compound	2,4-DNPH	Bisulfite	Tollens'	Fehling's	Iodoform
acetone					
heptanal					
benzaldehyde					
3-pentanone					
2-propanol					

3. How would you distinguish an aliphatic aldehyde from an aliphatic ketone by IR spectroscopy?

4. How would you use IR spectroscopy to tell apart an aliphatic aldehyde from an aromatic one? What chemical test would you use for the same purpose?

5. What properties should a good derivative have?

6. What precautions should be taken in handling the Tollens' reagent and the contents of the test tubes used for the Tollens' test?

7. Make a table with the toxicity/hazards of sodium bisulfite, iodine, hydrochloric acid, semicarbazide, and 2,4-dinitrophenylhydrazine.

In-lab

1. Make a table similar to the one in pre-lab 2 with all the chemical tests and compounds tested, including your unknown, in E20A.1 and E20A.3. Write + or −

for positive and negative results, respectively, and ± if your result is doubtful. Indicate N.T. (Not Tested) when necessary.

2. Report the melting point of the semicarbazone of *trans*-cinnamaldehyde.

3. Report the refractive index or your unknown and the melting point of its derivative.

4. Assign as many bands as possible in the IR of your unknown.

5. Report the identity of your unknown. Explain your reasoning in finding out its structure.

6. Interpret the ^{1}H-NMR and ^{13}C-NMR of your unknown.

Synthesis of trans-Cinnamic Acid

In this experiment you will prepare *trans*-cinnamic acid by a **Perkin condensation** between acetic anhydride and benzaldehyde using potassium acetate as a catalyst. You will reflux the reaction mixture for about one hour. The reaction should be performed in a dry environment because water and moisture decompose the reagent. Dry the glassware with a heat pistol (hair dryer) before use. Make sure that the potassium acetate is anhydrous. Keep this reagent in a tightly closed container.

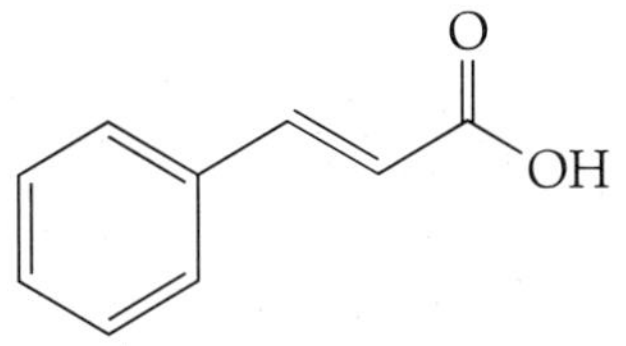

trans-cinnamic acid

At the end of the reflux period some benzaldehyde remains. You will separate the product from the unreacted starting material by acid-base extraction (section 14.4), purify it by recrystallization from mixed solvents (section 4.4), and finally characterize it by its melting point and IR spectrum. You will also interpret its ^{1}H- and ^{13}C-NMR spectra.

trans-Cinnamic acid obtained in this experiment can be used as the starting material to make a compound reminiscent of strawberry aroma: methyl *trans*-cinnamate. The preparation of this compound, an esterification reaction, is described in Experiment 22. You will need about 200 mg of *trans*-cinnamic acid to make methyl *trans*-cinnamate. Check with your instructor to see if you will perform that experiment.

PROCEDURE

> **Background reading:** 20.1–3; 14.4; 4.4
> **Estimated time:** 3 hours

E20B   SYNTHESIS OF *trans*-CINNAMIC ACID: PERKIN CONDENSATION

1. Assemble a reflux apparatus using a 15-mL round-bottom flask equipped with a microscale water-jacketed condenser, a drying tube filled with anhydrous calcium chloride, a stir bar, and a sand bath on a hot plate. The glassware should be perfectly dry. Place a thermometer, secured to a ringstand, in the sand bath (Fig. 20. 1).

2. In the round-bottom flask, weigh 1 mL of benzaldehyde, 0.6 g of potassium acetate (anhydrous), and deliver (in the hood) 1.4 mL of acetic anhydride.

3. Heat the system with stirring for 1 hour. The temperature of the sand bath should be between 170 and 190°C. In the meantime, obtain the IR spectrum of benzaldehyde using Teflon tape or NaCl windows (section 31.5).

4. Turn off the heat and let the system cool down by carefully removing it from the sand bath. Wait for 5 minutes and while still hot, slowly pour the contents of the flask into 40 mL of water in a 100 mL beaker.

Safety First

- Acetic anhydride is an irritant and a lachrymator. Avoid skin contact. Dispense this liquid in the hood.

- Benzaldehyde is a possible mutagen. Avoid skin contact.

- *tert*-Butyl methyl ether is a flammable solvent.

433

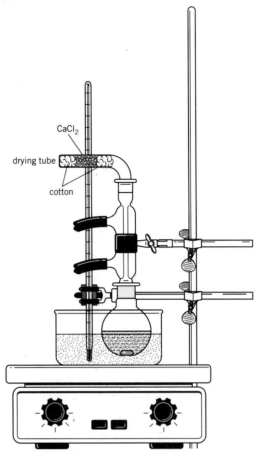

Figure 20.1 Microscale reflux apparatus for Perkin condensation.

Cleaning Up

- Dispose of the BME solution (step 6) in the waste container labeled "Cinnamic Acid-Liquid Waste."

- Dispose of the filtrate from step 8 in the container "Cinnamic Acid-Liquid Waste."

- Dispose of the mother liquor from the recrystallization of *trans*-cinnamic acid in the container labeled "Cinnamic Acid-Liquid Waste."

- Dispose of Pasteur pipets, filter papers, and capillary tubes in the "Cinnamic Acid Solid Waste" container.

- Turn in your *trans*-cinnamic acid in a labeled, capped vial to your instructor, or keep it until Exp. 22.

- Dispose of CDCl$_3$ solution in the container labeled "Chloroform Liquid Waste."

5. Add saturated sodium carbonate solution until the pH is 8–10. Bubbling will occur due to CO$_2$ formation. With the aid of a glass rod break up any precipitate that may appear.

6. After bubbling has subsided, transfer the liquid to a separatory funnel and extract twice with 10 mL of *tert*-butyl methyl ether (BME) each time. The organic layer will finally be discarded.

7. Transfer the aqueous layer to a 250-mL Erlenmeyer flask and acidify with 6 M HCl until the pH = 2. Cool down the system in an ice-water bath, and gently scratch the inner walls of the flask to obtain complete precipitation of *trans*-cinnamic acid.

8. Filter the solid using a Hirsch funnel. Wash it with cold water and dry it by leaving it on the filter with the vacuum on for 10–15 minutes. Weigh the crude product and calculate the % yield. Determine the melting point.

9. Recrystallize the product by the mixed solvent method using a methanol-water mixture. Heat water and methanol in two test tubes in a sand bath (place a glass rod or a boiling chip in each tube to prevent boil-over). Dissolve the product in a test tube using the minimum amount of hot methanol while heating in the sand bath and stirring with a microspatula. Add hot water dropwise. Keep adding hot water until the turbidity persists. At this point, add a few more drops of hot methanol to make the turbidity go away. Remove the test tube from the sand bath and let the product crystallize. Filter the product using a Hirsch funnel.

10. Weigh the product and determine its melting point. Calculate the % yield. Run the IR in nujol (section 31.5). If possible, obtain its ^{1}H-NMR in CDCl$_3$ (section 33.20). (*Caution: CDCl$_3$ is a possible carcinogen. Dispense this chemical in the hood.*)

Experiment 20B   REPORT

Pre-lab

1. Make a table showing the physical properties (MM, b.p. or m.p., density, solubilities, toxicity/hazards) of benzaldehyde, acetic anhydride, potassium acetate, and *trans*-cinnamic acid. Write in the same table the amount (volume and/or mass and mmoles) used for each compound in the preparation of cinnamic acid.

2. Write a chemical reaction for the synthesis of *trans*-cinnamic acid from benzaldehyde and acetic anhydride. Which is the nucleophilic species in the reaction? How is it formed? What role does potassium acetate play?

3. Draw a flowchart for the synthesis, purification, and characterization of *trans*-cinnamic acid.

4. Why is the *trans* isomer the preferred product of the Perkin condensation?

5. In the isolation of *trans*-cinnamic acid: Why do you add sodium carbonate (step 5)? Why do you extract with BME (step 6)? Why is the solution acidified with HCl (step 7)?

6. What would happen if instead of using the minimum amount of hot methanol, a large excess of this solvent were used in the recrystallization?

In-lab

1. Calculate the % yield of *trans*-cinnamic acid. Report its melting point (before and after recrystallization). Discuss any source of product loss.

2. Interpret the IR of *trans*-cinnamic acid and compare it with the one shown in Figure 20.2.

3. Interpret the IR of benzaldehyde. Which bands would you use to tell benzaldehyde and *trans*-cinnamic acid apart?

4. Interpret the ^{1}H-NMR of *trans*-cinnamic acid shown in Figure 20.3. Compare it with the one you obtained (if applicable). Calculate the expected chemical shifts and assign the NMR peaks to the corresponding hydrogens. Present your data in a table format.

5. How can you use ^{1}H-NMR to distinguish between *trans* and *cis* hydrogens?

6. Interpret the ^{13}C-NMR of *trans*-cinnamic acid shown in Figure 20.4.

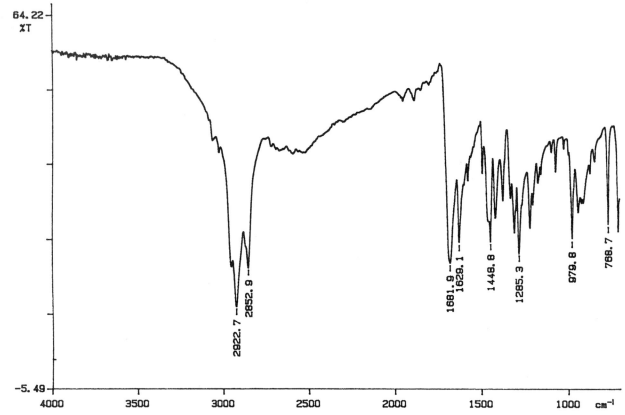

Figure 20.2   IR spectrum of *trans*-cinnamic acid (nujol).

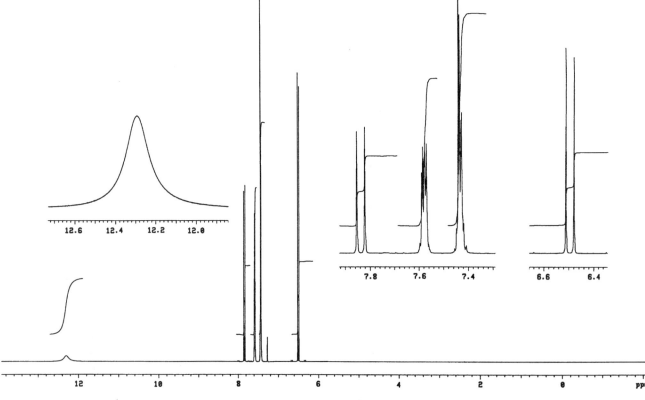

Figure 20.3   500-MHz ^{1}H-NMR spectrum of *trans*-cinnamic acid in CDCl$_3$.

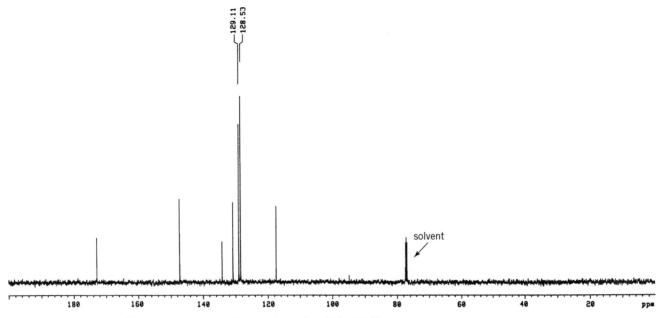

Figure 20.4   125.7-MHz ^{13}C-NMR spectrum of *trans*-cinnamic acid in CDCl$_3$.

UNIT 21

Oxidation-Reduction

21.1 OVERVIEW

Some of the most important transformations between functional groups involve oxidations and reductions. Before we consider the experimental details of this type of reaction, let's review some basic concepts.

In the most general sense an **oxidation** implies an electron loss and a **reduction** an electron gain. In order to decide whether an organic reaction is a reduction or an oxidation one should consider the oxidation states of the carbon atoms involved in the transformation. The oxidation state of a carbon atom is determined by the electronegativity of the atoms attached to it. Elements more electronegative than carbon (oxygen, nitrogen, halogens, etc.) pull the bond electrons toward themselves and leave the carbon atom with a positive charge density. Each bond between an electronegative element and a carbon atom contributes with $+1$ to the oxidation state of the carbon atom. On the other hand, elements more electropositive than carbon such as hydrogen and metals (Li, Na, Mg, etc.) have their electrons pulled by the carbon atom, and thus they contribute with -1 to the oxidation state of the carbon atom. Finally, the bond shared by two carbon atoms is not polarized, and the contribution to the oxidation state is 0. The examples below illustrate these concepts. The oxidation state of the carbon atom in methane is -4, in ethane -3, and in ethylene -2.

methane	ethane	ethylene
-4	-3	-2

The carbon atom bearing the functional group has an oxidation state of -1 in primary alcohols, $+1$ in aldehydes, $+2$ in ketones, and $+3$ in carboxylic acids, esters, and acid halides.

a primary alcohol	an aldehyde	a ketone	a carboxylic acid	an acid chloride
-1	$+1$	$+2$	$+3$	$+3$

An oxidation reaction increases the oxidation state of the carbon atom while a reduction diminishes it. A corollary of this is that, in general, an *increase in the number of hydrogens attached to a carbon is a reduction (a decrease in that number is an oxidation), while an increase in the number of oxygens, nitrogens, and halogens is an oxidation (a decrease is a reduction).* For example, the conversion of a primary alcohol into a carboxylic acid is an oxidation because two hydrogen atoms are replaced by oxygen; the reverse reaction is a reduction.

a primary alcohol   **a carboxylic acid**

There are countless methods and reagents to perform oxidations and reductions and a comprehensive treatment of the subject is beyond the purpose of this unit. Here, we will see the uses, applications, and limitations of some important reagents. A more thorough discussion of the issue can be found in References 1–4.

21.2 SELECTIVITY

Some of the reagents used for oxidations and reductions are very general and react with a wide range of functional groups; others are more selective. Let's now review some definitions about selectivity. A reaction is **chemoselective** if one type of functional group reacts preferentially in the presence of others. For example, the reaction of sodium bicarbonate with a carboxylic acid in the presence of a phenolic –OH group is chemoselective: only the –COOH group reacts with sodium bicarbonate because the phenolic –OH group is not acidic enough to react with such a weak base.

A reaction is **regioselective** when it can afford two or more constitutional isomers, but gives one preferentially. For example, the bromination of acetanilide gives 4-bromoacetanilide as the predominant product with negligible amounts of the other two possible isomers (2-bromo and 3-bromoacetanilide). This reaction is highly regioselective.

A reaction is **stereoselective** if it produces more of one stereoisomer than of the other, for example, *cis* versus *trans*, *E* versus *Z*, or *S* versus *R*. For instance, the elimination of 1-bromo-1,2-diphenylethane with a strong base is stereoselective because it gives *trans*-stilbene as the major product.

1-bromo-1,2-diphenylethane   *trans*-stilbene   *cis*-stilbene
                              major product   minor product

A reaction is **stereospecific** if different stereoisomeric substrates give different stereoisomeric products. For example, when (*R*)-2-bromobutane reacts with HO⁻ under S_N2 conditions, (*S*)-2-butanol is obtained. Conversely, (*S*)-2-bromobutane gives (*R*)-2-butanol. These reactions are stereospecific. We will see more examples of stereoselective and stereospecific reactions in Unit 28.

(*R*)-2-bromobutane   (*S*)-2-butanol

21.3 OXIDATIONS

Transition Metals in High Oxidation States

A list of reagents used to perform oxidation reactions is given in Table 21.1. A first glimpse reveals that many of the oxidizing agents are transition metals in high oxidation states such as Mn(VII) and Cr(VI). They all differ in oxidation power and selectivity.

One of the most important oxidation reactions in organic chemistry is the transformation of alcohols into carbonyl compounds. Different reagents are available to perform this task. For example, the oxidation of a primary alcohol with sodium dichromate leads to a carboxylic acid (Table 21.1, entry 1). An intermediate aldehyde forms but it is difficult to isolate and is quickly oxidized by the reagent. To stop the oxidation at the aldehyde stage, pyridinium chlorochromate (PCC), a milder oxidizing agent, should be used instead (Table 21.1, entry 2).

Although these reagents are very effective, and the reactions proceed with very high yields, chromium compounds must be handled with special precautions. Compounds of Cr(VI) (CrO_3 and $K_2Cr_2O_7$) are not only very corrosive, particularly in the presence of sulfuric acid, but they are also carcinogens. In addition, compounds of Cr(III), a byproduct of the reaction, are very toxic to fish and create a waste disposal problem. For these reasons, organic chemists found safer alternatives. Among them,

Table 21.1 **Most Common Oxidizing Agents and their Applications**

	Oxidizing agent	Main application	Comments
1.	CrO_3, H_2SO_4 chromium (VI) oxide or Na_2CrO_4, H_2SO_4 sodium chromate or $Na_2Cr_2O_7$, H_2SO_4 sodium dichromate	$R\!-\!CH_2OH \longrightarrow R\!-\!CO_2H$ $\underset{\displaystyle R-CH-R}{\overset{\displaystyle OH}{}} \longrightarrow \underset{\displaystyle R-C-R}{\overset{\displaystyle O}{}}$	The oxidation of 1° alcohols can be stopped at the aldehyde stage by distilling off the aldehyde as it forms. Only for aldehydes with boiling point below 100°C. A solution of CrO_3 and H_2SO_4 in water is called **Jones' reagent.** CrO_3 is highly toxic and a carcinogen. High waste disposal cost.
2.	[pyridinium ring with $N^+\!-\!H$] CrO_3Cl^- pyridinium chlorochromate (PCC)	$R\!-\!CH_2OH \longrightarrow R\!-\!CHO$	Oxidation of 1° alcohols stops at the aldehyde stage. Suspect carcinogen. High waste disposal cost.
3.	$KMnO_4$ potassium permanganate HO^-, cold	[alkene → 1,2-diol with OH OH] $R\!-\!C\!\equiv\!C\!-\!R \longrightarrow R\!-\!\overset{O}{\underset{}{C}}\!-\!\overset{O}{\underset{}{C}}\!-\!R$	Reactions must be performed under controlled conditions (basic pH, cold and dilute solutions) to avoid overoxidation.
4.	$KMnO_4$ potassium permanganate heat	[alkene → two C=O products] $R\!-\!CH_2OH \longrightarrow R\!-\!CO_2H$ $\underset{\displaystyle R-CH-R}{\overset{\displaystyle OH}{}} \longrightarrow \underset{\displaystyle R-C-R}{\overset{\displaystyle O}{}}$ [benzene-CHR_2 → benzene-CO_2H]	Nonselective oxidation. Many functional groups are oxidized (alkenes, alkynes, alcohols, aldehydes, etc.). The final products are usually carboxylic acids or ketones. Aromatic ring side chains with at least one benzylic hydrogen are oxidized to a carboxylic acid group.
5.	OsO_4 osmium tetraoxide	[alkene → 1,2-diol with OH OH]	Selective oxidation of alkenes. Reaction stops at the 1,2-diol. OsO_4 is highly toxic.
6.	MnO_2 manganese dioxide	$R\!-\!CH\!=\!CH\!-\!CH_2OH \longrightarrow$ $R\!-\!CH\!=\!CH\!-\!CHO$ $Ar\!-\!CH_2OH \longrightarrow Ar\!-\!CHO$	Selective oxidation of benzylic and allylic alcohols to aldehydes and ketones.
7.	HIO_4 periodic acid	[1,2-diol with OH OH] $\longrightarrow 2\ \overset{\displaystyle R}{\underset{\displaystyle R}{C}}\!=\!O$	Oxidation of 1,2-diols to aldehydes and ketones. R = alkyl, aryl, hydrogen.

(*continued*)

Table 21.1 (continued)

Oxidizing agent	Main application	Comments
8. $R-\underset{\underset{O}{\|\|}}{C}-OOH$ peroxyacids		Among the most common peroxyacids: Magnesium monoperoxyphthalate (MMPP)
9. O_3 ozone followed by Zn/AcOH		Aldehydes and ketones are formed
10. O_3 ozone followed by H_2O_2		Ketones and caroboxylic acids are formed. If both substituents on the alkene carbon are H, CO_2 is formed.
11. SeO_2 selenium dioxide		Oxidation of CH_2 (next to aromatic rings or C=O) to carbonyl. Oxidation of allylic C–H to C–OH.
12. Ag_2O/NH_3 $(Ag(NH_3)_2{}^+)$ Tollens' reagent	$R-CHO \longrightarrow R-CO_2H$	Mild oxidizing agent. Only aldehydes are oxidized to carboxylic acids
13. NaClO sodium hypochlorite (bleach)		Oxidation of secondary alcohols to ketones.

the use of sodium hypochlorite (bleach) in acetic acid is particularly convenient for the oxidation of secondary alcohols to ketones.

Sodium hypochlorite oxidizes secondary alcohols to the corresponding ketones (Table 21.1, entry 13). The reaction proceeds quickly and in good yields. The mechanism of the reaction has not been completely elucidated but it seems to involve the formation of an intermediate alkyl hypochlorite ester, which then decomposes to give the ketone and chloride anion. How the alkyl hypochlorite ester forms is still unclear.

an alkyl hypochlorite ester   a ketone

The byproduct of this reaction is sodium chloride, which does not pose environmental hazards. A potential drawback of this reagent is the presence of Cl_2, a potent irritant, in the hypochlorite solution. However, its concentration can be kept below physiologically detectable limits by working with dilute solutions.

Oxidation of Benzylic and Allylic Positions

The oxidation of benzylic and allylic alcohols to aldehydes and ketones can be achieved with a great degree of chemo and regioselectivity by the use of solid manganese dioxide, MnO_2 (Table 21.1, entry 6). The oxidizing power of this reagent depends on its particle size and degree of hydration. The oxidation reaction takes place on the surface of the oxide and probably involves the formation of radicals as intermediates. The byproduct of the reaction is a reduced form of manganese (MnO). Solvents such as petroleum ether, acetone, and methylene chloride, in which the oxide is totally insoluble, are normally used to carry out these oxidations. Saturated alcohols are also oxidized by MnO_2, but at a much slower rate than benzylic or allylic alcohols, making this reagent chemoselective for these types of alcohol. Primary benzylic and allylic alcohols are converted into aldehydes and secondary ones into ketones.

trans-cinnamyl alcohol   trans-cinnamaldehyde

Although manganese compounds are also environmental hazards, they are not as toxic as chromium compounds and are safer to handle.

Oxidation of Aromatic Side Chains

Alkyl groups attached to an aromatic ring can be oxidized to a carboxyl group by treatment with hot potassium permanganate, which is reduced to MnO_2 (Table 21.1, entry 4). The reaction has little selectivity, and in general, all side chains are oxidized regardless of length and degree of substitution, provided that there is at least one hydrogen atom on the carbon attached to the aromatic ring. tert-Butyl groups are not oxidized under these conditions.

21.4 PHASE TRANSFER CATALYSIS

Many of the reagents used for oxidation reactions are inorganic salts, such as $KMnO_4$ and $NaClO$, which are soluble in water but insoluble in organic solvents. On the other hand, most organic substrates, with the exception of small molecules with fewer than 6 carbon atoms, are insoluble in water but dissolve readily in organic solvents. How can an oxidation reaction be carried out if there is no good solvent for both the substrate and the oxidizing agent? One approach is to dissolve the inorganic salt in water and add this solution to a solution of the substrate in acetone. This is how Jones' reagent (a dilute aqueous solution of CrO_3 and H_2SO_4, Table 21.1, entry 1) is normally used. Another alternative is the use of **phase transfer catalysts**.

A phase transfer catalyst is a compound that facilitates the passage of inorganic salts from an aqueous solution into an *immiscible organic solvent*. The most common phase transfer catalysts are **quaternary alkylammonium salts**, such as **tetrabutyl-ammonium hydrogen sulfate** and **benzyltrimethylammonium chloride**; they are represented by Q^+X^-.

tetrabutylammonium hydrogen sulfate   benzyltrimethylammonium chloride

Quaternary alkylammonium salts are soluble in both water and organic solvents. They are soluble in water because they are ionic, and they are soluble in organic solvents because they have four nonpolar alkyl groups on the nitrogen atom. The nature of the anion (chloride, bromide, hydrogen sulfate, etc.) has little effect on the solubility of tetraalkylammonium salts.

If sodium hypochlorite, $NaClO$, is dissolved in water and the solution mixed with an immiscible organic solvent such as toluene or methylene chloride, virtually no $NaClO$ goes to the organic phase. If now a small amount of a tetraalkylammonium salt, Q^+X^-, is added, this salt partitions between both phases and some ClO^- passes from the aqueous to the organic phase by virtue of the following equilibrium that takes place in the aqueous phase.

$$Q^+X^- + Na^+ClO^- \rightleftharpoons Q^+ClO^- + Na^+X^-$$

The hypochlorite anion, ClO^-, and X^- trade places as the counter ion of the tetraalkylammonium cation, Q^+. Because the alkylammonium salts are soluble in water and organic solvents, the group Q^+ *carries* the hypochlorite anion from the aqueous phase to the organic phase where it can react with the organic substrate.

$$Q^+ \ ClO^- \rightleftharpoons Q^+ \ ClO^-$$
aqueous phase   organic phase

As the ClO⁻ reacts in the organic phase, more ClO⁻ is transported by the quaternary alkylammonium cation from the aqueous phase to reestablish the equilibrium.

Typically, the substrate is dissolved in methylene chloride, ethyl acetate, or some other organic solvent immiscible with water, and mixed with the oxidizing agent (KMnO₄, Na₂Cr₂O₇, NaClO, etc.) in water. A small amount (less than 10% of the stoichiometric amount) of the tetraalkylammonium salt is added and the mixture vigorously stirred to thoroughly mix both phases. The reaction is usually carried out at room temperature.

The use of phase transfer catalysts is not limited to oxidation reactions. Many other reactions that require the use of inorganic anions, such as nucleophilic substitutions and eliminations, can be dramatically accelerated by the use of tetraalkylammonium salts.

21.5 REDUCTIONS

There are three main ways to perform reductions in the laboratory: **catalytic hydrogenation**, reduction with **metal hydrides**, and reduction with **metals**. Catalytic hydrogenation uses hydrogen gas in the presence of a transition metal catalyst. It is one of the most powerful reductive methods affecting almost all functional groups with the exception of carboxylic acids. The metal is supported on an inert solid material such as carbon, alumina, or calcium sulfate. The nature of the catalyst determines the selectivity of the reaction. Temperature and pressure are also factors that affect the selectivity of the process. For example, the side chain of styrene (vinylbenzene) can be reduced to an ethyl group using Pd/C as a catalyst at low temperature and pressure. Under these conditions the aromatic ring remains intact. On the other hand, if the reduction is carried out in the presence of PtO₂ at elevated pressure, the aromatic ring is also reduced and ethylcyclohexane is obtained.

Catalytic hydrogenation is particularly useful to reduce alkenes and alkynes, which are difficult to reduce with other reducing agents. A drawback in the use of catalytic hydrogenation is the need for hydrogen gas, which is highly flammable and explosive.

The use of metal hydrides as reducing agents is the most widespread reduction technique in the laboratory. Two metal hydrides: **sodium borohydride**, NaBH₄, and **lithium aluminum hydride**, LiAlH₄, have a prominent place among all metal hydrides. Most aldehydes and ketones can be easily reduced to alcohols by treating them with the metal hydride followed by hydrolysis under mild acidic conditions.

Lithium aluminum hydride is a more potent reducing agent than $NaBH_4$. Most functional groups, with the exception of nonconjugated alkenes and alkynes, can be reduced with $LiAlH_4$. On the other hand, $NaBH_4$, a more selective reagent, is useful to reduce aldehydes and ketones to alcohols in the presence of other functional groups, such as carboxylic acids and esters. A list of the applications of different reducing agents is given in Table 21.2.

Table 21.2 **Useful Reducing Agents and their Applications**

Reduction	$NaBH_4$ in ethanol	$LiAlH_4$ in ether	H_2/metal catalyst
$R-CHO \longrightarrow R-CH_2OH$	+	+	+
$R-CO-R \longrightarrow R-\underset{\underset{OH}{\mid}}{CH}-R$	+	+	+
$R-COOH \longrightarrow R-CH_2OH$	−	+	−
$R-COOR' \longrightarrow RCH_2OH + R'OH$	−	+	+[a]
$R-CONR_2 \longrightarrow R-CH_2NR_2$	−	+[b]	+[a]
$R-COCl \longrightarrow R-CHO$	+[c]	+[d]	+
$R-CN \longrightarrow R-CH_2NH_2$	−	+	+
$R-NO_2 \longrightarrow R-NH_2$	−	+[e]	+
alkene reduction	−	−	+
epoxide reduction	−	+	+

[a]High temperature and pressure.
[b]Reduced to aldehyde if amide in excess.
[c]In THF
[d]Reduced to alcohol.
[e]Aromatic nitro groups reduced to azo compounds ($Ar-N=N-Ar$)

All four hydrogen atoms in $LiAlH_4$ and $NaBH_4$ are chemically active. In the reduction of ketones, for example, four carbonyl groups can be reduced with one molecule of $NaBH_4$. The reaction is shown below.

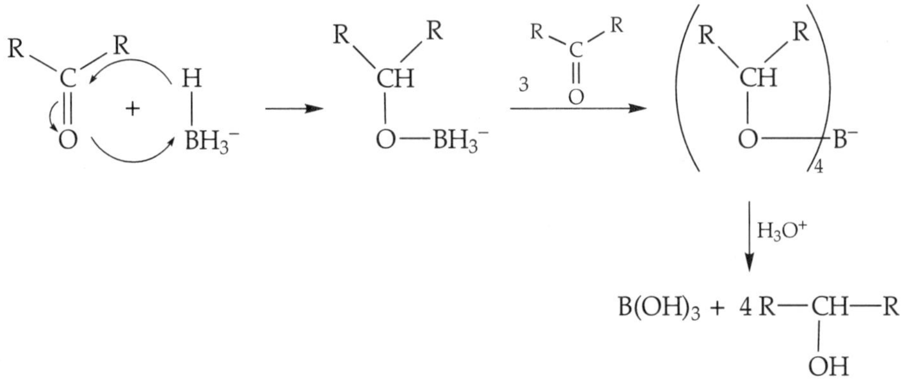

$$B(OH)_3 + 4\ R-\underset{\underset{OH}{\mid}}{CH}-R$$

LiAlH$_4$ is flammable and reacts violently with water liberating hydrogen gas.

$$\text{LiAlH}_4 + 4\,\text{H}_2\text{O} \longrightarrow \text{LiOH} + \text{Al(OH)}_3 + 4\,\text{H}_2$$

In working with LiAlH$_4$ all traces of moisture must be excluded from solvents and glassware, and all work should be done under an atmosphere of nitrogen to prevent fires. LiAlH$_4$ is used in aprotic solvents such as THF or diethyl ether.

Contrary to LiAlH$_4$, NaBH$_4$ can be used in water as a solvent. Aqueous solutions of NaBH$_4$ are very stable at high pH but decompose very quickly in the presence of acid. A disadvantage of using aqueous solutions is the limited solubility of most organic compounds in water. NaBH$_4$ is normally used in ethanol or methanol solutions because these solvents are relatively good solvents for organic compounds. When using methanol as a solvent, an excess of reducing agent should be used because methanol decomposes sodium borohydride rather quickly, producing hydrogen gas. The reagent is more stable in ethanol but its solubility in this solvent is lower (Table 21.3).

Table 21.3 Solubility of NaBH$_4$ and LiAlH$_4$ in Different Solvents at Room Temperature (g/100 mL)

Solvent	NaBH$_4$	LiAlH$_4$
water	55	reacts
methanol	16	reacts
ethanol	4	reacts
THF	0.1	13
diethyl ether	—	35–40

When methanol and ethanol are used as solvents, the final hydrolysis step with H$_3$O$^+$ is not necessary, because the solvent decomposes the tetraalkoxyboron compound to give the final products. This reaction is accelerated by heating.

$$\left(\begin{array}{c} \text{R} \diagdown \text{CH} \diagup \text{R} \\ | \\ \text{O} - \text{B}^- \end{array}\right)_4 \xrightarrow[\Delta]{\text{R'OH }(excess)} 4\,\text{R} - \underset{\overset{|}{\text{OH}}}{\text{CH}} - \text{R} + \text{B(OR')}_4^-$$

Finally, reduction with metals is useful for the transformation of specific functional groups. For instance, Na and Li dissolved in ammonia are used to reduce alkynes to alkenes (normally the *trans* isomer is obtained).

$$\text{R} - \text{C} \equiv \text{C} - \text{R} \xrightarrow{\text{Li/NH}_3} \begin{array}{c} \text{R} \diagdown \quad \diagup \text{H} \\ \text{C} = \text{C} \\ \text{H} \diagup \quad \diagdown \text{R} \end{array}$$

an alkyne   a *trans*-alkene

Heavy metals such as Fe, Zn, and Sn in acidic media are particularly useful to reduce aromatic nitro compounds to anilines. The mechanism of these reactions has not been fully elucidated. The metals are oxidized to a stable cationic form (Fe^{3+}, Zn^{2+}, Sn^{4+}).

nitrobenzene   aniline

When there are two or more nitro groups attached to the ring, the treatment with metals such as Fe and Zn under acidic conditions usually leads to the reduction of all the nitro groups. The reduction of only one nitro group can be achieved with sodium sulfide, (Na_2S), titanous sulfate ($TiSO_4$), or with copper in the presence of HBr. In the latter case copper is oxidized to cuprous bromide (Cu_2Br_2) and separates from solution.

BIBLIOGRAPHY

1. Reductions in Organic Chemistry. M. Hudlicky. 2nd ed. ACS, Washington, D. C., 1996.

2. Advanced Organic Chemistry. Reactions Mechanisms, and Structure. J. March. 4th ed. Wiley, New York, 1992.

3. Oxidations in Organic Chemistry. M. Hudlicky. American Chemical Society, Washington, 1990.

4. Reduction. Techniques and Applications in Organic Synthesis. R. L. Augustine, ed. Marcel Dekker, Inc. New York, 1968.

5. Sodium Borohydride Reduction of Conjugated Aldehydes and Ketones. M. R. Johnson, and B. Rickborn. *J. Chem. Ed.* , **35**, 1041–1045 (1970).

6. Conjugate and Nonconjugate Reduction with $LiAlH_4$ and $NaBH_4$. G. R. Meyer. *J. Chem. Ed.* , **58**, 628–630 (1981).

7. Reduction and Oxidation Tables. M. Hudlicky. *J. Chem. Ed.*, **54**, 100–106 (1977).

8. Reductions by Lithium Aluminum Hydride. W. G. Brown. Organic Reactions. Vol. VI. Wiley, New York, 1951.

9. The Design of Laboratory Experiments in the 1980's. A Case Study on the Oxidation of Alcohols with Household Bleach. J. R. Mohrig, D. M. Nienhuis, C. F. Linck, C. Van Zoeren, B. G. Fox, and P. G. Mahaffy. *J. Chem. Ed.* , **62**, 519–521 (1985).

10. Ketone Synthesis Using Household Bleach. R. A. Perkins, and F. Chau. *J. Chem. Ed.* , **59**, 981 (1982).

11. Oxidation of Cyclohexanol to Cyclohexanone by Sodium Hypochlorite. N. M. Zuczek and P. S. Furth. *J. Chem. Ed.* , **58**, 824 (1981).

EXERCISES

Problem 1

Calculate the oxidation state of the carbon atoms marked with an asterisk:

(a)

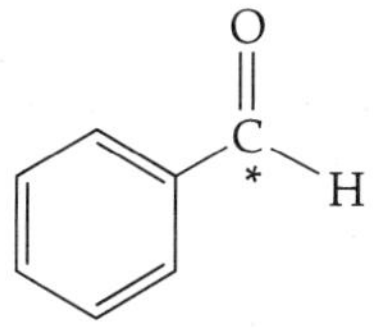

(b)

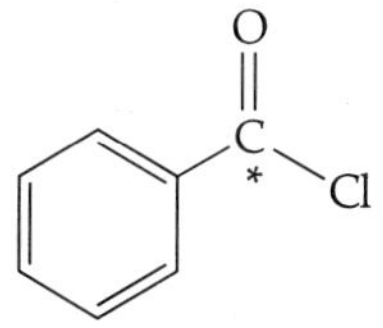

(c)

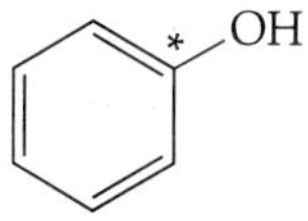

(d)

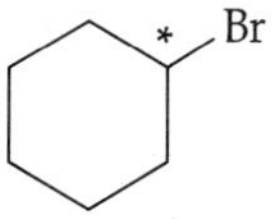

(e)

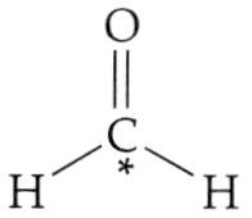

(f)

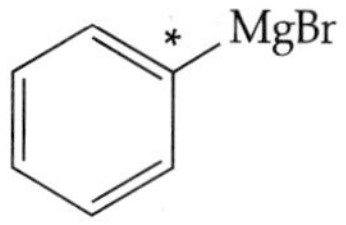

Problem 2

Classify each of the following transformations as oxidation, reduction, both, or neither.

(a)

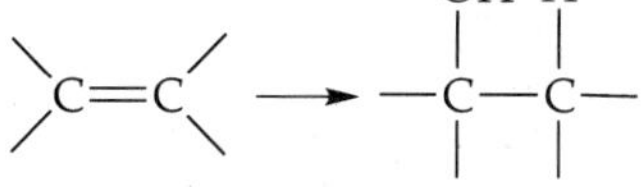

(b)

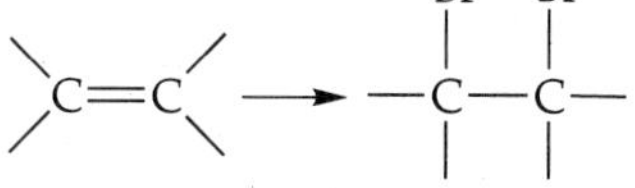

(c)

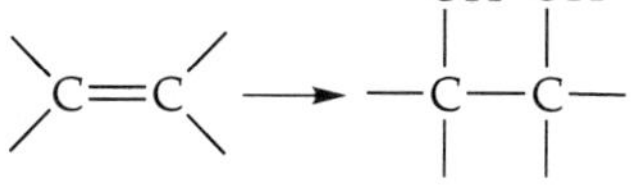

(d)

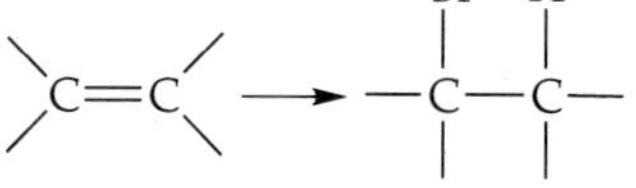

(e)

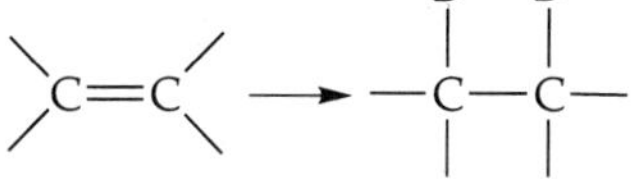

(f)

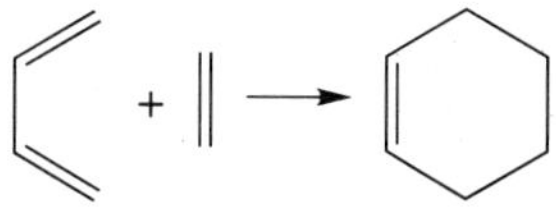

Problem 3

How would you perform the following transformations? Which reagents would you use? Check Tables 21.1 and 21.2.

(a)

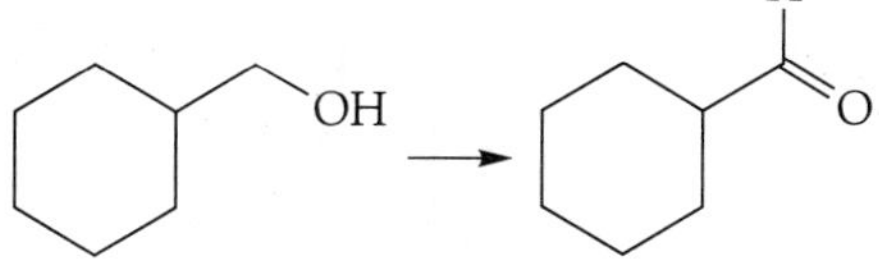

(b)

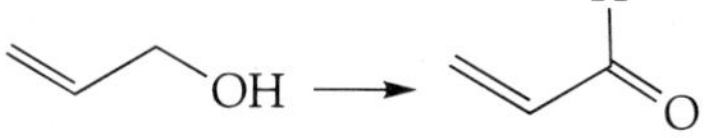

(c)

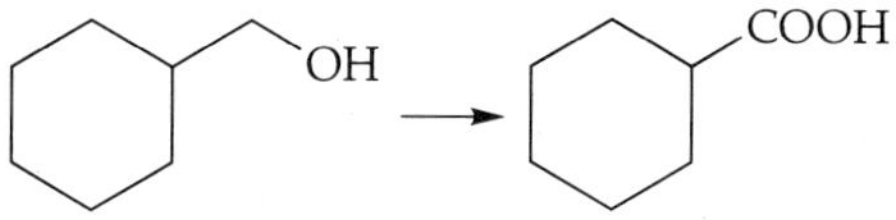

(d)

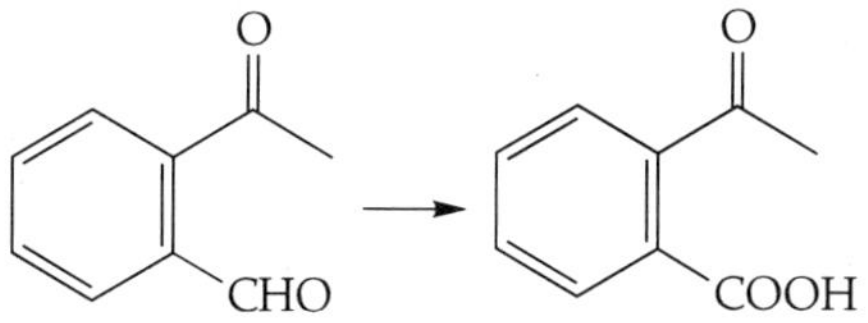

(e)

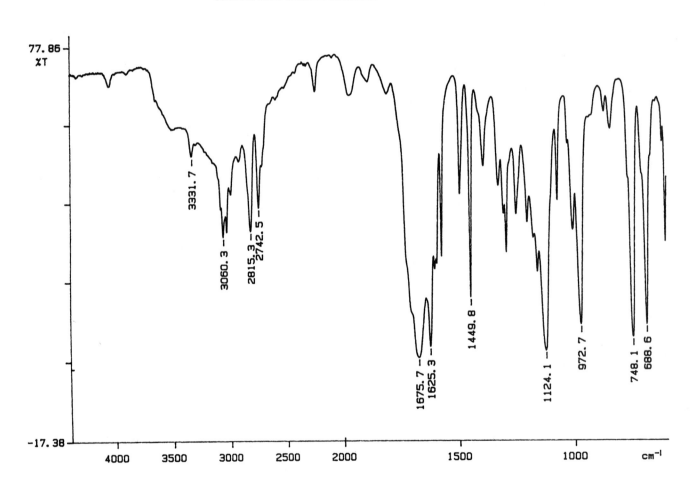

(f)

Problem 4

The partial reduction of 1-phenyl-1,2-propanedione can give two hydroxyketones. Write their structures and explain how you would distinguish both products by IR and ^{1}H-NMR.

Problem 5

(*Previous Synton: Problem 7, Unit 16.*)   Synton is preparing the reagents for the reduction of cinnamyl aldehyde, an experiment that his students will perform in the lab. He finds a bottle of *trans*-cinnamaldehyde that has a white solid deposit around the cap and inside the neck. He runs the IR of the aldehyde and obtains the spectrum shown in the figure. Do you think that the aldehyde is pure? If not, what is the most likely contaminant? Can you speculate how it formed? How can it be removed?

Oxidation-Reduction

In this experiment you will investigate several oxidation and reduction reactions. There are a total of four reductions and four oxidations. You are not expected to do all of them. Talk to your instructor to find out which ones will be available to you.

The first six reactions are organized in pairs. In each pair there are two compounds that can be interconverted by oxidation and reduction reactions. For example, in E21.1 we discuss the reduction of benzophenone to benzhydrol with sodium borohydride. In E21.2 the oxidation of benzhydrol to benzophenone with bleach (sodium hypochlorite) and a phase transfer catalyst is presented. E21.3 deals with the reduction of *trans*-cinnamaldehyde with sodium borohydride in methanol, while E21.4 considers the chemoselective oxidation of *trans*-cinnamyl alcohol to *trans*-cinnamaldehyde with manganese dioxide. In E21.5 you will investigate the reduction of camphor and determine whether borneol or isoborneol is the major product; this is a puzzle-like experiment. In E21.6 you will oxidize isoborneol with calcium hypochlorite to obtain camphor. In E21.7 you will carry out the side chain oxidation of *p*-toluic acid (*p*-methylbenzoic acid) to terephthalic acid by using potassium permanganate. In E21.8 you will perform the reduction of *m*-dinitrobenzene to *m*-nitroaniline.

The course of these reactions will be followed by either TLC or gas chromatography. You will determine when the reaction is complete and can be stopped. To this end, you will analyze standards of your starting material and product. If you team up with a classmate and choose an oxidation-reduction pair, the analysis will be easier. One student should do the oxidation while the other performs the corresponding reduction. Each student should work individually but both students should compare notes, especially in regard to the chromatographic analyses.

> **Important Note**
>
> In this experiment you and your classmates will be handling oxidizers and reducing agents. These two types of compounds should not be mixed or violent reactions may occur. Dispose of all waste following the cleaning-up instructions at the end of E21.4 and E21.8. Read carefully the label of each waste disposal container before you pour in your waste.

PROCEDURE

> **Background reading:** 21.1–5
> **Estimated time:** 2–3 hr for each reaction

E21.1   REDUCTION OF BENZOPHENONE

benzophenone   benzhydrol

In a 25-mL Erlenmeyer flask equipped with a magnetic stir bar, dissolve 250 mg of benzophenone in 5 mL of methanol. Add 0.13 g sodium borohydride little by little with a microspatula while the reaction mixture is stirred in an ice-water bath. This reaction produces flammable hydrogen gas and should be carried out in a well-ventilated fume hood.

> **Safety First**
>
> - Sodium borohydride is a flammable solid. It is decomposed by protic solvents and produces flammable hydrogen gas. Handle this chemical in the fume hood. It should be stored in plastic containers.
> - Benzophenone and benzhydrol are irritants. Methanol is toxic and flammable.
> - Ethyl acetate, petroleum ether, and *t*-butyl methyl ether are flammable.

After 10 minutes, remove a small aliquot (a few drops) with a Pasteur pipet and place it in a small test tube with 0.5 mL of water and 0.5 mL of ethyl acetate. Shake the test tube well and analyze the upper (ethyl acetate) layer by TLC using silica gel plates with fluorescence indicator and petroleum ether/t-butyl methyl ether (10:1) as a solvent. On the same plate, spot standard samples of benzophenone and benzhydrol. Visualize the plates with UV light. If the reaction is not complete (benzophenone is still present in the reaction mixture), continue the stirring for an additional 15-minute period and analyze the reaction mixture again by TLC. Keep repeating the analysis and stirring until the reaction is complete and no benzophenone spot is detected by TLC analysis.

Pour the liquid in a beaker with 20 mL of ice-water. Collect the solid by vacuum filtration. Recrystallize the product from petroleum ether and determine its melting point. Weigh the product and calculate the percentage yield of the synthesis.

Obtain the IR of the product and starting material using nujol (section 31.5). Compare the spectra.

Cleaning Up

- Dispose of the filtrate in the container labeled "Sodium Borohydride Waste."
- Dispose of the petroleum ether from the recrystallization step in the container labeled "Organic Solvents Waste."
- Dispose of the solvents from the TLC analysis in the container labeled "Organic Solvents Waste."
- Dispose of Pasteur pipets, filter papers, capillary tubes, and TLC plates in the container labeled "Oxidation-Reduction Solid Waste."
- Turn in your products in labeled screw-cap vials.

E21.2 OXIDATION OF BENZHYDROL

benzhydrol →[NaClO, Q⁺X⁻, ethyl acetate/water]→ benzophenone

In a 25-mL Erlenmeyer flask equipped with a magnetic stir bar, add 0.37 g of benzhydrol, 5 mL of commercial bleach (approximately 0.7 M NaClO), 5 mL of ethyl acetate, and 40 mg of tetrabutylammonium hydrogen sulfate. Secure the flask to a ringstand, stopper it, and stir the reaction mixture at high speed on a magnetic stirrer.

After about 15 minutes remove a small aliquot (a few drops) of the upper layer with a Pasteur pipet, transfer it to a small test tube, and analyze it by TLC using a silica gel plate with fluorescence indicator and petroleum ether/t-butyl methyl ether (10:1) as a solvent. On the same plate, spot standard samples of benzophenone and benzhydrol.

If the reaction is not complete and benzhydrol is still present in the reaction mixture, continue the stirring for another 15 minutes, and repeat the analysis. Keep repeating the analysis and the stirring until the reaction is complete and no benzhydrol is left in the reaction mixture.

Transfer the reaction mixture to a 16 × 125 mm screw-cap test tube and remove the lower aqueous layer with a Pasteur pipet. Wash the organic layer with 3 mL of a saturated aqueous solution of sodium chloride followed by a wash with 2 mL of water. Dry the organic layer over anhydrous magnesium sulfate. Gravity-filter it to separate the drying agent, collecting the filtrate in a pre-tared 25-mL round-bottom flask. Evaporate the solvent using a rota-vap with a boiling-water bath. Weigh the product. Recrystallize the product from hexanes, determine its melting point, and calculate its percentage yield.

Obtain the IR of the product and starting material using nujol (section 31.5). Compare the spectra (adapted from Ref. 1).

Cleaning Up

- Dispose of the solvents from the TLC analysis in the container labeled "Organic Solvents Waste."
- Dispose of the aqueous layers in the container "Hypochlorite Waste."
- Dispose of magnesium sulfate, Pasteur pipets, filter papers, capillary tubes, and TLC plates in the container labeled "Oxidation-Reduction Solid Waste."
- At the end of the lab period the instructor will empty the contents of the rota-vap traps into the container "Organic Solvents Waste."
- Turn in your products in labeled screw-cap vials.

E21.3   REDUCTION OF *trans*-CINNAMALDEHYDE

trans-cinnamaldehyde → (NaBH$_4$) → *trans*-cinnamyl alcohol

Safety First

- Sodium borohydride is a flammable solid. It is decomposed by protic solvents and produces flammable hydrogen gas. Handle this chemical in the fume hood. It should be stored in plastic containers.
- *trans*-Cinnamaldehyde and *trans*-cinnamyl alcohol are irritants. Methanol is toxic.
- Ethyl acetate, methanol, petroleum ether, and *t*-butyl methyl ether are flammable.

In a 15-mL round-bottom flask equipped with a magnetic stir bar, place 250 μL of *trans*-cinnamaldehyde and 4 mL of methanol. Secure the flask to a ringstand and place it in an ice-water bath. Add 0.13 g of sodium borohydride little by little with a microspatula. Stir the reaction mixture on a magnetic stirrer. This reaction produces flammable hydrogen gas and should be carried out in a well-ventilated fume hood.

After about 10 minutes take a small aliquot (a few drops) of the reaction mixture with a Pasteur pipet and transfer it to a small test tube with 0.5 mL of water and 0.5 mL of ethyl acetate. Shake well and analyze the ethyl acetate layer by TLC using silica gel plates with fluorescence indicator. Use petroleum ether/*t*-butyl methyl ether (10:1) as a solvent. On the same plate, spot standard samples of *trans*-cinnamaldehyde and *trans*-cinnamyl alcohol. If the reaction is not complete and the *trans*-cinnamaldehyde spot is still visible in the reaction mixture lane, continue the stirring for an additional 10-minute period, and repeat the TLC analysis.

Evaporate the methanol in a rota-vap. Add 2 mL of water to the residue and about 5 mL of *t*-butyl methyl ether (BME). Transfer the liquid to a 16 × 125 mm screw-cap test tube. Rinse the round-bottom flask with 1 mL of water and 2 mL of BME, and transfer the rinse to the screw-cap test tube. Cap the tube, shake, and vent. Remove the aqueous layer using a Pasteur pipet. Wash the organic layer with 2 mL of a saturated solution of NaCl in water. Dry the organic layer with anhydrous magnesium sulfate. Remove the drying agent by gravity filtration. Collect the filtrate in a pre-tared 25-mL round-bottom flask. Evaporate the solvent in a rota-vap.

Weigh the product and calculate the percentage yield. The product solidifies upon cooling.

Obtain the IR of the starting material and the product as neat liquids on NaCl plates (section 31.5).

Cleaning Up

- Dispose of the solvents from the TLC analysis in the container labeled "Organic Solvents Waste."
- Dispose of the aqueous layers in the container labeled "Sodium Borohydride Waste."
- Dispose of any magnesium sulfate, Pasteur pipets, filter papers, TLC plates, and capillary tubes in the container labeled "Oxidation-Reduction Solid Waste."
- At the end of the lab period the instructor will empty the contents of the rota-vap traps into the container "Organic Solvents Waste."
- Turn in your products in labeled screw-cap vials.

E21.4   OXIDATION OF *trans*-CINNAMYL ALCOHOL

Safety First

- Manganese dioxide is an oxidizer.
- *trans*-Cinnamaldehyde and *trans*-cinnamyl alcohol are irritants.
- Hexanes, petroleum ether, and *t*-butyl methyl ether are flammable.

---

trans-cinnamyl alcohol $\xrightarrow[\text{hexanes}]{\text{MnO}_2}$ *trans*-cinnamaldehyde

In a perfectly dry 15-mL round-bottom flask (to make sure the flask is dry use a piece of twisted paper towel), equipped with a stir bar, place 250 µL of *trans*-cinnamyl alcohol and 3 mL of hexanes. (If the alcohol is solidified, open the container and place it in a warm-water bath until the solid melts, m.p. 33°C.) Add 0.3 g of manganese dioxide, (activated, <5 micron) cap the flask with a septum, and insert a needle (with blunt tip) through the septum to vent the system. Stir the reaction on a magnetic stirrer for about 15 minutes.

At the end of this period take a few drops of the reaction mixture with a Pasteur pipet and dilute them in about 0.5 mL of hexanes. Filter the suspension using a filter made with a Pasteur pipet and a small cotton plug. Analyze this mixture by TLC using silica gel plates (with fluorescence indicator) and petroleum ether/*t*-butyl methyl ether (10 : 1) as a solvent. Spot on the same plate standard samples of *trans*-cinnamaldehyde and *trans*-cinnamyl alcohol dissolved in hexanes. If the reaction is not complete and the *trans*-cinnamic alcohol spot is still visible in the reaction mixture lane, continue the stirring for an additional 15 minutes, and repeat the analysis at the end of this period. Continue the reaction until all *trans*-cinnamyl alcohol is consumed according to the TLC analysis.

Add 2 mL of hexanes to the reaction mixture and filter the suspension using a Pasteur pipet with a cotton plug. Collect the liquid in a pre-tared 25-mL round-bottom flask and evaporate the solvent in a rota-vap. Weigh the product and calculate the percentage yield.

Obtain the IR of the starting material and the product as neat liquids using NaCl plates (section 31.5).

Cleaning Up

- Dispose of the solvents from the TLC analysis in the container labeled "Organic Solvents Waste."
- Dispose of the Pasteur pipets with manganese dioxide in the container labeled "Manganese Dioxide Waste."
- Dispose of Pasteur pipets, TLC plates, and capillary tubes in the container labeled "Oxidation-Reduction Solid Waste."
- At the end of the lab period the instructor will empty the contents of the rota-vap traps into the container "Organic Solvents Waste."
- Turn in your products in labeled screw-cap vials.

E21.5 REDUCTION OF CAMPHOR

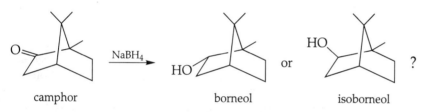

| camphor | | borneol | | isoborneol | |

In a 15-mL round bottom flask equipped with a stir bar, place 0.11 g of (1R)-camphor and add 2 mL of methanol. Secure the flask to a ringstand, place the flask in a water bath, and start the stirring on a magnetic stir plate. Add 0.14 g of sodium borohydride. This reaction produces flammable hydrogen gas and should be carried out in a well-ventilated fume hood.

After 20 minutes of reaction at room temperature, add 4 mL of water to the flask, and collect the white solid by vacuum filtration using a Hirsch funnel. Weigh the product. Obtain its melting point and calculate its percentage yield.

Obtain the IR of the product and the starting material using a nujol mull (section 31.5).

Obtain the ^{1}H-NMR of the product and the starting material using $CDCl_3$ as a solvent (section 33.20). Add a few drops of D_2O and rerun the spectrum. Compare with the ^{1}H-NMR spectrum of borneol and isoborneol from the literature and determine if the product is borneol or isoborneol.

Safety First

- Sodium borohydride is a flammable solid. It is decomposed by protic solvents and produces flammable hydrogen gas. Handle this chemical in the fume hood. It should be stored in plastic containers.
- Camphor and borneol are flammable solids. Borneol is toxic.
- $CDCl_3$ is a possible carcinogen. Handle this chemical in the fume hood.

Cleaning Up

- Dispose of the aqueous filtrate in the container labeled "Reduction of Camphor Liquid Waste."
- Dispose of Pasteur pipets, filter paper, and capillary tubes in the container labeled "Oxidation-Reduction Solid Waste."
- Dispose of $CDCl_3$ in the container labeled "$CDCl_3$-Waste."
- Turn in your products in labeled screw-cap vials.

E21.6 OXIDATION OF ISOBORNEOL

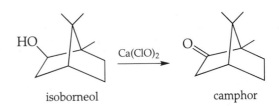

isoborneol   camphor

In a 25-mL Erlenmeyer flask equipped with a stir bar, place 0.3 g of isoborneol and 2.0 mL of a mixture of acetonitrile and acetic acid (3 : 2). Secure the flask to a ringstand and place it in an ice-water bath on a magnetic stirrer. Start the stirring and add dropwise, with the help of a Pasteur pipet, 3.0 mL of a solution of calcium hypochlorite in water (5% w/v). Stir the mixture for about 30 minutes.

In the meantime inject standards of isoborneol and camphor in a GC with a polar column such as Carbowax 20 M (135°C).

Remove a small aliquot (a few drops) and place it in a small test tube with *tert*-butyl methyl ether (BME) (0.5 mL) and water (0.5 mL). Shake well, separate the layers with a Pasteur pipet (section 5.4), and transfer the organic layer to a scintillation vial. Evaporate the solvent by blowing a gentle stream of nitrogen until a few drops remain. Perform this operation in the fume hood. Inject the sample in a gas chromatograph. Compare the GC trace from the reaction mixture with those from the standards. If the reaction did not reach completion, and a peak for isoborneol is observed, continue the stirring of the reaction mixture for another 30 minutes and repeat the analysis at the end of this period.

When the reaction is complete, transfer the reaction mixture to a screw-cap test tube, and add 3.5 mL of water and 3.5 mL of BME; cap, shake, and vent. Separate the layers with a Pasteur pipet. Save the organic layer and extract the aqueous layer twice with a 2.5 mL portion (each time) of BME. Wash the combined BME layer with 3 mL of 3% sodium hydroxide aqueous solution, and finally with 3 mL of saturated sodium chloride solution. Dry the organic layer over anhydrous magnesium sulfate. Remove the drying agent by gravity filtration, collecting the liquid in a pre-tared round-bottom flask. Evaporate the solvent in a rota-vap. Weigh the product and calculate its percentage yield. Determine its melting point.

Obtain the IR of the product and the starting material using nujol (section 31.5). Obtain the ¹H-NMR of the product and the starting material using CDCl₃ as a solvent (section 33.20; adapted from Refs. 2 and 3).

E21.7 SYNTHESIS OF TEREPHTHALIC ACID

COOH

$\xrightarrow[\text{2. } H_3O^+]{\overset{\text{1. } KMnO_4, HO^-}{heat}}$

COOH

p-toluic acid   terephthalic acid

In a 15-mL round-bottom flask equipped with a magnetic stir bar, a Claisen head, and a microscale reflux condenser, weigh 0.3 g of *p*-toluic acid (*p*-methylbenzoic acid), and add 3 mL of water and 0.21 g of sodium hydroxide. Cap the Claisen head with a septum and plastic screw cap as shown in Figure 21.1a. Heat the mixture to a boil directly on a hot plate at a medium setting. Turn off the heat momentarily. With the help of a syringe and a needle (with blunt tip) inserted in the septum (Fig. 21.1b), add little by little 6 mL of a solution of potassium permanganate (15% w/v) at a rate such that the boiling of the mixture is maintained without external heating. Remove the syringe and needle and reflux the mixture by heating for 30 minutes.

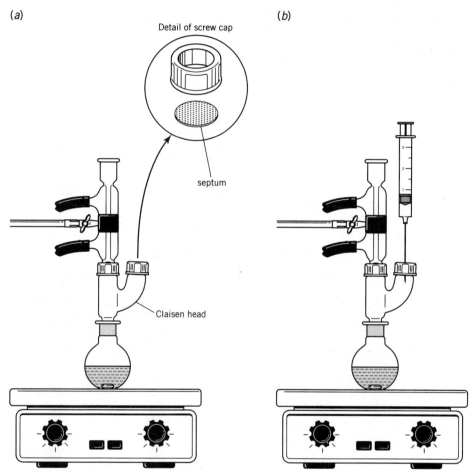

(a)   (b)

Detail of screw cap

septum

Claisen head

Figure 21.1   Reflux apparatus with Claisen head.

At the end of this period let the system cool down, remove the screw-cap from the Claisen head and with a Pasteur pipet add 1.2 mL of ethanol to destroy any excess of permanganate. Separate the manganese dioxide by vacuum filtration (avoid skin contact!) and wash it on the filter with 1.2 mL of hot water. Add 3 mL of 6 M hydrochloric acid to the flask to precipitate the product. Cool down in an ice-water bath and collect the product by vacuum filtration. Wash the product on the filter with small portions of cold water. Dry the product on a porous plate and weigh it. Calculate the percentage yield of the synthesis.

Do not attempt to determine the melting point because terephthalic acid sublimes at 300°C without melting. Obtain the IR of the product using a nujol mull (section 31.5). If possible, obtain the ^{1}H-NMR using DMSO-d$_6$ as a solvent (section 33.20; adapted from Ref .4).

Cleaning Up

- Dispose of the manganese dioxide in the container labeled "Manganese Dioxide Waste."
- Dispose of the aqueous filtrate in the container labeled "E21.7-E21.8 Liquid Waste."
- Dispose of the DMSO-d$_6$ in the container labeled "E21.7-E21.8 Liquid Waste."
- Dispose of Pasteur pipets and filter paper in the container labeled "Oxidation-Reduction Solid Waste."
- Turn in your products in labeled screw-cap vials.

E21.8   MONOREDUCTION OF *m*-DINITROBENZENE

In a 25-mL Erlenmeyer flask, weigh 0.4 g of *m*-dinitrobenzene. Add 4 mL of methanol, 1.8 mL of concentrated hydrobromic acid (48%, w/w), and a stir bar. Weigh on a piece of weighing paper 1.0 g of copper powder (200 mesh). Add the finely divided copper powder to the flask over a 15-minute period. Stir the mixture at room temperature for one extra hour.

Filter the suspension under vacuum using a Buchner funnel. Transfer the filtrate to a 50-mL Erlenmeyer flask and add 20 mL of 6 M aqueous HCl. Cool the mixture in an ice-water bath, stirring with a glass rod to crystallize the unreacted *m*-dinitrobenzene.

Remove the unreacted starting material by vacuum filtration. Transfer the filtrate to a 250-mL beaker and neutralize it with concentrated ammonium hydroxide. Perform this operation in the fume hood. Extract the aqueous layer twice with 15 mL of toluene each time. Dry the combined organic layer with anhydrous magnesium sulfate. Filter it and evaporate the solvent in a rota-vap using a pre-tared round-bottom flask. Weigh the product.

Recrystallize the product from water. Dry it on a piece of porous plate by repeatedly spreading and scraping the solid with a spatula. Weigh the product and determine its melting point.

Obtain the IR of the product using a nujol mull (section 31.5) and, if possible, its ^{1}H-NMR using CDCl$_3$ as a solvent (section 33.20, adapted from Ref. 5).

Cleaning Up

- Dispose of the cuprous bromide in the container labeled "Oxidation-Reduction Solid Waste."
- Dispose of the unreacted dinitrobenzene in the container labeled "Oxidation-Reduction Solid Waste."
- Dispose of the neutralized aqueous layer and the mother liquor from the recrystallization in the container labeled "E21.7-E21.8 Liquid Waste."
- Dispose of CDCl$_3$ in the container labeled "CDCl$_3$-Waste."
- Dispose of magnesium sulfate, Pasteur pipets, filter paper, and capillary tubes in the container labeled "Oxidation-Reduction Solid Waste."
- At the end of the lab period the instructor will empty the contents of the rota-vap traps into the container "Organic Solvents Waste."
- Turn in your products in labeled screw-cap vials.

BIBLIOGRAPHY

1. Phase Transfer Catalysis Applied to Oxidation, C. Amsterdamsky, *J. Chem. Ed.*, **73**, 92 (1996).

2. The Oxidation of Alcohols and Ethers Using Calcium Hypochlorite [Ca(ClO)$_2$]. S.O. Nwaukwa and P.M. Keehn, *Tetrahedron Lett.*, **23**, 35–38 (1982).

3. β-Cyclodextrin as Inverse Phase-Transfer Catalyst in Oxidations of Isoborneol, Borneol and Menthol by Calcium Hypochlorite in Aqueous Solution. R. Ravichandran and S. Divakar, *J. Molec. Catalysis*, **88**, L117–L120 (1994).

4. Terephthalic Acid. C.F. Koelsch. Organic Syntheses, Coll. Vol. III, 791. Wiley, New York, 1955.

5. Selective Reduction of Dinitroarenes. D. Palleros, P. MacCormack and N.S. Nudelman. *Anales Asoc. Quim. Arg.*, **70**, 155–160 (1982).

EXPERIMENT 21   REPORT

Pre-lab

For All Sections

1. Write chemical reactions for the transformations you will carry out and give a mechanistic account for each one.

2. Make a table with the physical properties (MM, m.p., b.p., density, toxicity/hazards, solubility, flammability) of reagents and products for the reactions that you will carry out.

3. For each reaction that you will carry out, outline the main differences that you expect to see in the IR spectra of starting material and product.

For Sections E21.1, E21.3, and E21.5

4. What are the hazards of working with sodium borohydride in methanol?

5. What types of compounds can be reduced with sodium borohydride? Which ones are unaffected?

6. What are the advantages and disadvantages of using methanol, ethanol, and water as a solvent for reductions with sodium borohydride?

For E21.2

7. Briefly explain how phase transfer catalysts work.

8. What is the advantage of using bleach as an oxidizing agent? What other oxidizing agents could be used to carry out the same transformation?

For E21.4

9. What types of compounds are oxidized with manganese dioxide?

10. At what approximate wavenumber do you expect to see the stretching of the carbonyl in the IR of the product?

For E21.5

11. What differences in the ^{1}H-NMR spectra of starting material and product do you expect to see? Why do you add D_2O?

For E21.6

12. What other oxidizing agents could be used in the oxidation of isoborneol to camphor? What is the advantage of using $Ca(ClO)_2$?

13. What differences in the ^{1}H-NMR of isoborneol and camphor do you expect to see?

For E21.7

14. What functional groups are oxidized by potassium permanganate?

15. Are *tert*-butyl side chains oxidized by potassium permanganate? Explain.

16. What differences in the ^{1}H-NMR of *p*-toluic acid and terephthalic acid do you expect to see?

For E21.8

17. What other reagents could be used for the monoreduction of polynitrocompounds?

18. How would you separate a mixture of *m*-nitroaniline and *m*-dinitrobenzene by acid-base extraction?

In-lab

For All Sections:
The Spectra of the Products are Shown in Figures 21.2–21.22.

1. Report the percentage yield and the m.p. (when applicable) of your product. Discuss your results.

2. Report the TLC or GC analysis (when applicable) and explain how you decided to stop the reaction.

3. Interpret the IR of the starting material and product. Present your data in a table format. Compare them and show that the reaction proceeded as planned. Compare the IR of the product with that of an authentic sample shown below (Figs. 21.2; 21.5; 21.8; 21.11; 21.14; 21.17; and 21.20). Discuss your data.

4. Interpret the ^{1}H-NMR of product and starting material (if applicable). Present your data in a table format. Compare the spectrum of your product with that of an authentic sample shown (Figs. 21.3; 21.6; 21.9; 21.12; 21.15; 21.18; and 21.21). In E21.5, use the ^{1}H-NMR of your product to determine whether it is borneol or isoborneol by comparing it with literature spectra (for example, *Aldrich Library of NMR spectra*). Justify your assignment.

5. Interpret the ^{13}C-NMR of your product shown below (Figs. 21.4; 21.7; 21.10; 21.13; 21.16; 21.19; and 21.22).

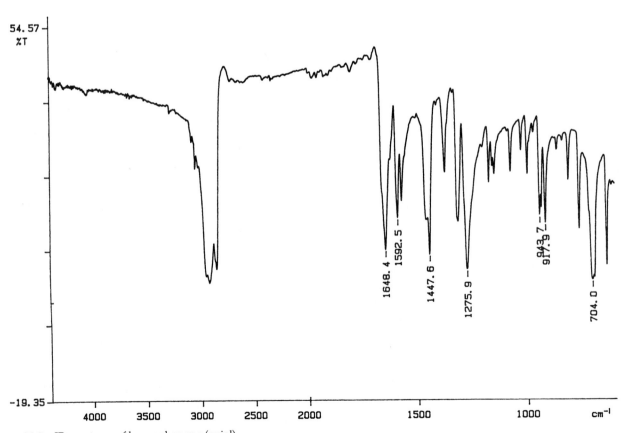

Figure 21.2 IR spectrum of benzophenone (nujol).

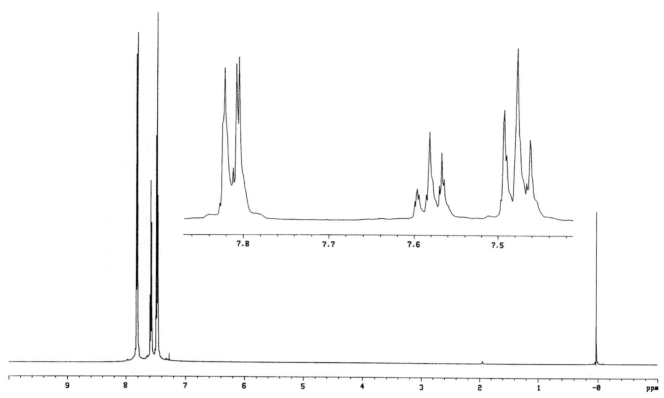

Figure 21.3   500-MHz ^{1}H-NMR spectrum of benzophenone in CDCl$_3$.

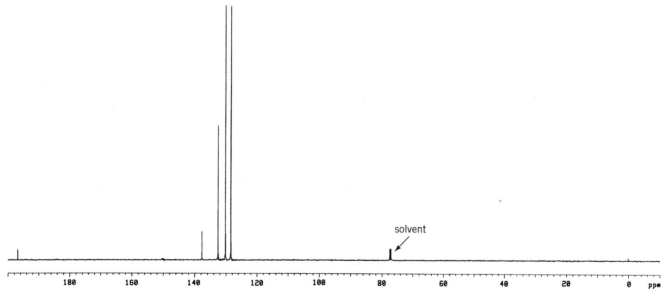

Figure 21.4   125.7-MHz ^{13}C-NMR spectrum of benzophenone in CDCl$_3$.

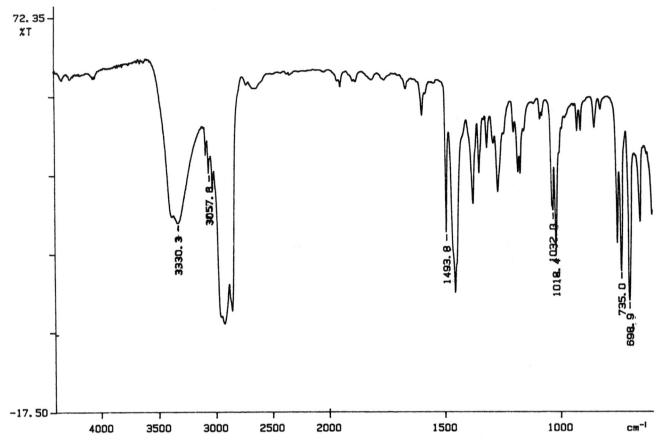

Figure 21.5   IR spectrum of benzhydrol (nujol).

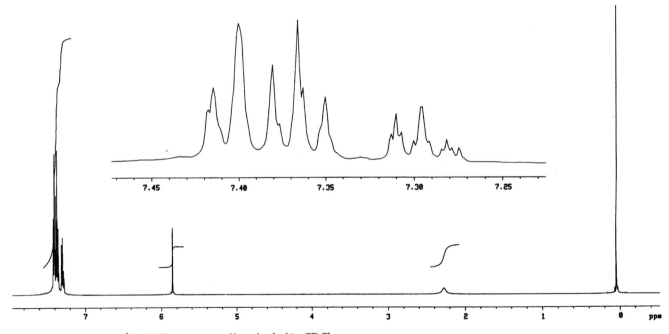

Figure 21.6   500-MHz ^{1}H-NMR spectrum of benzhydrol in CDCl$_3$.

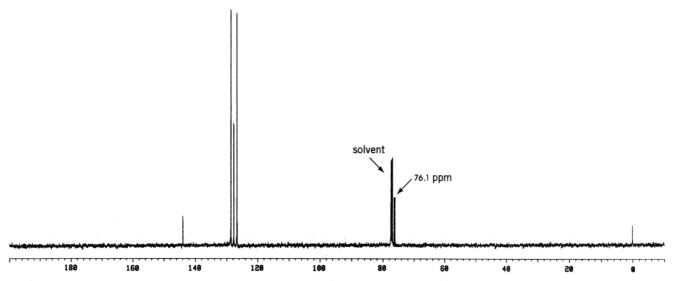

Figure 21.7   125.7-MHz ^{13}C-NMR spectrum of benzhydrol in CDCl$_3$.

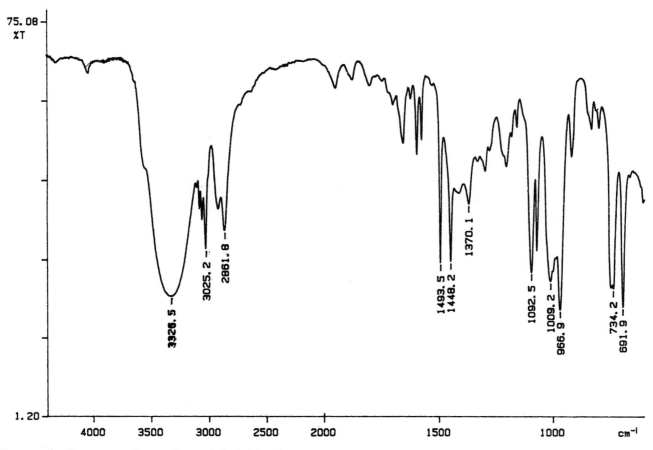

Figure 21.8   IR spectrum of *trans*-cinnamyl alcohol (neat).

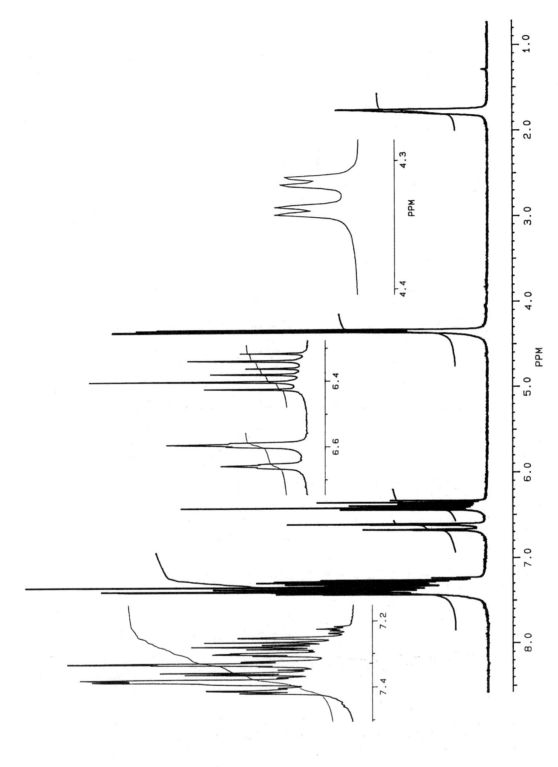

Figure 21.9 250-MHz ^{1}H-NMR spectrum of *trans*-cinnamyl alcohol in CDCl$_3$.

Figure 21.10   125.7-MHz ^{13}C-NMR spectrum of *trans*-cinnamyl alcohol in CDCl$_3$.

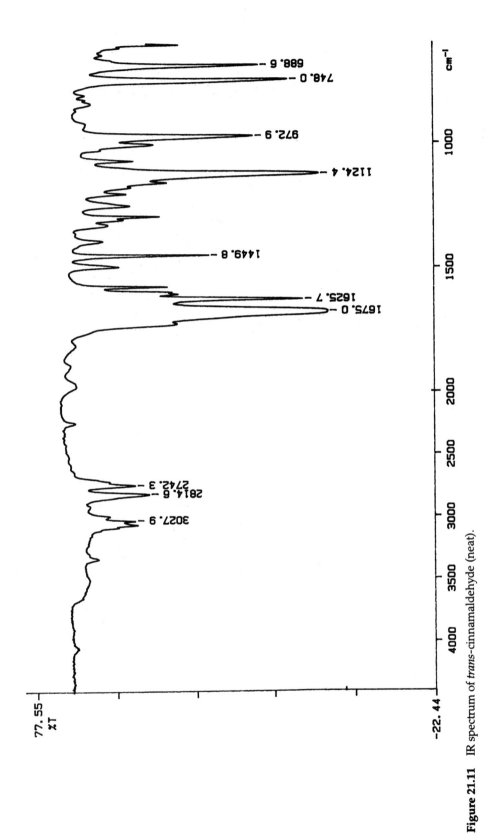

Figure 21.11 IR spectrum of *trans*-cinnamaldehyde (neat).

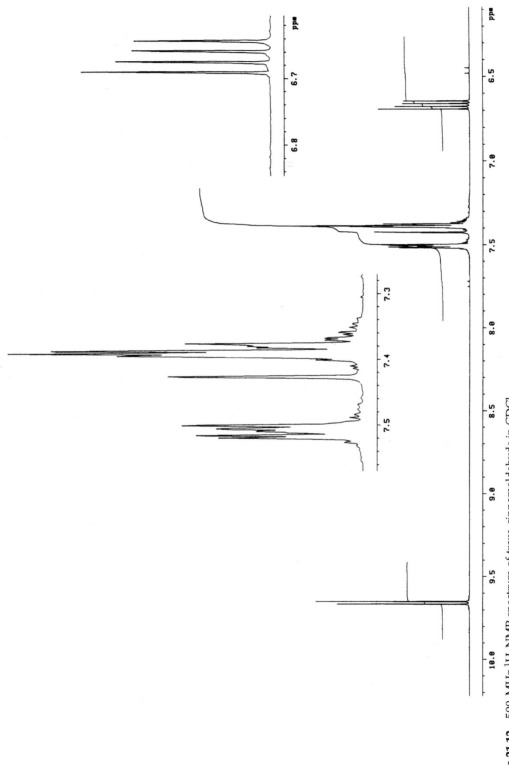

Figure 21.12   500-MHz ^{1}H-NMR spectrum of *trans*-cinnamaldehyde in CDCl$_3$.

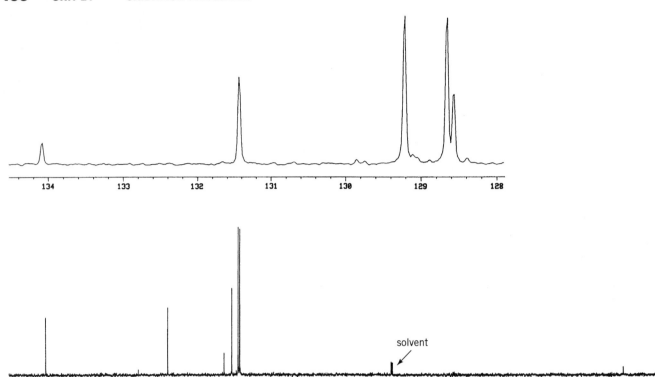

Figure 21.13 125.7-MHz ^{13}C-NMR spectrum of *trans*-cinnamaldehyde in CDCl$_3$.

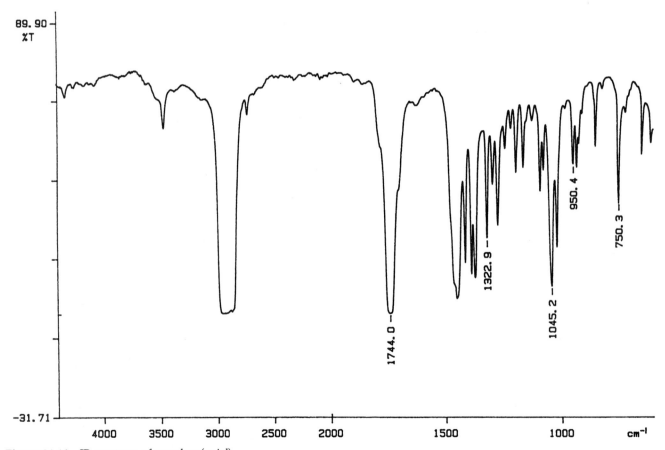

Figure 21.14 IR spectrum of camphor (nujol).

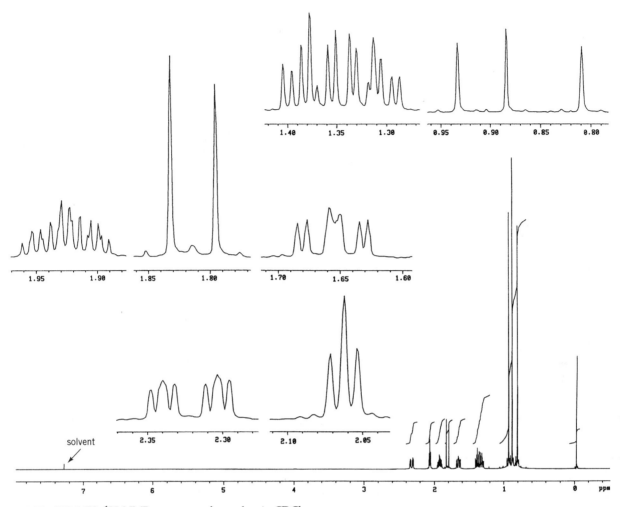

Figure 21.15   500-MHz ^{1}H-NMR spectrum of camphor in CDCl$_3$.

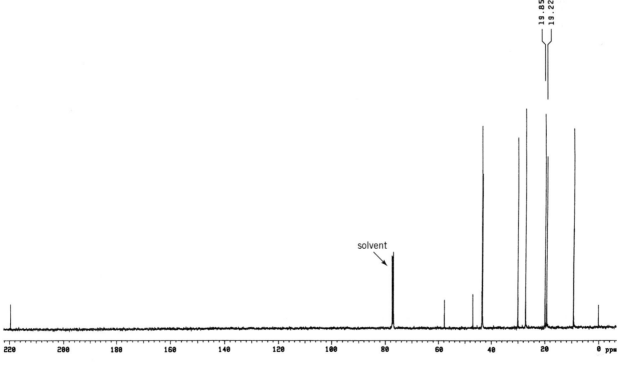

Figure 21.16   125.7-MHz ^{13}C-NMR spectrum of camphor in CDCl$_3$.

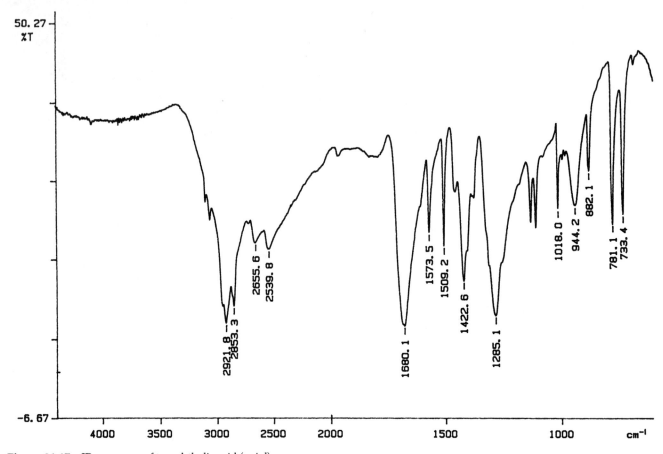

Figure 21.17 IR spectrum of terephthalic acid (nujol).

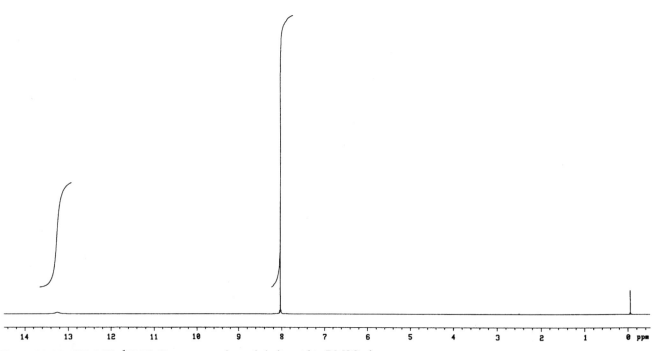

Figure 21.18 500-MHz ^{1}H-NMR spectrum of terephthalic acid in DMSO-d_6.

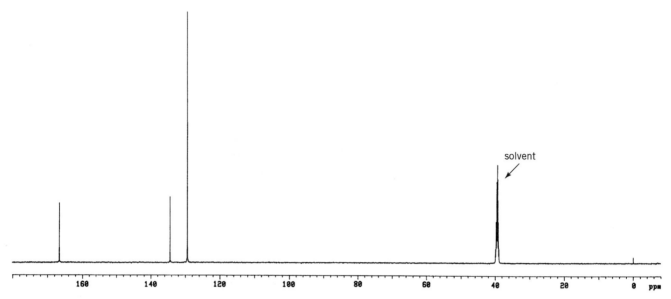

Figure 21.19   125.7-MHz ^{13}C-NMR spectrum of terephthalic acid in DMSO-d$_6$.

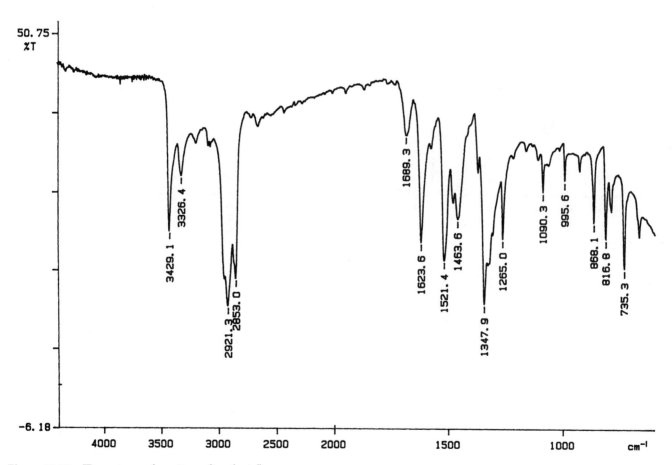

Figure 21.20   IR spectrum of *m*-nitroaniline (nujol).

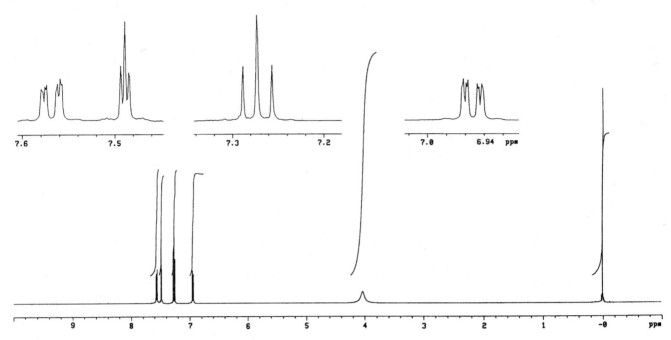

Figure 21.21   500-MHz ¹H-NMR spectrum of *m*-nitroaniline in CDCl₃.

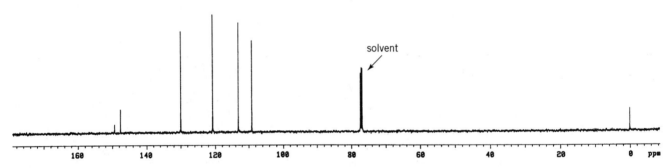

Figure 21.22   125.7-MHz ¹³C-NMR spectrum of *m*-nitroaniline in CDCl₃.

Esters

22.1 ESTERS IN NATURE AND SOCIETY

Carboxylic esters encompass a large family of organic compounds with broad applications in medicine, biology, and industry. Carboxylic esters are represented by the structure R–CO–OR′, in which R and R′ are alkyl or aryl groups.

$$\underset{R}{\overset{\displaystyle O}{\underset{\displaystyle}{\overset{\displaystyle \|}{C}}}}\underset{OR'}{}$$

Esters are widespread in nature. They occur naturally in plants and animals. Small esters, in combination with other volatile compounds, produce the pleasant aroma of fruits. In general, a symphony of chemicals is responsible for specific fruity fragrances; however, very often one single compound plays a leading role. For example, an artificial pineapple flavor contains more than twenty ingredients but ethyl butyrate is the major component. In the table below some examples of ester flavors and fragrances are shown.

Some Ester Flavors and Fragrances

n-Propyl acetate	$CH_3-CO-O-CH_2-CH_2-CH_3$	pears
n-Octyl acetate	$CH_3-CO-O-(CH_2)_7-CH_3$	oranges
Isoamyl acetate	$CH_3-CO-O-CH_2-CH_2-CH-(CH_3)_2$	banana
Ethyl butyrate	$CH_3-CH_2-CH_2-CO-O-CH_2-CH_3$	pineapple
n-Butyl acetate	$CH_3-CO-O-CH_2-CH_2-CH_2-CH_3$	apple
Methyl *trans*-cinnamate	$Ph-CH=CH-CO-O-CH_3$	strawberry

Some esters play an important role in insect communication. Isoamyl acetate, the main component of banana aroma, is also the alarm pheromone of the honeybee; 1-methylbutyl decanoate is the sex pheromone of the bagworm moth; and (Z)-6-dodecen-4-olide, a **lactone** or cyclic ester, is the "social scent" of the black-tailed deer.

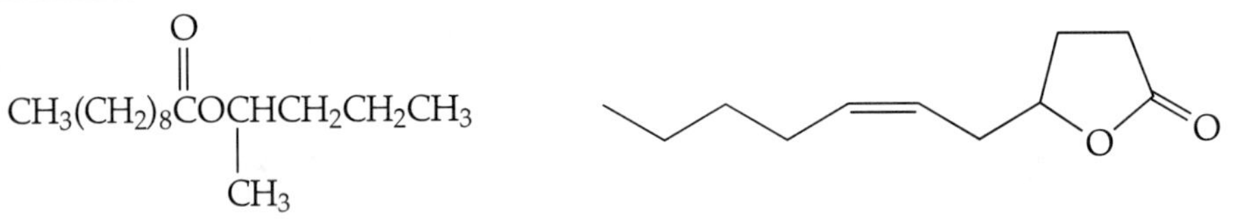

$$CH_3(CH_2)_8\overset{\displaystyle O}{\overset{\displaystyle \|}{C}}OCHCH_2CH_2CH_3$$

$$\underset{CH_3}{|}$$

1-methylbutyl decanoate
sex pheromone of the bagworm moth

(Z)-6-dodecen-4-olide
social scent of the black-tailed deer

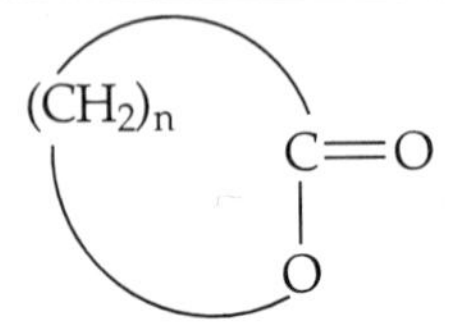

n = 21; 23; 25; 27; 29; 31; 33; 35; 37

Macrocylic **esters** found in the toxin of Armitermes.

Lactones are also found in the oily poisonous secretion of termites of the genus *Armitermes*. Macrocyclic lactones with 22 to 38 carbon atoms are used as a chemical weapon by the soldiers of the *Armitermes*. The soldiers apply the toxic oil to the wounds they make with their mandible and often get stuck to the victim and die after the attack.

Esters also have remarkable applications in everyday life. Plexiglas is a stiff, transparent plastic obtained by polymerization of methylmethacrylate. Dacron, a fiber used for fabrics, is a polyester obtained by polymerization of terephthalic acid and ethylene glycol. This polymer is also used to manufacture 2-L soda bottles. We will come back to the subject of polyesters in Unit 26.

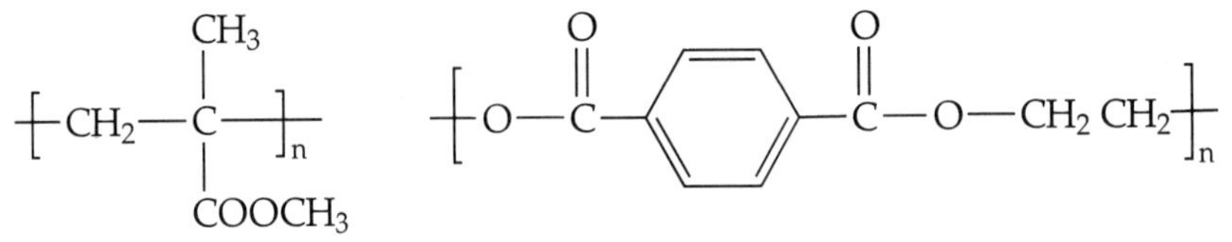

Poly(methyl methacrylate)
Plexiglas

Poly(ethylene terephthalate)
Dacron

Esters derived from *p*-aminobenzoic acid (PABA) have local anesthetic properties. Benzocaine, procaine, and butesin are chemicals used in medicinal preparations to alleviate pain caused by skin burns.

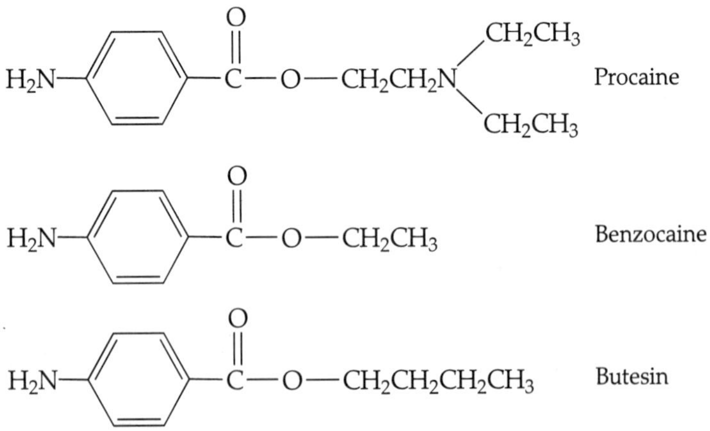

Procaine

Benzocaine

Butesin

22.2 PREPARATION OF ESTERS

Carboxylic esters are carboxylic acid derivatives. They can be obtained by heating a carboxylic acid with an alcohol in the presence of catalytic amounts of mineral acids such as sulfuric or hydrochloric acids, HB.

$$\left[\underset{\text{a carboxylic acid}}{\text{RCOOH}} \;+\; \underset{\text{an alcohol}}{\text{R'OH}} \xrightleftharpoons{\text{HB}} \underset{\text{an ester}}{\text{RCOOR'}} + H_2O \right]$$

This reaction is known as the **Fischer esterification**. It is a reversible reaction and therefore, its application in the preparation of esters is limited by the equilibrium constant, K_e.

$$\left[K_e = \frac{[\text{ester}][\text{water}]}{[\text{acid}][\text{alcohol}]} \right]$$

Typical values of K_e for the esterification of alcohols with unhindered carboxylic acids are in the range $K_e = 1\text{–}10$. The experimental K_e value for the reaction of acetic acid and ethanol is 3.8. Therefore, if we start with an equimolar mixture of acid and alcohol, after equilibrium is attained the yield of ester is only 66%, as the following calculation shows.

$$\underset{\text{acetic acid}}{\text{CH}_3\text{COOH}} + \underset{\text{ethanol}}{\text{CH}_3\text{CH}_2\text{OH}} \xrightleftharpoons{\text{HB}} \underset{\text{ethyl acetate}}{\text{CH}_3\text{COOCH}_2\text{CH}_3} + H_2O$$

start: 1 mole   1 mole   0

end: $(1-a)$ mole $(1-a)$ mole a mole

$$K_e = \frac{a^2}{(1-a)^2} = 3.8 \qquad \Longrightarrow \qquad a = \frac{\sqrt{K_e}}{1+\sqrt{K_e}} = 0.66$$

The yield can be considerably increased by using a large excess of one of the reactants (which can be recovered at the end of the reaction) or by constantly removing the products from the reaction mixture. Both approaches are used in the synthesis of esters.

In principle, the equilibrium constant does not depend on the catalyst concentration. It was found, however, that an increase in sulfuric acid concentration can result in higher yields. This is probably because of a change in the reaction medium. Sulfuric acid has a strong tendency to associate with water molecules, and thus it can shift the equilibrium toward products by removing water from the medium.

The mechanism of the Fischer esterification has been extensively studied and is presented in the scheme below where HB is the acid catalyst.

The Fischer esterification is only recommended with primary and secondary alcohols and unhindered carboxylic acids. Steric hindrance near the reaction center slows down the esterification.

22.3 ACETYLATION

An important reaction in the laboratory is the introduction of an acetyl group (CH_3CO-) in a molecule bearing an alcohol or phenol group. The acetylation reaction transforms these reactive compounds into their less-reactive acetates. The acetyl group is sometimes introduced to act as a temporary blocking agent that protects the hydroxyl group from unwanted reactions. The acetyl group can then be removed by hydrolysis. One of the most commonly used acetylation reagents is acetic anhydride. It easily reacts with phenols and alcohols to give acetates and acetic acid. The example below illustrates the esterification of phenol.

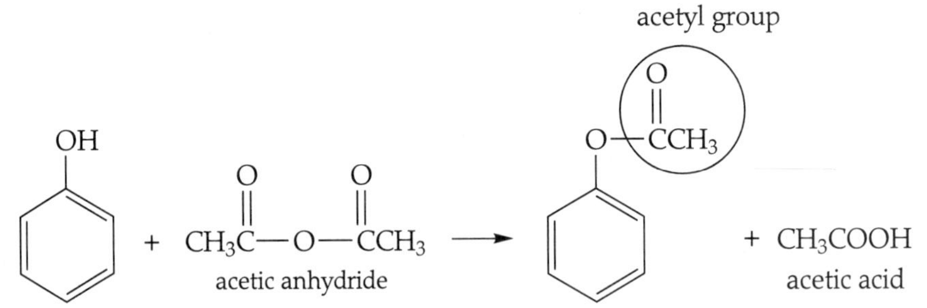

The mechanism is shown below:

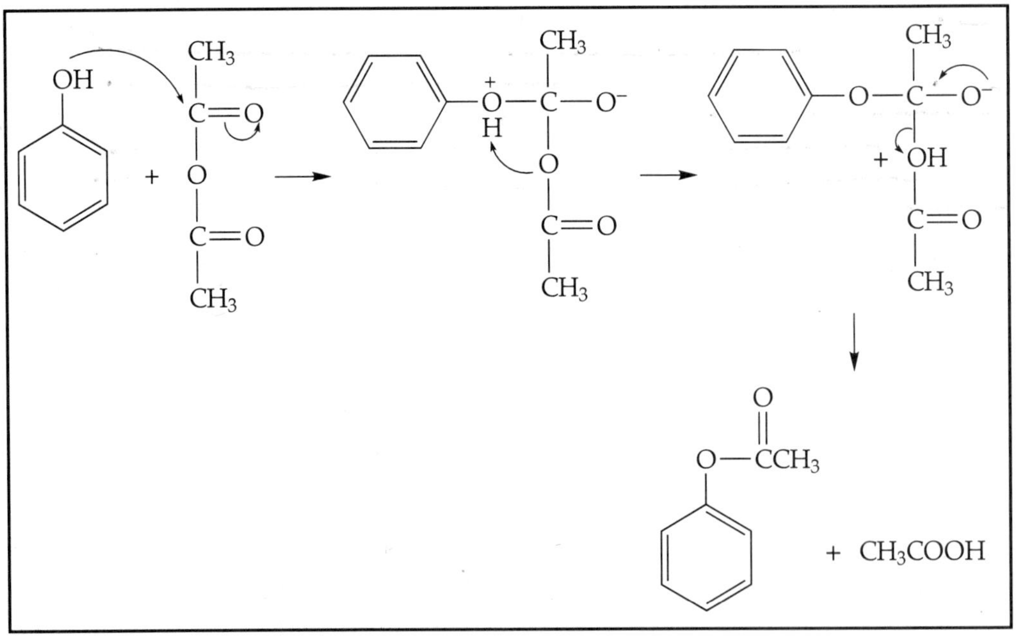

22.4 ESTER HYDROLYSIS

In the presence of water and catalytic amounts of mineral acid, esters are hydrolyzed to the corresponding carboxylic acid and alcohol (or phenol). The mechanism is the reverse of that for the Fischer esterification. Like the Fischer esterification, the acid-catalyzed hydrolysis of esters is reversible. From an experimental standpoint, more satisfactory results are obtained when the hydrolysis is performed in the presence of at least one equivalent of a strong base. The strong base neutralizes the carboxylic acid as it forms and pulls the reaction to completion. The overall reaction is irreversible. The basic hydrolysis of esters is called **saponification**:

$$R-\overset{\overset{\displaystyle O}{\|}}{C}-OR' \;+\; NaOH \;\xrightarrow[heat]{H_2O}\; R-\overset{\overset{\displaystyle O}{\|}}{C}-O^-Na^+ \;+\; R'OH$$

We will see this reaction in Experiment 22B, and will encounter it again in Unit 25 ("Lipids").

22.5 CHARACTERIZATION OF ESTERS

Hydroxamic Acid Test

Esters react with hydroxylamine (NH_2OH) under basic conditions to give **hydroxamic acids**. Hydroxamic acids form red-violet octahedral complexes with ferric chloride in acidic medium. This reaction constitutes a very sensitive test for esters. Carboxylic acids, with the exception of formic acid, give negative results. Other acid derivatives such as anhydrides, acid chlorides, and imides also react with hydroxylamine to give hydroxamic acids and, therefore, may interfere in the detection of esters by giving positive tests. Nitriles and amides need more drastic conditions to produce hydroxamic acids and thus do not interfere in the detection of esters. Some phenols and hydroxy acids also give red-violet colors in the presence of ferric chloride (without hydroxylamine added; section 15.3). If a phenol or hydroxy acid is suspected in the sample, a test in which hydroxylamine is omitted should be performed; if the reaction gives a red color in the absence of hydroxylamine, the compound is most likely a phenol or a hydroxy acid, and the test with hydroxylamine should not be performed.

$$\left[\; R-C\!\!\begin{array}{c}\nearrow O\\[4pt]\searrow OR'\end{array} \;+\; NH_2OH \;\longrightarrow\; R-C\!\!\begin{array}{c}\nearrow O\\[4pt]\searrow NH-OH\end{array} \;+\; R'OH\right.$$

a hydroxamic acid

$$\left.3\,R-C\!\!\begin{array}{c}\nearrow O\\[4pt]\searrow NH-OH\end{array} \;+\; FeCl_3 \;\longrightarrow\; \left[R-C\!\!\begin{array}{c}\nearrow \overset{\cdot\cdot}{O}:\text{-}\text{-}\text{-}\\[4pt]\searrow NH-O\end{array}\!\!\!Fe\right]_3 \;+\; 3\,HCl\;\right]$$

red-violet complex

22.6 IR AND NMR OF ESTERS AND CARBOXYLIC ACIDS

IR Spectra

Esters present a strong band around 1740 cm^{-1} due to C=O stretching. This band is very diagnostic because no other functional group absorbs at such a frequency. If the carbonyl is conjugated with an aromatic ring or a double bond, this band shows around 1720 cm^{-1}. Six- and five-membered lactones absorb at 1735 and 1770 cm^{-1}, respectively. Esters also present strong bands in the fingerprint region, between 1050 and 1320 cm^{-1} due to C−O−C stretching. These bands are not diagnostic for esters because other functional groups with C−O−C bonds (such as ethers) also absorb in the same region.

Carboxylic acids show a broad band between 2500 and 3500 cm^{-1} due to O−H stretching. In the solid, liquid, and even vapor state, carboxylic acids form dimers by H-bond association. The carbonyl group for such dimers absorbs around 1710 cm^{-1} for aliphatic carboxylic acids and at lower wavenumbers (1690 cm^{-1}) for aromatic and conjugated ones.

a carboxylic acid dimer

NMR Spectra

In ^{1}H-NMR, methyl, methylene, and methine groups attached to the oxygen end of the carboxyl group are deshielded and resonate at high chemical shifts (3.7–5.1 ppm). Practically no other groups (with the exception of alkyl phenyl ethers) resonate in this region. Methyl, methylene, and methine groups attached directly to the carbonyl end of the carboxyl group resonate at lower chemical shifts (2.0–2.5 ppm). ^{13}C-NMR signals show similar deshielding patterns.

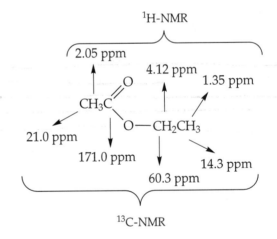

The most diagnostic peak in the ^{1}H-NMR spectra of carboxylic acids is the one of the acidic hydrogen, usually observed between 10 and 13 ppm. Only the carboxylic acid proton resonates in this region, but often it is not detected because of dissociation and rapid exchange.

The carbonyl carbon resonates in the range of 165–175 ppm for esters (aliphatic and aromatic), and around 170–180 ppm for aliphatic and aromatic carboxylic acids.

BIBLIOGRAPHY

1. Fruit Flavors. Biogenesis, Characterization, and Authentication. R.L. Rouseff and M.M. Leahy, ed. American Chemical Society, Washington, DC, 1995.

2. Common Fragrance and Flavor Materials. K. Bauer, D. Garbe, and H. Surburg. 2nd ed. VCH, Weinheim, 1990.

3. Esterification. E.G. Zey. Kirk-Othmer Encyclopedia of Chemical Technology. Vol. 9. 3rd ed. Wiley, New York, 1980.

4. Organic Esters. E.U. Elam. Kirk-Othmer Encyclopedia of Chemical Technology. Vol. 9. 3rd ed. Wiley, New York, 1980.

5. Synthetic Food. M. Pyke. John Murray, London, 1970.

6. Vogel's Textbook of Practical Organic Chemistry. A.I. Vogel, B.S. Furniss, A.J. Hannaford, P.W.G. Smith, and A.R. Tatchell. 5th ed. Longman, Harlow, UK, 1989.

7. Semimicro Qualitative Organic Analysis. N.D. Cheronis, J.B. Entrikin, and E.M. Hodnett. 3rd. ed. Interscience Pub., New York, 1965.

EXERCISES

Problem 1

The hydrolysis of ethyl acetate under acidic conditions is a reversible reaction. Using the equilibrium constant value $K_e = 3.8$ for the esterification reaction, calculate how many moles of acetic acid and ethanol will be produced after equilibrium is attained if: a) 1 mol of ethyl acetate is treated with 1 mol of water; b) 1 mol of ethyl acetate is treated with 100 mol of water.

Problem 2

Which IR bands would you use to distinguish a carboxylic acid from an ester? Which ones would you use to distinguish an alcohol from an ester?

Problem 3

Compound X ($C_8H_8O_2$) gives a red-violet color when reacted with hydroxylamine followed by ferric chloride. The IR of X shows bands at 1724, 750, and 700 cm^{-1}. Give a possible structure for X and briefly explain your reasoning.

Problem 4

(*Previous Synton: Problem 5, Unit 21*) Vanadia was impressed with Synton's chemical abilities. After several visits and countless phone calls, she finally persuaded him to put his knowledge to work and help her start her own aromatherapy business. Synton was reluctant at first because he didn't want to take any time off from his beloved research, but the offer made by Vanadia was irresistible and he finally gave in. He would spend only a few hours a week as the scientific consultant for the formulation of extracts, lotions, and potions and would get 40% of the profits! His first project is to decide whether to buy or to make some esters used in essence formulations. Synton made a table with the price (per liter) of the starting materials and products, along with densities and molecular masses. Synton wants to make the esters by the Fischer esterification and he is planning on using one of the reactants in a 10-fold molar excess. Here is a sample of Synton's table:

Compound	Molecular mass	Density g/mL	Price per liter
isoamyl acetate	130	0.876	114.8
isoamyl alcohol	88	0.809	35.80
acetic acid	60	1.05	15.23

(a) Which reactant, the acid or the alcohol, should Synton use in excess?

(b) If the equilibrium constant for ester formation is $K_e = 4$, and one reagent is used in a 10-fold molar excess over the other, is the preparation of the ester cost effective? The cost of other chemicals (catalysts, drying agent, etc.) plus labor and overhead can be estimated at $40 per liter of ester.

(c) Answer the same question if 80% of the reagent in excess is recycled at the end of the synthesis and the cost of recycling is $5 per liter of ester produced.

EXPERIMENT 22A

Preparation of Fruity Fragrances

In this experiment you will choose and prepare one of the following esters: methyl *trans*-cinnamate (a **strawberry** aroma component) from methanol and *trans*-cinnamic acid, or *n*-propyl acetate (a **pear** aroma component) from acetic acid and *n*-propanol, or isoamyl acetate (also known as isopentyl acetate or **banana oil**) from acetic acid and isoamyl alcohol. (Incidentally, isoamyl acetate is also the alarm pheromone of the honeybee and thus, it should be kept away from bee hives!) You will perform these syntheses at the microscale level by the Fischer esterification method. To separate the ester from the unreacted carboxylic acid you will use acid-base extraction (section 14.4) with an aqueous sodium bicarbonate solution. If you are making methyl *trans*-cinnamate, you will use *tert*-butyl methyl ether as the organic solvent for the extraction. In the preparation of *n*-propyl acetate and isoamyl acetate, on the other hand, no organic solvent is necessary for the acid-base extraction because these esters are liquids and separate from the aqueous solution as an immiscible layer.

You will purify methyl *trans*-cinnamate by recrystallization from a mixed solvent (section 4.4) and isoamyl acetate and *n*-propyl acetate by microscale distillation (section 6.8). You will characterize the product by GC, IR spectroscopy, and the hydroxamic acid test.

As starting material for the synthesis of methyl *trans*-cinnamate, you can use *trans*-cinnamic acid obtained in Experiment 20B. You will need 200 mg. If you do not have enough, supplement your product with commercial *trans*-cinnamic acid.

PROCEDURE

> **Background reading:** 22.1–3; 14.4; 6.8 or 4.4
> **Estimated time:** 3–4 hours

E22A.1 PREPARATION OF METHYL *trans*-CINNAMATE: A COMPONENT OF STRAWBERRY AROMA

trans-cinnamic acid methanol methyl *trans*-cinnamate

Synthesis

1. In a 15-mL round-bottom flask weigh 200 mg of *trans*-cinnamic acid. Add 3 mL of methanol and a magnetic stir bar. Swirl the contents to dissolve the solid and add 3 drops of concentrated sulfuric acid.

2. Attach a microscale reflux water-cooled condenser and heat directly on a hot plate at a medium setting for one hour (Fig. 22.1). In the meantime obtain the IR spectra of the starting materials (see Analysis, step 2).

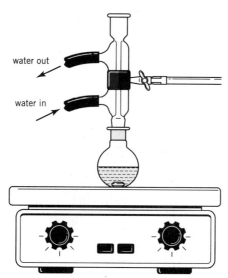

water out

water in

Figure 22.1   Reflux apparatus with 15-mL round-bottom flask.

3. Allow the system to cool down and transfer the liquid to a 16 × 125 mm screw-cap test tube labeled "organic." Wash the walls of the round-bottom flask with 5 mL of *tert*-butyl methyl ether (BME). Keep the wash in the round-bottom flask until step 5.

4. To the liquid in the test tube labeled "organic", slowly add 3 mL of 5% $NaHCO_3$ in 15% NaCl aqueous solution. Bubbling will occur because of CO_2 formation.

5. After bubbling subsides, add the BME wash from step 3. Cap the tube and invert it several times to mix the layers. Frequently vent the system to release the pressure by momentarily unscrewing the cap.

6. Let the layers settle and transfer the lower (aqueous) layer with a Pasteur pipet into another 16 × 125 mm screw-cap test tube labeled "aqueous." Keep the upper (organic) layer for step 8.

7. Extract the aqueous layer from the previous step with 3 mL of BME. Let the layers settle and remove the lower (aqueous) layer. Place this layer in a labeled test tube, keep it until the end of the experiment, and then discard it.

8. Consolidate both organic layers from steps 6 and 7 in the screw-cap test tube labeled "organic," and wash the combined layers with 3 mL of a saturated NaCl aqueous solution. After the layers settle, remove the lower aqueous layer, keep it until the end of the experiment, and then discard it.

9. Dry the combined organic layers by adding anhydrous $MgSO_4$ *little by little* with a *microspatula* until the solid no longer clumps together at the bottom of the tube but runs freely in the liquid.

10. Filter the suspension using a Pasteur pipet with a cotton plug (Fig. 3.9*a*) and collect the liquid in a dry, pre-tared round-bottom flask.

11. Evaporate the solvent in a rota-vap. You will obtain an oily residue that solidifies on ice. Weigh the flask with the product. The product is fairly pure and adequate for most applications. If necessary, you can purify it by recrystallization from a mixture of methanol-water (step 12).

12. *Recrystallization* (optional): Dissolve the solid in about 1.5 mL of warm methanol and transfer the solution to a 3-mL Craig tube. Heat the Craig tube in a sand bath, stirring with a microspatula to avoid boil-over. While heating, add water drop by drop. The product starts to precipitate but it quickly redissolves. Keep adding water until the turbidity persists. Add a few drops of methanol just to redissolve the solid while the solution is still hot. Remove from the heat and let the system cool down to ambient temperature and then place it in an ice-water bath. Gently scratch the walls of the tube with a glass rod. Centrifuge to separate the solid. Dry the solid on a porous plate. Determine its melting point.

Analysis

1. Perform the hydroxamic acid test on your product (E22A.3). See "Cleaning Up" at the end of E22A.3.

2. While you are refluxing the reaction mixture, obtain the IR spectra of *trans*-cinnamic acid (nujol mull) and methanol (neat) between NaCl plates (section 31.5).

3. Obtain the IR spectrum of the ester. Place a small amount of the solid in a test tube and warm it in a water bath to melt. Transfer a drop of the melted solid to a NaCl plate and immediately obtain its IR spectrum. If possible, obtain the ^{1}H-NMR of your product using $CDCl_3$ as a solvent (section 33.20). (*Caution: $CDCl_3$ is a possible carcinogen; dispense this chemical in the hood.*)

E22A.2 PREPARATION OF *n*-PROPYL ACETATE AND ISOAMYL ACETATE

Synthesis

1. Weigh approximately 10 mmoles of the desired alcohol (isoamyl alcohol or *n*-propanol) and 40 mmoles of glacial acetic in a pre-tared 15-mL round-bottom flask. Use a Pasteur pipet to transfer the liquids. Add 3 drops of concentrated sulfuric acid.

2. Add a magnetic stir bar and attach a microscale water-jacketed condenser (Fig. 22.1, E22A.1). Reflux for one hour by heating directly on a hot plate at a medium setting. In the meantime obtain the GC and IR spectra of the starting materials (see Analysis, step 2).

3. Remove from the heat and allow the mixture to cool to ambient temperature. Disassemble the apparatus and transfer the liquid to a 16 × 125 mm screw-cap test tube.

4. Rinse the round-bottom flask with 2 mL of 5% sodium bicarbonate in 15% sodium chloride solution. Slowly transfer the rinse to the screw-cap test tube. Stir the mixture with a microspatula until gas evolution (carbon dioxide) has subsided.

5. Cap the tube and invert it several times to mix the layers. Frequently vent the system to release the pressure by momentarily unscrewing the cap. Let the system settle for about 10 minutes.

6. Using a Pasteur pipet transfer the lower (aqueous) layer to a labeled test tube, keep it until the end of the experiment, and then discard it.

7. Wash the organic layer remaining in the screw-cap test tube twice with 1 mL (each time) of the sodium bicarbonate-sodium chloride solution. Invert and vent well in each wash.

8. Collect the aqueous washes in the same labeled test tube mentioned in step 6 (to be discarded).

9. Dry the organic layer by adding anhydrous $MgSO_4$ *little by little* with a *microspatula* until the solid no longer clumps together at the bottom of the tube but runs freely in the liquid. If you have more than 400 μL of product, proceed to the distillation (step 10); otherwise remove the liquid from the drying agent using a filter-Pasteur pipet (Fig. 3.9*b*) and transfer it to a pre-tared vial. Weigh the product.

10. Transfer the ester with a filter-Pasteur pipet to a 5-mL round-bottom flask. Assemble a distillation apparatus using a Hickman still, a boiling chip, a thermometer and an air condenser. Support the thermometer using a notched stopper resting on top of the condenser; this will make the removal of the thermometer very easy (Fig. 22.2). (If no 5-mL round-bottom flask is available use a 3-mL

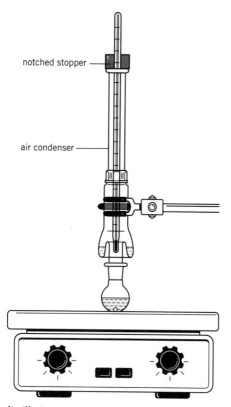

notched stopper

air condenser

Figure 22.2 Microscale distillation apparatus.

conical vial heated on a sand bath set at a temperature 40–50°C above the boiling point of the ester.)

11. Distill by heating directly on a hot plate at a medium setting until two or three drops of liquid remain in the flask. Record the temperature during the distillation. If the Hickman still is full before the distillation is complete, carefully remove the thermometer and condenser and empty the still with a Pasteur pipet. Reassemble the apparatus and continue the distillation.

12. Transfer the distillate to a tared vial. Weigh the product.

Analysis

1. Perform the hydroxamic acid test (E22A.3).

2. While you are refluxing the reaction mixture, obtain the IR spectrum (between NaCl plates) of the acid and alcohol used. Run GC of the starting materials using a column of medium polarity such as SF-96 at 90°C.

3. Run GC of the ester under the same conditions.

4. Obtain the IR spectrum of the ester using NaCl plates (section 31.5). If possible, obtain the ^{1}H-NMR of your product using $CDCl_3$ as a solvent (section 33.20). (*Caution: $CDCl_3$ is a possible carcinogen; dispense this chemical in the hood.*)

E22A.3   HYDROXAMIC ACID TEST

Note

In the event that you are investigating an unknown sample or you suspect the presence of a phenol, carry out the following test with ferric chloride before proceeding to the test for esters. Dissolve a drop (or 30 mg) of your sample in 1 mL of 95% ethanol and add 1 mL of 2 M aqueous hydrochloric acid followed by two drops of a 3% aqueous ferric chloride solution. If a violet, red, or orange color develops, your sample may be a phenol or a hydroxy acid and the following test for esters should not be performed.

Test for esters

Add one drop (or about 30 mg of solid) of your sample to 1 mL of 0.5 M hydroxylamine hydrochloride in 95% ethanol placed in a test tube. Add 0.2 mL of a 6 M sodium hydroxide aqueous solution dropwise and a boiling chip. Bring the mixture to a boil by heating in a water bath. Let the system cool down and add dropwise a 2 M aqueous solution of hydrochloric acid until the pH is 3–2. If cloudiness develops add 2 mL of 95% ethanol. Add 2 drops of 3% ferric chloride solution. A red-violet color is a positive test. Try the test with the synthesized ester and the starting materials (the acid and the alcohol). For comparison, also run the test on a pure sample of ethyl acetate.

Safety first

- Hydroxylamine, hydrochloric acid, and sodium hydroxide are corrosive and toxic.

Cleaning Up

- Dispose of the aqueous solutions from E22A.1 and E22A.2 and the hydroxamic acid solutions in the container labeled "Esters Liquid Waste."

- Dispose of the Pasteur pipets and capillary tubes in the container labeled "Esters Solid Waste."

- At the end of the experiment the instructor will empty the contents of the rota-vap traps into an "Esters Liquid Waste" bottle.

- Dispose of the $CDCl_3$ solution in the container labeled "Chloroform Waste."

- Turn in your ester in a labeled capped vial to your instructor.

EXPERIMENT 22A   REPORT

Pre-lab

1. Make a table with the physical properties including toxicity/hazards of reagents and products for the synthesis of the ester.

2. Write the chemical reaction for the Fischer esterification of the ester of your choice. Show the mechanism.

3. Make a flowchart for the preparation of the desired ester.

4. Why do you extract the reaction mixture with 5% sodium bicarbonate plus 15% sodium chloride solution? What role does each salt play?

5. For the preparation of methyl *trans*-cinnamate, calculate the number of mmoles of *trans*-cinnamic acid and methanol used. Determine the limiting reagent.

6. For the preparation of isoamyl acetate and *n*-propyl acetate, calculate the mass and volume of alcohol and acid that you will mix.

7. Consult your textbook and show another way to prepare the desired ester.

8. Calculate the theoretical yield of the synthesis in mg (assume that the reaction goes to completion).

9. What compounds interfere in the characterization of esters by the hydroxamic acid test?

In-lab

The spectra of the products are shown in Figures 22.3–22.12

1. Report the observed melting or boiling point of your ester and compare it with a literature value.

2. Calculate the percentage yield of the synthesis and discuss any source of product loss.

3. Discuss how you could increase the yield of your ester.

4. In the synthesis of *n*-propyl acetate and isoamyl acetate analyze your GC results.

5. Report and interpret your hydroxamic acid test results.

6. Assign as many bands as possible in the IR spectrum of the product. Compare the spectrum with the one of an authentic sample shown below (Figs. 22.3; 22.6; and 22.9).

7. Compare the IR of your product with those of the starting materials obtained in the lab.

8. Analyze the ^{1}H-NMR and ^{13}C-NMR of your product shown in the Figures below (Figs. 22.4; 22.5; 22.7; 22.8; 22.10; and 22.11).

9. Analyze the MS of your product shown in Figure 22.12 (optional).

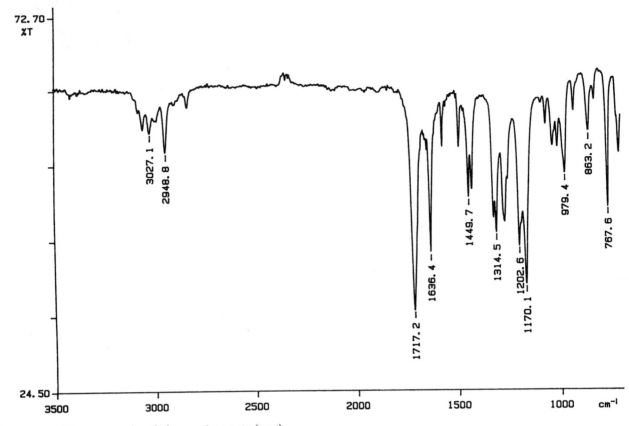

Figure 22.3 IR spectrum of methyl *trans*-cinnamate (neat).

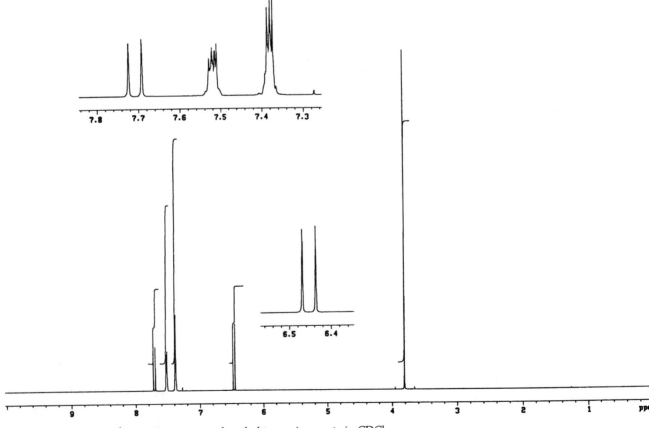

Figure 22.4 500-MHz ^{1}H-NMR spectrum of methyl *trans*-cinnamate in CDCl$_3$.

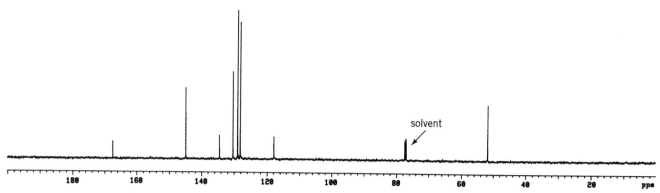

Figure 22.5   125.7-MHz ^{13}C-NMR spectrum of methyl *trans*-cinnamate in CDCl$_3$.

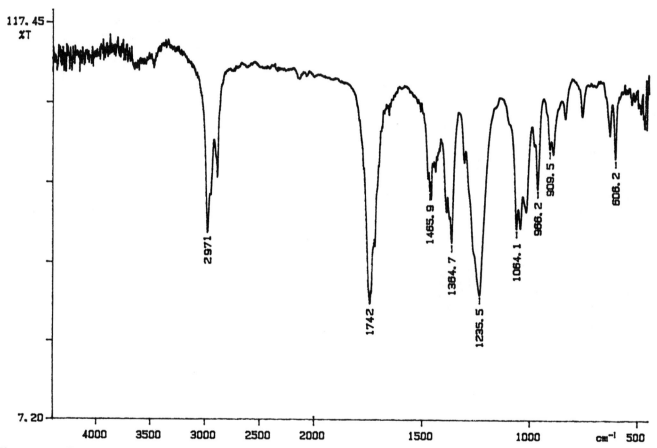

Figure 22.6   IR spectrum of *n*-propyl acetate (neat).

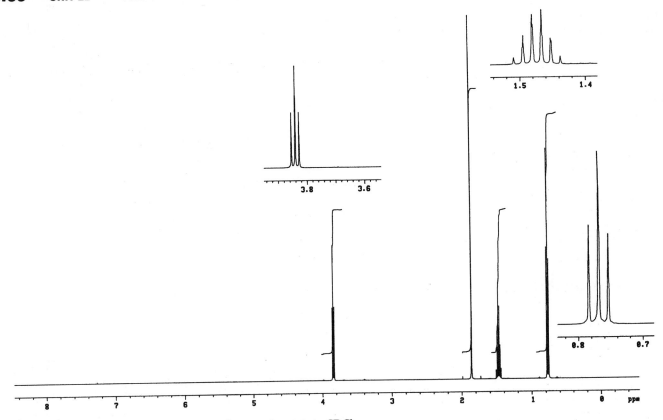

Figure 22.7   500-MHz ^{1}H-NMR spectrum of *n*-propyl acetate in CDCl$_3$.

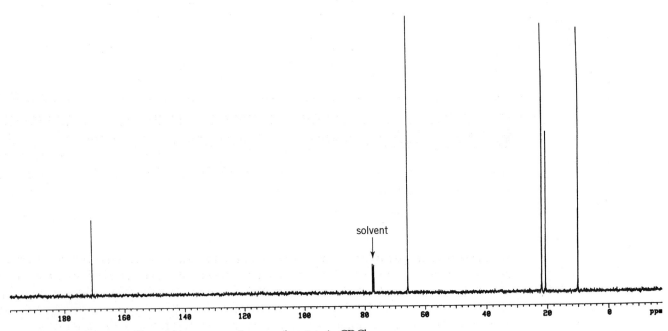

Figure 22.8   125.7-MHz ^{13}C-NMR spectrum of *n*-propyl acetate in CDCl$_3$.

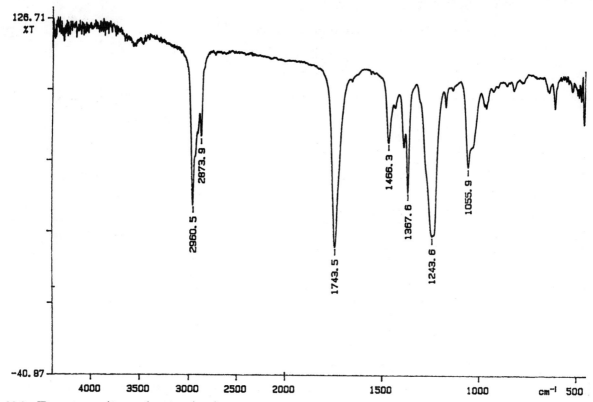

Figure 22.9   IR spectrum of isoamyl acetate (neat).

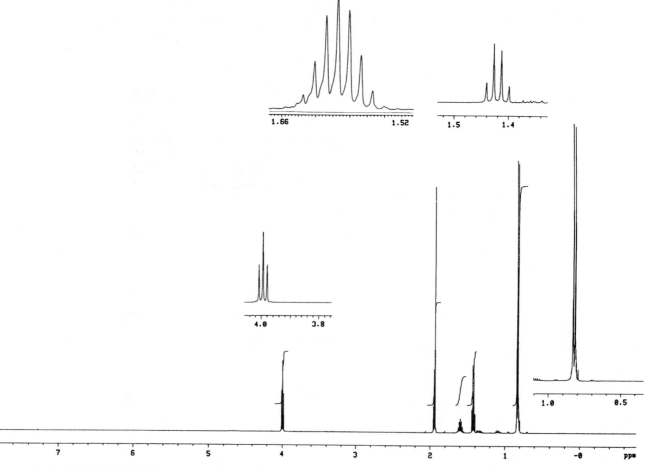

Figure 22.10   500-MHz ^{1}H-NMR spectrum of isoamyl acetate in CDCl$_3$.

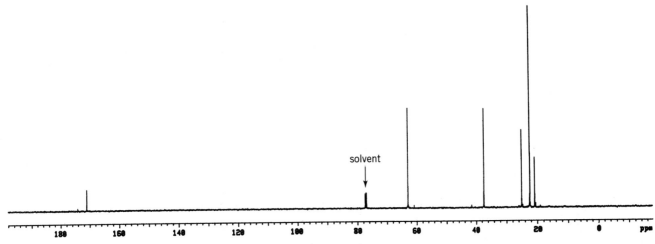

Figure 22.11 125.7-MHz ^{13}C-NMR spectrum of isoamyl acetate in CDCl$_3$.

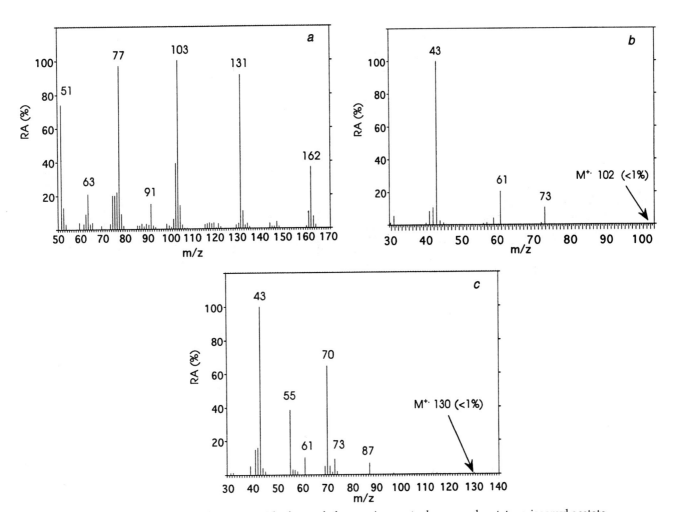

Figure 22.12 Mass spectra (electron impact, 70 eV) of a: methyl *trans*-cinnamate; b: n-propyl acetate; c: isoamyl acetate.

Transforming Bengay into Aspirin

E22B.1 OVERVIEW

In this experiment you will isolate methyl salicylate from a muscle pain relief ointment, such as Bengay, and transform it into aspirin by a series of reactions. Bengay (original formula) contains two active ingredients, methyl salicylate (18.3%) and menthol (16%), in a matrix of wax and other inactive compounds. Methyl salicylate is also known as **wintergreen oil** and has a sweet and minty smell. It is used in perfumery and as a flavoring agent in candy manufacture.

methyl salicylate   menthol

Methyl salicylate is a phenol and an ester. The ester group can be hydrolyzed to afford salicylic acid. In the presence of acetic anhydride, the phenol group of salicylic acid reacts, readily yielding aspirin.

The whole transformation from Bengay to aspirin encompasses several steps. The first step is the extraction of methyl salicylate and menthol from the ointment, followed by their separation. Whereas methyl salicylate and menthol are soluble in methanol, the inactive ingredients, such as wax, are not. The methanol extract is evaporated and the residue, consisting of methyl salicylate and menthol, is separated by **acid-base extraction**. Both menthol and methyl salicylate are insoluble in water, but methyl salicylate is soluble in basic aqueous solutions because the phenol group is dissociated at high pH. Menthol is a neutral compound (an alcohol) and does not react with bases. Thus, a mixture of the two can be separated by extraction with a sodium hydroxide solution. An outline of the separation is shown in the chart below.

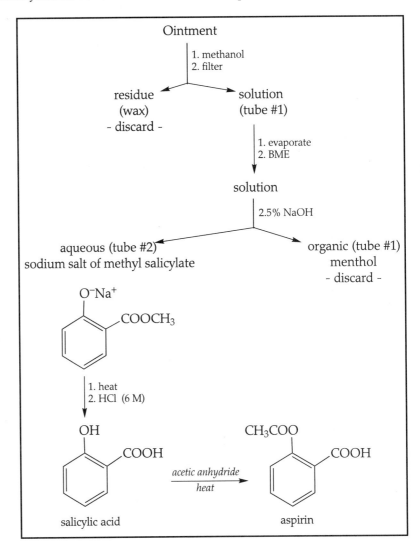

Flow chart for the separation of methyl salicylate from Bengay and its transformation to aspirin.

After the separation of methyl salicylate, you will perform its hydrolysis under basic conditions (saponification) followed by acidification with hydrochloric acid to obtain a white powder of salicylic acid. Without further purification you will react salicylic acid with acetic anhydride to obtain aspirin.

You will assess the success of the transformation by IR spectroscopy and melting points. Methyl salicylate, salicylic acid, and aspirin show significant differences in their IR spectra.

PROCEDURE

Estimated time: 3–5 hours
Background reading: 22.1–5; 14.4

E22B.2 SEPARATION OF METHYL SALICYLATE

Weigh 3.5 g of ointment (Bengay original formula) in a 50-mL beaker. Add 10 mL of methanol and break the paste against the walls of the beaker with the help of a spatula. Do this for about 5 minutes to ensure extraction of the active ingredients. Remove the paste with the spatula and let it dry over a pre-tared double paper towel in the fume hood. Weigh the paste on the paper after the methanol has evaporated.

Transfer the liquid in the beaker into a 25-mL round-bottom flask. Make sure that no residual paste is transferred along with the liquid. Evaporate the solvent in a rota-vap using a hot water bath. Little methanol should remain in the liquid, which should become a viscous residue. Stop the evaporation when no more solvent is collected in the rota-vap's cold trap.

Leave aside one or two drops of the residue for IR analysis; label this "original mixture." With the help of a Pasteur pipet transfer the rest of the liquid in the round-bottom flask to a 16 × 125 mm screw-cap tube (labeled tube "1"). Rinse the walls of the round-bottom flask twice with 3 mL of t-butyl methyl ether (BME) each time, and transfer the rinses to tube 1.

Add 2 mL of 2.5% aqueous NaOH solution, cap the tube, and invert it about 50 times; occasionally vent the tube. To avoid emulsions, do not shake the tube. Let the layers settle. Remove the lower layer with the help of a Pasteur pipet and transfer it to another 16 × 125 screw-cap tube. Label this test tube "2." If the layers are emulsified and do not separate, centrifuge the tube for about 2 minutes (balance the centrifuge!).

Extract the BME solution two more times with 2.5 mL of 2.5% NaOH solution each time, collecting the aqueous layers in test tube 2. Wash the combined aqueous layer in tube 2 with 2 mL of BME. With the help of a Pasteur pipet remove the lower aqueous layer and transfer it to a 15-mL round-bottom flask. Proceed to E22B.3. See "Cleaning Up" at the end of section E22B.5.

Safety First

- HCl and NaOH are corrosive and toxic. Handle them with care.
- *tert*-Butyl methyl ether and methanol are flammable.
- Menthol is an irritant. Avoid skin contact.
- Methyl salicylate is toxic.

E22B.3 HYDROLYSIS OF METHYL SALICYLATE

To the flask from the previous step add 2 mL of 2.5% aqueous NaOH solution and a small boiling chip. Lubricate the joint with a very small amount of grease. Attach a microscale water condenser and reflux the system directly on a hot plate at a medium setting for 25 minutes (Fig. 22.13).

Safety First

- HCl and NaOH are corrosive and toxic. Handle them with care.

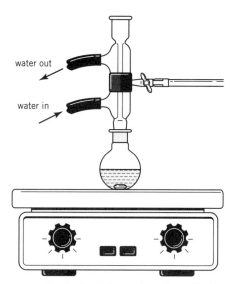

Figure 22.13 Reflux apparatus for the hydrolysis of methyl salicylate.

At the end of this period, turn the heat off and let the solution cool down to ambient temperature. With the help of a Pasteur pipet transfer the solution to a 50–100 mL beaker. Wash the walls of the flask with a little water and transfer the rinses to the beaker. Acidify the solution by adding 6 M HCl until the pH is clearly acidic (pH < 2); about 1–2 mL of acid will be necessary. As the acid is added, a white precipitate of salicylic acid forms.

Cool the beaker in an ice-water bath. Stir the liquid with a glass rod to ensure complete precipitation. Vacuum-filter the suspension using a Hirsch funnel. Let the solid dry on the filter with the vacuum on for about 5 minutes. Transfer the solid to a piece of porous plate and dry it by spreading it on the surface and pressing it with a spatula. Repeat this operation for about 5 minutes. The product should be dry before proceeding to the synthesis of aspirin. Weigh the product and leave a few crystals aside for IR and melting point analysis. Proceed to E22B.4. See "Cleaning Up" at the end of section E22B.5.

Safety First
• Acetic anhydride is a lachrymator. Dispense this chemical in the fume hood.

E22B.4   SYNTHESIS OF ASPIRIN

Fill approximately one-half of a small crystallizing dish (about 8 cm in diameter) with water and place it on a hot plate to boil. In the meantime, in a clean and **dry** 15-mL round-bottom flask weigh approximately 200 mg of salicylic acid from the previous step. Add 2 mL of acetic anhydride with the help of a pluringe (**fume hood!**). If you have less than 200 mg of salicylic acid, reduce the amount of acetic anhydride accordingly (see section 2.9).

Add a small boiling chip and attach a microscale water condenser. If you are using the condenser from the previous step, make sure that its inner walls are perfectly dry (moisture decomposes the acetic anhydride and decreases the yield of aspirin). Use a piece of paper towel to dry the condenser.

Place the round-bottom flask with condenser in the boiling-water bath and let the system react for about five minutes. At the end of this period add 1 mL of water through the top of the reflux condenser. Turn off the heat and let the system react for an additional five-minute period. This will destroy the excess acetic anhydride.

Carefully remove the water bath from the hot plate and let the system cool down to ambient temperature. With the help of a Pasteur pipet transfer the liquid to a 50–100 mL beaker. Wash the walls of the flask with 2 mL of water and transfer the wash to the beaker. Add 4 mL of water to the beaker and cool the system in an ice-water bath to crystallize the aspirin. Sometimes aspirin gives a supersaturated solution, which is difficult to crystallize. To induce precipitation of aspirin, gently scratch the walls of the beaker with a glass rod. After a few minutes aspirin will precipitate as a white mass. If this fails, you can induce the crystallization by "seeding" the supersaturated solution with a few crystals of pure aspirin. Stirring and scratching with the glass rod will finally afford the desired solid.

Vacuum-filter the aspirin using a Hirsch funnel. Let the solid dry on the filter with the vacuum on for about 5 minutes and then transfer it to a piece of porous plate. Dry the solid as you dried the salicylic acid. Weigh the product. Determine its melting point and run its IR spectrum.

Aspirin can be recrystallized from water. Approximately 3 mL of boiling water are needed to dissolve 100 mg of aspirin.

E22B.5   ANALYSIS

Ferric Chloride Test

Background reading: 15.3

Prepare three labeled test tubes with 1 mL of a 0.1% aqueous solution of ferric chloride. To two of them add a small amount (a few crystals) of either salicylic acid (from E22B.3) or aspirin (from E22B.4). Add a drop of water to the third one. Observe the change in color. A red-purple color is a positive test for phenols; yellow is considered negative.

IR Analysis

Run the IR of the "original mixture" directly on NaCl plates. Also obtain the IR of salicylic acid and aspirin using a nujol mull (section 31.5).

The success of the transformation can be followed by the IR spectra of the products. Because the original mixture contains methyl salicylate and menthol, its IR shows typical peaks for these two compounds. The most important bands expected for methyl salicylate, menthol, salicylic acid, and aspirin are indicated below.

Note that aspirin has two carbonyl stretching bands, one around 1750 cm^{-1}, which corresponds to the acetyl group, and another around 1690 cm^{-1}, which belongs to the carboxylic acid.

If possible, run the ^{1}H-NMR of aspirin using CDCl$_3$ as a solvent (section 33.20). (*Caution: CDCl$_3$ is a possible carcinogen; dispense this chemical in the hood.*)

> **Cleaning Up**
>
> - Dispose of the wax residue and the Pasteur pipets in the container labeled "Bengay Solid Waste."
> - At the end of the experiment the instructor will empty the contents of the rota-vap traps in the container labeled "Bengay Organic Solvent Waste."
> - Dispose of the organic layer from E22B.2 (menthol fraction) in the container labeled "Bengay Organic Solvent Waste."
> - Dispose of the filtrates from E22B.3 and E22B.4 in the container labeled "Bengay Aqueous Waste."
> - Dispose of the ferric chloride test solution in the container labeled "Bengay Aqueous Waste."
> - Dispose of the CDCl$_3$ NMR solution in the container labeled "Chloroform Waste."
> - Turn in your product in a capped, labeled vial to your instructor.

EXPERIMENT 22B REPORT

Pre-lab

1. Make a table showing the physical properties (molecular mass, m.p., b.p., solubility, refractive index, flammability [for solvents only], and toxicity/hazards) of methyl salicylate, menthol, methanol, *t*-butyl methyl ether, salicylic acid, aspirin, and acetic anhydride.

2. Classify methyl salicylate and menthol as acidic, basic, or neutral. Identify the functional groups present in each molecule and determine how they affect its acid-base properties.

3. Could you use $NaHCO_3$ instead of $NaOH$ for the separation of methyl salicylate from menthol? Briefly justify your answer.

4. Write a chemical equation for the reaction of methyl salicylate with 2.5% aqueous $NaOH$.

5. Write a chemical equation for the hydrolysis of methyl salicylate with sodium hydroxide. What is the general name for this type of reaction? Is the reaction reversible?

6. Why do you acidify with HCl at the end of the hydrolysis of methyl salicylate?

7. Write a chemical reaction for the acetylation of salicylic acid with acetic anhydride.

8. What would happen if the glassware is not dry? Explain with a chemical reaction.

9. What differences do you expect to see in the IR spectra of methyl salicylate, salicylic acid, and aspirin?

In-lab

1. Calculate the amount in mg of menthol and methyl salicylate present in 3.5 g of ointment, considering that the ointment contains 18.3% methyl salicylate and 16% menthol.

2. Report the mass of the original mixture after evaporation of methanol. Compare this mass with the amount of methyl salicylate and menthol that should be present in 3.5 g of ointment. Read the label of the ointment and speculate what other compound is likely to be present in the methanol extract.

3. Report the mass of salicylic acid and its melting point. How does the melting point compare with the literature value? Briefly discuss.

4. Report the mass of aspirin and its melting point. How does the melting point compare with the literature value? Briefly discuss.

5. Calculate the % yield in the synthesis of aspirin. Discuss your results.

6. Report and interpret your ferric chloride results.

7. Interpret the IR spectra of the original mixture (methyl salicylate plus menthol) and aspirin. Compare them with the ones shown in Figures 22.14 and 22.15.

8. Interpret your ^1H-NMR of aspirin. Compare it with the spectrum shown in E14B (Fig. 14.10).

9. Interpret the ^1H-NMR and ^{13}C-NMR of methyl salicylate shown in Figures 22.16 and 22.17.

10. What spectroscopic evidence for the formation of an intramolecular H-bond in methyl salicylate did you find?

11. (Advanced level) Interpret the mass spectrum of aspirin shown in Figure 22.18. The peak at m/z 138 corresponds to a loss of CH_2CO. Propose a mechanism for this fragmentation.

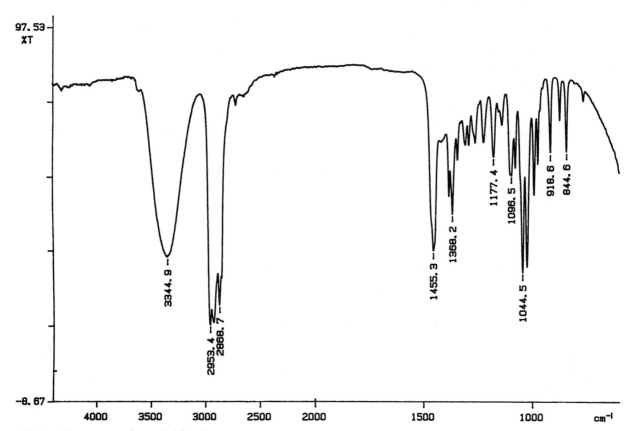

Figure 22.14 IR spectrum of menthol (neat).

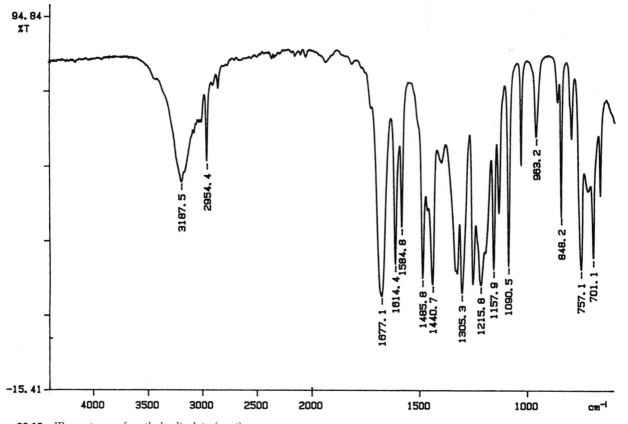

Figure 22.15 IR spectrum of methyl salicylate (neat).

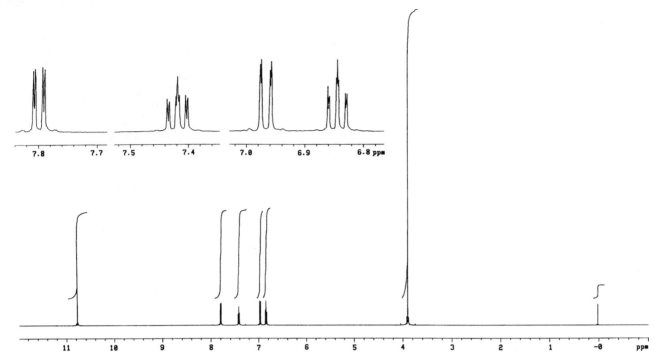

Figure 22.16   500-MHz ^{1}H-NMR spectrum of methyl salicylate in CDCl$_3$.

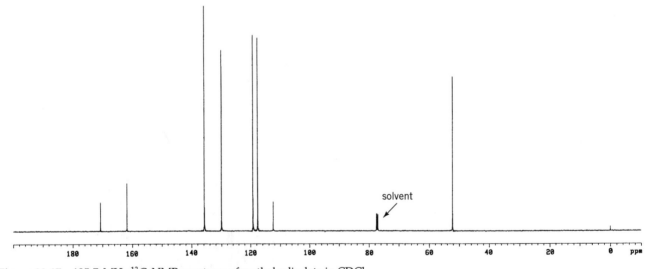

Figure 22.17   125.7-MHz ^{13}C-NMR spectrum of methyl salicylate in CDCl$_3$.

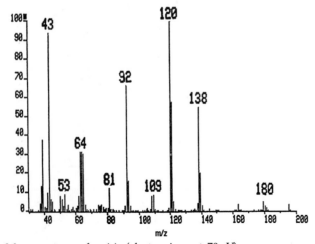

Figure 22.18   Mass spectrum of aspirin (electron impact, 70 eV).

Multistep Synthesis

23.1 OVERVIEW

Look at the molecule below. It represents a compound that you wouldn't want to wear as a perfume. It is the main component of the American cockroach sex attractant. It was discovered as early as 1952 but its isolation was elusive for many years. In 1974, chemists finally managed to isolate 200 μg (equivalent to about one-hundredth of a grain of rice) of periplanone-B from 75,000 cockroaches!

Periplanone-B
American cockroach sex attractant

The chemistry of natural products abounds in cases like this where a biologically active compound can be isolated from its source only in minuscule amounts. This scarcity creates serious medical and social problems as the price for some antibiotics, hormones, or antitumor agents are prohibitively high. Fortunately, organic synthesis can sometimes provide alternative production routes at a fraction of the isolation cost.

Organic synthesis is the science, craft, and art of building organic molecules. It is a science because we can organize, classify, and generalize our findings; it is a craft because it entails laborious work; and it is also an art because it is limited only by our own creativity. To illustrate this point, it can be mentioned that a group of Japanese chemists reported in 1987 the total synthesis of periplanone-B from limonene as starting material. The synthesis was accomplished in 28 steps with an overall yield of 0.5%. It took much less than a gram of limonene to produce 200 μg of periplanone-B, instead of 75,000 cockroaches.

(R)-(+)-limonene   Periplanone-B

28 steps
0.5% overall yield

What are the rules of organic synthesis? How can we build a molecule like periplanone-B from simpler materials? Given a certain final molecule, or a **target molecule**, how do we determine which compounds to use as raw material? How do we choose a given synthetic route? Although a comprehensive answer to all these questions will not be attempted here, this unit will get you acquainted with some fundamental concepts of organic synthesis with emphasis in its experimental facets.

23.2 MULTISTEP SYNTHESIS

To be successful in organic synthesis one should have a solid knowledge of organic reactions and an appreciation for the compounds that can be used as starting materials. One should also be familiar with stereochemistry and reaction mechanisms. With this in mind, let's suppose now that we want to prepare 2-methylcyclohexanone and we have at our disposal cyclohexanol as a reactant. How can we carry out this transformation?

cyclohexanol   2-methylcyclohexanone

A comparison of the target molecule, 2-methylcyclohexanone, with the starting material indicates that we *must add one carbon atom* to the cyclohexanol skeleton. We must also *change the functional group* from an alcohol to a ketone. There is no single reaction that accomplishes both tasks in one step and we should therefore perform a series of reactions. This is a typical example of a **multistep synthesis**.

As you may recall from the chemistry of alcohols and ketones, the hydroxyl group and the carbonyl group can be easily interconverted by redox reactions (Unit 21). For example, ketones can be transformed into alcohols by reduction with $NaBH_4$, and conversely, alcohols can be oxidized to ketones with CrO_3 or ClO^-.

To add a methyl group, one must resort to the C—C bond-forming reactions. Among them is the *alkylation of enolates*, which is very significant in this case because the target molecule is a ketone. It seems reasonable to perform the synthesis of 2-methylcyclohexanone from cyclohexanol according to the following route: oxidation of cyclohexanol to cyclohexanone, followed by the alkylation of the enolate of cyclohexanone with methyl iodide. The enolate of cyclohexanone is obtained by treatment with a strong base such as lithium diisopropylamide (LDA).

oxidation   alkylation

This example illustrates two of the most important types of chemical transformations used in organic synthesis: **functional group interconversions**, and **C—C bond-forming reactions**.

A summary of the most important functional group interconversions for aliphatic and aromatic compounds is given in Figures 23.1 and 23.2, respectively. Table 23.1 offers a list of the most relevant C—C bond-forming reactions. Reactions that transform functional groups and create new C—C bonds at the same time (for example, Grignard reactions) have been excluded from Figures 23.1 and 23.2, and have been incorporated into Table 23.1. These diagrams should contain no new information for you; they just summarize the most important reactions that you have learned in introductory organic chemistry. The diagrams are not comprehensive, and should be taken only as a guide to orient your thoughts when planning a synthesis.

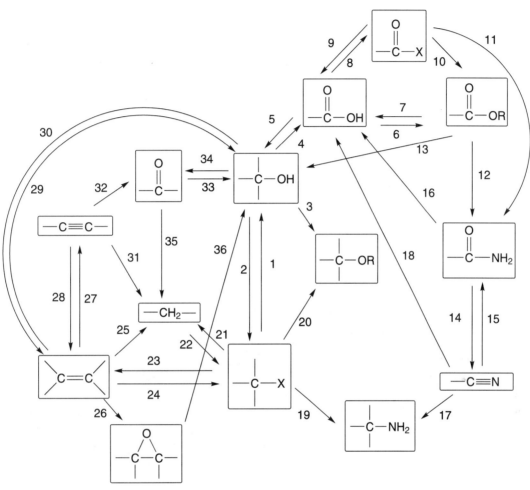

1. HO⁻
2. HX
3. i. NaH: ii. RX
4. CrO₃, H₃O⁺ (for primary alcohols only)
5. i. LiAlH₄ or BH₃; ii. H₃O⁺
6. ROH, acid catalyst
7. H₃O⁺ or NaOH
8. X=Cl: SOCl₂ or PCl₅ or PCl₃; X=Br: PBr₃ or PBr₅
9. H₃O⁺
10. ROH, pyridine
11. NH₃ excess
12. NH₃
13. i. LiAlH₄; ii. H₃O⁺
14. SOCl₂

15. H₂SO₄(c)
16. H₃O⁺ or NaOH
17. i. LiAlH₄, ether; ii. H₃O⁺
18. H₃O⁺ or NaOH
19. NH₃ excess
20. RO⁻
21. i. Mg, ether; ii. H₃O⁺
22. X₂, light or heat
23. Strong base
24. HX
25. H₂, Pd/C
26. R—CO₃H
27. i. X₂; ii. 2 eq. NaNH₂
28. H₂, Lindlar catalyst (gives *cis* alkenes); or Li, NH₃ (gives *trans* alkenes)

29. i. BH₃, THF; ii. H₂O₂, HO⁻; or i. Hg(OAc)₂; ii. NaBH₄
30. Acid catalyst, heat
31. 2 H₂, Pd/C
32. H₂O, HgSO₄, H₂SO₄; or i. disiamylborane; ii. H₂O₂, HO⁻
33. i. NaBH₄ or LiAlH₄; ii. H₃O⁺
34. 1° alcohols; PCC (pyridinium chlorochromate), CH₂Cl₂; 2° alcohols: PCC, CH₂Cl₂; or CrO₃, H₃O⁺; or ClO⁻, AcOH
35. H₂N—NH₂, KOH; or i. 2 RSH; ii. Raney Ni
36. i. LiAl H₄, ether; ii. H₃O⁺

Figure 23.1 Functional group transformations.

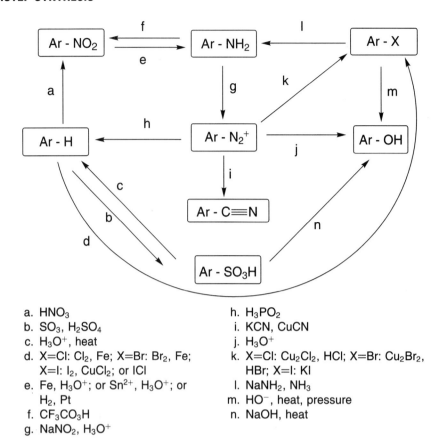

a. HNO₃
b. SO₃, H₂SO₄
c. H₃O⁺, heat
d. X=Cl: Cl₂, Fe; X=Br: Br₂, Fe;
   X=I: I₂, CuCl₂; or ICl
e. Fe, H₃O⁺; or Sn²⁺, H₃O⁺; or
   H₂, Pt
f. CF₃CO₃H
g. NaNO₂, H₃O⁺

h. H₃PO₂
i. KCN, CuCN
j. H₃O⁺
k. X=Cl: Cu₂Cl₂, HCl; X=Br: Cu₂Br₂,
   HBr; X=I: KI
l. NaNH₂, NH₃
m. HO⁻, heat, pressure
n. NaOH, heat

Figure 23.2   Functional group transformations for aromatic compounds.

23.3   RETROSYNTHETIC ANALYSIS

What we have done in the case of 2-methylcyclohexanone is a very simple **retrosynthetic analysis**. In a retrosynthetic analysis one compares the target molecule with the starting material and figures out the necessary transformations. In general, we take the target molecule and disconnect bonds, looking for fragments recognizable as possible starting materials. In doing so, it is crucial that we know a reaction that can reconnect the bonds. In a retrosynthetic analysis we operate backward; we start with the target molecule, and end up with the prospective starting materials. The figure below illustrates the retrosynthetic pathway for 2-methylcyclohexanone. The special arrow ⟹ is used to indicate a retrosynthetic step.

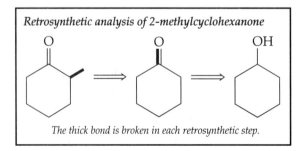

Retrosynthetic analysis of 2-methylcyclohexanone

The thick bond is broken in each retrosynthetic step.

Since the most useful reactions for organic synthesis take place by *polar mechanisms* (an important exception is the Diels–Alder reaction), when we break C–C bonds looking for possible starting materials, we should break them *heterolytically*. For example, in the case of 2-methylcyclohexanone, in breaking the bond between the

Table 23.1 **The Most Important C—C Bond-Forming Reactions**

⁻CN

+ R—X ⟶ R—CN

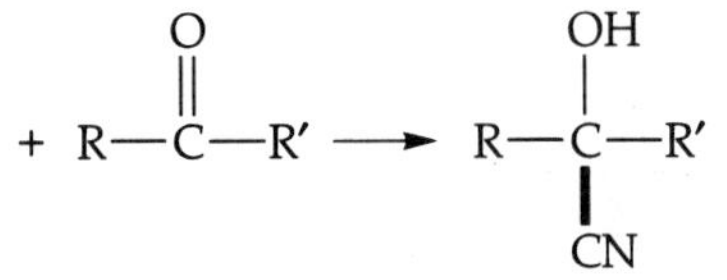

+ $\underset{\underset{\text{O}}{\parallel}}{\text{R—C—R'}}$ ⟶ $\underset{\underset{\text{CN}}{\mid}}{\overset{\overset{\text{OH}}{\mid}}{\text{R—C—R'}}}$

R—C≡C⁻

+ R'—X ⟶ R—C≡C—R'

R—Mg—X   Grignard reactions

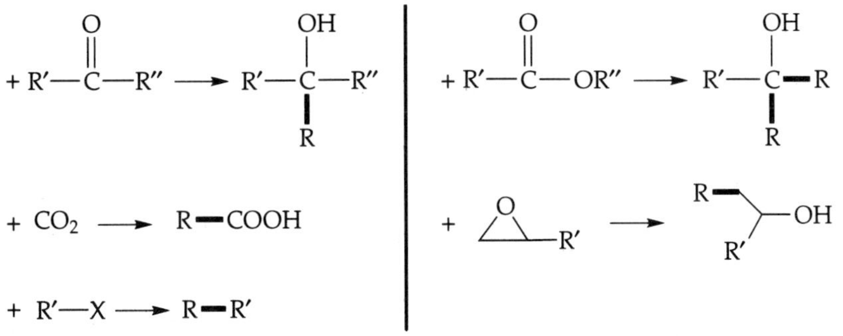

+ $\underset{\underset{\text{O}}{\parallel}}{\text{R'—C—R''}}$ ⟶ $\underset{\underset{\text{R}}{\mid}}{\overset{\overset{\text{OH}}{\mid}}{\text{R'—C—R''}}}$ | + $\underset{\underset{\text{O}}{\parallel}}{\text{R'—C—OR''}}$ ⟶ $\underset{\underset{\text{R}}{\mid}}{\overset{\overset{\text{OH}}{\mid}}{\text{R'—C—R}}}$

+ CO_2 ⟶ R—COOH | + epoxide-R' ⟶ R—CH(R')—CH(R')—OH

+ R'—X ⟶ R—R'

R—X   Alkylations

+ R'₂CuLi ⟶ R—R'

Friedel-Crafts alkylation

+ benzene $\xrightarrow{\text{Al Cl}_3}$ benzene-R

+ $\underset{\underset{\text{O}}{\parallel}}{\text{R'—C—CH—R''}}$ ⟶ $\underset{\underset{\text{R}}{\mid}}{\overset{\overset{\text{O}}{\parallel}}{\text{R'—C—CH—R''}}}$

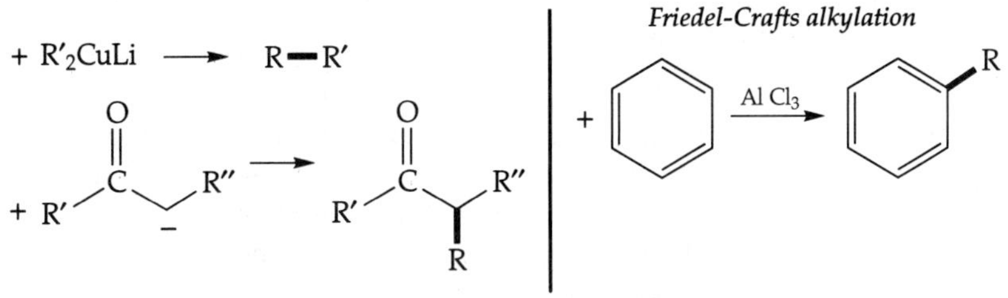

Friedel-Crafts acylation

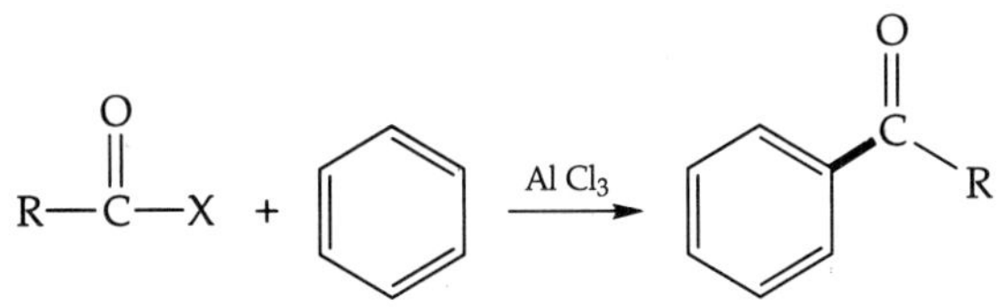

$\underset{\underset{\text{O}}{\parallel}}{\text{R—C—X}}$ + benzene $\xrightarrow{\text{Al Cl}_3}$ benzene-C(=O)-R

Diels-Alder reaction

diene + dienophile ⟶ cyclohexene

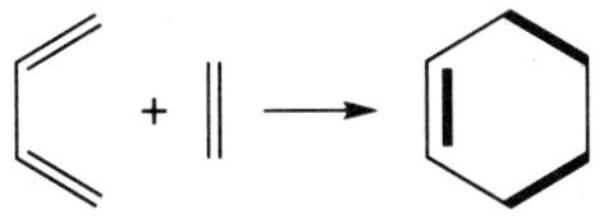

(continued)

Table 23.1 (continued)

Wittig reaction

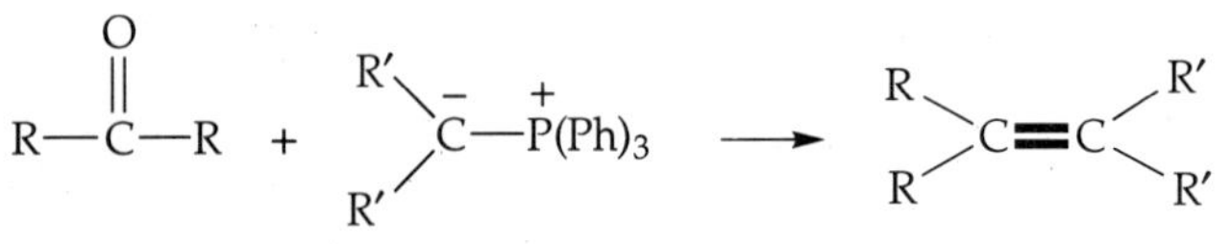

$$R-\overset{\overset{\displaystyle O}{\|}}{C}-R \;+\; \underset{R'}{\overset{R'}{\diagdown}}C\overset{+}{-}\overset{-}{P}(Ph)_3 \;\longrightarrow\; \underset{R}{\overset{R}{\diagup}}C\!\!=\!\!C\underset{R'}{\overset{R'}{\diagdown}}$$

Aldol condensation

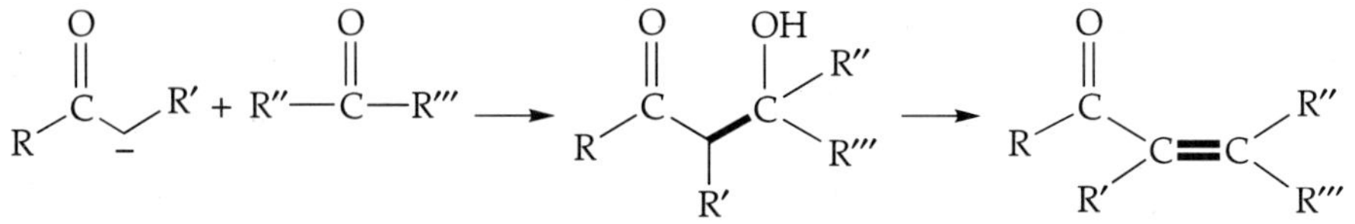

$$\underset{R}{\overset{\overset{\displaystyle O}{\|}}{C}}\diagdown R' \;+\; R''-\overset{\overset{\displaystyle O}{\|}}{C}-R''' \;\longrightarrow\; \underset{R}{\overset{\overset{\displaystyle O}{\|}}{C}}-\underset{R'}{\overset{OH}{\underset{|}{C}}}\overset{R''}{\diagdown}R''' \;\longrightarrow\; \underset{R}{\overset{\overset{\displaystyle O}{\|}}{C}}\underset{R'}{\overset{}{}}C\!\!=\!\!C\overset{R''}{\underset{R'''}{}}$$

Claisen condensation

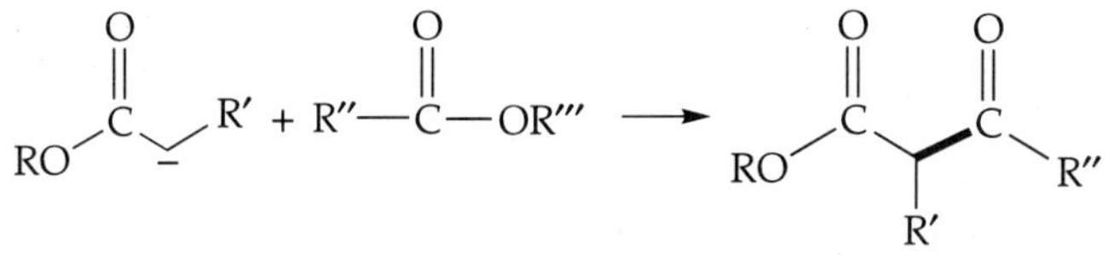

$$\underset{RO}{\overset{\overset{\displaystyle O}{\|}}{C}}\diagdown R' \;+\; R''-\overset{\overset{\displaystyle O}{\|}}{C}-OR''' \;\longrightarrow\; \underset{RO}{\overset{\overset{\displaystyle O}{\|}}{C}}\underset{R'}{\overset{}{}}\overset{\overset{\displaystyle O}{\|}}{C}\diagdown R''$$

Michael reaction

$$\underset{R}{\overset{\overset{\displaystyle O}{\|}}{C}}\diagdown W \;+\; \underset{\|}{\overset{\overset{\displaystyle O}{\|}}{C}}\diagdown R' \;\longrightarrow\; \underset{R}{\overset{\overset{\displaystyle O}{\|}}{C}}\underset{W}{\overset{}{}}\overset{\overset{\displaystyle O}{\|}}{C}\diagdown R'$$

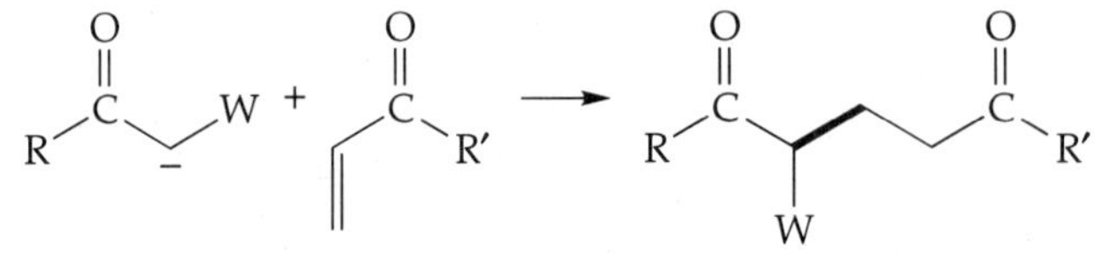

$$-W:-COR'';-COOR'';-CN;-NO_2$$

methyl group and the ring, we should assign positive or negative charges to the two disconnected carbon atoms. There are, in principle, two possibilities: the methyl group takes a positive charge and the ring a negative charge, or vice versa. The resulting fragments are called **synthons**. Synthons are not real molecules, but they can be directly correlated to true starting materials.

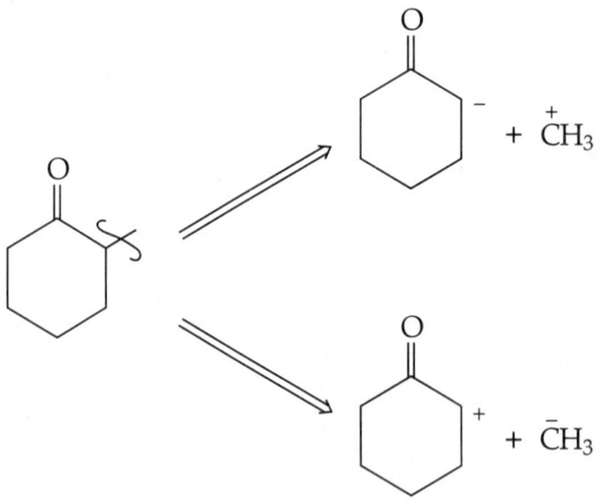

There are two ways of breaking the CH₃-ring bond, leading to four different synthons.

Let's consider the first possibility. A methyl group with a positive charge can be compared to a methyl halide. Methyl halides, because of the difference in electronegativity between the carbon and the halogen, have a *positive charge density on the carbon atom*. A methyl halide is the **synthetic equivalent** for the synthon $^+CH_3$.

$$\overset{+}{C}H_3 \quad \text{its synthetic equivalent is} \quad \overset{\delta+ \quad \delta-}{CH_3\text{---}X}$$

a methyl halide

A synthon with a negative charge on an α-carbon to a carbonyl has a synthetic equivalent in the enolate anion:

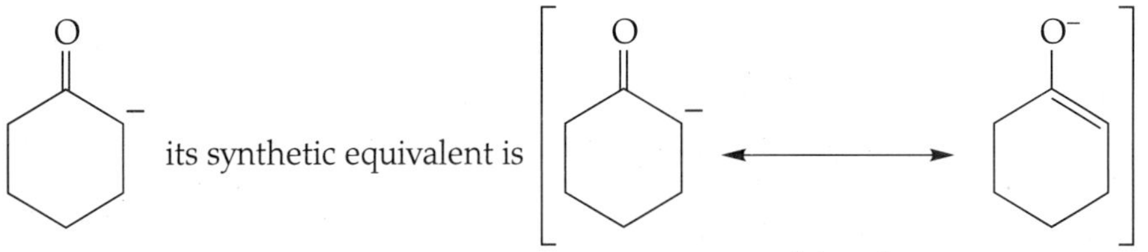

enolate anion

Let's consider now the other possibility in the retrosynthetic analysis of 2-methylcyclohexanone: a methyl group with a negative charge and a ring with a positive charge. The synthetic equivalent of an alkyl group with a negative charge is an organometallic reagent, such as a Grignard reagent. In Grignard reagents the magnesium atom is less electronegative than the carbon, leaving a *negative charge density on the carbon atom*.

$$\overset{-}{C}H_3 \quad \text{its synthetic equivalent is} \quad \overset{\delta- \quad \delta+}{CH_3\text{---}MgX}$$

a Grignard reagent

The synthetic equivalent of the ring with a positive charge on an α-carbon to the carbonyl group is an α-halocarbonyl compound. In Table 23.2 you can find the synthetic equivalents of some common synthons.

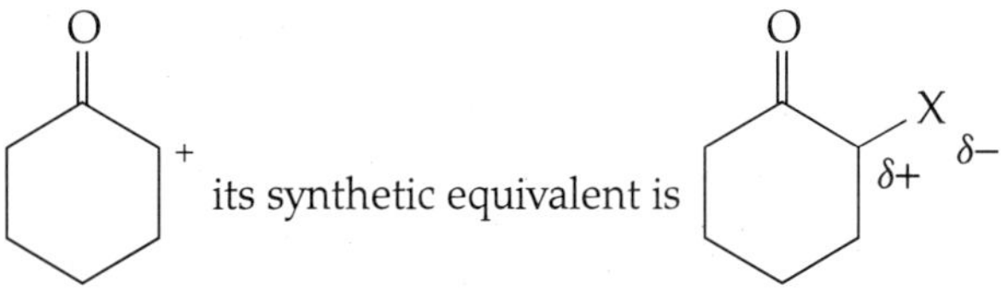

an α-haloketone

In summary, the retrosynthetic analysis of 2-methylcyclohexanone gives two possible synthetic routes: the alkylation of the enolate anion with a methyl halide, and the reaction of a Grignard reagent with an α-haloketone. Which one is better?

We know that alkylation of enolate anions is a common synthetic route that leads to products in good yields. On the other hand, the reaction of Grignard reagents with α-haloketones can produce more than one product. The nucleophilic Grignard reagent cannot only attack the α-carbon, leading to the desired product, but it can also add to the carbonyl group, producing an alcohol as shown below. Therefore, this reaction is not a good choice in the synthesis of 2-methylcyclohexanone. The best approach is the alkylation of the enolate, as already discussed in the previous section.

Table 23.2 Some Important Synthons and their Synthetic Equivalents

Synthon	Synthetic equivalent	Synthon	Synthetic equivalent
$R-\overset{R}{\underset{R}{C}}{}^{+}$	$R-\overset{R}{\underset{R}{C}}-X$	$R-\overset{R}{\underset{R}{C}}{}^{-}$	$R-\overset{R}{\underset{R}{C}}-MgX$; $\left(R-\overset{R}{\underset{R}{C}}\!\!\!-CuLi\right)_2$
$R-\overset{OH}{\underset{+}{C}}-R$	$\underset{R}{\overset{O}{\underset{\parallel}{C}}}\!\!R$	$R-\overset{OH}{\underset{-}{C}}-R$	(1,3-dithiane) $\underset{R}{\overset{S\ \ \ S}{C}}\!\!Li$
$R-\overset{O}{\underset{+}{C}}$	$\underset{R}{\overset{O}{\underset{\parallel}{C}}}\!\!X$	$R-\overset{O}{\underset{-}{C}}$	(1,3-dithiane) $\underset{R}{\overset{S\ \ \ S}{C}}\!\!Li$
$R-\overset{O}{\underset{\parallel}{C}}-\overset{R}{\underset{+}{C}}$	$R-\overset{O}{\underset{\parallel}{C}}-\overset{R}{\underset{X}{C}}$	$R-\overset{O}{\underset{\parallel}{C}}-\overset{R}{\underset{-}{C}}$	$R-\overset{O^-}{C}=\overset{R}{C}$
$R-\overset{OH}{C}-\overset{R}{\underset{+}{C}}$	(epoxide) $R\overset{O}{\triangle}R$		
$R-\overset{O}{\underset{\parallel}{C}}-\overset{+}{C}-R$	$R-\overset{O}{\underset{\parallel}{C}}-CH=CH-R$		

attack on α-C → desired product

+ CH₃MgX

attack on carbonyl → unwanted product

A retrosynthetic analysis is a systematic way to open the door to different synthetic strategies. It should be understood, however, that disconnecting bonds and finding the synthetic equivalents does not give us the answer to the best synthetic route. The best synthetic approach has to be decided based on our knowledge of organic reactions.

In the retrosynthetic analysis of any target molecule, one can try disconnecting all the bonds, one at a time, looking for prospective starting materials. This approach is, however, very time-consuming and impractical, especially for large molecules. Faster results are obtained if one starts the analysis by breaking C–C bonds *near* or *at* the functional groups. One should avoid disconnecting C−C bonds far away from functional groups because there are no important reactions for connecting them back. It is also a good strategy to disconnect the most reactive functional groups first. This means leaving the incorporation of these groups as the last steps of the synthesis. If very reactive groups are placed in the molecule early on during its synthesis, they have to be protected (section 22.3), otherwise chances are that they would react with some of the reagents needed in the intermediate steps and lead to unwanted products. For example, in the synthesis of periplanone-B, the epoxide rings and the carbonyl group were incorporated into the molecule in the very last steps of the synthesis.

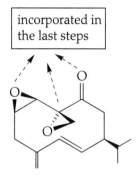

23.4 PLANNING A MULTISTEP SYNTHESIS

As we have already seen, sometimes there are two or more possible synthetic routes to make a desired target molecule. Deciding which one to follow depends on many factors that we must carefully consider before embarking on experimental work. Among these factors are:

1. Number of steps
2. Feasibility of each step
3. Yield of each step
4. Purity of intermediates and final product
5. Experimental conditions
6. Starting materials

We will separately consider each one of these factors, keeping in mind that they are interconnected.

- *Number of steps.* The number of steps must be kept to a minimum. Each step in a synthesis adds to the cost in chemicals and time, and decreases the overall yield. Even when a transformation takes place with quantitative yield, mechanical losses invariably lead to lower yields.

- *Feasibility of each step.* All the steps in a synthesis must be feasible. If one step fails to afford the desired product, the whole synthesis fails. If possible, the questionable step in a synthetic route must be at the beginning, so if it fails we haven't wasted much time or chemicals.

- *Yield of each step.* It is important that each step proceeds with the maximum yield possible. The overall percentage yield is determined by the yield of individual steps.

Let's suppose that the synthesis of a natural product X, begins with compound A and involves three steps with intermediates B and C. Let's also suppose that each step proceeds with a molar yield of 50%. In other words, the number of moles halves in each step. If we start with 20 mmol of A, we obtain 10 mmol of B, 5 of C, and 2.5 of the final product X.

$$A \xrightarrow{\text{50\%}} B \xrightarrow{\text{50\%}} C \xrightarrow{\text{50\%}} X$$

| 20 | 10 | 5 | 2.5 |
| mmol | mmol | mmol | mmol |

The overall yield is:

$$\% \text{ yield} = \frac{2.5 \text{ mmol}}{20 \text{ mmol}} \times 100 = 12.5\% \tag{1}$$

The figure 2.5 mmol was obtained by multiplying 20 mmol by 0.5 three times:

$$2.5 \text{ mmol} = 20 \text{ mmol} \times 0.5 \times 0.5 \times 0.5 \tag{2}$$

thus, substituting equation 2 into equation 1, we obtain:

$$\% \text{ yield} = \frac{20 \text{ mmol} \times 0.5 \times 0.5 \times 0.5}{20 \text{ mmol}} \times 100 = 0.5 \times 0.5 \times 0.5 \times 100$$

The initial number of millimoles, 20 mmol in this case, cancels out and the overall percentage yield is the product of the individual yields. In general:

$$\text{overall } \% \text{ yield} = Y_1 \times Y_2 \times \cdots \times Y_n \times 100 \tag{3}$$

where Y_1, Y_2, Y_n are the individual % yields (divided by 100) for each step.

- *Purity of intermediates and final product.* Ideally, all the steps should afford single products or products that can be easily separated. If possible, one should avoid steps that give mixtures of isomers or very similar products. Remember that compounds that closely resemble each other are usually difficult to separate.

- *Experimental conditions.* Other things being equal, one should choose the synthetic route that requires the simplest experimental conditions. Temperature, solvent, and the need to exclude air or moisture from the reaction mixture are all important factors that one must consider in deciding a synthesis. Many times we have no choice in these matters, but when possible we should avoid working under experimentally demanding conditions, such as inert atmospheres with total exclusion of air or moisture, high pressures or temperatures, or pyrophoric compounds (compounds that ignite spontaneously when exposed to air).

- *Starting materials.* Ultimately, a synthetic route is determined by the availability of the starting materials. In general, small organic compounds with no or very few stereogenic centers are commercially available in high degree of purity. One should always check different manufacturers' catalogs to find out which compounds can be bought directly.

23.5 LINEAR VERSUS CONVERGENT SYNTHESIS

There are two types of synthetic strategies: **linear synthesis** and **convergent synthesis**. The synthesis of 2-methylcyclohexanone from cyclohexanol discussed in section 23.2 is an example of a linear synthesis. In a linear synthesis the product of each step is the main starting material for the following step and the other chemicals needed are inorganic reagents or readily available organic compounds. We can represent a linear synthesis as follows:

A linear synthesis

$$A \longrightarrow B \longrightarrow C \longrightarrow X$$

In this sequence, the reagents as well as the side products of each step have been omitted. A, B, C, and X represent the compounds we are interested in (those bearing the desired carbon skeleton and functional groups).

On the other hand, in a **convergent synthesis**, the final product is obtained by the combination of two intermediates that have also been synthesized. In the following example, B is obtained from A, and D from C. The reaction of B and D then affords the desired product X.

A convergent synthesis

$$A \longrightarrow B$$
$$+ \longrightarrow X$$
$$C \longrightarrow D$$

One may wonder which approach is more effective from the standpoint of yield: linear or convergent synthesis? When the number of steps are the same and the percent yields for individual steps are similar, a convergent synthesis leads, in general, to a better overall *mass yield* than does a linear one. To understand this, let's consider the following examples.

Let's assume that a linear synthesis A $\longrightarrow$ B $\longrightarrow$ C $\longrightarrow$ X proceeds with a *mass percentage yield* of 50% in each step. If we start with 80 g of A, then we obtain 40 g of B, 20 g of C, and finally 10 g of product X.

$$\begin{array}{ccccccc} & 50\% & & 50\% & & 50\% & \\ A & \longrightarrow & B & \longrightarrow & C & \longrightarrow & X \\ 80\ g & & 40\ g & & 20\ g & & 10\ g \end{array}$$

Let's consider now the case of a convergent synthesis, as indicated below, also with three individual steps, each one with a mass percent yield of 50%. If we start with 40 g of each starting material A and C (also a total of 80 g as in the case of the linear synthesis) we end up with 20 g of final product X. This is twice as much as in the case of the linear synthesis.

$$\begin{array}{ccc} & 50\% & \\ A & \longrightarrow & B \\ 40\ g & & 20\ g \end{array}$$

$$\begin{array}{ccc} & 50\% & \\ + & \longrightarrow & X \\ & & 20\ g \end{array}$$

$$\begin{array}{ccc} & 50\% & \\ C & \longrightarrow & D \\ 40\ g & & 20\ g \end{array}$$

BIBLIOGRAPHY

1. Organic Synthesis. M.B. Smith. McGraw-Hill, New York, 1994.
2. The Logic of Chemical Synthesis. E.J. Corey and X-M. Cheng. Wiley, New York, 1989.
3. Organic Synthesis: The Disconnection Approach. S. Warren. Wiley, Chichester, 1982.
4. Principles of Organic Synthesis; 3rd ed. R.O.C. Norman and J.M. Coxon. Blackie Academic and Professional, London, 1993.
5. Guidebook to Organic Synthesis. R.K. Mackie and D.M. Smith. Longman, London, 1982.
6. Advanced Organic Chemistry; 4th ed. J. March. Wiley, New York, 1992.

7. Disconnect by the Numbers. A Beginner's Guide to Synthesis. M.B. Smith, *J. Chem. Ed,* **67,** 848–856 (1990).

8. Total Synthesis of (−)-Periplanone-B, Natural Major Sex-Excitant Pheromone of the American Cockroach, *Periplaneta americana.* T. Kitahara, M. Mori, and K. Mori, *Tetrahedron,* **43,** 2689–2699 (1987).

EXERCISES

Problem 1

Using Figure 23.1, find a synthetic route to convert a primary alcohol into a nitrile with the same number of carbons.

Problem 2

Using Figure 23.1, find a synthetic route to transform a nitrile into an alkane with the same number of carbons.

Problem 3

Using Figure 23.1 and 23.2, find a synthetic route to obtain benzoic acid from benzene.

Problem 4

Using Table 23.2, give synthetic equivalents for the following synthons:

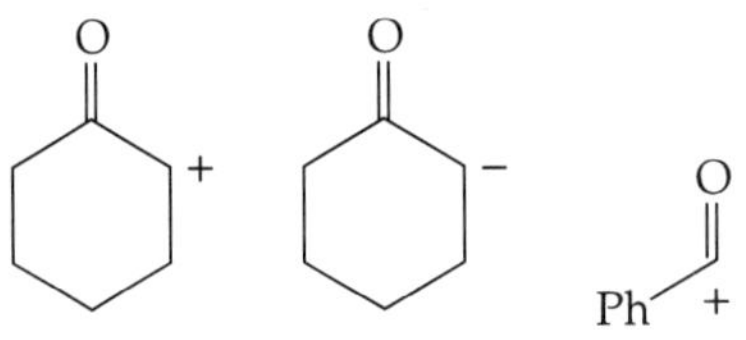

Problem 5

Do a retrosynthetic analysis of the following target molecule. Use only 4-carbon-atom compounds as starting materials.

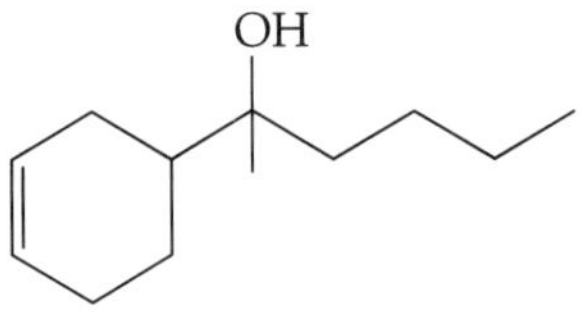

Problem 6

(a) A linear synthesis has 17 steps. What should be the average percentage yield of each step if the overall percentage yield is to be i) 95%; ii) 50%; and iii) 1%?

(b) Calculate the overall percentage yield of a linear synthesis of iv) 5 steps; v) 17 steps; and vi) 22 steps, if the average percentage yield of each step is 50%.

Synthesis of an Ant Alarm Pheromone: 2-Methyl-4-heptanone

E23A.1 OVERVIEW

In this experiment you will synthesize 2-methyl-4-heptanone using starting materials with four carbon atoms. 2-Methyl-4-heptanone is the alarm pheromone of two species of ant (*Tapinoma niggerimum* and *Tapinoma simrothi*) with habitat in the Mediterranean region.

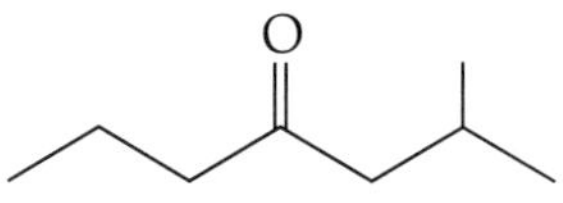

2-methyl-4-heptanone
an ant alarm pheromone

Before you go on, make a retrosynthetic analysis of 2-methyl-4-heptanone, using compounds with four or less carbon atoms as starting materials. Evaluate the feasibility of the different synthetic routes.

E23A.2 SYNTHETIC ROUTE

The synthetic route chosen to make 2-methyl-4-heptanone is based on the following retrosynthetic analysis:

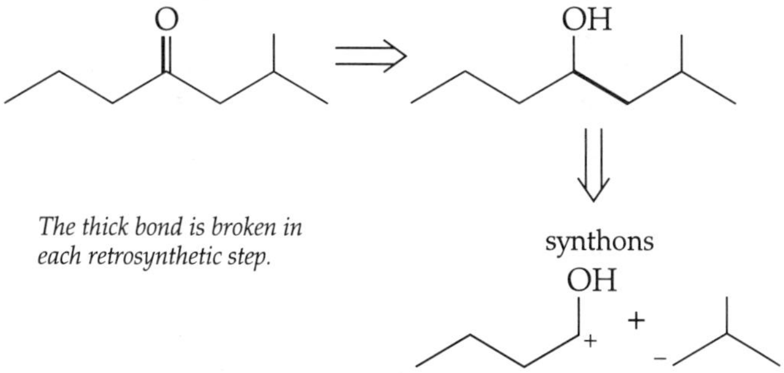

The thick bond is broken in each retrosynthetic step.

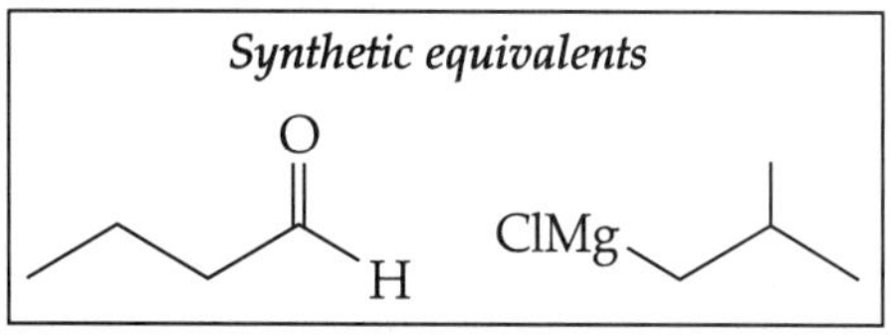

synthons

Synthetic equivalents

This retrosynthetic analysis leads to two convenient starting materials: butanal and the Grignard reagent from 1-chloro-2-methylpropane. In the first step, a Grignard reaction connects these two starting materials, leading to a secondary alcohol:

Pheromones are compounds produced and released to the environment by numerous species of animals and detected by members of the same species, on which they evoke specific behavioral or physiological responses. The term *pheromone* was first coined in 1959. It is derived from the Greek and it literally means "to bear excitement" (*pherein*, to bear; and *hormon*, excitement). Pheromones can be single compounds or a mixture of compounds acting in a synergistic way. They are active in minute quantities and elicit different types of responses ranging from sexual attraction to alarm and territory marking. Essential to the life of social insects such as honey bees and ants, pheromones are the most important means of communication between members of the community. Social insects possess a repertoire of pheromones that indicate the timing for work, procreation, brood rearing, fighting, and even ousting the dead from the colony. The use of pheromones for pest control is very promising. Pheromones can be used to lure specific insects to field traps where they are killed. This form of pest control is environmentally safer than the use of broad-spectrum insecticides, which kill destructive and beneficial insects alike. This field is still in its infancy.

First step: Grignard reaction

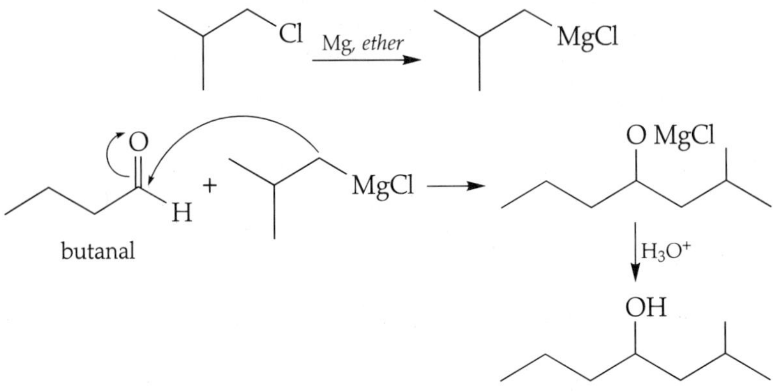

butanal

2-methyl-4-heptanol

In the second step, oxidation of the secondary alcohol produces the desired ketone:

Second step: Oxidation of a secondary alcohol

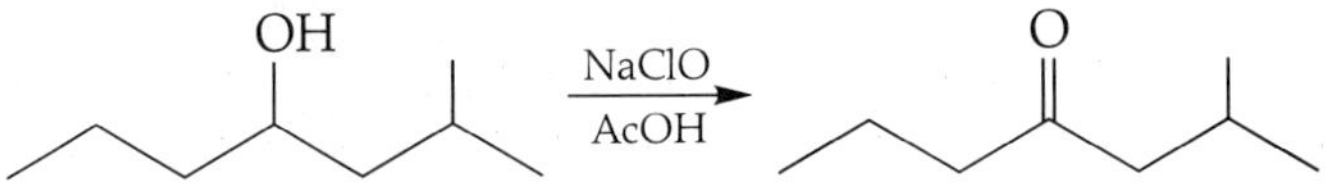

2-methyl-4-heptanone

Each individual reaction proceeds in high yield, making the synthesis of the target molecule, 2-methyl-4-heptanone, feasible. This experiment has been adapted from Reference 1.

E23A.3 GRIGNARD REACTIONS

Grignard reagents are among the most versatile compounds in organic chemistry. They react with a large number of electrophiles, leading to the formation of a vast array of functional groups such as primary, secondary, and tertiary alcohols; aldehydes; ketones; and carboxylic acids. Grignard reagents are obtained in the laboratory by the reaction of alkyl and aryl halides with magnesium in an ether solvent. The most widely used solvents are diethyl ether, commonly called just "ether," and tetrahydrofuran (THF).

Common solvents for Grignard reagents

Diethyl ether
("ether")

tetrahydrofuran
(THF)

The structure of the Grignard reagents has been the subject of numerous studies. The C–Mg bond is covalent, not ionic, with the expected polarization due to differences in electronegativity: the carbon atom has a negative charge density and the magnesium a positive charge density. The solvent plays an important role in the stability and structure of the Grignard reagents. In diethyl ether, Grignard reagents form complexes with two molecules of solvent:

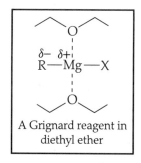

A Grignard reagent in
diethyl ether

In the preparation of the Grignard reagent, the reaction is initiated by adding the alkyl or aryl halide (iodide, bromide, or chloride) to a suspension of *activated magnesium* in ether. To ensure a rapid reaction, the magnesium should be grease-free and activated with a small amount of iodine, I_2. The role of iodine is not totally clear. It etches the surface of the metal making it more reactive, but in the process it forms small amounts of MgI_2 that may be a catalyst for the reaction.

$$R{-}X \xrightarrow[\text{small amount } l_2]{\text{Mg, ether}} R{-}Mg{-}X$$

$$X = Cl, Br, l$$

The addition of the alkyl or aryl halide should be done slowly to keep its concentration to a minimum and to avoid its reaction with the freshly formed Grignard reagent, a process that forms alkanes:

$$R{-}Mg{-}X + R{-}X \longrightarrow R{-}R + MgX_2$$

It is very important to exclude any trace of moisture from the reaction mixture when handling Grignard reagents because water decomposes them, producing alkanes. The solvent must be anhydrous and the reactants and glassware must be free of water.

$$R{-}Mg{-}X + H_2O \longrightarrow R{-}H + MgXOH$$

Because Grignard reagents react with oxygen, it is also recommended, although not indispensable, to handle them under nitrogen.

The mechanism of the Grignard addition to electrophiles is a matter of controversy. Impurities in the magnesium, such as iron, seem to have an important effect in the mechanism. The addition to aliphatic carbonyl compounds seems to involve a nucleophilic attack from the Grignard reagent in which the R group adds with its two electrons as shown below. For diaryl ketones (R′ and R″ aromatic), the mechanism seems to involve the formation of free radicals.

R′ and R″ aliphatic

The magnesium alkoxide is hydrolyzed by the addition of dilute HCl or H_2SO_4, producing the final alcohol. For tertiary alcohols, which easily undergo dehydration reactions in the presence of strong acids, NH_4Cl in water is used instead of HCl or H_2SO_4.

$$R''{-}\underset{\underset{R}{|}}{\overset{\overset{OMgX}{|}}{C}}{-}R' \xrightarrow{\text{HCl, H}_2\text{O}} R''{-}\underset{\underset{R}{|}}{\overset{\overset{OH}{|}}{C}}{-}R' + MgXCl$$

Grignard reagents react with all functional groups with the exception of tertiary amines, alkenes, alkanes, nonterminal alkynes, and ethers. This high reactivity makes Grignard reagents very versatile but, at the same time, limits their applications because they lead to a mixture of products when used with many polyfunctional molecules.

E23A.4   OXIDATION OF SECONDARY ALCOHOLS

The oxidation of secondary alcohols to ketones is a relatively easy transformation that can be performed in the laboratory using a variety of reagents, as discussed in section 21.3. In this experiment you will use sodium hypochlorite as an oxidizing agent. The overall reaction, whose mechanism is discussed in section 21.3, can be represented as follows:

$$R{-}\underset{\underset{R'}{|}}{\overset{\overset{OH}{|}}{C}}{-}H + NaClO \xrightarrow{\text{acetic acid}} R{-}\overset{\overset{O}{||}}{C}{-}R' + NaCl + H_2O$$

An important safety alert in the use of sodium hypochlorite is the presence of Cl_2 (a potent irritant) in the hypochlorite solution. Free chlorine concentration can be kept below physiologically detectable limits by working with dilute solutions of hypochlorite. At the end of the reaction, the excess hypochlorite can be destroyed with sodium thiosulfate, which reduces ClO^- to Cl^-:

$$2\,S_2O_3{}^{2-} + ClO^- + 2\,H_2O \longrightarrow S_4O_6{}^{2-} + Cl^- + 2\,HO^-$$

PROCEDURE

> **Background reading:** 23.1–5; E23A.1–4;
> 5.4; 6.6; 6.8; 6.10; 3.8
> **Estimated time:** 6 hours

E23A.5   SYNTHESIS OF 2-METHYL-4-HEPTANOL

Assemble the apparatus shown in Figure 23.3 using a 15-mL round-bottom flask, a microscale Claisen head, a microscale water-jacketed condenser, a drying tube with anhydrous calcium chloride, and a rubber septum. The flask should contain a stir bar and should rest on a hot plate. To remove last traces of moisture, heat the assembled apparatus by blowing hot air from a heating gun or hair dryer. Stop the heat, remove the rubber septum from the Claisen tube, and add 0.4 g of magnesium turnings followed by a few crystals of iodine. Immediately cap the Claisen head with the rubber septum.

In the meantime, prepare a solution of 1.1 mL of 1-chloro-2-methylpropane in 6 mL of diethyl ether in a perfectly dry 125-mL Erlenmeyer flask (that has been kept in the desiccator) equipped with a stopper. Draw the solution using a 12 mL syringe and a needle with a blunt tip. Insert the needle with the syringe in the right-side port of the Claisen head with the rubber septum (Fig. 23.3b).

Safety First

- Diethyl ether is very flammable and prone to form explosive peroxides on standing. Use a freshly opened container of diethyl ether. Do not use or store it beyond its expiration date.

- Acetic acid is corrosive and flammable.

- Methylene chloride and DCCl₃ are possible carcinogens. Handle them in a well-ventilated place.

Important Note

The glassware should be scrupulously dried. Place the glassware shown in Figure 23.3 (without the plastic caps and rubber septum) in an oven at 110°C for at least 30 minutes. Also dry in the oven two 125 mL Erlenmeyer flasks (the glassware can be put out to dry the night before the experiment). Remove the glassware from the oven and let it cool down in a desiccator. Commercial butanal contains water and must be distilled before used. The water is removed as an azeotrope that distills at 68°C; the fraction distilling between 72–75°C should be used. 1-Chloro-2-methylpropane should be dried over anhydrous sodium sulfate for at least 24 hours prior to use.

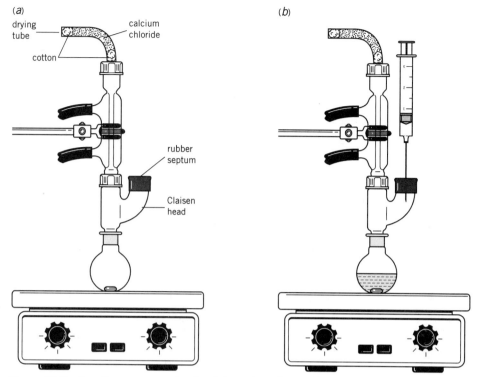

Figure 23.3 Reaction apparatus for the Grignard synthesis of 2-methyl-4-heptanone.

Heat the round-bottom flask with the hair dryer to sublime the iodine onto the magnesium turnings. Stop the heat and start adding the solution in the syringe dropwise to produce a gentle reflux. At the beginning of the reaction you should observe a light bubbling on the magnesium turnings. When the magnesium is totally covered by liquid, start the stirring. The addition of the liquid should take about 20 minutes. If the reflux stops, turn on the heat (low setting). After the addition of the liquid is complete, continue the reflux for about 20 minutes.

Stop the heat and cool down the reaction mixture to ambient temperature. Remove the syringe and load it with 2.7 mL of a freshly prepared solution of butanal in ether (0.6 mL of butanal in 2.1 mL of ether). Reposition the syringe and add the aldehyde solution *dropwise* to the Grignard solution while stirring. After the addition is complete reflux the system for about 20 minutes.

Stop the heat and cool down the reaction mixture to ambient temperature. Remove the syringe. Rinse the syringe with water and load it with 1.0 mL of water. Reposition the syringe and add the water dropwise while stirring. After the addition is complete, load the syringe with 3 mL of a 10% (w/v) aqueous HCl solution and add it dropwise. Separate most of the solution from the solid using a Pasteur pipet and transfer the liquid to a 16 × 125-mm screw-cap test tube. Add 0.5 mL of ether to the magnesium remaining in the round-bottom flask, and filter it using a Pasteur pipet with a cotton plug. Collect the filtrate in the same screw-cap test tube.

Separate the layers with the help of a Pasteur pipet. Transfer the lower layer to a container labeled "aqueous"; keep it until the end and then discard it. Wash the ether layer with 3 mL of a 10% aqueous NaOH solution. Separate the layers, collecting the lower layer in the container labeled "aqueous." Dry the organic layer with anhydrous MgSO₄. Remove the drying agent by gravity filtration using a microscale funnel (25 mm in diameter) and a small cotton plug. Collect the filtrate in a dry 15-mL pre-tared round-bottom flask and evaporate the solvent in a rota-vap. Weigh the product.

Remove a very small drop of product to obtain the IR spectrum of the neat liquid (section 31.5). Weigh the product again and use it in its totality for the next step.

Adjust the volumes of reagents according to the mass of your product (see section 2.9). The quantities indicated below are for 0.5 g of product.

E23A.6   SYNTHESIS OF 2-METHYL-4-HEPTANONE

Dissolve the liquid, 2-methyl-4-heptanol, from the previous step (0.50 g) contained in the 15-mL round-bottom flask in 2.5 mL of acetic acid. Assemble an apparatus as shown in Figure 23.3b but without the drying tube. Cool the round-bottom flask using a cold water bath. With the help of a 12-mL syringe add dropwise, while stirring, 3.75 mL of an aqueous 2.1 M sodium hypochlorite solution (approximately 12.7% [w/w] available chlorine; "swimming-pool" chlorine; the solution should be fresh). The addition should take about 10 minutes. Remove the water bath and stir the system at room temperature for about 1.5 hours.

With a Pasteur pipet transfer the liquid to a 125-mL separatory funnel. Wash the flask with 15 mL of water and transfer the water to the separatory funnel. Extract the aqueous layer twice with 15 mL of methylene chloride each time. Collect the aqueous layer in a 250-mL beaker labeled "aqueous." Extract the combined organic layer twice with 10 mL (each time) of a saturated aqueous solution of sodium bicarbonate. Perform these extractions carefully because CO_2 is produced. Frequently vent the funnel. Collect the aqueous layers in the beaker labeled "aqueous." Extract the organic layer twice with 10 mL (each time) of an aqueous 10% sodium thiosulfate solution. Collect the aqueous layers in the beaker labeled "aqueous." Dry the organic layer over magnesium sulfate. Remove the drying agent by gravity filtration using a small glass funnel and a cotton plug. Collect the filtrate in a pre-tared round-bottom flask and evaporate the solvent in a rota-vap. Weigh the product and calculate the percentage yield. If you obtain more than 300 mg of product you can purify the liquid by vacuum distillation (E23A.7); b.p. 40°C (10 Torr); 155°C (750 Torr). Obtain the IR spectrum of the neat liquid (section 31.5). Obtain its NMR in $CDCl_3$ (section 33.20).

E23A.7   VACUUM DISTILLATION

Assemble a microscale distillation apparatus as shown in Figure 23.4, using a 5-mL round-bottom flask, a Hickman still, and a vacuum adaptor. Use lengths of vacuum

---

Cleaning Up

- Dispose of the "aqueous" layers from E23A.5 and E23A.6 in the container labeled "Pheromones Liquid Waste."

- Dispose of the excess magnesium in the container labeled "Pheromones Magnesium Waste."

- Dispose of the Pasteur pipets and solid magnesium sulfate in the container "Pheromones Solid Waste."

- At the end of the experiment the instructor will empty the rota-vap traps and immediately dispose of the liquid as a hazardous waste.

- Dispose of the NMR solution in the container labeled "$CDCl_3$ Waste."

- Turn in your product to your instructor in a labeled vial.

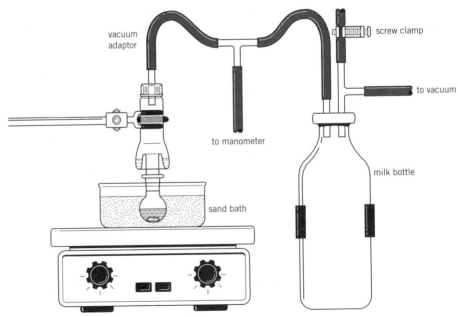

Figure 23.4 Microscale vacuum distillation apparatus.

tubing to connect the system as indicated in the figure; use regular latex tubing for the length with the screw clamp. This apparatus is suitable to distill very small amounts of liquids, but doesn't allow the determination of the temperature during the distillation.

To generate the vacuum you can use a vacuum pump, the house vacuum, or a water aspirator. Once the system is assembled and the manometer has been connected, begin the stirring of the liquid in the round-bottom flask. Make sure that the screw clamp is open and turn the vacuum on. Tighten the screw clamp and read the pressure on the manometer. Once the pressure has stabilized turn on the heat. The liquid should distill without bumping. If bumping occurs, increase the stirring speed. To stop the distillation, turn off the heat and open the screw clamp. Turn off the vacuum.

BIBLIOGRAPHY

1. The Synthesis of 2-Methyl-4-Heptanone. An Ant Alarm Pheromone. E.A. de Jong and B.L. Feringa. *J. Chem. Ed.*, **68**, 71–72 (1991).

2. The Sex Pheromones of the Gypsy Moth and the American Cockroach. W.F. Wood. *J. Chem. Ed.*, **59**, 35–36 (1982).

3. Will Pheromones Be the Next Generation of Pesticides? A. Shani. *J. Chem. Ed.*, **59**, 579–581 (1982).

4. Chemical Communication by Insects. W.R. Stine. *J. Chem. Ed.*, **63**, 603–606 (1986).

5. Chemical Ecology: Chemical Communication in Nature. W.F. Wood. *J. Chem. Ed.*, **60**, 531–539 (1983).

6. Ecology and Chemistry of Mammalian Pheromones. R.L. Brahmachary. *Endeavour*, **10**, 65–58 (1986).

7. Caste-Selective Pheromone Biosynthesis in Honeybees. E. Plettner, K.N. Slessor, M.L. Winston, and J.E. Oliver. *Science*, **271**, 1851–1853 (1996).

EXPERIMENT 23A   REPORT

Pre-lab

1. Make a table with the physical properties of reactants and products. Include toxicity/hazards and solvent flammability.

2. Make a flowchart for the preparation of 2-methyl-4-heptanone.

3. Write the chemical reactions involved in the synthesis.

4. What precautions should you take when working with Grignard reagents?

5. What are the potential hazards of working with diethyl ether?

6. Which of the following functional groups react with Grignard reagents: alkenes; terminal alkynes; nonterminal alkynes; alcohols; alkanes; tertiary amines; secondary amines?

7. After the addition of butanal to the Grignard reagent you add water and HCl. Why? What could you use instead of HCl?

8. After the oxidation with NaClO you extract the organic layer with a 5% sodium thiosulfate solution. Why? Answer with a chemical equation.

In-lab

1. Report the mass and percentage yield of 2-methyl-4-heptanol.

2. Interpret the IR of 2-methyl-4-heptanol.

3. How can you tell that the Grignard reaction was successful from your IR spectrum?

4. Report the mass and percentage yield of 2-methyl-4-heptanone. Discuss losses.

5. How can you tell that the oxidation was successful from the IR spectrum?

6. Interpret the IR and NMR of 2-methyl-4-heptanone and compare it with the ones shown in Figures 23.5–23.7.

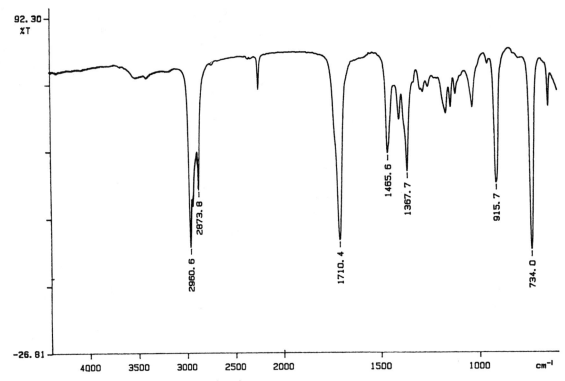

Figure 23.5   IR spectrum of 2-methyl-4-heptanone (neat).

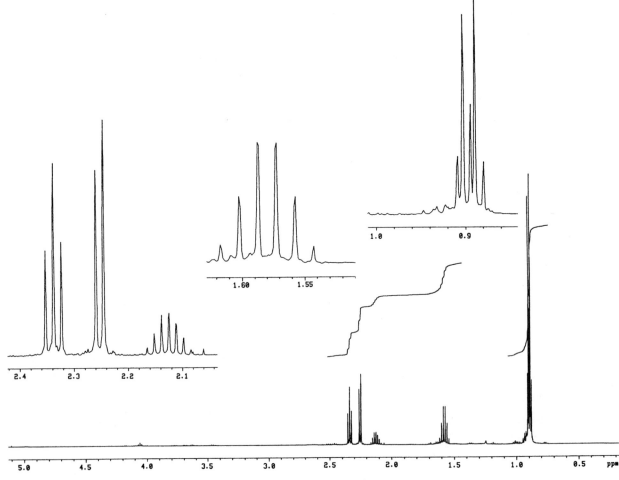

Figure 23.6   500-MHz ^{1}H-NMR spectrum of 2-methyl-4-heptanone in CDCl$_3$.

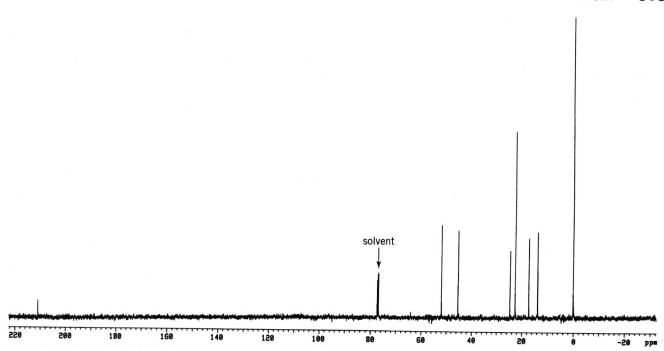

solvent

Figure 23.7   125.7-MHz ^{13}C-NMR spectrum of 2-methyl-4-heptanone in CDCl$_3$.

EXPERIMENT 23B

Synthesis of Ionones: An Open-Ended Experiment

E23B.1   IONONES

Ionones are unsaturated ketones responsible for the characteristic fragrance of violets. The most abundant isomers, α- and β-ionone, can be synthesized in the laboratory by a two-step process. In the undiluted form, the mixture of ionones has a smell that resembles cedarwood.

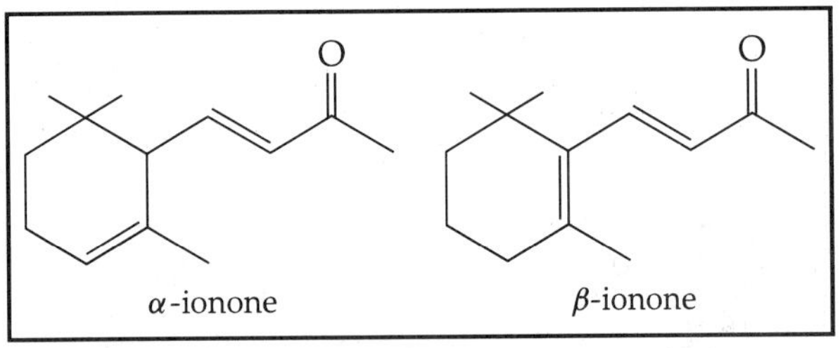

α-ionone   β-ionone

Perfumes from violets are difficult to obtain and have been highly appreciated since antiquity. An old method for obtaining this perfume consisted of embedding the petals of freshly cut violets between layers of animal fat. The aromatic oils were slowly absorbed by the fat that was then used as a hair cream or ointment. Napoleon was fascinated with the scent of violets, which was one of Josephine's favorite perfumes. When she died, Napoleon ordered violets to be planted by her grave. Before his exile to St. Helena, he visited her tomb, gathered some violets and concealed them in a locket that he wore until his death.

E23B.2   SYNTHETIC PATHWAY

The synthesis of ionones is an important industrial process. Ionones are not only used in perfumery, but they are also key intermediates in the manufacture of vitamin A. Both ionones can be prepared by the **aldol condensation** of citral with acetone followed by acid treatment. Before you read on, look at the ionones' structures and reason which bond can be formed by an aldol condensation (section 20.3).

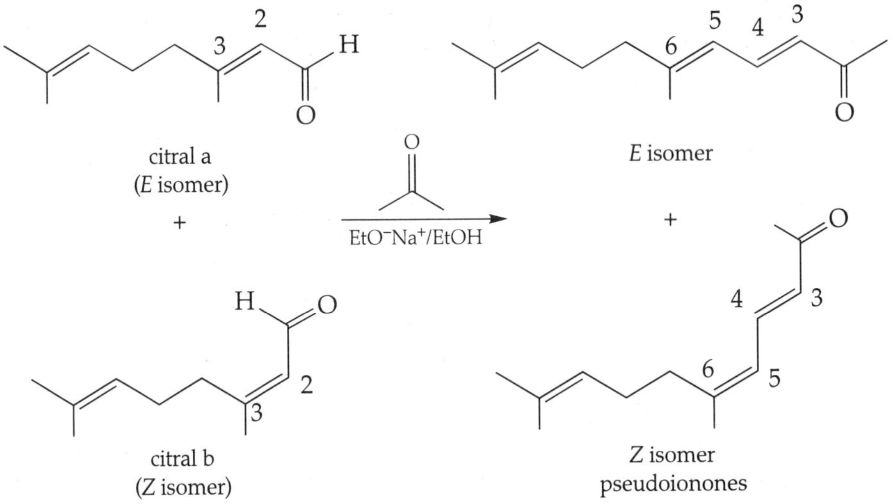

citral a
(E isomer)

citral b
(Z isomer)

EtO⁻Na⁺/EtOH

E isomer

Z isomer
pseudoionones

520

Citral is a mixture of the two *E-Z* stereoisomers of 3,7-dimethyl-2,6-octadienal. **Citral a** or **geranial** (*E* isomer) and **citral b** or **neral** (*Z* isomer) are the main components of the lemongrass oil. The aldol condensation of citral with acetone in the presence of sodium ethoxide gives a mixture of two products, **pseudoionones**, which differ only in the geometry of the double bond between C5 and C6.

Notice that in making the pseudoionones, the newly formed double bond between C3 and C4 has *trans* geometry.

The treatment of pseudoionones with acids induces their **cyclization** to α- and β-ionone. Depending on the *nature and concentration of the acid, the formation of one isomer is preferred over the other*. The products obtained in the presence of sulfuric or phosphoric acids are different; in each case mainly one isomer, either α- or β-ionone is obtained.

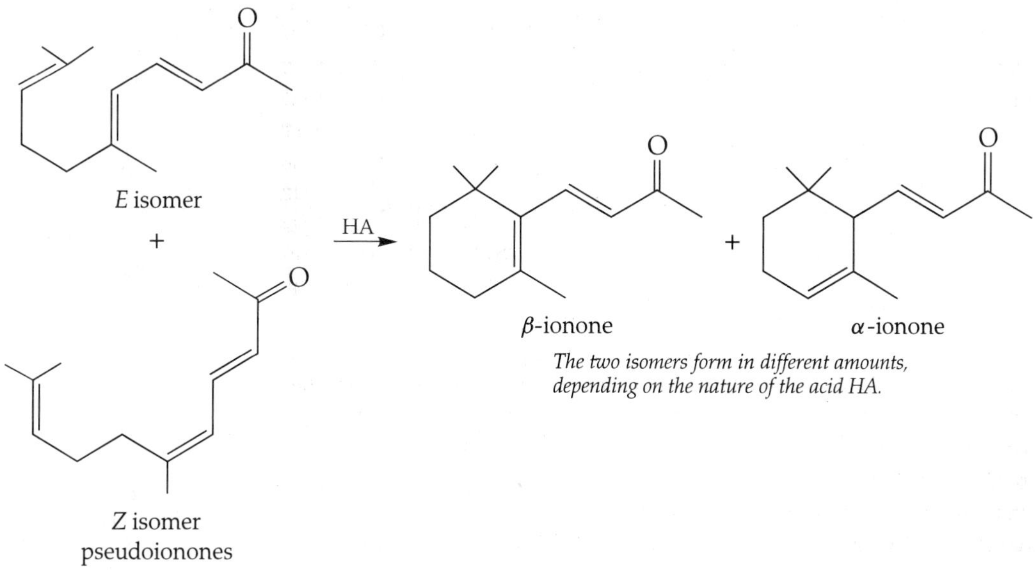

β-ionone   α-ionone

The two isomers form in different amounts, depending on the nature of the acid HA.

The first step in the cyclization of pseudoionones is the protonation of C9 by the acid catalyst. This is followed by an intramolecular attack from C5 to form a six-membered ring with a tertiary carbocation:

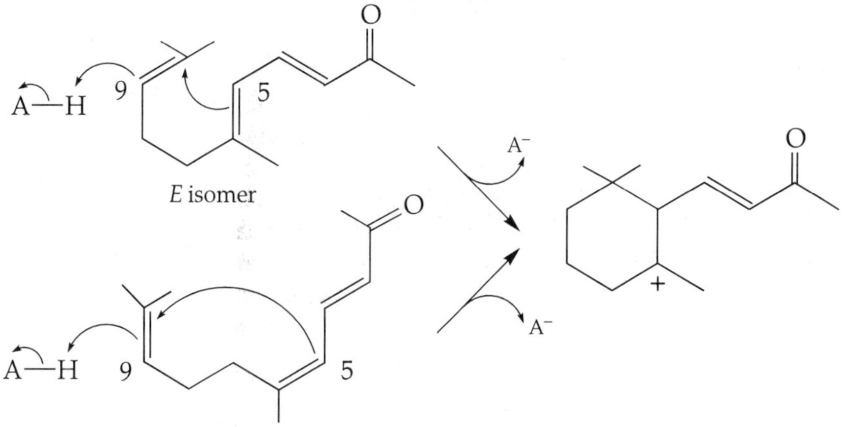

Both pseudoionones form the same carbocation intermediate. Then, the conjugated base of the acid catalyst removes a proton from the carbocation to form the final product. Depending on the acid catalyst used, the loss of the proton occurs from one of the two positions adjacent to the positive charge, giving either α- or β-ionone.

Why different acids give different isomers as the major product is still an unresolved matter.

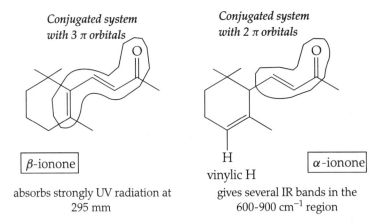

In this experiment you will choose either sulfuric or phosphoric acid to carry out the cyclization of pseudoionones, and determine which ionone is the predominant product in your reaction mixture. Both ionones have different physical properties, as explained below, and can be distinguished by a number of methods.

E23B.3   CHARACTERIZATION OF IONONES

α- and β-ionone can be distinguished by their IR and UV spectra, by their GC behavior, and also by their refractive indices. β-Ionone has two C—C double bonds conjugated with the carbonyl group, while α-ionone has only one. This structural difference results in different UV-visible spectra. β-Ionone has an absorption maximum at longer wavelengths than α-ionone. β-Ionone shows a strong band at 295 nm, which is absent in α-ionone; both isomers absorb at 227 nm.

Conjugated system with 3 π orbitals

Conjugated system with 2 π orbitals

β-ionone

α-ionone

H vinylic H

absorbs strongly UV radiation at 295 mm

gives several IR bands in the 600-900 cm^{-1} region

The IR spectra of both isomers differ in the C—H out-of-the-plane bending region (600–900 cm^{-1}). α-Ionone, because of its extra vinylic hydrogen, presents several bands in this region that are absent in the IR spectrum of β-ionone.

Both isomers have significantly different refractive indices that can be used to identify them. They can also be distinguished by their GC elution order. In columns of medium polarity they come out in order of increasing boiling points. Table 23.3 shows some of the distinctive features of the ionones and the starting materials.

Table 23.3 **Physical Properties of α- and β-Ionone and Related Compounds**

	α-Ionone	β-Ionone	Citral	Pseudoionones
n_D^{20}	1.4980	1.5200	1.4876	1.5332
b.p. (3 mm)	93–95°C	101–103°C	$\approx 75°C$	120–140°C
ε @ 295 nm	280	10,700	—	—
@ 227 nm	14,200	6,500		
(ethanol) (L mol^{-1}cm^{-1})				
IR bands in the 600–900 cm^{-1}	620;738; 800;827	—	842	—

PROCEDURE

> **Background reading:** 20.1–3; 23.1–3; 11.3; 32.1–5
> **Estimated time:** 6–8 hours (2 lab periods)
> Synthesis: 3–4 hours. Analysis: 3–4 hours

In this experiment you will prepare α- or β-ionone from citral. In the first lab period you will carry out the aldol condensation of citral with acetone followed by the cyclization of the pseudoionones with acids. For the cyclization step you will **choose one** of the two alternatives presented below: a sulfuric acid-acetic acid mixture or phosphoric acid. Each method leads to a different isomer as the major product. In the second lab period you will investigate which isomer is predominant under your cyclization conditions. To this end, you will characterize the final product mixture by a battery of methods: GC, UV spectroscopy, IR, and refractive index (Unit 11). To aid in the elucidation process, you will compare notes with one or more students who have chosen the other cyclization conditions. This experiment has been adapted from References 1–4.

> **Safety First**
>
> - Concentrated sulfuric acid is very corrosive; with the aid of a Pasteur pipet and a pluringe take just the amount of sulfuric acid needed. Do not remove the sulfuric acid bottle from the hood.
> - Acetic and phosphoric acids are corrosive and irritant.
> - *tert*- Butyl methyl ether is a flammable solvent.

E23B.4 PREPARATION OF PSEUDOIONONES

Synthesis

1. In a 25-mL Erlenmeyer flask equipped with a stir bar, weigh 1.5 g of citral (a mixture of citral a and b) and add 7.5 mL of acetone. Cool the mixture in an ice-salt bath placed on a magnetic stirrer, at approximately −8°C (outside temperature). See Figure 23.8.

2. With the aid of a Pasteur pipet add dropwise, over a period of 10 minutes, 1.5 mL of a 2.25 M sodium ethoxide solution in ethanol.

3. After the addition of the base is complete, keep stirring the reaction mixture for an additional 15-minute period.

4. Neutralize the reaction mixture by adding a 2 M solution of HCl in water to turn the solution permanently yellow (approximately 2.1 mL of HCl are needed).

5. Transfer the product to a 125-mL separatory funnel. Wash the Erlenmeyer flask with 10 mL of *t*-butyl methyl ether (BME) and transfer the wash to the separatory funnel.

6. Add 6 mL of water and extract the product into the organic solvent. Separate the layers. Label the aqueous layer "aqueous," and wash the BME layer with 12 mL of a 10% (w/v) solution of sodium chloride in water; collect this second aqueous layer in the "aqueous" container.

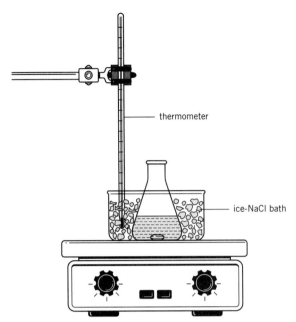

Figure 23.8   Setup for the synthesis of pseudoionones.

---

7. Dry the organic layer over anhydrous magnesium sulfate. Separate the drying agent by gravity filtration using fluted filter paper and collect the liquid in a dry, pre-tared round-bottom flask.

8. Evaporate the solvent in a rota-vap. Weigh the product and calculate the % yield of the synthesis of pseudoionones.

Analysis

1. Obtain the IR of citral and the mixture of products using NaCl plates (section 31.5).

2. Determine the refractive index of citral and pseudoionones (section 11.3).

3. Analyze the mixture of products by GC on a column of medium polarity such as SF-96 (suggested column temperature: 180°C).

E23B.5   CYCLIZATION WITH SULFURIC ACID-ACETIC ACID

Synthesis

1. In a 50-mL Erlenmeyer flask equipped with a stir bar, place 1.3 mL of glacial acetic acid and slowly add 1.7 mL of concentrated sulfuric acid (98% w/w). Cool the flask in an ice-water bath placed on a magnetic stirrer (Fig. 23.9).

2. Over a period of 15 minutes add, with the aid of a Pasteur pipet, 0.9 g of pseudoionones dropwise. As the pseudoionone mixture is added, the solution turns red-brown and becomes viscous. After the addition is complete, remove the reaction mixture from the ice-water bath and stir it at room temperature for 15 minutes.

3. To the reaction mixture, add a **mixture** of 20 mL of cold water and 4 mL of BME **at once**. Swirl the contents of the flask and transfer it to a 125-mL separatory funnel. Extract the product into the organic layer. Separate the layers and extract again the aqueous layer with 4 mL of BME.

4. Wash the combined BME layers twice with 8 mL (each time) of a sodium bicarbonate (5%, w/v) plus sodium chloride (10%, w/v) aqueous solution.

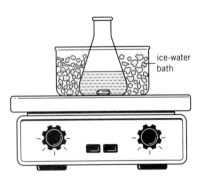

Figure 23.9   Setup for the cyclization with sulfuric acid-acetic acid.

5. Combine the aqueous layers, check the pH and, if necessary, adjust it using NaHCO$_3$ solution until it is basic. Wait until the bubbling has completely subsided before disposing.

6. Dry the organic layer over anhydrous magnesium sulfate. Filter the suspension and evaporate the solvent in a dry, pre-tared round-bottom flask by using a rota-vap. Calculate the % yield of the synthesis.

Analysis

1. Obtain the IR of the product using NaCl plates (section 31.5).

2. Determine the refractive indices of α- and β-ionone standards and that of the product.

3. Analyze the final product by GC using a column of medium polarity such as SF-96 (suggested column temperature: 180°C).

4. Analyze the product by UV-visible spectroscopy. Weigh approximately 5 mg of the product (with a precision of at least ±0.1 mg) in a 10-mL volumetric flask. Add 100% ethanol to the mark. Dilute this solution by dissolving 500 μL (measured with an automatic pipet) with 100% ethanol in a 10.0 mL volumetric flask. Obtain the UV-visible spectrum of this solution in the range 200–400 nm. Measure the absorbance at 227 and 295 nm. If the absorbance value is larger than 2, dilute the solution with ethanol and measure it again.

E23B.6   CYCLIZATION WITH PHOSPHORIC ACID

Synthesis

1. In a 25-mL Erlenmeyer flask equipped with a stir bar and immersed in a water bath at 30°C place 4.0 mL of concentrated phosphoric acid (85% w/w). See Figure 23.10.

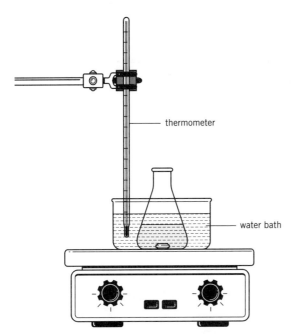

Figure 23.10   Setup for the cyclization with phosphoric acid.

Cleaning Up

- Dispose of the combined aqueous layers from E23B.5 (step 5) and E23B.6 (step 5) in the container labeled "Ionones Aqueous Waste."
- Dispose of the ethanolic solutions used for UV studies in the container labeled "Ionones Organic Waste."
- Dispose of Pasteur pipets and the filter paper with the drying agent in the container labeled "Ionones Solid Waste."
- At the end of the lab period the instructor will empty the contents of the rota-vap traps (BME).

2. With the aid of a Pasteur pipet add dropwise 0.9 g of pseudoionones over a period of 15 minutes while stirring. After the addition is complete keep the reaction mixture in the water bath for an additional 15 minute period with continuous stirring.

3. Add 20 mL of an aqueous 10% (w/v) sodium chloride solution and transfer the mixture to a 125-mL separatory funnel. Wash the flask with 10 mL of BME and transfer the wash to the separatory funnel. Shake the layers and separate them. Extract the aqueous layer again with 10 mL of BME. Combine the organic layers.

4. Wash the combined organic layers first with 10 mL of an aqueous solution containing 5% (w/v) sodium bicarbonate and 10% (w/v) sodium chloride, followed by 10 mL of a 10% sodium chloride solution in water.

5. Combine the aqueous layers, check the pH, and, if necessary, adjust it by using NaHCO₃ solution until it is basic. Wait until the bubbling has completely subsided before disposing.

6. Dry the organic layer over magnesium sulfate. Filter the suspension collecting the liquid in a dry, pre-tared round-bottom flask. Evaporate the solvent using a rota-vap. Weigh the product and calculate the % yield.

Analysis

Proceed as indicated in E23B.5.

BIBLIOGRAPHY

1. Synthesis of Pseudoionone Homologs and Related Compounds. W. Kimel, J.D. Surmatis. J. Weber, G.O. Chase, N.W. Sax, and A. Ofner. *J. Org. Chem.*, **22**, 1611–1618 (1957).

2. Note on the Preparation of β-Ionone. H.J.V. Krishna and B.N. Joshi. *J. Org. Chem.*, **22**, 224–226 (1957).

3. Condensation of Citral with Ketones and Synthesis of Some New Ionones. H. Hibbert and L.T. Cannon. *J. Amer. Chem. Soc.*, **46**, 119–130 (1924).

4. Cyclization of Pseudoionone by Acidic Reagents. E.E. Royals. *Ind. Engineer. Chem*, **38**, 546–548 (1946).

5. Novel Synthesis of Dienones and Enones from Propargyl Alcohols and Allyl Alcohols with 2,2-dimethoxypropane: Synthesis of Ionone and Irone. T. Ishihara, T. Kitahara, and M. Matsui. *Agr. Biol. Chem.*, **38**, 439–442 (1974).

EXPERIMENT 23B REPORT

Pre-lab

1. Make a table with the physical properties (molecular mass, b.p., density, solubility, and toxicity/hazards) of citral, α- and β-ionone, and sulfuric, acetic, and phosphoric acids. Include also the refractive index of α- and β-ionone, and the flammability of BME and acetone.

2. Make a flowchart for the synthesis of pseudoionones and α- and β-ionone.

3. Write a mechanism for the aldol condensation of geranial and neral with acetone. Why is the *trans* isomer around C3–C4 preferred?

4. Do you expect pseudoionones to have a UV maximum at a longer or shorter wavelength than citral?

5. What changes do you expect in the IR spectrum of pseudoionones as compared with citral?

6. Why does β-ionone absorb in the UV-visible region at longer wavelengths than α-ionone?

7. Why is sodium chloride used in the liquid-liquid extractions?

8. Using the Woodward–Fieser rules (section 32.3), calculate the λ_{max} of β-ionone. How does it compare with the experimental λ_{max} of 295 nm?

In-lab

The spectra of α- and β-ionone are shown in Figures 23.11–23.17.

1. Calculate the % yield for the syntheses of pseudoionones and ionones.
2. Using a table format, report the refractive index of citral, pseudoionones, and the ionones.
3. Based on the GC analysis, calculate the % of α- and β-ionone in your product mixture.
4. Compare the refractive index of the product with those of pure α-ionone and β-ionone.
5. Interpret the IR spectra of citral, pseudoionones, and the ionones. Note the bands that are important in assessing the structure of the compounds. Compare the ionones' IR spectra with those shown below. (Figs. 23.11 and 23.14).
6. Explain the differences in the IR spectra of α- and β-ionone. Which isomer is the predominant product according to the IR spectrum? How does it compare with the GC analysis and index of refraction determination?
7. Hypothesize why the cyclization with sulfuric or phosphoric acids leads to different products. Consider the strength of the acid in your hypothesis. What experiments would you carry out to check your proposal?
8. Calculate the % of both ionones in the product mixture based on the UV spectrum (Unit 32). In the equations below, A is the measured absorbance at the specified wavelengths, ε is the molar absorptivity, ℓ is the length of the cell (1 cm), and C_α and C_β are the molar concentrations of α- and β-ionone, respectively. The superscripts represent the wavelengths.

$$A^{295} = \varepsilon_\alpha^{295} \times C_\alpha \times \ell + \varepsilon_\beta^{295} \times C_\beta \times \ell \quad A^{227} = \varepsilon_\alpha^{227} \times C_\alpha \times \ell + \varepsilon_\beta^{227} \times C_\beta \times \ell$$

The molar absorptivities are given in Table 23.3. Substituting them we obtain:

$$A^{295} = 280 \times C_\alpha + 10{,}700 \times C_\beta$$
$$A^{227} = 14{,}200 \times C_\alpha + 6{,}500 \times C_\beta$$

This is a system of two equations with two unknowns (C_α and C_β). By solving it, the molar concentration of each isomer can be calculated (Section 32.4).

$$C_\beta = \frac{A^{227} \times 280 - A^{295} \times 14{,}200}{280 \times 6{,}500 - 14{,}200 \times 10{,}700}$$
$$C_\alpha = \frac{A^{295} - 10{,}700 \times C_\beta}{280}$$

How does this calculation compare with that from the GC analysis? Discuss the sources of error.

9. (Advanced level) Interpret the ^{1}H-NMR and ^{13}C-NMR spectra of α- and β-ionone shown in Figure 23.12; 23.13; 23.15; and 23.16.
10. (Advanced level) Interpret the mass spectra of α- and β-ionone shown in Figure 23.17. Compare the two spectra. Why is the m/z 177 peak so intense for β-ionone?

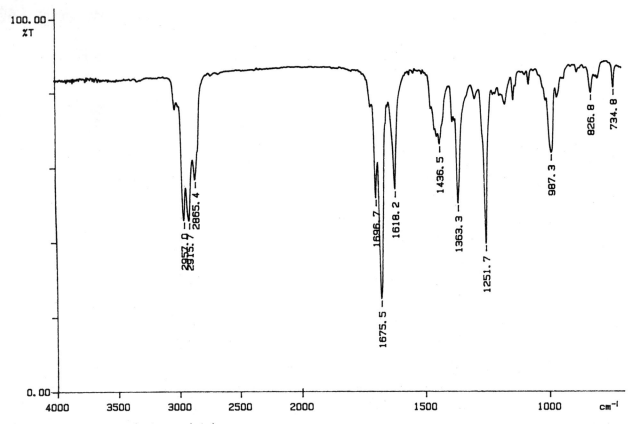

Figure 23.11 IR spectrum of α-ionone (neat).

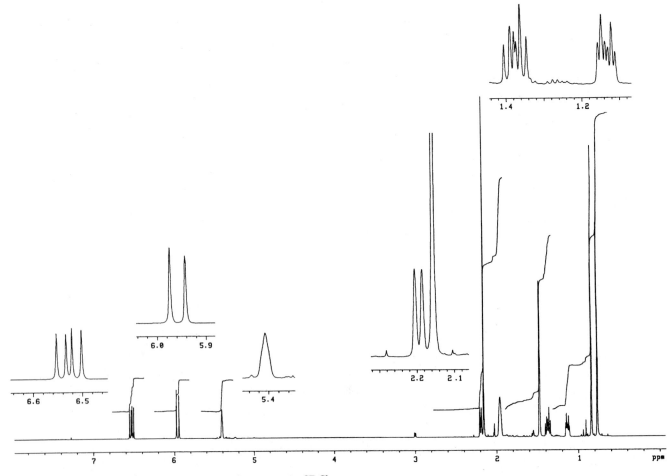

Figure 23.12 500-MHz ^{1}H-NMR spectrum of α-ionone in CDCl$_3$.

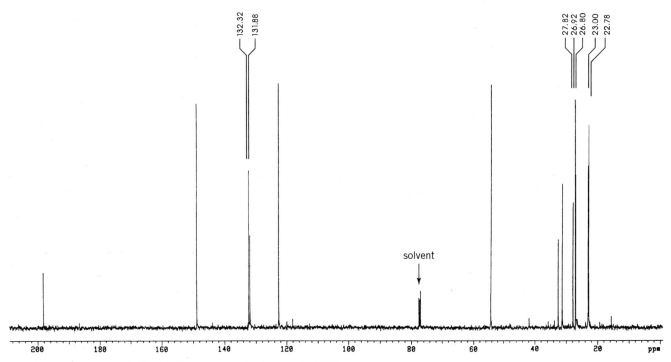

Figure 23.13   125.7-MHz ^{13}C-NMR spectrum of α-ionone in CDCl$_3$.

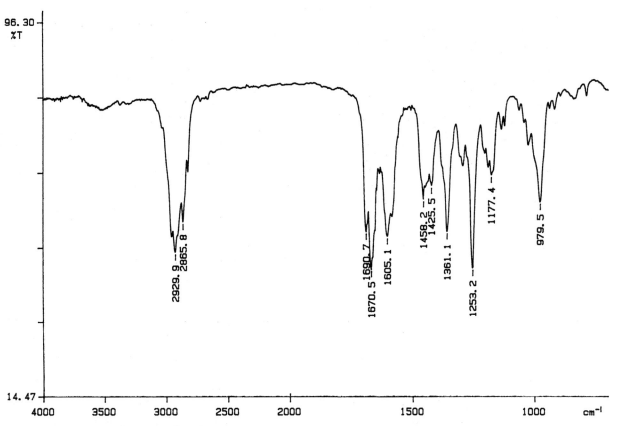

Figure 23.14   IR spectrum of β-ionone (neat).

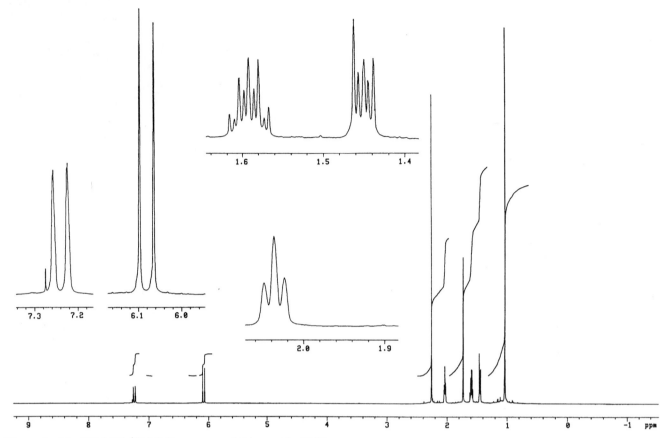

Figure 23.15 500-MHz ^{1}H-NMR spectrum of β-ionone in CDCl$_3$.

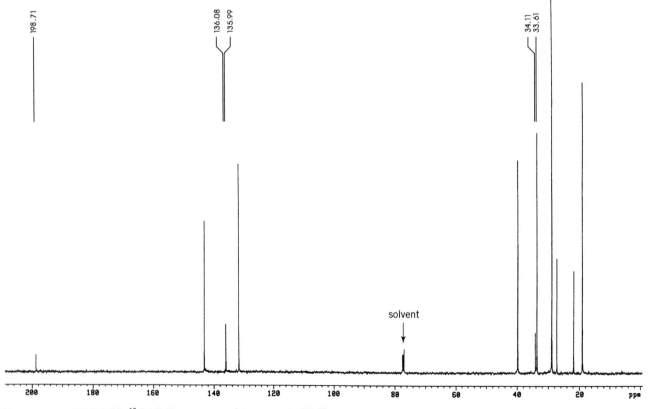

Figure 23.16 125.7-MHz ^{13}C-NMR spectrum of β-ionone in CDCl$_3$.

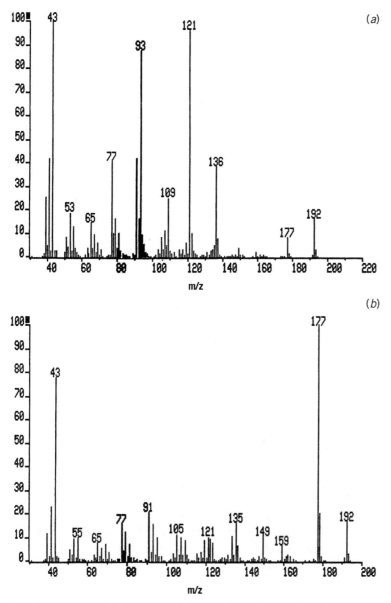

Figure 23.17   Mass spectra (electron impact, 70 eV) of *a*: α-ionone; *b*: β-ionone.

Molecules of Life

The secrets of life encoded in the nucleic acids DNA and RNA are translated by the cells' machinery in the biosynthesis of proteins, carbohydrates, and lipids. Organisms at the bottom of the food chain synthesize these compounds from carbon dioxide, water, nitrogen, and nitrates. As we ascend in the food chain, we find that more evolved organisms have lost their capacity to utilize such simple molecules as food sources; to sustain life they need carbohydrates, lipids, and proteins, which they obtain from other organisms. In this unit we will study proteins and carbohydrates. Their chemistry will be illustrated by the isolation and characterization of the main components of milk: casein and lactose.

24.1 PROTEINS

Proteins are macromolecules made up of α-amino acids linked together by amide or **peptide bonds**. They may have from 50 to more than 20,000 amino acids linked together with molecular masses in the range 5,000 to 2,000,000. The metabolism of proteins produces about 16.7 kJ (4 kcal) per gram of protein, a similar calorie output to that of carbohydrates. However, the use of proteins for energy production is a low-priority task. Proteins are the cells' last resource for energy production, used only when the intake of carbohydrates and lipids is not sufficient.

$$\begin{array}{c} O \\ \parallel \\ -C-NH- \end{array}$$

a peptide bond

Amino Acids

α-Amino acids are the building blocks of proteins. With a few exceptions, most of the biologically important amino acids have the same configuration around C2 (Cα) where the $-COOH$, $-NH_2$, and $-R$ groups are bound; this configuration is related to that of L-glyceraldehyde, and therefore they are referred to as L-amino acids.

an L-amino acid   L-glyceraldehyde

Only about 20 amino acids are found in proteins; they differ in the nature of the substituent R (Table 24.1). Amino acids (aa) behave as acids or bases depending on the pH. At low pH, the amino group of amino acids is protonated and the amino acid has a positive charge (aa$^+$). At higher pH values, the proton from the carboxyl group is lost, giving a doubly charged species (aa$^\pm$) called the **zwitterion** (from German: *zwitter*, hybrid; hermaphrodite). At even higher pH, the zwitterion loses a proton from the ammonium group, giving a negatively charged species (aa$^-$). The pH where the concentration of aa$^+$ equals the concentration of aa$^-$, and where the concentration of the zwitterion is maximum, is the **isoelectric point** of the amino acid, **pI** (Table 24.1). With some exceptions, the pI of most amino acids is in the range 5–7.

$$\overset{+}{H_3N}-CH-COOH \underset{H^+}{\overset{HO^-}{\rightleftharpoons}} \overset{+}{H_3N}-CH-COO^- \underset{H^+}{\overset{HO^-}{\rightleftharpoons}} H_2N-CH-COO^-$$

| low pH (aa$^+$) | zwitterion (aa$^\pm$) | high pH (aa$^-$) |

Table 24.1 **Side Chains of α-Amino Acids Occurring in Proteins**

R—	Name	Abbreviation	pI	Type
H—	glycine	Gly; G	5.97	nonpolar
CH$_3$—	alanine	Ala; A	6.00	nonpolar
CH$_3$—CH— CH$_3$	valine	Val; V	5.96	nonpolar
CH$_3$—CH—CH$_2$— CH$_3$	leucine	Leu; L	5.98	nonpolar
CH$_3$—CH$_2$—CH— CH$_3$	isoleucine	Ile; I	6.02	nonpolar
CH$_3$—S—CH$_2$—CH$_2$—	methionine	Met; M	5.74	nonpolar
$\left(^-OOC-CH-S-\right)_2$ $\overset{+}{NH_3}$	cystine^a	Cys; C	4.60	nonpolar
proline ring structure with $\overset{+}{N}$H and COO$^-$	proline^a	Pro; P	6.30	nonpolar
HS—CH$_2$—	cysteine	Cys; C	5.07	polar
HOCH$_2$—	serine	Ser; S	5.68	polar
CH$_3$—CH— OH	threonine	Thr; T	5.60	polar
$^-$OOC—CH$_2$—	aspartic acid	Asp; D	2.77	acidic

(continued)

Table 24.1   *(Continued)*

R—	Name	Abbreviation	pI	Type
$H_2N-\overset{\overset{O}{\|\|}}{C}-CH_2-$	asparagine	Asn; N	5.41	polar
$^-OOC-CH_2-CH_2-$	glutamic acid	Glu; E	3.22	acidic
$H_2N-\overset{\overset{O}{\|\|}}{C}-CH_2-CH_2-$	glutamine	Gln; Q	5.65	polar
$\overset{+}{H_3N}-CH_2-CH_2-CH_2-CH_2-$	lysine	Lys; K	9.74	basic
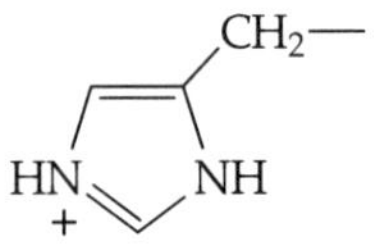 imidazole CH_2-	histidine	His; H	7.59	basic
$H_2N-\overset{\underset{\overset{+}{N}H_2}{\|\|}}{C}-NH-CH_2-CH_2-CH_2-$	arginine	Arg; R	10.76	basic
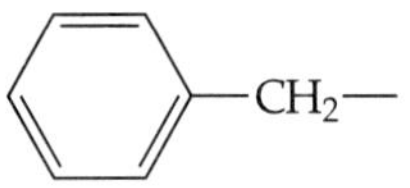 phenyl $-CH_2-$	phenylalanine	Phe; F	5.48	aromatic
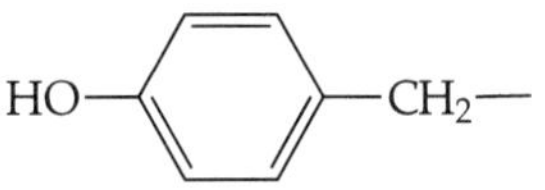 $HO-$ phenyl $-CH_2-$	tyrosine	Tyr; Y	5.66	aromatic
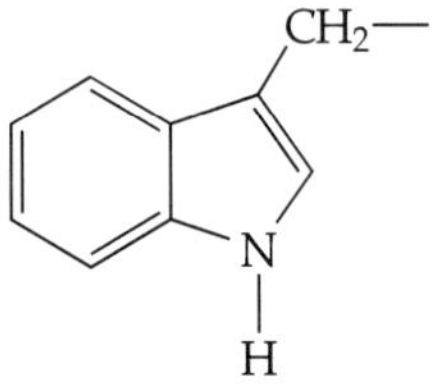 indolyl CH_2-	tryptophan	Trp; W	5.89	aromatic

[a]For proline and cystine the structures shown correspond to the whole amino acids instead of the R groups.

The solubility of amino acids in water is strongly affected by the pH. For most amino acids, the solubility is high at extreme pH values, where the amino acid is charged, and very low at the isoelectric point, where the amino acid has no net charge. This behavior is frequently used in the laboratory to separate and purify amino acids and proteins.

Amino acids can be classified according to the properties of their side chains, R. Lysine, histidine, and arginine are basic amino acids and bear a net positive charge at neutral pH. Aspartic acid and glutamic acid are acidic and have a net negative charge at neutral pH. Asparagine, glutamine, cysteine, threonine, and serine are polar amino acids without net charge. Alanine, valine, leucine, isoleucine, proline, cystine, and methionine are nonpolar, **hydrophobic**, amino acids. Phenylalanine, tyrosine, and tryptophan are aromatic amino acids. Amino acids are usually abbreviated by the first three letters of their names, for example, Lys for lysine and Phe for phenylalanine, or by a one-letter code, K for lysine and F for phenylalanine (Table 24.1).

Certain amino acids cannot be synthesized by the human body and must be obtained from the diet. These amino acids are called **essential amino acids** and they are lysine, threonine, valine, leucine, isoleucine, methionine, tryptophan, and phenylalanine. In children, arginine and histidine are also essential. The nutritional value of a protein is defined by its content of essential amino acids. Proteins derived from animals (eggs, meat, milk) are **nutritionally complete** (they contain all essential amino acids in adequate amounts), while vegetable proteins are more diverse. For example, corn proteins are particularly insufficient to supply the human body with all the essential amino acids, and should be supplemented with other proteins. Rice and potato proteins, on the other hand, are complete. The quality of the protein should be a major concern for populations whose livelihood largely depends on one crop.

By **hydrolysis**, proteins are broken down into smaller fragments called peptides or polypeptides. The reaction can be carried out in the laboratory with boiling 6 M HCl. It takes several hours to produce complete hydrolysis to the individual amino acids, and some amino acids, such as tryptophan, are destroyed in the process.

$$-NH-CH-\overset{\overset{\displaystyle O}{\|}}{C}-NH-\overset{\overset{\displaystyle R}{|}}{CH}-\overset{\overset{\displaystyle O}{\|}}{C}- \xrightarrow[\Delta]{H_3O^+} n\ H_3\overset{+}{N}-CH-\overset{\overset{\displaystyle O}{\|}}{C}-OH$$

$$\underset{R}{|} \qquad\qquad\qquad\qquad\qquad\qquad \underset{R}{|}$$

peptide α-amino acids

Protein Structure

The linear sequence of amino acids in a protein is called the **primary structure** and is defined by the genetic code. The amino acids in a protein are commonly referred to as **residues**. Peptides and proteins are written, by convention, with the residue with the free $-NH_2$ group (the **N-terminal residue**) on the left and the residue with the free $-COOH$ (the **C-terminal residue**) on the right; for example, the pentapeptide Arg-Lys-Gly-Ser-Ala has arginine as the N-terminal residue and alanine as the C-terminal one.

Amino acid residues within a protein are associated through H-bonds. This H-bond interaction may result in two different types of structures, **α-helix** (Fig. 24.1) and **β-sheet** (Fig. 24.2), called the **secondary structure** of the protein. Different elements of secondary structure interact with one another, conferring on the protein its stable fold or conformation. Interactions between positive and negative charges in different parts of the protein, disulfide bonds between two cysteine residues, and interactions between hydrophobic (nonpolar) amino acids are responsible for the tridimensional shape of the polypeptide chain or its **tertiary structure**. The nonpolar, hydrophobic groups such as leucine, isoleucine, and valine are usually buried in the inside of the protein to avoid contact with water molecules, while the polar and charged groups (threonine, lysine, aspartic acid, etc.) are on the outside, where they interact directly with water molecules (Fig. 24.3). Some proteins contain more than one polypeptide chain held together by H-bonds, ion-ion interactions, and hydrophobic interactions. This association, not present in all proteins, is known as the **quaternary structure**.

Protein folding is encoded in the primary structure of the protein. If the quaternary, tertiary, and secondary structures of a protein are disrupted by chemical or physical means without breaking any peptide bond, many proteins are able to fold back spontaneously into their original conformation once the source of disruption is eliminated. The natural conformation of a protein is called its **native state**. Changes from the native state caused by heat, chemicals, and pH usually result in a loss of biological activity. This process is called **denaturation** and takes place, for example, every time we cook or bake. Depending on the conditions, denaturation can be reversible or irreversible. Heat denaturation of a concentrated solution of several proteins (a situation that to some extent may mimic the process of cooking) usually results in the **unfolding** of native proteins followed by entanglement and aggregation

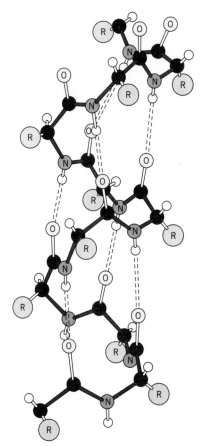

Figure 24.1 An α-helix.

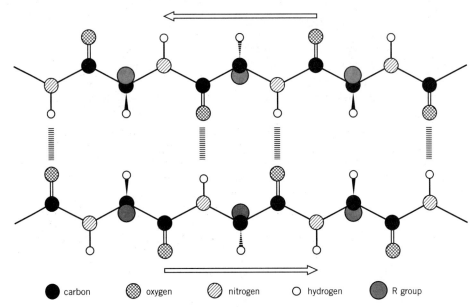

Figure 24.2   Representation of a β-sheet. The arrows point in the N-terminal to C-terminal direction.

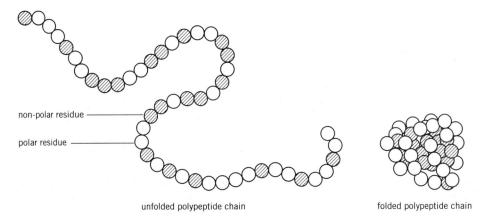

Figure 24.3   Representations of unfolded and folded polypeptide chains.

of agglomerated polypeptide chains, which results in precipitation; this process is irreversible (Fig. 24.4). On the other hand, denaturation and unfolding brought on by chemicals at low protein concentrations are usually reversible.

One of the chemicals most widely used to induce unfolding of polypeptide chains is **urea**. Most proteins lose their secondary and tertiary structures when treated with concentrated solutions of urea. The concentration of urea required to produce

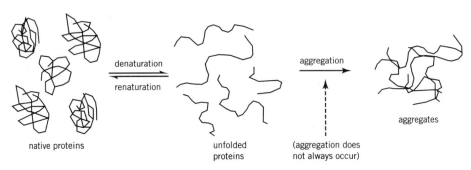

Figure 24.4   Protein denaturation and aggregation.

unfolding is a function of the protein stability. The more stable the protein, the higher the concentration of urea necessary to unfold the protein. Most proteins, however, are completely **unfolded** in 8 M urea. How urea unfolds the protein is a matter of speculation. It is believed that urea drives the unfolding of the protein by interacting with nonpolar, hydrophobic residues usually buried in the inside of the protein.

Types of Proteins

Proteins can be classified on the basis of their structure, function, or chemical composition. According to their structure they can be divided into two major categories: **globular** and **fibrous**. Globular proteins, as indicated by their name, are compact and have rather spherical shapes; they are relatively soluble in water. Examples of globular proteins are albumins, globulins, and all enzymes. Fibrous proteins usually have only one type of secondary structure, either α-helix or β-sheet. Examples of this type of protein are α-keratin, a protein made up exclusively of α-helices, which is the major constituent of hair, nails, and wool; collagen, also an all-α-helix protein, which is the constituent of connective tissue; and fibroin or β-keratin, which is made up of β-sheets and is the major constituent of silk.

According to their function, proteins can be classified into two groups: **structural** and **enzymes**. Enzymes are protein catalysts; they are ubiquitous in all cells where they perform countless tasks. Enzymes can be further subdivided into many categories depending on the reactions they catalyze (reductases, peroxidases, proteases, etc.). Structural proteins do not assist directly in chemical transformations; they are the building blocks of the cellular architecture. Most structural proteins are fibrous proteins.

Characterization of Proteins and Amino Acids

Coomassie Blue Staining In the laboratory, proteins are often separated by electrophoresis (a separation method similar to chromatography, but using an electric field) and the protein bands are visualized by binding to Coomassie Blue, an organic dye shown below. There are two different types, known as R and G, which differ only by two methyl groups. Sometimes the letters R or G are followed by the number 250, a number given by the manufacturers without structural meaning. The nature of the interaction between the dye and protein is not clearly understood, but binding seems to be proportional to the protein's basic-amino-acid content (lysine, arginine, and histidine).

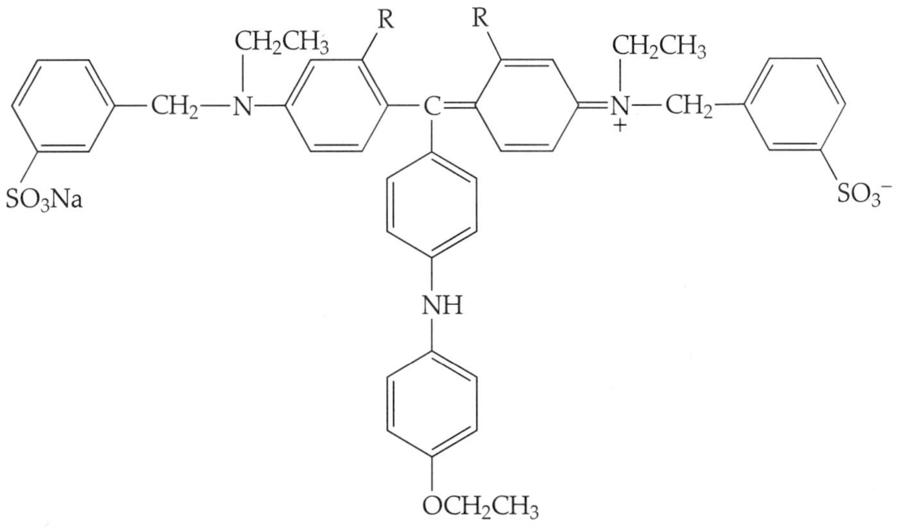

Coomassie Blue G   (R = CH₃)
Coomassie Blue R   (R = H)

Biuret Test   Proteins react with Cu^{2+} under alkaline conditions to give a blue-violet color, probably due to the formation of a coordination complex between cupric ions and peptide bonds.

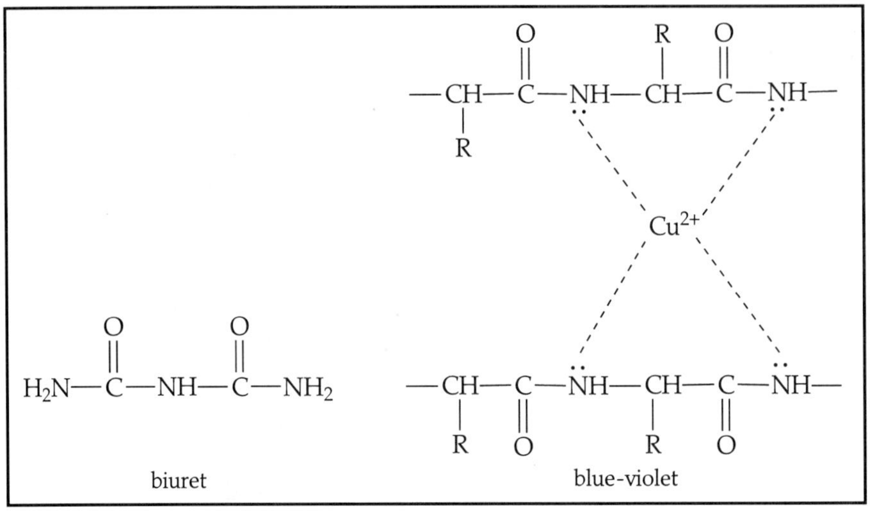

biuret   blue-violet

This test is named after a derivative of urea, **biuret**, that also gives a blue-violet color with Cu^{2+}. A tripeptide (two peptide bonds) is the shortest peptide that gives a positive test. Many chemicals can interfere in this reaction; some free amino acids and other compounds that coordinate with Cu^{2+} under the assay's conditions, for example, diamines, give a positive test. The reaction is important, however, for the quantitative determination of protein concentration by measuring the absorbance of the resulting blue solution.

Ninhydrin Test   The presence of free α-amino acids can be tested by several reactions; among them, the **ninhydrin test** is perhaps the most important one because of its application in the quantitative determination of amino acids. α-Amino acids react with ninhydrin to give a blue-violet condensation product with formation of CO_2. Proline gives a yellow-red color instead of blue.

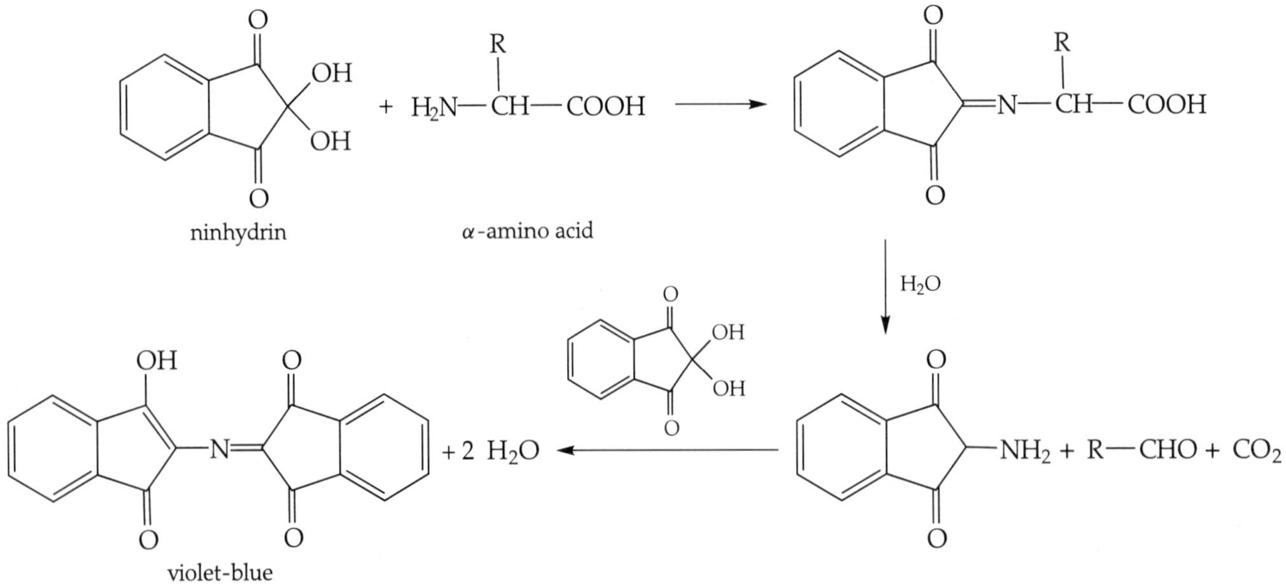

ninhydrin α-amino acid

violet-blue

24.2 CARBOHYDRATES

Carbohydrates are naturally occurring polyhydroxy aldehydes or polyhydroxy ketones. The name carbohydrates derives from the nineteenth-century concept of *hydrated carbons*, based on the fact that the molecular formula of some of the smallest members of the family is $C_6H_{12}O_6$ or $C_6(H_2O)_6$. They are often referred to as **sugars** because of the sweetness of some members of the family, especially glucose and sucrose. Some important carbohydrates are: glucose, a six-carbon polyhydroxy aldehyde or an **aldohexose**; ribose, a five-carbon polyhydroxy aldehyde, an **aldopentose**; and fructose, a six-carbon polyhydroxy ketone or a **ketohexose**.

Most naturally occurring sugars belong to the D-series. This means that the configuration of the stereogenic carbon farthest away from the carbonyl group (C5 in hexoses and C4 in pentoses) is similar to that of D-glyceraldehyde. In the Fischer projections of glucose, ribose, and fructose shown in Figure 24.5, the configuration of the carbon atom *next to the —CH₂OH group*, with the —OH to the right, indicates that these sugars belong to the D-series (Fig. 24.5).

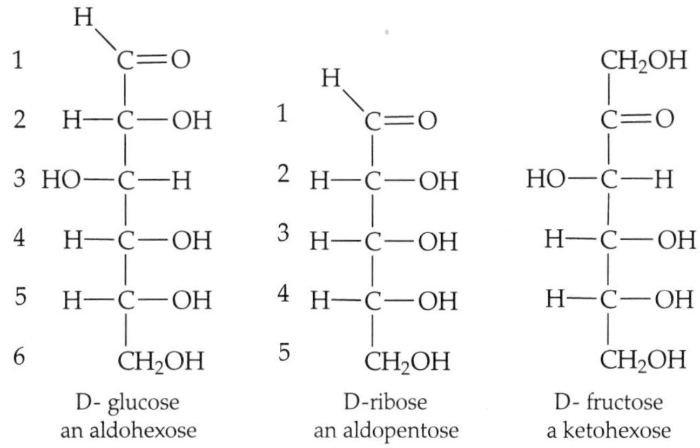

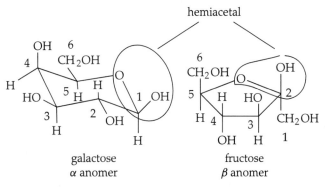

D- glucose
an aldohexose

D-ribose
an aldopentose

D- fructose
a ketohexose

Figure 24.5 Some common carbohydrates (Fischer projection).

Carbohydrates are also known as **saccharides** (from Latin; *saccharum*, sugar). Simple carbohydrates such as glucose, ribose, and fructose are called monosaccharides. Monosaccharides are not found in the aldehyde or ketone form, as shown in the Fischer projections, but rather they exist predominantly in cyclic configurations where one of the hydroxy groups, from C5 in hexoses and C4 in pentoses, has reacted with the carbonyl to form a cyclic **hemiacetal** (Fig. 24.6).

hemiacetal

galactose
α anomer

fructose
β anomer

Figure 24.6 Galactose and fructose in their hemiacetal forms.

The hemiacetalic OH group (C1 in the case of aldoses) can adopt two different orientations, giving rise to two stereoisomers called **anomers**. When the hemiacetalic HO− and the last CH_2OH group (C6 in hexoses, C5 in pentoses) are *cis* to one another, the anomer is called β, and when these groups are *trans* to one another, the anomer is called α (Fig. 24.7). Depending on the experimental conditions, one anomer or the other can be crystallized from solution.

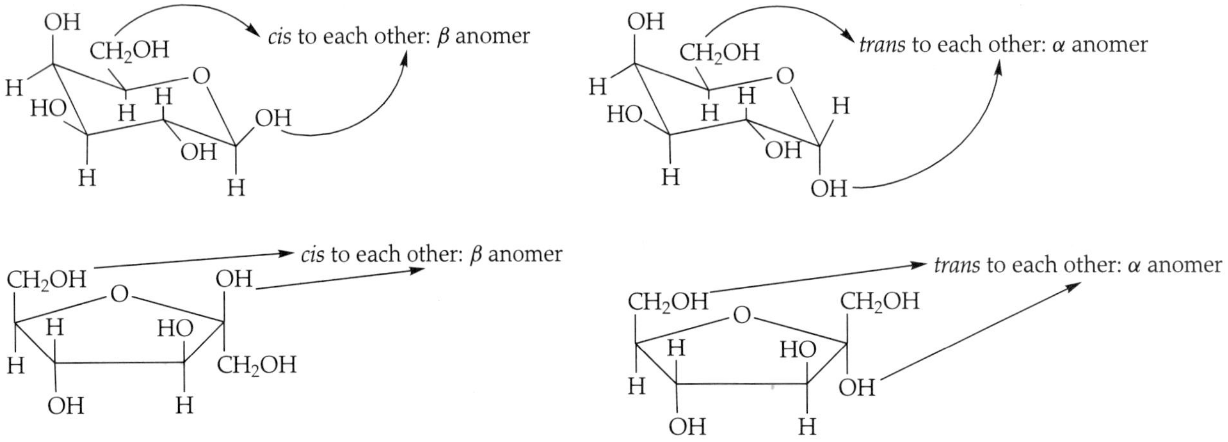

Figure 24.7 α and β anomers of galactose and fructose.

In solution one anomer is in equilibrium with the other. This **interconversion** process is called **mutarotation** and takes place because the hemiacetals are in equilibrium with the carbonyl (open chain) form.

Interconversion of glucose anomers

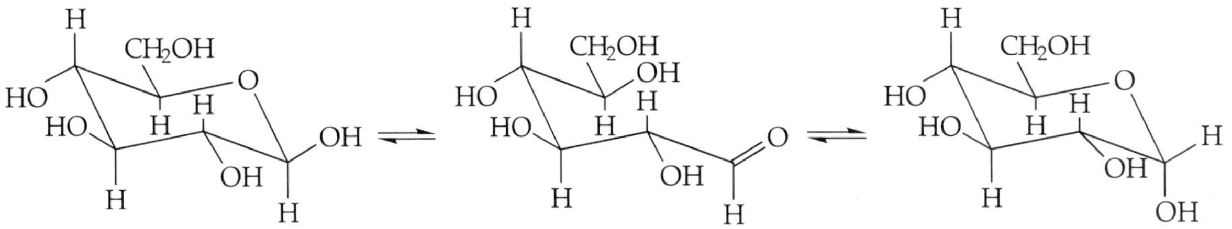

At equilibrium, the concentration of the carbonyl form is very low and one anomer is usually more abundant than the other. For example, in the case of glucose shown above, the β anomer is more stable than the α anomer, and therefore more abundant, because all the OH groups in the β anomer are equatorial.

The concentration of the anomers at equilibrium can be determined by polarimetry. In general, each pure anomer has a different specific rotation, which is different from the specific rotation of the mixture. For example, the specific rotations of the α and β anomers of glucose are $+102°$ $(g/mL)^{-1}$ dm^{-1} and $+18.7°$ $(g/mL)^{-1}$ dm^{-1}, respectively. The specific rotation of the equilibrium mixture is $+52.7$. If the α anomer is dissolved in water and the specific rotation measured as a function of time, it will change from $+102$ to the equilibrium value of $+52.7$. If the β anomer is dissolved in water, the specific rotation increases from $+18.7$ to $+52.7$. The concentrations of both anomers at equilibrium can be calculated with the help of Equation 11.7, rewritten here:

$$[\alpha]_T = [\alpha]_\alpha \times \frac{\%\alpha}{100} + [\alpha]_\beta \times \frac{\%\beta}{100} \tag{1}$$

where $[\alpha]_T$ is the specific rotation of the solution, $[\alpha]_\alpha$ and $[\alpha]_\beta$ are the specific rotations of the anomers, and $\%\alpha$ and $\%\beta$ are their percentages at equilibrium:

$$52.7 = 102 \times \frac{\%\alpha}{100} + 18.7 \times \frac{\%\beta}{100} \qquad (2)$$

Considering that $\%\alpha + \%\beta = 100$, the values for $\%\alpha$ and $\%\beta$ can be calculated. Substituting $\%\beta = 100 - \%\alpha$ in Equation 2:

$$52.7 = 102 \times \frac{\%\alpha}{100} + 18.7 \times \frac{(100 - \%\alpha)}{100}$$

Solving for $\%\alpha$ and then for $\%\beta$

$$\%\alpha = 41\% \implies \%\beta = 59\%$$

If the H atom from the anomeric $-$OH is replaced by an alkyl group, the carbohydrate is no longer a hemiacetal, it is an **acetal**. Acetals are more stable than hemiacetals. They do not exist in equilibrium with the carbonyl form, and do not mutarotate.

When monosaccharide units are linked together by an oxygen bridge, they form disaccharides, oligosaccharides, and polysaccharides. Some examples of disaccharides are shown in Figure 24.8. Bonds involving anomeric carbons are called **glycoside bonds**. Maltose is formed by two glucose molecules linked together by the C1 of one molecule (in α configuration) and C4 of the other; this glycoside bond is represented by $1\alpha \longrightarrow 4'$. Lactose has a $1,4'$-β glycoside bond (or $1\beta \longrightarrow 4'$). In sucrose, a disaccharide formed by the condensation of glucose and fructose, C1 of the α anomer of glucose is linked to C2 of the β-anomer of fructose.

Notice that carbons 1' of the maltose and the lactose molecules are in the hemiacetal form; therefore, these molecules are in equilibrium with the open chain form (the aldehyde form) and two anomers of maltose and lactose exist in solution. In Figure 24.8 only one anomer (α) is shown. Sucrose, on the other hand, has no hemiacetalic OH group because the anomeric carbons of glucose (C1) and fructose (C2) are linked together in a glycoside bond; therefore no mutarotation takes place in sucrose solutions.

Polysaccharides

Cellulose is one of the most widespread polysaccharides. It is the main constituent of wood and plant tissue. It is made up of glucose units linked together by $1,4'$-β-glycoside bonds. Cellulose cannot be digested by the human body, but its presence is essential in some foods, to which it adds bulk in the form of fiber. Cellulose is a linear polymer with a variable number of glucose units per molecule (3000–5000). Other important polysaccharides are **starch** and **glycogen**. Starch is the carbohydrate used for energy storage in plants and glycogen is used for the same purpose in animals.

Cellulose

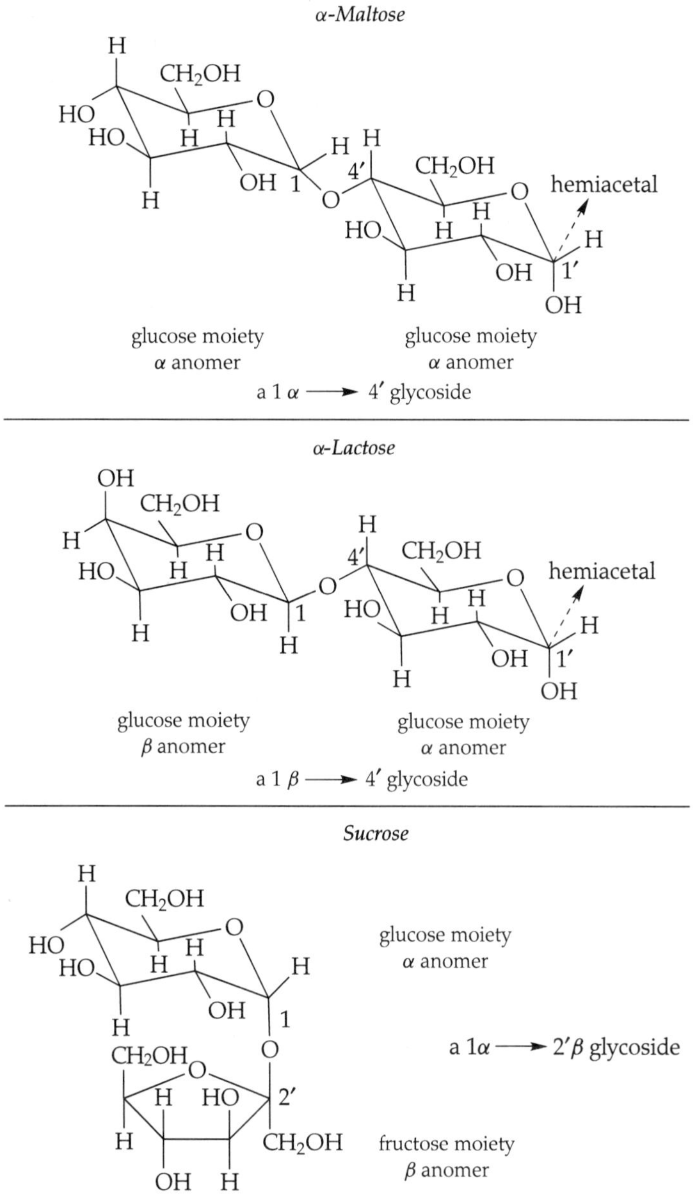

Figure 24.8   Some common disaccharides.

Characterization Reactions

Carbohydrates in the hemiacetal form are oxidized in the presence of Cu^{2+} to carboxylic acids while Cu^{2+} is reduced to Cu^+. Because of this redox capability such carbohydrates are called **reducing**. The reactive species is not the hemiacetal, but rather the aldehyde in equilibrium with the hemiacetal.

$$R{-}CHO + 2\,Cu^{2+} + 5\,HO^- \longrightarrow R{-}COO^- + Cu_2O + 3\,H_2O$$

The reaction shown above is performed in the presence of tartrate (a chelating agent that keeps the Cu^{2+} in solution and prevents its precipitation in the basic medium) and is known as **Fehling's test** (Unit 20). A modification of this reaction is carried out in acidic instead of basic conditions and is called **Barfoed's test**. This test is useful to distinguish reducing monosaccharides from reducing disaccharides, because the former react faster than the latter.

Ketoses, such as fructose, are also reducing sugars. The carbonyl on C2 migrates to C1 via keto-enol tautomerism especially under basic conditions, and gives an aldehyde that can be oxidized to a carboxylic acid.

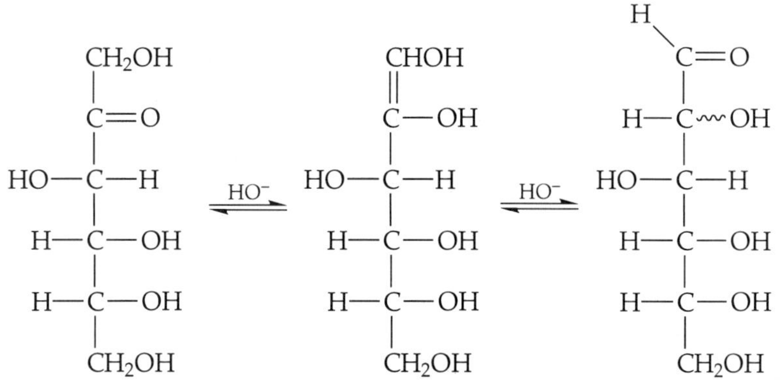

a ketose   enol form   an aldose

Carbohydrates with glycoside bonds cannot exist in the open chain form and are **nonreducing**. An example of a nonreducing carbohydrate is the disaccharide sucrose, where the anomeric carbons from both monosaccharides are linked together by a glycoside bond.

The hydrolysis of disaccharides gives monosaccharides. The reaction is catalyzed by acids.

Hydrolysis of lactose

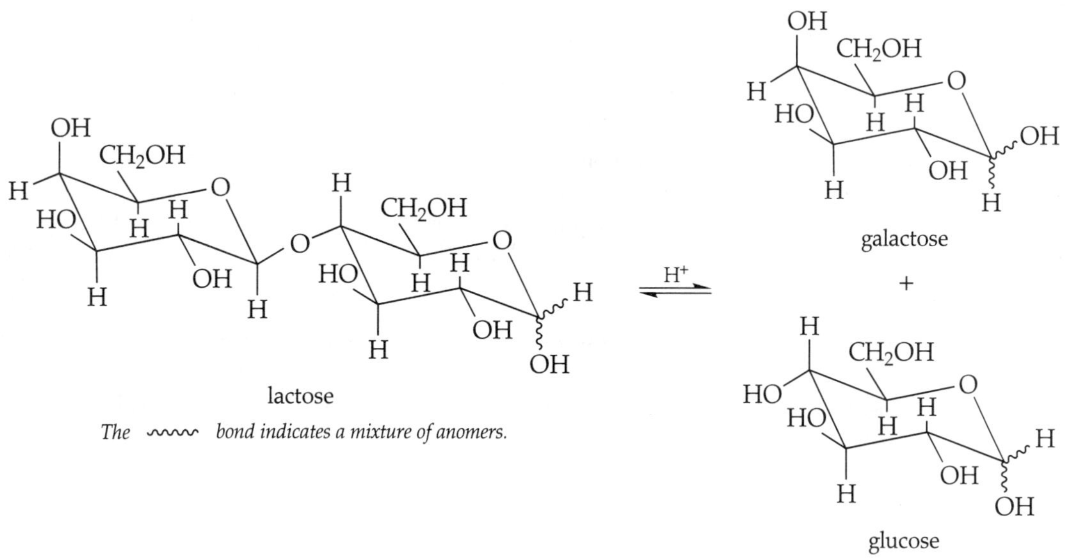

lactose

The ∿∿∿ *bond indicates a mixture of anomers.*

galactose

+

glucose

Analysis of Sugars by Chromatography

Sugars can be studied by several chromatographic techniques. Because of their low volatility and their tendency to decompose at high temperatures, carbohydrates must be derivatized before GC analysis. Trimethylsilyl derivatives where the hydrogens of the —OH groups have been substituted by —Si(CH$_3$)$_3$ are particularly useful. These derivatives are easily made and are volatile enough to be studied by GC.

$$R-OH \text{ (sugar)} + Cl-Si(CH_3)_3 \longrightarrow R-O-Si(CH_3)_3 + HCl$$

Carbohydrates can be directly studied without derivatization by HPLC using a variety of columns and mobile phases. They can also be analyzed by TLC on silica if the plates are deactivated before the run to reduce the power of the adsorbent. TLC plates with cellulose as the stationary phase (Unit 8) can also be used for the study of carbohydrates. Because carbohydrates are very polar molecules, very polar solvents are used in their chromatographic study. The solvents usually consist of a mixture of water, an alcohol, and an acid. After the run, the spots can be visualized with the aid of specific reagents. One of the most commonly used visualization reagents is a mixture of *p*-anisidine and phthalic acid. The chromatogram is sprayed with an ethanolic solution of these two compounds and then heated at 70–90°C for a few minutes; spots corresponding to aldoses produce a brown-reddish color. Ketoses and nonreducing sugars give a much weaker coloration.

p-anisidine   *o*-phthalic acid

BIBLIOGRAPHY

1. Proteins. Structures and Molecular Properties. T.E. Creighton, 2nd ed. Freeman, New York, 1993.

2. The Binding of Coomassie Brilliant Blue to Bovine Serum Albumin. J.L. Sohl and A.G. Splittgerber, *J. Chem. Ed.*, **68**, 262–264 (1991).

3. Studies and Critique of Amido Black 10B, Coomassie Blue R, and Fast Green FCF as Stains for Proteins after Polyacrylamide Gel Electrophoresis. C.M. Wilson, *Anal. Biochem.*, **96**, 263–278 (1979).

4. Why Does Coomassie Brilliant Blue R Interact Differently with Different Proteins? A Partial Answer. M. Tal, A. Silberstein, and E. Nusser, *J. Biol. Chem.*, **260**, 9976–9980 (1985).

5. Determination of Sugars on Paper Chromatograms with *p*-Anisidine Hydrochloride. J.B. Pridham, *Anal. Chem.*, **28**, 1967–1968 (1956).

6. Quantitative Paper Chromatography of Methylated Aldose Sugars. Improved Colorimetric Method Using Aniline Hydrogen Phthalate. W.C. Schaeffer and J.W. Cleve, *Anal. Chem.*, **28**, 1290–1293 (1956).

EXERCISES

Problem 1

Solubility values (g/100 mL) in water for some L-amino acids are (at 25°): Gly: 25.0; Ser: 25.0; Ala: 16.7; Val: 8.9; Ile: 4.1. Explain the observed trend in solubility.

Problem 2

Solubility values for leucine in water as a function of pH are given in the table below:

pH	Solubility at 25°C (g/100 mL)
2.1	6.9
2.5	4.4
3.0	3.0
3.5	2.7
6.0	2.3
8.3	2.7
9.1	3.5
9.4	4.4
9.9	6.7

Plot the solubility values against pH and explain the observed trend.

Problem 3

(*Previous Synton: Problem 4, Unit 22*) As part of his research as a graduate student, Synton was working on one of his favorite beverages, milk, from which he had to isolate a compound to prove (or disprove) a proposed metabolic route. In the process of optimizing the experimental conditions Synton separated the casein and hydrolyzed it to obtain the compound in question. He followed an old method described by H. Barnett in *J. Biol. Chem.*, **100**, 543–550 (1933). He hydrolyzed casein (100 g) by refluxing it with 300 mL of 20% HCl in water overnight. He concentrated the hydrolysate under reduced pressure to remove free HCl, dissolved the residue in 350 mL of hot water, and added 80 mL of 6 M NaOH in water. The final pH was about 3. He decolorized the solution with charcoal, filtered it, and cooled down the filtrate. Crystals of a compound (3.7 g) separated out quickly. Synton analyzed the product by physical and chemical methods. Its melting point was 340–344°C (with decomposition). The solid was sparingly soluble in water but it dissolved well under basic conditions. It gave a blue solution when treated with ninhydrin. Its ^{1}H-NMR (60 MHz), run in buffered D_2O, showed three doublets at 3.2, 6.9, and 7.2 ppm and a triplet at 4.4 ppm. The relative integration of these peaks was $2:2:2:1$. What is the structure of Synton's product?

Problem 4

Galactose exists in water solution as a mixture of α- and β-anomers. The specific rotation of a solution at equilibrium is $+80.2°$ $(g/mL)^{-1}dm^{-1}$. Calculate the % of each anomer in solution if their specific rotations are $[\alpha]_\alpha = +150.7°$ $(g/mL)^{-1}dm^{-1}$, and $[\alpha]_\beta = +52.8°$ $(g/mL)^{-1}dm^{-1}$.

Problem 5

(a) How would you distinguish maltose, galactose, and sucrose by chemical tests (Fehling's, Barfoed's)?

(b) How would you distinguish alanine, proline, and N-acetylglycine by the ninhydrin test?

EXPERIMENT 24

Chemistry of Milk

E24.1 COMPOSITION OF MILK

Milk is the most nutritionally complete food. It provides young mammals with all the nourishment they need, including water, during the first weeks after birth. The chemical composition of milk varies from species to species and depends on seasonal as well as environmental factors. The average percent composition of the milk from various mammals is shown in Table 24.2. It can be seen that the milk of mammals living in cold environments (polar bear, reindeer, and marine mammals) is richer in fats than the milk of animals from temperate climates. This is attributed to the higher caloric content of fats; the metabolism of fats produces 37 kJ/g (8.8 kcal/g) while the energy derived from carbohydrates and proteins is 16 and 17 kJ/g (3.8 kcal/g and 4.0 kcal/g), respectively. Also, the milk of mammals living in cold weather and marine environments is more concentrated because the need for water is not as critical as for animals in warmer climates.

Table 24.2 Average Percent Composition of Various Types of Milk

	Proteins	Fat	Carbohydrates	Minerals (ash)
cow	3.5	3.7	4.8	0.7
human	1.0	3.8	7.0	0.2
goat	2.9	4.5	4.1	0.8
sheep	5.5	7.4	4.8	1.0
horse	1.4	0.6	6.1	0.4
yak	5.8	6.5	4.6	0.9
llama	7.3	2.4	6.0	—
camel	3.6	4.5	5.0	0.7
rabbit	13.9	18.3	2.1	—
dog	7.9	12.9	3.1	—
cat	7.0	4.8	4.8	—
polar bear	10.9	33.1	0.3	—
reindeer	11.5	16.9	2.8	1.2
sealion	8.9	53.3	0.1	—
gray seal	11.2	53.2	0.7	—
blue whale	10.9	42.3	1.3	—

Cow's milk contains proteins, carbohydrates, fats, and also vitamins and minerals. Milk is rich in calcium, phosphorous, vitamins A, B_{12}, and riboflavin, but it is poor in three important nutrients, vitamins D, C, and iron. Vitamin D is added to whole and skimmed milk (fortified milk), but vitamin C and iron must be obtained from other food products. The proteins present in milk are rich in essential amino acids. One glass of cow's milk (250 mL) provides about 25% of the recommended daily allowance of protein; 30% of the riboflavin; 9% of the vitamin A; 35% of the calcium; and 25% of the phosphorous.

Whole, unprocessed milk is a suspension of fat in an aqueous milieu of sugars, proteins, and minerals. The fat, mainly triglycerides, is present in globules of 4–10 μm in diameter. If milk is left standing at room temperature, the fat globules rise to the surface and separate as cream. This problem is avoided by the **homogenization** process, in which whole milk is forced to pass through small holes under high pressure, reducing the size of the fat globules to about 1–2 μm in diameter. **Pasteurized milk** is milk treated at 71.7°C for at least 15 seconds. **Sterilization** involves treating the milk at higher temperatures (132–150°C) for variable periods of time (1–6 sec). Sterilized milk has a much longer shelf-life than pasteurized milk.

Milk Proteins

Casein accounts for 80% of the total protein content of bovine milk; the remaining 20% is primarily composed of α-lactalbumin and β-lactoglobulin. Casein is not a single protein but rather a conglomerate of at least four different types of polypeptide chains; α, β, κ, and γ, which associate to form highly hydrated spherical complexes, known as **casein micelles**, with a diameter between 30 and 300 nm (0.03–0.3 μm). The proportion of α, β, κ, and γ-casein is approximately 55, 25, 15, and 5%, respectively. Whereas α- and β-casein are insoluble in water, κ-casein is soluble in water because it is coupled to carbohydrate molecules, which increase its overall polarity. Some properties of the different components of casein are shown in Table 24.3.

Table 24.3 **Some Properties of the Components of Whole Bovine Casein**

Casein	MM (Da)	pI	% of total casein	Phosphate groups bound/unit	Ca^{2+} bound/unit
α	30,000	4.1	55	8	6–8
β	24,100	4.5	25	5	5–6
κ	20,000	3.7	15	1–2	1–2
γ	30,000	5.8–6.0	5	—	—

Casein exists as a complex with calcium, often called **calcium caseinate**, in which negatively charged parts of the protein (primarily phosphate groups) are bound to Ca^{2+}. The casein micelles are formed by the aggregation of submicelles with diameters of about 0.01 μm; submicelles contain a variable number of α, β, κ, and γ units. Submicelles rich in polar κ-casein are at the surface of the micelle to increase its solubility (Fig. 24.9). In bovine skimmed milk about 65% of the calcium and 55% of the phosphate are found within the casein micelle.

When the pH of milk is decreased by addition of acid or by natural acidification caused by bacteria, some of the Ca^{2+} ions are displaced from the caseinate by protons and precipitation of casein occurs (see "**Isolation of Casein**," E24.2.1). The precipitated casein is called the **curd**, and the liquid supernatant is the **whey**.

$$\text{casein-O} \overset{\overset{\textstyle O}{\|}}{\underset{\underset{\textstyle O^-}{|}}{P}} O^- \quad Ca^{2+} \quad + \ 2\,H^+ \longrightarrow \quad \text{casein-O} \overset{\overset{\textstyle O}{\|}}{\underset{\underset{\textstyle OH}{|}}{P}} OH \ + \ Ca^{2+}$$

calcium caseinate · insoluble casein

The curd obtained from low-fat milk by acid precipitation contains about 1.5% fat. The fat can be removed from the curd by washing it with ethyl ether or acetone. The whey contains lactose and two proteins, β-**lactoglobulin** and α-**lactalbumin**, which together account for 75% of the whey protein. Other proteins present in the whey are serum album, immunoglobulins, and a variety of enzymes. The whey proteins can be precipitated by partial evaporation. Lactose can then be obtained by addition of ethanol to the liquid remaining after the proteins have separated.

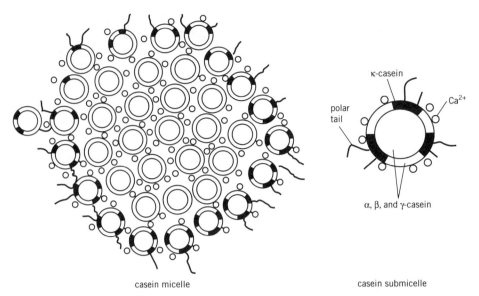

Figure 24.9 Casein micelle and submicelle.

β-Lactoglobulins are denatured by heat; they aggregate above 65°C at neutral pH. The function of β-lactoglobulins is not well understood, although they are believed to be involved in phosphate metabolism. They are known to associate with κ-casein in the casein micelle, playing an important role in the heat stability of milk.

The other important whey protein is α-**lactalbumin**. α-Lactalbumin catalyzes the synthesis of lactose from galactose and glucose. β-Lactoglobulin and α-lactalbumin dissolve readily in acids and stay in solution when casein precipitates by acidification of milk. They are denatured by heat and precipitate during the evaporation of the whey.

Carbohydrates in Milk

The main carbohydrate present in milk is **lactose**, a disaccharide formed by glucose and galactose; some 50 other carbohydrates are also present, although in very low concentrations. Lactose is six times less sweet than sucrose (table sugar); its concentration in cow's milk is in the range 4.4–5.2%. After casein has been precipitated and lactalbumin and lactoglobulin are removed by evaporation and filtration, the resulting liquid can be diluted with ethanol to precipitate lactose. Lactose, a very polar molecule, is soluble in water but only slightly soluble in ethanol, a less polar solvent. To maximize the yield of lactose, the whey should be neutralized after casein has been separated by the addition of acid; failure to neutralize the whey results in hydrolysis of the lactose catalyzed by the acid.

Lactose crystallizes from a water-ethanol solution at room temperature as the α **anomer**. The crystallization is a slow process that usually requires several days to complete. If the solution contains impurities, particularly fats, lactose may fail to crystallize. It should be pointed out that in aqueous solution, the β anomer of lactose is more stable than the α anomer and predominates (about 70% β, and 30% α anomer); however, it is α anomer that crystallizes. Crystallization of the β anomer can be induced from very concentrated solutions at temperatures higher than 95°C. α-Lactose precipitates as a monohydrate with melting point 201–202°C; the crystals lose their water of hydration at 120°C.

Lipids in Milk

The lipid content in milk depends strongly on the species, from a mere 0.6% in horses to more than 50% in marine mammals. Cow's milk contains 3.4–5.1% lipids, of which more than 95% are triglycerides (see Unit 25). Phospholipids account for less than 1%. Milk triglycerides contain more than a hundred different fatty acids; the most prominent ones are oleic (30%), palmitic (24%), stearic (13%), and myristic (9%). Saturated fatty acids make up about 60% of the fatty acids in cow's milk. Compared to other food products, milk is particularly rich in the shorter members of the fatty acid family: butyric (3%), caproic (2%), and caprilic (1%) acids. Bovine milk is relatively poor in cholesterol; a glass of milk contains about 30 mg of cholesterol, while a large egg has about 300 mg. Carotenoids and vitamins A, D, E, and K are associated with the milk lipids. Skimming milk, a process normally done by centrifugation, results in a substantial reduction of the fat content and loss of fat-soluble vitamins. In the United States, the Department of Agriculture requires that vitamins A and D be added to the milk at the end of the skimming process.

BIBLIOGRAPHY

1. Basic Food Chemistry. F.A. Lee. Avi Publishing Co.; 2nd ed. Westport, CT, 1983.
2. Characteristic of Edible Fluids of Animal Origin: Milk. H.E. Swaisgood in "Food Chemistry," 2nd ed., chapter 13; O.R. Fennema, ed. Marcel Dekker, Inc., New York, 1985.
3. Milk Proteins. Chemistry and Molecular Biology; vol. 1 and 2. H.A. McKenzie, ed. Academic Press, New York, 1971.
4. Applied Protein Chemistry. R.A. Grant, ed. Applied Science Pub., London, 1980.
5. Casein. P. Jollès in "Glycoproteins. Their Composition, Structure and Function;" vol. 5, chapter 7, section 3. A. Gottschalk ed. Elsevier, Amsterdam, 1972.
6. The Technology of Cheesemaking. J. Stadhouders and G. van den Berg. *Endeavour*, New Series, **12**, 107–112 (1988).

PROCEDURE

In this experiment you will isolate two of the most important components of cow milk: casein and lactose. The milk will be provided in the lab but, if you want, you can experiment with your favorite brand of nonfat milk.

This experiment will be conducted over two lab periods. In the first period you will isolate casein and characterize it by chemical tests. Also in the same period, you will come to the point in the lactose isolation where the lactose solution is left to crystallize (after charcoal filtration). It is crucial that you reach this point in the isolation of lactose in the first lab period because the crystallization of lactose is a slow process that takes several hours to complete.

In the second lab period you will finish the isolation of lactose and perform the chemical tests and the chromatographic analysis. Because the TLC run takes at least 2 hours, you should spot the TLC plate as soon as possible and leave it in the developing chamber until about 30 minutes before the end of the period.

Safety First
• Acetic acid is corrosive. Handle it with care (E24.2.1).
• Acetone is highly flammable (E24.2.1).
• Hydrochloric acid is corrosive and toxic. Handle it with care (E24.2.2; E24.2.5; E24.3.2).
• NaOH and Ca(OH)$_2$ are caustic. Handle them with care (E24.2.2; E24.2.5; E24.3.2; E24.3.3).
• Ammonium hydroxide irritates the respiratory tract. Dispense it in the hood (E24.2.5).
• Avoid skin contact with Coomassie Blue (E24.2.6). It produces unsightly stains.

E24.2   ISOLATION AND CHARACTERIZATION OF CASEIN

Background reading: 24.1; E24.1
Estimated time: Isolation: 1 hour.
Characterization: 2 hours

E24.2.1 Isolation of Casein

Warm 50 mL of nonfat milk in a 100-mL beaker placed in a water bath on a hot plate. Once the milk has reached a temperature of about 40–45°C, remove the beaker from the heat and, while stirring with a glass rod, add dropwise an aqueous solution of acetic acid (10%) until no further precipitation is observed; less than 2 mL is usually enough. Stir the suspension to break up the solid or **curd** (casein and any residual fat) into small pieces the size of rice grains. Separate the **curd** from the supernatant, **whey**, by vacuum filtration using a Buchner funnel and a disk of paper towel cut to fit the Buchner funnel (the paper should cover the holes without touching the walls of the funnel). Press the curd with a spatula against the filter to remove as much whey as possible. Transfer the liquid to a 250-mL beaker and, without delay, add 1.25 g of calcium carbonate and a magnetic stir bar and then proceed to **Isolation of Lactose** (E24.3.1). Reassemble the filtration apparatus and let the curd dry under vacuum for about 5–10 min, pressing occasionally with a spatula.

Transfer the curd from the Buchner funnel to a 100-mL beaker; add 15 mL of acetone; with the aid of a spatula, break the curd and stir the mixture to allow a thorough mixing with acetone. Acetone extracts any residual fat; it also removes absorbed water and helps in the drying process of casein. Separate the liquid from the solid by decantation, letting the system settle down and pouring the liquid into an Erlenmeyer flask. It does not matter if a few particles of curd pass along with the liquid; add a second 15-mL portion of acetone to the curd and repeat the extraction process. Assemble a filtration apparatus using a Buchner funnel and a disk of paper towel. Vacuum-filter the acetone solution contained in the Erlenmeyer flask first, and then the mixture of curd and acetone. Let the casein dry on the filter with the vacuum on for about 10–15 min, occasionally pressing with a spatula.

Transfer the casein to a piece of preweighed filter paper or paper towel and let the solid dry on the bench top; determine the weight of casein by difference. If the solid casein has a "plastic" texture with big granules, crush them using a mortar and pestle until they are no bigger than sand grains. Weigh out about one half of the mass of casein in a 25-mL beaker (this will be used to make casein glue; E24.2.7). Weigh out about one quarter to perform the tests indicated below (E24.2.2–6) and keep the rest (about one quarter) in your drawer until the following lab section, protected between two watch glasses. The following week, dry the casein in an oven at 70°C for about 15 min and determine its weight. Estimate the % of casein in milk. Take into account that about 3/4 of the casein was used to carry out the chemical tests listed below.

With some of the casein reserved for the chemical tests, make a saturated solution in water by suspending a microspatulaful of finely crushed casein powder in about 1 mL of water and perform the tests as indicated below (E24.2.3–6).

E24.2.2 Partial Hydrolysis of Casein

In a 15-mL round-bottom flask fitted with a microscale water condenser, suspend about 50 mg of casein in 2 mL of 6 M HCl, add a small boiling chip, and reflux in a sand bath on a hot plate for 40 min. In the meantime, proceed to the solubility tests (E24.2.5). At the end of this period, remove from the heat and add a microspatulaful of activated charcoal pellets (Norit RB1). Boil the system for a couple of minutes and then let it cool down to ambient temperature. Gravity-filter the suspension using a Pasteur pipet and a cotton plug (Fig. 3.9a). Neutralize the solution by adding a 6 M NaOH solution dropwise; final pH should be 7–9 (to determine the pH, do not dip the pH paper into the liquid, because the chemicals from the paper may leach into the solution and contaminate it; wet the tip of a glass rod with the solution to be tested and immediately touch a piece of pH paper). Use this hydrolysate for the ninhydrin test.

E24.2.3 Ninhydrin Test

To seven labeled test tubes containing 1 mL of an aqueous ninhydrin solution (0.5 % w/v) each, add one or two drops of the solution to be tested. Heat the test tubes in a water bath at 90–95°C for a couple of minutes. A positive test is the development of a violet-blue color. Interpret your results. The solutions to be tested, all aqueous, are: ovalbumin (2% w/v), gelatin (2%, w/v), glycine (2% w/v), N-acetylglycine (2% w/v), the saturated solution of casein from E24.2.1, the casein hydrolysate from E24.2.2, and water (blank).

E24.2.4 Biuret Test

To 1 mL of the solution to be tested in a test tube, add 0.5 mL of a 3 M NaOH solution, 2 mL of water, and two or three drops of a 2% solution of $CuSO_4$ in water. Solutions to be tested are the same as in Ninhydrin test. Record your observations.

E24.2.5 Solubility Tests

Label six test tubes and put in each one 1 mL of one of the following aqueous solutions: NaOH (3 M); HCl (3 M); HCl (concentrated); NH_4OH (concentrated); and urea (8 M); put 1 mL of water in the sixth tube. Add a microspatulaful of finely crushed casein powder (about 50 mg) to each test tube. Stir slowly and continuously with a glass rod for about 1–2 minutes, crushing the solid against the walls of the test tube; record your results. Keep these test solutions on the bench top for at least 30 min and observe any changes in solubility after this period. Save the solution in 8 M urea for the staining test (E24.2.6).

Using a Pasteur pipet, add the casein water solution to the 3 M HCl solution. Then add dropwise the 3 M NaOH solution followed by the HCl (c) solution. Finally add dropwise the NH_4OH (c) and urea solutions. Conduct this operation in the hood. Proceed to "Cleaning Up."

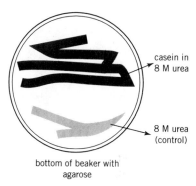

casein in 8 M urea

8 M urea (control)

bottom of beaker with agarose

Figure 24.10 Staining with Coomassie Blue.

E24.2.6 Coomassie Blue Staining

In a 100-mL beaker, weigh 200 mg of agarose. (Agarose is a complex polysaccharide obtained from sea weed and used as a gelled stationary phase in the study and separation of proteins and nucleic acids.) Add 20 mL of water and heat the suspension on a hot plate, stirring with a glass rod until the water boils and the agarose dissolves completely. Remove from the heat and let the agarose harden. In the meantime proceed to the other tests.

Heat the solution of casein in 8 M urea, saved from E24.2.5, in a sand bath to make sure that the casein is completely dissolved. On the hardened agarose, spread some of this solution in a semicircle by using a cotton swab. As a control, cover part of the other semicircle with 8 M urea solution without casein (Fig.24.10). Let these solutions penetrate the agarose for about 5 minutes.

Using a Pasteur pipet, cover the gel with a few milliliters of Coomassie Blue solution. Leave the staining solution in contact with the gel for about 2 minutes and then remove it with a Pasteur pipet. Add a few milliliters of the destaining solution (enough to cover the gel) and leave it for about 5 minutes. Remove the destaining solution and observe the results. The gel should take on a dark-blue color in the semicircle with casein. The rest of the gel should be pale violet.

Staining Solution A 0.2% (w/v) solution of Coomassie Brilliant Blue R250 in water-methanol (4:6) is diluted with an equal volume of a 20% solution of acetic acid in water right before use.

Destaining Solution Water-methanol-acetic acid (6:3:1).

E24.2.7 Casein Glue

Mix 1 part of casein with 2.5 parts of water (by weight); allow the mixture to stand at room temperature for 10 min. Slowly add approximately 0.1 parts of solid sodium hydroxide (each pellet weighs about 0.1–0.2 g; do not break the pellets to weigh out an exact amount; handle them with tweezers!) and 0.2 parts of calcium hydroxide while stirring the mixture. Transfer the glue to a jar and place a label "casein glue" on it using the glue just made. Try the glue on different pieces of cardboard and paper.

Cleaning Up

- Dispose of the acetone filtrate (E24.2.1) in the "Casein Organic Solvents Waste" container.

- Dispose of unused protein and amino acid solutions, the ninhydrin test solutions, the biuret test solutions, the solubility test solutions, and hydrolyzed casein solution (after neutralization) in the "Casein Aqueous Waste" container.

- Dispose of the Coomassie Blue solutions and rinses and destaining solutions in the container labeled "Coomassie- waste."

- Dispose of the filter papers, Pasteur pipets, and agarose gel in the container labeled "Casein Solid Waste".

- Coomassie Blue can be removed from glassware by rinsing with ethanol.

E24.3   ISOLATION AND CHARACTERIZATION OF LACTOSE

> **Background reading:** 24.2
> **Estimated time:** Isolation: 1 hour.
> Characterization: 1/2 hour.
> TLC: 2.5–3 hours

E24.3.1 Isolation of Lactose

After adding calcium carbonate to the whey (see under **"Isolation of Casein,"** E24.2.1), concentrate the liquid by evaporation on a hot plate for about 10 minutes with **magnetic stirring**. Let the mixture cool down and vacuum-filter it, using a Buchner funnel and a piece of filter paper. Concentrate the filtrate to a volume of about 15 mL by heating on a hot plate; the magnetic stirrer should be on during this operation to avoid bumping. At the end of this period, separate any precipitate by vacuum filtration. This removes the whey proteins. Transfer the filtrate to an Erlenmeyer flask and add 90 mL of 95% ethanol and about 0.5 g of powder activated charcoal (no pellets). Prepare a layer of filter aid (Celite) as described in the next paragraph. Heat the suspension on a hot plate **while constantly stirring with a glass rod**; when the system starts to boil, vacuum-filter the hot solution through a layer of filter aid on the Buchner funnel.

To prepare the layer of filter aid, suspend about 0.5 g of Celite in 10 mL of 95% ethanol in an Erlenmeyer flask; heat the suspension while stirring with a glass rod. Pour the suspension in a Buchner funnel fitted with a piece of filter paper. The filtrate, which should be clear without Celite particles, is disposed of in the proper container (see "Cleaning Up"). The Celite layer left on the funnel should be even, without cracks. To avoid cracks, do not allow the layer to dry out; once the liquid stops draining into the filtration flask, turn the vacuum off and use the Buchner funnel with the filter aid immediately.

Filter the charcoal suspension. If charcoal particles pass through the Celite layer, repeat the process. This filtration should be carried out while the suspension is still hot. The filtrate should be clear and colorless. Transfer the filtrate to a 125-mL Erlenmeyer flask, stopper it, and leave it in your drawer, labeled "lactose," until the next lab period to allow complete precipitation of the product. The precipitation of lactose may already start as the solution cools down. This often shows as turbidity due to small lactose particles.

During the following lab period, cool the lactose suspension in an ice-water bath, and with the help of a glass rod break loose the crystals adhered to the walls of the flask. Collect the solid by vacuum filtration on a Buchner funnel. Wash the crystals with a small portion of cold 95% ethanol. Let the solid dry on the filter with the vacuum on for about 10 min. Weigh the product and calculate its % in milk. Determine its melting point and compare it with literature values. Obtain the IR spectrum in a nujol mull (section 31.5).

Prepare a 5% solution of lactose in water (100 mg in 2 mL) and perform the following tests.

E24.3.2 Hydrolysis of Lactose

In two separate, labeled test tubes (13 × 100 mm) place about 1 mL of your 5% lactose solution and 1 mL of a 5% sucrose solution. Add one drop of HCl (c) to each test tube and heat them in a boiling water bath for about 10 min. Allow the reaction mixtures to cool and neutralize them by adding a 10% NaOH solution dropwise until it gives a slightly basic pH on pH paper; about five drops will be necessary. Use these solutions for the tests below.

E24.3.3 Fehling's Test

In six separate, labeled test tubes add 1 mL of Fehling's reagent (prepared by mixing equal volumes of Fehling's A and Fehling's B) and one drop of 5% solutions of lactose, glucose, galactose, sucrose, and hydrolyzed lactose and hydrolyzed sucrose from the previous experiment. Heat the test tubes in a boiling-water bath for about 5 min and record any change in color. A brown-reddish precipitate or the formation of a copper mirror is a positive test.

Fehling's Reagent Fehling's A: 34.5 g of pentahydrate copper sulfate dissolved in 500 mL of water; Fehling's B: 173 g of sodium potassium tartrate (Rochelle salt) dissolved in 500 mL of a 10% sodium hydroxide solution.

E24.3.4 Barfoed's Test

In six separate, labeled test tubes add 1 mL of Barfoed's reagent and one or two drops of the same solutions used for Fehling's test. Heat the test tubes in a boiling-water bath for about 10 min and record any change in color. Compare your results with those from the Fehling's test.

Barfoed's reagent (Cu^{2+}, in acidic conditions) reacts with disaccharides much more slowly than with reducing monosaccharides.

Reagent Cupric acetate (33.3 g) and acetic acid (4.5 mL) dissolved in 500 mL of water. The solution is filtered if necessary.

E24.3.5 TLC Analysis of Sugars

On a 5×10-cm silica TLC plate, spot one small drop of the following solutions using a capillary pipet: 5% galactose; 5% glucose; 5% lactose; hydrolyzed lactose. Develop the chromatogram using a mixture of *n*-butanol : acetic acid : ethyl ether : water (10 : 6 : 3 : 1). Let the chromatogram run for at least 2.5 hours. Let the plate evaporate in the fume hood. Visualize the chromatogram by spraying it with a solution of *p*-anisidine and phthalic acid in ethanol. Place the plate on a hot plate (at low-medium setting) in the fume hood or in a well-ventilated oven at 70°C for about 10 min or until the spots show on the plate.

> **Safety First**
>
> - The spraying of the TLC plate should be done in the hood.

Reagent 1.23 g of *p*-anisidine (4-methoxyaniline) and 1.66 g of *o*-phthalic acid (1,2-benzenedicarboxylic acid) are dissolved in 100 mL of 95% ethanol; this makes a 0.1 M solution in both reagents.

E24.3.6 Mutarotation of Lactose

Using mortar and pestle crush the lactose crystals to a very fine powder. Weigh 2.0 g of the solid and transfer it to a 25-mL volumetric flask.

The following operations should be conducted quickly. Have everything ready. Add distilled water to the mark, stopper the flask, and invert it several times to dissolve the lactose. If the lactose does not dissolve completely, warm the flask in a water bath at 50–60°C for about a minute. Start the timer. With the help of a funnel fill a 2-dm polarimeter tube.

> **Safety First**
>
> - Concentrated ammonium hydroxide is a corrosive lachrymator. Dispense this chemical in the fume hood.

Read the optical rotation (about 14° at the beginning). Read the optical rotation at 15–20 minute intervals for about 1–2 hours. Add two drops of concentrated ammonium hydroxide and read the equilibrium value. Ammonium hydroxide catalyzes the mutarotation and allows the system to reach equilibrium almost instantly.

After all the measurements have been taken, transfer the solution to a beaker and check the pH. If necessary, neutralize it with a few drops of dilute HCl to reach a final pH 6–8.

Plot optical rotation as a function of time. Estimate the optical rotation at time zero. Calculate the specific rotation at time zero and at equilibrium. In calculating the concentration of lactose, take into account that the solid you weighed is not 100% lactose; it is lactose monohydrate. Calculate the percentage of both anomers at equilibrium.

Cleaning Up

- Dispose of the calcium carbonate, charcoal, and Celite from the isolation of lactose, capillary tubes, Pasteur pipets, and capillary pipets in the container labeled "Lactose Solid Waste."
- Dispose of the ethanol solution from the isolation of lactose in the container labeled "Lactose Organic Solvents Waste."
- Dispose of hydrolyzed lactose and sucrose (after neutralization) in the container labeled "Lactose Aqueous Waste."
- Dispose of the Fehling and Barfoed's solutions in the container labeled "Lactose Copper Waste."
- TLC plates should be disposed of in the container labeled "Lactose Solid Waste."
- Dispose of the mutarotation solution in the container "Lactose Aqueous Waste."
- Turn in the casein and lactose in labeled vials to your instructor.

BIBLIOGRAPHY

1. Laboratory Investigations in Organic Chemistry. D.C. Eaton. McGraw-Hill, New York, 1989.
2. The Tools of Biochemistry. T.G. Cooper. Wiley, New York, 1977.

EXPERIMENT 24   REPORT

Pre-lab

1. Make a flowchart for the complete study of the chemistry of milk. Make a table with the physical properties of α-lactose, casein, acetone, acetic acid, urea, ammonium hydroxide, sodium hydroxide, and hydrochloric acid. Include toxicity/hazards and solvent flammability.
2. What are the main components of cow's milk? In what percentage are they present?
3. Why do you wash the curd with acetone?
4. Write a generic chemical reaction for the hydrolysis of peptide bonds.
5. Show the reaction of ninhydrin with glycine.
6. Formulate a chemical reaction for the biuret test.
7. Which of the following sugars are reducing: glucose, fructose, sucrose, lactose, galactose?
8. Write a chemical reaction for the hydrolysis of lactose.
9. In the isolation of lactose, after the separation of casein the whey is treated with calcium carbonate. Why is calcium carbonate added? Why must it be added without delay? Why is ethanol added?
10. In which step are the whey proteins separated?
11. How can β-lactose be crystallized from solution?
12. Give the chemical equation for the reaction of a reducing sugar with Fehling's reagent.
13. Why is the Fehling's reagent kept in two separate bottles (A and B)?

14. α-Lactose is dissolved in water and the optical rotation of the solution measured at different times. Immediately after the solution is made, the specific rotation is $+92.6°$ $(g/mL)^{-1}$ dm^{-1}; after 50 min: $+69°$ $(g/mL)^{-1}$ dm^{-1}; and after 22 hours (and longer): $+52.3°$ $(g/mL)^{-1}$ dm^{-1}. Explain the results and calculate the % of α- and β-lactose at equilibrium in the aqueous solution, knowing that the specific rotations are: $[\alpha]_\alpha = +92.6$ and $[\alpha]_\beta = +34.2°$ $(g/mL)^{-1}$ dm^{-1}.

In-lab

1. Report the % of casein and lactose in milk. Compare these numbers with those given in Table 24.2.

2. Interpret your ninhydrin-test and biuret results. Present your data in a table format. Did the casein hydrolysate give positive or negative tests? What conclusions can be reached about the hydrolysis?

3. Report the solubility behavior of casein. Present your data in a table format. Interpret your results.

4. Report and interpret the Coomassie Blue staining test results.

5. Report the melting point of lactose and compare it with literature values.

6. Interpret the IR spectrum of lactose and compare it with the one shown in Figure 24.11. Which band can be used as evidence for the precipitation of a *hydrate* of lactose? (Hint: consider the IR of liquid water shown in Fig. 31.19)

7. Present the results of the Fehling and Barfoed's tests in a table (each sugar a column; each test a row) with + and−. Interpret your results.

8. Discuss your TLC results. How do they confirm the results of the chemical tests?

9. Present and discuss your mutarotation results.

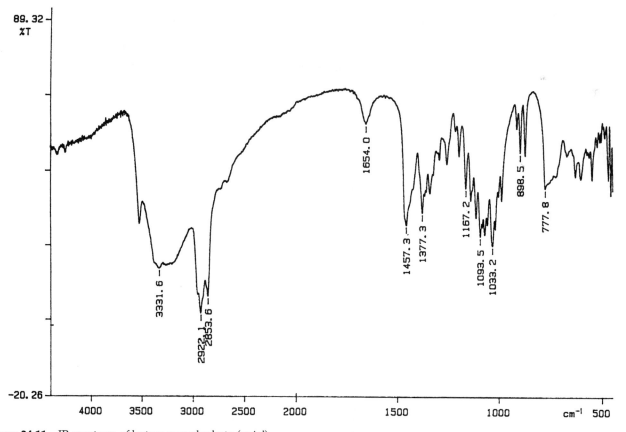

Figure 24.11 IR spectrum of lactose monohydrate (nujol).

Lipids

25.1 FATS AND OILS

Lipids are biological compounds insoluble in water but soluble in organic solvents. Unlike other families of organic compounds such as aldehydes, amines, or carbohydrates, lipids do not have a common functional group. What all lipids have in common is a dislike for water and a high affinity for organic solvents. The most important types of lipids are **fats, oils, waxes, phospholipids, terpenes**, and **steroids**. In this Unit we will concentrate on fats, oils, and steroids. Terpenes are discussed in Experiment 11A.

The main difference between **fats** and **oils** is that, at room temperature, fats are solids and oils are liquids. Fats and oils, also called **triglycerides** or **acylglycerols**, are esters of glycerol with carboxylic acids. The carboxylic acids found in natural fats and oils, called **fatty acids**, have an even number of carbon atoms ranging from 4 to 22 (Table 25.1). The three carboxylic acid residues in a triglyceride are usually different from each other, but triglycerides with only one type of carboxylic acid are not uncommon; for example, **trimyristin**, the major component of nutmeg fat, is a triglyceride with only myristic acid residues.

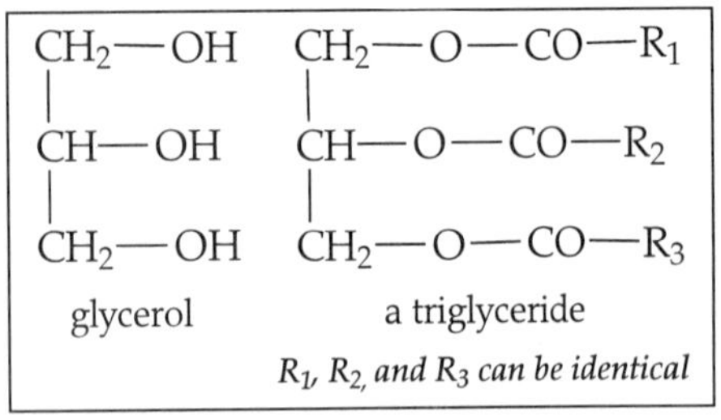

$$
\begin{array}{ll}
CH_2\!-\!OH & CH_2\!-\!O\!-\!CO\!-\!R_1 \\
| & | \\
CH\!-\!OH & CH\!-\!O\!-\!CO\!-\!R_2 \\
| & | \\
CH_2\!-\!OH & CH_2\!-\!O\!-\!CO\!-\!R_3
\end{array}
$$

glycerol   a triglyceride

R_1, R_2, and R_3 can be identical

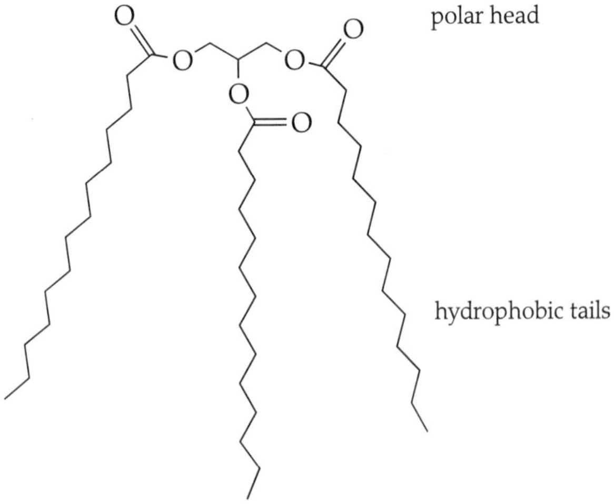

polar head

hydrophobic tails

trimyristin: trimyristylglycerol
a triglyceride present in nutmeg

Fats and oils are insoluble in water because they are **hydrophobic** (from Greek: *hydor*, water; *phobos*, fear). The long hydrocarbon chains of the fatty acid residues are nonpolar and repel water. Palmitic and stearic acids (Table 25.1) are saturated fatty

Table 25.1 **Some Common Natural Fatty Acids**

Number of carbons	Common name	Systematic name	Abbreviation	Structure	Melting point (°C)
			Saturated Fatty Acids		
12	lauric	dodecanoic	12:0	‿‿‿‿‿COOH	44
14	myristic	tetradecanoic	14:0	‿‿‿‿‿‿COOH	54
16	palmitic	hexadecanoic	16:0	‿‿‿‿‿‿‿COOH	63
18	stearic	octadecanoic	18:0	‿‿‿‿‿‿‿‿COOH	70
20	arachidic	eicosanoic	20:0	‿‿‿‿‿‿‿‿‿COOH	77
22	behenic	docosanoic	22:0	‿‿‿‿‿‿‿‿‿‿COOH	82
			Unsaturated Fatty Acids (all double bonds are *cis*)		
16	palmitoleic	9-hexadecenoic	16:1	‿‿‿═‿‿‿COOH	−0.5
18	oleic	9-octadecenoic	18:1	‿‿‿‿═‿‿‿COOH	4
18	ricinoleic	(*R*)-12-hydroxy-9-octadecenoic	18:1	‿‿‿ȮH═‿‿COOH	6
18	linoleic	9,12-octadecadienoic	18:2	‿‿═‿═‿‿COOH	−12
18	linolenic	9,12,15-octadecatrienoic	18:3	‿═‿═‿═‿COOH	−16
20	erucic	13-docosenoic	20:1	‿‿‿═‿‿‿‿‿COOH	34
20	arachidonic	5,8,11,14-eicosatetraenoic	20:4	═‿═‿═‿═‿COOH	−50

acids. Oleic and palmitoleic acids are **monounsaturated** (with one double bond) and linoleic and linolenic acids are **polyunsaturated** (with two or more double bonds). For naturally occurring unsaturated fatty acids, the double bonds are always *cis* and never conjugated.

Fatty acids are sometimes abbreviated by giving the total number of carbon atoms followed by the number of double bonds, as indicated in Table 25.1.

A closer inspection of Table 25.1 also reveals that unsaturated fatty acids have lower melting points than saturated ones with the same number of carbon atoms. The carbon-carbon double bond of unsaturated fatty acids bends the hydrocarbon chain, leading to a loose packing of molecules and a lower melting point. Fats are rich in saturated fatty acids with a tight chain packing, and have higher melting points, while oils, richer in loosely packed unsaturated chains, have lower melting points. The composition of some typical oils and fats is shown in Table 25.2.

Table 25.2 **Composition of Some Typical Oils and Fats**

	Iodine number	m.p. (°C)	Lauric	Myristic	Palmitic	Stearic	Palmitoleic	Oleic	Linoleic	Linolenic	Others
					VEGETABLE OILS AND FATS						
coconut	7–10	25	48	19	7	3		6	1		
olive	74–94	−6		1	8	2	1	80	5		
castor bean	81–91	−14			1			5	5		86% ricinoleic
peanut	83–100	3			8	4	1	60	20		
canola	94–106	−6		1	1	1		60	30		2% erucic acid
cottonseed	103–115	−1		1	23	1	1	28	44		
sesame	104–116	−6			9	4	1	45	40		
corn	116–130	−20		1	9	3	1	46	38		
sunflower	122–136	−18			6	2		30	60		
soybean	124–136	−16		1	9	3		25	53	7	
safflower	130–150	−15			3	3		19	75	3	
linseed	170–204	−24			6	3		19	24	47	
					ANIMAL OILS AND FATS						
butter	26–45	32	2	10	29	9	5	30	4		
lard	46–66	30		1	29	15	2	45	6		
whale	110–150	24		7	15	3	15	35	10		
fish (sardine)	120–190			7	13	1	10				70% unsat. C16–C22

25.2 CHARACTERIZATION OF OILS

The content of unsaturated fatty acid residues is a characteristic of each type of oil and is used to distinguish fats from oils. An assay that has been devised to measure the unsaturation content is the **iodine number test**. The carbon-carbon double bonds of unsaturated fatty acids react with iodine to give addition products.

$$R—CH{=}CH—CH_2—CH{=}CH—R' + 2\,I_2$$

unsaturated chain

$$R—\underset{I}{CH}—\underset{I}{CH}—CH_2—\underset{I}{CH}—\underset{I}{CH}—R'$$

The larger the number of double bonds, the larger the amount of iodine necessary to react with all of them. The iodine number is defined as the mass of iodine, in grams, that reacts with 100 g of fat or oil. In other words, the iodine number is the *percentage* of iodine absorbed by the fat or oil. Oils rich in unsaturated fatty acids such as linseed, soybean, and sunflower have high iodine numbers. On the other hand, butter, lard, and coconut oil are poor in unsaturated fatty acids and consequently have a low iodine number. The degree of unsaturation of fats and oils has important implications in our diets. Saturated fatty acids tend to form deposits on artery walls, and these deposits can cause cardiovascular disease. Physicians and nutritionists recommend reducing our fat intake and replacing oils rich in saturated fatty acids with others richer in monosaturated and polyunsaturated ones.

Peanut oil contains a small amount, about 5%, of arachidic acid residues (20:0). Many vegetable oils also contain arachidic acid but in smaller quantities. Sesame oil, for instance, has only about 1%. Arachidic acid is different from other saturated fatty acids, such as stearic and palmitic, in that it is insoluble in cold ethanol. This is the basis for the **turbidity test** or **Evers's test**. The oil is hydrolyzed to produce the free fatty acids, which are dissolved in warm ethanol. The clear solution is then cooled down slowly with stirring. The temperature at which turbidity first appears is a characteristic of each oil and is related to the content of arachidic acid. The higher the content of arachidic acid, the higher the turbidity temperature. Peanut oil produces turbidity at 39–40°C, sesame at 15°C, cottonseed at 13°C, corn at 7–14°C, and olive oil at 6–9°C.

(+)-sesamin

Some oils can be characterized by specific tests. For example, sesame oil gives a reddish color when treated with furfural in the presence of concentrated hydrochloric acid. Although the reaction, known as the **Villavecchia's test**, has not been elucidated, cyclic ethers present only in sesame oil, such as sesamin, seem to be responsible for color development.

25.3 UNCOMMON FATTY ACIDS AND RELATED COMPOUNDS

The most common fatty acids are devoid of other functional groups with the exception of double bonds. A small percentage of natural fatty acids, however, have oxygen atoms in the hydrocarbon chain in the form of epoxide rings or hydroxyl groups. Two such acids are vernolic and ricinoleic.

$$CH_3(CH_2)_5\overset{\underset{\displaystyle |}{OH}}{C}HCH_2CH = CH(CH_2)_7CO_2H \qquad CH_3(CH_2)_4\overset{\displaystyle O}{\overset{\displaystyle /\backslash}{C}HCHCH_2CH} = CH(CH_2)_7CO_2H$$

ricinoleic acid
found in castor oil

vernolic acid
found in seed oils of the genus *Vernonia*

Ricinoleic acid (12-hydroxyoleic acid) is the main fatty acid of castor oil (ricinus oil), accounting for 90% of its fatty acid content. Castor oil has been used since the time of the Pharaohs as a purgative. This application is in decline today but the oil is still used as a lubricant, in cosmetics, and as the raw material for the production of **azelaic acid** (nonanedioic acid). Azelaic acid has been approved by the FDA in 1995 as a prescription medicine for the treatment of acne.

Azelaic acid can be produced in the laboratory by oxidation of ricinoleic acid with potassium permanganate. The oxidizing agent cleaves the double bond, producing azelaic acid and 3-oxononanoic acid, which, under the reaction conditions, is decarboxylated to 2-octanone.

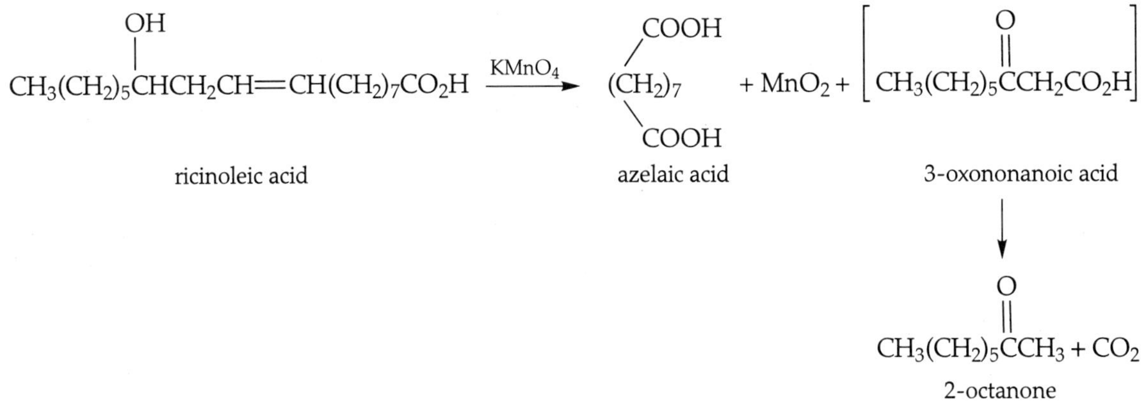

$$CH_3(CH_2)_5\underset{\overset{|}{OH}}{CH}CH_2CH{=}CH(CH_2)_7CO_2H \xrightarrow{KMnO_4} \underset{COOH}{\overset{COOH}{(CH_2)_7}} + MnO_2 + \left[CH_3(CH_2)_5\overset{\overset{O}{\parallel}}{C}CH_2CO_2H \right]$$

ricinoleic acid   azelaic acid   3-oxononanoic acid

$$CH_3(CH_2)_5\overset{\overset{O}{\parallel}}{C}CH_3 + CO_2$$

2-octanone

You will prepare azelaic acid from castor oil in E25.3.

25.4 SAPONIFICATION

Esters are hydrolyzed in the presence of strong bases, such as NaOH, to yield an alcohol and the salt of the carboxylic acid. This reaction is irreversible and is called **saponification**. Saponification, which literally means *making soap*, is one of the oldest chemical transformations of which we have record. It was used by the Babylonians in the manufacture of soaps from fats and oils. Triglycerides treated with concentrated NaOH give glycerol and the sodium salts of the fatty acids called **soap**.

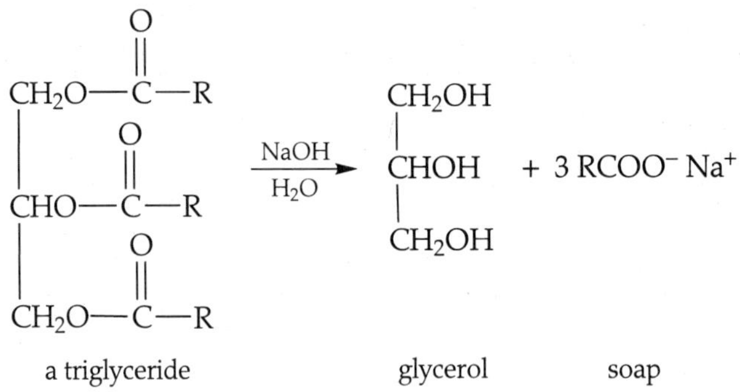

$$\begin{matrix} CH_2O{-}\overset{\overset{O}{\parallel}}{C}{-}R \\ | \\ CHO{-}\overset{\overset{O}{\parallel}}{C}{-}R \\ | \\ CH_2O{-}\overset{\overset{O}{\parallel}}{C}{-}R \end{matrix} \xrightarrow[H_2O]{NaOH} \begin{matrix} CH_2OH \\ | \\ CHOH \\ | \\ CH_2OH \end{matrix} + 3\,RCOO^-\,Na^+$$

a triglyceride   glycerol   soap

Sodium salts of fatty acids have a polar end that is **hydrophilic** (water-loving) and a long hydrocarbon chain that is hydrophobic and **lipophilic** (fat-loving). When dissolved in water, soap molecules associate in large aggregates, called **micelles**, with the hydrophilic ends on the outside surrounded by water molecules, and the nonpolar chains buried inside to avoid contact with water. These micelles confer cleansing properties to soap because they dissolve fats and oils inside their lipophilic cores (Fig. 25.1).

25.5 STEROIDS

Steroids encompass a large family of lipids; cholesterol is its most important member. Cholesterol is the precursor of steroid hormones and bile acids, which play vital roles in animal metabolism. All steroids contain the four-ring system shown below. The four rings are named A, B, C, and D. Most steroids have two methyl groups on

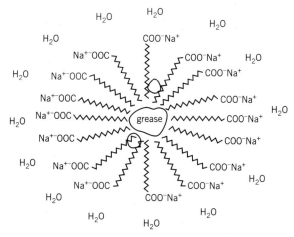

Figure 25.1 A soap micelle engulfing grease droplets.

carbons 10 and 13, which are called **angular methyl groups** and labeled 18 and 19. These two methyl groups are above the plane of the page and are represented by solid wedges.

The fusion between rings B and C is always *trans*. This means that the hydrogens on C8 and C9 are *trans* to each other. Rings C and D are also *trans*-fused. Rings A and B, on the other hand, can be *cis*- or *trans*-fused depending on the type of steroid. They are *trans*-fused in cholesterol and *cis*-fused in bile acids.

Rings A and B are trans-fused.

Rings A and B are cis-fused.

The structure of cholesterol is shown below. It has an HO— group in position 3, a double bond between carbons 5 and 6, and a side chain on carbon 17. Cholesterol has eight stereogenic carbons, indicated by asterisks in the figure, and thus there are $2^8 = 256$ possible stereoisomers. Only one isomer, to which we reserve the name cholesterol, is biologically active. The steroid ring system has two faces called α and β. Substituents on the same side as the angular methyl groups are on the **β-face**. Substituents opposite to the angular methyl groups are on the **α-face**.

Cholesterol

*stereogenic carbon

A typical adult human being has about 150–200 g of cholesterol in her or his body, most of it located in the brain, muscles, and the connective tissue, including blood. An average person produces about 1–1.5 g of cholesterol daily but needs only about 0.6–0.8 g to sustain life. In addition, a Westerner consumes about 0.5 g of cholesterol a day, mainly from egg yolks, meats, diary products, and shellfish. The excess cholesterol is mostly excreted but some accumulates in the body fat and also as plaque deposits on the walls of arteries. Over time, these deposits get calcified and become a permanent obstruction, narrowing the internal diameter of arteries and reducing blood flow. This disorder is called *atherosclerosis* and is the leading cause of heart disease.

Cholesterol has 27 carbons, 46 hydrogens, and one oxygen atom in a hydroxyl group. The presence of the polar OH group is not enough to counterbalance the effect of a large number of C—H bonds, and overall, cholesterol is nonpolar and insoluble in water. To be transported in the bloodstream cholesterol must be solubilized first. This is achieved by forming complexes with lipoproteins (proteins associated with fats) present in the plasma. Cholesterol carried by low-density lipoproteins (LDL-cholesterol) is sometimes called "bad cholesterol" because it is responsible for fatty deposits on arteries. On the other hand, cholesterol carried by high-density lipoproteins (HDL-cholesterol) helps to protect against heart disease and is called "good cholesterol."

25.6   BILE ACIDS

Bile is a greenish liver secretion that aids in the digestion of fats in the intestines. The main components of bile are carboxylic acids synthesized from cholesterol and called **bile acids**. Two important bile acids, **cholic** and **deoxycholic acids**, are shown on the next page. They have a carboxyl group on the side chain and several hydroxyl groups attached to the α face of the ring system. Contrary to most steroids, cholesterol included, rings A and B of bile acids are *cis*-fused.

The polar OH groups of bile acids are attached exclusively to the α face of the steroid ring system. This arrangement creates two faces with very different polarities: a polar α face and a nonpolar β face. As a result, bile acids can simultaneously associate with polar and nonpolar compounds, acting as *biological detergents* and keeping nonpolar molecules in solution. In combination with other compounds present in the bile, bile acids form micelles with nonpolar interiors where fats and cholesterol can be accommodated, and increase the solubility of these lipids in aqueous solutions.

25.7   INCLUSION COMPOUNDS

Urea (NH_2CONH_2) forms **inclusion compounds** with straight-chain hydrocarbons and saturated fatty acids. Inclusion compounds are noncovalent adducts between a

Bile acids

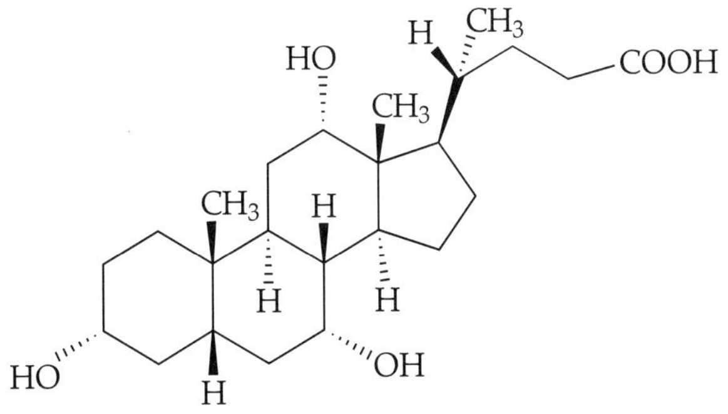

cholic acid

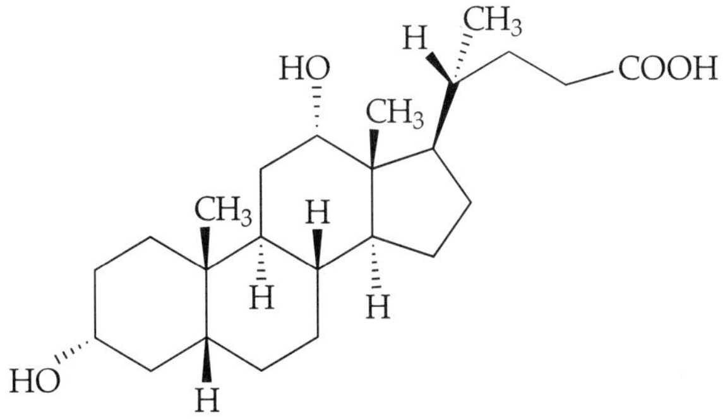

deoxycholic acid

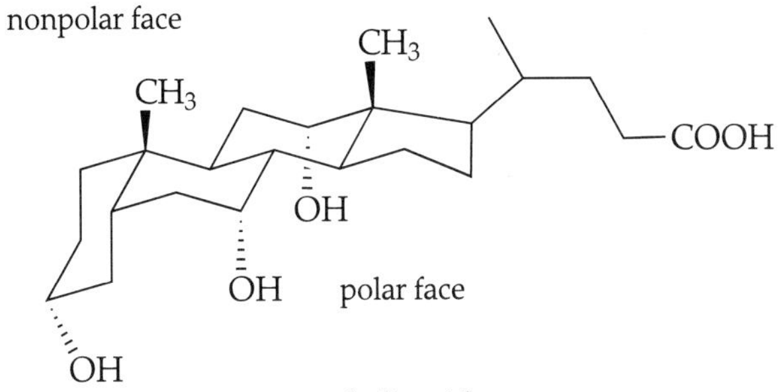

cholic acid

host (urea) and a **guest** (the hydrocarbon or fatty acid). The host molecules associate with each other by H-bonds, forming a lattice that leaves a cavity resembling a channel where the guest molecules fit tightly. The internal diameter of the urea channels is approximately 5 Å, big enough to accommodate linear hydrocarbons and saturated fatty acids, but too small to form stable adducts with branched and nonlinear compounds.

Urea inclusion compounds, called **clathrates**, are useful to separate straight-chain hydrocarbons from branched ones. They are also useful to characterize fatty acids that form stable adducts with dissociation temperatures considerably higher than the melting points of the free fatty acids. The inclusion compounds between urea and its guests are not stoichiometric. The number of urea molecules per guest molecule is variable and increases as the length of the guest molecule increases. Some examples of inclusion compounds between urea and fatty acids are shown in Table 25.3.

Table 25.3 **Inclusion Compounds Between Urea and Fatty Acids**

Acid	m.p. (°C)	Complex Dissociation Temp. (°C)	Urea: acid (molar)
capric (C10)	32	85	8.0
lauric (C12)	44	96	9.7
myristic (C14)	54	103	10.6
palmitic (C16)	63	114	12.8
stearic (C18)	70	126	14.2

Urea is not the only compound that can form inclusion compounds with other organic chemicals. Deoxycholic acid, a bile acid discussed in the previous section, also forms noncovalent adducts with a variety of compounds such as hydrocarbons, fatty acids, and alcohols. These adducts are called **choleic acids**.

There exist some differences between choleic acids and clathrates. Contrary to clathrates, which have a fractional number of host molecules per guest molecule, choleic acids are formed by an integral number of deoxycholic acid molecules (2, 4, 6, or 8) surrounding one guest molecule that sits in the center of a cavity, as the diagram in Figure 25.2 indicates.

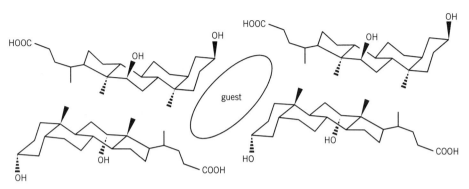

Figure 25.2   A choleic acid with four deoxycholic acid molecules and a guest.

Another difference between clathrates and choleic acids is that bile acids can form complexes with both polar and nonpolar compounds, while urea clathrates are formed only with long nonpolar compounds. Yet another difference is that the urea channels, where the guest molecules fit, collapse in the absence of the guest molecules while bile acids are able to associate themselves and form rigid structures even without a guest. This property is dramatically illustrated by deoxycholic acid. Below pH 6.8, this acid self-associates through H-bonds, forming helical structures that can be drawn from solution into long glassy fibers, as you will see in E25.6.

BIBLIOGRAPHY

1. Fatty Acid and Lipid Chemistry. F. Gunstone. Blackie Academic & Professional, London, 1996.
2. Lipids. Chemistry, Biochemistry, and Nutrition. J. F. Mead, R. B. Alfin-Slater, D. R. Howton, and G. Popják. Plenum Press, New York, 1986.
3. Natural Products. A laboratory Guide. R. Ikan, 2nd ed. Academic Press, San Diego, 1991.
4. Liver, Nutrition and Bile Acids. G. Galli and E. Bosisio, ed. Plenum Press, New York, 1985.
5. The Structure of Choleic Acids. W. C. Herndon. *J. Chem. Ed.*, **44**, 724–728 (1967).
6. Solid-State Photochemistry of Guest Aliphatic Ketones Inside the Channels of Host Deoxycholic and Apocholic Acids. R. Popovitz-Biro, C. P. Tang, H. C. Chang, M. Lahav, and L. Leiserowitz. *J. Am. Chem. Soc.*, **107**, 4043–4058 (1985).
7. Molecular Association of Organic Substances. L. N. Ferguson. *J. Chem. Ed.*, **31**, 628–630 (1954).

EXERCISES

Problem 1

Saponification of a triglyceride gave equimolar amounts of stearic, palmitic, and oleic acids. How many isomeric triglycerides, counting enantiomers, with these three different fatty acids are possible?

Problem 2

Triolein (trioleylglycerol) is the most abundant triglyceride of olive oil. Calculate its iodine number.

Problem 3

A vegetable oil has a melting point of $-6°C$ and an iodine number of 115. With the help of Table 25.2, determine the most likely origin of the oil.

Problem 4

Estimate the iodine number of a triglyceride that has 60% linoleic and 40% oleic acid.

Problem 5

(*Previous Synton: Problem 3, Unit 24*) During his internship in the forensic lab, Synton had to analyze a shipment of imported olive oil that had presumably been adulterated. His preliminary results showed that the main fatty acid present in the oil was oleic ($70 \pm 10\%$). The iodine number was 94 ± 2 and the melting point between 3 and $-9°C$. The turbidity temperature was 25°C. The oil gave a negative Villavecchia's test. Synton had to write a report about his preliminary findings, stating whether the oil had been adulterated and giving the names of possible adulterants. What do you think that Synton wrote in his report?

EXPERIMENT 25

Lipids

This experiment consists of several independent parts. In part E25.1 you will analyze some typical oils by chemical tests, and identify an unknown oil sample that may be peanut, sesame, or olive oil. In E25.2 you will isolate trimyristin, a triglyceride of myristic acid, from nutmeg. You will then proceed to the saponification of the trimyristin using sodium hydroxide and the identification of myristic acid by TLC analysis, and by its inclusion compound with urea. This part of the experiment has been adapted from References 4–6.

In part E25.3 you will prepare azelaic acid from castor oil. This involves the saponification of the oil to give ricinoleic acid followed by the oxidation of the acid with potassium permanganate. The whole procedure requires two laboratory periods, but while you are doing the saponification, which takes about 2 hours, you can do other parts of the experiment.

Parts E25.4, E25.5, and 25.6 are devoted to bile acids and inclusion compounds. In part E25.4 you will prepare the urea clathrate of a fatty acid. This part has been adapted from Reference 6. In part E25.5, you will qualitatively study the solubility of cholesterol in different solvents and in the presence of bile acids. Finally, in part E25.6 you will form the choleic acid of p-xylene and characterize it by its melting point. You will also form a complex of deoxycholic acid that can be drawn from solution as long and brittle fibers.

Consult with your instructor to determine which parts you will do.

E25.1 ANALYSIS OF OILS

> **Background reading:** 25.1 and 25.2
> **Estimated time:** 45 min.

Detection of Sesame Oil: Villavecchia's Test

Place 0.5 mL of the oil to be tested in a small test tube. Add 2 drops of a 2% (v/v) solution of furfural in 95% ethanol followed by 2 drops of concentrated sulfuric acid. Observe any change in color at the interface between the liquids. Sesame oil or oils containing 2% or more sesame oil develop a reddish color due to the presence of cyclic ethers, such as sesamin. Try this test with samples of sesame, peanut, olive, and your unknown.

Turbidity Temperature: Evers's Test

In a 15-mL round-bottom flask equipped with a water-jacketed reflux condenser and a stir bar, place 1 mL of the unknown oil and 5 mL of 1.5 M alcoholic KOH (in 95% ethanol). Lubricate the joint with a very small amount of grease. Boil the mixture directly on a hot plate at a medium setting for 10 min.

At the end of this period, let the system cool down and transfer the liquid with a Pasteur pipet to a 100-mL Erlenmeyer flask equipped with a stir bar. Add 50 mL of ethanol (70% v/v) and 0.8 mL of concentrated HCl and warm the solution on the hot plate to dissolve any precipitate that may have formed. After acidification, some oils give a slight opalescence that does not disappear by warming but it does not interfere in this test either. If you observe opalescence continue on with the test.

Place a thermometer in the Erlenmeyer flask. The bulb should be completely immersed in the liquid but it should not touch the bottom of the flask. There should be room for the stir bar to rotate freely. Cool the system in a water bath made with a beaker or another transparent piece of glassware (Fig. 25.3). Stir continuously and observe the temperature. It should fall at a rate of about 1°C per minute. Record the temperature at which a definite precipitate first appears. This is the turbidity temperature. To observe the formation of precipitate, look at the flask from the side. Peanut oil produces turbidity at 22–29°C; sesame and olive oil do not produce turbidity under these conditions.

Iodine Test

In a scintillation vial place 10 mL of the sample oil and add iodine tincture dropwise. Count the number of drops. After each addition swirl the vial to mix well. Stop the addition when the color of the iodine tincture persists. The iodine tincture does not completely dissolve in the oil but the iodine slowly reacts with the double bonds and the iodine color disappears. The higher the content of unsaturated fatty acid residues, the faster the oil reacts with iodine. Carry out this test with samples of olive oil, sesame oil, peanut oil, and your unknown. Check Table 25.2 and interpret your results.

Reagent Iodine tincture contains 2% (w/v) iodine and 2.4% (w/v) sodium iodide in a 47% (v/v) ethanol solution in water.

The analysis of oils has been adapted from References 1–3.

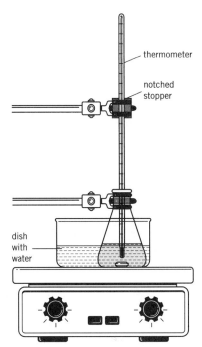

Figure 25.3 Setup for the turbidity test.

E25.2 TRIMYRISTIN FROM NUTMEG

> **Background reading:** 25.1; 25.4; 25.7
> **Estimated time:** 3 hours

Isolation

$$CH_2-O-CO-(CH_2)_{12}CH_3$$
$$CH-O-CO-(CH_2)_{12}CH_3$$
$$CH_2-O-CO-(CH_2)_{12}CH_3$$

In a 16 × 125 mm screw-cap test tube, weigh 1 g of finely powdered nutmeg and add 10 mL of *tert*-butyl methyl ether (BME). Cap the tube and shake the mixture for about 15 minutes. Vent frequently to release the pressure by momentarily unscrewing the cap.

Transfer the contents of the tube to a Pasteur pipet with a cotton plug, sand, and 1.5 cm of anhydrous sodium sulfate (Fig. 25.4). Collect the filtrate in a dry, pre-tared 25-mL round-bottom flask. Rinse the solid in the Pasteur pipet with 2 mL of BME and collect the liquid in the same round-bottom flask. Evaporate the solvent in a rota-vap. Determine the weight of the product.

Recrystallize the product from acetone. To this end, add about 2 mL of acetone to the round-bottom flask, heat the mixture on a sand bath to dissolve the solid, and with a Pasteur pipet transfer the solution to a small test tube. Let the system

Cleaning Up

- Dispose of the liquids from the Villavecchia's and Evers's tests in the container labeled "Lipids Organic Waste."

- Dispose of Pasteur pipets in the container "Lipids Solid Waste."

- Dispose of the iodine test waste in the container labeled "Lipids Iodine Waste." Do not mix with other waste.

Safety First

- Sodium hydroxide is toxic and caustic. Hydrochloric acid is highly toxic and corrosive. Handle them with care.

- Iodine is highly toxic and corrosive. Handle the iodine chamber in the fume hood.

- *t*-Butyl methyl ether is a flammable solvent.

cool down at ambient temperature and then in an ice-water bath. Collect the solid by vacuum filtration using a Hirsch funnel. Alternatively, you can recrystallize the product from 95% ethanol instead of acetone. Determine its melting point. Reserve a few milligrams of trimyristin for TLC and IR analysis and proceed to the saponification.

Saponification of Trimyristin

$$
\begin{array}{l}
CH_2-O-CO-(CH_2)_{12}CH_3 \\
| \\
CH-O-CO-(CH_2)_{12}CH_3 \quad + \quad 3\ NaOH \quad \xrightarrow{heat} \\
| \\
CH_2-O-CO-(CH_2)_{12}CH_3 \\
\qquad\qquad \text{trimyristin}
\end{array}
\qquad
\begin{array}{l}
CH_2-OH \\
| \\
CH-OH \quad + \quad 3\ CH_3(CH_2)_{12}COO^-Na^+ \\
| \\
CH_2-OH \qquad\qquad\quad \Big\downarrow 3\ HCl \\
\quad \text{glycerol} \\
\qquad\qquad 3\ CH_3(CH_2)_{12}COOH + 3\ NaCl \\
\qquad\qquad\qquad \text{myristic acid}
\end{array}
$$

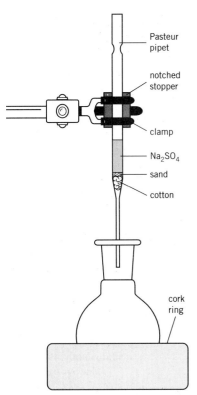

Pasteur pipet
notched stopper
clamp
Na_2SO_4
sand
cotton
cork ring

Figure 25.4 Pasteur pipet with anhydrous sodium sulfate for drying nutmeg extract.

In a 15-mL round bottom flask weigh 100 mg of trimyristin. Add 3.0 mL of a 6 M aqueous sodium hydroxide solution and 3.0 mL of 95% ethanol. Lubricate the joint with a very small amount of grease. Add a magnetic stir bar and attach a microscale water-jacketed condenser. Reflux directly on a hot plate at a medium setting for one hour.

Turn off the heat and allow the mixture to cool down. Pour the contents of the flask in a beaker with 30 mL of water and add 6 mL of concentrated hydrochloric acid. Cool down the mixture in an ice bath. Collect the white myristic acid that separates by using a Hirsch funnel. Wash the solid with 1 mL of cold water and let it dry on the filter. Dry the solid by repeatedly spreading it with a spatula over a piece of porous plate. Weigh the product and calculate the percentage yield. Determine its melting point and analyze it by TLC.

Analysis

Analyze trimyristin and myristic acid on silica gel TLC plates using toluene-ethanol (9 : 1) as a solvent. Detect the spots using an iodine chamber. Draw the plates in your lab notebook. Calculate the R_f values.

Obtain the IR spectrum of trimyristin. Using a sand bath, melt a small amount of the compound in a small test tube and quickly transfer the liquid to a NaCl plate (section 31.5). Obtain the IR of myristic acid using the same technique.

Obtain the inclusion complex of myristic acid with urea as described in E25.4 and determine its dissociation temperature. Compare it with the value from Table 25.3.

Cleaning Up

- Dispose of the Pasteur pipet used as a filter in the container labeled "Lipids Solid Waste."
- Dispose of the mother liquor from the recrystallization of trimyristin in the container labeled "Lipids Organic Waste."
- Dispose of the aqueous solution from the isolation of myristic acid in the container labeled "Lipids Aqueous Waste."
- Dispose of the TLC plates, Pasteur pipets, filter paper, and capillary tubes in the container labeled "Lipids Solid Waste."
- Turn in your myristic acid and myristic acid-urea complex in labeled vials to your instructor.

E25.3 PREPARATION OF AZELAIC ACID FROM CASTOR OIL

> **Background reading:** 25.1; 25.3, and 25.4
> **Estimated time:** 6 hours

$$CH_2-O-CO-(CH_2)_7CH\!=\!CHCH_2\overset{\overset{\displaystyle OH}{|}}{CH}(CH_2)_5CH_3$$

triricinoleylglycerol
(castor oil)

$$\xrightarrow[\text{2. } H_2SO_4]{\text{1. KOH, }\Delta}$$

glycerol + 3 ricinoleic acid

$$CH_3(CH_2)_5\overset{\overset{\displaystyle OH}{|}}{CH}CH_2CH\!=\!CH(CH_2)_7CO_2H$$

ricinoleic acid

$$\downarrow KMnO_4$$

azelaic acid $+ CH_3(CH_2)_5\overset{\overset{\displaystyle O}{\|}}{C}CH_3 + CO_2 + MnO_2$

2-octanone

Saponification

In a 15-mL round-bottom flask equipped with a magnetic stir bar, weigh 3.0 g of castor oil and add 8 mL of a 30% (w/w) solution of KOH in 95% ethanol. Lubricate the joint with a small amount of grease. Attach a microscale water-jacketed reflux condenser and boil the system for 2 hours by heating directly on a hot plate at a medium setting.

Turn off the heat and let the system cool down to room temperature. With the help of a Pasteur pipet, transfer the liquid to a 125-mL separatory funnel containing 80 mL of water. If the ethanol in the round-bottom flask has evaporated during the reflux period and a solid instead of a liquid is obtained after cooling, dissolve the solid with some of the water from the separatory funnel, and then transfer the solution to the separatory funnel.

Add enough 25% (w/w) aqueous sulfuric acid to make the solution acidic (about 2.5 mL may be needed). Cap the separatory funnel and invert it a few times to mix thoroughly (do not shake it vigorously). Add 20 mL of t-butyl methyl ether (BME) and invert the separatory funnel a few times and vent. Remove the lower aqueous layer and keep aside. Wash the upper BME layer with 5 mL of water. Collect the lower aqueous layer in the container set aside with the previous aqueous layer. Keep this aqueous phase until the end and then discard it as indicated in "Cleaning Up." Collect the BME layer in an Erlenmeyer flask and dry it with magnesium sulfate. Remove the drying agent by gravity filtration and collect the liquid in a pre-tared round-bottom flask. Evaporate the solvent in a rota-vap, weigh the product, and determine the percentage yield. Obtain the IR of the acid using NaCl plates (section 31.5).

Oxidation

In a 125-mL Erlenmeyer flask, place 2.8 g of potassium permanganate and 30 mL of water at 35°C. Add a magnetic stir bar and stir on a hot plate until the solid dissolves. The heat should be off.

Safety First

- Potassium hydroxide is toxic and caustic. Handle it with care.
- Sulfuric acid is very corrosive. Take only the amount needed.
- Potassium permanganate is an oxidant. Avoid skin contact.
- Azelaic acid is an irritant.
- t-Butyl methyl ether is a flammable solvent.

In a 50-mL Erlenmeyer flask weigh 1 g of ricinoleic acid from the previous step. (If you have less than 1 g, change the amounts given below accordingly, see section 2.9.) Add 7 mL of a 4% (w/w) aqueous solution of KOH and swirl to dissolve the acid.

Add the solution of ricinoleic acid in a single portion to the solution of potassium permanganate. Continue stirring the system for about 30 minutes. Add dropwise 7 mL of a 25% (w/w) aqueous solution of sulfuric acid.

Heat the system for about 5–10 minutes in a boiling-water bath to coagulate the solid manganese dioxide. Carefully vacuum-filter it while still hot using a Buchner funnel and a 125-mL filter flask. Wash the solid on the filter with 5 mL of hot water. Transfer the filtrate to a 250-mL beaker.

Place a stir bar in the beaker and heat on a hot plate while stirring to evaporate most of the water. When the volume has been reduced to about 20 mL, remove from the heat and let the system cool down at ambient temperature first and then in an ice-water bath.

Collect the crystals by vacuum filtration using a Hirsch funnel. Wash the solid with a little cold water. Let the solid dry on the filter with the vacuum on for about 10 minutes. Weigh the product, and calculate the percentage yield.

Reserve a small amount of the product to determine its melting point and recrystallize the rest from water. About 16 mL of water is needed per gram of azelaic acid. Adapted from Reference 7.

Analysis

Determine the melting point of crude and recrystallized azelaic acid. Obtain the IR spectrum of the recrystallized acid using a nujol mull (section 31.5). Compare it with the IR of ricinoleic acid.

E25.4   UREA COMPLEXES OF FATTY ACIDS

> **Background reading:** 25.7
> **Estimated time:** 30 min

Obtain an unknown fatty acid from your instructor and determine its melting point. The fatty acid is selected among those in Table 25.3.

In a 13 × 100 mm test tube weigh 150 mg of the unknown fatty acid and add 4.5 mL of a saturated solution of urea in methanol (about 16 g of urea per 100 mL of methanol). Warm the system in a sand bath while stirring with a glass rod. The acid should dissolve at this point. If the acid does not completely dissolve add a small amount (less than 0.5 mL) of isopropanol.

Cool down the system to room temperature. Separate the crystals by vacuum filtration using a Hirsch funnel. Determine the dissociation temperature of the complex. At the dissociation temperature a small amount of liquid exudes from the crystals, but melting does not occur. Identify the fatty acid using Table 25.3. Examine the crystals with a magnifying glass, and compare them with urea crystals.

E25.5   SOLUBILITY OF CHOLESTEROL

> **Background reading:** 25.5; 25.6
> **Estimated time:** 15 min

In small test tubes place 1 mL of the following solvents: water, acetone, toluene, ethanol, and a 0.5 M solution of sodium deoxycholate in water. Add a spatulaful of cholesterol to each test tube, shake well and observe the results. If cholesterol

---

Cleaning Up

- Dispose of the aqueous solution from the isolation of ricinoleic acid in the container labeled "Lipids Aqueous Waste."
- Dispose of the aqueous solution from the synthesis of azelaic acid and the mother liquor from the recrystallization of azelaic acid in the container labeled "Lipids Aqueous Waste."
- Dispose of the filter paper with the manganese oxide in the container labeled "Manganese Oxide Waste." Do not mix with other waste.
- Dispose of drying agent, filter papers, capillary tubes, and Pasteur pipets in the container labeled "Lipids Solid Waste."
- Turn in your azelaic acid in a labeled vial to your instructor.

Safety First

- Methanol and isopropanol are flammable.

Cleaning Up

- Dispose of the methanol solutions in the container labeled "Lipids Organic Waste."
- Dispose of Pasteur pipets, filter papers, and capillary tubes in the container "Lipids Solid Waste."
- Turn in your urea complex in a labeled vial to your instructor.

Safety First

- Acetone, toluene, and ethanol are flammable.

completely dissolves, keep adding the solid in small portions until the solution becomes saturated. For comparison, keep a record of how much cholesterol was added to each tube. Interpret your results.

E25.6 BILE ACIDS

> **Background reading:** 25.5 and 25.6
> **Estimated time:** 1 hour

Choleic Acid of *p*-Xylene

In a 13 × 100 mm test tube place 1 mL of *p*-xylene and heat the liquid in a sand bath while continuously stirring with a glass rod. Remove the tube from the bath and add 100 mg of deoxycholic acid. Add the minimum amount of 100% hot ethanol to dissolve the deoxycholic acid while heating in the sand bath.

Let the solution slowly cool down at room temperature. If crystals do not separate, concentrate the solution by evaporating some of the ethanol in a sand bath. Cool and collect the crystals by vacuum filtration using a Hirsch funnel. Wash the crystals in the funnel with a very small amount of cold ethanol. Let the crystals dry on the filter with the vacuum on for about 10 minutes. Weigh the product and determine its melting point (m.p. 183°C; adapted from Ref. 5).

Complex of Deoxycholic Acid

Method A   In a 50-mL beaker mix equal volumes (approximately 10 mL) of a 0.2 M aqueous solution of sodium deoxycholate and 0.2 M glycylglycine in water. Stir well with a glass rod. The viscosity of the solution starts to increase. Let it sit undisturbed for a while and then stir again until it becomes very viscous. With the help of a glass rod or tweezers, draw fibers from the bulk of the solution and let them dry in air.

Method B   To a 0.1 M solution of sodium deoxycholate in water add a 1 M aqueous solution of HCl little by little so that the pH falls just below 6.8. Stir well. The solution becomes very viscous. With the help of a glass rod or tweezers, draw fibers from the bulk of the solution and let them dry in air. The formation of the complex is reversed by raising the pH. Adding more HCl results in the precipitation of deoxycholic acid (adapted from Ref. 6).

BIBLIOGRAPHY

1. The Chemical Analysis of Foods. D. Pearson. Chemical Pub. Co. New York, 1977.
2. Chemical Analysis of Foods and Food Products. M. Jacobs. 3rd ed. D. Van Nostrand Co. Princeton, 1962.
3. The Extraordinary Chemistry of Ordinary Things. C. H. Snyder. 2nd ed. Wiley, New York, 1995.
4. Trimyristin from Nutmeg. F. Frank, T. Roberts, J. Snell, C. Yates, and J. Collins, *J. Chem. Ed.*, **48**, 255–256 (1971).
5. Isolation of Trimyristin and Cholesterol . M. M. Vestling, *J. Chem. Ed.*, **67**, 274–275 (1990).
6. Natural Products. A Laboratory Guide. R. Ikan, 2nd ed. Academic Press, 1991.
7. Azelaic Acid. J. W. Hill and W. L. McEwen. *Org. Synth. Coll.*, **2**, 53 (1950).
8. The Structure and Properties of Choleic Acids. R.G. Jesaitis and A. Krantz. *J. Chem. Ed.*, **48**, 137–138 (1971).
9. Formation of a Helical Steroid Complex. A. Rich, and D.M. Blow. *Nature*, **182**, 423–426 (1958).

EXPERIMENT 25   REPORT

Pre-lab

1. Make a table with the physical properties (MM, m.p., b.p., density, solubility, toxicity, flammability [for solvents only]) of the chemicals used.

2. Write the chemical reaction for the iodine test with oleic acid.

3. Write a flowchart for the isolation of myristic acid from nutmeg.

4. Write the chemical reaction for the saponification of trimyristin.

5. Write a flowchart for the synthesis of azelaic acid from castor oil. Write the chemical reactions involved in the process.

6. Briefly describe the structure of the complexes between urea and fatty acids. How are these complexes useful?

7. Explain why cholesterol is nonpolar despite its OH group.

8. What is the most important structural feature in the bile acid molecules? What role do bile acids play in vertebrates?

9. Briefly describe the structure of choleic acids.

10. Enumerate the differences between urea complexes and choleic acids.

In-lab

1. Report and discuss the results from the Villavecchia's, Evers's, and iodine tests. Present your data in a table format. Identify your unknown sample.

2. Report and discuss the % yield of trimyristin from nutmeg. Report its melting point and compare it with literature values.

3. Report and discuss the % yield and the melting point of myristic acid.

4. Report and interpret the TLC results of trimyristin and myristic acid. Draw the TLC plate in your lab notebook and indicate R_f values.

5. Interpret the IR of trimyristin and myristic acid. Compare them with those shown in Figures 25.5 and 25.6.

6. Interpret the ^{1}H-NMR and ^{13}C-NMR of myristic acid shown in Figures 25.7 and 25.8. (The hydrogen of the carboxyl group does not show on this ^{1}H-NMR).

7. Report the dissociation temperature of the urea complex of myristic acid. Compare it with the literature value.

8. Report the % yield of ricinoleic acid from castor oil.

9. Calculate the percentage yield of azelaic acid from ricinoleic acid. Report the melting point of azelaic acid and compare it with the literature value.

10. Interpret the IR spectrum of ricinoleic and azelaic acids. Compare your IR of azelaic acid with that of an authentic sample shown in Figure 25.9.

11. Interpret the NMR spectra of azelaic acid shown in Figures 25.10 and 25.11.

12. Report the dissociation temperature of the urea complex of the unknown fatty acid. Identify the acid.

13. Draw the crystal structures of the urea complex and urea as observed through the magnifying glass.

14. Report and discuss the solubility tests of cholesterol. Present your data in a table format. Assign ++++ to the highest solubility and + to the lowest. Did you observe any effect due to the presence of the bile acid?

15. Report the melting point of the choleic acid of *p*-xylene, and compare it with the literature value.

16. Discuss the results of your deoxycholic acid complex experiment.

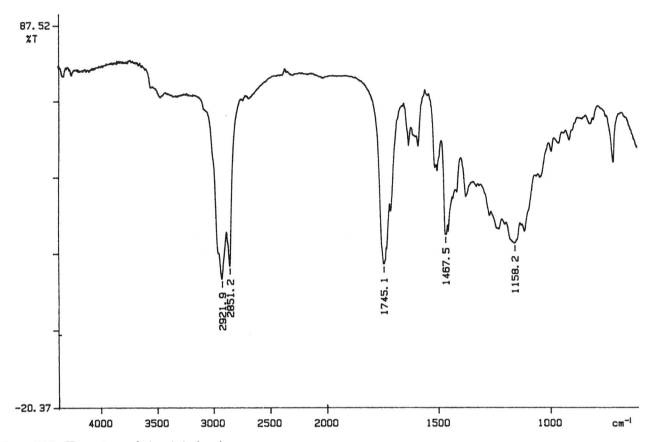

Figure 25.5   IR spectrum of trimyristin (neat).

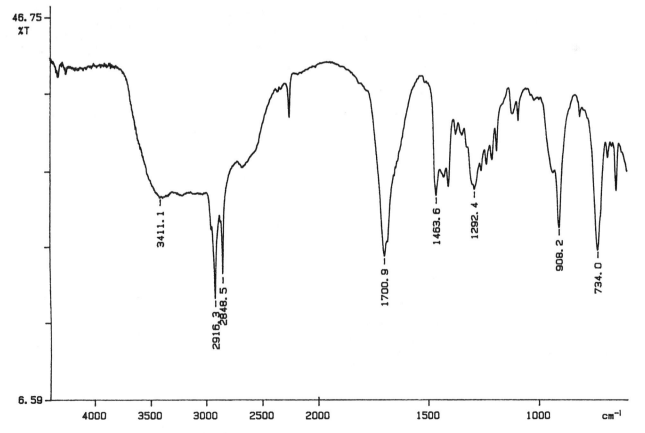

Figure 25.6   IR spectrum of myristic acid (neat).

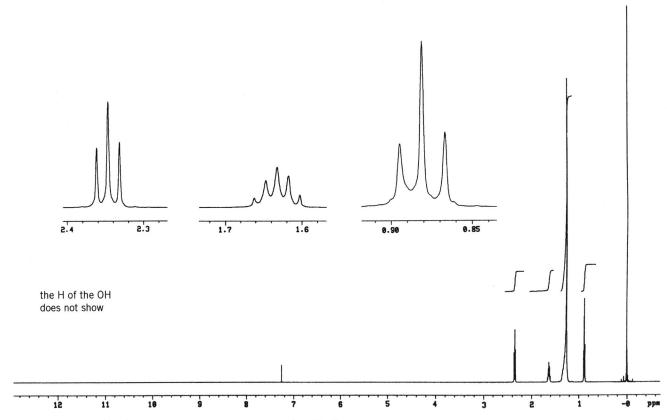

the H of the OH
does not show

Figure 25.7 500-MHz ^{1}H-NMR spectrum of myristic acid in CDCl$_3$.

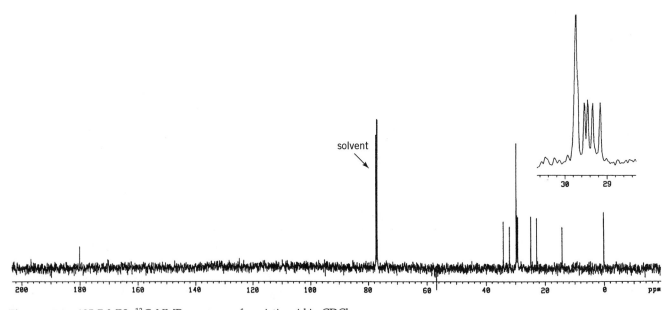

solvent

Figure 25.8 125.7-MHz ^{13}C-NMR spectrum of myristic acid in CDCl$_3$.

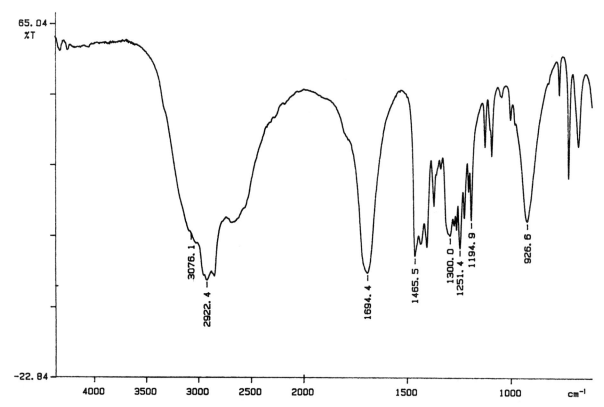

Figure 25.9   IR spectrum of azelaic acid (nujol).

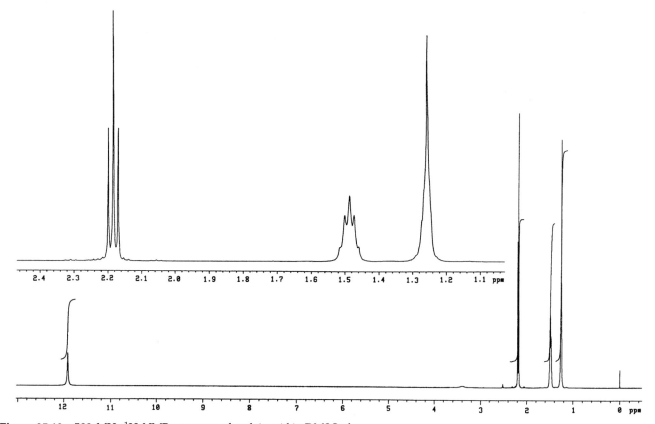

Figure 25.10   500-MHz ^{1}H-NMR spectrum of azelaic acid in DMSO-d_6.

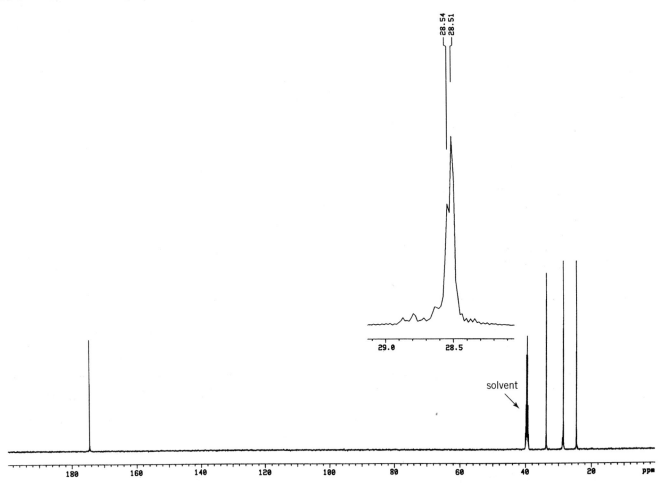

Figure 25.11   125.7-MHz ^{13}C-NMR spectrum of azelaic acid in DMSO-d$_6$.

Polymers

26.1 INTRODUCTION

Plastics are everywhere. Look around you and make an inventory of all the things made of plastic. Although the list is probably extensive, chances are that you have missed a thing or two; polyester, nylon, and other synthetic fibers are also plastics! In less than 100 years, these newcomers to our cultural landscape have dramatically changed how we live and die. Plastics provide us with sturdy football helmets, nonstick cookware, electrical insulators, dentures, blood-storage bags, and artificial arteries, just to mention a few items.

Plastics belong to a broader family of chemicals called **polymers**. The word polymer derives from Greek and literally means "many parts" (*poly*, many; *meros*, part). Polymers consist of **repetitive units** joined together like the links of a chain. A polymer molecule can easily contain 1000 or more of these repetitive units. Because of their large molecular size, polymers are also called **macromolecules**. Polymers with only a few repetitive units are referred to as **oligomers** (from Greek: *oligo*, few).

Although all plastics are polymers, not all polymers are plastics. In fact, the most widespread polymers are not man-made (synthetic) plastics but naturally occurring macromolecules such as cellulose, nucleic acids, and proteins. The scheme in Figure 26.1 shows the most important types of polymers with examples.

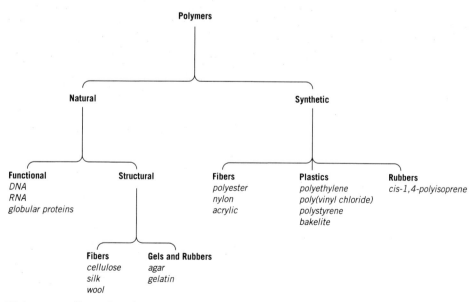

Figure 26.1 Examples of natural and synthetic polymers.

Table 26.1 shows examples of typical synthetic polymers and their applications. The raw materials to make them, called **monomers**, are also included. The repetitive units are shown in brackets. Some plastics are familiar to all of us, such as polyethylene, used in plastic bags; poly(vinyl chloride) or PVC, used for floor covering and pipes; nylon and polyester used as synthetic fibers; and Teflon, used for cookware. We will discuss the chemistry of these and other polymers in the following sections.

Table 26.1 Some Typical Polymers

Name	Repetitive unit	Monomer	Applications
high-density polyethylene (HDPE)	$-[CH_2-CH_2]-$	$CH_2{=}CH_2$	milk jugs, insulation, pipes
low-density polyethylene (LDPE)	$-[CH_2-CH_2]-$	$CH_2{=}CH_2$	packaging film, insulation, flexible bottles
phenol-aldehyde R=H (formaldehyde) R= (furfural)		$+ R-CHO$	electrical equipment, utensil handles, plywood adhesives
polyacrylonitrile	$-[CH_2-\underset{\underset{CN}{\mid}}{CH}]-$	$CH_2{=}\underset{\underset{CN}{\mid}}{CH}$	clothing, sports accessories
polyamides: nylon: 6.6 (x = 6; y = 4) 6.9 (x = 6; y = 7) 6.10 (x = 6; y = 8)	$-[NH-(CH_2)_x-NH-\overset{O}{\overset{\|}{C}}-(CH_2)_y-\overset{O}{\overset{\|}{C}}]-$	$H_2N-(CH_2)_x-NH_2$ + $HO_2C-(CH_2)_y-CO_2H$	clothing
cis-1,4-polyisoprene		$CH_2{=}\underset{}{\overset{CH_3}{\overset{\mid}{C}}}-CH{=}CH_2$	tires, footwear, sports accessories, insulation
poly(ethylene terephthalate) (PETE)		$HO_2C-\bigcirc-CO_2H$ + $HO-CH_2-CH_2-OH$	clothing, soda bottles
poly(methyl methacrylate) (PMMA)	$-[CH_2-\underset{\underset{CO_2CH_3}{\mid}}{\overset{\overset{CH_3}{\mid}}{C}}]-$	$CH_2{=}\underset{}{\overset{CH_3}{\overset{\mid}{C}}}-CO_2CH_3$	display signs, car tail-light lenses, dentures
polypropylene (PP)	$-[CH_2-\underset{\underset{CH_3}{\mid}}{CH}]-$	$CH_2{=}\underset{\underset{CH_3}{\mid}}{CH}$	carpeting, appliances, automobile parts

(continued)

Table 26.1 **(continued)**

Name	Repetitive unit	Monomer	Applications
polystyrene (PS)	$-\!\left[CH_2\!-\!CH\right]\!-$ (with phenyl group)	$CH_2\!=\!CH$ (with phenyl group)	packaging, cassette-tape boxes, drinking cups, insulation
poly(vinyl chloride) (PVC)	$-\!\left[CH_2\!-\!CH\right]\!-$ $\quad\quad\ \ \ \mid$ $\quad\quad\ \ \ Cl$	$CH_2\!=\!CH$ $\quad\quad\ \mid$ $\quad\quad\ Cl$	floor covering, cable insulation
poly(vinylidene chloride) (Saran)	$-\!\left[CH_2\!-\!CCl_2\right]\!-$	$CH_2\!=\!CCl_2$	package film
polytetrafluoro-ethylene (PTFE or Teflon)	$-\!\left[CF_2\!-\!CF_2\right]\!-$	$CF_2\!=\!CF_2$	nonstick coatings, stopcocks
polyurethanes n, m: 2; 4; 6, etc.	$\left[NH\!-\!(CH_2)_m\!-\!NH\!-\!\overset{O}{\overset{\|}{C}}\!-\!O\!-\!(CH_2)_n\!-\!O\!-\!\overset{O}{\overset{\|}{C}}\right]$	$O\!=\!C\!=\!N\!-\!(CH_2)_m\!-\!N\!=\!C\!=\!O$ $+$ $HO\!-\!(CH_2)_n\!-\!OH$	fibers, adhesives, foams

26.2 CHAIN-REACTION POLYMERIZATION

According to the reaction mechanism, there are two types of **polymerizations: chain-reaction polymerization** and **step-reaction polymerization**. The use of one or the other depends on the monomer employed.

As the name indicates, a chain-reaction polymerization occurs by a chain-reaction mechanism that consists of three steps: **initiation, propagation,** and **termination**. In the **initiation step**, an initiator molecule is decomposed by heat or light into two free radicals. One of the most commonly used initiators is **benzoyl peroxide**. The free radicals react with monomer molecules, usually vinylic compounds with structure $CH_2\!=\!CHY$, to give an adduct that is also a radical. These reactions are illustrated below. In vinylic monomers of the type $CH_2\!=\!CHY$, the most common $-Y$ groups are: $-CN$, $-C_6H_5$, $-Cl$, and $-CH_3$.

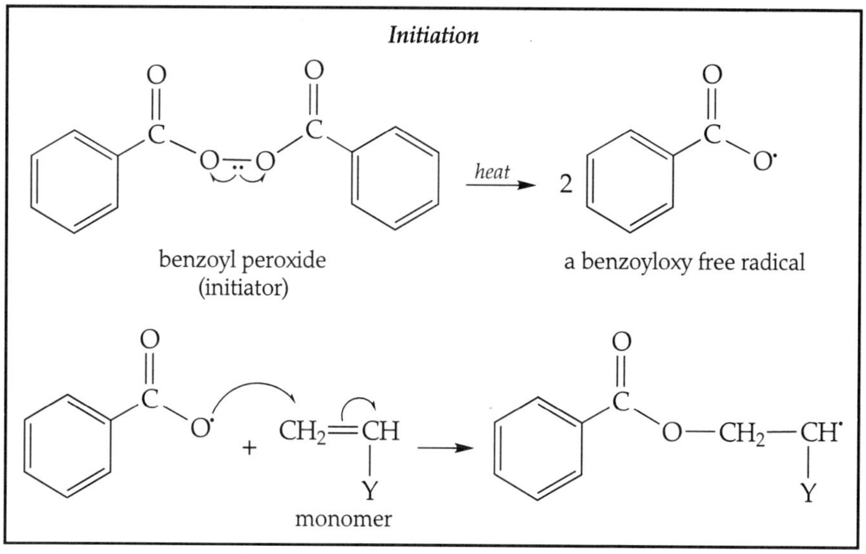

Initiation

benzoyl peroxide (initiator) → 2 a benzoyloxy free radical

monomer

In the **propagation step** the newly formed radical reacts with a monomer molecule yielding another radical. This process repeats itself many times, each monomer molecule adding at the end of the growing chain, one at a time. The growing polymeric radical attacks the monomer at the least substituted carbon of the double bond.

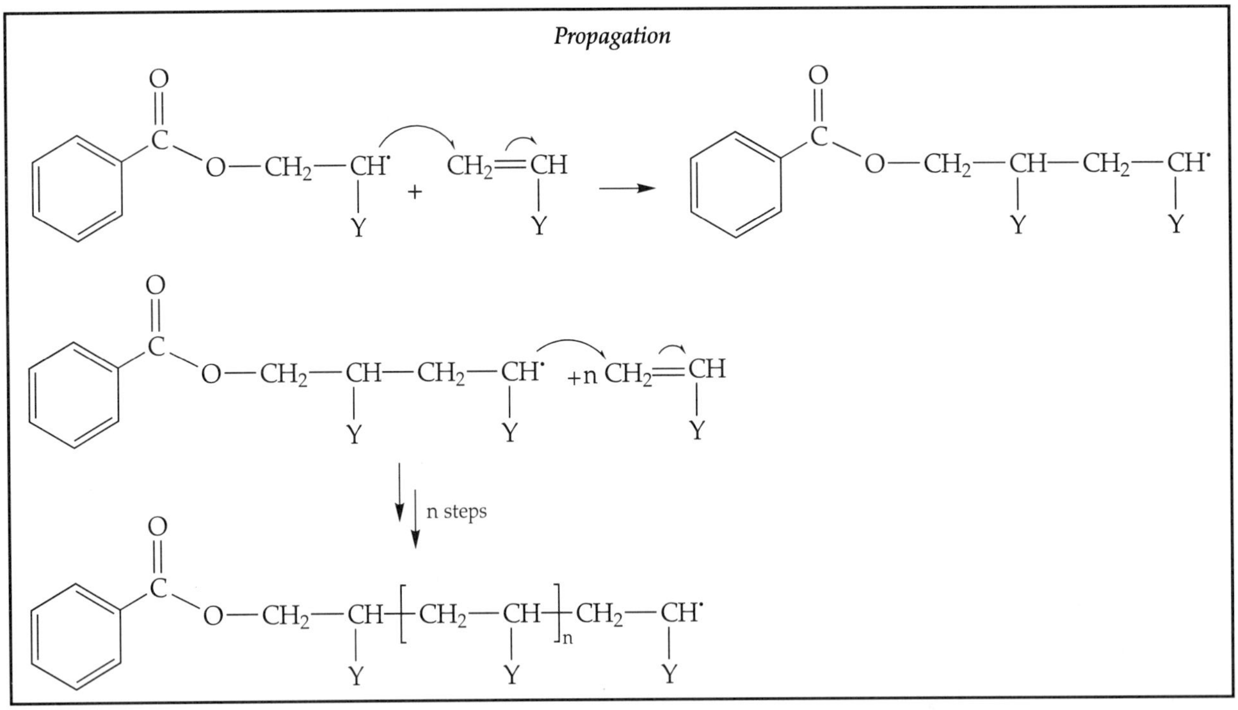

Propagation

The number of growing polymeric chains is determined by the initiator concentration, which is normally very low. As a result, only a limited number of chains form and they can grow to have very high molecular masses. The propagation step continues on until all the monomer molecules are consumed or until a **termination** event occurs.

In the **termination** step, the growing polymeric radical reacts with another radical in a **combination reaction** to give a nonradical molecule, or it undergoes a **disproportionation reaction** in which two radicals react, one loses a hydrogen atom and forms a carbon-carbon double bond, and the other gains the hydrogen atom, forming a C—H bond.

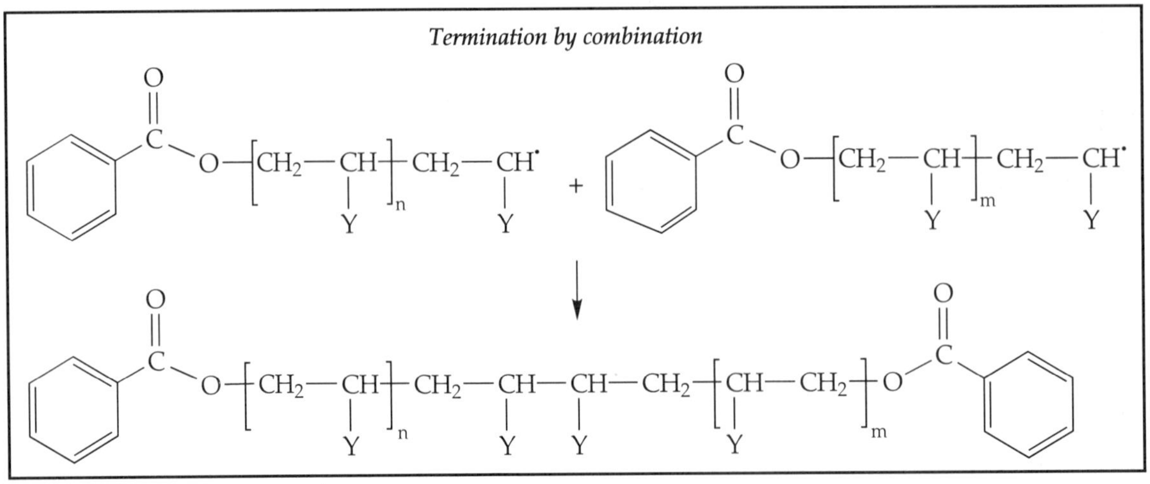

Termination by combination

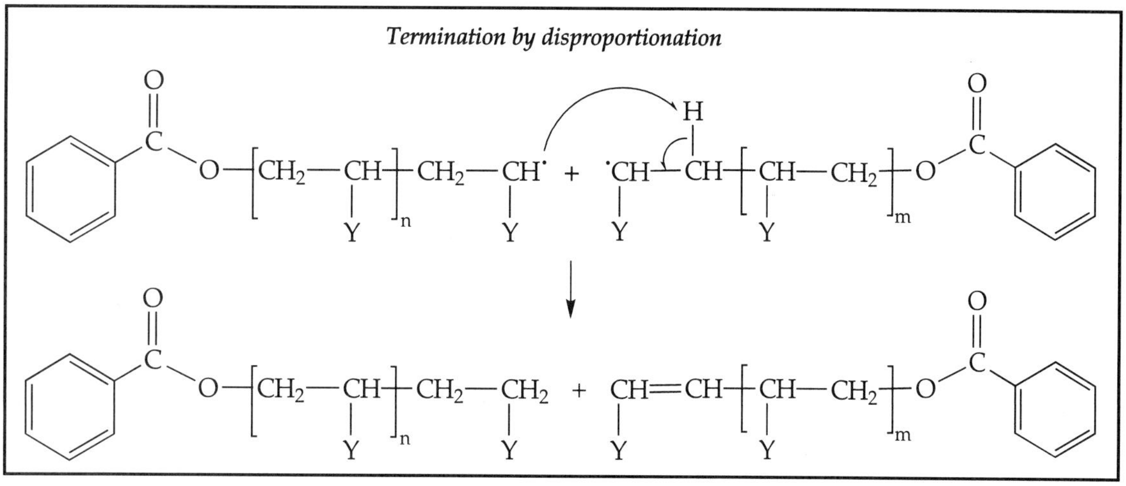

Termination by disproportionation

Polymers obtained by chain-reaction polymerization include **polyethylene (PE)**, **polypropylene (PP)**, **polystyrene (PS)**, **poly(methyl methacrylate) (PMMA)**, **poly(vinyl chloride) (PVC)**, and **polytetrafluoroethylene (PTFE**, or **Teflon**, registered mark of DuPont). Chain-reaction polymerizations are not limited to radical mechanisms. They can also take place by ionic mechanisms where a cation or an anion, instead of a free radical, initiates the reaction. We will not discuss those processes here.

26.3 STEP-REACTION POLYMERIZATION

Step-reaction polymerizations occur by a totally different mechanism. The monomers involved in this type of reaction must have at least two reactive functional groups. *Diols*, *amino acids*, and *hydroxy acids* are good examples of such monomers. The self-esterification of a hydroxy acid, 4-hydroxybenzoic acid, illustrates this point.

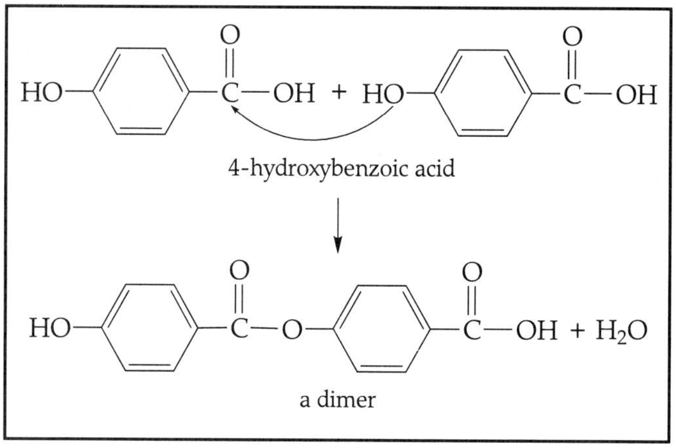

4-hydroxybenzoic acid

a dimer

In step-reaction polymerizations two monomer molecules react to form **dimers** (oligomers with two units), which then react with other monomer molecules, forming **trimers** (three units), **tetramers** (four units), and so on. Both ends of these molecules are reactive, and thus which one reacts is determined by the orientation of the molecules during collision.

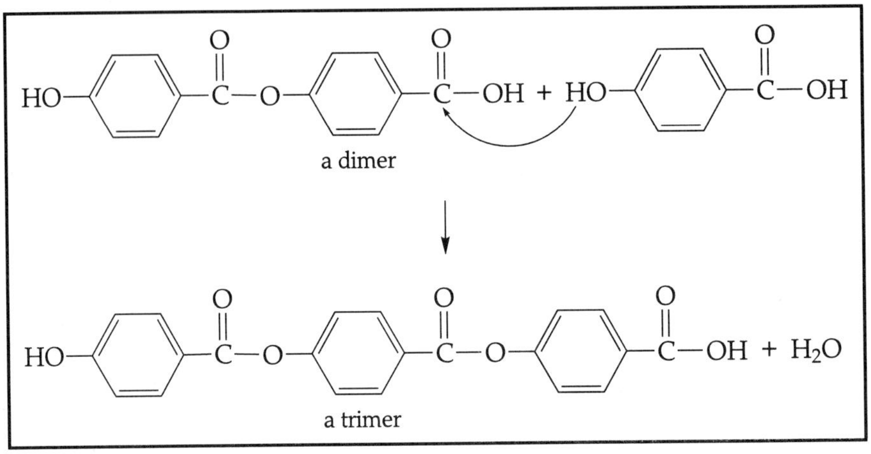

a dimer

↓

a trimer + H₂O

The oligomers also react with one another to form bigger and bigger molecules. In the example below, a tetramer reacts with a trimer to yield a heptamer.

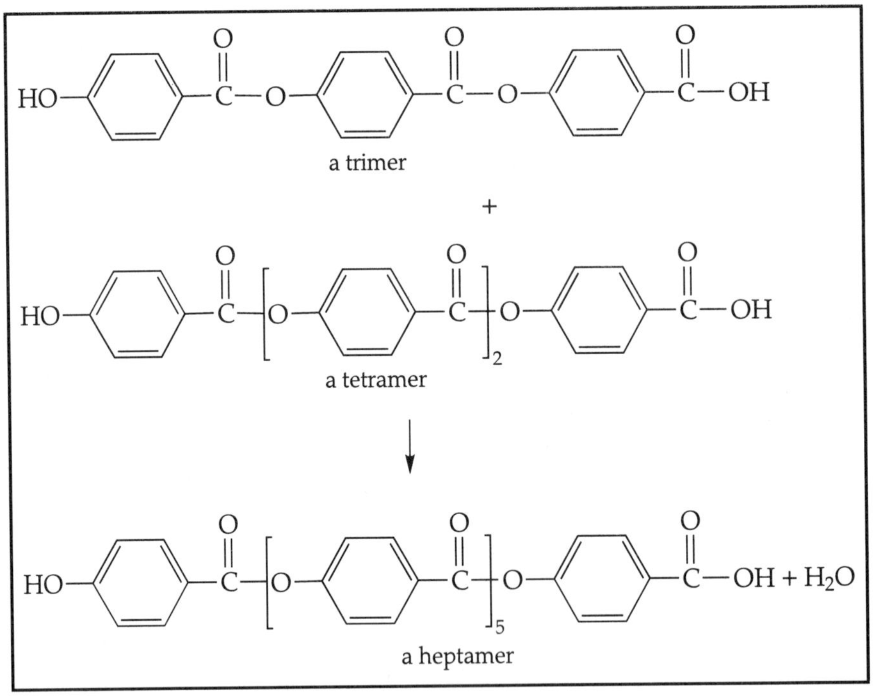

a trimer

+

a tetramer

↓

a heptamer + H₂O

Unlike chain-reaction polymerization, in step-reaction polymerization *there is no initiator to limit the number of growing chains* and, therefore, a large number of chains form from the beginning. These chains have to compete with each other for a limited number of monomer molecules and, thus, their growth is restricted. As a result, polymers obtained by step-reaction polymerizations are in general shorter than those obtained by chain-reaction polymerization.

Many polymers obtained by step-reaction polymerizations are formed by the reaction of two different monomers, usually a dicarboxylic acid and a diol, or a dicarboxylic acid and a diamine. The reaction of dicarboxylic acids with diols leads to **polyesters**, and the reaction with diamines to **polyamides** (also known as **nylons**). For instance, terephthalic acid reacts with ethylene glycol, forming **poly(ethylene terephthalate), (PETE)**, a type of polyester used in the manufacture of 2-L soda bottles. The reaction is shown below.

n HO—C(=O)—⟨benzene⟩—C(=O)—OH + n HO—CH₂—CH₂—OH

terephthalic acid   ethylene glycol

↓

HO—[C(=O)—⟨benzene⟩—C(=O)—O—CH₂—CH₂—O]ₙ—H + (2n–1) H₂O

poly(ethylene terephtahlate)

PETE

Because many (but not all) of the step-reaction polymerizations are condensation reactions (reactions where the loss of a small molecule such as H_2O or HCl occurs), these reactions are often called **condensation polymerizations**, and the resulting polymers, **condensed polymers**.

26.4 POLYMER STRUCTURE

Polymers can be classified as **linear**, **branched**, or **network** based on their three-dimensional structures. The three-dimensional structure dictates the polymer's physical properties, which in turn determine its uses and applications.

A linear polymer has each repetitive unit linked to no more than *two* repetitive units. In branched polymers, there are some repetitive units linked to *three or more* repetitive units, forming branches or ramifications. Network polymers have even more complex three-dimensional structures with many connections between branches. These polymers are also referred to as **crosslinked** (Fig. 26.2).

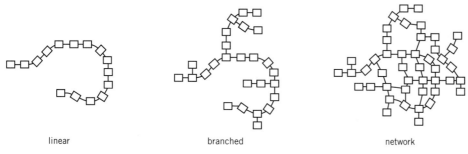

linear   branched   network

Figure 26.2   Linear, branched, and network polymers.

Physical properties such as viscosity, solubility, elasticity, and pliability depend on the interactions between polymeric chains. Linear polymers can easily form fibers through side-by-side association of individual molecules. These fibers can be manipulated with heat and cast into any desired shape. Such polymers are called **thermoplastics** and are used in numerous applications from windshields to clothing. Examples of thermoplastics are polyethylene, polyesters, and PVC. Contrary to linear polymers, network polymers have entangled chains and tremendous chain-chain interactions. These polymers are **infusible** (do not melt) and thus cannot be molded by heat. They are called **thermosetting** plastics and must be prepared in their final molds. Bakelite, a polymer made from phenol and formaldehyde and widely used as an electrical insulator, is a thermosetting plastic. Bakelite was developed by the chemist-entrepreneur Leo Baekeland in the early 1900s and it was the first synthetic polymer to be commercialized on a large scale.

Polymers do not have a well-defined molecular mass. During polymerization, a distribution of molecules with variable molecular mass is normally formed. The **average molecular mass**, rather than the molecular mass, is used to characterize polymers. Changing the experimental conditions of the polymerization usually leads to a change in the average molecular mass. For example, in chain-reaction polymerizations, an increase in the amount of initiator results in an increase in the total number of chains and a decrease in the average molecular mass.

A chemist can design polymers with specific physical properties by polymerizing two or more different monomers together. This process is called **copolymerization** and the resulting products, **copolymers**. Copolymerization is limited to structurally similar monomers such as two different vinylic monomers, two different alcohols, and so forth. The polymerization of styrene, for example, gives polystyrene, a hard plastic used to manufacture cassette-tape boxes and plastic cups. If styrene is copolymerized with 1,3-butadiene, a flexible product is obtained instead. This copolymer is a substitute for rubber and can be used for shoe soles and hoses.

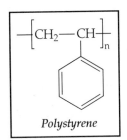

Polystyrene

26.5 SOME TYPICAL POLYMERS

Polystyrene (PS)

It is prepared by chain-reaction polymerization. Benzoyl peroxide is the initiator commonly used. If the polymerization is carried out without solvent, a block of polymer is obtained in a process called **bulk polymerization**. This technique is useful for embedding objects in a clear polymer matrix.

Polystyrene can also be obtained by **emulsion polymerization**, using water as a solvent. Because the monomer is not soluble in water, a detergent such as sodium dodecyl sulfate (SDS) is added as an emulsifier.

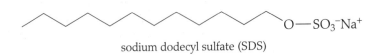

sodium dodecyl sulfate (SDS)

The polymerization takes place inside tiny droplets of monomer that are kept in solution by a surrounding layer of detergent molecules. These droplets are called micelles. In each micelle the detergent molecules are oriented with the ionic sulfate groups on the outside, interacting with surrounding water molecules. The nonpolar hydrocarbon chains are directed inward, avoiding contact with water and interacting instead with the nonpolar monomer and the growing polymeric chains (Fig. 26.3).

In emulsion polymerization, initiators soluble in water such as potassium persulfate are used. The initiator is decomposed in the aqueous solution by heat or light.

a polymerization reaction taking place inside a detergent micelle

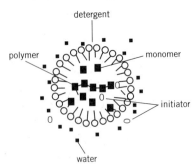

Figure 26.3 Detergent micelle in emulsion polymerization.

The free radicals then diffuse inside the micelles, where they start the polymerization.

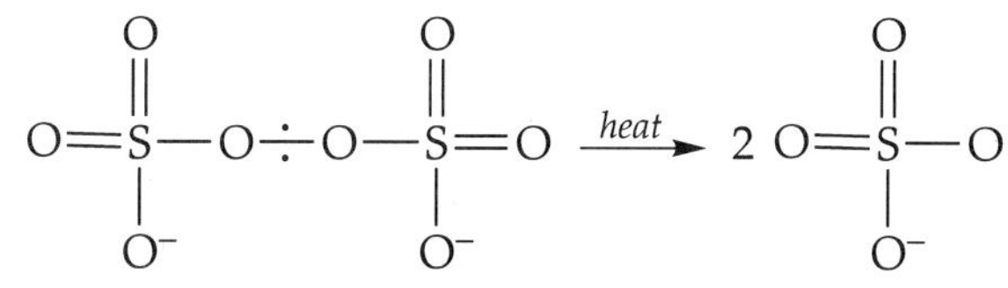

persulfate

Copolymerization of styrene with *p*-divinylbenzene, a compound with *two* reactive double bonds, leads to a crosslinked polymer. The degree of crosslinking can be controlled by the proportion of divinylbenzene in the copolymerization reaction.

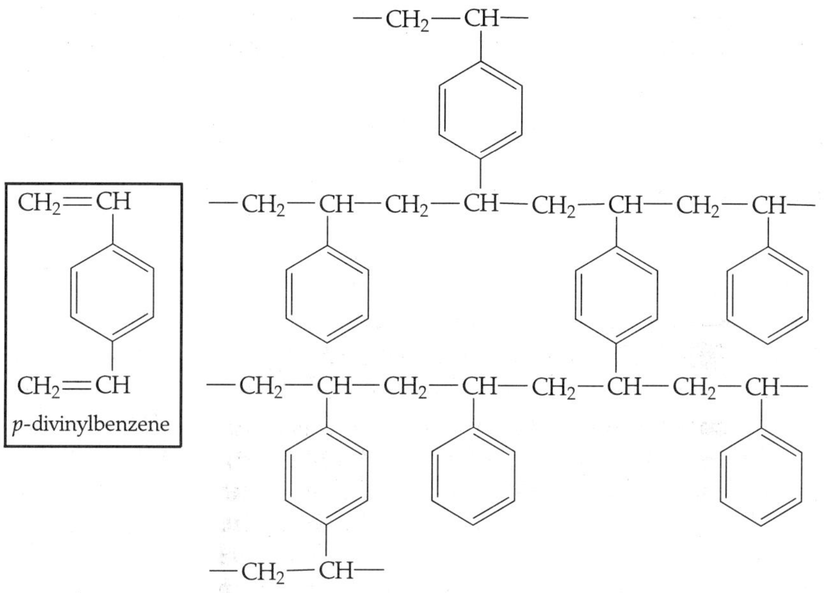

crosslinked polystyrene

Polystyrene is used to manufacture lighting panels, lenses, drinking cups, cassette-tape boxes, plant pots, and many other objects. It can also be obtained as polystyrene foam used for insulators and fast-food containers.

Poly(methyl Methacrylate) (PMMA)

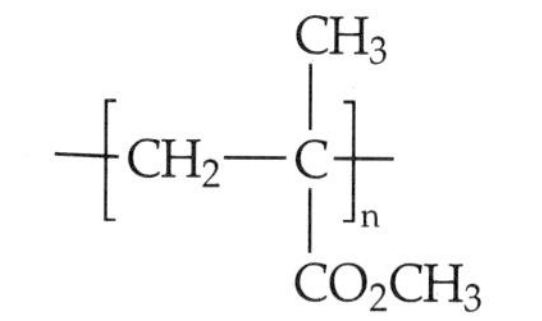

Plexiglas (trademark of Rohm and Haas) and Lucite (trademark of DuPont) are PMMA. The polymer is obtained by chain-reaction polymerization of methyl methacrylate. It can be used to embed objects and make castings by bulk polymerization. It can also be obtained as a powder by emulsion polymerization. This polymer has many applications in the manufacture of dentures and other dental prostheses.

poly(methyl methacrylate)

Nylons

Nylons are polyamides obtained by self-polymerization of amino acids and lactams (for example, nylon 6) or by the condensation of carboxylic acid derivatives with diamines (for example, nylon 6.6, 6.9, and 6.10). Nylon 6 is the polymerization product of ε-caprolactam; the number 6 refers to the number of carbons in the monomer.

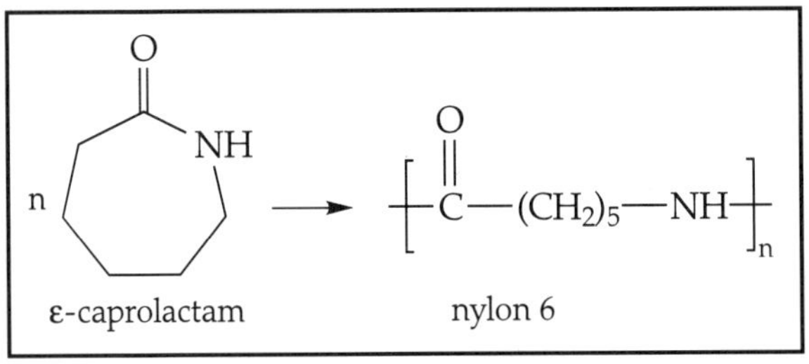

ε-caprolactam   nylon 6

Nylon 6.6, 6.9, and 6.10 are obtained by a step-reaction polymerization of a diamine with six carbon atoms (hexamethylene diamine) with diacyl chlorides having 6, 9 and 10 carbon atoms, respectively. The figures separated by a period indicate the number of carbon atoms in the diamine and in the diacid group, in that order. The reaction is shown below.

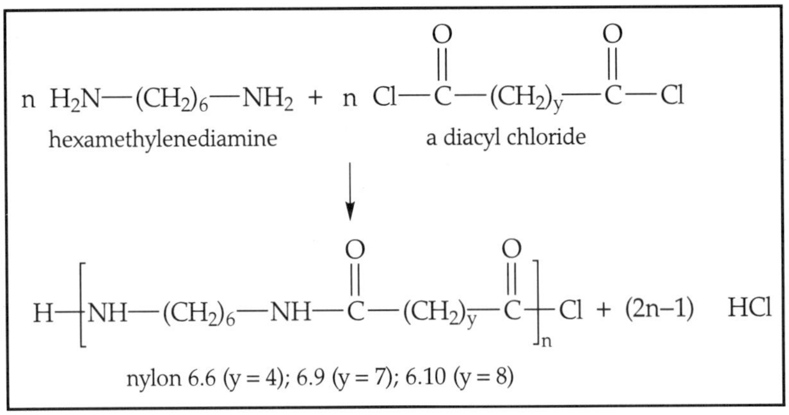

Nylon's most important application is in the production of fibers for clothing, rope, and carpets. Nylons behave as thermoplastics and can be melted and molded. They are used to make pipes, zippers, and wire insulation.

Phenol-Aldehyde

Aldehydes react with phenols, giving highly crosslinked polymers called **phenolic resins**. The first phenolic resin ever made was Bakelite. These resins with complex tridimensional structures are examples of thermosetting polymers. These polymers have applications as electrical insulators and also as adhesives in the manufacture of plywood, among many others (see box "All the Vermeers in the World"). The reaction is illustrated below with formaldehyde.

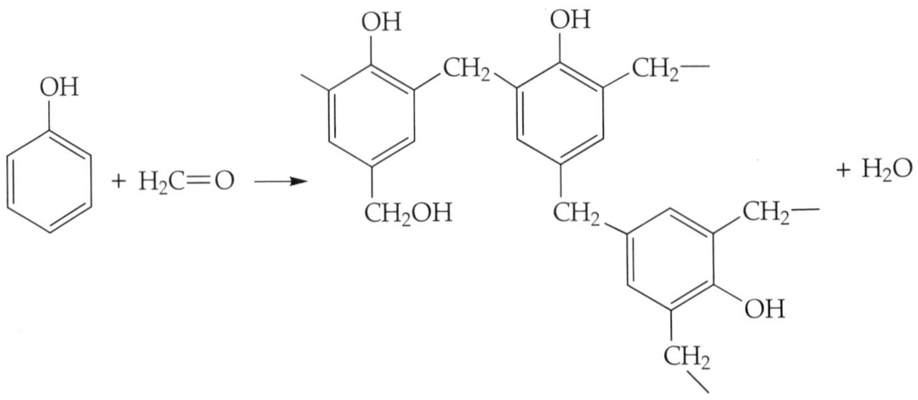

The reaction is complex and takes place in several steps. An electrophilic aromatic substitution on the phenol ring gives a hydroxymethylenephenol, which presumably loses water to give a reactive benzyl carbocation intermediate.

The benzyl carbocation reacts with other phenol and hydroxymethylenephenol molecules via electrophilic aromatic substitutions. The products of these reactions

phenolic resins

then react with more formaldehyde and phenol molecules to give a totally crosslinked resin. The substitutions always take place in the ortho and para positions with respect to the −OH group.

Cellulose Acetate

This is a semisynthetic polymer made from cellulose by acetylation of its hydroxyl groups. The resulting polymer, sometimes simply called **acetate**, is used for textile applications and also as a thermoplastic in the manufacture of photographic films and toys. Acetate belongs to a larger family of **cellulose polymers**. Among them we find **celluloid**, **cellophane**, and **rayon**. Celluloid was one of the first polymers to be commercialized. In the nineteenth century, cellulose treated with nitric acid (cellulose nitrate) was used to make billiard balls as a substitute for ivory but it proved to be too dangerous. The balls exploded on collision! The safety of the product was later improved by mixing it with camphor; the resulting plastic was called celluloid. Cellophane and rayon are different forms of reconstituted cellulose.

cellulose acetate
(It has a variable number of acetyl groups per unit.)

26.6   PLASTICIZERS

Most commercial polymers contain compounds added to alter their physical and mechanical properties. These substances, called **additives**, can be odorants, pigments, flame retardants, plasticizers, and so on. Plasticizers increase polymer flexibility; they are usually *small molecules that accommodate themselves between polymeric chains, pushing them apart and increasing their mobility*. Plasticizers are essential in the manufacture of PVC. They transform PVC, which in the pure state is a rigid and brittle polymer, into one of the most flexible and versatile plastics. Up to 50% of the weight of commercial PVC may be the plasticizer. The "new-car smell" and the greasy film that deposits on windshields is due to the plasticizer that slowly evaporates from the vinyl upholstery.

    One the most commonly used plasticizers is dioctyl phthalate (DOC), more properly called bis(2-ethylhexyl) phthalate.

bis(2-ethylhexyl) phthalate

26.7 APPLICATIONS AND RECYCLING

Besides countless applications in everyday life, polymers brought a remarkable revolution to medicine. It would be too lengthy to enumerate here their many uses in the medical sciences. Suffice it to say that PMMA is used as a cementing substance in orthopedic surgery; this polymer is used to attach steel prostheses to fractured bones. Polyesters and nylons are used to manufacture artificial arteries. Copolymers of methacrylates and methacrylic acid are used to make soft contact lenses, and are also being investigated as materials for artificial corneas. The impact of polymers in medicine has just begun. As many new polymers are developed and tested now, we will witness an explosion in their medical applications.

One of the drawbacks in the use of plastics is the waste they generate, or more appropriately, the public's perception of such waste. Buried in our sanitary landfills, in the dark, and deprived of oxygen to hasten its decomposition, *paper* is as resistant to degradation as plastics. However, people's outcry focuses on plastics, hardly on paper. This is not to say that one should not worry about the disposal of plastics; on the contrary, it is just a reminder that we must also fiercely undertake the recycling of what accounts for more than 40% of our waste: paper!

The recycling of plastics poses special challenges. There are many types of plastics with very different physical and chemical properties. Because some plastics are incompatible with others, they must be sorted before they can be recycled. Separation, identification, and classification of plastics is a costly process that must be fully considered to make the recycling process profitable. The codes in Table 26.2 have been devised for easy identification of the most common types of plastics. The code is usually imprinted on the bottom of the object.

Table 26.2 **Polymer Recycle Codes**

	Code	Name	Uses
♻1	PETE	poly(ethylene terephthalate)	soda bottles; clothing
♻2	HDPE	high-density polyethylene	toys; milk and water jugs
♻3	V	poly(vinyl chloride) (PVC)	pipes; flooring; bottles for cleaning materials
♻4	LDPE	low density polyethylene	plastic bags; squeeze bottles
♻5	PP	polypropylene	deli containers; microwaveable containers
♻6	PS	polystyrene	disposable cups; fast-food containers; insulating foam; cassette-tape boxes
♻7	Other	mixed polymers, poly(methyl methacrylate), phenolic resins, etc.	insulators; dentures, etc.

There are three main processes for the recycling of plastics: 1) melting, mixing, and reprocessing into new plastics; 2) pyrolysis to produce energy, oil, and gas; 3) pyrolysis or other chemical process to win the monomer back. Of all these methods, melting and reprocessing is the most widely employed. For example, PETE, used in soft-drink bottles, has a high scrap value and is recycled into fibers for carpets and fabrics.

26.8 IDENTIFICATION OF PLASTICS

The first step in the recycling of plastics is their classification and separation. This can be achieved by physical methods based on their density, solubility, and spectroscopic properties.

Density

Density determinations are very useful to *separate* and also to *identify* plastics. Some plastics such as Teflon, PETE, PS, and PMMA are denser than water while others such as PP and polyethylene are lighter (Table 26.3). Mixed scrap plastics can be separated by flotation in water or other solvents: denser plastics sink while lighter ones float. Flotation can also be used to identify unknown plastics by their density. We will see this in Experiment 26.9. Some plastics such as phenolic resins and PVC have no definite density. Their densities depend on their actual manufacture process. For instance, phenolic resins can be made with different proportions of phenol and aldehyde, resulting in lighter or heavier products. PVC also has variable density because the amount of plasticizer is not always the same.

Table 26.3 **Densities of Common Polymers**

Polymer	density (g/mL)
polypropylene	0.90–0.91
low-density polyethylene	0.92–0.94
high-density polyethylene	0.95–0.97
polystyrene	1.05–1.08
poly(methyl methacrylate)	1.18–1.24
cellulose acetate	1.27–1.34
poly(ethylene terephthalate)	1.39
poly(vinylidene chloride)	1.80
polytetrafluoroethylene	2.10–2.30

Solubility

Solubility tests are another quick method to identify plastics. Each polymer has a very unique solubility behavior. A plastic will dissolve in some solvents and will be insoluble in others. Polymers follow the adage "like-dissolves-like" and dissolve in solvents with which they bear chemical resemblance. For example, low-density polyethylene (LDPE) is a hydrocarbon polymer soluble in toluene (also a hydrocarbon), and insoluble in ethyl acetate (an ester), while cellulose triacetate (an ester) is insoluble in toluene but soluble in ethyl acetate. The student should be aware that there are exceptions to this rule. For example, PETE (an ester) is insoluble in ethyl acetate, and PS (a hydrocarbon) is soluble in this solvent. Nylons are insoluble in most organic solvents but they are solubilized, and eventually hydrolyzed, in acids. Teflon is a very inert compound and insoluble in most solvents. We will see the application of solubility tests for polymer identification in Experiment 26.9.

Infrared

Infrared provides one of the most useful methods for the identification of polymers. It is quick and reliable, and it can be easily automated. Modern recycling plants are equipped with infrared sensors to sort plastics wastestream. Table 26.4 lists some of the characteristic bands useful in polymer identification.

Table 26.4 **Characteristic IR Bands of Polymers**

Polymer	Characteristic IR bands
nylon 6.6	3300 (N—H st.)
	2930; 2850 (C—H st.)
	1640 (C=O st.)
polypropylene	2800–2900 several bands (C—H st.)
	1380 (C—H bend. —CH₃)
polyethylene (low and high density)	2920; 2855 (C—H st.)
	1480; 1460 (C—H bend.)
	720; 730 (C—H bend.)
polystyrene	2800–3100 several bands (C—H st.)
	1606 (C—C skeletal; aromatic ring)
	760; 700 (C—H bend)
poly(methyl methacrylate)	3000–2850 several bands (C—H st.)
	1740 (C=O st.)
	1150 (C—O—C st.)
cellulose triacetate	3500 (O—H st. from adsorbed water)
	2950 (C—H st. weak)
	1750 (C=O st.)
	1050 (C—O—C st.)
poly(ethylene terephthalate)	1721 (C=O st.)
poly(vinyl chloride)	3000–2800 several bands (C—H st.)
	700; 640 (C—Cl st.)
poly(vinylidene chloride)	3000–2800 several bands C—H st.)
	750 (C—Cl st.)
polytetrafluoroethylene	1222; 1160 (C—F st.)

BIBLIOGRAPHY

1. Polymer Chemistry. An Introduction. M.P. Stevens. 2nd ed. Oxford University Press, New York, 1990.

2. Polymers: Chemistry and Physics of Modern Materials. J.M.G. Cowie. 2nd ed. Blackie, Glasgow, 1991.

3. Emerging Technologies in Plastics Recycling. G.D. Andrews and P.M. Subramanian, ed. ACS, Washington, DC, 1992.

4. Introduction to Polymer Chemistry. F.W. Harris. *J. Chem. Ed.*, **58**, 837–843 (1981).

5. A Closer Look at Cotton, Rayon, and Polyester Fibers. T.M. Letcher and N.S. Lutseke. *J. Chem. Ed.*, **67**, 361–363 (1990).

6. Polymers Are Everywhere. R.B. Seymour. *J. Chem. Ed.*, **65**, 327–334 (1988).

7. Polymer Structure-Organic Aspects (Definitions). C.E. Carraher, Jr. and R.B. Seymour. *J. Chem. Ed.*, **65**, 314–318 (1988).

8. Chemistry of Coatings. J.R. Griffith. *J. Chem. Ed.*, **58**, 956–958 (1981).

9. The Fractal Nature of Polymer Conformations. C. Vijayan and M. Ravikumar. *J. Chem. Ed.*, **70**, 830–831 (1993).

10. Polymer Properties and Testing-Definitions. C.E. Carraher, Jr. and R.B. Seymour. *J. Chem. Ed.*, **64**, 866–867 (1987).

11. Chain Reaction Polymerization. J.E. McGrath. *J. Chem. Ed.*, **58**, 844–861 (1981).

12. Step Growth Polymerization. J.K. Stille. *J. Chem. Ed.*, **58**, 862–866 (1981).

13. Molecular Weight and Molecular Weight Distributions in Synthetic Polymers. T.C. Ward. *J. Chem. Ed.*, **58**, 867–879 (1981).

14. Fundamentals of Epoxy Formulation. B. Dewprashad and E.J. Eisenbraun. *J. Chem. Ed.*, **71**, 290–294 (1994).

15. Elastomers I. Natural Rubber. G.B. Kauffman, and R.B. Seymour. *J. Chem. Ed.*, **67**, 422–425 (1990).

16. Auxetic Polymers: A New Range of Materials. K.E. Evans. *Endeavour*, **15**, 170–174 (1991).

EXERCISES

Problem 1

What is the average number of units per polymeric chain when 9 mL of methyl methacrylate (density = 0.936 g/mL; MM = 100.12) is polymerized in bulk using 100 mg of benzoyl peroxide (MM = 242.23)? Assume that all the benzoyl peroxide is consumed to initiate polymer chains, all the monomer has reacted, and termination occurs exclusively by disproportionation.

Problem 2

In a 100 mL beaker, a chemist mixes 20 mL of a 2% (v/v) solution of sebacoyl chloride (density of monomer = 1.121 g/mL, MM = 239.14) in methylene chloride and adds 10 mL of a 2% (w/v) solution of hexamethylene diamine (MM = 116.21) in water with sodium carbonate. She obtains 200 mg of nylon rope. What type of nylon was produced (give the number)? What was the % yield of the synthesis?

Problem 3

You are running a plastic recycling plant. How can you separate caps from bodies of 2-L soda bottles? Come up with a plan feasible at a large scale. Caps are made of PE and bodies of PETE.

Problem 4

How can you tell polyethylene and polypropylene apart by IR spectroscopy?

Problem 5

(*Previous Synton: Problem 5, Unit 25*)   Synton volunteered to help with the science projects at a local grade school where Ester, his fiancee, teaches the fifth grade. After much thinking, he decided that for Earth Day he would teach the kids how to recognize and sort out plastics for recycling. In collecting the samples for his demonstration, Synton ran into some pieces without identification code: a plastic cup, a broken diskette storage box, and an old ruler. He performed some simple density tests to identify them. A piece of the plastic cup floated in dilute corn syrup (d = 1.29 g/mL) and sank in a sodium chloride aqueous solution (d = 1.09 g/mL). A small piece of the ruler floated in the same sodium chloride solution and sank in water (d = 1.00 g/mL). A piece of the diskette box floated in water and sank in a 1 : 1 ethanol-water solution (d = 0.94 g/mL). Assuming that the plastics are among those listed in Table 26.3, how did Synton classify them?

All the Vermeers in the World   One of the most fascinating cases of art forgery occurred in the heart of Europe in the days prior to the Second World War. This story of vengeance and audacity unfolds with a lot of ingenuity, a little chemistry, and two main characters from the Netherlands: Johannes Vermeer as the seventeenth-century artist, and Han van Meegeren as the forger. The former bequeathed us a small treasure of delicate and intimate paintings, most of them depicting people quietly enjoying themselves in some simple task such as writing a letter, playing music, or making lace. The latter was a little-known artist not very well treated by the critics, who rated his work as traditional and superficial. Too much of a reactionary to follow the artistic trends of his time, and deeply hurt by what he considered ill-spirited criticism, van Meegeren set off on a journey to destroy the credibility of the self-appointed judges of art. He would create a fake that would be regarded as the work of a master and admired as such. He would then reveal the true authorship and ridicule the art world in the process. This would be van Meegeren's revenge, and in this respect he was a true master. He amply succeeded. For his first serious forgery he bought an old painting from the seventeenth century, scraped off the paint, and used the canvas to create what the world later knew as Vermeer's *Supper at Emmaus*. He even saved the old nails and the stretcher to mount "his Vermeer" and give it a true old look. Instead of using the pigments commonly available in his day, van Meegeren obtained the same pigments that Vermeer used; lapis lazuli for the blues, cochineal and cinnabar for the reds, and white lead for the whites. He ground them by hand, as the old masters did, to make the particles appear of different sizes under close inspection. With all these precautions taken, van Meegeren was yet to face the dilemma of how to make a new painting look old. Old paintings are hardened and cracked and the paint does not dissolve easily in any of the usual solvents. One common test to determine whether a painting is old or not is to see if the paint dissolves readily. Van Meegeren knew that his painting would be seriously tested for authenticity and spent quite some time researching ways to age a new painting. After months of experimentation in his villa in the south of France, he finally devised a technique that only a brilliant chemist could have conceived. He mixed equal amounts of formaldehyde and phenol, and before each stroke he dipped his brush first into this mixture and then into the paint. After the painting was finished, he baked the canvas in an oven at about 100°C. The high temperature hastened the polymerization of the phenol-formaldehyde mixture, which hardened into a firm resin. To produce the cracks typical of old paintings, van Meegeren then rolled the canvas around a tube. It was not difficult for him to fool the Dutch art dealers and the painting was sold for the equivalent of $300,000 to the Boymans Museum in Rotterdam. What makes this case even more enthralling is the fact that *Supper at Emmaus*, as well as five more fakes produced later, little resembles Vermeer's style but has instead the unmistakable imprint of van Meegeren's early work, the one looked down upon by the critics. The way he shaded the eyes, mouths, and noses gave the human figures a theatrical and sickly expression, totally foreign to the work of Vermeer. However, a critic of the time called this painting a "glorious work of Vermeer," a "masterpiece . . . of the highest art, the highest beauty." This

Vermeer, "Girl with a Pearl Earring," ca. 1660–65, Mauritshuis, The Hague, The Netherlands.

Van Meegeren, "Last Supper at Emmaus," 1937, Collection of Museum Böijmans Van Beuningen, Rotterdam, The Netherlands.

must have been van Meegeren's triumphant moment. The critics had come to "appreciate in the faker those very qualities they have previously denied in the painter" (Ref. 2). In 1947, ten years after he sold "his first Vermeer," van Meegeren stood trial for forgery. He was found guilty.

Bibliography

1. What Is Art History? M. Roskill, 2nd ed. University of Massachusetts Press, Amherst, 1989.

2. Vermeer. The Complete Paintings. A. Blankert and M. Villa. Granada, London, 1981.

3. The Master Forger. The Story of Han van Meegeren. J. Godley. Wilfred Funk, Inc., New York, 1951.

4. Van Meegeren: Master Forger. Lord Kilbracken. Charles Scribner's Sons, New York, 1967.

Synthesis and Analysis of Polymers

E26.1 OVERVIEW

In this experiment you will have the chance to synthesize several polymers. You are not expected to do all the different parts of the experiment. Consult with your instructor to find out which polymers you will prepare. In E26.2 and E26.3 you will obtain polystyrene (PS) and poly(methyl methacrylate) (PMMA) by emulsion polymerization in water. In both cases, you will use potassium persulfate as an initiator and a detergent, sodium dodecyl sulfate (SDS), as an emulsifier. Polymers obtained by emulsion polymerization are powdery. You will transform these powders into films, by dissolving the polymer in a good solvent and allowing the solvent to slowly evaporate. The emulsion polymerizations described in E26.2 and E26.3 take about one hour of reflux; in the meantime you can perform other parts of the experiment.

In E26.4 you will carry out the bulk copolymerization of styrene and divinylbenzene with and without a plasticizer (dioctyl phthalate), using benzoyl peroxide as an initiator. You will use different amounts of the plasticizer to assess its effect on the strength of the polymer. You will compare this polymer with PS obtained in E26.2.

In E26.5 you will perform the bulk polymerization of methyl methacrylate with benzoyl peroxide as an initiator. You can embed small objects (dry flowers, sea shells, coins, etc.) in PMMA and make a paper weight and other ornaments. Talk to your instructor to know how much monomer will be available to you (this will limit the size of your object) and let your imagination fly! You will compare this type of PMMA with that obtained in E26.3.

In E26.6 you will make cellulose triacetate from cotton by acetylation of the glucose units with acetic anhydride. This is the lengthier part of the experiment. It will show you how to "dissolve" and transform cotton. In E26.7 you will make an infusible, thermosetting polymer (a phenolic resin) by the reaction of phenol and furfural. In E26.8 you will prepare nylon 6.6 in the form of a nylon rope. You will also transform this rope into a powder.

Finally, in E26.9 you will analyze the polymers you made and other plastics by a battery of methods including solubility, density, melting, and IR spectroscopy. You can bring your own samples for study and classification. Identification and sorting of polymers is an essential step for their recycling. As you will see in this experiment, never try to classify a plastic by its appearance. The word "plastic," after all, is derived from the Greek "*plastikos*" meaning "fit for molding." The references at the end of the experiments indicate the sources from which the experimental procedures were adapted.

The monomers used in this experiment must be distilled before use to free them from polymerization inhibitors added by the manufacturer to prolong their shelf life. The monomers will be already distilled and ready to use, but if eventually you need to distill them, a simple distillation will be enough. The distillation should be done at a fast rate to minimize the "cooking" of the monomer, which leads to polymerization and loss of material.

PROCEDURE

Before doing E26.2–5, read the safety alert.

Safety First

- Styrene, *p*-divinylbenzene, and dioctyl phthalate are irritants. Styrene is a possible carcinogen. Handle them **only** in the fume hood.

- Potassium persulfate is a powerful oxidant. Keep the container well closed in a cool place.

- Methyl methacrylate is a severe irritant. Handle this chemical only in the hood.

- Benzoyl peroxide is a powerful oxidant. Never heat the solid; it may explode. Do not handle the solid with a metal spatula. Use a piece of cardboard to transfer the solid. Keep the solid well closed in a plastic bottle in the refrigerator when not in use.

595

E26.2   EMULSION POLYMERIZATION OF STYRENE WITH PERSULFATE

Background reading: 26.1; 26.2; 26.5
Estimated time: 1.5 hr

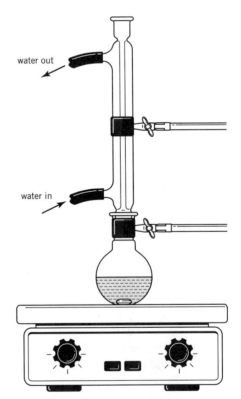

Perform this synthesis in a **well-ventilated fume hood**. In a 50-mL round-bottom flask equipped with a magnetic stir bar and a reflux condenser (Fig. 26.4), place 4 mL of an aqueous solution containing sodium dihydrogen phosphate (NaH_2PO_4) (0.01 M) and sodium dodecyl sulfate (SDS; 1.25%, w/v). Add 1 mL of a freshly prepared aqueous solution of potassium persulfate ($K_2S_2O_8$) (1%, w/v), and 2.5 mL of freshly distilled styrene. Use a new Pasteur pipet to dispense each solution. Place a small amount of silicon grease (the size of a rice grain) on the condenser joint and spread it by reattaching the round-bottom flask and rotating it.

Figure 26.4   Reflux apparatus for emulsion polymerization.

Heat the mixture directly on a hot plate at a low-medium setting making sure that the *liquid is refluxing without foaming in the condenser*. Stir and reflux for 1 hour.

At the end of this period, turn off the heat and add 5 mL of water from the top of the condenser. Let the system cool down with continuous stirring. Disassemble the reflux apparatus momentarily and make sure that the stir bar is free to rotate. Carefully dislodge any chunks of polymer attached to the walls using a spatula. Add little by little 0.5 mL of a 10% (w/v) solution of aluminum potassium sulfate (AlK(SO$_4$)$_2$ · 12H$_2$O) to completely precipitate the polymer. Reattach the condenser and stir the system for a couple of minutes.

Vacuum-filter the solid using a large Buchner funnel fitted with fast filter paper (coarse porosity, for example, Whatman 4). Wash the solid with two 10-mL portions of water, followed by two 10-mL portions of methanol. Let it dry on the filter for about 15 minutes, occasionally applying pressure with a spatula. Transfer the polymer to a 150-mL beaker labeled with your name and the name of the polymer. Cover it loosely with a piece of filter paper or paper towel fastened to the glass with a small piece of tape, and let the polymer dry for at least 12 hours in an oven set at 50°C. Weigh the product and determine its yield (disregard the contribution of the initiator). At the end of the experiment, the condenser and round-bottom flask can be cleaned with a little acetone (Ref. 1).

Preparation of a Polystyrene Film

In a test tube dissolve 0.1 g of powder polystyrene from the previous experiment (after it has been in the oven for at least 30 minutes) with 1.5 mL of ethyl acetate. Filter the solution using a Pasteur pipet with a small cotton plug, and collect the filtrate in a 50-mL beaker labeled with your name and the polymer's name. Let the solvent evaporate in a well-ventilated hood at room temperature (it will take several hours). The evaporation may be hastened by heating in a water bath. To remove the film from the bottom of the beaker, add a few milliliters of water and heat in a boiling-water bath for a couple of minutes. Add cold water and remove the polymer film with a spatula. Dry it between pieces of paper towel. Obtain its IR spectrum (see E26.9). Analyze the film by solubility and melting tests. See "Cleaning Up" at the end of E26.5.

E26.3   EMULSION POLYMERIZATION OF METHYL METHACRYLATE

> **Background reading:** 26.1; 26.2; 26.5
> **Estimated time:** 1.5 hr
> Read **Safety First** on p. 595

$$n \; CH_2{=}\underset{\underset{CO_2CH_3... }{}}{\overset{\overset{CH_3}{|}}{C}}{-}CO_2CH_3 \xrightarrow[\text{SDS}]{\text{K}_2\text{S}_2\text{O}_8} \left[CH_2{-}\underset{CO_2CH_3}{\overset{CH_3}{C}} \right]_n$$

methyl methacrylate   poly(methyl methacrylate)
PMMA

Follow the same procedure as for styrene (E26.2). Replace styrene by 2.5 mL of freshly distilled methyl methacrylate.

Preparation of a Poly(methyl Methacrylate) Film

In a 50-mL dry Erlenmeyer flask weigh approximately 0.1 g of poly(methyl methacrylate) from the previous part (after it has been in the oven for at least 30 minutes). Add 5 mL of acetone and mix to dissolve the solid. Remove any insoluble particles by gravity filtration, using a small cotton plug and a Pasteur pipet. Collect the filtrate in a 50–150 mL beaker labeled with your name and the polymer's name. Let the solvent evaporate in a well-ventilated fume hood at room temperature (it will take several hours). The evaporation may be hastened by heating in a water bath. Once the solvent has completely evaporated, remove the film with a spatula and obtain its IR spectrum (see E26.9). Analyze the polymer by solubility and melting tests.

E26.4   BULK COPOLYMERIZATION OF STYRENE AND DIVINYLBENZENE: EFFECT OF A PLASTICIZER

> **Background reading:** 26.1; 26.2; 26.4–6
> **Estimated time:** 0.5 hr
> Read **Safety First** on p. 595

styrene   p-divinylbenzene

crosslinked polystyrene

The following operation should be **conducted in a well-ventilated hood**. In a 50-mL Erlenmeyer flask weigh 10 g of freshly distilled styrene, using a pluringe. Add 2.5 mL of freshly distilled divinylbenzene with a pluringe with a new Pasteur-pipet tip. Add 0.13 g of benzoyl peroxide to the mixture and swirl the liquid to dissolve. Place 3.0; 2.5; 2.0; 1.5; and 1.0 mL of the styrene-divinylbenzene mixture in five clean, dry, and labeled test tubes (13 × 100 mm). Using a pluringe with the Pasteur pipet tip broken off (to break off the tip, gently etch the glass with a file and wrap the pipet in several layers of paper towels to protect your hands) add 0; 0.5; 1.0; 1.5; and 2.0 mL of dioctyl phthalate (DOC), in that order, to the test tubes, so the total volume in all of the test tubes is approximately 3 mL. Using a clean Pasteur pipet, mix the contents of the test tubes well. Do this by drawing the liquid into the pipet several times (dioctyl phthalate is very viscous).

Place the tubes in a beaker labeled with your name and the name of the polymer, add a little water to the beaker, and heat it in a water bath at about 70–80°C for about 15 minutes or until the liquid becomes viscous. Cap the tubes loosely with cork stoppers, and let them polymerize in a water bath at 30°C for the rest of the lab section and then transfer them to a well-ventilated oven set at 30°C for at least 24 hours.

To remove the polymers, check with a spatula that they have hardened and then wrap the tubes, one at a time, in several layers of paper towel or in a cloth, and carefully smash them with a hammer. Remove the plastic rods with tweezers and place them in a beaker with water to remove small pieces of glass. Dry them, determine their densities, and analyze their flexibility. Do **not** do a melting test (E26.9) on these samples (they may give off foul-smelling vapors as they melt), Ref. 2. See "Cleaning Up" at the end of E26.5.

E26.5 BULK POLYMERIZATION OF METHYL METHACRYLATE

> **Background reading:** 26.1; 26.2; 26.4; 26.5
> **Estimated time:** 0.5 hr
> Read **Safety First** on p. 595

methyl methacrylate → poly(methyl methacrylate) PMMA

Conduct the following operations **in a well-ventilated hood**. In a 25-mL Erlenmeyer flask place about 5 mL of freshly distilled methyl methacrylate. Add 0.05 g of benzoyl peroxide, and swirl the liquid to dissolve the solid. Transfer the liquid to a clean and dry test tube (13 × 100 mm). Place it in a beaker labeled with your name and the name of the polymer, add a little water to the beaker, and heat the system in a water bath at 70–80°C for about 15 minutes or until the liquid becomes viscous. You can also add small objects to make incrustations. To make bigger objects check with your instructor to know how much monomer is available to you. You can also use glass or metal containers, instead of a test tube, sprayed with a little cooking oil, such as Pam, to prevent adhesion to the plastic. Cap the tube loosely with a cork stopper, and let it polymerize in a water bath at 30°C for the rest of the lab section and then transfer it to a well-ventilated oven set at 30°C for at least 24 hours.

Before you remove the polymer make sure that it has hardened by poking it with a spatula. Wrap the tube in several layers of paper towels or in a cloth and carefully smash it with a hammer. Remove the plastic rod with tweezers, and place it in a beaker with water to remove small pieces of glass. Dry it and determine its density (Ref. 3).

> ### Cleaning Up
>
> - Dispose of the filtrates from the emulsion polymerizations (E26.2 and E26.3) in the container labeled "Polymers Liquid Waste."
>
> - Dispose of the filter papers, paper towels, and Pasteur pipets in the container labeled "Polymers Solid Waste."
>
> - Dispose of any unused portion of monomers (styrene, p-divinylbenzene, methyl methacrylate) and dioctyl phthalate in the "Polymers Liquid Waste" container.
>
> - To clean the glassware from the emulsion polymerizations, rinse with a little acetone and dispose of the rinses in the "Polymers Liquid Waste" container.
>
> - To dispose of the polymers, see "Cleaning Up" at the end of the analysis (E26.9).

E26.6 CELLULOSE TRIACETATE

> **Background reading:** 26.1; 26.5
> **Estimated time:** 2 hr

cellulose + 3 acetic anhydride → cellulose triacetate + 3 CH$_3$COOH acetic acid

Ac-: CH$_3$—C

In a 50-mL Erlenmeyer flask weigh 0.5 g of cotton. Add 3 mL of glacial acetic acid and stir the cotton with a glass rod to wet it as homogeneously as possible. Add 1 drop of concentrated sulfuric acid and a stir bar. Stopper the flask with a cork and place it in a water bath kept at 60–70°C. After approximately 10 minutes add 4 mL of acetic anhydride and continue heating at the same temperature for about 30 minutes. Stir with the magnetic stirrer while heating. At the end of this period the cotton will be completely dissolved and acetylated.

Add 4 mL of an 80% (v/v) aqueous solution of acetic acid and continue the heating for an additional 10-minute period. This will hydrolyze the excess acetic anhydride. Slowly add 10-mL of water to the flask. The cellulose acetate precipitates at this point. Stir the mixture with a glass rod. Transfer the thick liquid to two centrifuge tubes using a Pasteur pipet with the tip broken off (to break off the tip, gently etch the glass with a file and wrap the pipet in several layers of paper towels to protect your hands). Balance the centrifuge tubes by making sure that the *total mass* of both tubes is the same. Place them opposite to each other in the centrifuge rotor, and centrifuge them for about 5 minutes at 3000 rpm. Remove the supernatant with a Pasteur pipet and place it in a 250-mL beaker, add about 8–10 mL of water to each tube. Stir well with a glass rod, balance the tubes, and centrifuge them again for 5 minutes at about 3000 rpm. Remove the supernatant and place it in the beaker. Wash the polymer by adding to each tube 8–10 mL of a 5% aqueous solution of sodium bicarbonate. Add this solution slowly, stirring with a glass rod after each addition to avoid spill-over due to CO_2 bubbling. Balance the tubes and centrifuge them as before. Separate the supernatant and collect it in the beaker. Repeat the wash with 8–10 mL of water. Balance and centrifuge the tubes again. Remove the supernatant and collect it in the beaker. Check the pH of the aqueous solution in the beaker and adjust it with sodium bicarbonate solution, if necessary, until it is basic. Dispose of it as indicated in "Cleaning Up."

With the help of about 10 mL of acetone (for each tube), transfer the suspension to a large Buchner funnel fitted with fast filter paper (coarse porosity, for example, Whatman 4). Wash the walls of the centrifuge tube with a little acetone and transfer the washes to the funnel. Filter the suspension under vacuum. Let it sit in the funnel for about 10 minutes with the vacuum on, occasionally applying pressure with a spatula. Place the polymer in a labeled beaker (your name and polymer's name), cover it with a piece of filter paper, and dry it in an oven at 50°C for at least 12 hours. Weigh the product and calculate the % yield, assuming that the cellulose was completely acetylated (three —OH groups per glucose unit). See "Cleaning Up" at the end of E26.8 (Ref. 4).

Preparation of a Cellulose Acetate Film

Weigh 0.1 g of dry cellulose triacetate (after it has been in the oven for at least 30 minutes) in a 50-mL dry Erlenmeyer flask. Add 5 mL of glacial acetic acid and warm it up in a water bath. Remove any insoluble particles by gravity filtration using a cotton plug and a Pasteur pipet. Collect the liquid in a 150-mL beaker labeled with your name and the name of the polymer. Let the solvent slowly evaporate in a well-ventilated fume hood. Once the solvent has completely evaporated (it will take a few hours at room temperature, but the evaporation my be hastened by heating in a water bath) remove the film of cellulose with a spatula and obtain its IR spectrum. If the film is firmly stuck to the bottom of the beaker, add about 5 mL of water, and heat for about one minute in a boiling-water bath; pour out the liquid and immediately add 5 mL of cold water. Remove the film with a spatula; dry it between pieces of paper towel before obtaining its IR spectrum (see E26.9). Analyze the polymer by melting and solubility tests. If the cellulose acetate used to make the film was not sufficiently dry, a powder rather than a film may be obtained. In that case obtain the IR spectrum using a nujol mull.

E26.7   PHENOLIC RESINS

Background reading: 26.1; 26.4; 26.5
Estimated time: 0.5 hr

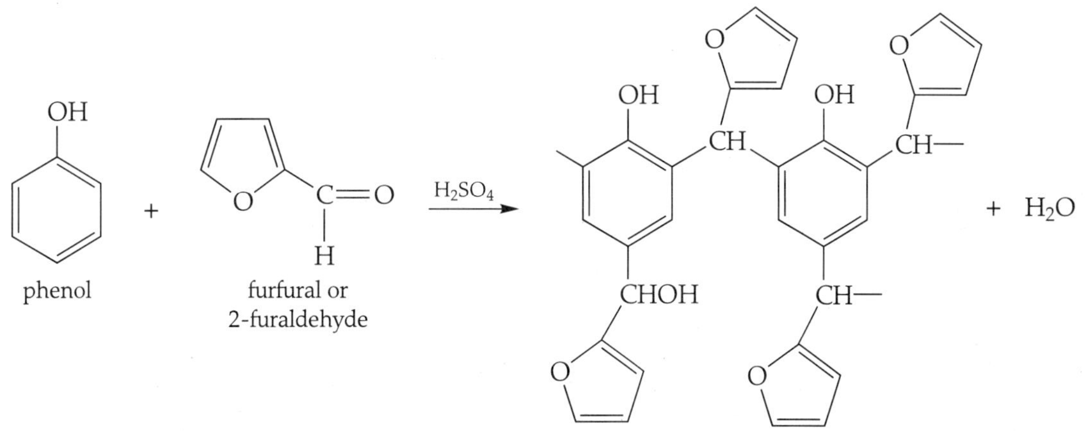

In a medium-size test tube, place 1 mL of freshly distilled furfural; add 1 g of phenol and two drops of concentrated sulfuric acid. Stir the tube with a glass rod and cap it. Leave it at room temperature for at least 24 hours. Break the polymeric product with a spatula. (If the polymer is too stiff and cannot be easily broken, wrap the test tube in several layers of paper towel or in a cloth and carefully smash it with a hammer; handle the pieces with tweezers and place them in a test tube). Wash the polymer pieces in the test tube with 2 mL of 95% ethanol, followed by 2 mL of 10% aqueous NaOH, then 2 mL of water, and finally 2 mL of 95% ethanol. In each wash, remove the liquid with the help of a Pasteur pipet. Analyze the polymer by solubility and melting tests. See "Cleaning Up" at the end of E26.8 (Ref. 3).

Safety First

- Phenol and furfural are toxic and corrosive. Avoid skin contact. Wear gloves when handling them. Remove gloves immediately.

- Sulfuric acid is corrosive. It may cause burns. Do not remove the sulfuric acid bottle from the hood. Dispense only the amount you need.

E26.8   PREPARATION OF NYLON 6.6: THE NYLON ROPE TRICK

Background reading: 26.1; 26.3; 26.5
Estimated time: 0.5 hr

Safety First

- Adipoyl chloride is corrosive and a lachrymator. Handle this chemical only in the fume hood. Avoid skin contact.

- Hexamethylene diamine and formic acid are corrosive. Handle them with care.

- Methylene chloride is a possible carcinogen.

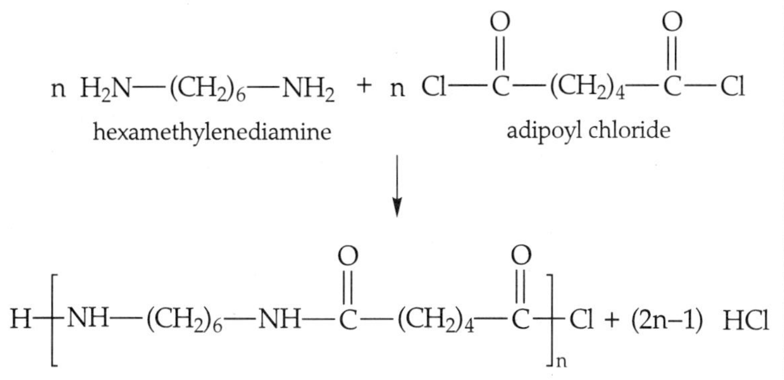

nylon 6.6

Perform this experiment in a **well-ventilated fume hood**. In a 100-mL beaker pour 20 mL of a solution of adipoyl chloride in methylene chloride (2% v/v). Carefully pour on top of this organic layer 10 mL of an aqueous solution of hexamethylenediamine (2%, w/v) containing sodium carbonate (4%, w/v). With the aid of tweezers pull the film that forms at the interface between the layers and coil it around a glass rod held horizontally about 5 inches above the beaker. With your fingers spin the glass rod to collect the nylon that keeps continuously forming at the interface (Fig. 26.5). Wash the nylon rope with water, followed by a 50% solution of ethanol in water. Dry and weigh the polymer. Analyze the polymer by solubility and melting tests (Refs. 5 and 6).

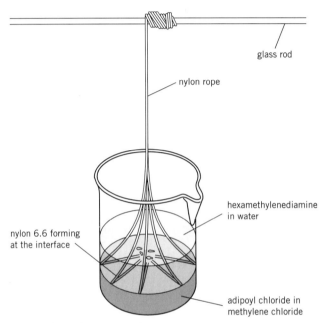

glass rod

nylon rope

hexamethylenediamine in water

nylon 6.6 forming at the interface

adipoyl chloride in methylene chloride

Figure 26.5   Setup for the nylon rope trick.

Cleaning Up

- Dispose of the supernatant and filtrate from E26.6 in the container labeled "Polymers Liquid Waste."

- Dispose of the washes from E26.7 in the container labeled "Polymers Liquid Waste."

- Dispose of unused liquid from E26.8 in the container labeled "Polymers Liquid Waste."

- Dispose of Pasteur pipets and paper towels in the "Polymers Solid Waste" container.

- To dispose of the polymers see "Cleaning Up" at the end of the analysis (E26.9).

Preparation of Nylon Powder

In a 13 × 100 mm test tube weigh about 50–100 mg of nylon 6.6 from the previous part. Dissolve the polymer by adding 1 mL of a 96% (w/w) aqueous solution of formic acid. Transfer the liquid to a watch-glass and let it evaporate overnight in a well-ventilated hood. The watch-glass should be labeled with your name and the polymer's name. If you obtained a powdery solid, prepare a nujol mull and obtain its IR. If you obtained a film instead, use it directly to run its IR spectrum (see E26.9).

E26.9   ANALYSIS OF POLYMERS

> **Background reading:** 26.1; 26.7; 26.8
> **Estimated time:** 1–3 hr

In this part of the experiment you will use a battery of methods to study samples of known polymers (such as PP, PE, PS, PMMA, PTFE, PETE, nylon, etc.) and also those that you have synthesized. You will employ solubility, density, melting, and IR spectroscopy to characterize these polymers. You will also perform the analysis and identification of two unknown polymer samples. You can also bring your own plastic samples for study.

You should bear in mind that not all types of plastics can be studied by these tests as described here. For example, the solubility and density tests are intended for films or blocks, not for powders. The melting test can be applied to powders, films, and blocks. IR analysis is easier to perform on films but the films should be very thin to give good spectra. You can also obtain the IR of powders by using nujol mulls. Some block polymers, such as PTFE, PETE, HDPE, and phenolic resins, which are difficult to melt or dissolve, require special treatment before IR spectra can be obtained. You will not be concerned with them.

Identification of Polymers by Solubility

To identify polymers by solubility you will follow the flowchart shown in Figure 26.6. Start by checking the solubility of the polymer in toluene. If the polymer is soluble, follow the branch on the right side; if it is insoluble, follow the branch on the left side. You should use a fresh sample of polymer in each step. For example, let's suppose that you try the solubility of an unknown polymer in toluene and you find out that it is soluble. Next, you try its solubility in ethyl acetate using a fresh sample of polymer. If the polymer dissolves, then try its solubility in 96% (w/w) formic acid with a fresh sample. If it dissolves in this solvent your unknown is likely to be poly(methyl methacrylate). You can use this solubility scheme to identify the most commonly used polymers such as PTFE, nylon, PETE, cellulose acetate, saran (poly(vinylidene chloride)), HDPE, LDPE, PP, PS, and PMMA.

Safety First
• Formic and acetic acids are corrosive. Handle them with care.
• Toluene and ethyl acetate are flammable.
• DMF is an irritant. Handle it only in the fume hood.
• DMSO is an irritant. Avoid skin contact.
• Perform this experiment in the fume hood.

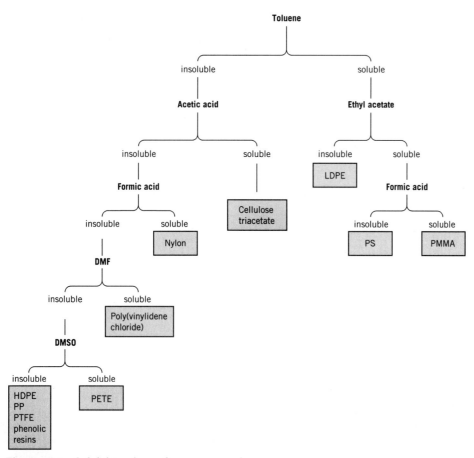

Figure 26.6 Solubility scheme for common polymers.

The solubility of a polymer in a given solvent depends on the physical state of the polymer. Powders dissolve much more quickly than solid blocks or films. This solubility scheme has been developed using polymer blocks and films, and is intended to be applied for the same type of polymers. You may obtain different results if you try these tests on powders.

Place about 50 mg of the polymer (a block the size of a rice grain or a 0.5 × 0.5 cm film) in a 13 × 100 mm labeled test tube and add 1 mL of the desired solvent. Stir the system with a thin glass rod or a microspatula and observe whether the polymer dissolves at room temperature. If it doesn't dissolve, warm up the system. When using toluene, ethyl acetate, acetic acid, or formic acid, heat the test tube for about 3 minutes in a bath of boiling water (in the **fume hood**!). For N, N-dimethylformamide (DMF) and dimethylsulfoxide (DMSO) heat the test tube in a sand bath at about 150–170°C while stirring (in the **fume hood**!). Consider the polymer soluble if it dissolves completely or if it gives a turbid solution. When heating HDPE and PP in DMSO, they *melt* but *do not dissolve*; PTFE and phenolic resins, on the other hand, don't melt and don't dissolve in hot DMSO. After determining the solubility in a given solvent, proceed to the next step in the flowchart as already explained.

Try these solubility tests with samples of nylon 6.6 (if available from E26.8), cellulose triacetate film (if available from E26.6), phenolic resin (E26.7), PS film (E26.2), PMMA film (E26.3), and at least two of the following: PETE (2-L soda bottles), PS (cassette-tape boxes, some disposable cups), poly(vinylidene chloride) saran wrap, PP (some deli containers), LDPE (some polyethylene bags), HDPE (some water and milk jugs), and teflon tape. Also, you will characterize and identify two unknown polymers.

At the end of the experiment collect all of the acetic acid and formic acid solutions in a beaker. Dilute the mixture with an equal volume of water and, using a Pasteur pipet, slowly add a 10% NaOH solution until the pH is basic. See "Cleaning Up."

Identification of Plastics by Density

You will estimate the density of polymers by placing small samples of plastics in a series of solvents of known density and observing whether the polymer floats or sinks. If the polymer sinks, you will try its buoyancy in the next solvent of higher density. You will start with the solvent of lower density (ethanol-water, 4 : 1) and stop when you find a solvent in which your sample floats. The density of the polymer will be in between the densities of the two solvents where the transition between sinking and floating occurs. You will use 150-mL jars containing the solvents listed in Table 26.5.

Table 26.5 **Solvents for the Determination of Polymer Density**

Solvent	Density (g/mL)	Solvent	Density (g/mL)
ethanol-water (4 : 1) (v/v)	0.86	10% NaCl (w/w)	1.06
ethanol-water (10 : 7) (v/v)	0.91	dilute light corn syrup (4 : 1) (w/w)	1.29
ethanol-water (1 : 1) (v/v)	0.92	light corn syrup	1.41
water	1.00		

After adding your sample (the size of a nickel for films, and the size of a peanut or bigger for blocks), cap the jar and invert it a few times. Corn syrup is very viscous; to sink your sample push it with a glass rod and observe if it rises. Some samples rise to the surface very slowly. Use a fresh sample of polymer for each solvent (except for the block samples of PS and PMMA obtained in E26.4 and E26.5, respectively, which must be reused because you will have only one). Remove your samples with tweezers at the end. Characterize the following polymers: PMMA from (E26.5), PS (E26.4), and at least two of the following: PETE (2-L soda bottles), PS (cassette-tape boxes, some disposable cups), poly(vinylidene chloride) (saran wrap), PP (some deli containers), LDPE (some polyethylene bags), HDPE (some water and milk jugs), and Teflon tape. Also, you will characterize and identify two unknown polymers. Compare your results with the densities listed in Table 26.3.

The solutions will be prepared for you to use, but if you need to make them, follow these directions. **Ethanol-water (4:1) (v/v)**: mix 4 volumes of 95% ethanol with 1 volume of water. For the other two ethanol-water solutions, mix volumes of 95% ethanol and water as indicated by the ratios 10:7 and 1:1. **10% NaCl (w/w)**: weigh 10 g sodium chloride and 90 g of water in a beaker. **Dilute corn syrup**: mix 100 g of syrup (light corn syrup) with 25 mL of water. To determine the density of these solutions, fill a pre-tared 10-mL volumetric flask to the mark, and determine the mass of the solution. Divide the mass (g) by the volume (mL), Ref. 7.

Melting of Plastics

Most plastics do not have a well-defined melting point. They melt at different temperatures depending on whether they are crystalline or amorphous solids. Some plastics such as high- and low-density polyethylene melt at relatively low temperatures (120–130°C); others, such as PTFE, have very high melting points (>330°C), while others do not melt at all, for example, phenolic resins.

You will compare the ease of melting of different plastics by placing them on a piece of aluminum foil on a hot plate and gradually heating them. You will also place a sand bath with a thermometer on the hot plate and will read the temperature as the plastics melt. These readings are not the melting points because the temperature across the hot plate is not homogeneous but will give you, nonetheless, a relative order of melting.

Conduct this experiment in the fume hood. Cut a piece of aluminum foil (15 × 10 cm) and place it on a hot plate. Place a sand bath with a well-immersed thermometer secured by a clamp to a ring stand. Place small pieces, the size of a lentil, of the following polymers about 2 cm apart from each other: HDPE, LDPE, PP, PMMA, PS, PETE, nylon 6.6 rope, PTFE, cellulose triacetate, saran, phenolic resins, and your unknowns. Heat at a medium-high setting and observe which samples melt. Read the temperature in the sand bath as the samples melt. Turn off the heat when the temperature has reached 210°C.

IR Analysis of Polymers

Obtain the IR spectra of polyethylene (plastic bag), poly(vinylidene chloride) (saran), the cellulose acetate as a film or as a nujol mull (E27.6), PMMA film (E27.3), PS film (E27.1), and nylon 6.6 as a film or as a nujol mull (E27.8). In obtaining the IR of films, position the film in the path of the IR light to obtain peaks that are on scale. Some films are too thick in the middle to give good spectra but produce satisfactory ones when the *edges* are irradiated instead.

Cleaning Up

- Dispose of the solvents from the solubility tests in the container labeled "Polymers-Liquid Waste."

- Dispose of Pasteur pipets in the "Polymers Solid Waste" container.

- Dispose of the aluminum foil from the melt test in the trash.

- Remove any pieces of polymers from the solvent jars used for the density tests. These jars will be used by other students.

- Dispose of the block PS with dioctyl phthalate in the "Polymers Solid Waste" container.

- You can keep the block PMMA and PS (without dioctyl phthalate).

- Turn in to your instructor leftovers of all the other polymers in labeled vials.

BIBLIOGRAPHY

1. Preparative Methods of Polymer Chemistry. W.R. Sorenson and T.W. Campbell. 2nd ed. Interscience, New York, 1968.

2. Introducing Plastics in the Laboratory. Synthesis of a Plasticizer, Dioctyl phthalate and, Evaluation of its Effects on the Physical Properties of Polystyrenes. A. Caspar, J. Gillois, G. Guillerm, M. Savignac, and L. Vo-Quang. *J. Chem. Ed.*, **63**, 811–812 (1986).

3. Classroom Demonstration of Polymer Principles. F. Rodriguez, L.J. Mathias, J. Kroschwitz, and C.E. Carraher, Jr. *J. Chem. Ed.*, **64**, 886–888 (1987).

4. Polymer Preparations in the Laboratory. G.M. Lampman, D.W. Ford, W.R. Hale, A. Pinkers, and C.G. Sewell. *J. Chem. Ed.*, **56**, 626–628 (1979).

5. The Nylon Rope Trick. P.W. Morgan and S.L. Kwolek. *J. Chem. Ed.*, **36**, 182–184 (1959).

6. An Alternative Procedure for the Nylon Rope Trick. G.C. East and S. Hassell. *J. Chem. Ed.*, **60**, 69 (1983).

7. Method for Separating or Identifying Plastics. K.E. Kolb and D.K. Kolb. *J. Chem. Ed.*, **68**, 348 (1991).

8. Laboratory Preparation for Macromolecular Chemistry. E.L. McCaffery. McGraw-Hill, New York, 1970.

9. Classroom Demonstrations of Polymer Principles. IV. Mechanical Properties. F. Rodriguez. *J. Chem. Ed.*, **67**, 784–788 (1990).

10. Rapid Identification of Thermoplastic Polymers. H. Cloutier and R.E. Prud'homme. *J. Chem. Ed.*, **62**, 815–819 (1985).

EXPERIMENT 26   REPORT

Pre-lab

1. Make a table with the physical properties of all reagents used. Specify toxicity/hazards and flammability. Calculate the number of mmoles used.

2. Write chemical reactions for the emulsion polymerization of styrene and methyl methacrylate. What role does potassium persulfate play? Could it be replaced by benzoyl peroxide?

3. In the emulsion polymerization of styrene and methyl methacrylate, why do you add sodium dodecyl sulfate?

4. Do you anticipate any changes in the length of the polymeric chains if the concentration of potassium persulfate is increased?

5. What role does dioctyl phthalate play in the polymerization of styrene?

6. In the bulk polymerization of styrene and methyl methacrylate, why do you use benzoyl peroxide? Could you use potassium persulfate instead?

7. In the copolymerization of styrene with divinylbenzene, what effect on the 3-D structure of the polymer does divinylbenzene have?

8. What precautions should you take in handling benzoyl peroxide?

9. Formulate a chemical reaction for the synthesis of cellulose triacetate from cellulose. Make a flowchart for this synthesis.

10. Write a chemical reaction for the synthesis of the furfural-phenol resin.

11. Briefly explain the meaning of thermosetting and thermoplastic. Reason why aldehyde-phenol resins are thermosetting.

12. Write a chemical reaction for the synthesis of nylon 6.6. What natural polymers do nylons resemble?

13. A solid sample of an unknown plastic was insoluble in toluene, acetic acid, formic acid, DMF, and DMSO. The plastic sank in an ethanol-water (4:1) solution but floated in an ethanol-water (10:7) solution. What is the most likely composition of the polymer?

14. A film of an unknown polymer was soluble in toluene and ethyl acetate. The IR of the polymer showed a strong band at 1740 cm^{-1}. What is the most likely composition of the polymer?

In-lab

1. Why did you add aluminum potassium sulfate in the emulsion polymerizations of styrene and methyl methacrylate?

2. Calculate the % yield in the synthesis of PS and PMMA by emulsion polymerization. Disregard the contribution of the initiator.

3. Do you think that polymers have a well-defined molecular mass? Speculate how you could control the average molecular mass of polystyrene obtained by emulsion polymerization.

4. Discuss the effects in the molecular structure of poly(methyl methacrylate) that an increase in the amount of benzoyl peroxide would have.

5. Did you observe any effect due to the addition of dioctyl phthalate in the properties of PS? Briefly explain.

6. What percentage increase in the mass of cellulose do you expect to see if all three OH groups of each glucose unit are acetylated? (Ignore the glucose units at the end of the chain, which may have more than three acetyl groups; their contribution to the whole molecule is negligible.)

7. Calculate the maximum amount of cellulose triacetate that you can expect in your synthesis (theoretical yield). Report the actual yield and calculate the % yield of the synthesis.

8. In the synthesis of nylon 6.6 calculate the number of mmoles of both reagents used. Which compound is the limiting reagent? Calculate the theoretical yield in grams. To do this you do not need to know the molecular mass of the polymer. Just calculate the molecular mass of the repetitive unit of the polymer and multiply this number by the number of moles of limiting reagent.

9. Report the actual yield and calculate the % yield in the nylon 6.6 synthesis.

10. Report and discuss your solubility, density, and melting tests.

11. Interpret the IR spectra and compare them with those shown in Figures 26.7–26.13.

12. Identify your unknowns. Explain your reasoning.

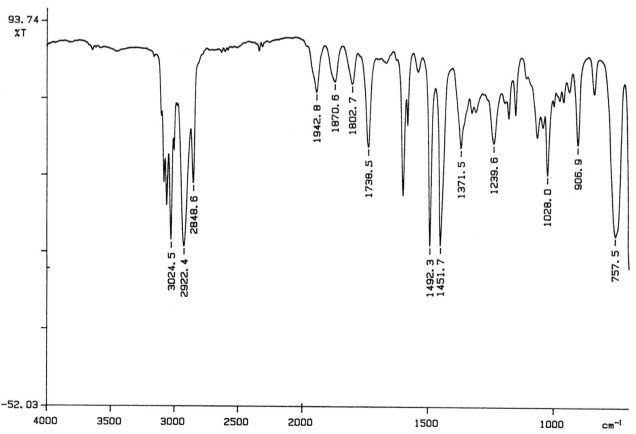

Figure 26.7 IR spectrum of polystyrene (film).

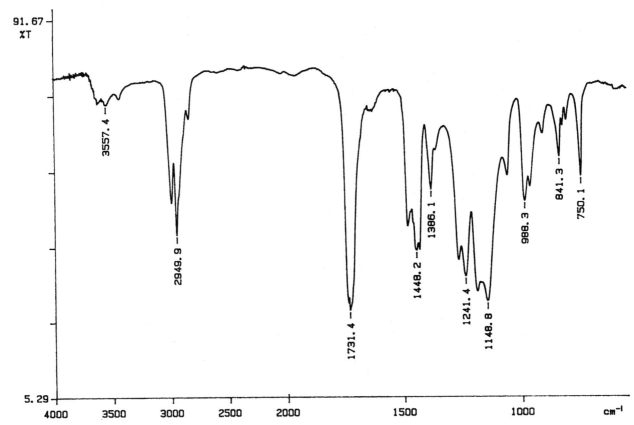

Figure 26.8 IR spectrum of poly(methyl methacrylate) (film).

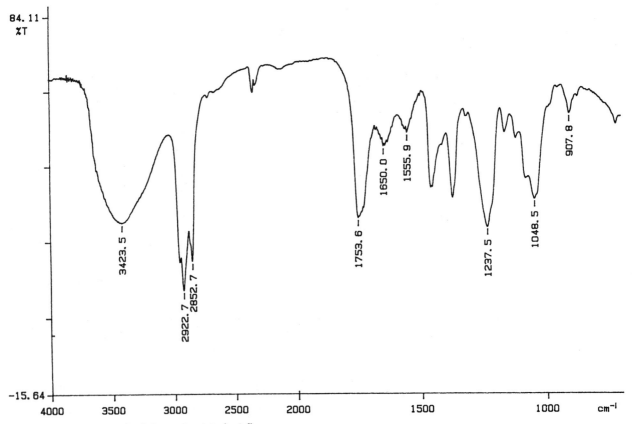

Figure 26.9 IR spectrum of cellulose triacetate (nujol).

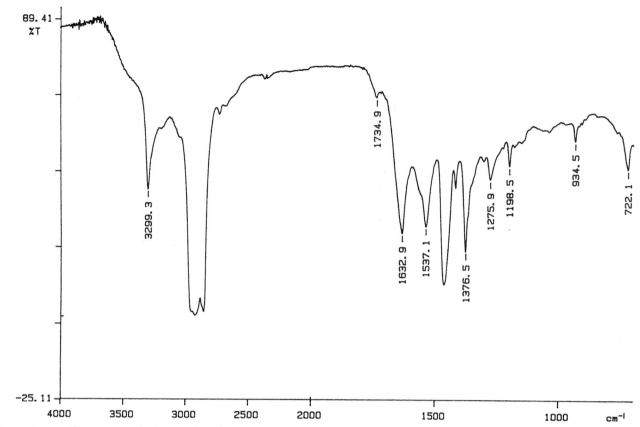

Figure 26.10   IR spectrum of nylon 6.6 (nujol).

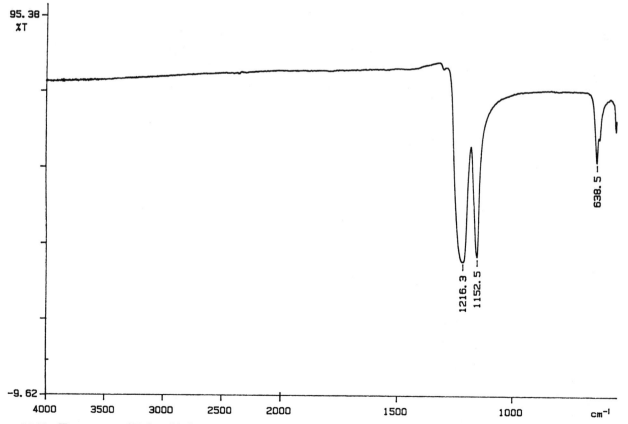

Figure 26.11   IR spectrum of Teflon (film).

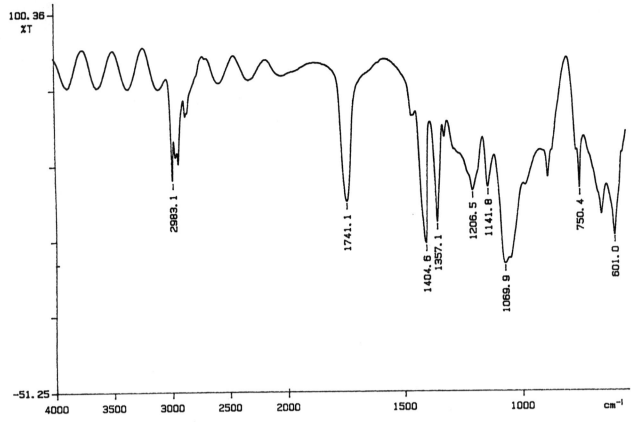

Figure 26.12 IR spectrum of saran (film).

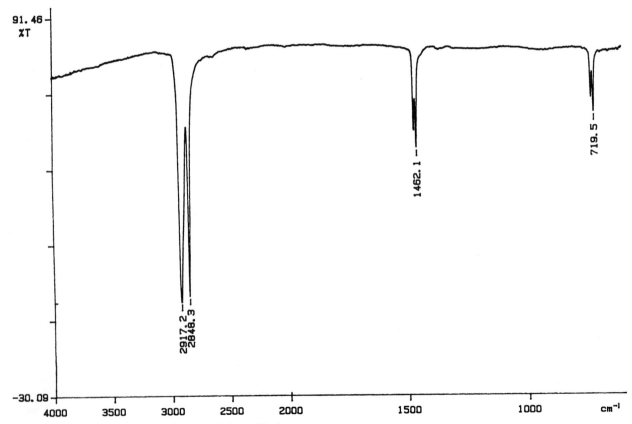

Figure 26.13 IR spectrum low-density polyethylene (film.)

Dyes and Pigments

O might it be, that Readers find delight
In this Work that to the living is so opportune.
Set apart are Purple, Yellow, and how to brown,
To color in Wine, and faded shades,
The green, the blues, and scarlets
And those that carry the emblem of fortune.

"The Plictho, Instructions in the Art of the Dyers Which Teaches the Dyeing of Woolen Cloths, Linens, Cottons, and Silk by the Great Art as Well as by the Common," by Gioanventura Rosetti, 1548. Translated by S.M. Edelstein and H.C. Borghetty.

27.1 DYES, PIGMENTS, AND COLORS

Dyes and pigments are colored compounds used to change the appearance of objects. Nature produces them to make flowers attractive, to tell predators to back off, to catch the sunlight and transform it into chemical energy, and also for no apparent reason when at the onset of autumn tree leaves explode in reds, browns, and yellows. We have learned to use natural coloring substances from a very early time, as cave paintings and ceramic artifacts testify. It has been only in the last one hundred years that we have discovered how to make our own dyestuffs. The creation of new colors and their applications in the textile and printing industries were the engine that, in the middle of the nineteenth century, propelled organic chemistry to the foreground of scientific research (see "The Color of Success," Unit 20, p. 424).

The difference between dyes and pigments lies in their solubilities. **Dyes** are soluble in the medium in which they are applied, normally water, while **pigments** are insoluble. This distinction is operational rather than structural. As we will see, the same compound can be regarded as a dye or a pigment depending on its mode of use.

Dyes can be classified according to their *structures* and also based on their *mode of application* to fibers. According to their structural differences, dyes can be classified into the following categories: **azo**, **cationic**, **anthraquinone**, and **indigo**. There are also many dyes that do not fall into any of these structural groups. Depending on their mode of application, dyes can be grouped into the following types: **direct**, **mordant**, **ingrain**, **vat**, **disperse**, **reactive**, and **solvent**.

In this unit we will concentrate on the synthesis of dyes and their use in the manufacture of textiles. We will first discuss different structural types of dyes (sections 27.2–27.7) and then their mode of application to fibers (sections 27.8 and 27.9). Before we go on, let's consider briefly the nature of color.

Objects appear colored because they absorb certain wavelengths of the visible spectrum (400–750 nm) and reflect others. The color that we observe is that of the reflected light. The colors of the spectrum can be arranged in a **color rosette** where they are displayed clockwise in order of decreasing wavelength; each color is diametrically opposed to its **complementary**: blue to orange, red to green, and

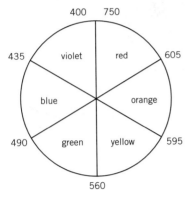

Figure 27.1 The color rosette. The numbers represent wavelengths in nm.

violet to yellow (Fig. 27.1). When a color is absorbed by an object, its complementary color is observed. Thus, when something appears red it is because the object absorbs green light, and conversely, when an object appears green, it absorbs red light.

> We take colors for granted. We do not hesitate to call a tree green, the ocean blue, and a salmon steak pink. But it has not always been like this. The Greeks in the times of Homer did not have a name for blue, and yellows and pale greens were lumped together in one single word: *chloros*. The language of the Pomo people, natives of central California, has words for only three colors: black, white, and red. Studies of about 100 languages from all over the world show that the perception of colors is evolutionary and its development follows similar patterns in distant cultures. It was found that all languages have words for black and white. If the language has a third word for colors (not all languages do), that word always corresponds to red. If there is a fourth word, it corresponds to either yellow or green. If there is a fifth one, then the language has words for both yellow and green. Only if the language has six words for colors is there one for blue. If there are seven terms, there is always one for brown. If the language has eight or more color words, then there are words for pink, purple, orange, and gray in no particular order.

27.2   AZO DYES

Azo dyes encompass the largest and most important family of dyes. They contain an azo group, $-N=N-$, linking two aromatic rings.

> *Azobenzene*
>
> $N=N$
>
> *Azo dyes are derivatives of azobenzene.*

Because of their extended conjugated pi-orbital systems these aromatic azo compounds absorb in the visible region of the electromagnetic spectrum and thus are deeply colored. The relationship between conjugation and light absorption is discussed in Unit 32. For now, it will suffice to say that the more extended the conjugated system (the larger the number of pi electrons), the longer the wavelength of the absorbed radiation. Most azo compounds absorb radiation in the violet-green range and therefore they have a yellow-red color (Fig. 27.1). The degree of conjugation and the nature of the substituents attached to the aromatic rings are important factors in determining the color of the azo compound.

To be useful as *dyes*, azo compounds must be *soluble* in water. This can be achieved by having polar and ionic groups attached to the aromatic rings. Favorite groups for this purpose are the sodium salts of sulfonic ($-SO_3Na$) and carboxylic acids ($-COONa$). Because of the presence of these acid groups, these dyes are also called *acidic* or *anionic dyes*. Shown below are some examples of anionic azo dyes. Notice that Direct Blue 67 has two azo groups linking three aromatic systems; because of its extended conjugation this dye absorbs visible light of long wavelength, orange, and thus it has a blue color (Fig. 27.1).

Soluble azo dyes

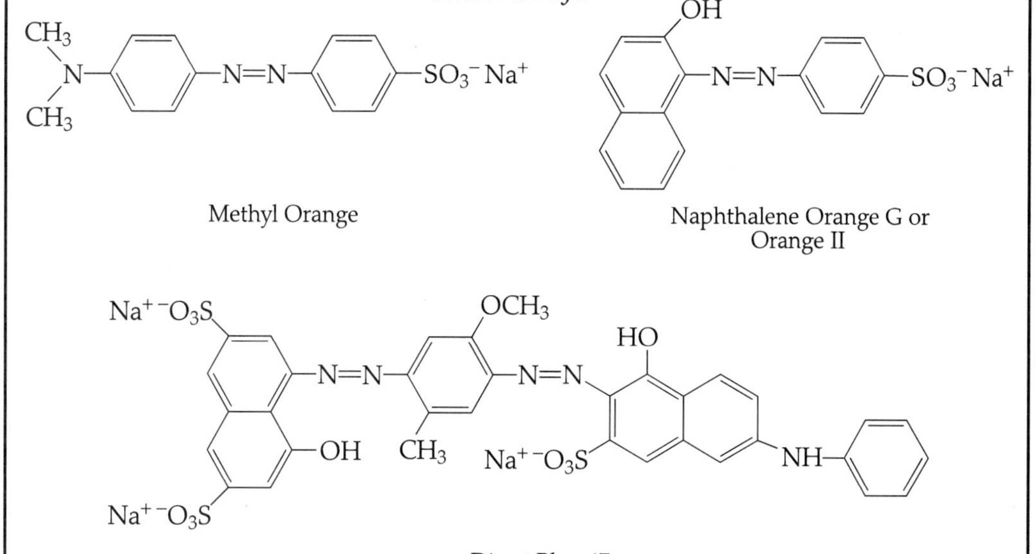

Methyl Orange

Naphthalene Orange G or Orange II

Direct Blue 67

Azo dyes without ionic groups are insoluble in water. These insoluble compounds can be used as *pigments* or, in solution with a suitable solvent, as *solvent dyes* (discussed in section 27.8). A typical example of insoluble azo dyes is shown below.

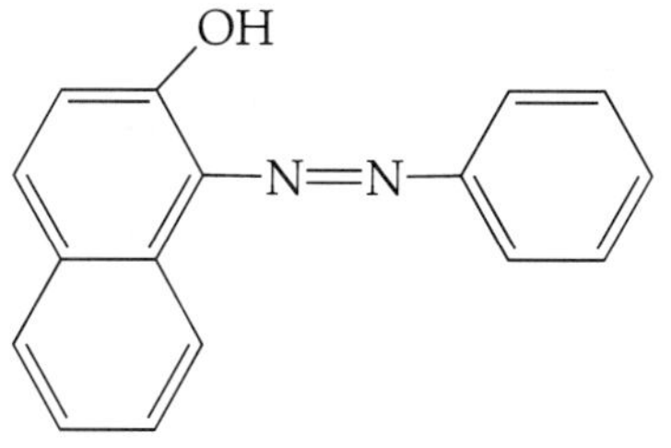

Solvent Yellow 14

27.3 SYNTHESIS OF AZO DYES

Azo dyes can be prepared in two easy laboratory steps. In the *first step*, an aromatic amine is transformed into a **diazonium salt** by the action of nitrous acid obtained by mixing sodium nitrite and a mineral acid. This reaction is called **diazotization**. In the *second step*, the diazonium salt is coupled to an aromatic compound, usually an aniline or a phenol, to yield an **aromatic azo compound**. The example below shows the coupling reaction of benzenediazonium chloride with phenol.

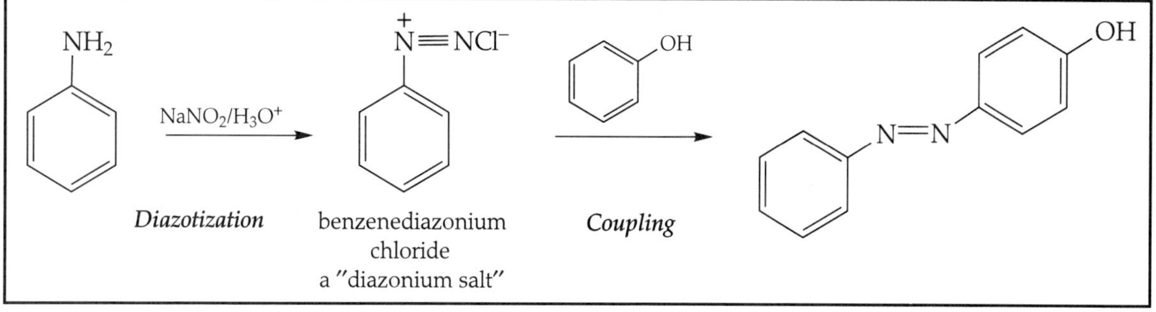

Diazotization   benzenediazonium chloride a "diazonium salt"   *Coupling*

Diazotization reactions are usually performed at low temperatures (0–5°C) to avoid the decomposition of the diazonium salts, which are unstable at higher temperatures. These reactions are normally carried out with a slight molar excess of sodium nitrite (about 10% excess), and a 3-fold molar excess of mineral acid. One mole of mineral acid is used to form nitrous acid from sodium nitrite, a second mole is needed for the actual diazotization reaction, and the third mole is necessary to keep any unreacted aniline protonated and avoid its coupling with the diazonium salt as it forms. The reaction mechanism for diazotization is complex and we won't be concerned with it here. Since some diazonium salts are explosive when dry, they are not isolated as solids, but used directly in the solution where they were formed.

benzenediazonium
chloride

The coupling reaction between diazonium salts and aromatic compounds takes place by an electrophilic aromatic substitution mechanism. The diazonium salt is the electrophile attacked by the aromatic ring. Only phenols, anilines, and other aromatic compounds with electron-donating substituents ($-NR_2$, $-OH$, $-OR$) are good substrates for coupling to diazonium salts. The coupling reaction normally occurs para to the electron-donating group. When the para position is already occupied, ortho substitution takes place. The example below illustrates the coupling of a diazonium salt with N,N-dimethylaniline.

Naphthols (and naphthylamines) are widely used in the manufacture of azo dyes. 1-Naphthols (and 1-naphthylamines) react with diazonium salts at the 4 position if it is free. If this position is occupied by another substituent, then coupling takes place in position 2. On the other hand, 2-naphthols (and 2-naphthylamines) couple exclusively at position 1; no coupling at the other ortho position (position 3) takes place (why? see Problem 1).

Azo coupling is strongly affected by pH. There is usually an optimum pH range in which to perform the coupling reaction. In general, coupling is not favorable above pH 10 because diazonium salts undergo undesirable side reactions. The recommended pH range for the coupling of phenols is 9–10 because phenoxide ions, obtained by dissociation of phenols at pH 9 or higher (Unit 14), are much more strongly activated for coupling than undissociated phenols. The optimum pH range for the coupling of

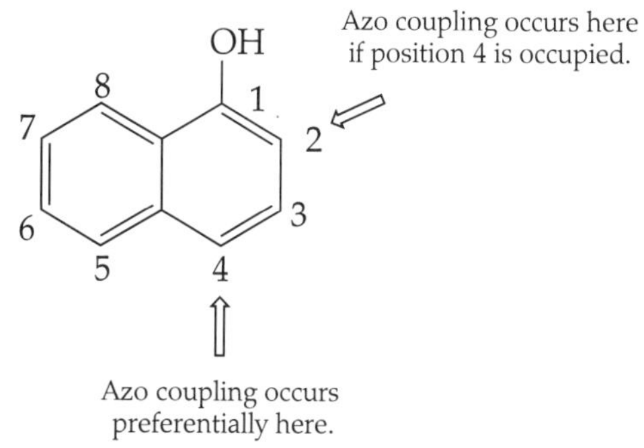

Azo coupling
occurs here only.

Azo coupling occurs here
if position 4 is occupied.

Azo coupling occurs
preferentially here.

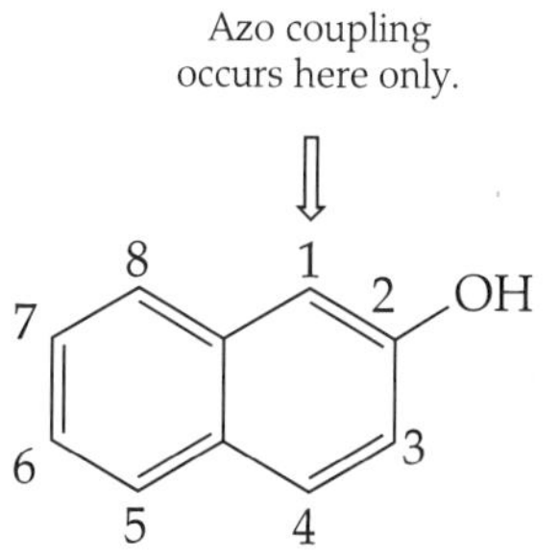

aromatic amines, on the other hand, is 5–10. This follows because the free amino group ($-NH_2$) is an activator for the coupling reaction, whereas the ammonium group ($-NH_3^+$), obtained by protonation below pH 5, is not.

This difference in the optimum pH range for amines and phenols allows selective coupling when both $-OH$ and $-NH_2$ groups are present in the same molecule. For example, the compound shown below reacts with diazonium salts at different positions of the naphthalene ring, depending on the pH. Under mild acidic conditions, pH 5–7, the $-OH$ group is not dissociated and the free amino group ($-NH_2$), a stronger activator than the $-OH$ group, directs the substitution. Coupling takes place in the ring where the amino group is located. At basic pH (pH 9–10), on the other hand, substitution occurs mainly in the ring that has the $-OH$ group because the $-O^-$ ion is a stronger activator than the amino group.

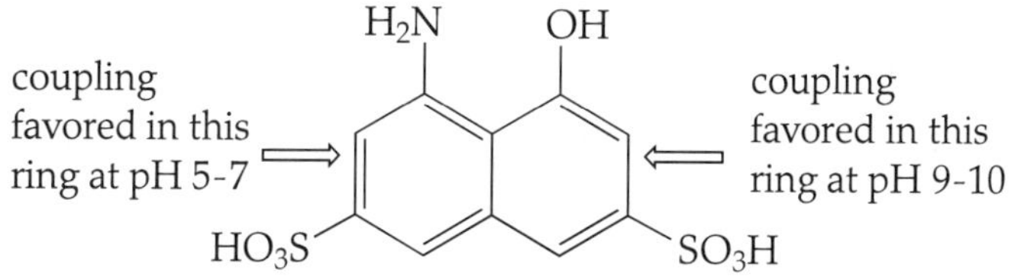

coupling
favored in this
ring at pH 5-7

coupling
favored in this
ring at pH 9-10

The coupling reaction between the diazonium salt and the activated aromatic ring can take place directly on the fiber. For instance, if cotton is immersed in a solution of 2-naphthol at pH 10, removed from the liquid, and then treated with a solution of the diazonium salt of *p*-nitroaniline, a deep red color called American Flag Red is obtained. This method is called **ingrain dyeing** and is particularly useful for applying azo dyes insoluble in water.

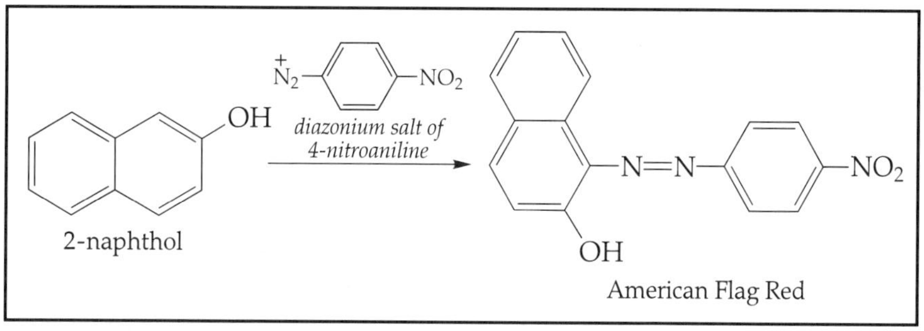

2-naphthol

diazonium salt of
4-nitroaniline

American Flag Red

27.4 CATIONIC DYES

There are various chemical classes of cationic dyes, the most important being the **derivatives of triphenylmethane**, such as Methyl Violet and Malachite Green.

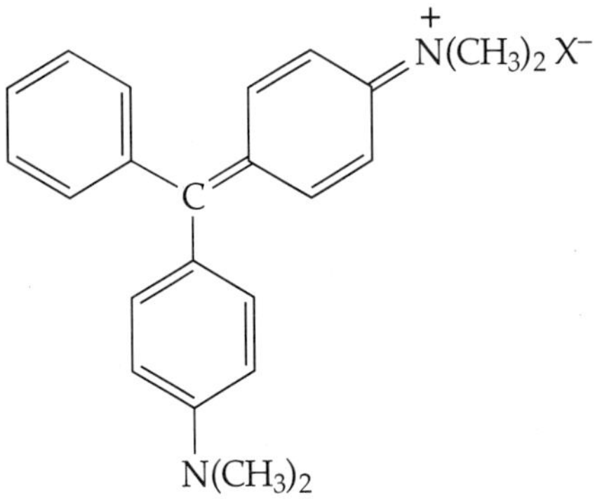

Malachite Green

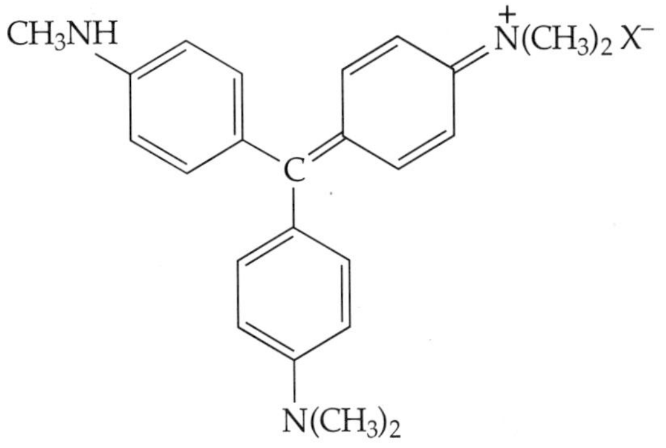

Methyl Violet

In triphenylmethane dyes, three aromatic rings are directly attached to a central sp^2-hybridized carbon atom. At least two of the rings have a dialkylamino group ($-NR_2$) para to the central carbon. These molecules are totally conjugated and have the positive charge delocalized among all three aromatic groups. The extensive conjugation makes these molecules colored. Because of the presence of the amino groups, these dyes are also called *basic dyes*.

Cationic dyes have exceptional brilliance and have been used extensively in the past to dye silk and wool but today they have been replaced with other dyes more resistant to light. They are employed in the manufacture of paper, inks, cosmetics, and pH indicators. You have already seen an example of cationic dyes in Unit 24, Coomassie Blue.

27.5   ANTHRAQUINONE DYES

These dyes of natural origin were used for centuries to dye cotton and leather with beautiful red hues. The most important dye of this class is **alizarin**, which is the main component of the dyestuff obtained from the roots of the madder plant *Rubia tinctorum*. These dyes are applied to fabrics in the presence of metal ions such as Al^{3+}, Fe^{3+}, Sn^{2+}, and Cr^{3+}, which keep the dye molecules anchored to the fiber through the formation of coordination complexes. In these complexes, the metal ion is at the center and the dye and the fiber molecules are bound as ligands. When alizarin is used to color cotton, a red hue is obtained if the metal ion is Al^{3+} or Sn^{2+}, a deep violet shade if it is Fe^{2+}, and a brown-black if Fe^{3+} is used instead. Compounds used to fix dyes to fibers are called **mordants**. Metal ions are the most important type of mordants. Tannins (phenolic compounds derived from plants; you have seen them in Experiment 5) are another type.

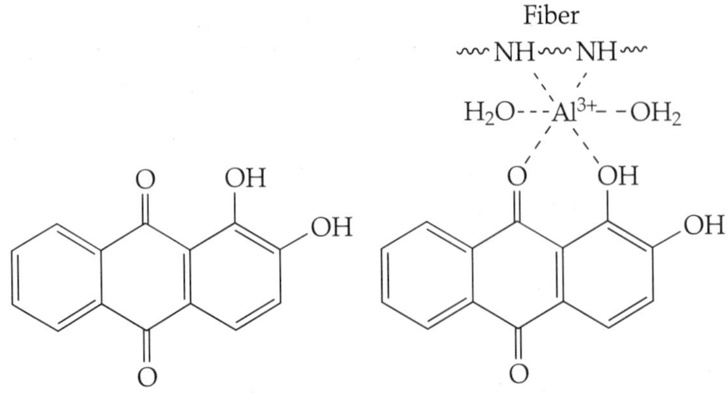

Alizarin   allizarin with mordant Al^{3+}

27.6 INDIGO DYES

The use of **indigo**, the dye of blue jeans, goes back at least 4000 years. The pigment was obtained from several indigenous plants from India and was introduced into the Middle East by Phoenician merchants. From there its use spread around the Mediterranean region and the rest of Europe. Indigo and indigo derivatives give blue-purple colors. According to their mode of application these dyes belong to the category of *vat dyes*.

Vat dyes are insoluble in water but they become soluble when reduced in the presence of a base. The reduced dye, called the *leuco* form because it is usually colorless (from Greek: *leuco*, white), is soluble in water and is applied onto the fiber by immersion. Upon drying and exposure to atmospheric oxygen, the dye is reoxidized and acquires its original color. Blue indigo is insoluble in water but its leuco form is water-soluble. Both forms are shown below.

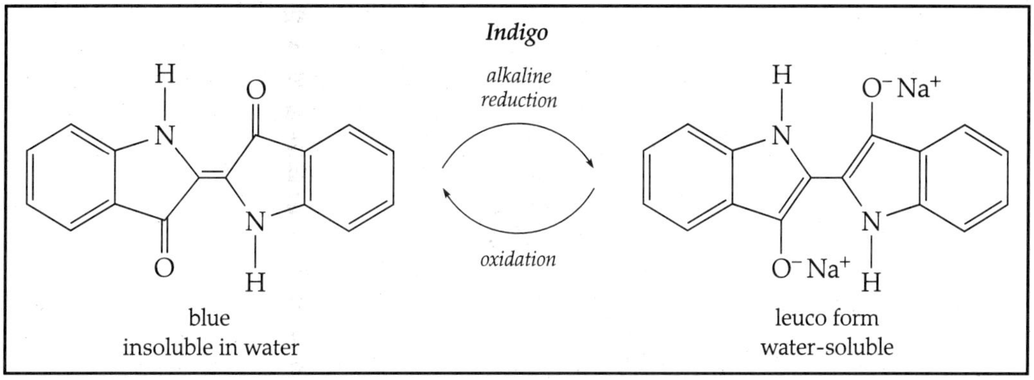

Indigo

alkaline reduction

oxidation

blue
insoluble in water

leuco form
water-soluble

The leuco form of indigo is obtained by reduction with sodium dithionite ($Na_2S_2O_4$) under basic conditions. Dithionite is oxidized to sulfite in the process.

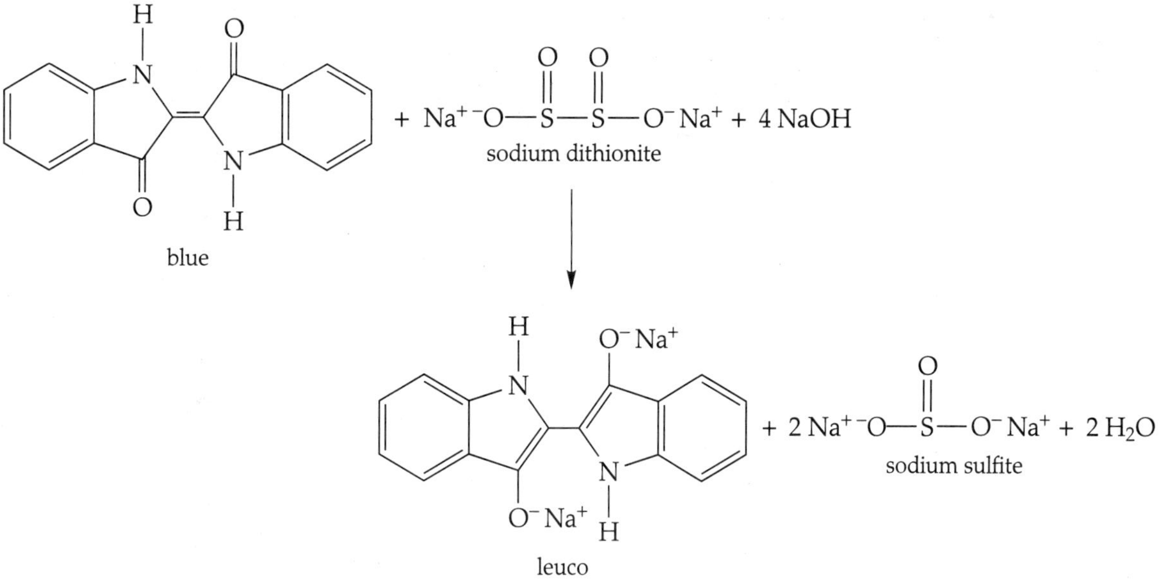

blue

+ sodium dithionite + 4 NaOH

leuco

+ 2 sodium sulfite + 2 H_2O

After applying the leuco form to the fiber by immersion, exposure to air causes oxidation; blue indigo reforms and deposits in the matrix of the fibers giving them color.

In the laboratory, indigo can be obtained by the condensation of *o*-nitrobenzalde-hyde and acetone under basic conditions. The reaction takes place in several steps and involves many intermediates that have not been fully characterized. The first step is an aldol condensation between the aldehyde and acetone, followed by a cyclization and loss of water. Finally, a series of cleavages and condensations takes place and indigo forms. The overall reaction is completed in a matter of minutes.

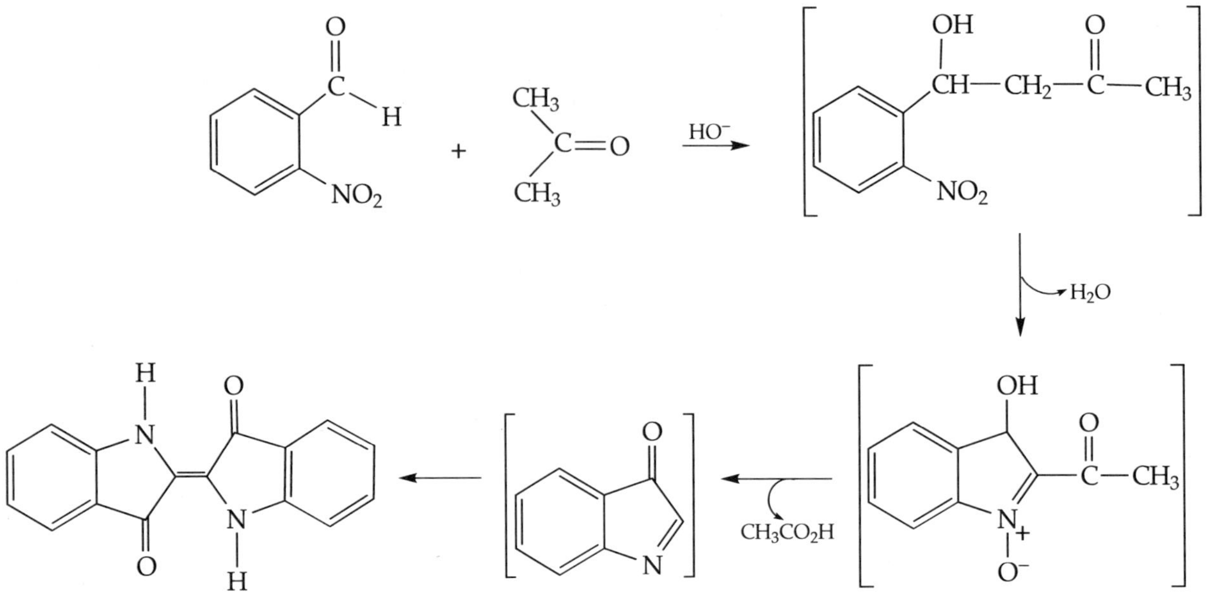

indigo

27.7 OTHER DYES

There are many other types of dyes that do not belong to the structural types mentioned above. Indanthrene Scarlet GG is among them. This is a *vat dye* that can be obtained by the reaction of 1,4,5,8-naphthalenetetracarboxylic acid (I) with *o*-phenylenediamine (II):

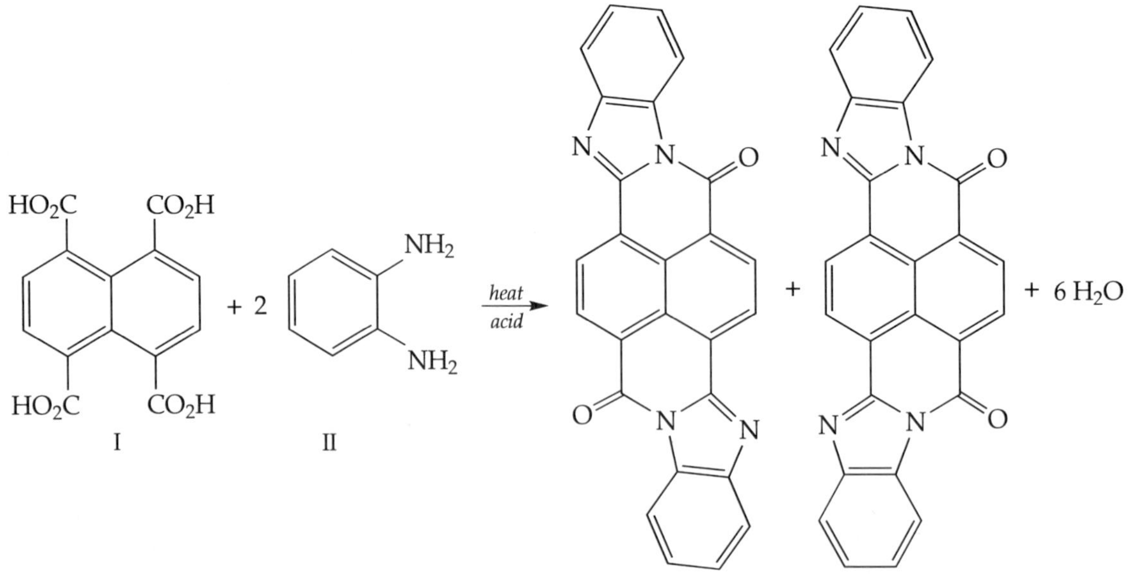

Indanthrene Scarlet GG

The product, Indanthrene Scarlet GG, is a mixture of two isomers (III and IV), each one with a different hue. Isomer III is called Indanthrene Brilliant Orange GR and isomer IV is called Indanthrene Bordeaux RR. Reduction with sodium dithionite yields the soluble form of the dyes (see E27.4), which can be directly applied to cotton. Reoxidation in the presence of air yields the original color.

27.8 MODE OF APPLICATION

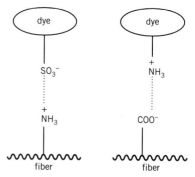

Figure 27.2 Direct dyes: ionic interactions between fibers and dye molecules.

We have already touched on how to apply some dyes, such as vat and ingrain, when discussing specific structural types. Let's now see how dyes are classified, in general, according to their mode of application to fibers.

Direct Dyes

Dyes of this type are soluble in water, at least in a given pH range. They have polar groups such as azo ($-N=N-$), sulfonate ($-SO_3Na$), hydroxyl ($-OH$), and amino ($-NH_2$), which not only increase their solubility in water but also help anchor the dye molecule to the fiber through ionic interactions and H-bonds (Fig. 27.2).

These dyes are used in aqueous solutions to color fibers by simple immersion. The fabric is soaked in the coloring solution and the dye is absorbed by the fibers. Some azo and cationic dyes belong to this category.

Mordant Dyes

Dyes in this category are anchored to the fiber with the aid of metal ions or tannins. Aluminum, tin, iron, and chromium salts are used for this purpose. The dye and the fiber act as ligands, forming a complex with the metal ion in its center. The nature of the metal ion determines the hue obtained. Anthraquinone dyes belong to this category.

Ingrain Dyes

The dye is produced directly on the fiber by a coupling reaction between two colorless compounds, such as a naphthol and a diazonium salt, which are applied to the fabric sequentially. They react with each other to form insoluble particles that become firmly attached to the fiber. Ingrain dyes belong to the chemical class of azo dyes (section 27.2); for this reason they are often called **azoic dyes**.

Vat Dyes

These dyes are insoluble in water but they become soluble when reduced in the presence of a base. The reduced dye, called the *leuco* form, is applied to the fiber by immersion. Exposure to atmospheric oxygen causes reoxidation of the dye, which regains its original color. Indigo is a classical example of a vat dye. Indanthrene dyes also belong to this group. These dyes are mainly used with cotton.

Disperse Dyes

These dyes are insoluble in water but they are "soluble" in the fibers. The fiber is soaked in a suspension of the finely ground dye in water. The dye migrates inside the fiber where it becomes adsorbed. Disperse dyes are very important in dyeing synthetic fibers, especially acetate, polyesters, and nylons.

Reactive Dyes

As their name suggests, dyes in this category are attached to the fiber by a chemical reaction. A covalent bond links the dye molecule to the fiber. These dyes are especially useful in dyeing cotton and other cellulosic fibers. A reactive chlorotriazinyl group links the fiber to the rest of the dye molecule.

$$\text{Dye - NH} \underset{\text{chlorotriazinyl group}}{\fbox{triazine with Cl}} + \text{HO - cellulose} \longrightarrow \text{Dye - NH} \fbox{triazine with O-cellulose} + \text{HCl}$$

Solvent Dyes

These are soluble in organic solvents. Their solutions are used for diverse purposes such as wood staining, ball-pen inks, and food, candle, and soap coloring. They are not normally used to dye fibers. Azo, anthraquinone and cationic dyes may belong to this category.

27.9 DYEING

To understand the process of dyeing we must consider the chemical nature of fibers and fabrics. Fibers can be **natural** or **artificial**. Natural fibers are either **cellulosic**, if they are derived from plants, or **proteins**, if they are of animal origin. **Cotton** and **flax** are examples of cellulosic fibers. The main component of these fibers is cellulose, a polymer of glucose containing an average of 3000 units of glucose per molecule. Cellulose is a highly polar compound because of the presence of free hydroxyl groups. Table 27.1 shows a list of the most important types of natural and synthetic fibers along with the types of dyes used to color them. Also included are the effects of acids and alkalis on fibers.

Silk and wool are examples of animal fibers. **Silk** is made of fibroin, a protein rich in the amino acids glycine, alanine, serine, and tyrosine. Three-dimensionally, fibroin adopts the structure of a β-sheet (Unit 24). The main component of **wool** is α-keratin, a helical protein rich in charged amino acids such as arginine and glutamic acid. It is also rich in cystine, an amino acid with a disulfide bond ($-S-S-$), which is essential in maintaining the helical structure of α-keratin. The presence of arginine and glutamic acid with their ionic side chains makes α-keratin very easy to dye because these ionic groups act as anchors for dye molecules (Fig. 27.3). Contrary to wool keratin, silk fibroin lacks ionic amino acids and is much more difficult to dye.

Among all of the artificial fibers available to us today, nylons and polyesters are the most widespread, followed by acrylics (Unit 26). From a chemical point of view **nylons** are similar to proteins because they are both polyamides; however, nylon molecules are much more homogeneous than proteins. While a protein molecule can have up to twenty-one different types of amino acids as building blocks, a nylon molecule has only one type of building block (Table 27.1).

The most important **polyester** used as a fiber is poly(ethylene terephthalate), PETE (Unit 26), also used in the manufacture of 2-L soda bottles. **Acrylics** are fibers whose building blocks are derived from acrylonitrile. **Rayon** is reconstituted and purified cellulose. **Acetate** is an ester of cellulose in which the free hydroxyl groups have been reacted to form acetate groups (see E26.6). There are three free hydroxyl groups per glucose unit in the cellulose molecule; if all three hydroxyl groups are acetylated, the fiber is called **triacetate**.

Different fibers subjected to the same dyeing process produce different color shades because each type of fiber interacts with the dye molecules in a unique way. Fibers with an abundance of polar groups, such as cotton and wool, are easier to dye than others with only a few polar residues, such as silk, polyesters, acetates, and acrylics. In general, synthetic fibers are less absorbent than natural ones and require special methods for color application. For example, polyesters are generally dyed using disperse dyes at high pressure and temperature.

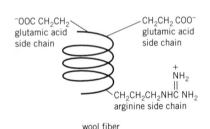

$^-$OOC CH$_2$CH$_2$
glutamic acid
side chain

CH$_2$CH$_2$ COO$^-$
glutamic acid
side chain

$\overset{+}{\underset{\parallel}{\text{NH}_2}}$
CH$_2$CH$_2$CH$_2$NHC NH$_2$
arginine side chain

wool fiber

Figure 27.3 A keratin α-helix, showing the ionic side chains of arginine and glutamic acid.

Table 27.1 **Natural and Synthetic Fibers and Their Mode of Dyeing**

Fiber	Unit	pH effect on fiber	Dyes used
cotton		resistant to alkalis attacked by acids	direct, ingrain, vat, mordant, reactive
flax		resistant to alkalis attacked by acids	ingrain, vat, mordant, reactive
wool		attacked by alkalis moderately resistant to acids	direct, mordant, reactive
silk		attacked by alkalis and acids	direct, mordant, reactive
nylon	$m = 6$: $n = 4$; 7; 8; etc.	resistant to alkalis attacked by acids	direct, disperse
polyester		resistant to weak alkalis, resistant to acids	direct, disperse
acetate		resistant to weak alkalis and acids	direct, ingrain, disperse
rayon		resistant to alkalis resistant to weak acids	direct, ingrain, vat, mordant, reactive
acrylic		resistant to weak alkalis and acids	direct, vat, disperse

A good dye should be resistant to degradation once it is applied to the fiber; this resistance is called **fastness**. The most common causes of chemical degradation are exposure to light, air, and water. Fastness not only depends on the molecular structure of the dye, but also on the mode of application and type of fiber used. It is *the interaction between the fiber and the dye molecule that ultimately determines dye fastness.* In general, the stronger the interaction, the longer the dye will stay put on the fiber. Azo dyes have good fastness to washing, bleach, and light when applied as ingrain dyes to cotton. They have good light fastness but poor fastness to washing when applied as direct dyes to cotton and wool. Triphenylmethane dyes have poor light fastness.

BIBLIOGRAPHY

1. Synthetic Dyes in Biology, Medicine and Chemistry. E. Gurr. Academic Press, London, 1971.

2. Azo and Diazo Chemistry. Aliphatic and Aromatic Compounds. H. Zollinger. Interscience Pub. New York, 1961.

3. Color Chemistry. Synthesis, Properties and Applications of Organic Dyes and Pigments. H. Zollinger. 2nd ed. VCH, Weinheim, 1991.

4. The Analytical Chemistry of Synthetic Dyes. K. Venkataraman, ed. Wiley, New York, 1977.

5. Colour Chemistry. R.L.M. Allen. Appleton-Century-Crofts, New York, 1971.

6. Aromatic Diazo Compounds. K.H. Saunders and R.L.M. Allen. 3rd ed. Edward Arnold, London, 1985.

7. Carcinogenicity and Metabolism of Azo Dyes, Especially Those Derived from Benzidine. M. Boeniger. NIOSH Technical Report. U. S. Department of Health and Human Services, 1980.

8. Basic Color Terms. Their Universality and Evolution. B. Berlin and P. Kay. University of California Press, Berkeley and Los Angeles, 1969.

9. The Plictho. Instructions in the Art of the Dyers Which Teaches the Dyeing of Woolen Cloths, Linens, Cottons, and Silk by the Great Art as Well as by the Common. G. Rosetti. Translation by S.M. Edelstein and H.C. Borghetty. M.I.T. Press, Cambridge, Massachusetts, 1969.

10. A History of Dyed Textiles. S. Robinson. M.I.T. Press, Cambridge, Massachusetts, 1969.

11. The Chemistry of Plant and Animal Dyes. M. Séquin-Frey. *J. Chem. Ed.*, **58**, 301–305 (1981).

12. Indigo. W.C. Fernelius and E.E. Renfrew. *J. Chem. Ed.*, **60**, 633–634 (1983).

13. Azo Dyes. R.V. Stick, M. Mocerino, and D.A. Franz. *J. Chem. Ed.*, **73**, 540–541 (1996).

14. Chemistry and Artists' Pigments. I.S. Butler and R.J. Furbacher. *J. Chem. Ed.*, **62**, 334–336 (1985).

15. Why Objects Appear as They Do. T.B. Brill. *J. Chem. Ed.*, **57**, 259–263 (1980).

16. Chemistry and Artists' Colors. M.V. Orna. *J. Chem. Ed.*, **57**, 264–269 (1980).

17. A Short History of the Chemistry of Painting. H.G. Friedstein. *J. Chem. Ed.*, **58**, 291–295 (1981).

18. A World of Color, Investigating the Chemistry of Vat Dyes. D.N. Epp. *J. Chem. Ed.*, **72**, 726–727 (1995).

19. The Chemistry of Fabric Reactive Dyes. M.C. Bonneau. *J. Chem. Ed.*, **72**, 724–725 (1995).

EXERCISES

Problem 1

In the coupling of 2-naphthol with diazonium salts, the reaction takes place preferentially in position 1 of the naphthol. Why is there no coupling in position 3?

Problem 2

When a sample of azobenzene is irradiated with visible light, an isomer forms. Both isomers have very different physical properties (see table below). When the irradiated isomer is stored in the dark, the original azobenzene reforms. Explain these observations. What are the implications of these results in dye chemistry?

	m.p. (°C)	dipole moment, μ (D)	$n \longrightarrow \pi$ λ_{max} (ε)	$\pi \longrightarrow \pi$ λ_{max} (ε)
original azobenzene	68	0	450 nm (463 L mol^{-1} cm^{-1})	330 nm (17,000 L mol^{-1} cm^{-1})
isomer	71	3	430 nm (1,500 L mol^{-1} cm^{-1})	280 nm (5,100 L mol^{-1} cm^{-1})

Problem 3

Aniline was treated with sodium nitrite and one equivalent of hydrochloric acid. Instead of a clear solution of the diazonium salt, a yellow precipitate was obtained. Explain this observation. What went wrong?

Problem 4

How would you make the following dyes using diazotization and coupling reactions?

Cl Pigment Red 49

Naphthalene Black 12B

Problem 5

Compound X ($C_{19}H_{16}O$) shows two singlets in its ^{1}H-NMR spectrum at 2.83 and 7.27 ppm with relative intensities 1 : 15, respectively. X dissolves in ethanol, ethyl ether, and glacial acetic acid and gives colorless solutions. When X is dissolved in concentrated sulfuric acid it gives a bright yellow solution. Give a structure for X and explain its color behavior.

Problem 6

Read "Color Additives" in the box and speculate what structural element common to all certified color additives, except Citrus Red No. 2, makes them relatively safe for human consumption.

Color Additives We eat with our eyes as well as our palates. Making food appealing to the eye is an art as old as our civilization. The Romans used *defrutum*, concentrated must of red grapes, for color and flavor enhancement. They even made a natural version of sherbet or shave ice by pouring red *defrutum* over a cup of white snow. Color additives are compounds that we add to foods, drugs, or cosmetics to impart color. In the United States color additives are regulated by the Federal Food, Drug, and Cosmetic (FD&C) Act and controlled by the FDA. There are nine certified color additives and about twenty exempt from certification approved for use. Certifiable color additives are man-made coloring products while those exempt from certification are additives of natural origin derived from vegetables, animals, and minerals. Both types of additives must meet rigorous safety standards before they are approved for use. Some are approved for limited use only; for example, Citrus Red No. 2 is restricted for coloring the skin of oranges not intended for further processing, and Orange B for imparting color to the surfaces of sausages. The list in the table below shows the nine certifiable color additives and their most important applications. Certifiable color additives used in food must be listed by their names on the label. Among the color additives exempt from certification are beet powder, β-carotene, carrot oil, caramel color, paprika, and titanium dioxide.

Bibliography

1. Food Color Facts. FDA/IFIC Brochure, Jan. 1993; http://vm.cfsan.fda.gov.

2. Basic Food Chemistry. F.A. Lee. 2nd ed. Avi, Westport, 1983.

Certified Color Additives in the United States

Name	Hue	Common food uses	Structure
FD&C Blue No. 1 Brilliant Blue FCF	bright blue	fruits, juices, baked goods, nonalcoholic beverages, candies	
FD&C Blue No. 2 Indigotine	royal blue	ice cream, candies, chewing gum	
FD&C Green No. 3 Fast Green FCF	sea green	fruits, juices, baked goods, nonalcoholic beverages	
FD&C Red No. 40 Allura Red AC	orange red	fruits, juices, nonalcoholic beverages, baked goods, candies	
FD&C Red No. 3 Erythrosine	cherry red	meat products, baked goods, fruits and juices, candies	

(continued)

Certified Color Additives in the United States (continued)

Name	Hue	Common food uses	Structure
FD&C Yellow No. 5 Tartrazine	lemon yellow	fruits, juices, candies, baked goods	
FD&C Yellow No. 6 Sunset Yellow	orange	nonalcoholic beverages, baked goods, frostings, gelatins.	
Orange B	orange	meat products, sausages	
Citrus Red No. 2	orange	skin of oranges	

EXPERIMENT 27

Colored Chemistry

In this experiment you will prepare several synthetic dyes and use them to dye fabrics. To fully appreciate how the same dye produces different colors on different fibers you will use a multifiber fabric. A typical multifiber fabric has cotton, wool, acrylic, nylon, polyester, and acetate swatches sewn together as indicated in Figure 27.4.[1] If a multifiber fabric is not available, use a piece of white cloth (3 × 10 cm) or a cotton ball.

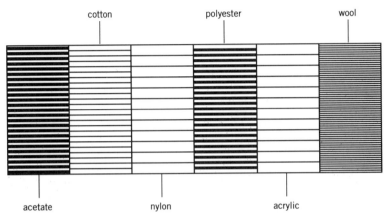

Figure 27.4   Multifiber fabric 10A from Testfabrics Inc.

Section E27.1 focuses on direct dyes. You will prepare Orange II, Magneson II, or Solochrome Orange M by coupling diazonium salts with naphthols or phenols. Choose one of them, synthesize it, and use it to dye a piece of fabric. Solochrome Orange M can also be used as a mordant dye (section E27.2). At the end of this section you will use two other direct dyes, Malachite Green and Eosin Y, to dye fabrics. These dyes are already prepared.

In section E27.2 you will work with mordant dyes. You will use commercial alizarin and the dyes Solochrome Orange M and Magneson II made in the previous section. You will use a series of metal ions such as Al^{3+}, Cu^{2+}, and Fe^{2+} as mordants.

In section E27.3 you will make the ingrain dyes: American Flag Red and Easter Purple. In section E27.4 you will prepare two vat dyes: indigo and Indanthrene Scarlet GG. You will reduce them with sodium dithionite to obtain their leuco forms, which you will apply to fabrics. Air oxidation produces the final colors. Finally, in section E27.5 you will try the fastness of the dyes to washing and bleach.

This experiment has many parts and you may not do all of them. Follow your instructor's directions concerning what parts to do. Students should work on different dyes so all dyes are prepared by the entire class and a final comparison among them can be made. Most dyes have self-describing names, others such as Magneson II, Solochrome Orange M, Eosin, and alizarin deserve some comment. Magneson II gives beautiful earth tones when applied directly to most fabrics. Solochrome Orange M produces ocher and beige colors. Alizarin gives a purple color when used directly, and reddish colors with Al^{3+} and Sn^{2+} as mordants. Eosin gives orange-red colors.

---

[1] They are available from Kontes (800-223-7150) and Testfabrics Inc. (570-603-0432).

Record any change in color during the preparation of dyes and during dyeing. Be very careful when handling the solid dyes and their solutions. Wear gloves, otherwise they may give you unsightly stains. Avoid skin contact with these chemicals. Handle the fabrics only with tweezers. Insert the paper clip in the acetate swatch before dyeing. This will help you in identifying the swatches after dyeing and also in handling them. The toxicity of many of these dyes has not been investigated. The references at the end of the experiments indicate the sources from which the experimental procedures have been adapted.

PROCEDURE

E27.1 DIRECT DYES

Note: If in the synthesis of the dyes you obtain a pasty solution that is difficult to pour, you can ease the transferring by adding a little water.

Orange II

Background reading: 27.2; 27.3; 27.8; 27.9
Estimated time: 45 min.

Safety First

- 4-Nitroaniline is a highly toxic compound. Handle it with care.
- Sodium hydroxide is caustic. Avoid skin contact.
- Hydrochloric acid is highly toxic and corrosive. Handle it with care.
- Sodium nitrite is a toxic oxidizer.
- 2-Naphthol, 1-naphthol, salicylic acid, sulfanilic acid, and alizarin are irritants. 1-Naphthol and Malachite Green are toxic.
- Diazonium salts are explosive in the solid state. Wash your glassware immediately after use.
- Wear gloves when handling the dyes.

Synthesis

Diazonium Salt Preparation   In a 13 × 100 mm test tube place 0.49 g of sulfanilic acid, 0.13 g of sodium carbonate, and 5 mL of water. Warm it in a water bath to obtain a clear solution. Remove it from the water bath and add 0.2 g of sodium nitrite dissolved in 0.5 mL of water. With the help of a Pasteur pipet add this solution dropwise to 0.53 mL of concentrated hydrochloric acid and 3 g of ice contained in an 18 × 150 mm test tube. Place the tube in an ice-water bath. The diazonium salt of sulfanilic acid precipitates. The suspension is used in the next step.

Coupling   In a 25 mL Erlenmeyer flask place 0.38 g of 2-naphthol and add 2 mL of a 2.5 M aqueous solution of sodium hydroxide. Place the system in an ice-water bath. With the help of a Pasteur pipet with the tip broken off (see note) add little by little the suspension of diazobenzene sulfonate prepared in the previous step. Stir with a glass rod after each addition. Observe and record any change in color. Let the system stand for about 10 minutes on ice with occasional stirring. In the meantime, weigh 1 g of sodium chloride. Heat the suspension on a hot plate until the solid dissolves. Add the sodium chloride and continue heating to dissolve the salt. Cool down the solution at room temperature first, and then in an ice-water bath. Filter off the solid by vacuum filtration using a Hirsch funnel. Wash the solid on the filter with 2 mL of a saturated sodium chloride aqueous solution and dry. Weigh the product (it contains about 20% sodium chloride; Ref. 1).

Note. To break the tip of a Pasteur pipet, gently etch the glass with a file and wrap the pipet in several layers of paper towel to protect your hands.

Dyeing   Dissolve 50 mg of the solid dye in 20 mL of water. Immerse a strip of fabric (3 × 10 cm) or a cotton ball and boil for about 5 minutes. Using tweezers remove it from the heat. Rinse the fabric well with tap water and pat dry with paper towels.

Magneson II

> **Background reading:** 27.2; 27.3; 27.8; 27.9
> **Estimated time:** 45 min.
> **Read Safety First** on p. 627

4- nitroaniline   Magneson II

Synthesis

Diazonium Salt Preparation   In a 13 × 100 mm test tube, mix 0.70 g of 4-nitroaniline, 0.38 g of sodium nitrite, and 1.5 mL of water to make a homogeneous suspension. With the help of a Pasteur pipet with the tip broken off (see note above) add the slurry to a mixture of concentrated hydrochloric acid (1.5 mL) and ice-water (1.5 mL) contained in a 13 × 100 mm test tube and placed in an ice-water bath. Stir with a glass rod during the addition and occasionally thereafter for about 10 minutes. Remove any solid particles by gravity filtration using a small glass funnel (2.5 cm in diameter) and a small cotton plug. Collect the filtrate in another test tube.

Coupling   Dissolve 0.74 g of 1-naphthol in 10 mL of a 2.5 M sodium hydroxide aqueous solution in a 25-mL Erlenmeyer flask and place it in an ice-water bath. Slowly and with continuous stirring add the diazonium salt solution from the previous step. Observe and record any color change. Let it stand on ice for about 10 minutes with occasional stirring. Slowly add concentrated hydrochloric acid to obtain pH 3–4 (it takes about 1.5 mL). Add 1 g of sodium chloride and heat to a boil on a hot plate. Remove from the heat. Let the system cool down at room temperature first and then in an ice-water bath. Vacuum-filter the solid using a small Buchner funnel. Wash the solid on the filter with 2–5 mL of water and let it dry. Weigh the product (Refs. 1 and 2).

Dyeing   In a 100-mL beaker place 50 mg of the solid dye; add 20 mL of water and 3 mL of a 2.5 M sodium hydroxide aqueous solution. Heat on a hot plate while stirring with a glass rod. Add a strip (3 × 10 cm) of fabric or a cotton ball and boil for about 3 minutes. Remove from the heat using tweezers and wash the fabric with tap water. Pat dry between paper towels. Let the solution cool down at room temperature and immerse another strip of fabric for about 5 minutes. Wash and dry as before. Different fabrics take on very different colors that are also affected by the temperature of the dyeing bath.

Solochrome Orange M

> **Background reading:** 27.2; 27.3; 27.8; 27.9
> **Estimated time:** 45 min.
> **Read Safety First** on p. 627

4-nitroaniline → Solochrome Orange M

Synthesis

Diazonium Salt Preparation Prepare the diazonium salt of 4-nitroaniline as indicated in the synthesis of Magneson II (first step).

Coupling In a 25-mL Erlenmeyer flask dissolve 0.68 g of salicylic acid in 10 mL of 2.5 M sodium hydroxide aqueous solution and place it in an ice-water bath. Add the diazonium salt solution little by little while stirring with a glass rod. Observe and record any color change. Let the reaction mixture sit on ice for another 10 minutes with occasional stirring. Add 1 g of sodium chloride and heat on a hot plate until the suspension boils. Cool it down at room temperature and then on ice. Filter off the solid by vacuum filtration using a Buchner funnel. Wash with a little water and dry. Weigh the product.

Dyeing Use the same technique described for dyeing with Magneson II but add 1.5 mL instead of 3 mL of 2.5 M sodium hydroxide aqueous solution.

Malachite Green and Eosin Y

> **Background reading:** 27.4; 27.8; 27.9
> **Estimated time:** 15 min.
> **Read Safety First** on p. 627

Cleaning Up

- Rinse the Pasteur pipets used with diazonium salts with the liquid from the dyeing baths and dispose of them in the "Dyes Solid Waste Container."
- Wash the glassware used with diazonium salts immediately.
- Dispose of the mother liquors and the liquids from the dyeing baths in the "Dyes Liquid Waste" container.
- Dispose of filter papers in the "Dyes Solid Waste" container.
- Dispose of paper towels in the "Dyes Solid Waste" container.
- Turn in your dyes to your instructor in labeled vials.

Malachite Green   Eosin Y

Dyeing In a 100-mL beaker dissolve 20 mg of Malachite Green or Eosin Y with 20 mL of water. Add a strip of fabric (3 × 10 cm) or a cotton ball and bring the system to a boil on a hot plate. Boil for about 3 minutes. Using tweezers remove from the heat and wash the fabric with tap water. Pat dry between paper towels.

E27.2 MORDANT DYEING

Background reading: 27.5; 27.8; 27.9
Estimated time: 45 min.
Read Safety First on p. 627

Mordant Treatment

In 100-mL beakers, boil strips of fabric (or cotton balls) for about 5 minutes in 20 mL of 0.1 M aqueous solutions of the following salts: copper (II) sulfate, aluminum potassium sulfate, and iron (II) sulfate. Remove the strips from the heat. Wash each strip with 20 mL of water, collecting the wash in a beaker (dispose of the washes as hazardous waste because they contain heavy metals). Pat dry between paper towels and then dry with a heat pistol or hair dryer. Use the fabrics to dye with alizarin, Magneson II, and Solochrome Orange M as described below.

Dyeing with Alizarin In a 100-mL beaker place 20 mg of alizarin, and 20 mL of a 0.5% sodium bicarbonate aqueous solution. Heat on a hot plate to dissolve the dye and bring to a boil. Add a strip of fabric (3 × 10 cm) or cotton ball and boil for about 5 minutes. Remove from the heat with tweezers. Wash with tap water and pat dry with paper towels. Try with strips treated with iron (II) sulfate, and aluminum potassium sulfate, and an untreated fabric for comparison.

Cleaning Up

- Dispose of the liquids from the mordant and dyeing baths and the mordant washes in the "Dyes Liquid Waste" container.
- Dispose of paper towels in the container labeled "Dyes Solid Waste."

Safety First

- 4-Nitroaniline is a highly toxic compound. Handle it with care.
- Sodium hydroxide is caustic. Avoid skin contact.
- Hydrochloric acid is highly toxic and corrosive. Handle it with care.
- Sodium nitrite is a toxic oxidizer.
- 2-Naphthol and 8-anilino-1-naphthalene sulfonic acid are irritants.
- Diazonium salts are explosive in the solid state. Wash your glassware immediately after use.
- Wear gloves when handling the dyes.

Alizarin

Dyeing with Magneson II and Solochrome Orange M Follow the same procedure indicated for direct dyeing with these dyes in section E27.1, but use fabric strips pretreated with mordants (copper (II) sulfate, aluminum potassium sulfate, and iron (II) sulfate). For comparison, also use a strip without mordant treatment.

E27.3 INGRAIN DYES

Background reading: 27.2; 27.3; 27.8; 27.9
Estimated time: 45 min.

Diazonium Salt Preparation

4-nitroaniline

In a 13 × 100 mm test tube mix 0.35 g of 4-nitroaniline, 0.19 g of sodium nitrite, and 0.75 mL of water to make a homogeneous suspension. Using a Pasteur pipet with the tip broken off (see note in E27.1, p. 627) add the slurry in small portions to a mixture of concentrated hydrochloric acid (0.75 mL) and ice-water (0.75 mL) contained in a 13 × 100 mm test tube placed in an ice-water bath. Stir with a glass rod during the addition and occasionally for about 10 minutes. Remove any solid particles by gravity filtration using a small glass funnel (2.5 cm in diameter) and a small cotton plug. Collect the filtrate in a 100-mL beaker and dilute the diazonium salt solution with 30 mL of water. Use the dilute solution to produce American Flag Red and Easter Purple as indicated below (Ref. 2).

American Flag Red   In a 250-mL beaker dissolve 0.2 g of 2-naphthol in 35 mL of water and add 5 mL of a 2.5 M sodium hydroxide aqueous solution. Warm on a hot plate until the solid dissolves. Immerse a strip of fabric or a cotton ball for about 1–2 minutes, stirring occasionally. Remove the fabric using tweezers and pat dry on paper towels. Immerse the fabric pretreated with 2-naphthol in the dilute diazonium salt solution of 4-nitroaniline prepared in the previous step. Let it stand at room temperature for about 5 minutes. Remove the strip from the bath and wash well. Dry on paper towels.

2-naphthol

diazonium salt of 4-nitroaniline

American Flag Red

Easter Purple   In a 100-mL beaker, dissolve 100 mg of 8-anilino-1-naphthalenesulfonic acid ammonium salt (ANS) in 35 mL of water. Immerse a strip of fabric for about 1–2 minutes with occasional stirring. Remove the fabric using tweezers and pat dry on paper towels. Immerse the fabric in the dilute solution of 4-nitroaniline diazonium salt previously prepared and let it stand at room temperature for a few seconds. With the aid of tweezers, remove the strip from the bath and wash well. Dry on paper towels.

ANS

diazonium salt of 4-nitroaniline

Easter Purple

Slowly transfer the remaining diazonium salt solution to the 250-mL beaker with the 2-naphthol solution from **American Flag Red**. Slowly add the remaining solution of ANS. See "Cleaning Up."

Cleaning Up

- Rinse the Pasteur pipets used with diazonium salts with the mixed solutions and dispose of them in the "Dyes Solid Waste" container.
- Dispose of the mixed solutions in the "Dyes Liquid Waste" container.
- Dispose of paper towels in the container "Dyes Solid Waste."

Safety First

- *o*-Phenylenediamine is toxic and a cancer suspect agent. Handle it with care.
- Phosphoric acid is corrosive.
- Sodium hydroxide is caustic. Avoid skin contact.
- *o*-Nitrobenzaldehyde is an irritant.
- Sodium dithionite is a flammable solid.
- Bleach is an oxidizer.
- Wear gloves when handling the dyes.

E27.4 VAT DYES

Indanthrene Scarlet GG

Background reading: 27.7; 27.8; 27.9
Estimated time: 45 min.

Synthesis Place 50 mg of 1,4,5,8-naphthalenetetracarboxylic acid and 50 mg of *o*-phenylenediamine in a 13 × 100 mm test tube. Stir with a microspatula to homogenize the mixture. Add 0.2 mL of concentrated phosphoric acid (85% w/w) and heat in a sand bath at 190°C for 15 minutes. Stir the system with a glass rod. Let the system cool down and add 7 mL of water. Stir to make a homogeneous suspension. Centrifuge at 2000–3000 rpm for about 3 minutes (balance the centrifuge!). Remove the supernatant with a Pasteur pipet and place it in a 100-mL beaker. Repeat the wash with another 7 mL portion of water and place the supernatant in the same beaker. Keep the product for the next step. Before disposing of the liquid in the beaker, slowly add with a Pasteur pipet a 10% NaOH aqueous solution until the pH is basic (Ref. 3).

Dyeing Perform this operation in the fume hood. Transfer the solid to a 100-mL beaker with the help of 6 mL of a 2.5 M sodium hydroxide aqueous solution. Add 10 mL of a *freshly prepared* 10% (w/v) aqueous solution of sodium dithionite and heat to boil on a hot plate with constant stirring. Add more sodium dithionite solution, if necessary, to completely dissolve the solid. Add 15 mL of water, bring to a boil, and add a strip of fabric or a cotton ball. Boil for 3 minutes; using tweezers remove the beaker from the heat, rinse the fabric with tap water, and let it dry.

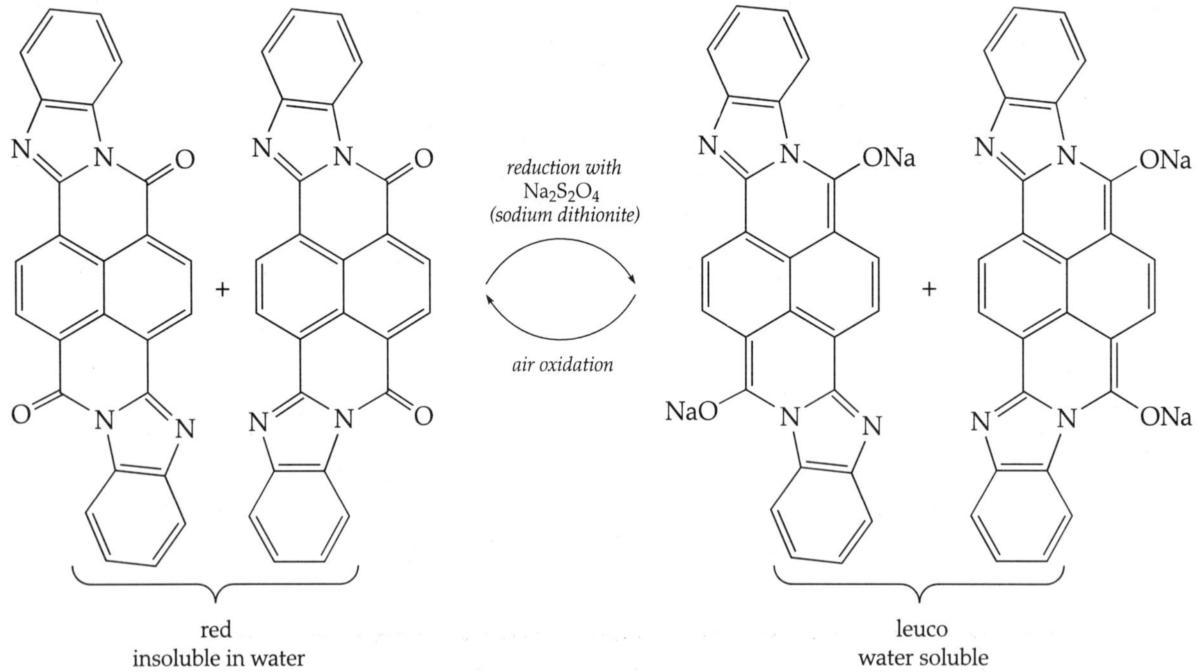
reduction with
$Na_2S_2O_4$
(sodium dithionite)
air oxidation
red
insoluble in water
leuco
water soluble

Indanthrene Scarlet GG

Indigo

Background reading: 27.6; 27.8; 27.9
Estimated time: 45 min.

Synthesis In a 13 × 100 mm test tube place 100 mg of *o*-nitrobenzaldehyde, 1 mL of acetone, and 1 mL of water. Stir the suspension with a glass rod and add 1 mL of a 2.5 M solution of sodium hydroxide dropwise. Blue indigo starts to form immediately. Let the reaction mixture stand for about 5–10 minutes (the reaction is exothermic) and then cool it in an ice-water bath. Collect the solid by vacuum filtration using a Hirsch funnel. Wash the solid on the filter with 2 mL of water followed by 2 mL of 95% ethanol. Weigh the product (Ref. 4).

Dyeing Perform this operation in the fume hood. Place 50 mg of indigo in a 100-mL beaker and add 5 mL of a 2.5 M sodium hydroxide solution. Boil on a hot plate while stirring with a glass rod. Add 7 mL of a freshly made solution of sodium dithionite (10% w/v) in water. Boil and observe any color change. If the blue color persists add more sodium dithionite until the solid dissolves completely and the solution turns clear yellow. Add 25 mL of water and bring the solution to a boil. Add a strip of fabric (3 × 10 cm) or a cotton ball and boil for 3 minutes. Using tweezers remove from the heat. Rinse well with tap water and let it dry.

> **Cleaning Up**
>
> - Dispose of the supernatants and liquids from the dyeing baths in the "Dyes Liquid Waste" container.
> - Dispose of filter papers in the "Dyes Solid Waste" container.
> - Dispose of paper towels in the trash.
> - Turn in your products to your instructor in labeled vials.

reduction with $Na_2S_2O_4$ *(sodium dithionite)*

air oxidation

blue indigo
insoluble in water

leuco form
water-soluble

E27.5   DYE FASTNESS

> **Background reading:** 27.8; 27.9
> **Estimated time:** 15 min.
> **Read Safety First** on p. 632

Cut each strip of dyed fabric into three narrower strips (if you used a cotton ball divide it into three equal parts). Keep one part as a control and immerse another in a water bath with detergent, and warm it at about 50–60°C for about 10 minutes with occasional stirring. Immerse the other part in a bleach solution (5 mL of bleach in 100 mL of water) and keep it at room temperature for 5 minutes. Remove the strips from their respective baths. Rinse with water and dry. Compare with the control.

> **Cleaning Up**
>
> - Dispose of the liquid from the detergent baths in the container labeled "Dyes Liquid Waste."
> - Dispose of the bleach solution in the container labeled "Bleach Dyes Waste."
> - Dispose of paper towels in the "Dyes Solid Waste" container.

BIBLIOGRAPHY

1. Vogel's Textbook of Practical Organic Chemistry. A.I. Vogel, B. S. Furniss, A.J. Hannaford, P.W.G. Smith, and A.R. Tatchell. 5th ed. pp. 920–952. Longman, Harlow, UK, 1989.

2. Aromatic Diazo Compounds. K.H. Saunders and R.L.M. Allen. 3rd ed. Edward Arnold, London, 1985.

3. The Preparation of Indanthrene Scarlet GG. W.B. Lutz, *J. Chem. Ed.*, **67**, 71 (1990).

4. A Microscale Synthesis of Indigo: Vat Dyeing. J.R. McKee and M. Zanger, *J. Chem. Ed.*. **68**, A242–A244 (1991).

EXPERIMENT 27   REPORT

Pre-lab

1. Make a table with the physical properties (including toxicity/hazards) of the reagents used. Look up the dyes in the *Aldrich Catalog* and in the *Merck Index* (not all of them will be found) and make a table with their physical properties. Synonyms: Orange II: 4-(2-hydroxy-1-naphthylazo)benzenesulfonic acid, sodium salt. Magneson II: 4-(4-nitrophenylazo)-1-naphthol. Solochrome Orange M: Mordant Orange 1, Alizarin Yellow R, and 5-(4-nitrophenylazo)salicylic acid. American Flag Red: Para Red, and 1-(4-nitrophenylazo)-2-naphthol.

2. What precautions should be taken when handling diazonium salts?

3. Consider the dye Orange II. Look up the structures of different fibers in Table 27.1, and predict which ones would color strongly and which ones weakly.

4. Write the resonance forms for Malachite Green.

5. Account for the difference in color between blue indigo and the leuco form.

6. How would you prepare FD&C Red No. 40 (one of the nine certifiable color additives approved by the FDA for foods, drugs, and cosmetics) by diazotization and coupling reactions? Answer with chemical reactions.

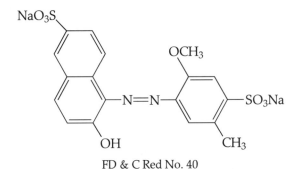

FD & C Red No. 40

7. Explain in a few words the meaning of: mordant, vat, ingrain, and direct dyes. Give one example of each type.

8. What are the advantages of ingrain dyeing over direct dyeing with azo dyes?

In-lab

1. Formulate the chemical reactions involved in the syntheses.

2. Give the mechanism for the coupling reactions of the azo dyes.

3. What is the recommended molar excess of HCl in the diazotization reactions? Why is this excess necessary?

4. Calculate the percentage yields of the syntheses. Discuss any source of error. Were the dyes perfectly dry before weighing?

5. Report any color change during the syntheses and dyeing.

6. In the coupling reactions of the diazonium salts with the naphthols, a basic medium is used. Why?

7. What role do Al^{3+}, Cu^{2+}, and Fe^{2+} play in mordant dyeing?

8. Attach your strips or cotton balls to your lab report (let them dry well first). If you used a multifiber fabric explain any trend in the dyeing capacity of different fibers. In general, which fiber is easier to dye: wool or polyester, nylon or acrylic?

9. Report and discuss your fastness results.

UNIT 28

Bioorganic Chemistry

28.1 ENZYMES: BIOLOGICAL CATALYSTS

Forget for a moment the hot plate, the boiling chips, and the reflux apparatus. The same reactions that we often perform at high temperatures and in organic solvents can take place at room temperature and in water. How is this possible? The answer is called **enzymes**.

Enzymes are a special type of proteins that accelerate the rates of chemical transformations. They lower the reaction energy barrier by stabilizing the transition state (Fig. 28.1). Enzymes are recovered intact at the end of the reaction and do not change the equilibrium constant; they are reaction **catalysts**.

Enzymes are widespread in nature. They are the true workers of life; most of the chemical reactions that take place inside the cells are catalyzed by enzymes. The word enzyme (from Greek: *en*, in; *zyme*, yeast) was coined by Willy Kühne in 1878 to emphasize that *something* present *in yeast* was responsible for yeast's fermentation power.

Despite some early efforts, the use of enzymes and microorganisms in organic synthesis did not gain popularity until recently. Today it is recognized that many organic reactions, which are difficult to perform by conventional methods, can be successfully carried out with the help of enzymes and microbes. Active research in this area has created the fast-growing field of **bioorganic chemistry**.

What are the reasons for using enzymes in organic synthesis? Perhaps the most important one is their **specificity**. Enzymes act on specific molecules, called **substrates**, which they recognize from a large pool of compounds. If the substrate has two similar functional groups (such as two hydroxyl groups in different locations), enzymes usually catalyze the transformation of only one of the them, or in other words, they are **regiospecific** (Unit 21). If the reaction yields a mixture of stereoisomers, enzymes generally catalyze the formation of one of them and not the other, and in this sense they are **stereospecific**. One of the major challenges in organic synthesis is obtaining specific stereoisomers in a pure form. This is particularly important in the manufacture of pharmaceuticals, where often only one stereoisomer has biological activity. To this end, the manufacturer must often do lengthy separations that not only decrease the overall yield but also increase the cost. In this respect the use of enzymes and microorganisms has proved invaluable because they generally give pure stereoisomers.

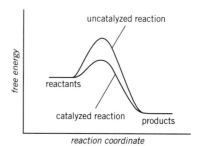

Figure 28.1 Energy diagrams for an uncatalyzed and a catalyzed reaction. The transition state for the catalyzed reaction is lower in energy, and thus the reaction is faster.

28.2 HOW ENZYMES WORK

Enzymes catalyze reactions by binding to their substrates. Within an enzyme molecule there is a certain region or pocket, called the **active site**, that binds specifically to its substrates, forming an **enzyme-substrate complex**. This complex then decomposes to give products and the intact enzyme, which is free to bind to new substrate molecules and initiate another round of catalysis.

The binding between the enzyme and its substrate takes place through H-bonds, and electrostatic and hydrophobic interactions. Enzymes are made of chiral L-amino acids (Unit 24) and, therefore, their active sites lack symmetry and are also chiral. Active sites are very specific in recognizing their substrates. Only substrates complimentary in shape to the active site bind. This is referred to as the **lock-and-key model** (Fig. 28.2) because only substrates with the right geometry (the key) fit into the active site (the lock). This model was realized by the brilliant chemist Emil Fischer as early as 1894 and its most important tenets are still valid. Experimental evidence accumulated in the last forty years indicates that in the process of binding, the enzyme as well as the substrate undergo some conformational changes that make the binding tighter.

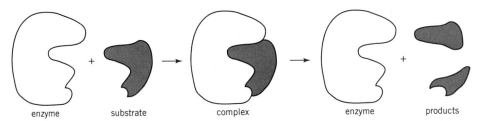

Figure 28.2 The lock-and-key model for enzyme catalysis.

There are more than 3000 known enzymes and many others waiting to be discovered. They catalyze all sorts of reactions from simple hydrolysis to the transferring of functional groups between substrates. Enzyme nomenclature is a knotty subject and we will not discuss it here in detail. For our purposes, it will suffice to know that the enzyme's name generally refers to the reaction it catalyzes or to the substrate it binds, and ends with the suffix *ase*. For example, the **hydrolases** are a large group of enzymes that catalyze hydrolysis reactions. Some hydrolases are called **proteases** because they catalyze the hydrolysis of proteins and polypeptides.

$$\text{〜NH—CH—C—NH—CH—C〜 + H_2O} \xrightarrow{\text{a protease}} \text{H_3N^+—CH—C—O^- + H_3N^+—CH—C—O^-}$$

a polypeptide   amino acids

Some enzymes require the presence of small organic molecules, called **cofactors**, to be fully active. Cofactors can be *transiently* or *permanently* bound to the enzyme. When they are transiently bound they receive the name of **coenzyme**. Vitamins of the group B are all essential coenzymes needed for vital processes. Most of the enzymes involved in oxidation-reduction processes, called **oxidoreductases**, require coenzymes to operate. Two of the most common coenzymes are **nicotinamide adenine dinucleotide** (NAD^+) and **nicotinamide adenine phosphate dinucleotide** ($NADP^+$), shown in Figure 28.3.

These coenzymes participate *actively* in redox reactions where they are chemically transformed. In this regard, they can be considered a second substrate. For example, the oxidation of ethanol to acetaldehyde that takes place in the liver is catalyzed by **alcohol dehydrogenase**, an enzyme that needs NAD^+ as a coenzyme. In the process of oxidizing ethanol, NAD^+ is reduced to NADH. The reduction reaction takes place in the nicotinamide ring as shown below.

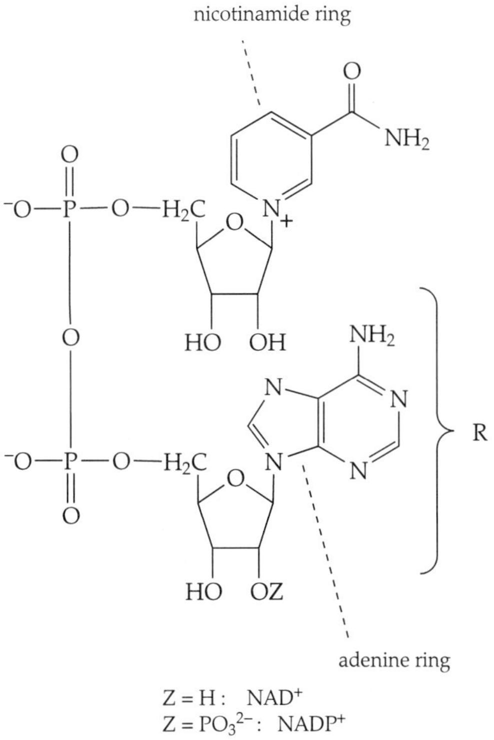

Figure 28.3   Structures of NAD$^+$ and NADP$^+$.

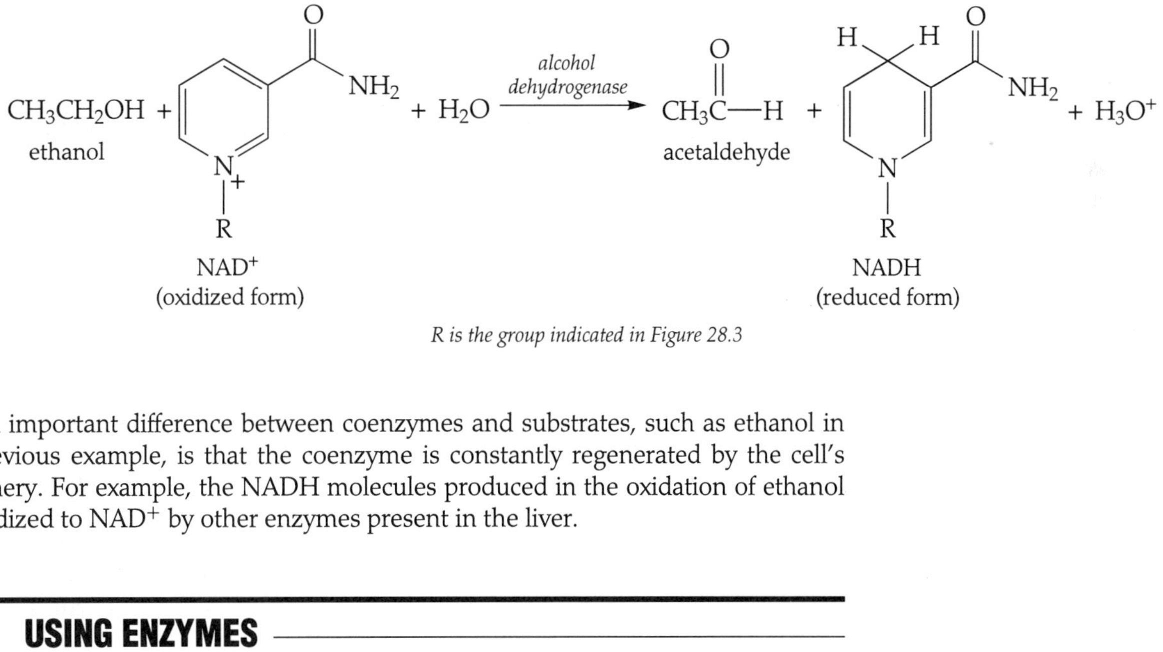

An important difference between coenzymes and substrates, such as ethanol in the previous example, is that the coenzyme is constantly regenerated by the cell's machinery. For example, the NADH molecules produced in the oxidation of ethanol are oxidized to NAD$^+$ by other enzymes present in the liver.

28.3   USING ENZYMES

There are two important methodologies for the incorporation of enzymes in organic reactions: *as pure chemicals* or *as a part of whole organisms*. The use of pure enzymes is restricted to those enzymes that can be isolated and purified with relatively high yields. As new technologies facilitate the identification, isolation, and purification of enzymes, the use of pure enzymes is gaining new ground in research as well as in industry.

The other approach, the use of whole cells, offers an inexpensive alternative because no isolation of enzymes is necessary. Among many types of bacteria and fungi used to produce specific transformations, **baker's yeast** (a class of fungi) is most widely applied. Several factors account for this. *Baker's yeast is cheap, readily available, grows easily, and can be used to perform an extensive variety of reactions.* Reductions are among the most important reactions performed with baker's yeast; ketones, β-ketoesters, double bonds, and nitro groups can all be reduced using baker's yeast. (For an account of the timeless use of yeast see the box "Yeast: The Midwife of Beer and Bread.")

Besides its lower cost, the use of whole microorganisms offers an additional advantage over isolated enzymes. If the enzyme needs coenzymes such as NAD^+ or $NADP^+$, these compounds do not need to be added to the reaction mixture because they are provided by the cells, which constantly regenerate them.

A potential drawback in the use of microorganisms is that there are hundreds of enzymes in the intact cell and therefore a single substrate could, in principle, undergo many different reactions, yielding more than one product. Fortunately, through a careful control of the experimental conditions we can often perform clean transformations. The substrate's structure and the conditions chosen to carry out the experiment ultimately determine the course of the reaction and the type of product obtained. Temperature, pH, presence of nutrients, and substrate concentration are among the most important experimental factors that can change the course of a reaction. For example, the reduction of 1-phenyl-1,2-propanedione with baker's yeast at pH 5 in the presence of sucrose as yeast's nutrient yields mainly (S)-(−)-2-hydroxy-1-phenyl-1-propanone, while the reduction of the same substrate without pH control gives (1R,2S)-1-phenyl-1,2-propanediol. We will come back to this in Experiment 28.

(S)-2-hydroxy-1-phenyl-1-propanone

1-phenyl-1,2-propanedione

pH 5

no pH control

(1R,2S)-1-phenyl-1,2-propanediol

The use of enzymes and microorganisms has important limitations. First, there are reactions that we perform in test tubes that are not known to occur in nature, and therefore no enzymes are available to catalyze them. Second, enzymes operate in water and, in principle, this limits their application to water-soluble substrates. Also, the volumes involved in working with enzymes are usually large and therefore the workups may be cumbersome. Finally, some enzymatic transformations give more than one product, making its isolation difficult.

Intensive research in bioorganic chemistry is trying to find ways to overcome these limitations. For example, in 1995 a group of Japanese researchers discovered an enzyme that catalyzes Diels–Alder cycloadditions, a type of reaction that for many years was thought not to occur in nature. Also, it is now known that some enzymes and microorganisms are still functional in nonpolar organic solvents (such as hexane or toluene) provided a small amount of water is present. This opens up a new window of possibilities because nonpolar substrates, which are insoluble in water but dissolve well in nonpolar organic solvents, can then be transformed with the help of enzymes and microorganisms.

28.4 PROCHIRAL MOLECULES

Certain achiral molecules react to give enantiomeric products. These molecules are called **prochiral**. For example, ketones with two different alkyl (or aryl) groups attached to the carbonyl are prochiral because when they undergo nucleophilic additions two enantiomers can form, depending on which side of the carbonyl is attacked. Let's call one of the alkyl groups, the bulkier, "L" (for Large) and the other one "S" (for Small). This lack of symmetry around the carbonyl creates two different faces for the carbonyl; one face has L on the right side, and the other face has L on the left (Fig. 28.4).

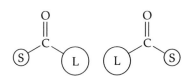

Figure 28.4 The two faces of asymmetrical carbonyl compounds.

Depending on which face is attacked by the nucleophile Nu⁻, two enantiomers can form as Figure 28.5 shows.

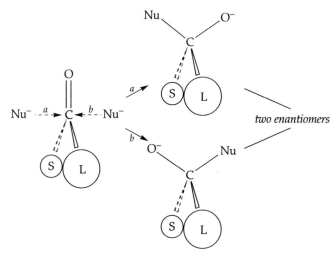

Figure 28.5 Nucleophilic attack to a carbonyl. Both modes of attack (*a* and *b*) are equally probable in an achiral environment and a racemic mixture results.

In an achiral environment both faces are attacked with equal probability and the product is a racemic mixture. A different situation arises, however, when the carbonyl compound and the nucleophile are bound to an enzyme. Because the enzyme's active site is chiral, the carbonyl compound and the nucleophile can bind in only one orientation, which is determined by the complementarity of shapes, and therefore, the attack is stereospecific. The nucleophile attacks only one of the carbonyl faces and only one enantiomer forms (Fig. 28.6).

The two faces of asymmetric carbonyl compounds can be named following rules similar to those used for stereogenic carbons. The three substituents on the trigonal carbonyl carbon are prioritized following the Cahn–Ingold–Prelog rules. In the case of aldehydes and ketones, the oxygen of the carbonyl has the highest priority. If the priority order decreases in a clockwise manner when the viewer faces the molecule, that face is called the **re face** (pronounced "ree"). The other face, where the priority decreases counter-clockwise is called the **si face** (pronounced "sigh"). For aldehydes and ketones with simple alkyl groups attached to the carbonyl, the bulkier group, L, is usually the one with the highest priority. This is illustrated in the example with butanone shown in Figure 28.7.

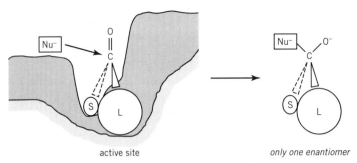

Figure 28.6   Nucleophilic attack in an enzyme's active site. Only one face is attacked by the nucleophile and only one enantiomer forms.

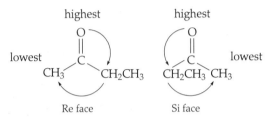

Figure 28.7   The two faces of butanone.

28.5   REDUCTION OF KETONES

In the 1960s, V. Prelog studied the reduction of asymmetric ketones by microorganisms and found that usually only one stereoisomer was produced. *He observed that the hydride anion (the nucleophile in a reduction reaction) is always added to the face that has the bulkier group (L) on the right when the carbonyl is pointing up.*

This mode of attack is equivalent to that shown in Figure 28.6 and is known as **Prelog's rule**.

Yeast: The Midwife of Beer and Bread   Biotechnology is not a twentieth century invention. The use of microorganisms to transform raw materials into useful products, one of biotechnology's aims, is at least 8000 years old. The Babylonians and Sumerians used yeast to produce beer. Of course, they didn't know that a microorganism was responsible for the miraculous metamorphosis of porridge into a palatable concoction, but they liked the effects of the drink and the custom continued on. The Sumerian clay tablet shown below dates from the twenty-first century B.C. and is the equivalent of a modern shopping list. Among the items requested are oil, 1 fish, 1 string of onions, and three different types of barley: for bread, for ordinary beer, and for first-class beer. Beermaking was mostly in the hands of women, who not only made it for the household but also for sale. Being a bartender in ancient Mesopotamia wasn't easier than in today's watering holes. In 1750 B.C. legislation was passed punishing bartenders with death if they served beer to outlaws. Bartenders who overcharged were to be tossed into the river! By the year 4000 B.C. the

Egyptians were already familiar with the brewing process. They used wheat and barley and perhaps some flavoring herbs like coriander for making beer. Recent archeological investigations of Egyptian food remains show that their brewing process was far more elaborate than previously thought, and not too different from our modern methods. Using scanning electron microscopy, Dr. Delwen Samuel was able to study the shape of the starch particles found in 3500-year-old vessels. Starch granules are very sensitive to water and heat, changing their morphology during the cooking process. It is believed now that in ancient Egypt the grains were allowed to germinate and were then heated. This converts the starch into smaller sugar molecules, which are a good food source for yeast. This blend, the equivalent of today's malt, was then mixed with unheated sprouted grains, which carried the yeast responsible for the fermentation process. In fermentation, small sugars such as glucose are transformed into ethanol and carbon dioxide. In Egypt's high temperatures, the fermentation was completed in a few days. Beer was not only an intoxicating drink for the Egyptians, it was also a source of nourishment. Thicker than modern brews, Egyptian beer was one of the few sources of vitamin B_{12}. The other source was leavened bread, a staple food in ancient Egypt. It is believed that the Egyptians invented leavened bread by allowing the dough to ferment before baking it. During cooking ethanol evaporates and carbon dioxide escapes, giving the bread its spongy and supple texture. Bread and beer were used as a sort of currency by the Egyptians. Peasants received two pints of beer and three loaves of bread a day.

Sumerian clay tablet, ca. 2030 B.C.,
University of California
Santa Cruz, Special Collections.

Egyptian bakers,
detail of an
Egyptian bas-relief.

Bibliography

1. Ancient Inventions. P. James and N. Thorpe. Ballantine Books, New York, 1994.

2. Investigation of Ancient Egyptian Baking and Brewing by Correlative Microscopy. D. Samuel. *Science*, **273**, 488–490 (1996).

3. Six Thousand Years of Bread. Its Holy and Unholy History. H. E. Jacob. Doubleday, Garden City, New York, 1944.

When the groups attached to the carbonyl are simple alkyl groups, L has higher priority than S, and the face with L on the right (carbonyl pointing up) is the **re face**. The reduction product obtained in such cases is the S enantiomer. It should be noticed, however, that if the substituents on the carbonyl are more complex than simple alkyl groups (groups with heteroatoms, aromatic rings, etc.) the L substituent may not necessarily have higher priority than S.

In the reduction of carbonyl compounds carried out with enzymes and microorganisms, the sources of hydride are the cofactors NADH or NADPH. With the enzyme's help, the reduced form of the nicotinamide ring transfers one of the hydrogens (in the form of a hydride) from position 4 of the ring to the carbonyl carbon to give an alcohol.

| NADH or NADPH (reduced form) | a carbonyl compound | | NAD⁺ or NADP⁺ (oxidized form) | an alcohol |

28.6 ENANTIOMERIC EXCESS

When a chemical transformation that creates a stereogenic center affords unequal amounts of both enantiomers, it is customary to determine the **enantiomeric excess** of the synthesis, *e.e.* The enantiomeric excess is the difference in yield between the two enantiomers divided by the total yield, and multiplied by 100 (Eq. 1). In Equation 1, e_1 and e_2 are the yields in each enantiomer.

$$e.e. = \frac{(e_1 - e_2)}{e_1 + e_2} \times 100 \tag{1}$$

To keep the *e.e.* a positive figure, e_1 refers to the enantiomer with larger yield. If a reaction gives 80% of one enantiomer and 20% of the other, the enantiomeric excess is 60%. If a reaction affords a racemic mixture, the enantiomeric excess is zero.

An easy way to determine the enantiomeric excess experimentally is by measuring the optical rotation of the mixture of products. If the specific rotation for the pure enantiomers is known, then the e.e. is calculated as:

$$e.e. = \frac{\text{measured specific rotation of mixture}}{\text{specific rotation of pure enantiomer}} \times 100$$

BIBLIOGRAPHY

1. Asymmetric Synthesis. R.A. Aitken and S.N. Kilényi, ed. Blackie Academic and Professional, London, 1992.

2. Specification of the Stereospecificity of Some Oxido-reductases by Diamond Lattice Sections. V. Prelog. *Pure and Applied Chem.*, **9**, 119 (1964).

3. Preparation of (−)-(1R,2S)-1-phenylpropane-1,2-diol by Fermenting Baker's Yeast Reduction of 1-phenyl-1,2-propanedione. E.C.S. Brenelli, P.J.S. Moran, and J.A.R. Rodrigues. *Synth. Comm.*, **20**, 261–266 (1990).

4. Application of Biocatalysts in Organic Synthesis. R. Azerad. *Bull. Soc. Chim. Fr.*, **132**, 17–51 (1995).

5. Baker's Yeast Mediated Transformations in Organic Chemistry. R. Czuk and B.I. Glänzer. *Chem. Rev.*, **91**, 49–97 (1991).

6. Baker's Yeast as a Reagent in Organic Synthesis. S. Servi. *Synthesis*, 1–25 (1990).

7. Enzymatic Activity Catalyzing Exo-Selective Diels-Alder Reaction in Solanapyrone Biosynthesis. H. Oikawa, K. Katayama, Y. Suzuki, and A. Ichihara. *J. Chem. Soc. Chem. Comm.*, 1321–1322 (1995).

8. Enzymes in Organic Synthesis: Oxidoreductions. J-M. Fang, C-H. Lin, C.W. Bradshaw, and C-H. Wong. *J. Chem. Soc. Perkin Trans. I*, 967–978 (1995).

9. The Use of Organic Solvent Systems in the Yeast Mediated Reduction of Ethyl Acetoacetate. L.Y. Jayasinghe, D. Kodituwakku, A.J. Smallridge, and M.A. Trewhella. *Bull. Chem. Soc. Jpn.*, **67**, 2528–2531 (1994).

Problem 1

EXERCISES

(a) Determine if the face shown for each ketone is *re* or *si*.

acetophenone   2-methylcyclohexanone

(b) Draw the reduction product (addition of H⁻) of acetophenone if Prelog's rule is obeyed. Name the product as *R* or *S*.

(c) Draw the reaction product of acetophenone with ethyl magnesium bromide. Assume that the Grignard reagent attacks the ketone following Prelog's rule. What is the stereochemistry of the product, *R* or *S*?

Problem 2

The reduction of the following compounds mediated by yeast affords in each case one stereoisomer as the major product. Do the reactions follow Prelog's rule? Briefly discuss.

Problem 3

The first reaction shown in Problem 2 proceeds with an enantiomeric excess of 95%. What is the percentage of each enantiomer in the product mixture?

Problem 4

The reduction of acetophenone (see Problem 1) mediated by yeast gives (S)-1-phenylethanol as the major product. In the presence of glucose as a yeast nutrient the % yield of the synthesis is only 16% and the specific rotation of the product is $-32.3°$ $(g/mL)^{-1}$ dm^{-1}. In the absence of glucose the reduction gives a 45% yield and the specific rotation of the product is $-40.5°$ $(g/mL)^{-1}$ dm^{-1}.

(a) Is Prelog's rule obeyed?

(b) Calculate the enantiomeric excess under both experimental conditions. The specific rotation of pure (S)-1-phenylethanol is $-44°$ $(g/mL)^{-1}$ dm^{-1}.

(c) In both cases calculate the % of each enantiomer.

Problem 5

The reduction shown below gives four stereoisomers with the indicated percentages. Calculate the enantiomeric excess for each pair of enantiomers.

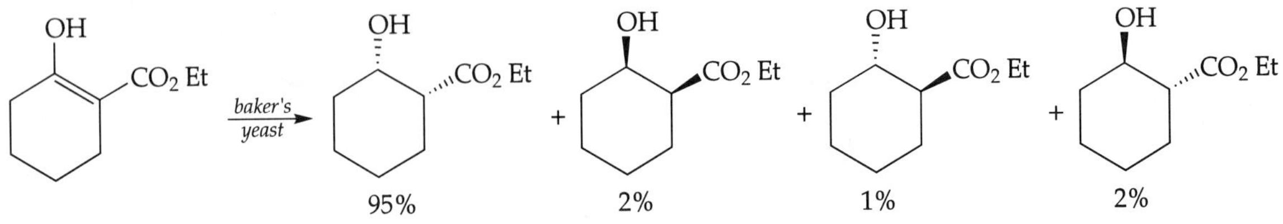

Asymmetric Synthesis with Baker's Yeast: An Open-Ended Experiment

E28.1 OVERVIEW

In this experiment you will carry out the asymmetric reduction of 1-phenyl-1,2-propanedione using baker's yeast (*Saccharomyces cerevisiae*). This reaction leads to the formation of 1-phenyl-1,2-propanediol as the final product. Because the diol has two stereogenic centers, four different stereoisomers are possible. However, the reaction proceeds in good yield (60–75%) and gives one isomer, $(-)$-$(1R,2S)$-1-phenyl-1,2-propanediol, in high enantiomeric excess (>98%).

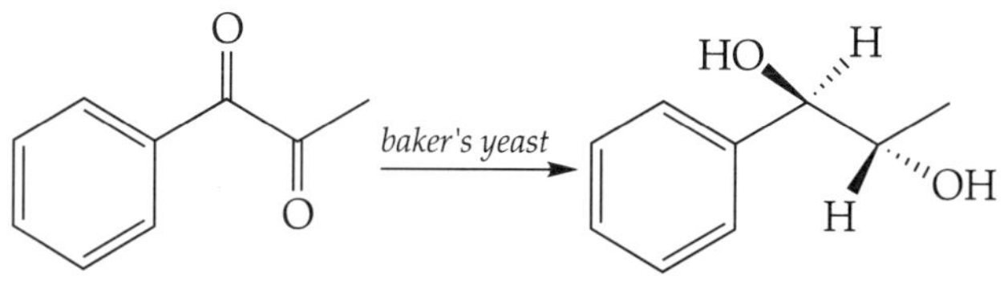

1-phenyl-1,2-
propanedione

baker's yeast

$(1R,2S)$-1-phenyl-1,2-
propanediol

How does this reaction take place? Does it proceed in one single step in which both carbonyl groups are reduced at once, as shown in the equation above, or does it occur in stages with the reduction of one carbonyl group at a time?

If this second hypothesis is correct and the reaction occurs in stages, we should observe an intermediate α-hydroxyketone, called an α-**ketol**, which is then reduced to the final diol.

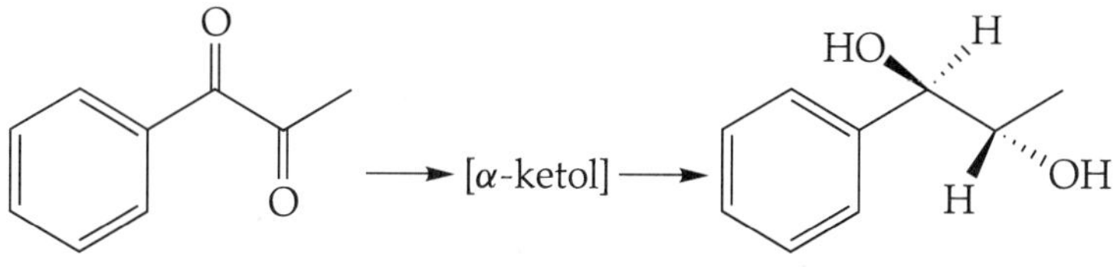

1-phenyl-1,2-
propanedione

$[\alpha\text{-ketol}]$

$(1R,2S)$-1-phenyl-1,2-
propanediol

If this hypothesis is right, two different α-ketols are, in principle, possible, depending on which carbonyl group is reduced first.

Possible α-ketols

(S)-2-hyroxy-1-phenyl-1-propanone (R)-1-hydroxy-1-phenyl-2-propanone

Do both α-ketols form, or is the reaction regiospecific with only one α-ketol as the intermediate? If this is the case, which α-ketol forms? This is an open-ended experiment and these are some of the questions that you will investigate.

You will study the course of the reaction and determine the nature of the intermediate compounds, if they form. You will carry out this experiment with a partner. Both students will follow the course of the reaction by taking small aliquots approximately every 15 minutes and analyzing them by TLC. One student will stop the whole reaction after it has proceeded to completion and no more starting material is left (approximately 90 minutes); this student will isolate the final product (1R,2S)-1-phenyl-1,2-propanediol. The other student will stop the whole reaction at an early time (after about 50 minutes) and analyze the products formed up to that point. Each student will perform the reaction *separately*, but both members of the team will compare notes during and after the experiment. You will use a combination of analytical tools to investigate the course of the reaction: TLC, GC, IR, and ^{1}H-NMR.

It is not yet known which enzymes are involved in the reduction of the diketone to the diol. In 1990, a group of German chemists isolated two enzymes from yeast that reduce α-diketones. These enzymes, called diacetyl reductases, may be involved in the reduction of 1-phenyl-1,2-propanedione. These enzymes need the reduced form of NADP$^+$ (NADPH) as a coenzyme to perform the reduction of the diketone. Other reductases such as glycerol dehydrogenase may also be involved. These enzymes probably fulfill a detoxification process in yeast, ridding the cell of toxins such as diketones and oxoaldehydes, formed as byproducts of the yeast's natural metabolism.

The procedure outlined below has been adapted from References 2 and 3. It uses baker's yeast in an *unbuffered aqueous solution*, and yields the product already mentioned ((1R,2S)-1-phenyl-1,2-propanediol) in high yield and enantiomeric purity. As with most reactions facilitated by yeast, this reduction is sensitive to pH and the presence of nutrients (sucrose, glucose, etc.). For example, at pH 5 and with sucrose added (different conditions from the ones used in our experiment), the major product is (−)-(S)-2-hydroxy-1-phenyl-1-propanone (Ref. 4).

E28.2   INVESTIGATING THE COURSE OF THE REACTION

You will follow the reaction by TLC and GC analyses and determine whether intermediate compounds are involved in the transformation. You will *remove small aliquots from the reaction mixture approximately every 15 minutes* and analyze them by TLC. If the reduction takes place in one step, without the formation of intermediate α-ketols, you will see no more than two spots on the TLC plates: one corresponding to the starting material, the diketone, and the other to the product, the diol. If, on the other hand, the α-ketol is an intermediate, you will see three spots on the TLC plates: one for the diketone, one for the diol, and one for the α-ketol. In either case, an aliquot taken from the reaction mixture toward the end of the reaction, after approximately 90 minutes, will show mainly one spot corresponding to the final product. You will confirm your TLC results by performing GC analysis of your reaction mixture and your partner's.

You and your partner will obtain the IR spectrum, and, if possible, the ^{1}H-NMR spectrum of the compounds present in the reaction mixture after 50 and 90 minutes. In the diagram below, you will find spectroscopic information about the diketone, the diol, and both possible α-ketols. Use this information to interpret the spectra of the starting material, the product, and possible intermediates. Keep in mind that the spectra obtained after 50 minutes of reaction corresponds to a mixture of starting material, product, and, possibly, intermediate α-ketols.

With all the chromatographic and spectroscopic information gathered by you and your partner you will be able to determine the course of the reaction and identify the intermediate α-ketol, if present.

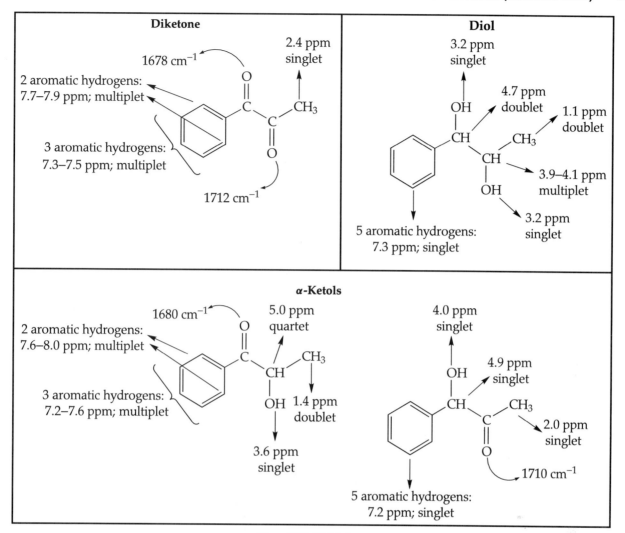

IR and ^{1}H-NMR data.

E28.3   CONFORMATIONAL ANALYSIS: DETERMINING WHETHER THE DIOL IS *THREO* OR *ERYTHRO* (Advanced Level)

The final product 1-phenyl-1,2-propanediol has two stereogenic centers, and therefore, there are four possible stereoisomers grouped in two pairs of enantiomers. Out of these four possible isomers, the reduction with baker's yeast is stereospecific and gives only (1R,2S)-1-phenyl-1,2-propanediol, as already discussed.

To confirm the absolute configuration of the diol you could measure its optical rotation and compare it with literature values. To measure the optical rotation in the laboratory, however, you would need a large sample or a very sensitive polarimeter. Either option is normally beyond what most undergraduate laboratories can offer. Therefore, unless instructed otherwise, you will not study the absolute configuration of the diol. You will instead determine whether the product is *erythro* or *threo*. To this effect you will perform a ^{1}H-NMR conformational analysis.

In the Fischer projections shown below you will find the four possible stereoisomers of 1-phenyl-1,2-propanediol. The two enantiomers that have the same groups on the same side (both —OH groups on one side, and both hydrogens on the other side) are called the *erythro* enantiomers. The enantiomers with similar groups on opposite sides are called *threo* enantiomers. Thus, as we can see in the projections below, the actual product of the reaction, the 1R,2S stereoisomer, is one of the two erythro isomers.

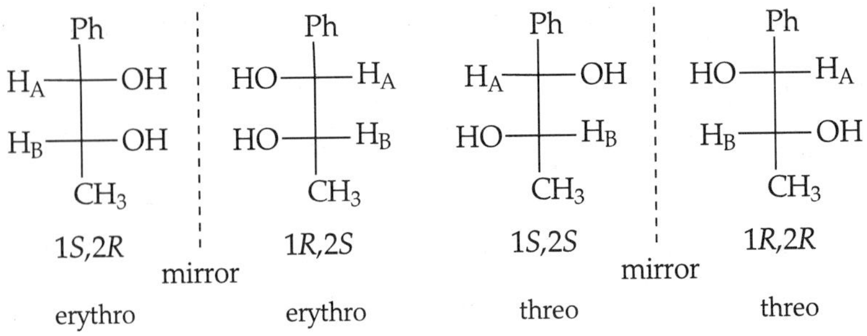

1S,2R	1R,2S	1S,2S	1R,2R
	mirror		mirror
erythro	erythro	threo	threo

The ^{1}H-NMR spectra of the two erythro isomers are indistinguishable in an achiral solvent because these two isomers are enantiomers (section 33.19). Likewise, the spectra of both threo enantiomers are identical, but different from those of the erythro isomers. In general, the spectra of the erythro and threo isomers need not be identical because they are diastereomers. In our case, the ^{1}H-NMR spectra of the erythro and threo diastereomers of 1-phenyl-1,2-propanediol are similar, but different. The *most important difference* is the coupling constant between H_A and H_B, J_{AB}. In the erythro isomers $J_{AB} = 4$ Hz, while in the threo isomers $J_{AB} = 7.5$ Hz. Therefore, by measuring the coupling constant J_{AB} you can determine whether the product is threo or erythro. The difference in J_{AB} values arises from differences in the stability of the conformers, as discussed below.

Newman projections for the three staggered conformers of the 1S, 2R isomer (erythro) are shown in the figure below. The two conformers with the hydroxyl groups gauche (A and C) are more stable (and therefore more abundant) because they can form intramolecular H-bonds. Conformer B, on the other hand, has both −OH groups anti with respect to each other and cannot form intramolecular H-bonds. As a result, conformer B is less stable than A and C and is found only in small amounts at equilibrium.

1S,2R (erythro)

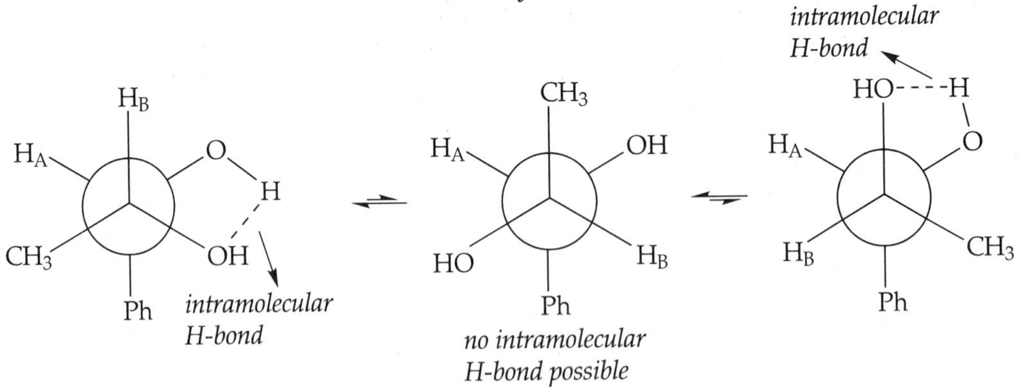

A	**B**	**C**
abundant conformer	minor conformer	abundant conformer
H_A and H_B gauche:	H_A and H_B anti:	H_A and H_B gauche:
$J_{AB} \sim 3$ Hz	$J_{AB} \sim 10$ Hz	$J_{AB} \sim 3$ Hz

In conformers A and C, H_A and H_B are gauche with respect to each other. The *coupling constant between gauche hydrogen atoms is approximately 3 Hz*. In the minor conformer (B), on the other hand, H_A and H_B are anti with respect to each other. The *coupling constant between anti hydrogens is approximately 10 Hz* (we should recall that the coupling constant between vicinal hydrogens is a function of the dihedral angle between them (section 33.12); this angle is 60° when the hydrogens are gauche and

180° when they are anti with respect to each other). The observed coupling constant is the weighted average of the coupling constant for each conformer, where the weight factor is the percentage of the conformer at equilibrium. Because the most abundant conformers have $J_{AB} \approx 3$ Hz, and the minor conformer has $J_{AB} \approx 10$ Hz, the observed coupling constant for the 1*S*,2*R* isomer is very close to 3 Hz ($J_{AB} = 4$ Hz). Now, do a similar conformational analysis for the other erythro enantiomer (1*R*,2*S*) and convince yourself that J_{AB} should also be close to 3 Hz.

Similar conformational analyses for the *threo* isomers (1*S*,2*S* and 1*R*,2*R*) indicate that the observed coupling constant should be an average of 10 and 3 Hz with similar weight for both. The observed value is 7.5 Hz. The analysis for the 1*S*,2*S* isomer is shown below.

1S,2S (threo)

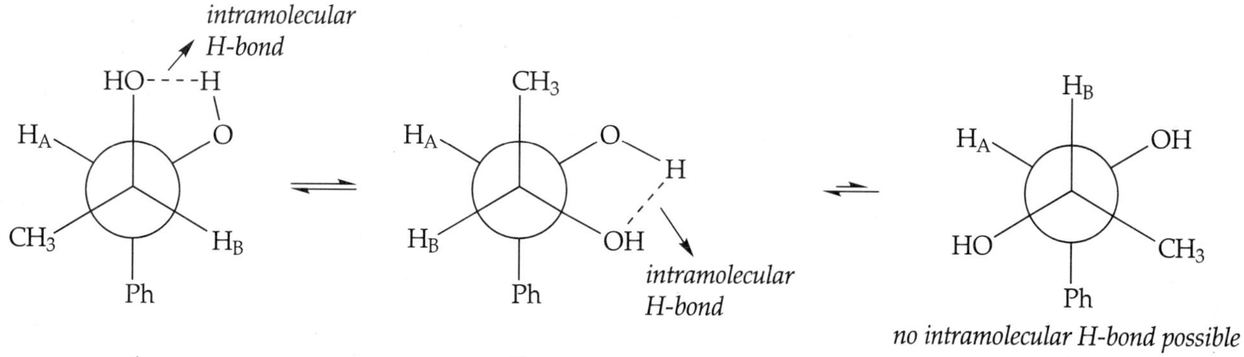
intramolecular
H-bond
HO---H
O
H_A
CH_3
H_B
Ph
CH_3
H_A
O
H
H_B
OH
Ph
intramolecular
H-bond
H_B
H_A
OH
HO
CH_3
Ph
no intramolecular H-bond possible

A	B	C
abundant conformer	abundant conformer	minor conformer
H_A and H_B anti:	H_A and H_B gauche:	H_A and H_B gauche:
J_{AB} ~ 10 Hz	J_{AB} ~ 3 Hz	J_{AB} ~ 3 Hz

You will measure J_{AB} for the diol obtained in the reaction with baker's yeast and confirm that it is an erythro isomer. To this end, you will measure the coupling constant for the doublet around 4.7 ppm (H_A) in the spectrum of the diol. You could also measure the same coupling constant using the signal for H_B (around 3.9–4.1 ppm), but because H_B gives a multiplet (H_B is also coupled to the adjacent methyl group) determining J_{AB} from this signal is more difficult. To find out how to measure coupling constants, see section 33.12.

If you have access to a very sensitive polarimeter, you can also measure the optical rotation of the diol and confirm that it is (−)-(1*R*,2*S*)-1-phenyl-1,2-propanediol. The specific rotation for the 1*R*,2*S* isomer is $-40°(g/mL)^{-1}dm^{-1}$, and $+40°(g/mL)^{-1}dm^{-1}$ for its enantiomer.

PROCEDURE

Background reading: 28.1–28.4; E28.1–E28.3
Estimated time: synthesis: 2–3 hr; Analysis: 1–2 hr

Safety First

- *tert*-Butyl methyl ether is flammable. Handle it with care.
- When releasing the pressure, point the separatory funnel stem away from your face and others.
- $CDCl_3$ is a possible carcinogen. Handle it with care. Deliver the liquid in the hood.

E28.4 REDUCTION OF 1-PHENYL-1,2-PROPANEDIONE WITH BAKER'S YEAST

In a 250-mL Erlenmeyer flask weigh 10 g of freeze-dried baker's yeast. Add 35 mL of deionized water and stir the mixture to obtain a smooth suspension. Add a magnetic stir bar and place the flask on a stir plate set at a medium speed. Add 230 μL of 1-phenyl-1,2-propanedione with the aid of an automatic pipet. Take aliquots of approximately 0.2 mL immediately after addition of the diketone, and every fifteen minutes thereafter. Analyze them by TLC as indicated in E28.5. In the meantime, between TLC runs, obtain the IR and NMR spectra of the starting diketone.

After 50 or 90 minutes of reaction at room temperature (depending on whether you are investigating the intermediate or the final product), transfer the reaction mixture to a 125-mL separatory funnel. Add 50 mL of *tert*-butyl methyl ether (BME); invert and vent the separatory funnel. Keep inverting the separatory funnel for a period of 5 minutes. Do not shake it because the layers will emulsify. Separate the layers, collecting the lower aqueous layer in the original Erlenmeyer flask. The interface between the layers will not be well defined and will show some emulsion. The interface can be made clearer by applying a gentle rotational motion to the separatory funnel in a vertical position while holding it from the top and lifting it from the ring (Fig. 28.8). Try to drain most of the emulsion with the lower aqueous layer. Collect the upper layer in a dry 250-mL Erlenmeyer flask, labeled "Organic," and do not worry if some of the emulsion passes along with the organic layer.

Transfer the aqueous layer back to the separatory funnel, add 35 mL of BME, and extract it as described above. Separate the layers, collecting the BME layer in the Erlenmeyer flask labeled "Organic." Repeat the extraction of the aqueous layer with another 35-mL portion of fresh BME.

To the combined organic layers add enough anhydrous magnesium sulfate so it runs freely in the liquid (approximately 4 g). After about 5 minutes, separate the drying agent by gravity filtration using a piece of fluted filter paper. Collect the liquid in a pre-tared **and dry** 150-mL round-bottom flask. Evaporate the solvent in a rota-vap using a warm-water bath. Remove final traces of solvent and water by blowing a gentle stream of nitrogen through the flask (this is particularly important for NMR analysis). Weigh the flask with the oily product and calculate its mass by difference.

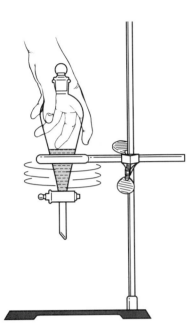

Figure 28.8 Swirling the separatory funnel helps break emulsions.

E28.5   ANALYSIS

TLC Analysis

Immediately after the addition of the diketone and every fifteen minutes thereafter, with the help of a Pasteur pipet remove aliquots of the reaction mixture of approximately 0.2 mL. Transfer the aliquot to a screw-cap test tube and add 0.4 mL of BME. Shake and vent immediately. Repeat this operation for about 30 seconds. Spot the upper layer on a silica gel plate with fluorescent indicator (F254) (about 8–10 applications). Develop the plate using a mixture of cyclohexane-BME (6:4). The R_f values for the diol, α-ketol, and diketone increase in that order, as expected from their relative polarities. Visualize the spots using a UV lamp (254 nm). The diol shows as a much fainter spot than the diketone. Under these conditions the two possible α-ketols run with similar R_f.

GC Analysis

Inject the oily residue obtained in E28.4 in a GC column of medium polarity such as SF-96 equilibrated at 175°C (suggested injection volume 0.2 μL). Also inject the starting diketone. To compare the retention times, the analysis of the diketone and the reaction product should be done the same day on the same GC instrument.

The diketone, the α-ketols, and the diol elute in the expected order ($t_{\text{Rdiketone}} < t_{\text{R}\alpha-\text{ketol}} < t_{\text{Rdiol}}$). Under the experimental conditions, the two possible α-ketols may elute as a single and broad peak. The diol also gives a broad peak. Calculate the areas under the peaks and estimate the % of each compound in the mixture.

IR Analysis

Obtain the IR spectrum (700–4000 cm^{-1} region) of the residue obtained in E28.4 between NaCl plates. Also obtain the IR of the starting diketone under the same conditions (section 31.5).

^{1}H-NMR Analysis

Dissolve approximately 50–100 mg (2–6 drops) of the oil obtained in E28.4 in 0.5 mL of CDCl$_3$ containing 1% TMS. Obtain the spectrum in the 0–10 ppm range (section 33.20).

If you see a broad peak around 3 ppm, it is probably due to water. In this case, remove the water by drying the product again. To this end, transfer the NMR sample to the flask containing the rest of the oil and add 15–20 mL of BME. Add a spatulaful of anhydrous magnesium sulfate and let it sit for about 5 minutes. Filter and evaporate the solvent. Obtain a new NMR spectrum.

An unexpected peak around 1.2 ppm may be due to traces of BME. This can be confirmed by comparing with the ^{1}H-NMR of BME. You can eliminate this peak by exhaustive evaporation of the solvent in the rota-vap, and obtaining a new spectrum of the sample after evaporation.

To identify the −OH peaks of your product, add two drops of D$_2$O to the NMR solution in CDCl$_3$, shake well, and obtain the spectrum of the product again. Compare with the spectrum obtained before deuterium exchange.

Also obtain the ^{1}H-NMR spectrum of the starting diketone in CDCl$_3$. Compare it with the spectrum of the product.

If you are performing the conformational analysis (advanced level), measure and interpret the coupling constant between H$_A$ and H$_B$ as indicated in E28.3.

Cleaning Up

- Dispose of the aqueous layer with yeast in the container labeled "Yeast Liquid Waste."
- Dispose of the magnesium sulfate, filter paper, Pasteur and capillary pipets in the container labeled "Yeast Solid Waste."
- Dispose of CDCl$_3$ in the container labeled "Chloroform Waste."
- Turn in your product in a labeled screw-cap vial.

BIBLIOGRAPHY

1. Purification and Properties of Two Oxidoreductases Catalyzing the Enantioselective Reduction of Diacetyl and Other Diketones from Baker's Yeast. J. Heidlas and R. Tressl. *Eur. J. Biochem.*, **188**, 165–174 (1990).
2. Baker's Yeast Reduction of α-Diketones. P. Besse, J. Bolte, and H. Veschambre. *J. Chem. Ed.*, **72**, 277–278 (1995).
3. Preparation of (−)-(1R,2S)-1-phenylpropane-1,2-diol by Fermenting Baker's Yeast Reduction of 1-phenyl-1,2-propanedione. E.C.S. Brenelli, P.J.S. Moran, and J.A.R. Rodrigues. *Synth. Comm.*, **20**, 261–266 (1990).
4. Baker's Yeast Reduction of 1,2-Diketones. Preparation of Pure (S)-(-)-2-Hydroxy-1-phenyl-1-propanone. R. Chenevert and S. Thiboutot. *Chem. Lett.*, 1191–1192 (1988).
5. Aluminum Hydride Reduction of α-Ketols. II. Additional Evidence for Conformational Flexibility in the Transition State. S.B. Bowlus and J.A. Katzenellenbogen. *J. Org. Chem.*, **39**, 3309–3314 (1974).

EXPERIMENT 28   REPORT

Pre-lab

1. List the physical properties (including toxicity/hazards) of 1-phenyl-1,2-propanedione and solvents. Calculate the molecular mass of the diol. In a pure form, *erythro*-1-phenyl-1,2-propanediol has a melting point of 89–91°C (the melting point for the threo isomers is 51–53°C). See Reference 5.
2. Draw and name all possible stereoisomers for 1-phenyl-1,2-propanediol. Indicate the relationship (enantiomers, diastereomers) between them.

3. Consider the spectroscopic data shown on page 647 and discuss what distinctive features of the IR and ^{1}H-NMR spectra of α-ketols you would use to tell the two α-ketols apart.

4. Consider ^{1}H-NMR data shown on page 647 and discuss how you can distinguish between the final diol and the intermediate α-ketols.

5. Briefly justify the TLC R_f order for the diketone, α-ketols, and diol.

6. Name the faces (re or si) of both carbonyl groups in the structures of 1-phenyl-1,2-propanedione shown below.

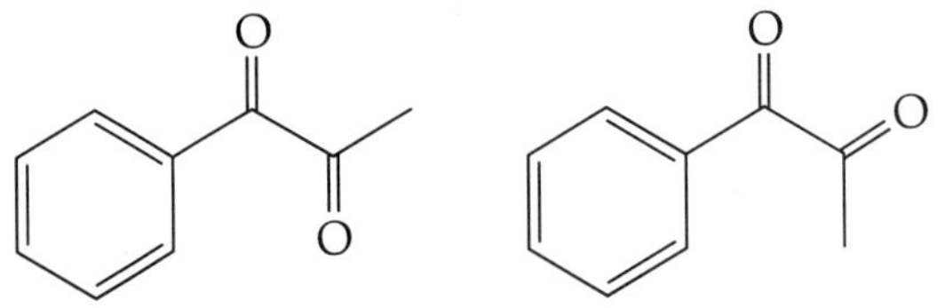

7. The product of the reduction of 1-phenyl-1,2-propanedione is $(1R,2S)$-1-phenyl-1,2-propanediol. Is Prelog's rule of enzymatic reduction obeyed in the reduction of each carbonyl group? Explain with diagrams.

8. Explain why $J_{AB} = 4$ Hz for the erythro enantiomers, while $J_{AB} = 7.5$ Hz for the threo enantiomers. Show your conformational analysis for the $1R,2S$ and $1R,2R$ isomers (advanced level).

In-lab

When reporting results, always report yours and your partner's.

1. Calculate the % yield of the synthesis.

2. Report the TLC results. Indicate the composition of the reaction mixture at different times. Make a table showing the R_f values for different spots. Are the spots in the order predicted in pre-lab 5?

3. Report the GC results. Make a table with the retention times and areas for the peaks observed. Identify the peaks.

4. Report and interpret the IR and ^{1}H-NMR spectra of starting material, final product and reaction mixture after 50 minutes.

5. Did the reaction take place in one step? Did it occur via α-ketols? Explain your reasoning.

6. Based on your ^{1}H-NMR spectrum, is your final diol erythro or threo? Briefly explain (advanced level).

7. Propose other experiments to continue and refine this investigation. For example, if an intermediate forms, how would you go about isolating it in a pure form?

8. What other aspects of this reaction would you like to investigate?

Molecules of Heredity

29.1 NUCLEOSIDES AND NUCLEOTIDES

Life is encoded in the molecular structure of **deoxyribonucleic acid** (DNA), which contains all the information for self-replication, development, and survival of the organism. No less important in the life cycle is **ribonucleic acid** (RNA), which translates the information contained in DNA into the cellular production of chemicals. Both DNA and RNA are polymers with repetitive units made of a sugar molecule, a heterocyclic base, and a phosphate group that links the units together. The sugar in RNA is D-ribose, and in DNA it is D-2'-deoxyribose (the prime is used to number the sugar positions and distinguish them from those of the base). The base is attached to position 1' of the sugar by a β-glycosidic bond.

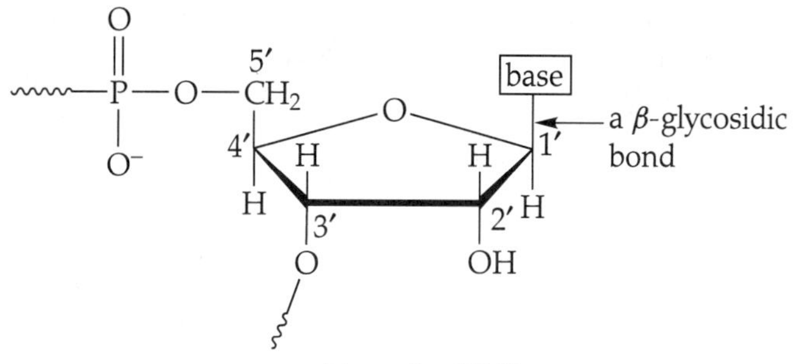

repetitive unit of RNA

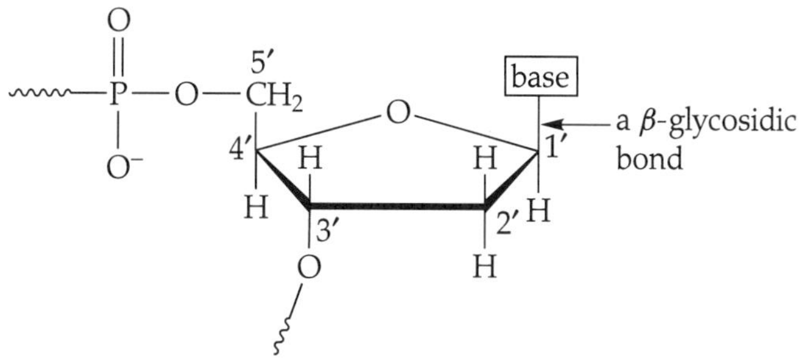

repetitive unit of DNA

The heterocyclic bases found in DNA and RNA are substituted **purines** and **pyrimidines**. The purine bases are **adenine** (A) and **guanine** (G) and the pyrimidine bases are **cytosine** (C) and **uracil** (U) in RNA, and cytosine and **thymine** (T) in DNA.

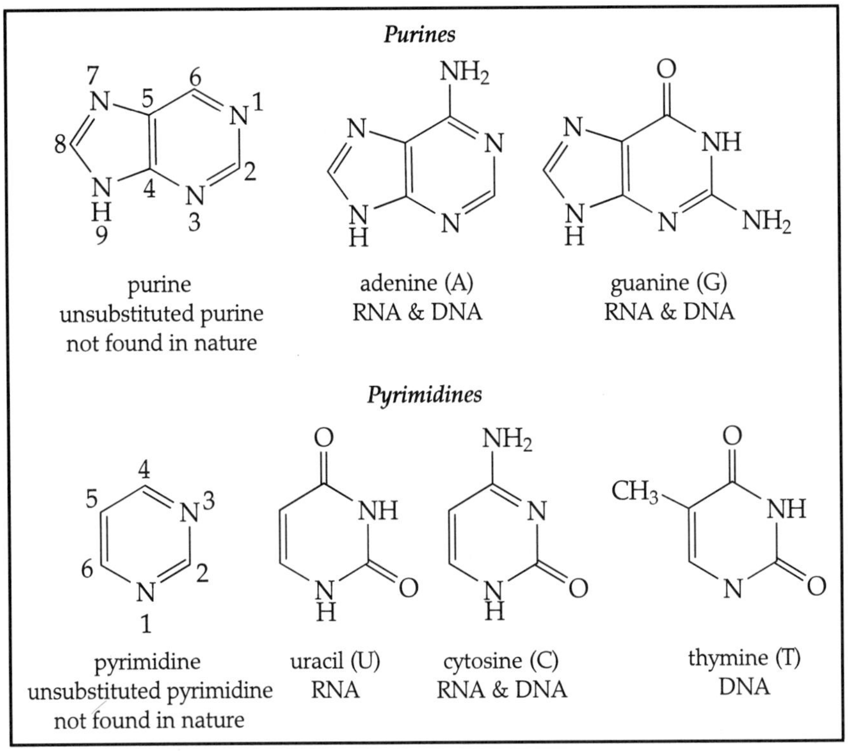

Purines

purine
unsubstituted purine
not found in nature

adenine (A)
RNA & DNA

guanine (G)
RNA & DNA

Pyrimidines

pyrimidine
unsubstituted pyrimidine
not found in nature

uracil (U)
RNA

cytosine (C)
RNA & DNA

thymine (T)
DNA

The purines are attached to the sugar residue through the nitrogen at position 9 and the pyrimidines through nitrogen 1.

The phosphate groups in DNA and RNA, called phosphodiesters, link position 5' of one sugar residue with position 3' of the next sugar. The phosphate groups are acidic with a pK_a of about 1.2 and thus they are dissociated under physiological conditions (pH near neutrality). The negative charge of the phosphodiester groups is balanced by association with metal cations, especially Mg^{2+}.

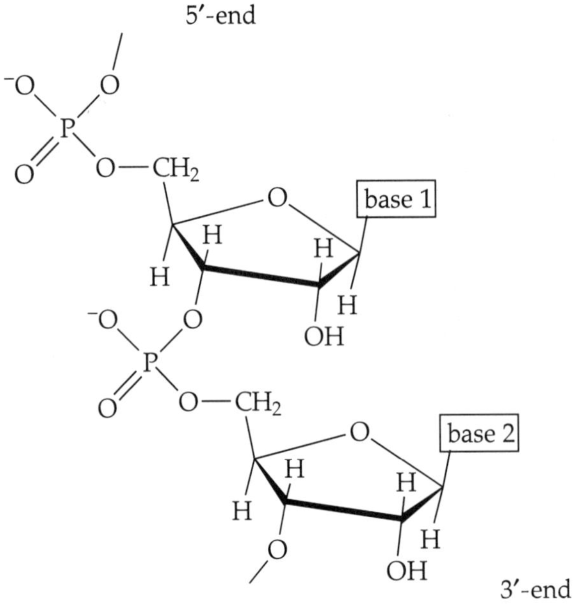

5'-end

base 1

base 2

3'-end

Hydrolysis of RNA and DNA cleaves the phosphodiester groups and produces **nucleotides** and **deoxynucleotides**, respectively. Nucleotides and deoxynucleotides are formed by one unit of ribose or deoxyribose, one of a purine or pyrimidine base, and one phosphate group usually attached to position 5' of the sugar. The structures below show examples of a nucleotide and a deoxynucleotide with adenine as the base.

Example of a nucleotide

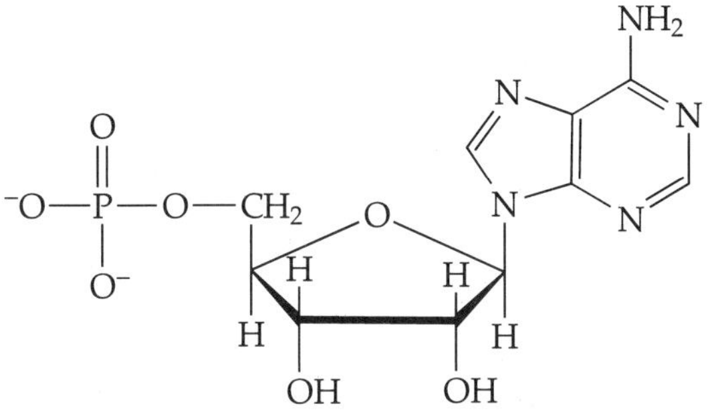

Example of a deoxynucleotide

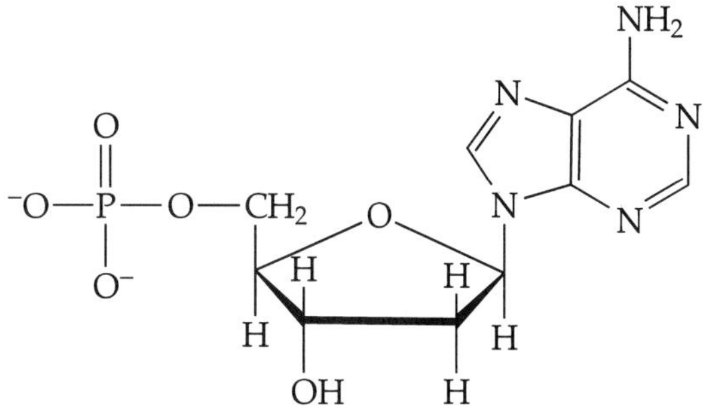

adenosine 5′-phosphate

2′-deoxyadenosine 5′-phosphate

Removal of the phosphate group from nucleotides and deoxynucleotides produces **nucleosides** and **deoxynucleosides**, respectively. The nucleosides from RNA are adenosine, guanosine, cytidine, and uridine. The deoxynucleosides from DNA are 2′-deoxyadenosine, 2′-deoxyguanosine, 2′-deoxycytidine, and 2′-deoxythymidine.

Nucleosides (from RNA)

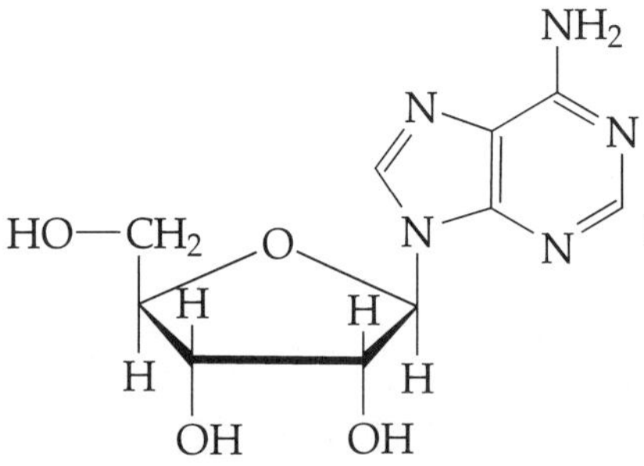

adenosine

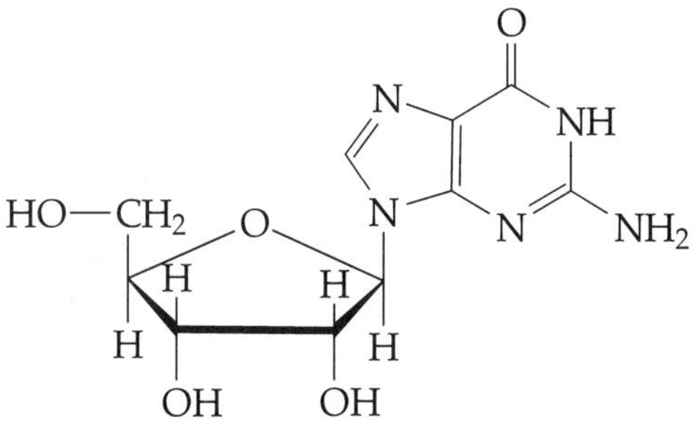

guanosine

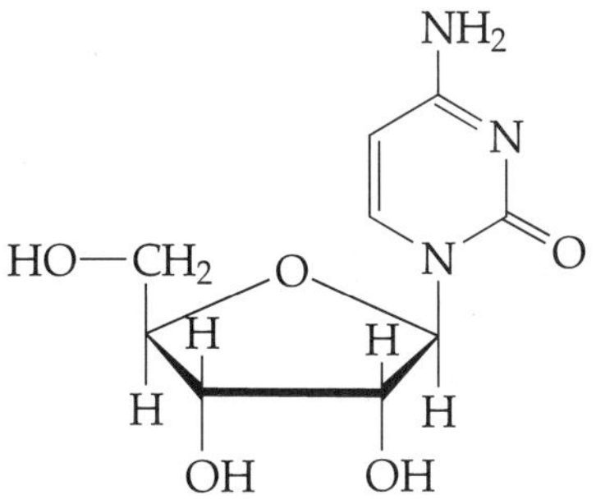

cytidine

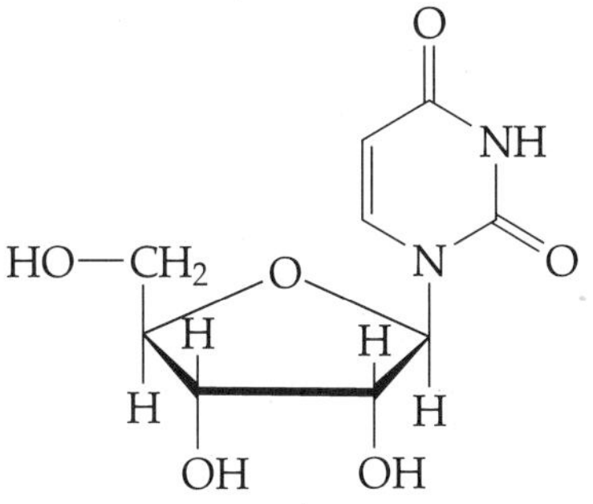

uridine

2′-deoxynucleosides (from DNA)

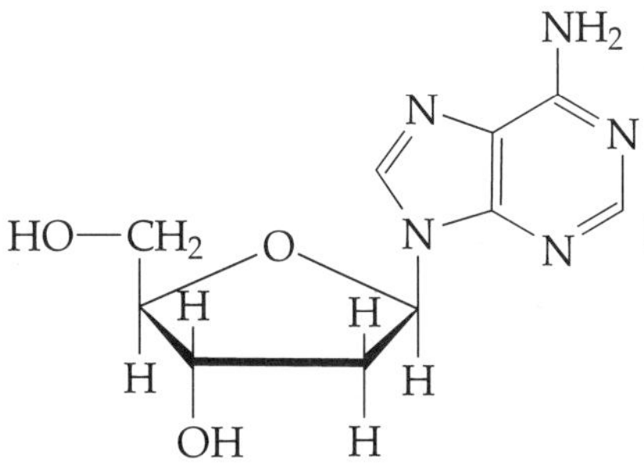

2′-deoxyadenosine

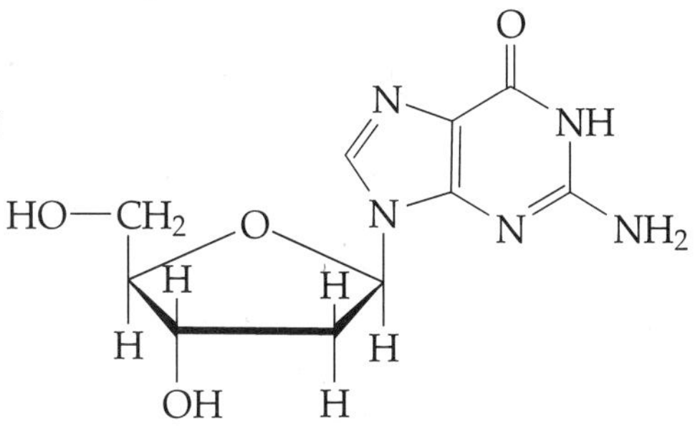

2′-deoxyguanosine

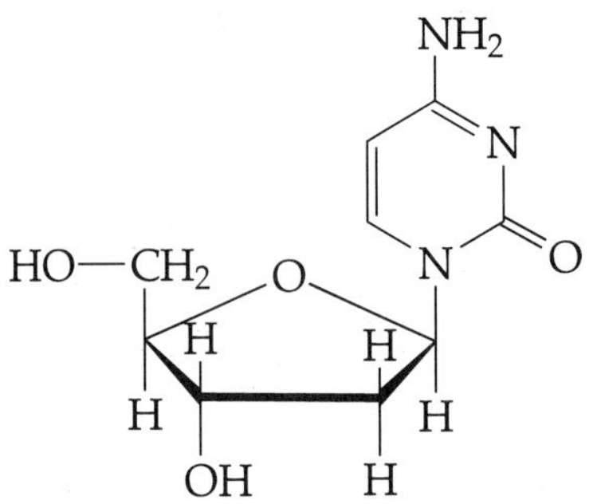

2'-deoxycytidine 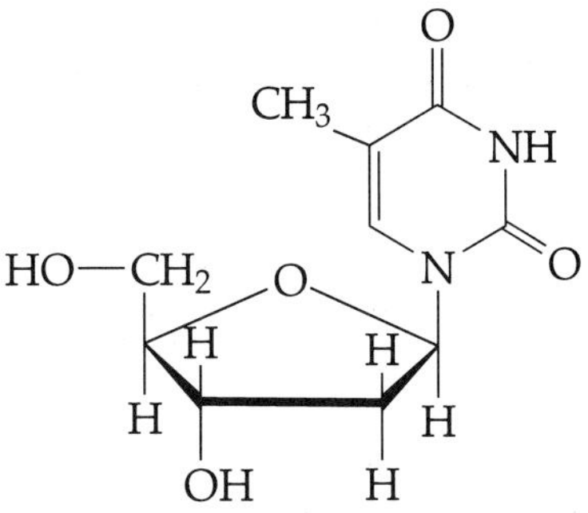   2'-deoxythymidine

29.2   BASE PAIRING

The sequence in which the bases are aligned in the polymeric chain of DNA carries the blueprint that distinguishes one life form from another. The DNA sequence is usually indicated by the base initials starting at the 5'-end. For example:

GGAAGTACAGCTCAGTCT

5'-end   3'-end

DNA exists as a double-stranded structure with the bases in one strand H-bonded to the bases in the other strand. The pairing of the bases is very specific. Adenine (A) is always H-bonded to thymine (T) and guanine (G) to cytosine (C). The two strands are antiparallel to each other, with the 5'-end of one strand facing the 3'-end of the other. For the DNA sequence shown above, the pairing with its complementary strand gives the following double-stranded DNA:

5'-end GGAAGTACAGCTCAGTCT 3'-end
       | | | | | | | | | | | | | | | | | |
3'-end CCTTCATGTCGAGTCAGA 5'-end

The H-bonds between bases are very specific. The A-T pair shares two H-bonds, and the C—G pair shares three, as indicated below. This is called Watson–Crick base pairing in honor of James Watson and Francis Crick, who in 1953 elucidated the three-dimensional structure of DNA based on X-ray data collected by Rosalind Franklin and Maurice Wilkins.

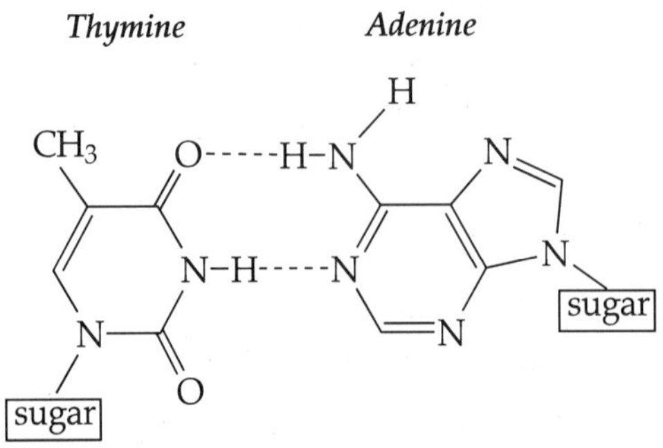

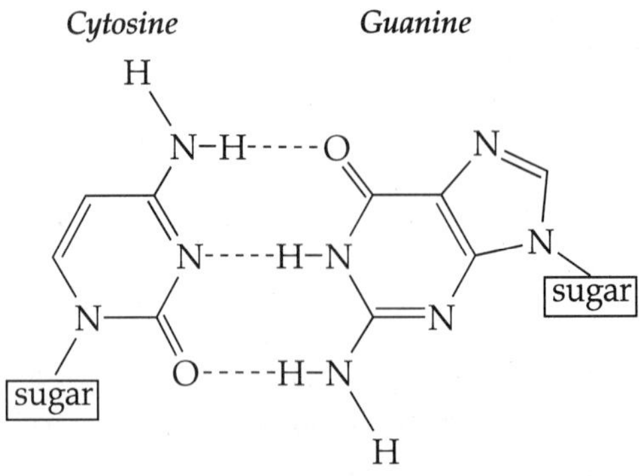

The two strands of DNA are coiled in a double helix with the base pairs in the center and the sugar and phosphate groups on the outside. This arrangement is thermodynamically favorable because the relatively nonpolar bases are on the inside

interacting with each other away from water molecules, while the polar phosphate groups on the outside can interact with water and other polar molecules in the medium.

29.3 PURINE AND PYRIMIDINE BASES

The purine and pyrimidine bases found in DNA and RNA can exist in different tautomeric forms. For example, adenine exists in the amino and the imino forms in equilibrium with one another. The same is true for cytosine. In both cases the equilibrium favors the amino form, which is the only tautomer found in nucleic acids.

Adenine

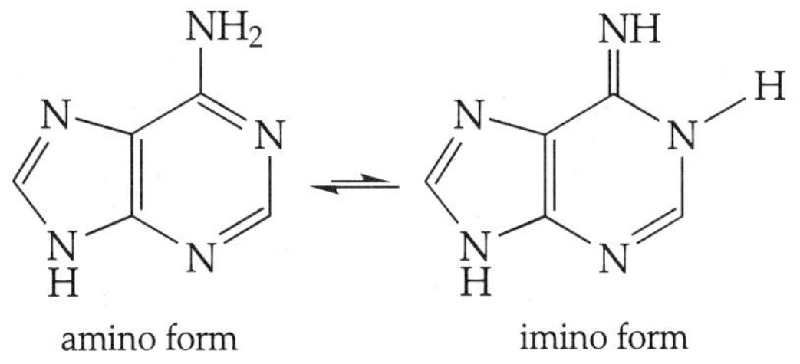

amino form   imino form

Cytosine

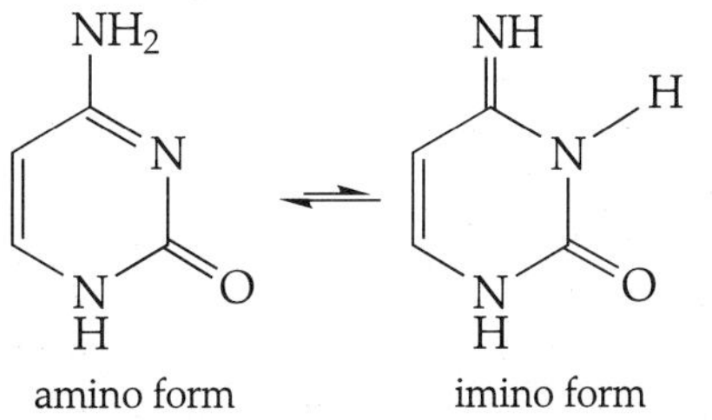

amino form   imino form

In the case of guanine, uracil, and thymine, the keto form is in tautomeric equilibrium with the enol form. The keto form is more stable than the enol form and is the major constituent at equilibrium. The less-stable tautomeric forms are present at equilibrium in very small amounts that usually do not exceed 0.01%. This is also true when the bases are bound, forming part of DNA and RNA.

Guanine

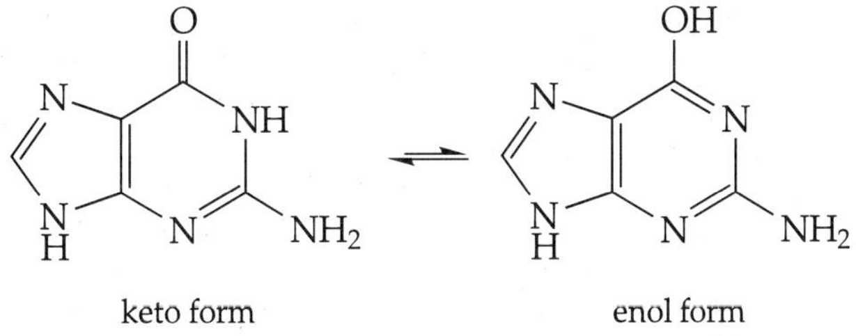

keto form   enol form

The purine and pyrimidine bases are aromatic. This is perhaps obvious in the case of adenine (in the amino form) and guanine (in the enol form) because they have three alternating "double bonds" in a six-membered ring. What may not be evident however, is that all five bases in all the tautomeric forms are aromatic. Let's consider, for example, cytosine. A close inspection reveals that in one of the resonance structures, the carbonyl is polarized with the oxygen atom bearing a negative charge and the carbon atom a positive charge (see below). The total number of π electrons in the ring is six: three are contributed by the carbon atoms at positions 4, 5, and 6, one by the nitrogen at position 3, and two by the nitrogen at position 1. With a total of six electrons, this ring is aromatic according to Huckel's rule ($4n + 2$; $n = 1$). We can carry out a similar analysis for all the other purine and pyrimidine bases in any of their tautomeric forms.

Cytosine

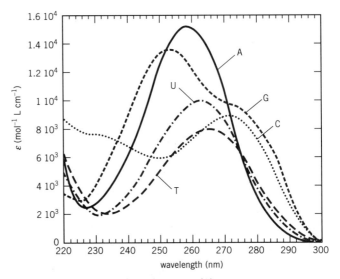

Because of the extended π–system of the purine and pyrimidine bases, nucleosides, nucleotides, and nucleic acids absorb ultraviolet radiation. They show maxima around 185, 205, and 260 nm due to $\pi \longrightarrow \pi^*$ transitions (see Section 32.1). The free base and the corresponding nucleoside and nucleotide all absorb in the same region and have nearly the same molar absorptivities and λ_{max}. The absorption of UV light around 260 nm is responsible for UV-induced DNA damage that may result in cancer and other genetic aberrations. The UV spectra of nucleosides in the 220–300 region are shown in Figure 29.1.

Figure 29.1   UV spectra of nucleosides adenosine (A); guanosine (G); cytidine (C); uridine (U); and free base thymine (T). Concentration $4-9 \times 10^{-5}$ M; pH 7.2. Adapted from Reference 8.

29.4   CHEMICAL TRANSFORMATIONS

The base residues as well as the sugar and phosphate groups of DNA and RNA are susceptible to chemical transformations that may lead to genetic damage. Chemicals normally present in the cell, environmental insults such as ultraviolet and ionizing radiation (X-rays, γ-rays), and exogenous chemicals all can produce modifications of the purine and pyrimidine bases, causing mutations that may lead to cancer.

The purine and pyrimidine rings can be attacked by electrophiles and nucleophiles alike. On one hand, the heterocyclic nature of these compounds with the electron-withdrawing nitrogen atoms in the ring skeleton leaves the carbon atoms adjacent to the nitrogens with a positive charge density and makes them the target for nucleophilic reagents such as water and amines. On the other hand, the nitrogen and the oxygen atoms in or attached to the rings are nucleophilic and react with electrophiles.

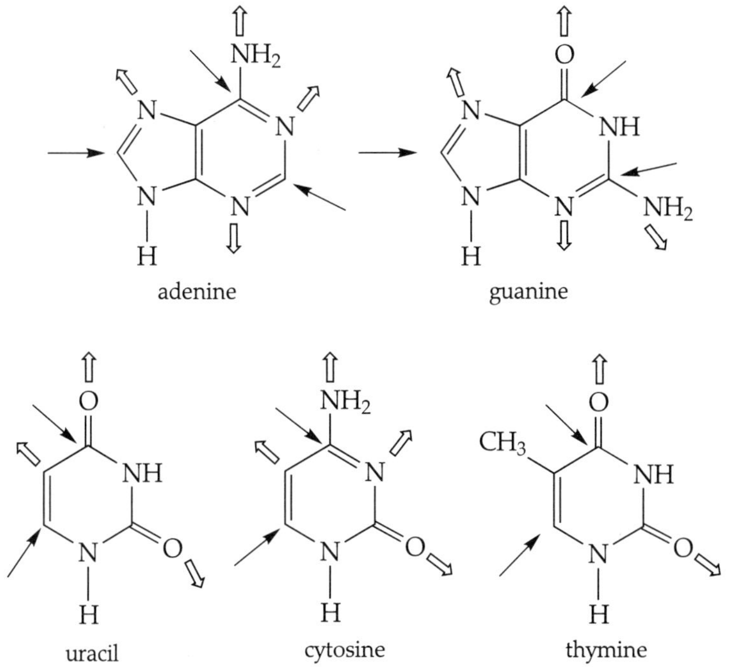

adenine

guanine

uracil

cytosine

thymine

→ indicates an electrophilic center (attacked by nucleophiles).
⇒ indicates a nucleophilic center (attacked by electrophiles).

An example of *nucleophilic attack* is the reaction of cytidine with water. This reaction leads to a **deamination** product where the $-NH_2$ is replaced by an oxygen atom. The product is uridine.

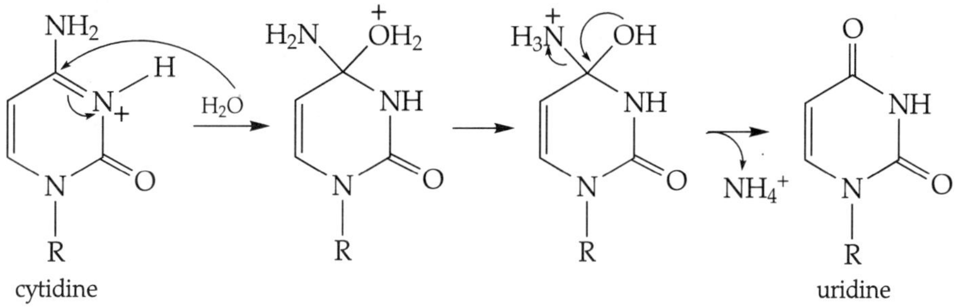

cytidine

uridine

This reaction takes place under physiological conditions at a low rate but it can be accelerated by physical and chemical factors.

Deamination of cytidine also occurs as a consequence of reaction with nitrous acid. In this case the mechanism probably involves the formation of a diazonium group ($-N_2^+$) from the $-NH_2$ group, followed by a substitution reaction with water.

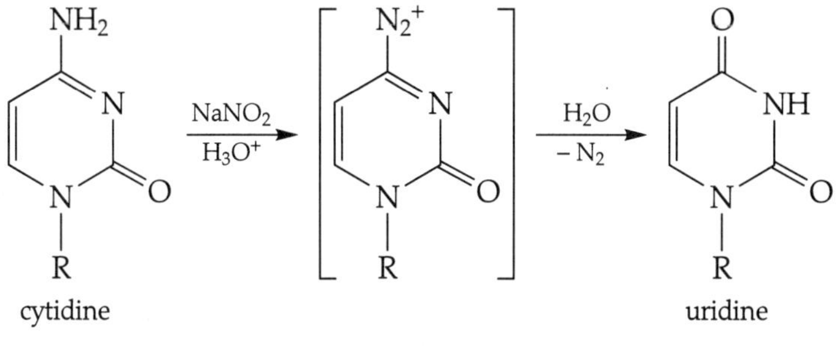

cytidine

uridine

The ease of transformation of cytidine into uridine explains why DNA contains thymine instead of uracil. DNA has a built-in repair mechanism that recognizes nonnative bases and replaces them with the correct ones. This mechanism ensures the fidelity of DNA as the repository of genetic information. When a cytosine residue of DNA is deaminated to a uracil residue, this is recognized by DNA as an abnormal base and eliminated from its sequence before it leads to genetic aberrations. However, if uracil were a native base of DNA, the built-in mechanism in DNA wouldn't be able to distinguish between the native and nonnative uracil, and the mutation wouldn't be corrected. Having thymine instead of uracil makes it possible for DNA to recognize any uracil as a mutation and remove it from its sequence.

An example of *electrophilic attack* on the heterocyclic bases is the reaction with alkylating agents. Compounds such as methyl bromide (used as a pesticide), dimethyl sulfate, nitrosamines (present in burnt meats), and mustard gas are damaging to DNA because they transfer alkyl groups to the nitrogen and oxygen atoms of the bases, and preclude their correct pairing.

Alkylating Agents

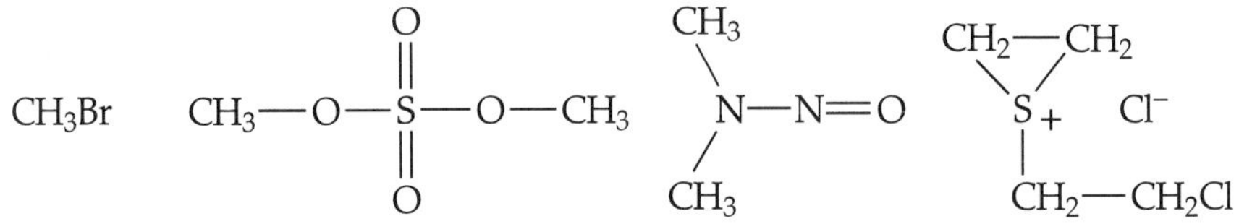

methyl bromide   dimethyl sulfate   dimethylnitrosamine   mustard gas

Usually one of the nitrogen atoms of the purine and pyrimidine rings is more reactive than the others and is preferentially alkylated. Which position is alkylated depends, among other factors, on the alkylating agent and whether the base is free in solution or forming part of a nucleic acid. Especially reactive with alkylating agents are nitrogens 1, 3, and 7 of adenine, nitrogens 3 and 7 of guanine, and oxygen 4 of thymine. The following example shows the reaction of free adenosine with dimethyl sulfate giving 1-methyladenosine, which is almost the only alkylated product when the reaction is carried out in solution. However, when DNA is alkylated with dimethyl sulfate, the adenine residue is preferentially alkylated in position 3 instead of 1.

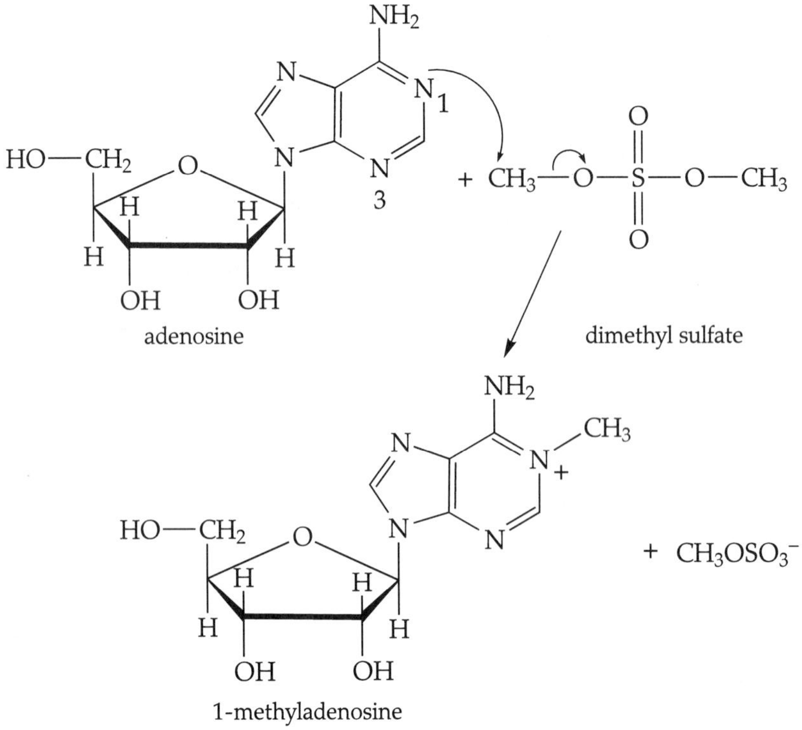

adenosine   dimethyl sulfate

1-methyladenosine

BIBLIOGRAPHY

1. Basic Principles in Nucleic Acid Chemistry. P.O.P. Ts'o, ed. Vol. II. Academic Press, New York, 1974.

2. Chemistry of Nucleosides and Nucleotides. L.B. Townsend. Vol. 1. Plenum Press, New York, 1988.

3. Bioorganic Chemistry: Nucleic Acids. S.M. Hecht. Oxford University Press, New York, 1996.

4. Molecular Biology of Mutagens and Carcinogens. B. Singer and D. Grunberger. Plenum Press, New York, 1983.

5. DNA Repair and Mutagenesis. E.C. Friedberg, G.C. Walker, and W. Siede. ASM Press, Washington, D. C., 1995.

6. Principles of Nucleic Acid Structure. W. Saenger. Springer-Verlag, New York, 1984.

7. The Tautomerism of Heterocycles. J. Elguero, C. Marzin, A.R. Katritzky, and P. Linda, ed. Academic Press, New York, 1976.

8. Absorption Spectra in the Ultraviolet and Visible Region. L. Lang, ed. Academic Press, New York, 1961.

EXERCISES

Problem 1

Explain why uracil is aromatic. Draw resonance structures.

Problem 2

Draw structures for three different enol forms of uracil.

Problem 3

Write the complementary strand of the following DNA fragment:

5'-ATTATACCAGGATGATAC-3'

Problem 4

How would you distinguish adenine and guanine by UV spectroscopy if you could measure the absorbance at only two wavelengths of your choice? What wavelengths would you choose?

Problem 5

Consider Figure 29.1 and decide what wavelength you would choose to monitor the deamination of cytosine into uracil.

Problem 6

(*Previous Synton: Problem 5, Unit 26*) Synton and Ester are getting married. They received the most unusual gift from Ester's parents, an HPLC system. And as Vanadia said, it wasn't very romantic but it was certainly practical. It would be of great help in their company, "Lotions & Potions, Inc.," which has been thriving. Synton is setting the instrument for the first time. He runs a mixture of standards sent by the manufacturer consisting of adenosine, guanosine, uridine, and cytidine. Which compound has the largest response factor (area/concentration) at 254 nm? Which one the lowest? What would happen with the response factors if Synton sets the detector at 240 nm?

EXPERIMENT 29

Analysis of Nucleosides

E29.1 OVERVIEW OF THE EXPERIMENT

High-performance liquid chromatography (HPLC) is a powerful tool for the investigation of biomolecules. Most biomolecules such as sugars, proteins, and nucleic acids are nonvolatile and cannot be directly analyzed by gas chromatography. For such systems, HPLC, which does not require sample vaporization, is very useful.

In the first part of this experiment you will analyze an RNA digest, which is a mixture of nucleosides, and will determine the composition of the RNA. HPLC using a phased-reversed column can separate the four different nucleosides in the mixture. Measurement of the peak areas and comparison with those of standard mixtures will give you the quantitative composition of the mixture. Mixtures like these are obtained in the molecular biology laboratory by hydrolysis of RNA samples.

In the second part of the experiment you will study the transformation of cytidine into uridine produced by nitrous acid. This type of reaction can cause mutations when it happens in DNA or RNA. HPLC is an effective tool to investigate the course of this transformation.

For the first part of the experiment you will use the standard curve method to analyze the mixture of nucleosides. Read the standard curve method in the HPLC unit (section 10.13). Several standard mixtures of nucleosides will be already prepared and ready to use. These mixtures contain adenosine, guanosine, uridine, and cytidine in known concentrations. Solutions of the individual nucleosides will also be available to you.

In the second part, you will follow the reaction of cytidine with nitrous acid by HPLC. You will mix cytidine with sodium nitrite in the presence of acetic acid and follow the course of the reaction by taking samples at different times and injecting them into the HPLC. You will confirm that the product is uridine by comparing the elution volume of the product with that of a uridine standard. A plot of the product peak area versus time will allow you to estimate the reaction rate.

PROCEDURE

Safety First

- Methanol is toxic and flammable.
- Sodium nitrite is toxic.

Background reading: 10.1–10; 10.13–14; 29.1–3; 30.3; 32.4
Estimated time: E29.2: 3–4 hr. E29.3: 2–3 hr

E29.2 ANALYSIS OF AN RNA DIGEST

HPLC conditions
Column: C18; particle size: 5 μm; dimensions: 250 × 4.6 mm
Solvent: 92.8% water; 7.2% acetonitrile; 10 mM trifluoroacetic acid/triethylamine; pH 5.7.
Flow rate: 1 mL/min
Detection: 254 nm
Chart speed: 2.5 cm/min
Injection loop: 20 μL

You will inject several standard mixtures containing the four nucleosides found in RNA: adenosine, guanosine, uridine, and cytidine. You will determine the peak areas by a triangulation method, and for each nucleoside you will plot the peak areas versus the concentrations in the standard solutions. The slope of the curve (area vs. concentration) is the response factor. The HPLC is equipped with an ultraviolet detector that monitors the absorbance of the peaks at 254 nm; therefore, the response factor is directly proportional to the nucleoside molar absorptivity at 254 nm (see sections 30.3 and 32.4). To facilitate your work, you will team up; each student will analyze only one standard solution (at least in duplicate) and share the results with the rest. At least four different standard solutions should be analyzed by the entire class to determine the response factors. To identify each peak in the mixture of standards, each nucleoside should be individually injected and its elution volume determined. Each student will individually analyze an unknown mixture of nucleosides and determine its composition. Your solutions should be particle free before injecting into the HPLC. Centrifuge, if necessary, to separate solid particles.

Set the absorbance range of the detector at an intermediate reading (0.5) and make sure that the system produces a steady baseline. An absorbance setting of 0.5 means that absorbance values in the range 0–0.5 will be on scale, and absorbance values larger than 0.5 will be off scale.

Stock Solutions

Four stock solutions, one for each of the four nucleosides adenosine, guanosine, cytidine, and uridine, will be provided to you. These solutions are made in the same HPLC solvent used for the runs. Each solution is approximately 0.1 mg/mL and ready to be injected. The recommended absorbance setting is 0.5. Inject them one by one to determine the elution volume for each nucleoside. To obtain reproducible results wash the injection loop, *with the injection valve in the load position*, at least 3 times with about 100 µL (each time) of the sample. Load 25 µL of sample, inject it (Fig. 10.5) and mark the beginning of the run. The loop always delivers the same amount of sample to the column (20 µL).

The peaks should be on scale. If they are not, change the detector sensitivity accordingly. Decrease the absorbance range (0.2; 0.1; 0.05; etc.) to obtain bigger peaks, or increase the range (1.0; etc.) to get smaller peaks. Repeat the injections until you find the optimum absorbance range where all the peaks are on scale. Determine the elution volume for each nucleoside. Repeat the injection and take the average of the elution volumes.

Standard Solutions

Make four standard solutions by mixing known volumes of the stock solutions, so that the concentrations for each nucleoside in all four solutions are different. The volumes for standard solutions 1 and 2 are given in Table 29.1, along with the dilution factors. The dilution factor is the volume of each stock solution divided by the total volume of the standard solution (4 mL). For standard solutions 3 and 4, choose volumes of adenosine, guanosine, cytidine, and uridine stock solutions that are different from the ones already used for each nucleoside. The total volume should be 4 mL. Calculate the dilution factor for each nucleoside in each solution. The product of the dilution factor times the concentration of the stock solution gives the concentration of the nucleoside in the standard solution:

$$[\text{nucleoside}]_{\text{standard}} = [\text{nucleoside}]_{\text{stock}} \times \text{dilution factor}$$

Inject the standard solutions following the same procedure as that for the stock solutions. All the standard solutions should be analyzed **with the same absorbance setting** and all the peaks should be on scale. Measure the areas and determine the elution volumes. Inject each standard mixture in duplicate and take the average values.

Table 29.1 **Volumes and Dilution Factors for Standard Solutions**

	Adenosine	Guanosine	Cytidine	Uridine	Standard solution
volume of stock solution (mL)	1.00	1.00	1.00	1.00	
dilution factor	1/4	1/4	1/4	1/4	1
volume of stock solution (mL)	1.50	0.50	1.50	0.50	
dilution factor	1.5/4	0.5/4	1.5/4	0.5/4	2
volume of stock solution (mL)					
dilution factor					3
volume of stock solution (mL)					
dilution factor					4

For each nucleoside, plot the area of the peak as a function of the nucleoside concentration in the standard mixture. You should obtain a straight line that passes through the origin. The slope of this line is the response factor.

Unknown

Obtain your unknown RNA digest mixture and weigh approximately 5 mg (0.0050 g) using an analytical balance and a piece of weighing paper. Transfer the solid quantitatively to a 50.0-mL volumetric flask, add the elution solvent to the mark, and shake well to dissolve. Run your unknown as you ran the standard mixtures. Identify all the nucleosides in your mixture and measure the area for each peak and determine its concentration using the response factor:

$$\text{concentration} = \frac{\text{area}}{\text{response factor}}$$

E29.3 DEAMINATION OF CYTIDINE TO URIDINE

cytidine → uridine (NaNO$_2$ / AcOH)

You and a classmate will study the reaction at two different temperatures: room temperature and 30°C.

For each reaction, place in a scintillation vial (or in a small test tube with a cork stopper) 2.0 mL of a freshly prepared solution of cytidine (2.0 mg/mL) in acetate buffer (pH 3.8; 3.6 M) and add 1.00 mL of a solution of sodium nitrite in water (4 M), and mix well. Remove 10-µL portions at 10-minute intervals for the reaction at room temperature and at 5-minute intervals for the reaction at 30°C. Quench the aliquots in 990 µL of HPLC solvent contained in a snap-cap plastic vial and inject them into the HPLC. Follow the reaction at 30°C for 1 hour and the one at room temperature for 1.5 hr.

Compare the retention times of the peaks with those of cytidine and uridine. You will also see a peak for sodium nitrite. To determine where sodium nitrite elutes, mix 1 mL of the 4 M sodium nitrite solution with 2 mL of acetate buffer (pH 3.8 , 3.6 M), take a 10-µL aliquot of this solution, dilute it with 990 µL of HPLC solvent, and inject it into the HPLC. The recommended setting for the detector is 0.1.

Using the height of the peaks as a measure for their concentration, calculate the percentage of cytidine and uridine in the reaction mixture. (In this case the height of the peaks, instead of their areas, can be used because the widths of the cytidine and uridine peaks are nearly the same.) Plot the percentage of cytidine against time and estimate the time when 50% of the cytidine was converted to uracil. This is the half-life of the reaction ($\tau_{1/2}$) (section 18.5). An estimate of the reaction rate can be obtained with the following formula:

$$\text{reaction rate: } k = (\ln 2)/\tau_{1/2} = 0.69/\tau_{1/2}$$

Compare the reaction rates at both temperatures. A plot of $\ln k$ versus the inverse of the absolute temperature ($1/T$ (K)) will allow you to estimate the reaction rate at physiological temperature (37°C).

> **Cleaning Up**
>
> - Dispose of the solvent from the HPLC waste and all the solutions used in E29.1, E29.2, and E29.3 in the container labeled "Nucleosides Liquid Waste."

BIBLIOGRAPHY

1. HPLC for Undergraduate Introductory Laboratory. S.A. Van Arman and M.W. Thomsen. *J. Chem Ed.*, **74**, 49–50 (1997).

2. The Reaction of Ribonucleoside with Nitrous Acid. Side Products and Kinetics. R. Shapiro and S.H. Pohl. *Biochemistry*, **7**, 448–455 (1968).

EXPERIMENT 29   REPORT

Pre-lab

1. Make a table with the physical properties of adenosine, guanosine, cytidine, and uridine (MM, mp, solubility, molar absorptivities [see below], etc.)

2. Explain why adenine, guanine, cytosine, and uracil are aromatic. Draw resonance structures for each compound.

3. The molar absorptivities of adenosine, guanosine, cytidine, and uridine at pH 7.2 and different wavelengths are:

Nucleoside	ε (mol^{-1}L cm^{-1}) at 254 nm	ε (mol^{-1}L cm^{-1}) at λ_{max}
adenosine	14,400	258 nm: 15,200
guanosine	13,600 (λ_{max})	274 nm: 9,500 (shoulder)
cytidine	6,300	272 nm: 8,900
uridine	8,700	262 nm: 10,000

Which compound would have a larger response factor (area/concentration) at 254 nm?

4. In the analysis of nucleosides, the peak for guanosine is almost full scale when the absorbance range is 2.0 and the eluent is monitored at 254 nm. Would the same peak be on scale if the absorbance range is 1.0 and the detection is done at 274 nm? Explain. (Hint, use the molar absorptivities given in the table above.)

5. Speculate why uridine is not a native base of DNA.

6. What type of reaction is the deamination of cytidine with nitrous acid?

In-lab

1. Calculate the HPLC response factors for adenosine, guanosine, cytidine, and uridine at 254 nm. Present your data in a table format. Show calculations.

2. In the same table report the elution volumes for the nucleosides.

3. Report the qualitative and quantitative composition of your RNA digest. Discuss any source of error. Show calculations.

4. Interpret your deamination of cytidine HPLC data.

5. From the plot % cytidine versus time, estimate the half-life and the reaction rate for the deamination reaction at room temperature and 30°C.

6. Using the ln k versus $1/T$ plot (where T is the absolute temperature), estimate the reaction rate at 37°C.

SECTION 3

Spectroscopy

Absorption Spectroscopy

This unit is an introduction to the absorption spectroscopies treated in more detail in Units 31–33 (infrared, ultraviolet-visible, and nuclear magnetic resonance). Here we will review some of the concepts common to all of them, namely the nature of light and the absorption process.

30.1 THE NATURE OF LIGHT

What we call light is a narrow range of the **electromagnetic spectrum** to which our retina is sensitive. The nature of light has fascinated philosophers, artists, and scientists for centuries. At the beginning of the twenty-first century its nature is still elusive. Electromagnetic radiation displays dual behavior: **particle** and **wave**.

As a wave, electromagnetic radiation can be characterized by its wavelength (λ) and frequency (ν). Wavelength is the distance between two consecutive crests (or troughs), see Figure 30.1. Wavelength and frequency are related by the speed of light (c), whose value in a vacuum is $c = 2.99792 \times 10^8$ m/s (Eq. 1):

$$\lambda \times \nu = c \qquad (1)$$

The units of wavelength are those of length: meter, **m**; centimeter, **cm**; micron, **μm** (1 μm = 10^{-6} m); nanometer, **nm** (1 nm = 10^{-9} m); Ångström, **Å** (1 Å = 10^{-10} m), and so forth. The unit of frequency is 1/second = s^{-1} = 1 hertz = 1 Hz. The speed of light decreases as radiation passes from a vacuum to any other medium.

The electromagnetic spectrum can be divided into arbitrary regions of wavelengths of which **ultraviolet** (UV), **visible**, and **infrared radiation** (IR) encompass a narrow range (Fig. 30.2). According to Equation 1, as the wavelength increases, the frequency decreases because the speed of light is constant. UV radiation, for example, has higher frequency than IR radiation and smaller wavelength.

Besides frequency and wavelength, another parameter is often used to characterize radiation: the **wavenumber**. By definition, the wavenumber ($\bar{\nu}$) is the reciprocal of the wavelength (Eq. 2). The most commonly used unit for wavenumber is cm^{-1}, and is extensively used in IR spectroscopy.

$$\bar{\nu} = \frac{1}{\lambda} \qquad (2)$$

Combining Equations 1 and 2, Equation 3 is obtained:

$$\bar{\nu} = \frac{\nu}{c} \qquad (3)$$

which indicates that frequency and wavenumber are directly proportional.

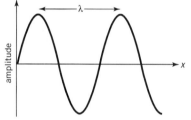

Figure 30.1 Wave-like description of light.

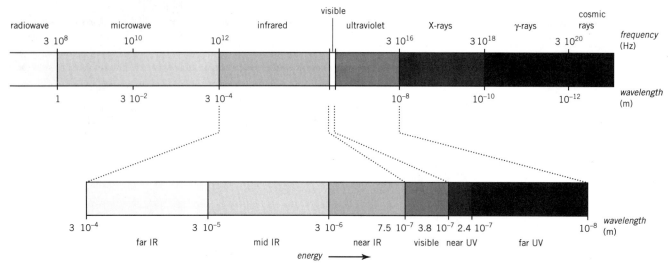

Figure 30.2 Regions of the electromagnetic spectrum.

According to the corpuscular description of light, electromagnetic radiation behaves like massless particles called **photons**, whose energy, E, is quantized and directly proportional to the frequency (Eq. 4). The proportionality constant, h, is a universal constant called Planck's constant $(h = 6.6242 \times 10^{-27}$ erg s), in honor of Max Planck, a German physicist and one of the founders of quantum mechanics.

$$\boxed{E = h \times \nu} \tag{4}$$

Combining Equations 1 and 4, Equation 5 is obtained:

$$E = h \times \frac{c}{\lambda} \tag{5}$$

Equations 4 and 5 indicate that *the higher the frequency of the electromagnetic radiation (or the shorter the wavelength), the higher the energy.*

30.2 INTERACTION BETWEEN ELECTROMAGNETIC RADIATION AND MATTER

Can we get a suntan on a cloudy day? Can we feel the heat? Can we get a suntan through a closed window when it is sunny outside? Can we feel the heat? The answers to these questions are: Yes, No, No, Yes.

The tanning of our skin is the result of the interaction between UV radiation and skin pigments. The heat sensation is due, in part, to the infrared radiation present in sunlight. The sunlight filtered by water in the clouds contains most of the UV radiation and lacks the IR component. This is why we can get a suntan but do not feel the heat when it is cloudy. On the other hand, the sunlight filtered through a window lacks the UV component but still contains IR radiation. We cannot get a suntan but we feel the heat. These observations indicate that there must be something intrinsic to glass and water that accounts for their disparate behavior as light filters (Fig. 30.3).

To understand the interaction between radiation and matter, we must descend to the molecular level and consider atomic motions. The motion of polyatomic molecules is very complex because all the atoms are moving simultaneously. However, the overall motion can be decomposed into three simple components: **translation, rotation,** and

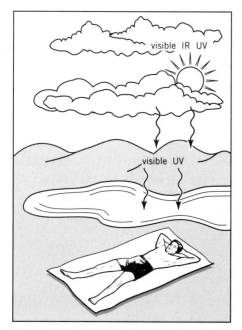 

Figure 30.3   *Left*: IR radiation is absorbed by clouds. *Right*: UV radiation is absorbed by glass.

vibration. Translation refers to the movement of the whole molecule in space from a point A to a point B. Rotation refers to the spinning of the molecule around the x, y, and z axes. Vibration describes the relative motions of atoms within the molecule. Because of these different contributions to the overall motion, a molecule's total kinetic energy can be divided into three components: translational, rotational, and vibrational.

In addition to kinetic contributions, potential energy also adds to a molecule's total energy. The potential energy of a molecule is provided by the bond energy or electronic energy, or in other words, "the glue" that keeps the atoms together. Therefore, the total energy (E_T) of a molecule is the sum of the **electronic** (E_e), **vibrational** (E_v), **rotational** (E_r), and **translational energy** (E_t) (Eq. 6):

$$E_T = E_e + E_v + E_r + E_t \tag{6}$$

The largest contribution to the total energy is that of electronic energy (on the order of hundreds of kJ/mol); the second-largest contributor is E_v (on the order of a few kJ/mol); E_r and E_t are both on the order of a few tenths of 1 kJ/mol.

With the exception of translational energy, the energy components are quantized. *Electronic, vibrational, and rotational energy can be taken or given by a molecule only in discrete amounts or **quanta** of energy.* A simplified representation of molecular energy levels is given in Figure 30.4.

For each electronic energy level, there are several vibrational levels and each one of them is comprised of numerous rotational levels. The energy difference between electronic levels is comparable to bond energies (200–400 kJ/mol) and much larger than the difference between vibrational levels (4–20 kJ/mol); rotational levels within the same vibrational level are only marginally different in energy. At room temperature, most molecules are in the first vibrational level of the first electronic level; this is called the **ground state**. Molecules at room temperature are found in all of the rotational levels of the ground state because these levels are very close in energy. The number of molecules in each energy level is called the **population** of the level. The population can be changed by absorption of a quantum of energy provided by electromagnetic radiation. Photons of the right energy (and therefore of the right frequency, Eq. 4) can provide the exact amount of energy to send molecules from a lower energy state

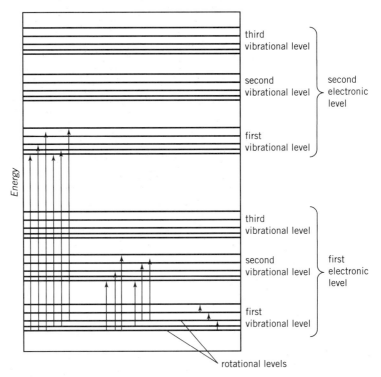

Figure 30.4 Energy level diagram for a polyatomic molecule.

(electronic, vibrational, or rotational) to a higher energy state, called an **excited state** (Eq. 7). This process is called a transition or excitation.

$$\Delta E = \underbrace{E_{\text{excited}} - E_{\text{ground}}}_{\substack{\Delta \text{ energy for} \\ \text{transition}}} = \underbrace{h \times \nu}_{\substack{\text{energy of} \\ \text{photon}}} \tag{7}$$

Radiation of high energy (high frequency) is required to produce electronic excitations. These electronic transitions can be induced by ultraviolet-visible radiation and X-rays. Vibrational transitions require less energy than the electronic excitations and therefore are produced by radiation of lower frequency (larger wavelength) than UV radiation such as IR radiation. Rotational transitions, demanding even less energy than vibrational transitions, can be produced by microwave radiation (less energetic than IR; Fig. 30.2). The study of the absorption of UV, IR, and microwave radiation is the subject of different branches of spectroscopy.

As already mentioned, all the rotational levels of the ground state are very close in energy and are populated at room temperature. As a result of this, an electronic transition can start at any rotational level of the ground state and finish, in principle, at any rotational level of any vibrational level of the second electronic state (Fig. 30.4). Because there are many possible initial and final states, the ΔE of an electronic transition is not unique but spans over a relatively wide range. Consequently, radiation of a wide frequency range is absorbed.

30.3 ABSORPTION SPECTROSCOPY

When a beam of electromagnetic radiation passes through a sample, photons within a certain frequency range are absorbed. This absorption of energy, as discussed above, ultimately causes electronic, vibrational, or rotational transitions. As a result, the power (power = energy/time) of the radiation is decreased (Fig. 30.5). If the power of the incident radiation is P_0, the power of the transmitted radiation, or throughput, is P, where $P \leq P_0$. The ratio between P and P_0 is called **transmittance**, T (Eq. 8):

$$T = \frac{P}{P_0} \tag{8}$$

When no photons are absorbed by the sample, the power of the radiation is not diminished, $P = P_0$, and $T = 1$; on the other hand, when total absorption occurs, the throughput is zero, and $T = 0$. Transmittance is usually expressed as a percentage: $\%T = 100 \times T$. By definition, $-\log T$ is called **absorbance**, A, (Eq. 9):

$$A = -\log T \tag{9}$$

Combining Equations 8 and 9:

$$A = -\log \frac{P}{P_0} = \log \frac{P_0}{P} \tag{10}$$

Absorbance values range between $A = 0$ ($T = 1$) and $A = \infty$ ($T = 0$).

It has been found that the absorbance increases linearly with the thickness of the sample (pathlength), ℓ, and the concentration of the molecules that absorb radiation, C (Eq. 11). The proportionality constant, ε, is called the **molar extinction coefficient** or **molar absorptivity**.

$$\boxed{A = \varepsilon \times C \times \ell} \tag{11}$$

Absorbance (also called optical density in old literature) has no units; the molar absorptivity has units of $M^{-1}\,cm^{-1}$ or $L\,mol^{-1}\,cm^{-1}$. Equation 11 is known as **Lambert–Beer's law**.

The molar absorptivity of a compound measures its ability to absorb radiation of a given frequency. It strongly depends on the frequency and, to a lesser extent, on the solvent and the temperature of the sample. A plot of absorbance, transmittance, or molar absorptivity as a function of wavelength, frequency, or wavenumber is called an **absorption spectrum**. Figure 30.6 shows a diagrammatic representation of typical UV and IR spectra for hypothetical compounds. The UV spectra generally show broad bands because, as discussed earlier, the ΔE of an electronic transition is not unique.

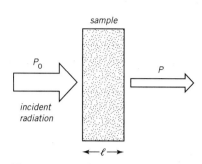

Figure 30.5 Absorption of electromagnetic radiation.

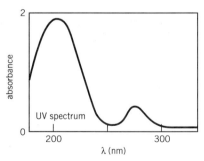
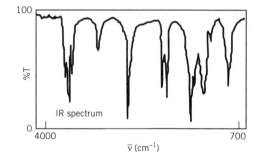

Figure 30.6 Typical UV and IR spectra.

BIBLIOGRAPHY

1. Quantum Chemistry. I.N. Levine. Allyn and Bacon, Boston, 1974.
2. Quantum Mechanics in Chemistry. M.W. Hanna. 3rd ed. Benjamin/Cummings, Menlo Park, CA. 1981.
3. Physical Chemistry. G.W. Castellan. 3rd ed. Addison-Wesley Pub. Co., Reading, MA, 1983.
4. Introduction to Classical and Modern Optics. J.R. Meyer-Arendt. Prentice-Hall, Englewood Cliffs, N.J., 1984.
5. A Unifying Approach to Absorption Spectroscopy at the Undergraduate Level. R.S. Macomber, *J. Chem. Ed.*, **74**, 65–67 (1997).

EXERCISES

Problem 1

Making use of Equations 1–5 and the following conversion factors and constants, complete the table below.

$h = 6.624 \ 10^{-34}$ J s; $N_A = 6.022 \ 10^{23}$ mol^{-1}; 1 cal = 4.184 J

Relate each row to a region of the electromagnetic spectrum.

λ (nm)	λ (Å)	ν (Hz)	$\bar{\nu}$ (cm^{-1})	E (kcal/mol)	region
280					
			1700		
		$3 \ 10^{10}$			

Problem 2

The absorbance at 450 nm of a solution of β-carotene in petroleum ether is $A_{450} = 1.06$. If the molar absorptivity is 127,000 M^{-1} cm^{-1} and the pathlength is 1 cm, calculate the concentration of β-carotene in solution.

Problem 3

The absorbance at 280 nm of a protein solution in water (0.1% w/v) is $A_{280}^{0.1\%} = 1.2$ using a cell of 1.0 cm pathlength. If the molecular mass of the protein is 70,000 g/mol (70 kDa), calculate the molar absorptivity at 280 nm in M^{-1} cm^{-1}.

Problem 4

The % transmittance of an IR band at 1700 cm^{-1} is 5% when a 20 μm pathlength is used. The concentration of the compound is 0.1% (w/v) in carbon tetrachloride. If the molecular mass of the compound is 230 g/mol, calculate the molar absorptivity in M^{-1} cm^{-1}.

Infrared Spectroscopy

Infrared (**IR**) spectroscopy was one of the most powerful tools for the elucidation of chemical structures in the 1950s and 1960s. With the explosive development of nuclear magnetic resonance (NMR) in the last 30 years, IR has been relegated to a less prominent position. Although the wealth of information provided by NMR overshadows IR spectroscopy, IR spectra are still routinely run, partly because it is fast and inexpensive, and partly because it can provide direct information about the presence (or absence) of certain functional groups in a way that no other spectroscopic technique can. With a simple look at the IR spectrum, a trained chemist can tell what functional groups are likely to be present in the molecule. In a few words, IR spectroscopy is the *spectroscopy of functional groups*.

The interaction between infrared radiation and matter causes vibrational excitation through the absorption of energy quanta (photons). Most organic molecules absorb IR radiation in the wavenumber region $700-4000$ cm^{-1}. As mentioned in the previous unit, a record of transmittance or absorbance as a function of the wavenumber in the IR region constitutes an IR spectrum. To understand the physical process behind IR spectroscopy we should take a close look at the vibration of molecules.

31.1 MOLECULAR VIBRATIONS

For the sake of simplicity let us consider a diatomic molecule such as H—Cl. In HCl, as in any other molecule, the atoms are constantly moving along the bond axis; the distance between H and Cl oscillates around an equilibrium value, r_e, the *bond length* (Fig. 31.1). This process is called a **vibration** and can be studied in detail with the help of classical and quantum mechanics. A rigorous treatment of such a system is beyond the scope of this unit. We will present, however, some of its most important features and applications to IR spectroscopy.

At first approximation, a diatomic molecule can be treated like two balls attached to the ends of a spring (Fig. 31.1). The frequency at which a diatomic molecule vibrates, ν_0, is given by Equation 1:

$$\nu_0 = \frac{1}{2\pi}\sqrt{\frac{k}{m_r}} \tag{1}$$

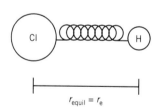

Figure 31.1 HCl molecule as an oscillator.

where k is the **force constant** of the bond and m_r is the **reduced mass** of the two atoms. The force constant is directly related to the energy of the bond. In general, as the bond energy increases, so does the force constant; for example, a C=C bond has a larger force constant than a C—C bond. Some typical force constant values are given in Table 31.1. The reduced mass, m_r, for a system of two particles of mass m_1 and m_2

Table 31.1 **Force Constants of Some Typical Bonds**

Bond	Force constant (N/cm)	Bond	Force constant (N/cm)
$C(sp^3)-H$	4.8	$C=O$	12.1
$C(sp^2)-H$	5.1	$C-N$	5.8
$C(sp)-H$	5.9	$C=N$	10.5
$N-H$	6.3	$C\equiv N$	18.0
$O-H$	7.6	$C-F$	5.7
$C-C$	4.5	$C-Cl$	3.65
$C=C$	9.6	$C-Br$	3.15
$C\equiv C$	15.5	$C-I$	2.65
$C-O$	5.1	$S-H$	4.2

is defined in Equation 2:

$$m_r = \frac{m_1 \times m_2}{m_1 + m_2} \tag{2}$$

Quantum mechanics shows that the total vibrational energy is quantized with energy levels given by Equation 3

$$E_n = \left(n + \frac{1}{2}\right) h \times v_0 \tag{3}$$

where E_n is the energy of the *nth vibrational level*, n is the vibrational quantum number ($n = 0, 1, 2, 3 \dots$ etc.), h is Planck's constant, and v_0 is the **frequency of the vibration** as defined in Equation 1. Most molecules at room temperature are in the first vibrational level ($n = 0$) and their vibrational energy is:

$$E_0 = \frac{1}{2} h \times v_0$$

When molecules absorb IR radiation they are promoted to the next energy level E_1:

$$E_1 = \frac{3}{2} h \times v_0$$

The gain in energy is hv_0, as Equation 4 shows:

$$\Delta E = E_1 - E_0 = \frac{3}{2} h \times v_0 - \frac{1}{2} h \times v_0 = h \times v_0 \tag{4}$$

The absorption of energy occurs only if the energy of the photon is equal to the difference in energy between the vibrational states E_1 and E_0, (Eq. 5):

$$\Delta E = \underbrace{E_1 - E_0 = h \times v_0}_{\substack{\Delta \text{ energy for} \\ \text{transition}}} = \underbrace{h \times v}_{\substack{\text{energy of} \\ \text{photon}}} \implies \boxed{v_0 = v} \tag{5}$$

Equation 5 indicates that to cause a vibrational excitation, *the frequency of the electromagnetic radiation (v) should be identical to the vibrational frequency of the bond (v_0).* Electromagnetic radiation in the wavenumber range 600–4000 cm^{-1} (λ: 16.6–2.5 µm; v: $1.8 \times 10^{13} - 12 \times 10^{13}$ Hz) excites the vibrational modes of most organic molecules.

Unlike the vibration of the HCl molecule, where there is only one frequency, the vibration of polyatomic molecules is very complex because all the atoms are simultaneously moving. However, the complex vibration can be regarded as the combination of the vibrations of individual bonds. In principle, the frequency of a bond vibration is independent of the rest of the molecule. For example, a carbonyl group always vibrates in the same wavenumber range ($1650-1800$ cm^{-1}) regardless of the functional group to which it belongs (aldehyde, ketone, ester, etc.); for most alkenes, the C=C double bond vibrates around 1600 cm^{-1}.

As we will see in more detail later, the environment in which a given bond is located (the nature of neighboring atoms, steric constraints, and so on) has a minor but *noticeable* effect on the vibrational frequency of the bond. This effect is of great importance to organic chemists. *These subtle but predictable changes in frequency due to the environment allow us to get valuable structural information from IR spectra.*

It should be noticed that in the absorption of IR radiation, the frequency of the vibration doesn't change. In the excited state, the molecule still vibrates at the same frequency as in the ground state. The energy of the absorbed photon is used to increase the amplitude of the vibration, not its frequency. Following an analogy discussed by R.S. Macomber (Ref. 7), the absorption of IR radiation can be compared to somebody pushing a swing. To keep the swing moving, the frequency of the push (light) has to be equal to the frequency of the swing (vibration) and the energy given in the push is used to increase the amplitude of the swing, not its frequency.

31.2 STRETCHING AND BENDING VIBRATIONS

Let's now analyze one of the simplest polyatomic molecules: H_2O. This molecule has three main vibrations as shown in Figure 31.2. Two vibrations are described by the elongation and shortening of the O—H bonds (Fig. 31.2a and b). This type of vibration where a bond length is altered is called **stretching**. Stretching vibrations are abbreviated by "st." The O—H stretchings can be in phase or out of phase. In the in-phase vibration both H atoms move away or toward the O atom simultaneously, while in the out-of-phase vibration one H atom moves away while the other moves toward the O atom. These vibrations are called **symmetrical** and **asymmetrical** stretching, respectively. The third vibration of the water molecule is described by a pendular motion of the hydrogen atoms (Fig. 31.2c). This vibration represents a kind of vibration that occurs only in polyatomic molecules: *the distortion of a bond angle.* This kind of vibration is called **bending** or **deformation** and is abbreviated by "d" or "δ."

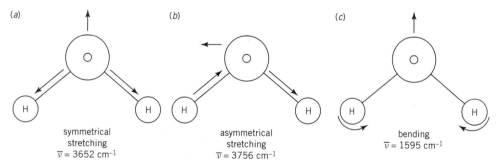

(a) symmetrical stretching $\bar{v} = 3652$ cm^{-1}

(b) asymmetrical stretching $\bar{v} = 3756$ cm^{-1}

(c) bending $\bar{v} = 1595$ cm^{-1}

Figure 31.2 The three vibrations of the water molecule.

It should be noticed that in the vibrations depicted in Figure 31.2 the O atom also moves. The motion of the O atom is shown by a smaller arrow to indicate that O moves less than H due to its larger mass. In true vibrations, as those shown in Figure 31.2, the molecule as a whole does not move.

In IR spectra of polyatomic molecules, besides the bands corresponding to the predicted vibrational transitions such as stretching and bending, there are others of much lesser intensity called **overtones** and **combination bands**. Overtones are bands that appear at a frequency that is approximately twice or three times the vibrational frequency of some fundamental band. For example, in the case of water there is an overtone at approximately twice the wavenumber of the bending vibration: $3152 \text{ cm}^{-1} \approx 2 \times \bar{\nu} = 2 \times 1595 \text{ cm}^{-1}$. Combination bands appear at a frequency that is a sum of fundamental frequencies; these bands are due to a single photon simultaneously exciting two vibrations.

Bending vibrations involve smaller force constants than stretching vibrations between the same atoms and, thus, they occur at lower frequency and require less energy to be excited. When asymmetrical and symmetrical vibrations are possible, the asymmetrical mode always occurs at higher frequencies, as the vibration of water illustrates.

The stretching and bending vibrations of a methylene group in an alkyl chain are shown in Figure 31.3. There are two stretching vibrations, symmetrical and asymmetrical, with the asymmetrical stretching (2925 cm^{-1}) at higher frequencies than the symmetrical (2850 cm^{-1}). There are four different bendings: **scissoring, rocking, wagging**, and **twisting**. In scissoring and rocking, both H's move in the plane of the paper; in wagging and twisting they move in a direction perpendicular to the paper (denoted with $+$ and $-$ signs). Wagging and twisting are called out-of-plane bendings.

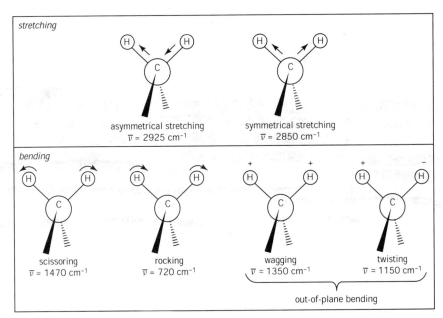

Figure 31.3 Stretching and bending vibrations of the methylene group.

Another kind of **out-of-plane bending** involves H atoms bound to C=C and aromatic rings. Figure 31.4 shows a C—H out-of-plane bending vibration for a generic alkene. The hydrogen atom moves above and below the plane defined by the double bond. As we will see, this kind of out-of-plane bending is very useful in obtaining structural information about alkenes and aromatic compounds.

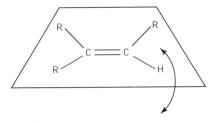

Figure 31.4 Out-of-plane bending.

31.3 IR AND DIPOLE MOMENT

As we discussed in Unit 30, the intensity of an IR band (or more properly its absorbance) depends directly on sample concentration, pathlength, and the molar absorptivity at the frequency under consideration (Lambert–Beer's law; Eq. 30.11). For vibrational transitions, the molar absorptivity is a function of *the change in the dipole moment associated with the vibration*; the larger the change in the dipole moment, the stronger the absorption. Let's consider a heteronuclear diatomic molecule such as HCl or CO. Due to a difference in electronegativity between the intervening atoms there is a net charge density on each atom. Positive and negative charges (q) separated by a given distance (r) generate a **dipole moment**, μ (Eq. 6; see also section 2.1). This is illustrated with H–Cl in Figure 31.5. The dipole moment is a vector aligned with the bond axis and normally represented pointing from the positive to the negative end. In molecules with a center of symmetry, the overall dipole moment is zero because the dipole moments associated with each bond cancel out as Figure 31.5 shows for the case of CO_2.

$$\mu = q \times r \tag{6}$$

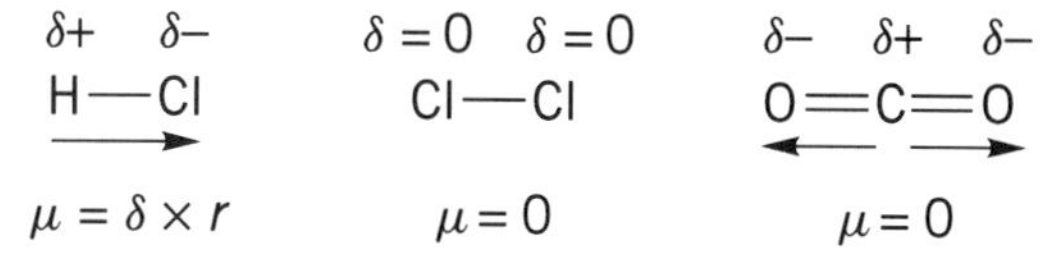

$$\underset{\mu = \delta \times r}{\overset{\overset{\delta+ \quad \delta-}{H—Cl}}{\longrightarrow}} \qquad \underset{\mu = 0}{\overset{\delta = 0 \quad \delta = 0}{Cl—Cl}} \qquad \underset{\mu = 0}{\overset{\delta- \quad \delta+ \quad \delta-}{O=C=O}}$$

Figure 31.5 Dipole moments of some simple molecules.

When a molecule like H–Cl vibrates along its molecular axis, the distance r separating the charges oscillates, causing a periodical variation in the dipole moment. Quantum mechanics shows that only vibrations accompanied by a change in dipole moment give bands in the IR spectrum. These vibrations are called **IR active**. Homonuclear diatomic molecules such as Cl_2, N_2, and H_2 do not absorb in the IR region because their vibrations do not involve a change in the dipole moment (Fig. 31.5).

For polyatomic molecules, the situation is a bit more complicated. The molecule may have an overall dipole moment of zero but still absorb IR radiation. To illustrate this point let us consider the case of CO_2, which has two stretching vibrations, symmetrical and asymmetrical, as illustrated in Figure 31.6. The molecule also has bending vibrations that we will ignore for now. The symmetrical stretching (Fig. 31.6a) does not result in a net change of the dipole moment of the molecule, because the changes associated with the vibration of each C=O bond cancel out. On the other hand, during the asymmetrical stretching (Fig. 31.6b) there is a net change in the dipole moment; the asymmetrical stretching gives a band at 2330 cm^{-1}.

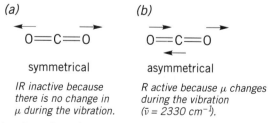

(a) symmetrical

IR inactive because there is no change in μ during the vibration.

(b) asymmetrical

R active because μ changes during the vibration ($\bar{\nu} = 2330$ cm^{-1}).

Figure 31.6 Vibrational modes of CO_2: a) symmetrical stretching, inactive in the IR; b) asymmetrical stretching, absorbs at 2330 cm^{-1}.

The carbonyl group (C=O) gives IR bands of high intensity because of the large dipole moment of the carbon-oxygen bond. On the other hand, carbon-carbon stretching bands (C–C and C=C) are usually of low intensity because carbon-carbon bonds are not highly polarized.

31.4 REGIONS OF THE IR SPECTRUM

Figure 31.7 shows the wavenumber range in which most organic chemicals absorb IR radiation. For practical purposes, the IR spectrum can be divided into three regions: 700 to 1000 cm^{-1}, 1000 to 1500 cm^{-1}, and 1500 to 4000 cm^{-1}.

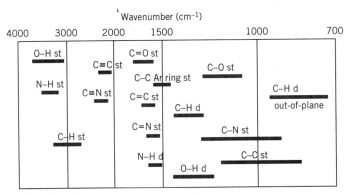

Figure 31.7 Infrared spectral regions and types of vibrations.

- **1500–4000 cm^{-1}:** This region of the IR spectrum gives very valuable information about functional groups. It can be subdivided into three subregions. Between 1500 and 1800 cm^{-1} is where most double bonds absorb (C=C, C=O, C=N); this subregion is very useful in the study of carbonyl compounds. Between 2100 and 2300 cm^{-1} triple bonds (C≡C and C≡N) have bands. Finally, between 2700 and 4000 cm^{-1} stretchings involving H atoms are found (C–H, N–H and O–H); this subregion gives valuable information in the study of alcohols, amines, amides, and carboxylic acids.
- **1000–1500 cm^{-1}:** This part of the IR spectrum is called the **fingerprint region**. It is generally very rich in bands because several types of stretching (C–C, C–O, C–N) and bending (C–H, N–H, O–H) vibrations occur in this wavenumber range. Although the information in this region is unique for each compound (like a fingerprint) and can be used to identify chemicals by comparison with a library of IR spectra, its interpretation is rather difficult and of little interest to most organic chemists. One of the few bands from this region that has some diagnostic value is the C–O–C stretching of ethers and esters (around 1200 cm^{-1}).
- **700–1000 cm^{-1}:** This region is particularly useful in the study of alkenes and aromatic compounds. Most out-of-plane bending vibrations give signals in this wavenumber range. Bands from this region give information about different types of substitutions on aromatic rings. For example, monosubstituted phenyl rings have two bands (around 700 and 750 cm^{-1}), while ortho disubstituted phenyl rings have only one band (around 750 cm^{-1}).

Figure 31.8 shows the IR of o-nitrophenol where the O–H st (3267 cm^{-1}), C–H st (3085 cm^{-1}), C–C st (arom. ring) (1620 cm^{-1}), C–N st (871 cm^{-1}), and C–H d out-of-plane (748 cm^{-1}) can be observed.

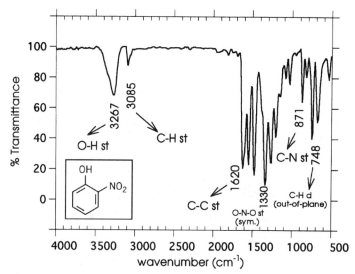

Figure 31.8 IR spectrum of *o*-nitrophenol.

31.5 EXPERIMENTAL ASPECTS

Sample Cells

IR spectra can be run with gas, liquid, and solid samples. In the organic chemistry laboratory, solid and liquid samples are routinely studied by IR. The analysis of gases and vapors requires special handling and is infrequently performed in organic chemistry teaching labs.

Materials such as glass, quartz, and plastics absorb some IR radiation and therefore are limited in their application as sample holders, called **cells** or **windows**. Some inorganic salts are transparent in the range of IR radiation commonly used to run IR spectra ($700-4000$ cm^{-1}), and are employed to manufacture the cells and other optical components of the IR spectrometer. Among the salts with the widest transparency range are NaCl, AgCl, CaF$_2$, and BaF$_2$. Due to its relatively low cost, sodium chloride is perhaps the most widespread material used for IR windows. Because of its high solubility in water, special precautions should be taken when handling NaCl windows. Aqueous solutions cannot be analyzed with NaCl cells, which should be washed only with organic solvents. Sodium chloride plates are brittle and should be handled with care.

Figure 31.9 shows a typical IR sample cell. The spacer is used only for quantitative analysis when a precise pathlength is required. The thickness of the spacer, generally

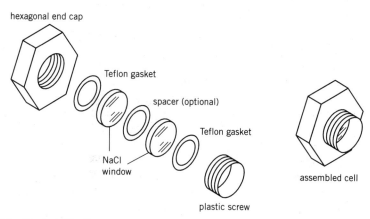

Figure 31.9 Typical IR cell.

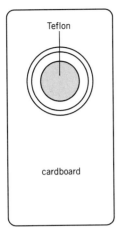

Figure 31.10   Disposable IR card.

made of Teflon, ranges between 0.006 and 1 mm. For qualitative studies, a drop of liquid is spread between the NaCl windows (without the spacer), and the windows placed in the hexagonal end cap between two Teflon gaskets. The assembly is secured in place with the plastic screw that prevents it from falling. The screw should not be excessively tightened because breakage of the brittle NaCl plates may occur.

Teflon (polytetrafluoroethylene) and polyethylene, with only a few well-characterized absorptions in the IR region, are very good materials to hold samples; thin films of Teflon or polyethylene are used for this purpose. **Disposable IR cards** in which a thin film of the polymer covers a hole on a piece of cardboard (Fig. 31.10), are now commercially available. Alternatively, they can be easily manufactured in the laboratory for a fraction of the cost of premade cards.

Homemade disposable IR cards can be prepared with Teflon tape and a piece of cardboard, metal, or plastic. The Teflon tape is first secured to the plate with a piece of adhesive tape. While avoiding touching the plate, a second piece of adhesive tape is applied on the Teflon tape so that about 5 mm of Teflon remains uncovered (Fig. 31.11*a*). The second piece of adhesive tape is then pulled downward to stretch the Teflon tape and cover the hole. When the desired length is reached, the adhesive tape is lowered to make contact with the plate (Fig. 31.11*b*). With most commercially available Teflon tapes, a stretching ratio (final uncovered length/initial uncovered length) of at least 10 is recommended; smaller ratios result in thicker Teflon films, which do not transmit enough IR radiation. The spectrum of Teflon is shown in Figure 31.12*a*.

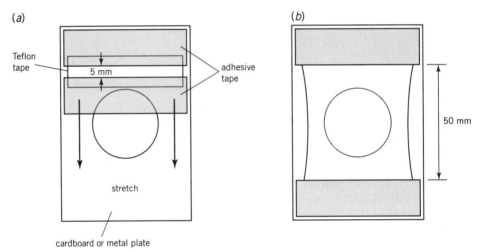

Figure 31.11   Homemade IR card: *a*) before stretching; 5 mm of Teflon tape are uncovered, *b*) after stretching; 50 mm of Teflon tape are uncovered. Stretching ratio = 50/5 = 10.

Preparing the Sample to Run the IR Spectrum

Liquid Samples   Liquid compounds are very easy to analyze because they do not need to be dissolved prior to running the spectrum. A small drop of liquid (5–30 μL) is all it takes to obtain good IR spectra. The liquid is spread between two NaCl plates and the IR is obtained. IR spectra of liquids obtained in this fashion are labeled "liquid film" or "neat," indicating that the IR spectrum of the neat liquid was run with no solvent added. For some special purposes, it may be necessary to dissolve the liquid sample before determining the IR spectrum. In such cases a solvent that does not absorb IR radiation in the region studied should be chosen. Most organic solvents absorb in one or more regions of the IR spectrum and therefore their application is limited. Carbon tetrachloride does not absorb significantly above 1600 cm^{-1}; thus it is suited for the study of the high-frequency region of the IR spectrum (1600–4000 cm^{-1}).

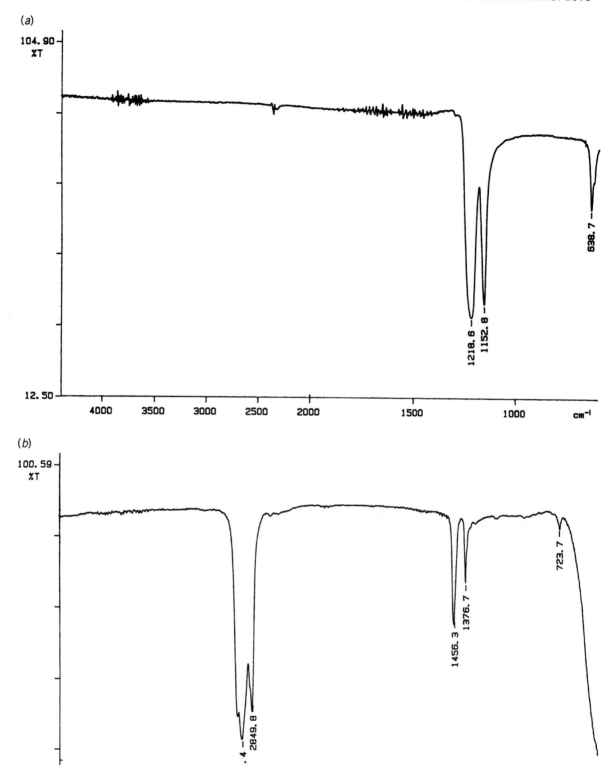

Figure 31.12 *a*) IR spectrum of Teflon. *b*) IR spectrum of nujol.

Tetrahydrofuran does not absorb in the region 1550–1900 cm^{-1}, which makes it an excellent solvent to study carbonyl absorption (1650–1800 cm^{-1}).

Solid Samples   Solids can be studied by IR in numerous ways. A solution can be made in a suitable solvent and the IR of the solution determined. In most cases, a 0.5–2% solution gives bands of adequate intensity. The same limitations already discussed for liquid samples apply to choosing a solvent for solids. The most common way of determining the IR of solids is as a fine dispersion in **nujol**, called a **mull**. Nujol is a mixture of high-boiling-point aliphatic hydrocarbons. It is not a good solvent for most organic compounds and is used as a **dispersant** for the finely divided solid particles. The particle size should be smaller than the wavelength of the IR radiation (<3 μm or 0.003 mm) to avoid **light scattering**. Light scattering distorts the shape of the spectrum, giving a concave baseline especially at high wavenumbers.

Making the mull: Place about 5–10 mg of the solid sample in a mortar and add a drop of nujol. Grind the mixture for several minutes using the pestle. As the suspension gets spread over the mortar during the grinding process, gather it in the center of the mortar using a rubber policeman and continue the grinding until a smooth dispersion is obtained. Finally, take a small amount of the nujol mull with the rubber policeman and spread it between two NaCl plates. Obtain the spectrum.

Nujol absorbs at 2921, 2850, 1456, 1377, and 724 cm^{-1} due to C—H stretching and bending vibrations. Because almost all organic compounds have C—H bonds and absorb in these regions, C—H bands are of little importance for structure elucidation, and thus the nujol absorption does not interfere with the analysis. The spectrum of nujol is shown in Figure 31.12*b*.

Another technique used to study solid samples involves the preparation of a **pellet** with KBr. KBr is transparent in the frequency region used in IR spectroscopy. The solid sample (2–5 mg) is finely ground with 100 mg of KBr with the aid of a mortar and pestle; the optimum concentration range is 0.2–1%. The mixture is then placed in a die and pressed with the help two wrenches, a torque wrench or a pneumatic press. This gives a disk or pellet that can be studied directly using the die barrel as a holder (Fig. 31.13). Alternatively, the pellet is removed from the die and placed between two NaCl windows. The die should be immediately washed with distilled water because KBr is highly corrosive. A die consisting of two bolts and a barrel is shown in Figure 31.13.

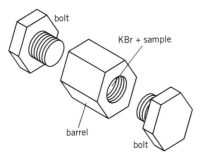

Figure 31.13   Die for the preparation of KBr pellets.

Solid samples with low melting points can also be studied as **cast films**. The solid sample is melted on a NaCl plate to form a thin solid film upon cooling to room temperature. An alternative technique consists in dissolving the sample in a volatile solvent such as acetone, applying a few drops of the solution on the window, and letting the solvent evaporate. A problem with cast films is that sometimes the solid deposits as large particles, producing light scattering.

In handling samples for IR spectroscopy special care should be taken to avoid moisture. Water not only absorbs IR radiation but also attacks the optical components of most IR instruments. Solvents and dispersants (nujol, KBr) should be anhydrous and kept in a desiccator.

Using IR Cards   Teflon cards are especially useful when FT-IR instruments (section 31.9) are used. They provide an easy way to run the IR of liquid and solid samples when only qualitative determinations are required. To obtain good IR spectra, the sample should penetrate the Teflon film. If the sample precipitates on top of the Teflon film, part of the radiation is lost due to scattering. To obtain a good IR spectrum, a drop (15–30 μL) of a 5–10% solution of the compound is applied on the Teflon film and the solvent is allowed to evaporate. Good solvents for this purpose are methanol, ethyl acetate, toluene, methylene chloride, and ethyl ether. Acetone and water should be avoided because they do not "wet" Teflon and therefore do not allow sample penetration into the polymer fibers. Saturated solutions should not be used because evaporation of the solvent and precipitation of the sample are faster than penetration into the Teflon, and result in big solid particles depositing on the surface of the tape. Better results are obtained if several applications of a dilute solution are made; the solvent should be allowed to evaporate between applications. Direct application of pure liquid samples has to be done with a pipet or a syringe that allows the delivery of no more than 10–15 µL of sample; the liquid should be uniformly spread over the Teflon tape. Delivering pure liquids with a Pasteur pipet is not recommended because it usually results in overloading. For more information about the uses of homemade cards see Reference 6.

31.6 INTERPRETING IR SPECTRA

Detailed information on IR absorption bands is provided in Table 31.2. Use this table to interpret and assign IR bands. A description of the most characteristic bands is given below.

Absorption of the Carbonyl

The carbonyl can be studied by IR spectroscopy like no other functional group. It gives a strong band, which is remarkably sensitive to structural effects, in the region 1650–1870 cm^{-1}. Given the widespread occurrence of the carbonyl group in organic compounds, IR spectroscopy provides a fast and economical way for studying them.

Aliphatic aldehydes absorb around 1725 cm^{-1}. Substituents directly attached to the carbonyl have a strong influence on its frequency. Electronegative elements with powerful inductive effects, such as oxygen and halogens, increase the double bond character of the carbonyl because they destabilize the resonance form with a positive charge density on the carbon. This results in higher frequencies for esters, acid anhydrides, and acid halides than for aldehydes. In the case of acid anhydrides, the two carbonyls can vibrate in phase or out of phase, giving rise to two bands around 1820 and 1760 cm^{-1}.

Substituents with strong resonance effects, such as the nitrogen of amides, decrease the double bond character of the carbonyl, and lower its frequency.

Table 31.2 **IR Characteristic Frequencies (± 15 cm^{-1})**

Aliphatic hydrocarbons

$-CH_3$	2960	st C–H
	2870	st C–H
	1460	d C–H (asym)
	1380	d C–H (sym); doublet between 1370 and 1385 for $(CH_3)_2CH-$
$-CH_2-$	2925	st C–H
	2850	st C–H
	1470	d C–H
	720–725	$-(CH_2)_n-$ rock when $n \geq 4$
	740–770	$-(CH_2)_n-$ rock when $n < 4$

Alkenes

$=CH_2$	3085	st C–H (asym)
	2980	st C–H (sym)
$=CH-$	3025	st C–H
$C=C$	1640–1680	st C=C
$R-CH=CH_2$	900–915	
	985–1000	
$R_2C=CH_2$	880–900	
RHC=CHR (trans)	960–990	d C–H out of plane
RHC=CHR (cis)	665–730	
$R_2C=CHR$	800–850	

Aromatics

$=CH-$	3020–3080	st C–H; several bands
	1580–1630	st C–C; phenyl ring skeletal;
	1470–1530	often doublets
mono-substituted	690–705 (s)	
	730–775 (s)	
1,2-disubstituted	730–770 (s)	
1,3-disubstituted	680–725 (m)	
	750–810 (s)	
	860–900 (m)	
1,4-disubstituted	790–860 (s)	d C–H out of plane
1,2,4-trisubstituted	800–860	
	860–900	
1,2,3-trisubstituted	685–720	
	770–810	
1,3,5-trisubstituted	675–730	
	800–850	
	850–900	
1,2,3,4-tetrasubst.	790–860	
1,2,4,5-tetrasubst. &		
1,2,3,5-tetrasubst. &	850–900	
pentasubstituted		

Alkynes

$\equiv$C–H	3250–3350	st C–H
$-C\equiv C-$ (terminal)	2100–2150	st C$\equiv$C
$-C\equiv C-$ (central)	2180–2260	st C$\equiv$C

Nitriles

$-C\equiv N$	2220–2260	st C$\equiv$N

(continued)

Table 31.2 (continued)

Alcohols and Phenols

R—OH and ArOH (monomeric)	3600–3650	st O—H
R—OH and ArOH (associated)	3200–3500	st O—H
R—CH$_2$—OH	1000–1075	
R$_2$CH—OH	1000–1125	st C—O
R$_3$C—OH	1100–1125	
Ar—OH	1180–1260	

Ethers

R—O—R	1000–1310	st C—O—C (asym)

Amines

R—NH$_2$ (primary)	3400–3500	st N—H (two bands)
	1560–1640	d N—H
	700–850	d N—H (doublet)
R$_2$NH (secondary)	3310–3450	st N—H (one band)
	1490–1580	d N—H
	700–850	d N—H

Ammonium salts

R—NH$_3^+$ & R$_2$NH$_2^+$	2700–3000	st N—H (broad)
R$_3$NH$^+$	2250–2700	st N—H

Ketones

R—CO—R (aliph.)	1715	
R—CO—CH=CH—R	1675	
α, β-unsaturated		
R—CO—Ar (aromatic)	1690	st C=O
cyclobutanone	1775	
cyclopentanone	1745	
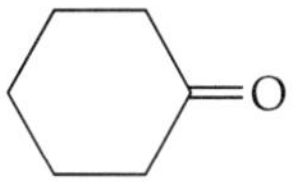 cyclohexanone	1715	

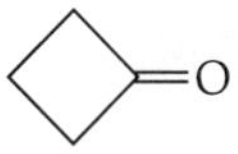

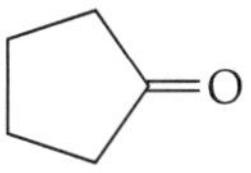

Aldehydes

R—CHO	2720 & 2820	st C—H & overtone d C—H (Fermi resonance)
R—CHO (aliphatic)	1720–1740	
R—CH=CH—CHO	1680	st C=O
Ar—CHO	1700	

Carboxylic acids

R—COOH	2500–3500	st O—H (broad)
R—COOH (dimer)	1710	
Ar—COOH (dimer)	1690	st C=O
R—COOH (monomer)	1740–1800	
R—COO$^-$	1580	st O—C—O (asym)
	1425	st O—C—O (sym)

(continued)

Table 31.2   (continued)

Esters & Lactones

R—COOR (aliph)	1740	
Ar—COOR	1720	
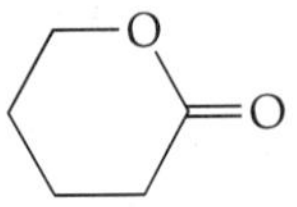	1735	st C=O
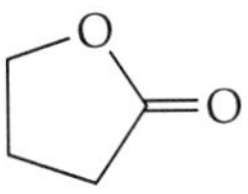	1770	
R—COOR & Ar—COOR	1050–1320	st C—O—C (two bands)

Acyl halides

R—CO—X	1750–1810	st C=O

Acid anhydrides

R—CO—O—CO—R	1730–1850	st C=O (two bands)

Amides & Lactams

R—CO—NH$_2$ (free)	3400–3500	st N—H (two bands)
	1690	st C=O (amide I)
	1610	d N—H (amide II)
R—CO—NH$_2$ (associated)	3100–3300	st N—H (several bands)
	1650	st C=O (amide I)
	1640	d N—H (amide II)
R—CO—NH—R (free)	3400–3500	st N—H
	1680	st C=O (amide I)
	1530	d N—H (amide II)
	1260	st C—N + d N—H (amide III)
R—CO—NH—R (assoc.)	3300–3500	st N—H (several bands)
(e.g.: proteins)	1660	st C=O (amide I)
	1550	d N—H (amide II)
	1300	st C—N + d N—H (amide III)
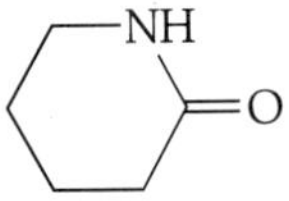 (free)	1660	st C=O

Nitrocompounds

Ar—NO$_2$	1520–1550	st O—N—O (asym)
	1320–1390	st O—N—O (sym)
	870	st C—N
R—NO$_2$	1540–1580	st O—N—O (asym)
	1350–1390	st O—N—O (sym)

Thiols

R—S—H	2550–2600	st S—H

Alkyl & aryl halides

R—F and Ar—F	1000–1400	st C—F
R—Cl and Ar—Cl	<600–840	st C—Cl
R—Br and Ar—Br	<700	st C—Br
R—l and Ar—I	<600	st C—l

st, stretching; d, deformation; s, strong; m, medium.

Double bonds and aromatic rings in conjugation with the carbonyl lower its frequency because they also decrease the double bond character of the carbonyl.

$$1680 \text{ cm}^{-1}$$

Absorption of the Most Common Functional Groups

Alkanes   Alkanes show bands due to C–H stretching (2850–2950 cm^{-1}) and bending (1380–1470 cm^{-1} and 720–790 cm^{-1}). All these bands are of little diagnostic value because they are present in the IR spectra of most organic chemicals. A doublet around 1380 cm^{-1} may be an indication of an isopropyl or *t*-butyl group.

Alkenes   Three groups of bands can be used to characterize alkenes: C–H stretching, C=C stretching, and C–H out-of-plane bending. C–H stretchings produce bands in the range 2970–3080 cm^{-1}. These bands occur at slightly higher wavenumbers than C–H stretching bands of alkanes. Because of the sp^2 hybridization of the C atom, C–H bonds of alkenes have higher s character than the C–H bond of alkanes and this translates into a larger force constant for alkene C–H bonds (see Table 31.1). The C=C stretching shows in the range 1640–1670 cm^{-1}. In most alkenes, C=C stretching gives rise to bands of only medium and low intensity. Conjugation lowers the absorption frequency; conjugated polyenes give a broad band between 1580–1660 cm^{-1}. Perhaps the most important bands for structure elucidation purposes are those resulting from out-of-plane bending, which occur between 670 and 990 cm^{-1} and give rise to characteristic patterns indicative of the substitution on the double bond. For example, *trans*-disubstituted alkenes give one band around 960–990 cm^{-1}, whereas *cis*-disubstituted alkenes give one band at lower frequencies (665–730 cm^{-1}).

Aromatics.   Aromatic groups give bands in the same regions as alkenes. C–H stretching occurs at 3000 cm^{-1} (3030–3050 cm^{-1}). The phenyl ring (as well as other aromatic rings) gives rise to a number of bands (2–4) in the 1450–1600 cm^{-1} region due to vibrations of the carbon skeleton. Substitution on aromatic rings can be determined by the out-of-plane bending bands between 690 and 900 cm^{-1}. Between 1700 and 2000 cm^{-1} a series of weak bands (overtones and combination bands of C–H bendings) can be used to identify different types of substitution patterns. These bands are shown in Figure 31.14.

Alkynes   This type of functional group can be characterized using two types of bands. The C–H stretching of terminal alkynes gives a strong band around 3300 cm^{-1}. The C≡C stretching gives a band in the range 2100–2150 cm^{-1} for terminal alkynes, and between 2180 and 2260 cm^{-1} for central alkynes; these bands are usually weak.

Alkyl Halides   C–Cl bonds give strong bands between 600 and 800 cm^{-1}. C–F stretching takes place at higher frequencies (1000–1400 cm^{-1}) while C–Br and C–I absorb at lower frequency values, below 700 and 600 cm^{-1}, respectively.

Alcohols and Phenols   The most important band for these functional groups originates in the O–H stretching. The position of this band is very sensitive to experimental conditions such as sample concentration and the nature of the solvent. These two factors greatly influence the formation of hydrogen bonds and change the frequency of the O–H vibration. Free *monomeric alcohols*, such as in the vapor phase or in very dilute solutions, absorb around 3600–3650 cm^{-1}. At high concentrations, the O–H stretching bands absorb between 3200 and 3500 cm^{-1} (this is discussed

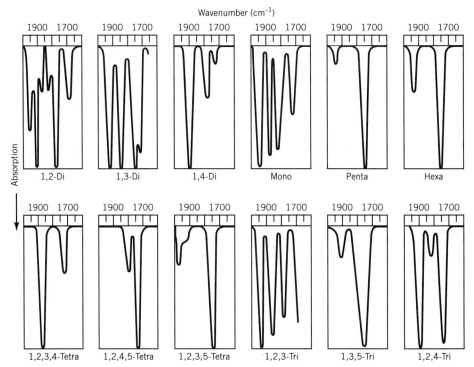

Figure 31.14   Overtones and combination bands of substituted benzenes (from D.W. Mayo, R.M. Pike, and P.K. Trumper, *Microscale Organic Laboratory*, 3rd ed. New York: Wiley, 1994).

in section 31.8). C—O stretchings and O—H bendings have limited diagnostic value because they appear in a very busy region of the IR spectrum (1000–1200 cm^{-1}).

Amines   Characteristic bands for amines arise from the N—H stretching. Similar to alcohols, the position of the N—H band is also sensitive to hydrogen bond formation. Primary amines (aliphatic as well as aromatic) give rise to two bands in the range 3400–3500 cm^{-1} due to N—H stretching. Secondary amines give a single band in the range 3300–3450 cm^{-1}. Association through H-bonds lowers the absorption frequency by 100–200 cm^{-1}. Other important vibrations are N—H bendings. In-plane N—H bendings give a band in the range 1560–1640 cm^{-1} for primary amines, and in the range 1500–1600 cm^{-1} for secondary amines. C—N stretching vibrations occurring in the range 1000–1400 cm^{-1} are of very limited diagnostic value.

Ethers   The only band that can be used to characterize ethers is the C—O—C stretching. This absorption takes place in the fingerprint region of the IR spectrum and therefore is not always easy to identify; however, in most cases it is a rather strong band around 1100 cm^{-1}. Aromatic ethers give this band at higher frequencies (1200–1300 cm^{-1}).

Aldehydes   Aldehydes can be characterized by two bands: one results from C—H stretching of the carbonyl group and the other originates in the C=O stretching. The frequency of C—H stretching is in the range 2700–2800 cm^{-1}; it coincides with an overtone of the C—H deformation band. When a coincidence like this happens, the intensity of the band is anomalously enhanced or the band shows as a doublet. This phenomenon is known as **Fermi resonance**. In the case of aldehydes, the C—H stretching band usually appears as a doublet. The carbonyl stretching produces a band in the range 1670–1740 cm^{-1}; it shows as a very strong band. Aliphatic aldehydes absorb between 1720 and 1740 cm^{-1}. Aromatic and α, β-unsaturated aldehydes have the C=O stretching band below 1700 cm^{-1} (1680–1700 cm^{-1}). Figure 31.15 shows the IR spectrum of *o*-chlorobenzaldehyde. The following bands are easily assignable: 3070 cm^{-1} (C$_{(Ar)}$-H st); 2868 and 2753 cm^{-1} (Fermi resonance: C—H st

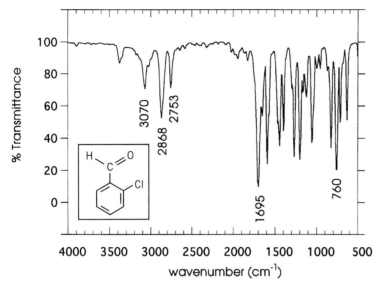

Figure 31.15 IR spectrum of *o*-chlorobenzaldehyde (vapor).

of aldehyde + C−H d overtone); 1695 cm^{-1} (C=O st arom. ald.); 760 cm^{-1} (C−H d out-of-plane, ortho disubstituted benzene ring).

Ketones The most important band in the IR of ketones is from the C=O stretching (1670–1850 cm^{-1}). Aliphatic ketones absorb around 1715 cm^{-1}. Aromatic and α, β-unsaturated ketones give bands in the range 1660–1690 cm^{-1}. The C=O stretching is sensitive to ring strain. Cyclohexanone (nonstrained ring) absorbs at 1715 cm^{-1} and cyclopentanone and cyclobutanone at 1745 and 1775 cm^{-1}, respectively. Electronegative elements (such as O, Cl, Br, I) on the α carbon to the carbonyl (−CO−C−X), increase the wavenumber of the C=O absorption (up to 50 cm^{-1}).

Carboxylic Acids This functional group is easily characterized by IR spectroscopy. Due to very strong H bonding, most carboxylic acids have a very broad O−H stretching band in the region 2500–3500 cm^{-1}. The C=O band, usually strong, can be found in the 1710–1750 cm^{-1} region. Under normal conditions, carboxylic acids are associated through H-bonds (which lower the frequency of the carbonyl) and absorb around 1710 cm^{-1}. At very low concentrations they are free in solution and absorb around 1750 cm^{-1}. As in the case of aldehydes and ketones, conjugation also decreases the absorption frequency of the carbonyl. Salts of carboxylic acids (**carboxylates**) have the carbon-oxygen stretching band in the range 1550–1620 cm^{-1}.

Esters The carbonyl group of esters absorbs in the range 1720–1840 cm^{-1}. Most aliphatic esters give the C=O stretching band around 1740 cm^{-1}. Conjugation with C=C decreases the frequency of the C=O vibration; aromatic and α, β-unsaturated esters absorb in the vicinity of 1720 cm^{-1}. Lactones (cyclic esters) with three or four C-atom rings absorb at higher frequencies (1770–1840 cm^{-1}) because of the ring strain.

Acid Anhydrides Most acid anhydrides give two strong C=O bands in the 1730–1850 cm^{-1} region. Conjugation with double bonds decreases the absorption frequency. If the anhydride forms part of a strained ring, the C=O bands move to higher wavenumbers (1780–1870 cm^{-1}).

Amides Amides have N−H as well as C=O bands. Primary amides (R−CO−NH$_2$) give two N−H bands in the range 3200–3500 cm^{-1}; the position of these bands depends on the degree of self-association (H-bonds). Secondary amides give only one band in the same region. The C=O of amides absorb in the region 1650–1690 cm^{-1}; this is called the Amide I band. Near the Amide I band is the Amide II band, which is due to a N—H bending (1530–1640 cm^{-1}). Amide III bands (C—N stretching mixed with N—H bending) are in the range 1200–1300 cm^{-1}. All these bands (especially Amide I) are very important in the study of protein structure.

Acyl Halides The C=O absorption in these halogenated compounds takes place at relatively high wavenumbers due to the presence of the electronegative halogen atom. Acyl chlorides absorb around 1800 cm^{-1}.

Nitriles This functional group gives a sharp band centered around 2220–2260 cm^{-1} due to C≡N stretching. Conjugation with double bonds decreases their absorption frequency.

Nitro Compounds Aromatic nitro compounds give bands at 1340 and 1530 cm^{-1}, arising from O—N—O stretching vibrations (symmetrical and asymmetrical, respectively). The frequency of C—N stretching is around 870 cm^{-1}.

Thiols The frequency of the S—H stretching band is between 2550 and 2600 cm^{-1}. This band is not as strong as the O—H band of alcohols.

Guidelines for the Interpretation of IR Spectra

If the structure of the chemical is known:

1. Identify the functional groups in the molecule: double bonds, aromatic rings, alcohol, aldehyde, carboxylic acid, and so on.

2. Find the characteristic frequencies (wavenumbers in cm^{-1}) for the functional groups present in the molecule using Table 31.2.

3. Match these frequencies with the peaks observed in the IR spectrum. In assigning IR bands, keep in mind that:

    (a) Not all bands of the IR spectrum are easily assignable.

    (b) The region between 1000 and 1500 cm^{-1} is difficult to interpret unless the compound under consideration has a very simple structure.

    (c) Deviations from the values reported in the tables may be observed (±15 cm^{-1}).

4. Do not assign peaks due to solvent, nujol, Teflon, or any other material used to run the IR spectrum. Check Figure 31.12 to find out where these compounds absorb.

5. If you are interpreting an IR from the literature and it has the legend "neat," it means that the spectrum of the neat liquid was run (no solvent added); "melt" means that the solid sample was melted on the IR window (no solvent added); "KBr" means that a KBr pellet was used.

6. Report the observed frequencies, the corresponding values from tables, and the type of vibration (stretching or bending). Remember that "st" and "d" are the usual abbreviations for stretching and bending (or deformation), respectively. For an example, see Table 31.3 in the next section.

If the structure of the chemical is unknown:

1. IR will shed light on the functional groups present in the molecule. Find the most prominent peaks in the IR of the unknown. Focus primarily in the region between 1600 and 4000 cm^{-1} where most functional groups absorb. See if they correspond to any of the characteristic frequencies given in Table 31.2. Make a list of possible functional groups present in the compound. Use this information in choosing other ways to test the identity of your unknown (chemical tests, other spectroscopic techniques, derivatives, etc.).

2. Follow points 3, 4, and 6 above.

3. The use of computerized library searches has become commonplace. Collections of a few thousand IR spectra are available from different sources as databases that can be input into a personal computer. In the database each spectrum is characterized by

a set of frequencies (the most important peaks) and their corresponding intensities (absorbance). The IR of the unknown, given as a set of frequencies and intensities, is entered into the program and the program searches for the best match within the database. A list of possible compounds with a margin of confidence is provided as output. This type of analysis is very important for routine studies of illegal drugs and pollutants, and for quality control purposes. Its application in the research lab is limited because it can positively identify only chemicals whose IR spectra have already been reported and form part of the database.

Examples of the interpretation of IR spectra are presented in the following case studies.

31.7   CASE STUDIES

These case studies are analyzed with the help of Table 31.2 and Figure 31.14.

1. 3-Ethyltoluene

Figure 31.16 shows the IR spectrum of 3-ethyltoluene. This aromatic hydrocarbon with two side chains has C–H bands above and below 3000 cm^{-1} (3030 cm^{-1}; 2950, and 2880 cm^{-1}) corresponding to $C(sp^2)$-H and $C(sp^3)$-H stretching vibrations, respectively. The three small bands (combination bands) between 1700 and 2000 cm^{-1} are characteristic of a meta-disubstituted aromatic ring (Fig. 31.14). The band at 1611 cm^{-1} corresponds to carbon-carbon stretching of the aromatic ring (skeletal stretching vibrations). The two bands around 1475 cm^{-1} can be assigned, in principle, as C–H bending of the methylene and methyl groups (1460 cm^{-1}) and skeletal C–C stretching vibrations of the aromatic ring (1425–1525 cm^{-1}). The bands between 1000 and 1300 cm^{-1} are of no practical importance. The three bands at 700, 780, and 883 cm^{-1} are C–H out-of-plane bendings of the aromatic ring; they agree with the expected bands for a meta-disubstituted phenyl ring (Table 31.2). The assignments are summarized in Table 31.3.

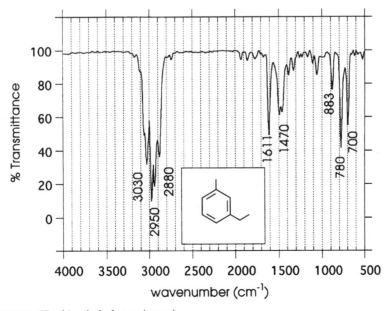

Figure 31.16   IR of 3-ethyltoluene (vapor).

Table 31.3 **IR Band Assignment for 3-Ethyltoluene**

Observed wavenumber (cm^{-1})	Expected (from table) (cm^{-1})	Assignment
3030	3020–3080	C—H st (arom)
2950 & 2880	2850–2960	C—H st (aliph)
1930; 1850 & 1760	2000–1700 three bands	combination bands (meta)
1611 & 1470	1600 & 1500	C—C skeletal st (arom)
883	860–900 ⎫	
780	750–810 ⎬	C—H d out-of-plane (meta)
700	680–725 ⎭	

2. 3-Nonanone

The IR spectrum of 3-nonanone is shown in Figure 31.17. The most relevant feature in the IR of this ketone is the carbonyl stretching band at 1712 cm^{-1}. Other bands present in the spectrum are those of C—H stretching and C—H bending. These bands have little diagnostic value. The assignments are summarized in Table 31.4.

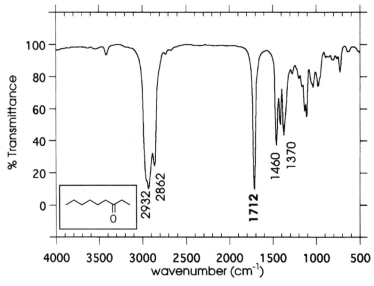

Figure 31.17   IR of 3-nonanone (neat).

Table 31.4 **IR Band Assignment for 3-Nonanone**

Observed wavenumber (cm^{-1})	Expected (from table) (cm^{-1})	Assignment
2932 & 2862	2850–2960	C—H st (aliph)
1712	1715	C=O st (ketones)
1460	1460	C—H d (asym)
1370	1380	C—H d (sym)

3. An Unknown

The IR of a compound that contains only C, H, and O was determined in the vapor phase and is shown in Figure 31.18. The band at 3657 cm^{-1} indicates the presence of an O—H group. This band could, in principle, also be assigned to a N—H bond, but this possibility is ruled out on the basis of the atomic composition of the unknown.

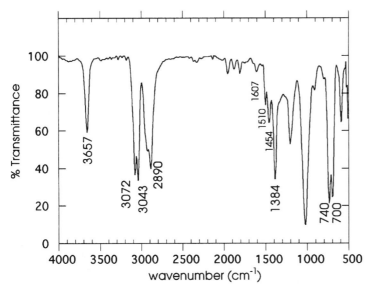

Figure 31.18 IR of an unknown (vapor).

The strong band around 1000 cm^{-1} confirms the presence of a C–O bond. The strong bands between 3000 and 3100 cm^{-1} and the small bands between 1600 and 2000 cm^{-1} suggest the presence of an aromatic ring. Bands between 2800 and 3000 cm^{-1} suggest that aliphatic C–H stretching is also present. The two bands around 700–750 cm^{-1} are in agreement with a monosubstituted aromatic ring. The information is gathered in Table 31.5.

Table 31.5 **IR Band Assignment for an Unknown Sample**

Observed wavenumber (cm^{-1})	Suggests	Expected (from table) (cm^{-1})
3657	O–H (free) st	3600–3650
3072 & 3043	C–H st (arom)	3080
2890	C–H st (aliph)	2850–2960
1700–2000	combination bands (arom)	1700–2000
1607 & 1510	C–C st (skeletal; arom)	1600 & 1500
1020	C–O st alcohol	1000–1125
740	C–H d out-of-plane (mono)	730–775
700	C–H d out-of-plane (mono)	690–705

A simple compound compatible with all this information is benzyl alcohol: $PhCH_2OH$. Compounds such as phenyl ethanol ($Ph–CH_2–CH_2–OH$) and other alcohols with a phenyl ring attached to an aliphatic chain are also in agreement with the IR spectrum. Further analysis (NMR, MS, etc.) is needed to determine the actual structure.

31.8 CONCENTRATION AND SOLVENT EFFECTS ON IR

IR spectra of vapors and gases show a high degree of resolution. Rotational levels, within each vibrational level (section 30.2), are well separated in energy and give rise to clusters of well-resolved peaks for a single vibrational transition. The IR of water vapor is shown in Figure 31.19a. It can be observed that the expected bands at 1595, 3652, and 3756 cm^{-1} (as discussed earlier; Fig. 31.2) are composed of many well-defined peaks. This level of resolution is absent in the IR spectrum of liquid water (Fig. 31.19b). Because of intermolecular interactions in the condensed phase, the rotational energy levels in liquids are "smeared" over a range of energy rather than having unique values. As a result, vibrational transitions do not occur at discrete frequency values but are spread over a wide range. It can also be observed in the IR spectrum of liquid water that the O—H stretching bands show at lower frequencies than in the vapor phase. This effect is also due to interactions, such as H-bonds, in the condensed phase.

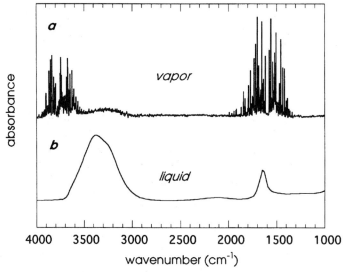

Figure 31.19 IR spectra of water: a) vapor; b) liquid.

IR spectroscopy provides an excellent means for the study of H-bonds. The self-association of alcohols and amines, among other functional groups, can be followed

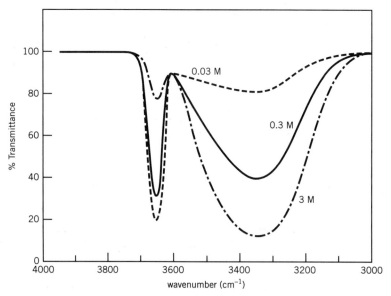

Figure 31.20 Diagrammatic IR spectrum of an alcohol in carbon tetrachloride solution in the O—H stretching region at different concentrations.

by the change in intensity of the characteristic bands of the free and associated species. The H-bond association of two alcohol molecules is depicted below. The formation of associated species depends strongly on the alcohol concentration. At low alcohol concentrations only free molecules are in solution. As the concentration increases, the equilibrium is shifted to the right, and the concentration of associated species increases.

$$2 \ R{-}O{-}H \ \rightleftharpoons \ R{-}O{-}H\cdots O{-}R$$
$$|$$
$$H$$

Figure 31.20 shows the high-frequency region of the IR spectra of an aliphatic alcohol in carbon tetrachloride solutions of different concentrations. At low concentrations (0.03 M) the alcohol is free and shows the O—H stretching at 3650 cm^{-1}. In contrast, the associated alcohol has the O—H band at lower wavenumbers (3350 cm^{-1}). This band becomes more and more intense as the concentration increases, at the expense of the band for the free alcohol at 3650 cm^{-1}.

31.9 INSTRUMENTATION

IR spectrometers can be divided into two main types: **dispersive** and **interferometric** instruments. Both kinds are commercially available; dispersive instruments, the first to be developed, are being replaced by more sophisticated and stable interferometric spectrometers.

Dispersive Instruments

A very simple diagram of a dispersive instrument is shown in Figure 31.21. Most dispersive spectrometers are **double-beam** instruments: radiation from the IR source is split into two beams, one passing through the sample cell and the other going through a reference cell containing solvent or just air. This arrangement compensates for unwanted drift in input energy and detector stability. Typical **IR sources** are incandescent materials heated to 1000–2000 K; at these temperatures they emit radiation centered in the IR range. Once the radiation passes through the sample it reaches the **monochromator** (from Greek: *mono*, one; and *chroma*, color). The monochromator is a **prism** or **grating** that disperses the IR radiation into its components of different frequencies.

Figure 31.21 Diagram of a dispersive IR spectrometer.

The monochromator is mounted on a platform that can be rotated to shine IR radiation of different frequencies on the detector. Very old instruments use prisms as monochromators, but they have been replaced by gratings of higher resolution. **Detectors** for IR spectrometers are thermosensitive: they respond to IR radiation by producing a change in temperature, which in turn causes an electrical imbalance in a circuit. Thermocouples, thermistors, temperature-sensitive capacitors, and gas thermometers are used as IR detectors.

Fourier-Transform Spectrometers

Fourier-transform (FT) spectrometers operate under a different principle than dispersive instruments. Whereas in dispersive instruments a narrow range of frequency, ideally **monochromatic radiation**, reaches the detector at any given time, in FT-IR spectrometers a broadband of IR frequencies reaches the detector all at once. The heart of the FT-IR spectrometer is the Michelson interferometer (Fig. 31.22).

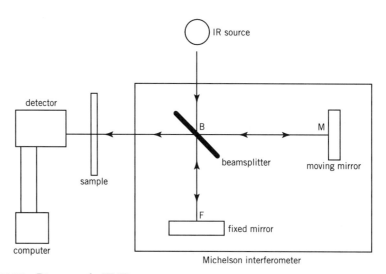

Figure 31.22 Diagram of a FT-IR spectrometer.

Radiation emitted by the source impinges on the beamsplitter and it is split into two beams of about equal intensity. One beam is reflected by the fixed mirror, and the other by a mirror that moves at a constant speed. Both beams recombine at the beamsplitter. Depending on the position of the moving mirror, the distance traveled by the beam that reflects on it can be equal to or smaller or greater than the distance traveled by the beam that reflects on the fixed mirror. This difference in the distance traveled by the beams gives rise to a difference in phase between the light waves when they recombine at the beamsplitter. This results in constructive or destructive interference. For each IR frequency that reaches the beamsplitter, an interference pattern in which the amplitude of the wave is a function of the position of the moving mirror is produced. The amplitude is given by a cosine wave, cos x, where x is the difference in the distances BM and BF (Fig. 31.22). When BM = BF, $x = 0$ and thus cos $x = 1$. At this position of the moving mirror the cosine wave has maximum amplitude (Fig. 31.23).

Each IR frequency is modulated into a cosine wave with a frequency f that depends on the speed of the moving mirror (v_m) and the wavenumber of the radiation ($\bar{v}$), (Eq. 7):

$$f = 2v_m\bar{v} \tag{7}$$

For example, IR radiation of $\bar{v} = 2000$ cm^{-1} is modulated with a frequency $f = 6000$ Hz if the mirror moves at a speed of 1.5 cm/s.

When several cosine waves are combined together, an **interferogram** emerges. The combination of three cosine waves, each one representing a modulated IR frequency, is shown in Figure 31.23.

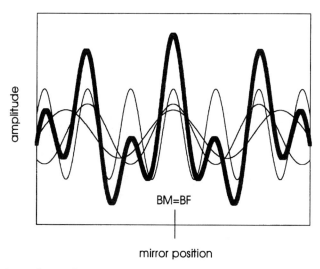

Figure 31.23 A simple interferogram. Three cosine waves and their addition (thick line). Each cosine wave represents a modulated IR frequency.

The modulated IR frequencies interact with the sample, which absorbs part of the radiation and decreases its intensity (amplitude). The radiation then reaches the detector where the intensities for different moving mirror positions are computed. This generates an interferogram as shown in Figure 31.23. The interferogram contains all the information of the IR spectrum: *an intensity associated with each IR frequency*. For easy interpretation, the interferogram must be decomposed into the component frequencies. Such conversion is called a **Fourier transform**. Figure 31.24 shows the Fourier transform (FT) of the simple interferogram of three frequencies shown in Figure 31.23. Fourier transform of an interferogram containing several hundred frequency components can be handled only with computers. A real interferogram and its corresponding Fourier transform, the IR spectrum, are shown in Figures 31.25 and 31.26. We will encounter Fourier transforms again in NMR spectroscopy.

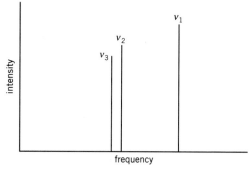

Figure 31.24 Fourier transform of the three cosine waves represented in Figure 31.23.

Fourier transform spectrometers have several advantages over dispersive instruments:

1. The optics of FT-IR instruments are very simple and easy to maintain.
2. A broad band of IR radiation is scanned at once and spectra can be obtained in a matter of seconds. With dispersive instruments, only a narrow range of frequency

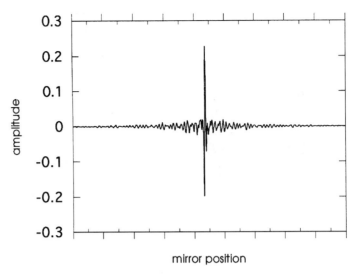

Figure 31.25 An interferogram for carvone.

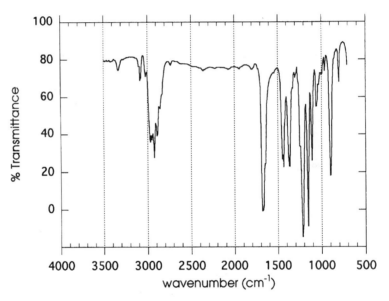

Figure 31.26 IR spectrum of carvone (Teflon).

is scanned at any given time. If it takes 0.5 sec to scan a range of 4 cm^{-1} and we want to scan from 500 to 4500 cm^{-1}, it will take: $0.5 \times (4500 - 500)/4 \text{ sec} = 500 \text{ sec} \approx 8 \text{ min}$ to scan the whole spectrum with a dispersive instrument. With an FT-IR system, between 200–1500 complete scans can be performed in the same period. In most cases, only one scan is necessary to obtain a good spectrum.

3. With FT-IR systems, multiple scans can be performed and added in a reasonable time to produce final spectra in which most of the irregularities in the baseline, called noise, are canceled out. Contrary to real signals, which are always positive, noise is random and can be positive or negative. The addition of successive scans results in noise cancellation and signal reinforcement. This results in a higher sensitivity or less material used to obtain a good spectrum.

4. Dispersive instruments make use of slits to control the monochromaticity of the radiation reaching the detector. These slits greatly reduce the throughput of energy. Since there is no monochromator in FT-IR systems, no slits are necessary and more energy reaches the detector. This also results in a higher sensitivity.

5. Dispersive instruments require external wavenumber calibration. After the IR of the compound of interest is run, a standard sample of polystyrene is normally used to calibrate the chart paper using well-known polystyrene IR bands. In FT-IR instruments, an internal HeNe laser provides automatic calibration.

A few disadvantages of the FT-IR systems can be mentioned:

1. Higher cost.

2. The Michelson interferometer used in FT-IR spectrometers is sensitive to vibrations. Ideally the instrument should lay on an antishock table.

3. The vast majority of FT-IR instruments are single-beam and thus, a spectrum of the solvent, nujol, KBr, Teflon, and so on called a **"blank"** or **"background"** must be run before each sample. Any small variation in the atmospheric contents of water vapor or CO_2 between the background and the run for the sample results in peaks for these compounds.

BIBLIOGRAPHY

1. Infrared Absorption Spectroscopy. K. Nakanishi and P. Solomon. 2nd ed. Holden-Day, San Francisco, 1977.

2. Advances in Applied Fourier Transform Infrared Spectroscopy. M.W. Mackenzie, ed. Wiley, Chichester, New York, 1988.

3. Organic Structural Analysis. J.B. Lambert, H.F. Shurvell, D.A. Lightmer, and R.G. Cooks. Prentice Hall, Upper Saddle River, N. J., 1998.

4. Organic Chemistry. An Experimental Approach. J.S. Swinehart. Appleton-Century-Crofts, New York, 1969.

5. Chemical, Biological and Industrial Applications of Infrared Spectroscopy. J.R. Durig, ed. Wiley, Chichester, New York, 1985.

6. Teflon Tape as a Sample Support for IR Spectroscopy. K.A. Oberg and D.R. Palleros. *J. Chem. Ed.*, **72**, 857–859 (1995).

7. A Unifying Approach to Absorption Spectroscopy at the Undergraduate Level. R.S. Macomber. *J. Chem. Ed.*, **74**, 65–67 (1997).

EXERCISES

Problem 1

Assuming identical force constants for $^{12}C–H$ and $^{12}C–D$ stretching,

(a) Calculate the isotopic effect $(\bar{v}_{C–H}/\bar{v}_{C–D})$ caused by deuteration.

(b) Compare the predicted effect with the observed values for chloroform: $Cl_3C–H$: 3020 cm^{-1} and $Cl_3C–D$: 2250 cm^{-1}.

Problem 2

Using force constant values provided in Table 31.1, calculate the wavenumbers for the following stretching vibrations: a) N–H; b) O–H; c) C=O; d) C–F; e) C–C; f) C=C.

Problem 3

Which of the following vibrations are active in the IR?

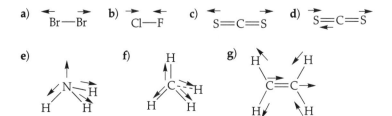

Problem 4

Classify the following vibrations as stretching or bending.

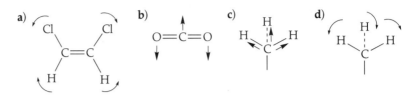

Problem 5

Assign a C=O st wavenumber (1–5) to the compounds (a–e) shown below.

Wavenumbers

    (1) 1675 cm^{-1};   **(2)** 1690 cm^{-1};   **(3)** 1715 cm^{-1};   **(4)** 1745 cm^{-1};   **(5)** 1770 cm^{-1}

Compounds

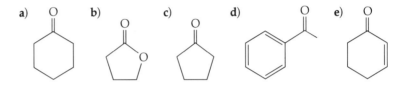

Problem 6

Assign as many bands as possible in the IR of limonene shown in the figure below.

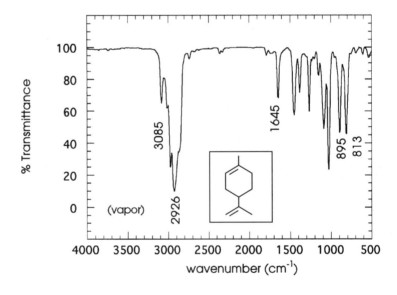

Problem 7

Assign C—H out-of-plane vibration wavenumbers (1–4) to the compounds listed below (a–d):

Wavenumbers

    (1) 886 cm^{-1};   **(2)** 695 cm^{-1};   **(3)** 992 and 911 cm^{-1};   **(4)** 964 cm^{-1}

Compounds

    (a) 2-methyl-1-butene;   **(b)** 1-pentene;   **(c)** *trans*-2-pentene;   **(d)** *cis*-2-pentene.

Problem 8

(a) Compound X (C_2Cl_3N) gives rise to the IR shown in the figure. Propose a structure for X and assign as many bands as possible.

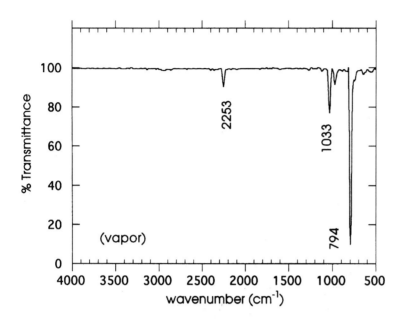

(b) Compound Y (C_6H_6O) gives rise to the IR shown in the figure. Propose a structure for Y and assign as many bands as possible.

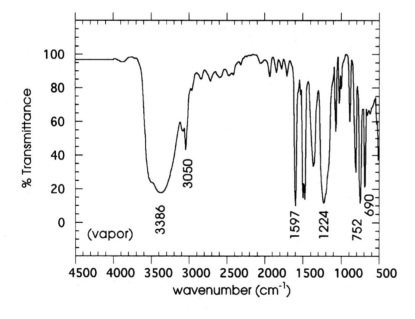

Problem 9

The IR spectrum shown in the figure below corresponds to one of the following isomers: 1,4-diethylbenzene; *tert*-butylbenzene; 1,2-diethylbenzene; 1,3-diethylbenzene. Analyze the IR spectrum and identify the compound that originated it.

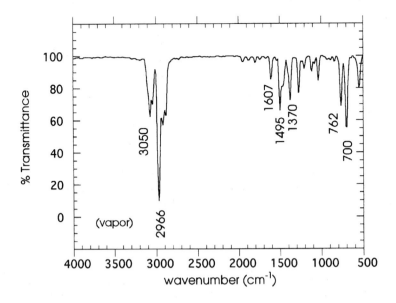

Problem 10

The IR spectra of four isomers A–D of molecular formula C_7H_7Cl are shown in the figures below. Analyze the spectra and give a structure for compounds A–D in agreement with the spectroscopic data. Pay special attention to the 1700–2000 cm^{-1} region.

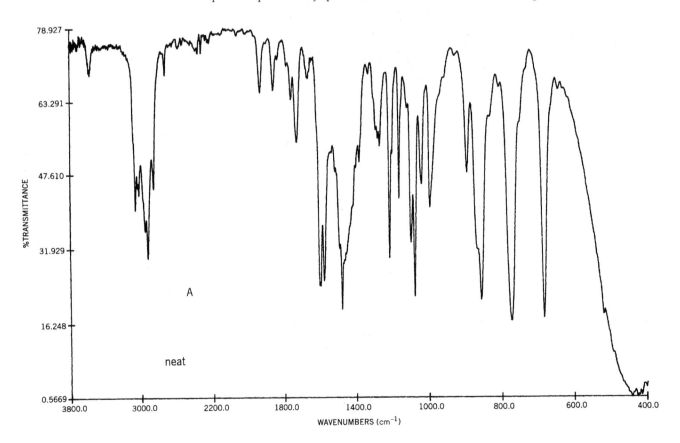

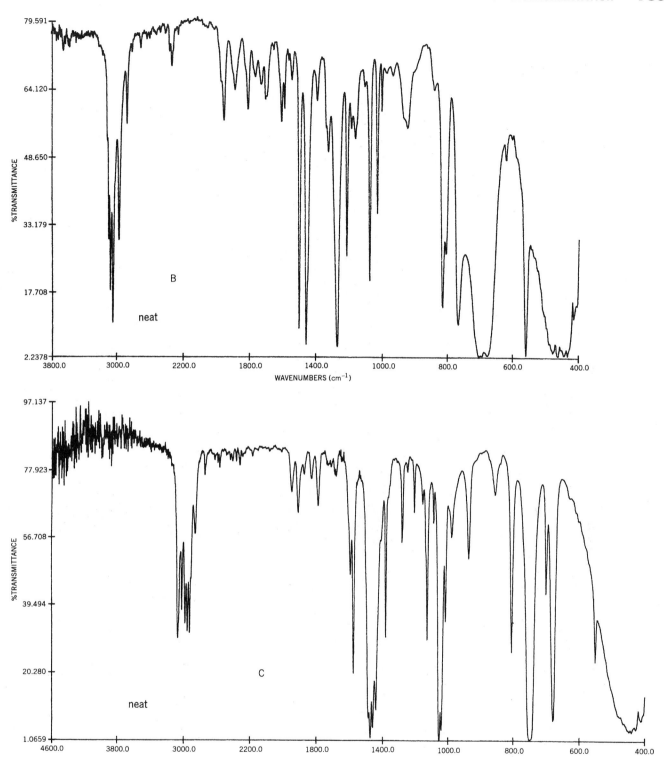

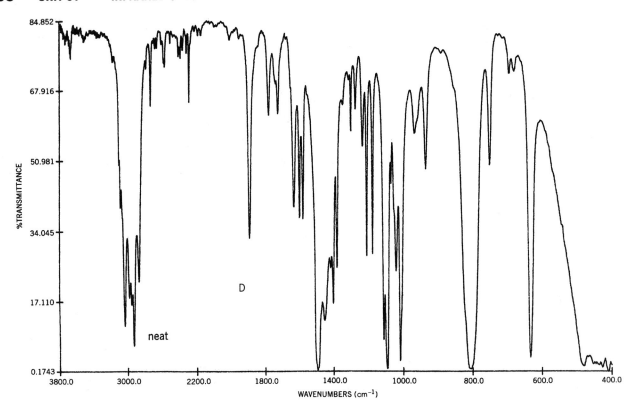

D

neat

Ultraviolet-Visible Spectroscopy

Ultraviolet-visible (**UV-vis**) spectroscopy is very useful in the study of molecules with double and triple bonds (C=C; C=O; C=N; C≡C; C≡N), especially those with conjugated systems and aromatic rings. Although UV-vis cannot be compared to IR and NMR, which give a wealth of structural information, it is nonetheless a very important tool in the organic chemistry laboratory with applications in the determination of concentrations, kinetic measurements, and the detection of samples eluting from chromatographic columns.

32.1 ELECTRONIC TRANSITIONS

As you may recall from Unit 30, when ultraviolet-visible radiation (UV: $\lambda = 190-380$ nm; visible: $\lambda = 380-750$ nm) interacts with matter, it produces electronic transitions by promoting electrons from the ground state to a higher energy level. The energy level of electrons is determined by the molecular orbital they occupy. There are two types of molecular orbitals, sigma (σ) and pi (π), with their corresponding antibonding counterparts, sigma star (σ^*) and pi star (π^*). When a molecule contains heteroatoms, such as N, O, and halogens, there are also nonbonding orbitals (n) where the heteroatom's lone electron pairs reside. To be absorbed by a molecule, the energy of the photon ($E = h\nu = hc/\lambda$; Unit 30) has to be equal to the energy difference between the ground and the excited state. Only photons of certain energy, and thus certain wavelengths, are absorbed by a sample. A general diagram of a molecule's energy levels is shown in Figure 32.1.

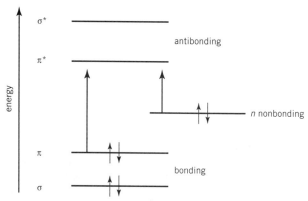

Figure 32.1 Electronic energy diagram of a hypothetical molecule showing electronic transitions.

Electrons can be promoted from a π to a π^* orbital (a $\pi \longrightarrow \pi^*$ transition) and also from an n to a π^* orbital ($n \longrightarrow \pi^*$ transition). Other transitions such as $\sigma \longrightarrow \sigma^*$ and $n \longrightarrow \sigma^*$ are also possible but require higher energy and occur beyond the UV-vis range. From Figure 32.1, you can see that the transition $n \longrightarrow \pi^*$ requires less energy than the $\pi \longrightarrow \pi^*$ transition and thus takes place at longer wavelengths (remember that the longer the wavelength, the lower the energy; Equation 30.5).

The usual range of UV-vis spectroscopy is between 200 and 700 nm. Below 200 nm, oxygen absorbs radiation and interferes with the measurements. To obtain UV spectra below 200 nm, the spectrometer should be purged with nitrogen or operated under very low pressure.

For isolated double bonds, the $\pi \longrightarrow \pi^*$ transition occurs below 200 nm and has no practical applications. For conjugated double bonds, on the other hand, the $\pi \longrightarrow \pi^*$ transition occurs at longer wavelengths, often in the visible range, giving rise to colored compounds. The reason for this is explained below.

Molecules with multiple double bonds have as many bonding pi orbitals as the number of double bonds. They also have an equal number of antibonding pi orbitals. For example, 1,3-butadiene has two bonding pi orbitals of different energy, and two antibonding pi orbitals. In the ground state, only the bonding orbitals are occupied. The occupied orbital with the highest energy is called **HOMO (highest occupied molecular orbital)**, and the unoccupied orbital with the lowest energy is called **LUMO (lowest unoccupied molecular orbital)**.

One of the most important transitions in UV-vis spectroscopy is from HOMO to LUMO. These transitions involve the smallest energy gap between bonding and antibonding orbitals and require low energy photons. It can be demonstrated with the help of quantum mechanics that, in general, the larger the number of conjugated double bonds, the smaller the energy gap between HOMO and LUMO, and the longer the wavelength of the absorbed light (Fig. 32.2).

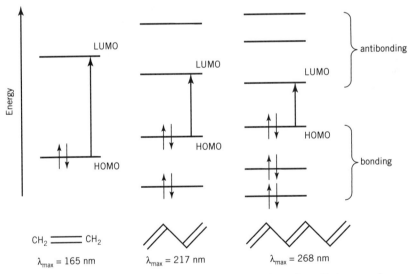

Figure 32.2 Pi system and HOMO and LUMO energy gap for ethylene and conjugated polyenes.

Polyenes with eight or more conjugated double bonds absorb in the visible portion of the spectrum and are colored. The observed color is complementary of the absorbed one. For example, if a compound absorbs blue light it will appear orange. This is the case of β-carotene, the orange pigment found in carrots and other vegetables (Unit 8), which is a polyene with 11 conjugated double bonds that absorbs blue light around 455 nm.

A typical UV-vis spectrum is presented in Figure 32.3. The wavelengths of maximum absorbance are called the λ_{max} (pronounced *lambda max*). The UV-vis spectrum in Figure 32.3 corresponds to 4-methyl-3-penten-2-one, a conjugated ketone, and presents the $\pi \longrightarrow \pi^*$ transition with λ_{max} 231 nm ($\varepsilon = 12{,}600$ L mol^{-1} cm^{-1}) and the $n \longrightarrow \pi^*$ transition with λ_{max} 321 nm ($\varepsilon = 40$ L mol^{-1} cm^{-1}). In general, the $\pi \longrightarrow \pi^*$ transitions are more favorable than the $n \longrightarrow \pi^*$ transitions and this is reflected by higher absorbances. As a comparison, it is worth mentioning that acetone (CH_3COCH_3), a nonconjugated ketone, presents both the $\pi \longrightarrow \pi^*$ and the $n \longrightarrow \pi^*$ transitions at shorter wavelengths (190 nm and 277 nm, respectively).

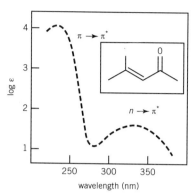

Figure 32.3 UV spectrum of 4-methyl-3-penten-2-one in heptane.

32.2 CHROMOPHORES AND AUXOCHROMES

A group of atoms that absorbs UV-visible radiation is called a **chromophore** (from Greek: *color bearer*). Examples of chromophores are groups with double and triple bonds such as C=C, C≡C, C=N, C≡N, C=O, N=O, and systems with conjugated double bonds including aromatic rings. Compounds with no double bonds such as alkanes, cycloalkanes, aliphatic alcohols, aliphatic ethers, and alkyl chlorides absorb UV radiation below 195 nm because they can only undergo high energy transitions such as $\sigma \longrightarrow \sigma^*$ and $n \longrightarrow \sigma^*$. These compounds are not usually studied by UV-vis spectroscopy.

An **auxochrome** is a group of atoms that does not absorb UV-visible radiation itself but changes the λ_{max} and absorption intensity of chromophores. Examples of auxochromes are the amino ($—\overset{..}{N}H_2$) and the hydroxyl group ($—\overset{..}{\underset{..}{O}}H$). When these groups are bonded to benzene, for example, they increase the λ_{max} and the molar absorptivity of the HOMO–LUMO transition. The spectral changes are due to changes in the energy levels of the ring's pi orbitals as a result of the orbital overlap with the unshared electron pairs of the auxochrome. The amino group, for example, has a free electron pair on the nitrogen atom that is delocalized by resonance with the aromatic ring. This conjugation has the effect of increasing the λ_{max} of aminobenzene (aniline) as compared to benzene. Aniline absorbs at 280 nm, 26 nm higher than benzene. In contrast, the ammonium group, with no free-electron pairs, does not interact with the ring's orbitals and has little effect on the absorption of UV light.

		NH$_2$	NH$_3{}^+$	OH	O$^-$
λ_{max} (nm)	254	280	254	270	287
ε (L mol^{-1} cm^{-1})	204	1430	160	1450	2600

An increase in the λ_{max} of a particular band is called a **bathochromic shift** or a **red shift**. On the other hand, a decrease in the λ_{max} is called a **hypsochromic shift** or a **blue shift**. Solvents are known to cause red and blue shifts. An increase in the polarity of the solvent usually results in a red shift of the $\pi \longrightarrow \pi^*$ transitions, and a blue shift of the $n \longrightarrow \pi^*$ transitions. This is diagrammatically shown in Figure 32.4. Other factors such as pH, salt concentration, temperature, and sample concentration also influence the UV-visible spectra of organic compounds.

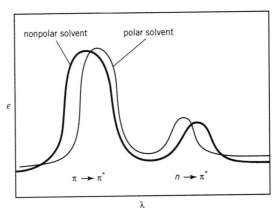

Figure 32.4 Solvent effect on the UV-visible spectrum of a hypothetical compound.

32.3 STRUCTURAL EFFECTS ON UV-VISIBLE SPECTRA: WOODWARD–FIESER RULES

The λ_{max} of the $\pi \longrightarrow \pi^*$ transition for conjugated polyenes can be calculated with the help of the rules developed by R.B. Woodward and Mary and Louis Fieser. These rules are applicable to systems with fewer than five conjugated double bonds, and are shown in Table 32.1.

Table 32.1 Woodward–Fieser Rules for Conjugated Polyenes

Base value for parent diene	217 nm
Two conjugated double bonds inside the same ring (homoannular double bonds)	+36 nm
Each additional conjugated double bond	+30 nm
Any exocyclic double bond	+5 nm
Each alkyl group on the conjugated system	+5 nm
Each Cl or Br on the conjugated system	+5 nm
Each OR group on the conjugated system	+6 nm
Each NR$_2$ group on the conjugated system	+60 nm
Solvent correction	0 nm

The application of the Woodward–Fieser rules is illustrated with the two examples below:

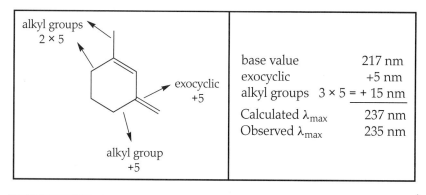

base value	217 nm
exocyclic	+5 nm
alkyl groups 3 × 5 =	+ 15 nm
Calculated λ_{max}	237 nm
Observed λ_{max}	235 nm

base value	217 nm
homoannular	+36 nm
alkyl groups 3 × 5 =	+ 15 nm
Calculated λ_{max}	268 nm
Observed λ_{max}	263 nm

The λ_{max} of the $\pi \longrightarrow \pi^*$ transition for conjugated carbonyl compounds can also be calculated by the Woodward–Fieser rules (Table 32.2).

These rules are illustrated with the conjugated ketone and the conjugated carboxylic acid shown below.

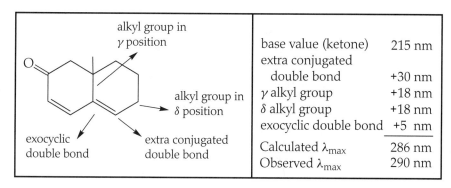

base value (ketone)	215 nm
extra conjugated	
double bond	+30 nm
γ alkyl group	+18 nm
δ alkyl group	+18 nm
exocyclic double bond	+5 nm
Calculated λ_{max}	286 nm
Observed λ_{max}	290 nm

base value	
(carboxylic acid)	193 nm
α alkyl group	+10 nm
β alkyl group	+12 nm
Calculated λ_{max}	215 nm
Observed λ_{max}	217 nm

32.4 APPLICATIONS OF UV-VISIBLE SPECTROSCOPY

UV-visible spectroscopy is useful in the study and identification of conjugated systems and aromatic molecules. One of the most important applications of UV-visible

Table 32.2 **Woodward–Fieser Rules for Conjugated Carbonyl Compounds**

Base value for		
	aldehyde (R = H)	210 nm
	ketone (R = alkyl)	215 nm
	carboxylic acid, ester (R = OH; O-alkyl)	193 nm

Each additional conjugated double bond	+30 nm
Two conjugated double bonds inside the same ring (homoannular double bonds)	+36 nm
Any exocyclic double bond	+5 nm

Each of the following groups in positions α, β, γ, or δ of the conjugated system increases the λ_{max} by ... (nm)

	α	β	γ	δ
Alkyl group	10	12	18	18
OH	35	30	—	50
O-alkyl	35	30	17	31
OCOCH$_3$	6	6	6	6
Cl	15	12	—	—
Br	25	30	—	—
NH$_2$	—	95	—	—

Solvent correction

methanol, ethanol	0 nm
water	+8 nm
chloroform	−1 nm
diethyl ether	−7 nm
hexane	−11 nm

spectroscopy is in the study of sample concentrations by application of Lambert-Beer's law, Equation 1 (see also section 30.3):

$$A = \varepsilon \times C \times l \tag{1}$$

where A is the absorbance at a given wavelength, ε is the molar absorptivity at that wavelength, C is the sample concentration (in mol/L), and ℓ is the pathlength of the **cuvette** or sample container in cm.

If the molar absorptivity of the sample is known, the concentration can be determined by measuring the sample's absorbance. To determine the molar absorptivity, several solutions of known molar concentration must be prepared and their absorbances measured at the desired wavelength. A plot of absorbance versus concentration gives, according to Equation 1, points that can be joined by a straight line with slope $\varepsilon \times \ell$; if a 1-cm cuvette is used, $\ell = 1$ cm, the slope gives the molar absorptivity (Fig. 32.5). If instead of a straight line the data show that the absorbance values level off at high concentrations, then the Lambert–Beer's law is not obeyed at high concentrations, and Equation 1 should not be used in that concentration range. In such cases, the plot of absorbance versus concentration (with a curve joining the experimental points) can be used to determine the concentrations. These plots are called **standard curves** (Fig. 32.5b).

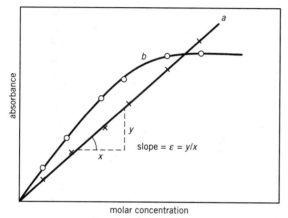

Figure 32.5 Absorbance versus concentration plot ($\ell = 1$ cm); a) a sample that obeys Lambert–Beer's law in the entire concentration range; b) a sample that deviates from Lambert–Beer's law at high concentrations.

In general, the absorbance readings should be in the range 0–2. Absorbance values higher than 2 have a larger intrinsic error and are likely to deviate from Lambert–Beer's law. Solvents commonly used for UV-visible spectroscopy are those with little or no absorbance in the wavelength range under investigation. In the visible region, 340–700 nm, water, and most organic solvents can be used without problems. In the 210–340 nm region methanol, isopropanol, n-hexane, cyclohexane, acetonitrile, and water can be used, and between 190 and 210 nm, only water, n-hexane, and acetonitrile are recommended. The cuvettes used in the UV region must be made of quartz because glass absorbs UV light. In contrast, glass and even plastic cuvettes can be used in the visible portion of the spectrum.

The absorbance of a multicomponent sample is the sum of the absorbances of the individual components. The absorbance, A, of a mixture of two compounds M and N, is given by Equation 2 where ε_M and ε_N are the molar absorptivities of M and N, respectively, and [M] and [N] are their molar concentrations:

$$A = \varepsilon_M \times [M] \times \ell + \varepsilon_N \times [N] \times \ell \tag{2}$$

This equation can be used to calculate [M] and [N] if one measures the absorbance at two different wavelengths, λ_1 and λ_2, and knows the molar absorptivities for both compounds at both wavelengths. Rewriting Equation 2 at each wavelength, a system of two equations with two unknowns, [M] and [N], is generated. This system can be solved with standard algebraic methods:

$$A^{\lambda_1} = \varepsilon_M^{\lambda_1} \times [M] \times \ell + \varepsilon_N^{\lambda_1} \times [N] \times \ell \tag{3}$$

$$A^{\lambda_2} = \varepsilon_M^{\lambda_2} \times [M] \times \ell + \varepsilon_N^{\lambda_2} \times [N] \times \ell \tag{4}$$

Rearranging Equation 4 we obtain:

$$[M] = \frac{A^{\lambda_2} - \varepsilon_N^{\lambda_2} \times [N] \times \ell}{\varepsilon_M^{\lambda_2} \times \ell} \tag{5}$$

Substituting Equation 5 into Equation 3, and solving for [N] Equation 6 is obtained:

$$[N] = \frac{A^{\lambda_1} \times \varepsilon_M^{\lambda_2} - A^{\lambda_2} \times \varepsilon_M^{\lambda_1}}{(\varepsilon_M^{\lambda_2} \times \varepsilon_N^{\lambda_1} - \varepsilon_M^{\lambda_1} \times \varepsilon_N^{\lambda_2}) \times \ell} \tag{6}$$

By measuring A^{λ_1} and A^{λ_2} and knowing $\varepsilon_M^{\lambda_1}$, $\varepsilon_N^{\lambda_1}$, $\varepsilon_M^{\lambda_2}$ and $\varepsilon_N^{\lambda_2}$, [M] and [N] can be calculated. This type of analysis can be extended to mixtures with more than two components on condition that the absorbance of the sample is measured at as many wavelengths as the number of components to be determined. Because of the errors associated with each measurement, the method usually has little reliability for samples with more than three components.

Another important application of UV-vis spectroscopy is the study of reaction rates. If there is a difference in the spectra of reactants and products, the course of a reaction can be investigated by following the absorbance of the reaction mixture as a function of time. This application is treated in Units 18 and 19. UV-vis spectroscopy is also useful to detect compounds as they elute from chromatographic columns. A discussion of this application is offered in section 10.8 and Experiments 10 and 29.

32.5 SPECTROPHOTOMETERS

Most **spectrophotometers** (also called **spectrometers**) operate by a similar principle. The radiation coming from a lamp is dispersed into its constituent wavelengths by a **monochromator**, such as a diffraction grating or a prism; the resulting monochromatic radiation passes through the sample and then reaches the detector. By comparing the light intensity transmitted by the sample with that transmitted by the solvent alone, the absorbance of the solute is computed.

Spectronic 20

One of the most popular spectrometers for teaching purposes is the Spectronic 20 manufactured by Spectronic Instruments (Rochester, New York). This simple and sturdy spectrophotometer is useful in the visible region of the spectrum (340–600 nm) and can be modified to operate in the near IR, but it is not appropriate to study the ultraviolet region. A diagram of the Spectronic 20 optical system is shown in Figure 32.6.

White light from a tungsten lamp is focused on a diffraction grating and decomposed into individual wavelengths. The desired wavelength is selected with the wavelength cam that moves the grating. The light passes through the exit slit and reaches the sample and the phototube detector, which converts photons into electrical current. The shutter prevents light from reaching the detector when the sample is removed from the sample compartment. The light control allows the setting of the 0 and 100% transmittance. A diagram of the Spectronic 20, model D, with its different parts is shown in Figure 32.7.

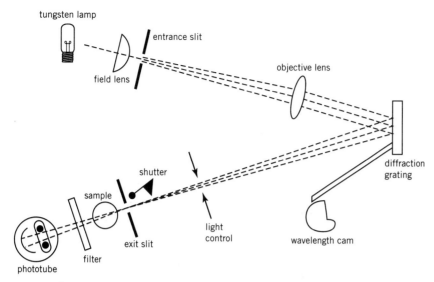

Figure 32.6 Spectronic 20 optical system (courtesy of Spectronic Instruments).

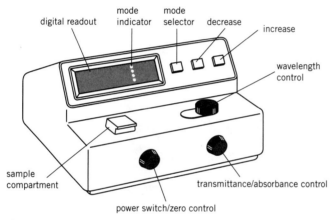

Figure 32.7 The Spectronic 20D (courtesy of Spectronic Instruments).

Operation Select the desired wavelength with the wavelength control knob. Read the wavelength on the left-hand side of the digital display. Set the mode to "transmittance" by depressing the mode selector until the "transmittance" indicator is lit. Without sample in the sample compartment, and with the lid down, adjust the zero control so that the reading on the digital display is 0% T. Insert the cuvette with the solvent in which the sample is prepared (blank) and adjust the "absorbance/transmittance control" until the reading is 100% T. Remove the blank and place the sample in the sample compartment. Read its % T or absorbance on the digital display. The Spectronic 20D also has a "factor" and a "concentration" mode that transform the absorbance readings into concentration units; the increase and decrease buttons are used only with these two modes. An older version of this spectrophotometer, Spectronic 20, does not have a digital display, the wavelength is read on a dial next to the wavelength control knob, and the transmittance or absorbance is read on an analog display. Despite these small differences, the operation of both spectrometers is the same.

The Spectronic 20 is useful to measure the absorbance at a determined wavelength but it is poorly suited for obtaining whole spectra because the operator must read the absorbance at each individual wavelength separately.

Diode-Array Spectrometers

This type of instrument has been developed in the last 15 years, and has revolutionized the way UV-visible spectra are obtained. With **diode-array spectrometers** it is possible to obtain high-quality spectra in a few seconds. They are especially useful in kinetic studies, as detectors for HPLC systems, and whenever a rapid acquisition of data is crucial.

Diode-array spectrometers have very simple optics (Fig. 32.8). The UV-visible radiation is generated by a lamp such as a low-pressure deuterium lamp, which emits radiation in the region 190–820 nm. The light is focused on the sample and then reaches a grating where it is dispersed into the individual component wavelengths before falling into the diode array. The diode array is a collection of photodiodes (photodetectors), normally more than 300, each one assigned to a particular wavelength and placed in sequence one next to another. The wavelengths assigned to the individual photodiodes are 1 or 2 nm apart, depending on the resolution of the instrument. All the light received by an individual photodiode is interpreted to be of that photodiode's assigned wavelength. The instrument is controlled by a computer equipped with software to display and analyze the data.

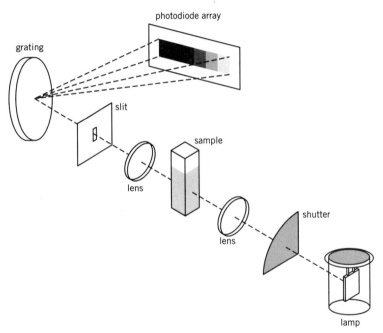

Figure 32.8 Optical system for a diode-array spectrometer (courtesy of Hewlett-Packard).

Contrary to conventional spectrometers where each wavelength is scanned individually, in a diode-array system all wavelengths are measured at once, which results in very fast data acquisition.

BIBLIOGRAPHY

1. Theory and Applications of Ultraviolet Spectroscopy. H.H. Jaffé and M. Orchin. Wiley, New York, 1962.

2. UV-Vis Spectroscopy and its Applications. H.-H. Perkampus. Springer-Verlag, Berlin, 1992.

3. UV-VIS Atlas of Organic Compounds. H.-H. Perkampus, 2nd ed. VCH, New York, 1992.

4. Computer Methods in UV, Visible, and IR Spectroscopy. W.O. George and H.A. Willis, ed. Royal Society of Chemistry, Cambridge (England), 1990.

5. UV Spectroscopy: Techniques, Instrumentation, Data Handling. B.J. Clark, T. Frost, and M.A. Russell, ed. Chapman & Hall, London, 1993.

6. Ultraviolet and Visible Spectroscopy. Chemical Applications C.N.R. Rao. 3rd ed. Butterworths, London, 1974.

7. Ultraviolet and Visible Spectroscopy. M.J.K. Thomas. 2nd ed. Wiley, Chichester, 1996.

EXERCISES

Problem 1

Which of these compounds would you expect to have the longest λ_{max}? and the shortest? Briefly justify.

Problem 2

A chemist was studying the $n \longrightarrow \pi^*$ transition of acetone in different solvents. He measured the λ_{max} in each solvent and noticed that it changed from solvent to solvent in the expected manner. He wrote down the λ_{max} values: 265 nm, 270 nm, 272 nm, 277 nm, and 279 nm, but he did not write the names of the solvents next to the λ_{max}. The solvents he used were: water, chloroform, ethanol, hexane, and methanol. Could you assign the λ_{max} to the corresponding solvent?

Problem 3

Calculate the λ_{max} of the following compounds using the Woodward–Fieser rules.

Problem 4

A mixture of α- and β-ionone (compounds responsible for the smell of violets, prepared in Experiment 23B) is analyzed by UV-visible spectroscopy.

The absorbance of the mixture at 227 and 295 nm is 0.808 and 0.175, respectively. If the molar absorptivities for α-ionone at 227 and 295 are 14,200 and 280 L mol^{-1} cm^{-1}, respectively, and for β-ionone at the same wavelengths are 6,500 and 10,700 L mol^{-1} cm^{-1}, respectively, calculate the concentration of each component in the mixture.

Problem 5

The molar absorptivity of 4-methyl-2-nitrophenol (MM 153.14 g/mol) was determined at its $\lambda_{max} = 360\,nm$, in a mixture of ethanol and concentrated HCl (9 : 1; v/v). Two stock solutions (I and II) were made and, from each one, three diluted solutions were prepared in the same solvent; the dilution factors are indicated in the table. The concentrations of solutions I and II were 4.18 mg/50 mL and 3.21 mg/50 mL, respectively. The absorbance values, A, measured using a 1-cm pathlength cuvette, are given in the table.

	Solution I			Solution II		
	Concen-tration (g/L)	Concen-tration (mol/L)	A (360 nm)	Concen-tration (g/L)	Concen-tration (mol/L)	A (360 nm)
stock solution			1.750			1.302
dilution factor 2/5			0.703			0.527
dilution factor 1/5			0.355			0.268
dilution factor 1/10			0.172			0.136

(a) Calculate the concentrations and complete the table.

(b) Plot A versus molar concentration. Is Lambert–Beer's law obeyed?

(c) Calculate the molar absorptivity at 360 nm.

Problem 6

(a) What changes would you expect to see in the UV-visible spectrum of 4-nitrophenol at pH 1 and pH 12? Explain with a chemical reaction.

(b) The UV-visible spectrum of 6-methyl-2-nitrophenol shows $\lambda_{max} = 360\,nm$ ($\varepsilon = 3{,}350\,L\,mol^{-1}\,cm^{-1}$) in a toluene solution. In the presence of a large molar excess of cyclohexylamine the λ_{max} shifts to 436 nm. Explain this observation.

Nuclear Magnetic Resonance

Nuclear magnetic resonance (**NMR**) is the most important tool for structure elucidation in all areas of chemistry. In a matter of minutes and using a relatively small amount of sample a chemist can obtain the NMR spectrum, which is like a map showing the connectivity between atoms. The NMR phenomenon in molecules was first described in 1946 by two independent groups, one from Harvard led by Edward Purcell, and the other from Stanford led by Felix Bloch. Both scientists were awarded the Nobel Prize in Physics in 1952. The use of NMR grew rapidly and the technique was already an important tool for structure elucidation in the 1960s. However, it was then limited to the study of the hydrogen nucleus (**^{1}H-NMR**). Technological advances, especially in the area of microcomputers in the late 1970s, made possible the routine study of other nuclei such as ^{13}C (**^{13}C-NMR**). Today, almost any element of the periodic table can be studied by nuclear magnetic resonance. Here, we will only discuss ^{1}H-NMR and ^{13}C-NMR because they are the most widely used NMR techniques in organic chemistry.

Before we go on, let's anticipate the kind of information offered by NMR. ^{1}H-NMR gives information about how many different types of hydrogens a molecule has, and how many hydrogens of each type are present; it also tells about the environments of the hydrogens (for example, is a hydrogen bound to oxygen or carbon?) and the connectivity between them (is a methyl group attached to a methylene or a methine?). How to obtain this information, and more, is the subject of this unit.

The technique of **Magnetic Resonance Imaging (MRI)**, amply used as a diagnostic tool in medicine, is based on the same principles as NMR. It should be pointed out that in nuclear magnetic resonance, the term *nuclear* has no relation to ionizing radiation or any other high-energy process that the lay public has learned to associate with the word "nuclear." On the contrary, NMR is a very mild and gentle process that is nuclear in only one sense: it involves the atomic nucleus.

33.1 NUCLEAR SPIN

Certain atomic nuclei such as ^{1}H and ^{13}C possess angular momentum, or **spin**, which can be interpreted as if the nuclei were rotating about their axes. Because moving charges generate a magnetic field, the positively charged spinning nuclei behave as small magnets. These nuclei have a **magnetic moment**, orient themselves in the presence of an external magnetic field, and show nuclear magnetic resonance phenomena. Not all nuclei behave like magnets, however. Those with even atomic mass and even atomic number, such as ^{12}C and ^{16}O, have no spin, are magnetically inactive, and do not show nuclear magnetic resonance.

Nuclear spin is characterized by the **nuclear spin quantum number**, I, which can take the values 0, 1/2, 1, 3/2, 2, and so forth. If $I = 0$, the nucleus has no spin. The nuclear spin quantum number and the natural abundance of selected nuclei are shown in Table 33.1.

Table 33.1 **Nuclear Spin Quantum Numbers and Natural Abundance of Selected Nuclei**

Isotope	I	Natural abundance	Isotope	I	Natural abundance
^1H	1/2	99.985	^{18}O	0	0.200
^2H	1	0.015	^{19}F	1/2	100
^{12}C	0	98.90	^{31}P	1/2	100
^{13}C	1/2	1.10	^{32}S	0	95.03
^{14}N	1	99.635	^{33}S	3/2	0.75
^{15}N	1/2	0.367	^{34}S	0	4.21
^{16}O	0	99.762	^{35}Cl	3/2	75.77
^{17}O	5/2	0.038	^{37}Cl	3/2	24.23

In the presence of an **external magnetic field** (B_0) nuclei with magnetic moment ($I \neq 0$) orient themselves. The total number of possible orientations for the **magnetic moment vector**, μ, is equal to ($2I + 1$) where I is the nuclear spin quantum number. Thus, for nuclei such as ^1H and ^{13}C, with $I = 1/2$, there are two possible orientations. According to the rules of quantum mechanics, the magnetic moment vector for these nuclei can adopt only two possible angles from the z axis (which coincides with the direction of B_0): $\Theta = 55°$ and $\Theta = 125°$ (Fig. 33.1). The angle ϕ as shown in Figure 33.1 can take any value. These two orientations have different energy and represent two spin states. The state with the magnetic moment vector pointing up is called the α state. This state is said to be aligned with the magnetic field B_0 because the z component of the magnetic moment, μ_z, is parallel to the magnetic field. The other orientation, with the magnetic moment vector pointing down, is called the β state, and is said to be against the field.

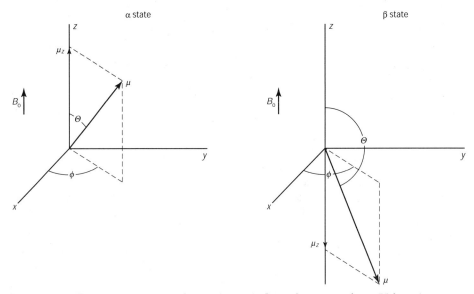

Figure 33.1 The two spin states for nuclei with $I = 1/2$: α state ($\Theta = 55°$) and β state ($\Theta = 125°$). The angle ϕ is not fixed.

The α state is lower in energy than the β state; the energy difference between the states, ΔE, depends on the strength of the applied magnetic field (Eq. 1):

$$\Delta E = E_\beta - E_\alpha = k \times B_0 \tag{1}$$

In Equation 1, k is a proportionality constant that depends on the nucleus in question; it has different values for ^1H and ^{13}C. The energy gap between the states is diagrammatically represented in Figure 33.2. The existence of these two energy states sets the condition for NMR spectroscopy. Nuclei in the lower energy state can be promoted to the higher energy state by absorption of a photon of the right frequency ν (remember that the energy of a photon is $h\nu$; h is Planck's constant).

$$h \times \nu = k \times B_0 \tag{2}$$

$$\Longrightarrow \boxed{\nu = \frac{k}{h} \times B_0} \tag{3}$$

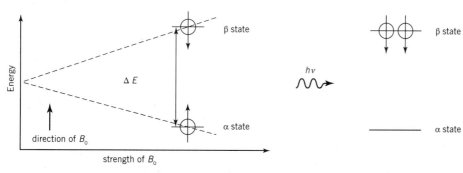

Figure 33.2 Spin flip from α to β by absorption of radiation.

The energy difference between the two spin states is normally very small, and thus electromagnetic radiation of low energy (low frequency) is needed for the transition. The frequency lies in the range 6–700 MHz within the radiowave portion of the electromagnetic spectrum. As can be seen from Equation 3, the exact frequency required to flip the spins from α to β, called the **resonance frequency**, depends on the strength of the magnetic field, B_0, and the constant k, which has a characteristic value for each type of nucleus. Since different nuclei have different k values, they resonate at different frequencies in a constant B_0 field. For example, a molecule such as chloroacetic acid has four magnetically active nuclei: ^1H, ^{13}C, ^{35}Cl and ^{37}Cl; ^{16}O, which accounts for more than 99.7% of natural oxygen, is not magnetically active (we will ignore ^{17}O ($I = 5/2$) because of its low natural abundance of only 0.038%). Thus, a hypothetical scan of the NMR spectrum of chloroacetic acid would show four absorptions. Using a magnet of $B_0 = 21,150$ gauss, the nuclei resonate at the frequencies shown in Figure 33.3.

The spectrum shown in Figure 33.3 is not, however, a typical NMR spectrum. In real NMR experiments, instead of scanning a wide range of frequencies to see the resonance of different nuclei, a small frequency region where only one nucleus (^1H, ^{13}C, etc.) resonates is explored at a time. We will come back to this in section 33.4.

33.2 THE RESONANCE PHENOMENON: A CLOSER LOOK

To understand how modern NMR instruments operate we should complement the simple quantum mechanical description of the resonance process discussed above

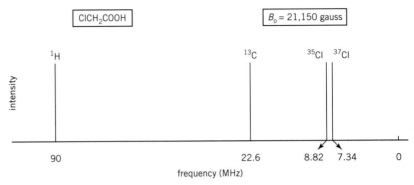

Figure 33.3 The hypothetical NMR of chloroacetic acid.

with a vector picture. A detailed analysis of the process is, however, beyond the scope of this book and the student is referred to References 1, 4, 7, and 8 for a more comprehensive treatment of the subject. To simplify our analysis we will restrict our discussion, for now, to ^{1}H-NMR. The concepts can be easily generalized to explain ^{13}C-NMR as well.

When nuclei orient in a magnetic field they don't do it "faithfully" like a compass in the Earth's magnetic field, but rather they **precess** about the magnetic field at the fixed angle Θ (remember that Θ can be 55° or 125°). To visualize nuclear precession, think of the rotation about the vertical position that a spinning top acquires when tilted (Fig. 33.4a). The frequency of the nuclear precession, ν_0, is called the **Larmor frequency** and is given by Equation 4 where μ, h and I have their usual meanings. The precession is represented in Figure 33.4b by the rotation of the magnetic moment μ about the magnetic field B_0.

$$\nu_0 = \frac{\mu}{h \times I} \times B_0 \qquad (4)$$

According to Equation 4, the stronger the magnetic field, the faster the hydrogen nuclei precess. Hydrogens in the same chemical environment sense the same magnetic field and precess at the same frequency. Molecules with two or more types of hydrogens have two or more Larmor frequencies.

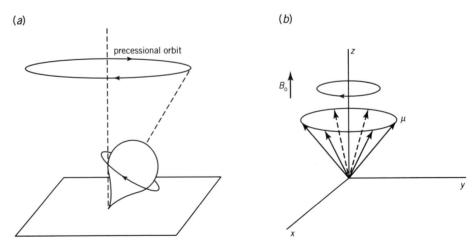

Figure 33.4 *a*) The precession of a spinning top about the Earth's gravitational field; *b*) the precession of the nuclear magnetic moment about the applied magnetic field B_0. Only nuclei in the α state are shown.

Since the angle ϕ is not fixed (Fig. 33.1), the magnetic moment vector μ can be at any value of ϕ, and at any given time, each individual nucleus of a sample has a different value for ϕ. Since there is a very large number of nuclei, the components in the $x-y$ plane of the individual magnetic moments cancel each other out. As a result, the overall magnetic moment for a group of equivalent nuclei only has a component in the z direction. Because there are more nuclei in the α state than in the β state, the **net magnetization (M)** for a system of equivalent nuclei is in the direction of $+z$ (Fig. 33.5).

To flip the spins, radiofrequency is given to the sample by a transmitter coil (Fig. 33.6). This radiofrequency is applied in such a way that its magnetic component, B_1, is in the plane $x-y$ and rotates in the same direction as the individual magnetic moments of the sample (Fig. 33.6a). When B_1 *rotates at the same frequency as the Larmor frequency of the nuclei, absorption of energy takes place and some of the nuclei flip spin from* α *to* β. This causes a change in the orientation of the net magnetization M, which is no longer aligned with the $+z$ axis and is tilted toward the $x-y$ plane (Fig. 33.6b). The final position of the magnetization vector depends on the strength of the field B_1 and the length of time for which the radiofrequency is applied. Normally, the oscillating field B_1 is applied for a very short period as a **pulse of radiation**. The angle θ through which the magnetization M is tipped from the z axis can be calculated with Equation 5:

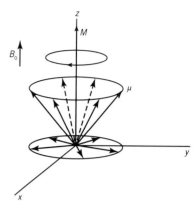

Figure 33.5 The magnetic moments precess about the applied magnetic field B_0. The individual components in the $x-y$ plane cancel each other out. The net magnetization is in the $+z$ axis.

$$\theta = 2\pi k \times B_1 \times t_p \tag{5}$$

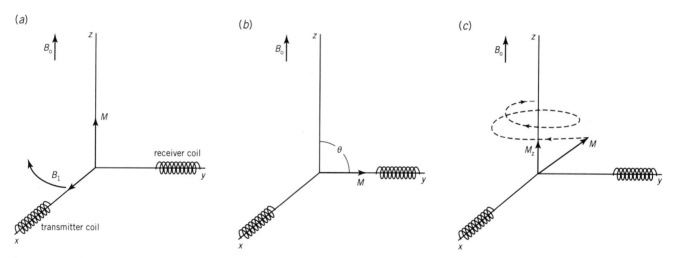

Figure 33.6 *a*) The rotating B_1 is generated by the transmitter coil; *b*) a $\pi/2$ pulse is applied, and the magnetization is tipped to the $x-y$ plane; *c*) the magnetization precess about B_0. The signal is measured in the receiver coil.

where k and B_1 have their usual meanings and t_p is the duration of the pulse in seconds. In a typical experiment, the duration of the pulse is such that the tip angle θ is $\pi/2$. This is called a 90° or a $\pi/2$ pulse (Fig. 33.6b).

At the end of the pulse, when B_1 is no longer applied, the net magnetization M tries to establish its equilibrium position and precesses about B_0 in a decreasing spiral, (Fig. 33.6c). The projection of M in the $x-y$ plane generates an oscillating magnetization in that plane, which is measured by a receiver coil. The process of restoring the magnetization to its equilibrium value is called **relaxation**. During relaxation, nuclei in the higher energy level return to the lower level by giving away the excess energy.

33.3 OBTAINING THE NMR SPECTRUM

In this section we will discuss how the NMR spectrometer works. We will delay until section 33.20 the description of how to prepare the samples for NMR analysis.

There are two main ways of obtaining NMR spectra. Old or very simple instruments operate by the principle of **continuous wave** (**CW**). Modern instruments use the **pulse Fourier transform** technique (**PFT** or **FT**). A simplified diagram of an NMR spectrometer is shown in Figure 33.7.

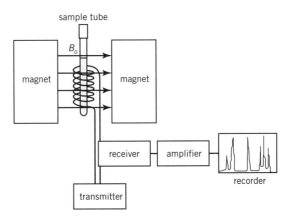

Figure 33.7   A simplified diagram of an NMR spectrometer.

The sample is placed in a slender tube and spun inside a magnet. The spinning of the sample averages out inhomogeneities in the applied magnetic field B_0. A coil wrapped around the tube delivers the radiofrequency radiation (transmitter coil) and in some instruments the same coil also acts as the receiver or detector. The magnetic component of the radiofrequency, B_1, is perpendicular to B_0. In the continuous wave technique, either the magnetic field B_0 is swept while the radiofrequency is kept constant, or the magnetic field is kept constant while the radiofrequency is varied. Most modern CW instruments use the latter mode. In these instruments, the magnetic field B_0 is held constant and the radiofrequency is varied from higher to lower values. When the applied radiofrequency equals the Larmor frequency of the nuclei in the sample, absorption of energy occurs and this is detected as magnetization in the receiver, which produces a voltage that is amplified and sent to the recorder. In the continuous wave mode, each type of hydrogen is excited individually. It takes a few minutes to obtain the whole NMR spectrum.

Modern NMR spectrometers operate by the pulse Fourier transform technique. This mode of obtaining the NMR is very fast compared with the CW mode and makes possible the study of nuclei with low natural abundance, such as ^{13}C. Instead of exciting each type of hydrogen (or carbon) individually as in CW, in the PFT technique all the nuclei are excited simultaneously with a short pulse of radiation that contains all the radiofrequencies of interest. Normally it is a $\pi/2$ pulse that tips the magnetization to the horizontal plane. If there is only one type of magnetic nuclei in the sample, all precessing at the same Larmor frequency, the receiver coil detects an oscillating signal that corresponds to the relaxation of the magnetization in the $x-y$ plane. This signal is called a **free induction decay (FID)** and contains all the information of the NMR spectrum, albeit in a cryptic way. Figure 33.8a shows a simple FID for a sample with only one Larmor frequency, and Figure 33.8b shows the FID of a more typical sample with several Larmor frequencies.

Before it can be interpreted as an NMR spectrum, the FID has to be decoded using a Fourier transform algorithm. A Fourier transform is a mathematical manipulation

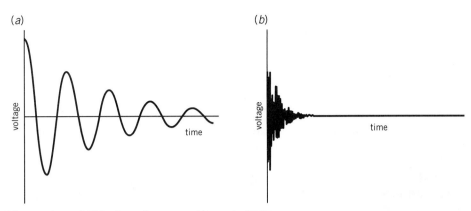

Figure 33.8   *a*) FID of one frequency; *b*) a typical FID.

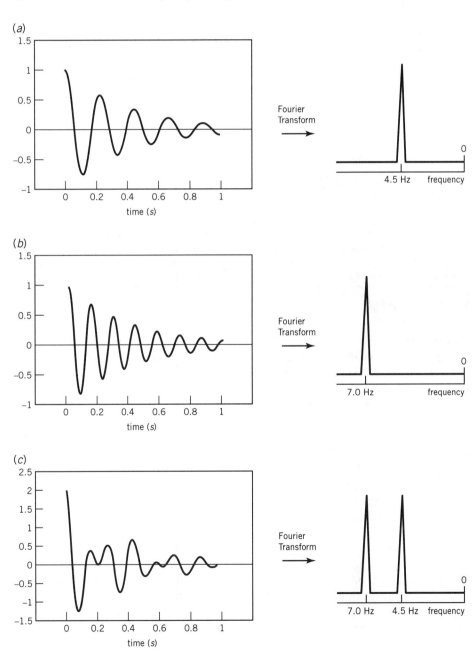

Figure 33.9   Fourier transform. The FID is Fourier-transformed to the corresponding frequencies: *a*) and *b*) FIDs with only one frequency (4.5 and 7.0 Hz); *c*) FIDs for *a* and *b* added up.

that decomposes a complex oscillating signal into its individual components. For example, when we hear a musical piece, our brain decodes the complex sound into the different components and allows us to distinguish the piano from the trumpet. The Fourier transform converts an oscillating signal of voltage versus time into a signal in the frequency domain, or in other words, it extracts the frequencies hidden in the FID and displays them separately (Fig. 33.9).

The big advantage of PFT versus CW is the rapid acquisition of spectra as a result of simultaneously exciting all the nuclei. In a few seconds a good-quality NMR spectrum can be obtained. This allows the study of nuclei with low natural abundance, such as ^{13}C. To obtain a good spectrum of such nuclei, the whole spectrum is run several times and the result of each run is stored in the memory of a computer. The individual spectra are added and the sum spectrum is displayed. This technique improves the sensitivity because it increases the **signal-to-noise ratio**. Noise, represented by the jagged baseline of the spectrum (Fig. 33.10), is a random process and can take positive or negative values. A true signal, on the other hand, is always positive. After adding many spectra, the noise tends to cancel out while the true signals get reinforced in each addition and augment in size.

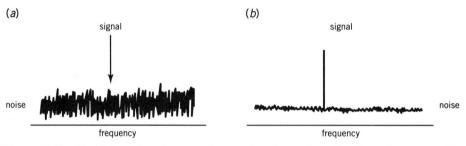

Figure 33.10 Signal-to-noise: *a*) a case of a poor signal-to-noise ratio after only one scan. *b*) improved signal-to-noise ratio after adding 400 scans.

33.4 ^{1}H-NMR

As was mentioned at the end of section 33.1, the radiofrequency is never swept in a wide range that covers all possible magnetically active nuclei in a molecule. Instead, a very narrow range around the resonance frequency of ^{1}H or ^{13}C is studied. For example, to obtain ^{1}H-NMR spectra using a magnet of $B_0 = 21{,}150$ gauss, the frequency is swept around 90 MHz, for example, from 90,000,000 Hz to 90,001,000 Hz. To obtain the ^{13}C-NMR spectrum in the same magnetic field, the frequency is swept around 22.6 MHz, also in a very narrow range. An instrument with a 21,150 gauss magnet, where ^{1}H resonates at 90 MHz, is called a 90-MHz spectrometer. The resonance frequency of ^{1}H is called the **operational frequency** of the instrument.

The ^{1}H-NMR spectrum of chloroacetic acid shows that the signal observed around 90 MHz for ^{1}H (Fig. 33.3) resolves into two signals very close together, one at 90,000,470 Hz and the other at 90,000,900 Hz (Fig. 33.11). Why are there two signals instead of one?

The molecule of chloroacetic acid has two types of hydrogens: the hydrogens of the methylene group, and the hydrogen of the acid. These hydrogens are chemically different because they are in different environments and do not sense exactly the same magnetic field, even though they are immersed in the same applied field B_0. The actual field sensed by each hydrogen is the result of the applied magnetic field B_0 and the local fields generated by electrons and other atoms in the surroundings. Hydrogens in different chemical environments precess at different Larmor frequencies, and thus resonate at different frequencies. These hydrogens are **nonequivalent**. ^{1}H-NMR allows us to distinguish them. *The ^{1}H-NMR spectrum shows, in principle, as many*

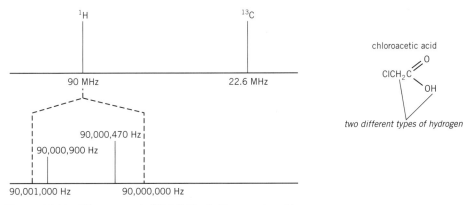

Figure 33.11 Diagrammatic ^{1}H-NMR of chloroacetic acid.

signals as the number of nonequivalent hydrogens. A few examples follow; we will come back to this in section 33.8.

One signal	*Two signals*	*Three signals*

33.5 ELECTRONIC SHIELDING

Hydrogen nuclei in organic molecules are covalently bound to other nuclei and surrounded by the electron clouds of the intervening bonds. In the presence of an applied magnetic field, the electrons flow around the nucleus and generate a small induced magnetic field, which, like any induced magnetic field, opposes the applied field (Lenz law). The induced magnetic field is called B_{local} (Fig. 33.12). Thus, the effective magnetic field sensed by the nucleus, B_{eff}, is smaller than the applied magnetic field B_0.

$$B_{eff} = B_0 - B_{local} \tag{6}$$

This effect caused by the electron flow prevents the nuclei from sensing the applied magnetic field B_0 full strength. The nuclei are "shielded" by the electron cloud. Electron-donating groups near the hydrogen nucleus increase the electron density around the nucleus and intensify the **shielding effect**. Electron withdrawing groups, on the other hand, pull the electrons toward themselves and away from the hydrogen and decrease shielding. These groups are said to be deshielding or to have a **deshielding effect**.

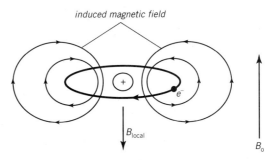

Figure 33.12   Electronic shielding.

What are the consequences of shielding? Since the effective magnetic field sensed by each nucleus is the resultant of the applied field and the local fields, nuclei in different environments will resonate at different frequencies. Instead of Equation 3, the following equation, where B_{eff} is different for different nuclei, applies:

$$ \nu = \frac{k}{h} \times B_{eff} $$

Substituting B_{eff} by Equation 6:

$$ \nu = \frac{k}{h} \times (B_0 - B_{local}) \qquad (7) $$

In the continuous wave mode, as you may remember, there are two ways of running NMR: keeping B_0 constant and sweeping the radiofrequency ν (newer instruments), or keeping the radiofrequency ν constant, and sweeping B_0 (old instruments). If the spectrum is recorded at constant B_0 by sweeping the radiofrequency, nuclei **shielded** by a large B_{local} will resonate at **low frequency** values. **Deshielded** nuclei, on the other hand, with weak B_{local} will resonate at **higher frequencies**. If the spectrum is obtained by keeping the frequency ν constant and varying the field B_0 (like in the old days), then Equation 7 reveals that an increase in B_{local} (an increase in shielding) must be matched by an increase in B_0 to keep the frequency ν constant. Under these conditions, **shielded** nuclei resonate at high B_0 fields, or **upfield**, while **deshielded** nuclei resonate at low fields, or **downfield**. Although the method of sweeping the magnetic field has become obsolete, the names *upfield* as a synonym of *shielded*, and *downfield* as a synonym of *deshielded* have survived until now.

Let's suppose now that we mix methane, methyl chloride, dichloromethane, and chloroform and record the ^{1}H-NMR of the mixture. Since each compound has only one type of hydrogen, we expect to see four peaks in the spectrum. This is illustrated in Figure 33.13.

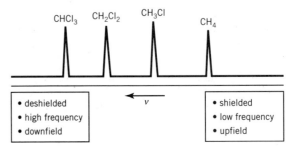

Figure 33.13   Deshielding caused by chlorine atoms.

As the number of chlorine atoms increases, the electron density around the hydrogens diminishes because of the electronegativity of chlorine. As a result, the deshielding increases and the hydrogens resonate at higher frequencies. Similar to IR and UV-visible spectra, the NMR spectrum is displayed with the frequency increasing from right to left.

The hybridization of the carbon atom, the presence of aromatic rings, and the formation of H-bonds are factors that also affect the resonance frequency. We will discuss these effects in later sections.

33.6 THE CHEMICAL SHIFT

The measurement of absolute frequencies is difficult. To circumvent this problem frequencies are measured using the resonance frequency of a standard compound as a reference. The compound is called the **internal standard** and is added to the sample before running the NMR. The standard universally used is **tetramethylsilane (TMS)**. TMS is an ideal reference compound because it has only one type of hydrogen and thus it shows one peak in the NMR. It resonates at lower frequency than most organic compounds, and thus it normally shows at the right end of the spectrum.

Let's suppose that we mix acetone, water, and TMS and obtain the ^{1}H-NMR of the solution using a 14.1 kgauss magnet where ^{1}H resonates around 60 MHz. The spectrum shows three sharp peaks, because each compound has only one type of hydrogen. The peaks for acetone and water are found at 126 Hz and 165 Hz from the TMS peak, respectively (Fig. 33.14). If the spectrum is obtained using a more powerful instrument with a 58.7 kgauss magnet, where the operational frequency is 250 MHz, the peaks for acetone and water show at 525 Hz and 687.5 Hz, respectively. Diagrams of both spectra are shown in Figure 33.14.

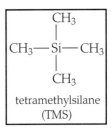

tetramethylsilane
(TMS)

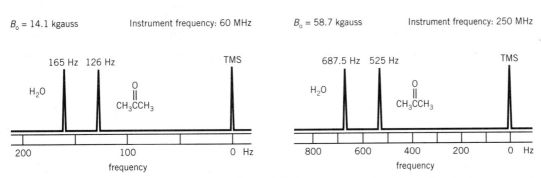

Figure 33.14 Resonance frequencies in Hz from TMS for acetone and water at $B_0 = 14.1$ kgauss and $B_0 = 58.7$ kgauss.

As the examples in Figure 33.14 illustrate, the frequency difference from TMS depends on the operational frequency of the instrument. The higher the operational frequency, the larger the frequency difference from TMS. However, the ratio of the frequency difference from TMS and the operational frequency is a constant for each compound. This ratio is independent of the strength of the magnetic field and is called the **chemical shift**:

$$\text{chemical shift} = \frac{\text{separation from TMS in Hz}}{\text{instrument frequency in MHz}} \qquad (8)$$

The chemical shift is represented by δ. The chemical shift for acetone is:

$$\delta = \frac{126 \text{ Hz}}{60 \text{ MHz}} = \frac{525 \text{ Hz}}{250 \text{ MHz}} = 2.1 \frac{\text{Hz}}{10^6 \text{ Hz}} = 2.1 \ 10^{-6} = 2.1 \text{ ppm}$$

The factor 10^{-6} that results from dividing Hz by MHz is abbreviated ppm. Notice that the chemical shift has no units.

The chemical shift for water is:

$$\delta = \frac{165 \text{ Hz}}{60 \text{ MHz}} = \frac{687.5 \text{ Hz}}{250 \text{ MHz}} = 2.75 \text{ ppm}$$

The chemical shift depends on the electronegativity of the atom to which the hydrogen is attached (C, N, O, S, P, etc.), the electronegativity of nearby groups, the hybridization of the carbon atom, the proximity of double bonds and aromatic rings, the formation of H-bonds, sample concentration, the solvent, and the temperature. The chemical shift range for most hydrogens is between 0 and 12 ppm. The diagram in Figure 33.15 shows the typical chemical shift range for hydrogens in different environments.

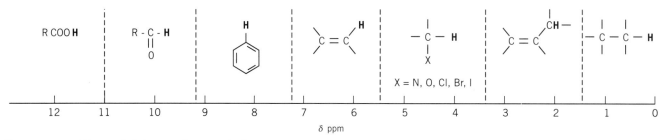

Figure 33.15 Chemical shift range for some hydrogen types.

33.7 EFFECT OF PI ELECTRONS

In the presence of an applied magnetic field, groups with pi orbitals such as carbon-carbon double and triple bonds, carbonyls, and aromatic rings produce an induced magnetic field that depends on the orientation of the molecule relative to the magnetic field. Let's consider first the case of benzene. When benzene is perpendicular to the magnetic field, the pi electrons circulate inside the ring and generate a ring current and an induced magnetic field that *opposes* the applied field *in the center of the ring*. In the *periphery of the ring*, on the other hand, the induced field *points in the same direction* as the applied field and reinforces it (Fig. 33.16). As a result, a deshielding region around the plane of the ring and a shielding cone right above and below the center of the ring are created. The aromatic hydrogens, which are located in the plane of the ring, sense an effective magnetic field larger than B_0 and thus resonate at higher frequencies. Aromatic hydrogens normally resonate between 7 and 9 ppm.

According to Figure 33.16, hydrogens located above or below the ring should be shielded and would resonate at lower chemical shifts than hydrogens in the plane of the ring. This is illustrated with the chemical shifts of 1,6-methano[10]annulene. The methylene hydrogens located above the ring resonate below the TMS signal, at -0.51 ppm, while the aromatic hydrogens resonate around 7 ppm. Hydrogens with resonance below TMS are uncommon in organic molecules and are an indication of an unusual molecular geometry. Some examples of these types of hydrogens are discussed in Experiment 19.

It should be noticed that aromatic molecules in solution, or in the liquid state, tumble very fast and do not adopt any preferential orientation in a magnetic field. Still, the effect of the magnetic field on aromatic molecules is noticeable because at any given time there is a number of molecules perpendicular to the field generating an induced magnetic field.

Similar effects are observed with other planar pi systems such as the carbon-carbon and carbon-oxygen double bonds. There is a deshielding region in the plane of

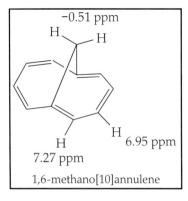

-0.51 ppm

7.27 ppm   6.95 ppm

1,6-methano[10]annulene

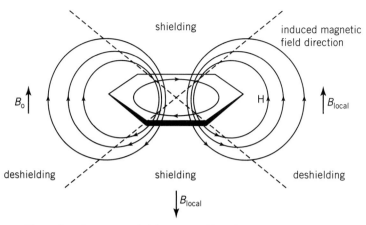

Figure 33.16   The induced magnetic field generated by a benzene ring in a magnetic field.

the double bond, and a shielding cone right above and below it (Fig. 33.17). As a result of this, vinyl hydrogens resonate around 5.5 and 7 ppm. Aldehyde hydrogens resonate at even higher chemical shifts (9–11 ppm) because of the electron-withdrawing effect of the oxygen atom.

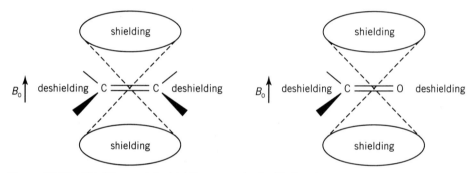

Figure 33.17   Shielding and deshielding zones in double bonds.

Finally, let's consider acetylenic systems. When the molecule is parallel to the applied magnetic field there is an induced electron flow around the molecular axis that creates an induced magnetic field that opposes the applied field along the molecular axis, and reinforces it elsewhere (Fig. 33.18). As a result, acetylenic hydrogens are relatively shielded and resonate at lower chemical shifts (2–3 ppm) than vinyl hydrogens.

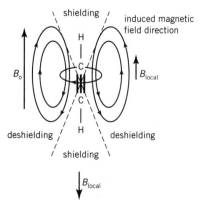

Figure 33.18   Shielding and deshielding zones in acetylenes.

33.8 HYDROGEN EQUIVALENCE: A CLOSER LOOK

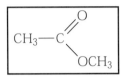

As was stated at the end of section 33.4, the number of signals that one expects to see in a ^{1}H-NMR is equal to the number of nonequivalent hydrogens. Let's consider, for example, the case of methyl acetate. This compound has two types of hydrogens: the methyl group attached to the carbonyl, and the methyl group attached to the oxygen.

The ^{1}H-NMR of methyl acetate shows two peaks (Fig. 33.19). The peak at lower chemical shift corresponds to the methyl group attached to the carbonyl, and the one at higher chemical shift belongs to the more deshielded methyl group, the one bound to the oxygen atom.

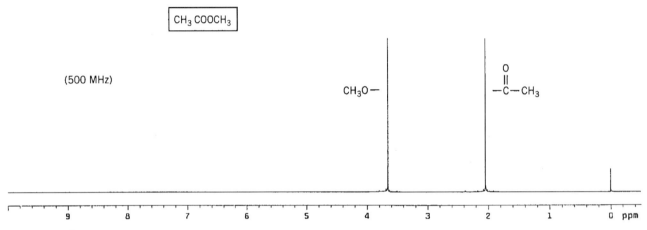

Figure 33.19 ^{1}H-NMR of methyl acetate.

Now let's consider the case of benzyl acetate. There are five nonequivalent types of hydrogens, A, B, C, D, and E, and therefore we expect to see five different peaks in the ^{1}H-NMR. The spectrum is shown in Figure 33.20. Contrary to our expectations it shows only three peaks, at 2.1, 5.1, and 7.4 ppm. What happened?

Benzyl acetate, like all monosubstituted benzene rings, has three different types of aromatic hydrogens (C, D, and E). Although these hydrogens are chemically different and thus nonequivalent, in the case of benzyl acetate they are in a similar magnetic environment and resonate at the same frequency. This unexpected coincidence is called **fortuitous equivalence**. Usually, the aromatic hydrogens of alkyl-substituted phenyl rings resonate at the same frequency because the electronic effects of the alkyl groups are not very strong. On the other hand, phenyl rings with strong electron-donating or electron-withdrawing substituents (for example, $-OH$, $-COR$, etc.) show the aromatic hydrogens at different frequencies.

benzyl acetate

33.9 INTEGRALS

An NMR spectrum not only gives different signals for different hydrogens, but it also tells if the signal corresponds to one, two, three, or more hydrogens. The area under each ^{1}H-NMR peak is proportional to the number of hydrogens that gave rise to the peak. After the ^{1}H-NMR is recorded, the operator can easily obtain the peak integration, which is normally displayed as an **integration line**. The integration line is a trace that runs above the peaks. Wherever the spectrum shows a steady and horizontal baseline, the integration line is also steady and horizontal; wherever there is a peak, the integration line shows a jump with a height that is directly proportional to the area of the peak, or in other words, to the number of hydrogens. The ^{1}H-NMR of benzyl acetate with integration line is shown in Figure 33.21.

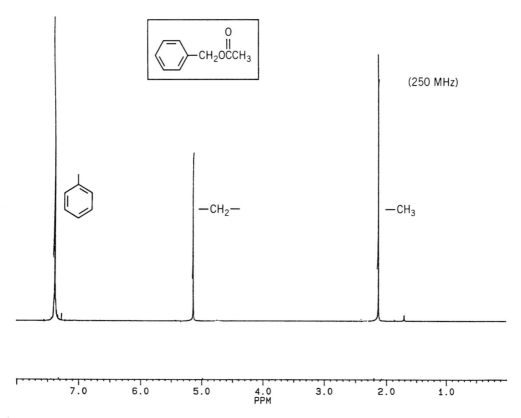

Figure 33.20 ¹H-NMR of benzyl acetate.

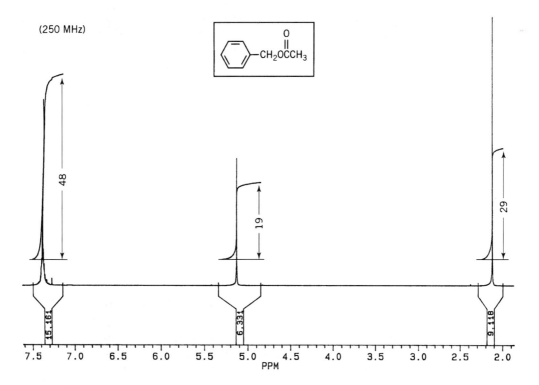

Figure 33.21 ¹H-NMR of benzyl acetate with integration lines. The distances between arrows are in mm.

Modern instruments not only show the integration line, but also give an integration number (in arbitrary units) for each peak. This number is usually shown between brackets to indicate the beginning and the end of the integration (Fig. 33.21). For benzyl acetate the integration numbers are 15.161, 6.331, and 9.118 from left to right and in arbitrary units. The sum of these numbers, or **total integration number**, I_T, is 30.61. The ratio of the integration number over the total integration number represents the proportion of hydrogens under the peak. For example, for the peak around 7.4 ppm, that ratio is 15.161/30.61 = 0.495. Since the total number of hydrogens (H_T) in benzyl acetate is 10, the number of hydrogens under that peak is 0.495 × 10 = 4.95, which is rounded off to 5. Because of intrinsic error in the integration process, the number of hydrogens calculated from the integration numbers may be fractional. They should be rounded off to the closest integer.

If the total number of hydrogens in the molecule is known, the number of hydrogens under a given peak, #H, can be calculated with the following equation:

$$\#H = \frac{I}{I_T} \times H_T \tag{9}$$

where I is the integration number for the peak, I_T is the total integration number, and H_T is the total number of hydrogens in the molecule. Applying this equation to the other two peaks of the benzyl acetate spectrum gives the following numbers. For the peak at 5.1 ppm:

$$\#H = \frac{6.331}{30.61} \times 10 = 2.07$$

which is rounded off to 2. For the peak around 2.1 ppm:

$$\#H = \frac{9.118}{30.61} \times 10 = 2.98$$

which is rounded off to 3.

When the integration numbers are not provided, we should measure the heights of the integration lines as indicated in Figure 33.21. The heights (in mm) for the three peaks are 48, 19, and 29. These are now our integration numbers and we should carry out the same calculations as before. In this case the total integration number is $I_T = 48 + 19 + 29 = 96$. The calculation of the number of hydrogens for the peak around 7.4 ppm gives the following result:

$$\#H = \frac{48}{96} \times 10 = 5.0$$

which agrees with the previous calculation using the integration numbers provided in brackets.

When the total number of hydrogens is not known, the integration numbers give the relative number of hydrogens under the peaks. If one didn't know that the compound that originated the spectrum of Figure 33.21 was benzyl acetate, the integration numbers of 15.161, 6.331, and 9.118, which can be approximated to 15, 6, and 9, would reveal that the *relative number* of hydrogens was 15 : 6 : 9, or (after dividing by 3) 5 : 2 : 3.

33.10 SPIN-SPIN SPLITTING

If the number of nonequivalent hydrogens and the integration were the only data derived from the ^{1}H-NMR spectrum, the technique would still be immensely useful for structure elucidation. However, there is an added piece of information that can be obtained from the NMR and this is the *number of hydrogens near a group of equivalent hydrogens*. How is this possible?

Let's consider the case of 1,1,2-trichloroethane. This molecule has two sets of hydrogens that we will call H_A and H_B. In its ^{1}H-NMR we would, in principle, expect to see two peaks. The signal for H_B is expected at a higher chemical shift than that for H_A, because H_B is next to two chlorine atoms while both H_A are next to only one. The ^{1}H-NMR (Fig. 33.22) shows, however, five peaks clustered in two groups. The integration reveals that the two peaks around 3.95 ppm integrate for two hydrogens, and the cluster of three peaks around 5.77 integrates for one hydrogen. Based solely on the integration, the peaks at 3.95 ppm can be assigned to H_A, and the peaks at 5.77 ppm to H_B. This assignment also agrees with the expected chemical shift order.

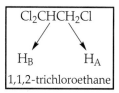

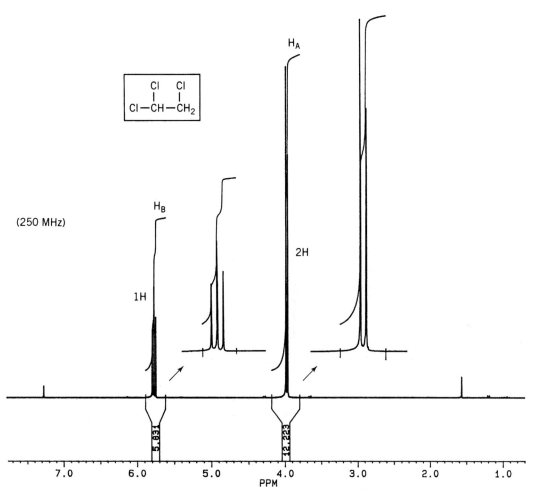

Figure 33.22 ^{1}H-NMR of 1,1,2-trichloroethane.

Why is the signal for H_A a group of two peaks, and the signal for H_B a group of three? The number of peaks for a given signal, called its **multiplicity**, depends on the *number of hydrogens on the adjacent carbon atoms*. A rigorous treatment of this phenomenon requires quantum mechanics and will not be attempted here. A simple and intuitive account of the multiplicity will be provided instead. Let's first consider H_A. The signal for H_A is a group of two peaks, or a **doublet**, because of the presence of *one* H_B on the next carbon. H_B can adopt two possible orientations, α and β, and each spin orientation has an effect on the effective magnetic field sensed by H_A. Depending on whether H_B is in the α or β state, the effective field sensed by H_A will be slightly different. Consequently, H_A shows two transitions at two slightly different frequencies (Fig. 33.23).

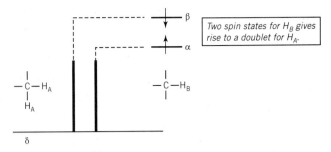

Figure 33.23 The doublet for the methylene hydrogens caused by the methine hydrogen.

On the other hand, H_B shows as a group of three peaks, or a **triplet**, because the *two* hydrogen atoms on the adjacent carbon, H_A, can adopt *three spin states*, and each one affects the magnetic field sensed by H_B. The three states are: a low energy state with both H_A in the α state ($\alpha\alpha$), a high energy state with both H_A in the β state ($\beta\beta$), and two intermediate states of equal energy with one H_A in the α state and the other H_A in the β state ($\alpha\beta$ and $\beta\alpha$) (Fig. 33.24).

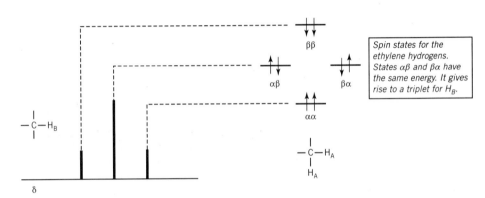

Figure 33.24 The triplet for the methine hydrogen caused by the methylene hydrogens.

The middle line of the triplet is twice as intense as the outer lines because it arises from two spin configurations of both H_A, $\alpha\beta$ and $\beta\alpha$.

The splitting of the signals is called **spin-spin splitting** and reflects the **coupling** of a hydrogen's spin state with its neighbor's. In general, the multiplicity of a signal is determined by the ***n* + 1 rule**, where *n* is the number of neighboring hydrogens. In the previous example, the methylene hydrogens, H_A, have only one neighbor, $n = 1$, and give rise to a doublet ($n + 1 = 1 + 1 = 2$). The methine hydrogen, H_B, on the other hand, has two neighbors, $n = 2$, and gives rise to a triplet ($n + 1 = 2 + 1 = 3$).

Cl H_A

The methine hydrogen has   Cl—C—C—Cl   *The two methylene hydrogens have one*
two neighbors and gives rise   *neighbor, and give rise to a doublet:*
to a triplet: H_B H_A *n* + 1 = 1 + 1 = 2
n + 1 = 2 + 1 = 3

Let's consider the case of ethyl iodide. There are two types of hydrogens in this molecule: the three hydrogens of the methyl group, and the two hydrogens of the methylene. We expect to see two signals with relative integration $3:2$. We also expect to see the methylene at higher chemical shift than the methyl group because the methylene is directly attached to the electron-withdrawing iodine atom. The multiplicity of the methyl group is a triplet because it has two neighbors ($n + 1 = 2 + 1 = 3$). The methylene group gives rise to four lines, or a **quartet**, because it has three neighbors ($n + 1 = 3 + 1 = 4$). The spectrum of ethyl iodide is shown in Figure 33.25.

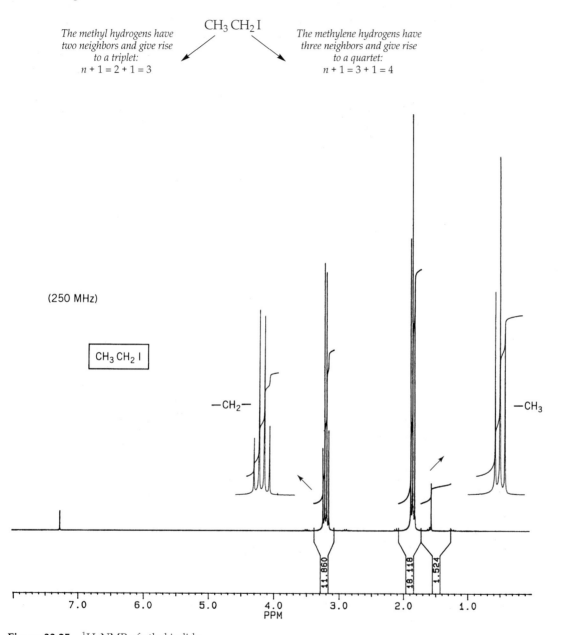

The methyl hydrogens have two neighbors and give rise to a triplet:
$n + 1 = 2 + 1 = 3$

$$CH_3\,CH_2\,I$$

The methylene hydrogens have three neighbors and give rise to a quartet:
$n + 1 = 3 + 1 = 4$

(250 MHz)

CH$_3$ CH$_2$ I

—CH$_2$—

—CH$_3$

Figure 33.25 ^{1}H-NMR of ethyl iodide.

The relative intensity of the lines of a quartet is $1:3:3:1$, because these are the numbers of possible configurations in the spin states of the methyl hydrogens (Fig. 33.26).

Let's consider the case of 2-chloropropane. The six hydrogens of the two methyl groups are equivalent and give a doublet because they have one neighbor, the methine hydrogen ($n + 1 = 1 + 1 = 2$). The methine hydrogen sees six neighboring hydrogens and gives rise to seven lines, or a **septet**. The spectrum is shown in Figure 33.27.

The methyl hydrogens have one neighbor and give rise to a doublet:
$n + 1 = 1 + 1 = 2$

$$\begin{array}{c} CH_3 \quad CH_3 \\ \diagdown\; CH \;\diagup \\ | \\ Cl \end{array}$$

The methine hydrogen has six neighbors and gives rise to a septet:
$n + 1 = 6 + 1 = 7$

2-chloropropane

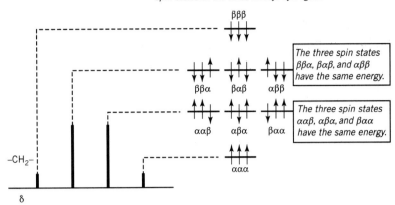

Figure 33.26   The splitting of the methylene signal by the methyl group.

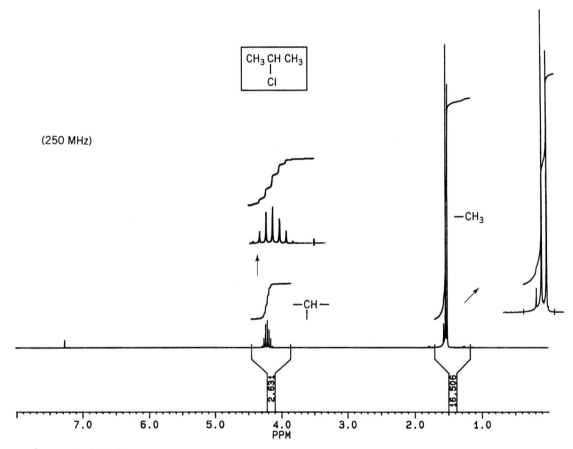

Figure 33.27   ¹H-NMR of 2-chloropropane.

The relative intensity of the lines of doublets, triplets, quartets, and so on can be predicted by Pascal's triangle, where each number is the sum of the two right above (Fig. 33.28).

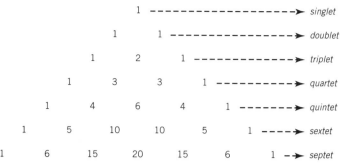

Figure 33.28 Pascal's triangle: each number is the sum of the two numbers right above. The numbers give the relative intensities of the lines of doublets, triplets, and so on.

As you can see, the relative intensity of the lines of a quartet according to the Pascal triangle is 1:3:3:1. Notice that as the multiplicity increases, the outer lines become negligible in relation to the central ones. For example, the outer lines of a septet are difficult to observe because their relative intensity is only one-twentieth that of the central one.

33.11 TYPICAL COUPLING PATTERNS

The coupling pattern of some typical groups is shown in Figure 33.29. It is assumed that no other hydrogens, except for those shown, participate in the coupling. Notice that the relative intensities of the multiplets deviate slightly from those predicted by the Pascal's triangle. Let's consider, for example, Figure 33.29b, where the lines of the doublet have been labeled a and b, and the lines of the quartet, c, d, e, and f. In the doublet, line b is taller than a, and in the quartet, line c is taller than f, and d taller than e. This behavior is called the **roof effect** and can be generalized by saying that in each multiplet, the peaks closer to the other multiplet are slightly taller than their symmetrical counterparts. The roof effect is very useful to match up multiplets that are coupled to each other, especially when there are several multiplets in the spectrum. Two sets of hydrogens coupled to one another always show the roof effect (see Problem 4).

It should be noticed that coupling is a reciprocal process. If we see a multiplet that belongs to a set of hydrogens (H_A), then we must see another multiplet, somewhere else in the spectrum, that belongs to another set of hydrogens (H_B) coupled to the first.

33.12 THE COUPLING CONSTANT

The distance between the lines of a multiplet is called the **coupling constant** and is represented by J. Within a triplet, a quartet, or a quintet the peaks are equally spaced and the separation between peaks is the coupling constant. The coupling constant is always measured in Hertz, instead of chemical shift units, because its value in Hertz does not depend on the operational frequency of the instrument. Whether the NMR is obtained at 60 or 600 MHz, the separation between the lines of a given multiplet is always the same in Hz. To measure the coupling constant, the distance between the peaks is measured in chemical shift units, and then multiplied by the operational frequency of the instrument. An example of how the coupling constant is measured is

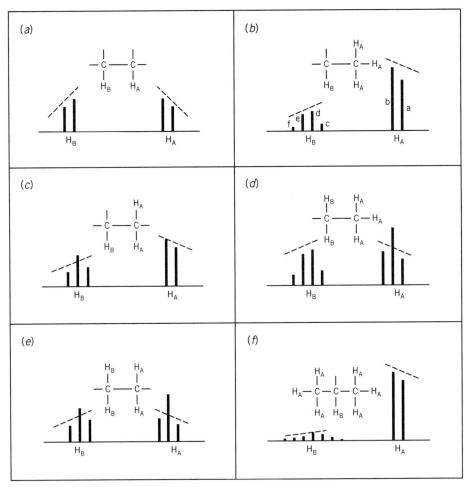

Figure 33.29 Some typical coupling patterns. The dotted lines show the roof effect.

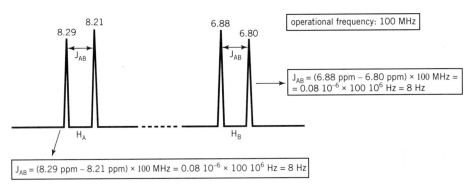

Figure 33.30 Measuring the coupling constant in two doublets.

given in Figure 33.30, where the doublets were obtained using a 100-MHz instrument. The coupling constant measured in both multiplets is the same.

The coupling between nuclei occurs through the intervening bonds. The electrons transmit the magnetic information from one nucleus to the other. When this happens between hydrogens on adjacent carbons, the coupling is called **vicinal**. Vicinal coupling takes place between hydrogens separated by three bonds (H−C−C−H) and is the most commonly observed coupling in ^{1}H-NMR. There is also **geminal coupling** between hydrogens attached to the same carbon atom (H−C−H). Geminal coupling is observed in alkenes and ring systems if both hydrogens are nonequivalent. Finally, there is **long range coupling** that takes place between hydrogens separated by four or more bonds (H—C—C—C—H). This type of coupling occurs when there is a pi-system between the carbon atoms connecting the hydrogens, or when the molecule has a rigid ring system and the atoms H—C—C—C—H are in a fixed conformation in the shape of a W (see Table 33.2).

Table 33.2 Some Typical Coupling Constant Values.

Geminal	Vicinal	Long range

The strength of the coupling decreases as the number of intervening bonds increases; usually long-range coupling constants are smaller than vicinal or geminal ones. Some typical coupling constant values are gathered in Table 33.2.

The coupling constant depends on geometrical factors such as bond length, bond angle, and carbon hybridization. The electronegativity of the substituents directly attached to the carbon atoms also influences the value of the coupling constant. The coupling between vicinal hydrogens depends strongly on the **dihedral angle**, ϕ (Fig. 33.31).

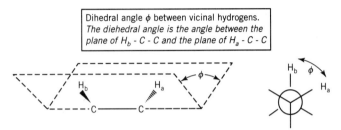

Figure 33.31   Dihedral angle between vicinal hydrogens.

The relationship between dihedral angle and J value is given by the **Karplus equation**:

$$J = -0.28 + 8.5 \times \cos^2 \phi, \qquad \text{for } 0° \leq \phi \leq 90° \text{ and}$$
$$J = -0.28 + 9.5 \times \cos^2 \phi, \qquad \text{for } 90° \leq \phi \leq 180°$$

The Karplus equation is represented in Figure 33.32. It can be observed that the coupling constant goes through a minimum for dihedral angles around 90° and it reaches maximum values for dihedral angles near 0 and 180°. Thus, by measuring the coupling constant J, valuable information about a molecule's geometry can be deduced.

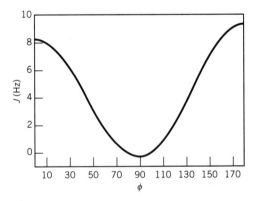

Figure 33.32   Karplus plot: J vs. ϕ.

33.13   COUPLING: A CLOSER LOOK

Equivalent Hydrogens Do Not Split

Equivalent hydrogens, attached to the same carbon or different carbons, give a singlet as long as they are not coupled to other nuclei. Some examples of molecules with equivalent hydrogens are given below. It is a common misconception to think that because spin-spin splitting is not observed between equivalent hydrogens, these hydrogens are not coupled to each other. Equivalent hydrogens are in fact coupled to each other but their coupling does not result in splitting. We will come back to this later in this section (under **Strongly Coupled Systems**). Special methods should be used to measure the coupling constant between equivalent hydrogens (see Ref. 2).

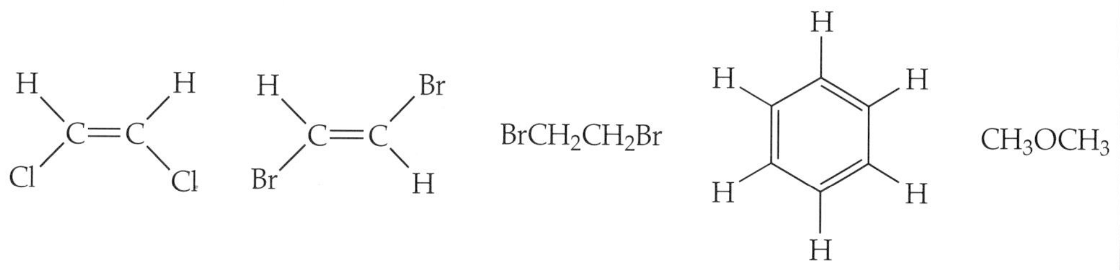

Molecules with all equivalent hydrogens. Each one gives a singlet.

Cyclohexane Ring System

An interesting case of hydrogen equivalence is that of cyclohexane. The cyclohexane molecule in the chair conformation has two types of hydrogens: six axial and six equatorial. There are four different coupling constants involved: one geminal coupling, J_{gem}, and three vicinal coupling constants J_{aa}, J_{ea}, and J_{ee}, and thus one should expect a complex spectrum for this molecule (Fig. 33.33). However, the ^{1}H-NMR of cyclohexane obtained at room temperature shows only one singlet. This is the result of a rapid interconversion between the two chairs, which causes all the hydrogens to experience the same average environment and behave as if they were equivalent. If the spectrum is obtained at lower temperatures where the rate of interconversion is much slower, then it is possible to see separate signals for the axial and equatorial hydrogens.

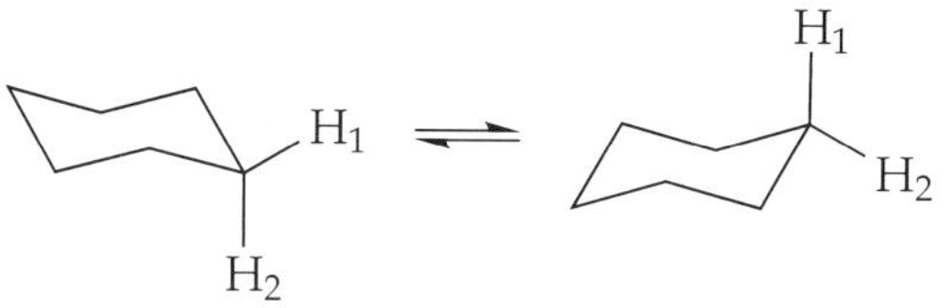

H_1 equatorial (left) becomes axial (right), and H_2 axial (left) becomes equatorial (right). Both hydrogens see the same average environment.

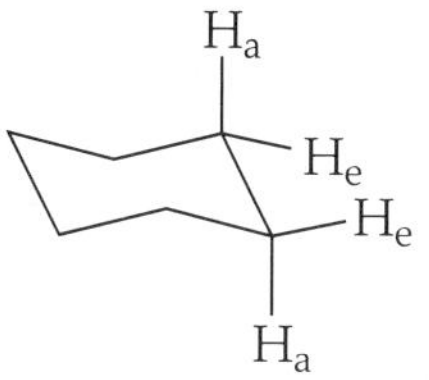

J_{gem} = 13 Hz (11–14 Hz)

vicinal coupling:
- J_{ee} = 3 Hz (2–5 Hz)
- J_{ae} = 4 Hz (2–6 Hz)
- J_{aa} = 13 Hz (7–13 Hz)

Figure 33.33 Coupling constants in cyclohexane. The values in parenthesis are for substituted cyclohexane rings.

Signals With More Than One Coupling Constant

So far we have considered the coupling between only two groups of hydrogens. In many organic molecules, however, a set of hydrogens is simultaneously coupled to two or more sets of hydrogens. Let's consider, for example, the case of *trans*-cinnamyl alcohol, a product used in perfumery and prepared in Unit 21. This compound has seven different types of hydrogens as indicated by the letters A–G. Hydrogens B, C, and D are the ones that interest us now. Let's ignore for now the aromatic hydrogens E–G and the alcohol hydrogen, A.

H_D is coupled to H_C with a vicinal coupling constant, J_{CD}, and also to H_B through long-range coupling, J_{BD}. H_D is also coupled through long-range coupling to the aromatic hydrogens, but the coupling between hydrogens in a side chain and hydrogens in the aromatic ring is not observed under normal conditions (the coupling constant is too small) and we will ignore it. H_C is coupled to H_D with J_{CD} and also to H_B (J_{BC}). Finally, H_B is coupled to H_C and H_D with coupling constants J_{BC} and J_{BD}. The coupling between hydrogen B and the alcohol hydrogen H_A is not observed for reasons that we will discuss in section 33.15. The experimental values for the coupling constants are $J_{BC} = 6$ Hz, $J_{BD} = 1.4$ Hz, and $J_{CD} = 16$ Hz (notice that these values agree with those given in Table 33.2).

Let's first consider H_B. What multiplicity do we expect for this hydrogen coupled to H_D and H_C with two different coupling constants? To answer this question let's first pretend that H_B was coupled only to H_C. In such a case we would see that the signal for H_B, according to the $n + 1$ rule, splits into a doublet with a separation equal to J_{BC}. Now, because H_B is also coupled to H_D, each line of the doublet splits, in turn, into a doublet with a separation equal to J_{BD} (which is smaller than J_{BC}) (Fig. 33.34a). This multiple splitting gives a total of four lines with similar intensities, called a **double doublet**. It does not matter which coupling is considered first; if the coupling between H_B and H_D is drawn first, a doublet with a J_{BD} is obtained; because H_B is also coupled to H_C, each line of this doublet splits into another doublet with a separation J_{BC}. The final result is the same double doublet as obtained before (Fig. 33.34b).

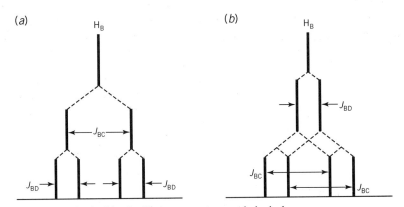

Figure 33.34 Double doublet for H_B in *trans*-cinnamyl alcohol.

Now let's consider the splitting of H_C. This hydrogen is coupled to H_D with a large coupling constant ($J_{CD} = 16$ Hz) and to H_B with $J_{BC} = 6$ Hz. Let's consider first the coupling with H_D. Because there is only one H_D, the signal for H_C is split, in principle, into a doublet according to the $n + 1$ rule. Since H_C is also coupled to H_B, and there are two H_B, each line of the doublet is split into a triplet according to the $n + 1$ rule ($2 + 1 = 3$) (Fig. 33.35a). This signal is called a **double triplet**. If the splitting is analyzed by first considering the coupling between H_C and H_B, and then the coupling between H_C and H_D, the final result is the same and a double triplet is obtained (Fig. 33.35b).

Finally, let's consider the splitting of H_D. H_D is coupled to H_C and H_B with coupling constants J_{CD} and J_{BD}. Because there is only one H_C, the signal for H_D, in principle, splits into a doublet. Each line of the doublet splits, in turn, into a triplet because H_D is also coupled to two H_B. The final multiplicity is a double triplet (Fig. 33.36).

(a)   (b)

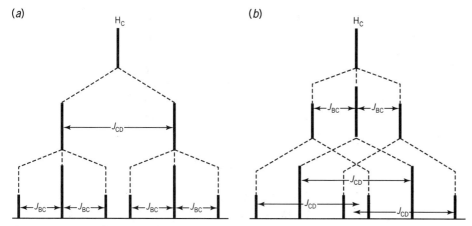

Figure 33.35 Double triplet for H_C in *trans*-cinnamyl alcohol.

(a)   (b)

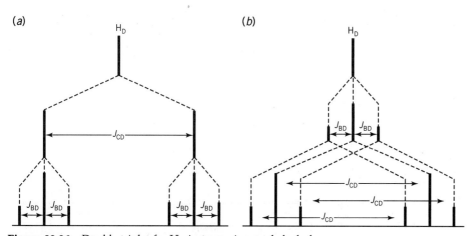

Figure 33.36 Double triplet for H_D in *trans*-cinnamyl alcohol.

The spectrum of cinnamyl alcohol is shown in Figure 33.37. The double doublet for H_B is observed at 4.32 ppm. The double triplet for H_C is observed at 6.37 ppm, and the double triplet for H_D shows at 6.62 ppm (notice that these triplets are not very well resolved). The aromatic hydrogens give a complex signal between 7.2 and 7.4 ppm. The alcohol hydrogen shows as a broad singlet at 1.74 ppm.

In general, if there are three groups of hydrogens, H_A, H_M, and H_X coupled to each other with three different coupling constants J_{AM}, J_{AX}, and J_{MX} and the number of hydrogens in each group is n_A, n_M, and n_X, respectively, then the signal for H_A has a number of peaks equal to $(n_M + 1) \times (n_X + 1)$; the signal for H_M has $(n_A + 1) \times (n_X + 1)$ peaks, and the signal for H_X has $(n_A + 1) \times (n_M + 1)$ peaks (Fig. 33.38).

What happens to the spin-spin splitting of the coupled hydrogens if two of the coupling constants are the same? To answer this question, let's go back to Figure 33.34 and assume, for the sake of argument, that $J_{BC} = J_{BD}$. In such a case the double doublet for H_B collapses into a triplet. If we assume, also for the sake of argument, that the two couplings constants involved in Figure 33.35, J_{CD} and J_{BC} are identical, then the double triplet collapses into a quartet (Fig. 33.39). In general, when a set of hydrogens is coupled to two different sets of hydrogens with the same coupling constant, the multiplicity for the first set of hydrogens is simply given by the $n + 1$ rule, *where n is the total number of hydrogens coupled to that set*. Let's consider, for example, the case of *n*-butyl bromide. There are four types of hydrogens: H_A, H_M, H_T, and H_X. The methyl group, H_A, sees two equivalent hydrogens, H_M, and gives a triplet. The methylene next to the bromine, H_X, sees two equivalent

$$\boxed{\begin{array}{l} H_X \ \ H_T \ \ H_M \ \ H_A \\ Br\ CH_2CH_2CH_2CH_3 \end{array}}$$

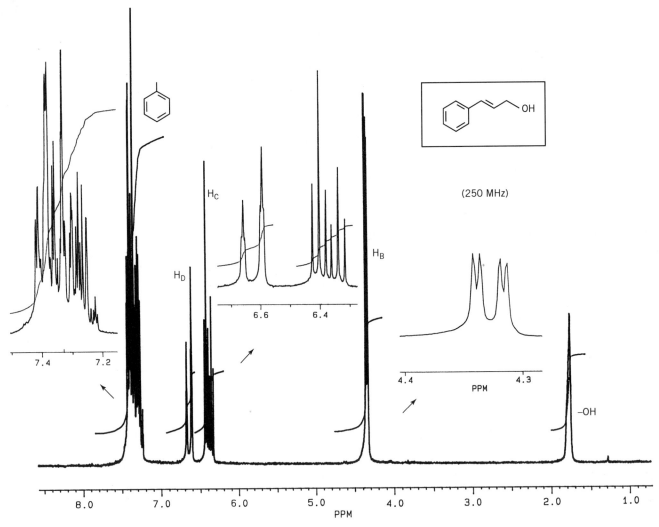

Figure 33.37 ^{1}H-NMR of *trans*-cinnamyl alcohol.

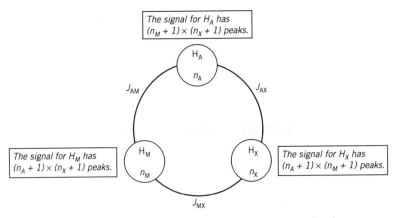

Figure 33.38 A system of three groups of hydrogens coupled to each other with three different coupling constants.

hydrogens, H_T, and also shows as a triplet. The methylene represented by H_M has two different neighbors, H_T and H_A, and, in principle, it should show a multiplicity of 12 peaks:

$$(n_T + 1) \times (n_A + 1) = (2 + 1) \times (3 + 1) = 12$$

(a)   (b)

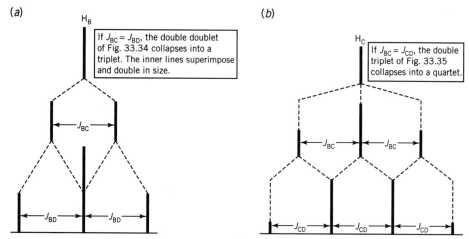

Figure 33.39   The collapse of a double doublet into a triplet, and a double triplet into a quartet.

However, the coupling constants J_{AM} and J_{MT} are almost identical and H_M shows as a sextet ($n_T + n_A + 1 = 2 + 3 + 1 = 6$). Using a similar argument it can be predicted that the observed multiplicity for H_T is a quintet ($n_M + n_X + 1 = 2 + 2 + 1 = 5$) instead of a multiplet with nine peaks, $(n_M + 1) \times (n_X + 1) = (2 + 1) \times (2 + 1) = 9$, because $J_{MT} = J_{TX}$.

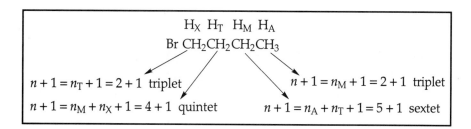

The spectrum of *n*-butyl bromide is shown in Figure 33.40. The triplet for H_A shows at 0.92 ppm, the sextet for H_M at 1.45 ppm, the quintet for H_T at 1.82 ppm, and the triplet for H_X at 3.40 ppm.

Because the coupling constants between different alkyl hydrogens are very similar, the *multiplicity of a set of equivalent hydrogens in an alkyl chain is normally n + 1, where n is the total number of hydrogens bound to all the adjacent carbons.*

Differences Between a Double Doublet and a Quartet

What is the difference between a double doublet and a quartet? First of all, there is only one coupling constant in a quartet while there are two different coupling constants in a double doublet. One of the coupling constants of a double doublet is measured as the distance between lines 1–2 or 3–4, J_1 in Figure 33.41, and the other coupling constant is measured as the distance separating lines 1–3 or 2–4, J_2 in Figure 33.41. In a quartet, the distance between lines 2–3 is equal to the distances between 1–2 and 3–4 (and is the *J* value); in a double doublet, on the other hand, the distance 2–3 is determined by the difference between the two coupling constants, and, in principle, can have any value.

A second difference between a double doublet and a quartet is that the four lines of a quartet show a relative intensity of $1 : 3 : 3 : 1$, as Pascal's triangle predicts, while in a double doublet all the peaks are, in principle, equally intense.

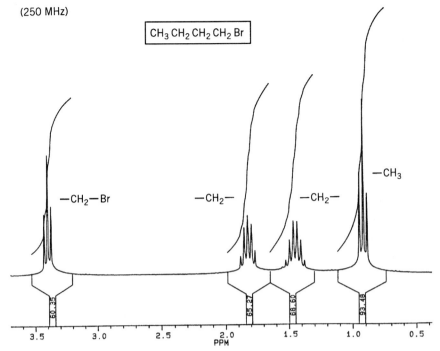

Figure 33.40 ^{1}H-NMR of *n*-butyl bromide.

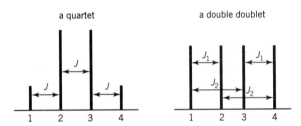

Figure 33.41   A quartet versus a double doublet.

Strongly Coupled Systems

The rules explained so far about multiplicity apply when the difference between the resonance frequencies of two sets of hydrogens H_A and H_B, Δv, is large compared to the coupling constant between them ($\Delta v/J > 10$). Systems where $\Delta v/J > 10$ are called **first order** and follow the $n + 1$ rule with intensities predicted by Pascal's triangle. For systems with Δv similar in magnitude to J, the splitting does not necessarily follow the $n + 1$ rule and the intensities deviate considerably from those expected from Pascal's triangle. These systems are said to be strongly coupled and often give complex signals of difficult analysis.

An example of how the $\Delta v/J$ ratio affects the appearance of the spectrum is illustrated in Figure 33.42 with the coupling of H_A and H_B. The coupling constant is 10 Hz in all cases, and the frequency difference between H_A and H_B is varied from 110 Hz (case *a*) to 0 Hz (case *e*). As the two hydrogens come closer together, the outer lines of the doublets become smaller. When $\Delta v = 0$ (Fig. 33.42*e*), both hydrogens are equivalent, the outer lines of the doublets have zero intensity, and splitting is not observed.

Complex spectra with strongly coupled hydrogens can be considerably simplified by working at a higher operational frequency. Multiplets that are difficult to interpret because they do not obey the $n + 1$ rule or do not show the typical intensities

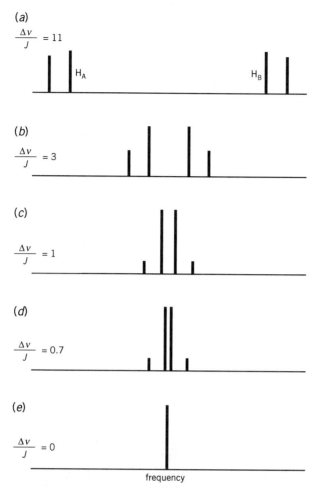

Figure 33.42 System of two hydrogens A and B for different values of the $\Delta v/J$ ratio.

as predicted by Pascal's triangle may become first-order systems at higher operational frequencies. This happens because as the operational frequency increases, so does the frequency difference between peaks (remember Fig. 33.14). The coupling constant, on the other hand, is independent of the operational frequency and remains unchanged. For example, if a system has $\Delta v/J = 2$ at 60 MHz, and therefore it is not a first-order system, it will become first order at 600 MHz, where the Δv increases ten times ($\Delta v/J = 20$). An example of a complex system at 60 MHz that becomes a first-order one at 500 MHz is provided by the aromatic hydrogens of aspirin (Fig. 33.43). We will discuss the splitting of aromatic hydrogens in section 33.18.

Naming Spin Systems

It is customary to label different types of hydrogens in a molecule with the letters A, B, C, and so on. Here we have made extensive use of this labeling system but we have not associated any special meaning to the letters. However, it is sometimes useful to label the hydrogens with letters that reflect how different or similar their chemical shifts are. Let's see how we do it.

A system of two coupled hydrogens with similar (but different) chemical shifts is usually called AB. If, on the other hand, the hydrogens resonate at very different frequencies (the ratio $\Delta v/J$ is larger than 10) the system is called AX. The choice of the letters A and X, well separated in the alphabet, reminds us that the hydrogens

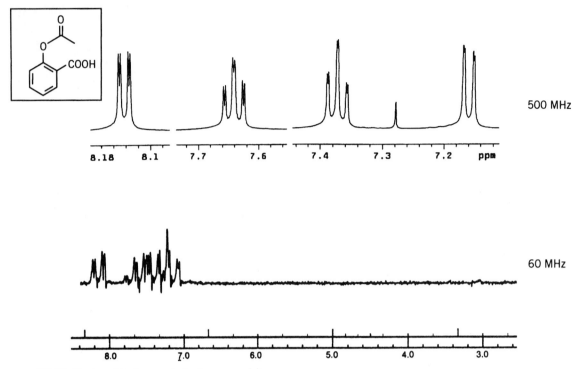

Figure 33.43 ^{1}H-NMR of aspirin (aromatic hydrogens only).

are also well separated. Equivalent hydrogens are labeled with the same letter. With this in mind, we can call the system represented in Figure 33.42*a*: AX; the systems represented in Figure 33.42*b-d*: AB; and the system in Figure 33.42*e*: A$_2$.

This type of notation can be extended to a system of three or more coupled hydrogens. A system of three very different hydrogens is usually named AMX. If two hydrogens are close to each other but separated from a third, the system is called ABX. If we reexamine Figure 33.43, we can call the system of four aromatic hydrogens at 500 MHz an AMTX system because four well-separated signals are observed. At 60 MHz, on the other hand, the same four aromatic hydrogens give a much more complex signal, with only one hydrogen (the signal at 8.1 ppm) separated from the rest. This system is called ABCX.

This notation is useful in the prediction and interpretation of coupling patterns. For example, once we know the coupling pattern of an AMX system, we can apply it to any molecule that contains such a system. We will come back to AMX systems in section 33.18.

33.14 CHEMICAL SHIFT CORRELATIONS

Alkyl Hydrogens

The effects of substituents on the chemical shifts can be predicted with the help of tables compiled from the actual spectra of hundreds of compounds. These tables, called **correlation tables**, play an essential role in structure elucidation because they allow us to translate chemical shifts into structural information. The chemical shifts of the hydrogens of **methyl groups** attached to different substituents are given in Table 33.3. Tables 33.3–14 have been compiled from Refs. 2–6; 15, and 16.

Table 33.3 **Chemical Shifts of Methyl Groups Attached to Different Substituents**

	Substituent	δ		Substituent	δ
	$-R$ (alkyl)	0.9		$-NH_2$	2.5
	$-C=C$	1.6		$-NH-C_6H_5$	2.8
C	$-C\equiv C$	1.8		$-NH-CO-R$	2.8
	$-C_6H_5$	2.3	**N**	$-\overset{+}{N}R_3$	3.2
	$-CO-NR_2$	2.0		$-\overset{+}{N}H_3$	2.6
	$-CO-OR$	2.0		$-NO_2$	4.3
C=O	$-CO-OH$	2.1	**S**	$-SH$	2.0
	$-CO-R$	2.1		$-SR$	2.1
	$-CO-H$	2.2		$-OR$	3.2
	$-CO-C_6H_5$	2.6		$-OH$	3.4
	$-C\equiv N$	2.2	**O**	$-O-C_6H_5$	3.7
	$-I$	2.2		$-O-CO-R$	3.7
	$-Br$	2.7		$-O-CO-C_6H_5$	3.9
X	$-Cl$	3.1			
	$-F$	4.3			

R stands for any alkyl group

Example of the use of Table 33.3

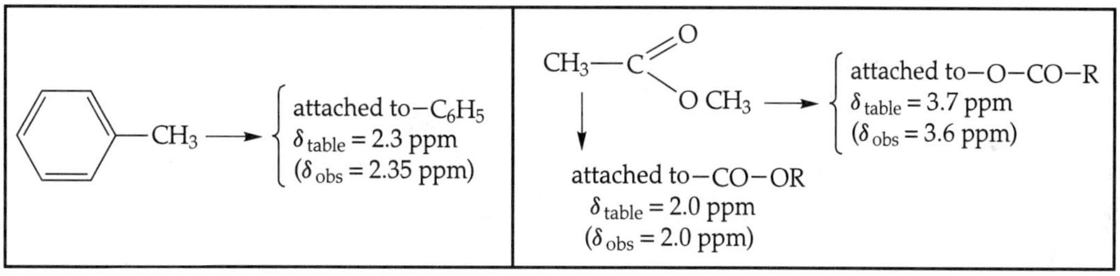

Examples of how to use Table 33.3 are also given. Deviations of up to ±0.2 ppm can be expected between the calculated values and the experimental ones.

Table 33.4 is used to calculate the chemical shifts of **methylene** and **methine** hydrogens. This table gives the shielding constants for different substituents, σ, which should be added to the base value for methylene (1.25 ppm) and methine (1.50 ppm). Examples of the use of Table 33.4 are shown below. If the substituent we are looking for is not listed in the table, we use the shielding constant of a structurally similar substituent. For example, the value for a phenyl ring ($\sigma = 1.3$) can be also used as an estimate for the shielding constant of any substituted phenyl ring. The typical error of this table is ±0.3 ppm.

Vinyl Hydrogens

The chemical shift of a vinyl hydrogen depends on the substituents attached to the double bond and their relative positions with respect to that hydrogen. To the base value of 5.28 ppm (chemical shift of ethylene), the shielding constant of the substituents should be added (Table 33.5).

Table 33.4 **Shielding Constants of Substituents Attached to Methylene and Methine Groups**

X—CH$_2$—Y			X—CH—Y │ Z		
$\delta(CH_2) = 1.25$ ppm $+ \sigma_X + \sigma_Y$			$\delta(CH) = 1.50$ ppm $+ \sigma_X + \sigma_Y + \sigma_Z$		
	Substituent	σ		Substituent	σ
	—R (alkyl)	0.0		—NH$_2$; —NHR; —NR$_2$	1.0
	—C=C	0.8	**N**	—NH—C$_6$H$_5$	1.8
C	—C≡C	0.9		—NH—CO—R	2.1
	—C$_6$H$_5$	1.3		—NO$_2$	3.2
	—CO—OR	0.7	**S**	—SH; —SR	1.3
	—CO—OH	0.8			
	—CO—NR$_2$	1.0		—OR	1.9
C=O	—CO—R; —CO—H	1.2	**O**	—OH	2.1
	—CO—C$_6$H$_5$	1.6		—O—C$_6$H$_5$	2.7
				—O—CO—R	2.7
	—C≡N	1.2		—O—CO—C$_6$H$_5$	2.9
	—I	1.4			
X	—Br	1.9			
	—Cl	2.0			

R is any alkyl group.

Example of the use of Table 33.4

Methylene

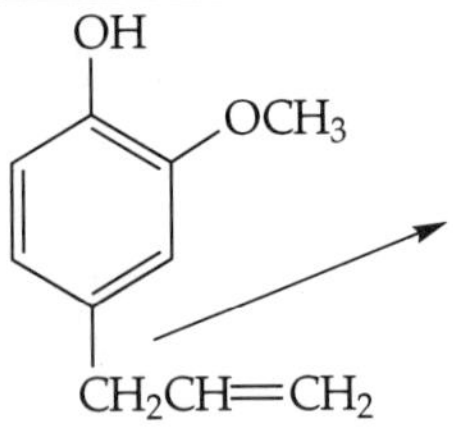

	δ table	
base	1.25 ppm	
—C$_6$H$_5$	1.3 ppm	
—C=C	0.8 ppm	
total	3.35 ppm	
	$\delta_{obs} = 3.39$ ppm	

	δ table	
base	1.25 ppm	
—CO—OR	0.7 ppm	
—Cl	2.0 ppm	
total	3.95 ppm	
	$\delta_{obs} = 4.10$ ppm	

ClCH$_2$COOCH$_3$

Methine

OH
│
CH$_3$ CH—COOCH$_3$

	δ table	
base	1.50 ppm	
—CO—OR	0.7 ppm	
—OH	2.1 ppm	
total	4.3 ppm	
	$\delta_{obs} = 4.22$ ppm	

	δ table	
base	1.50 ppm	
3—C$_6$H$_5$	3 x 1.3 ppm	
total	5.4 ppm	
	$\delta_{obs} = 5.55$ ppm	

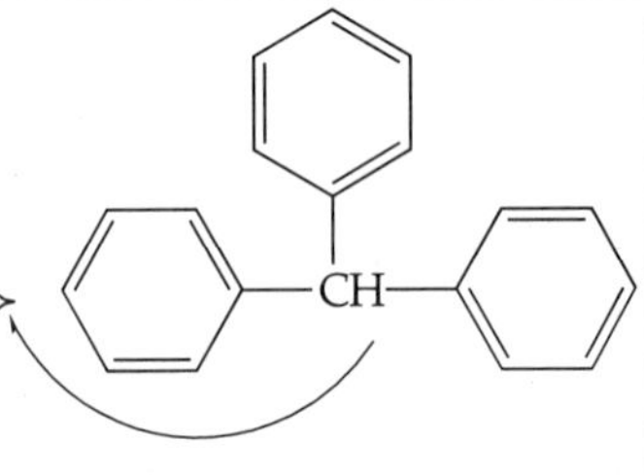

Table 33.5 **Shielding Constants of Substituents Attached to Alkenes**

$$\delta_H = 5.28 \text{ ppm} + \sigma_{Zgem} + \sigma_{Zcis} + \sigma_{Ztrans}$$

	Substituent	σ_{gem}	σ_{cis}	σ_{trans}		Substituent	σ_{gem}	σ_{cis}	σ_{trans}
	—H	0	0	0		—C≡N	0.23	0.78	0.58
C	—R (alkyl)	0.44	−0.26	−0.29		—I	1.14	0.81	0.88
	—CH₂—C₆H₅	1.05	−0.29	−0.32		—Br	1.04	0.40	0.55
	—CH₂OR	0.67	−0.02	−0.07	**X**	—Cl	1.00	0.19	0.03
	—C=C	0.98	−0.04	−0.21		—F	1.54	−0.40	−1.02
	—C=C (conj.)ᵃ	1.26	0.08	−0.01		—NH₂; —NHR; —NR₂	0.69	−1.19	−1.31
	—C≡C	0.50	0.35	0.10		—NH—C₆H₅	2.30	−0.73	−0.81
	—C₆H₅	1.43	0.39	0.06	**N**	—NH—CO—R	2.08	−0.57	−0.72
	—CO—NR₂	1.37	0.93	0.35		—NO₂	1.87	1.32	0.62
	—CO—OR	0.84	1.15	0.56	**S**	—SH; —SR	1.00	−0.24	−0.04
C=O	—CO—OR (conj.)ᵃ	0.68	1.02	0.33		—OR	1.18	−1.06	−1.28
	—CO—OH	1.00	1.35	0.74	**O**	—OC₆H₅	1.14	−0.65	−1.05
	—CO—OH (conj.)ᵃ	0.69	0.97	0.39		—O—CO—R	2.09	−0.40	−0.67
	—CO—R	1.10	1.13	0.81					
	—CO—H	1.03	0.97	1.21					

ᵃFor systems where the alkene, the substituent, or both are conjugated.
R stands for any alkyl group.

An example of how to use Table 33.5 is given below.

Example of the use of Table 33.5

Aromatic Hydrogens

The chemical shifts of aromatic hydrogens can be calculated with the help of Table 33.6. The effect of the substituent on the chemical shift depends on the relative position of the substituent with respect to the hydrogen. The base value is 7.26 ppm (chemical shift of benzene). The effects are additive; if there is more than one substituent on the ring, each one will contribute to the chemical shift of each hydrogen. The error between the calculated values and the experimental ones is typically less than 0.3 ppm, but if there are three or more substituents on the ring, larger errors can be expected.

Table 33.6 **Shielding Constants for Substituents Attached to Phenyl Rings**

$$\delta_{H_A} = 7.26 \text{ ppm} + \sigma_{Zortho}$$
$$\delta_{H_B} = 7.26 \text{ ppm} + \sigma_{Zmeta}$$
$$\delta_{H_C} = 7.26 \text{ ppm} + \sigma_{Zpara}$$

	Substituent	σ_{ortho}	σ_{meta}	σ_{para}		Substituent	σ_{ortho}	σ_{meta}	σ_{para}
	−H	0	0	0		−I	0.39	−0.21	0
						−Br	0.18	−0.08	−0.04
C	−R (alkyl)	−0.15	−0.08	−0.19	**X**	−Cl	0.03	−0.02	−0.09
	−CF$_3$	0.32	0.14	0.20		−F	−0.26	0	−0.20
	−CH$_2$OH	−0.07	−0.07	−0.07					
	−C=C	0.06	−0.03	−0.10		−NH$_2$; −NHR;	−0.78	−0.24	−0.67
	−C≡C	0.15	−0.02	−0.01		−NR$_2$	−0.66	−0.18	−0.67
	−C$_6$H$_5$	0.37	0.20	0.10	**N**	−N$^+$R$_3$	0.69	0.36	0.31
						−NH−CO−R	0.12	−0.07	−0.28
	−CO−NR$_2$	0.61	0.10	0.17		−N=N−C$_6$H$_5$	0.67	0.20	0.20
	−CO−OR	0.70	0.10	0.20		−NO$_2$	0.95	0.26	0.38
	−CO−OH	0.85	0.18	0.27					
C=O	−CO−X (X=Cl; Br)	0.82	0.22	0.37	**S**	−SH; −SR	−0.08	−0.13	−0.23
	−CO−R	0.62	0.13	0.20					
	−CO−H	0.56	0.22	0.29		−OH	−0.56	−0.12	−0.45
					O	−OR	−0.47	−0.09	−0.44
	−C≡N	0.36	0.18	0.28		−O−C$_6$H$_5$	−0.29	−0.05	−0.23
						−O−CO−R	−0.25	0.03	−0.13

R stands for any alkyl group.

Example of the use of Table 33.6

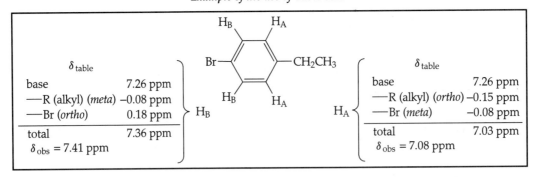

δ table	
base	7.26 ppm
──R (alkyl) (*meta*)	−0.08 ppm
──Br (*ortho*)	0.18 ppm
total	7.36 ppm

$\delta_{obs} = 7.41$ ppm

δ table	
base	7.26 ppm
──R (alkyl) (*ortho*)	−0.15 ppm
──Br (*meta*)	−0.08 ppm
total	7.03 ppm

$\delta_{obs} = 7.08$ ppm

Hydrogens Attached to Other Unsaturated Carbons

As we discussed in section 33.7 and have shown in Tables 33.5 and 33.6, hydrogens on carbon-carbon double bonds and aromatic systems are deshielded and resonate in the range 5–8 ppm. Hydrogens attached to carbonyls and carbon-nitrogen double bonds are also deshielded and resonate in the range 6–11 ppm. Hydrogens on sp carbons are relatively shielded and resonate at lower chemical shifts (1.5–3 ppm). A summary of chemical shifts is given in Table 33.7.

Table 33.7 **Chemical Shift of Hydrogens on Unsaturated Carbons**

	Compound	δ	Compound	δ
alkynes	$R-C\equiv C-H$	1.7–1.8	Formates: $RO-CO-H$	8–8.2
	$Ar-C\equiv C-H$	2.8–3.2	Formamides: $R_2N-CO-H$	8–8.7
	$R^*-C\equiv C-H$	2.7–3.1	Imines: $R-N=CRH$	6–8
	$R-C\equiv C-C\equiv C-H$	1.8–2.0	Alkenes: $C=C-H$	Table 33.5
aldehydes	aliphatic: $R-CO-H$	9–10	Aromatics: $Ar-H$	Table 33.6
	aromatic: $Ar-CO-H$	9.6–10.5		

Ar is any aryl group; R* has at least a double bond in conjugation.

Hydrogens Attached to Heteroatoms

Hydrogens of alcohols resonate around 0.5 ppm when the molecules are not associated through H-bonds (monomeric form), a situation most likely to be encountered at very low concentrations. Under the experimental conditions in which most spectra are obtained, however, alcohols are H-bonded and resonate between 0.5 and 5 ppm depending on the solvent, the concentration, and the temperature. Phenols and amines behave in a similar fashion.

a carboxylic acid dimer

The hydrogen of carboxylic acids resonates between 10 and 13 ppm. Unlike alcohols and phenols, its chemical shift is not strongly affected by the concentration because carboxylic acids exist as dimers even at very low concentrations. The chemical shifts of hydrogens attached to heteroatoms are gathered in Table 33.8.

Table 33.8 **Chemical Shifts of Hydrogens Attached to Heteroatoms**

	Compound	δ		Compound	δ
	alcohols: $R-OH$ (monomeric; low concentration)	0.5	**SH**	thiols: $R-SH$; $Ar-SH$	1–4
	alcohols: $R-OH$ (H-bonded; higher concentration)	0.5–5		aliphatic amines $R-NH_2$; R_2NH	0.5–4
	phenols: $Ar-OH$ (monomeric; low concentration)	4.5		aromatic amines $ArNH_2$; $ArNHR$; Ar_2NH	2.5–5
OH	phenols: $Ar-OH$ (H-bonded; higher concentration)	4.5–9	**NH**	ammonium salts: R_3NH^+	7–9
	nitrophenols $O_2N-C_6H_4-OH$	9–11		primary amides $RCONH_2$; $ArCONH_2$	5–6.5
	carboxylic acids $R-COOH$	10–13		secondary amides $RCONHR$; $ArCONHR$	6–8.5
	oximes: $R_2C=N-OH$	7–11		secondary amides $RCONHAr$; $ArCONHAr$	7.5–9.5

The values are strongly affected by solvent and concentration. Ar stands for any aryl group.

33.15   COUPLING OF HYDROGENS ATTACHED TO HETEROATOMS

Hydrogens bonded to oxygen and nitrogen *do not normally show coupling* to other hydrogens. Let's consider the case of ethanol (Fig. 33.44). Its ^{1}H-NMR shows three signals: a triplet at 1.2 ppm that corresponds to the methyl group, a quartet at 3.6 ppm that belongs to the methylene, and a singlet at 2.6 ppm that corresponds to the hydroxyl group.

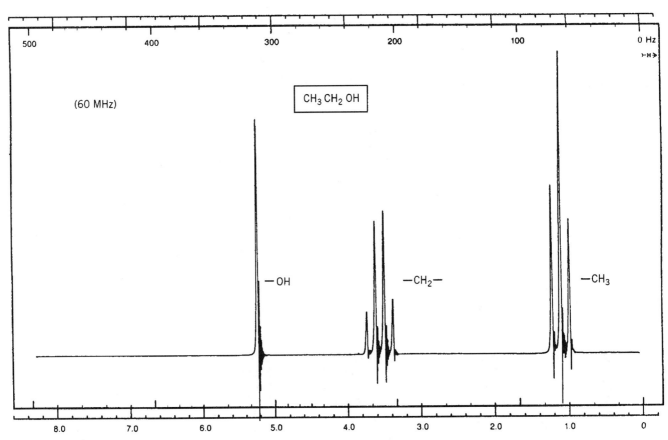

Figure 33.44 ^{1}H-NMR of ethanol.

The lack of spin-spin splitting between the hydrogen of the alcohol and the hydrogens of the methylene is due to a rapid exchange of the −OH hydrogen between different ethanol molecules. This exchange process is accelerated by traces of acid or base.

coupling not observed

$$R—CH_2OH + R—CH_2OH \rightleftharpoons R—CH_2OH + R—CH_2OH$$

hydrogens are exchanged between molecules

To observe the coupling of hydrogens bonded to heteroatoms with other hydrogens, the sample must be scrupulously pure, free of acids and bases. Under those conditions the spectrum of ethanol still shows a triplet at 1.2 ppm for the methyl group (Fig. 33.45), but it now shows a triplet for the HO− because the −OH hydrogen is coupled to the methylene group. The signal for the methylene becomes a multiplet

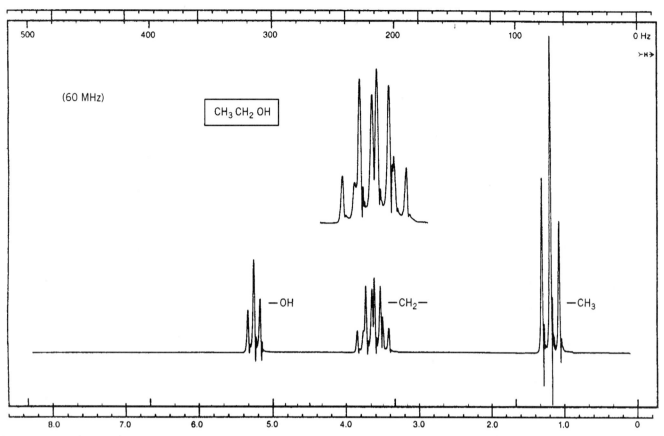

Figure 33.45 ^{1}H-NMR of ethanol showing coupling with the OH group.

with eight lines because the methylene is coupled to the methyl group and the HO group with two different coupling constants $((3 + 1) \times (1 + 1) = 8)$.

When dimethylsulfoxide (DMSO) is used as a solvent (totally deuterated DMSO is used in ^{1}H-NMR; section 33.20), the hydrogen of alcohols shows spin-spin splitting with its neighbors. This arises because DMSO forms strong H-bonds with the hydroxyl group and slows down the rate of hydrogen exchange between alcohol molecules.

$$R-CH_2OH + O{=}S\begin{matrix} CD_3 \\ CD_3 \end{matrix} \rightleftharpoons R-CH_2OH{\cdots}O{=}S\begin{matrix} CD_3 \\ CD_3 \end{matrix}$$

coupling observed between
these hydrogens

The exchange reaction of hydrogens attached to heteroatoms can be used to obtain structural information about the sample. In the presence of deuterium oxide (D_2O) the hydrogens of alcohols and amines, for example, undergo an exchange reaction with the deuterium:

$$R{-}OH + D_2O \rightleftharpoons R{-}OD + HDO$$

$$R{-}NH_2 + D_2O \rightleftharpoons R{-}NHD + HDO$$

$$\Big\updownarrow D_2O$$

$$R{-}ND_2 + HDO$$

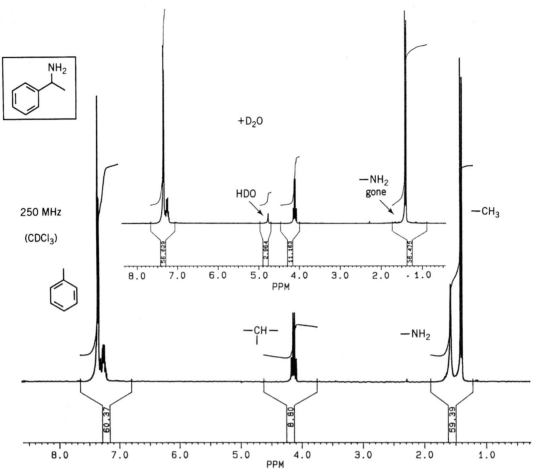

Figure 33.46 ¹H-NMR of 1-phenylethylamine. The inset shows the spectrum in the presence of D₂O. The peak for —NH₂ a has disappeared.

Deuterium does not resonate at the same frequency as hydrogen. Thus, a sample that has been treated with D₂O will lack the peaks of the hydrogens attached to oxygen and nitrogen. To locate these peaks in the spectrum, the spectrum is first run in a solvent such as CDCl₃, then a few drops of D₂O are added to the NMR tube and the tube is shaken. The exchange process is very fast for alcohols, phenols, carboxylic acids, and amines and is complete in a matter of minutes. The spectrum is obtained again and compared with the one run without D₂O. The peaks that have disappeared correspond to hydrogens attached to oxygen or nitrogen. Spectra obtained in the presence of D₂O always show a peak for HDO around 4.65 ppm (Table 33.14, p. 771).

For some compounds, such as amides and certain amines, the exchange rate may be very slow. To speed up the exchange process, DCl dissolved in D₂O is used. Figure 33.46 shows the spectrum of 1-phenylethylamine before and after D₂O exchange.

33.16 ¹³C-NMR

Even though the natural abundance of ¹³C is only 1.1%, ¹³C-NMR spectra are routinely obtained. As was discussed in section 33.3, the problem of low ¹³C abundance is overcome by using a number of pulses, ranging from a few to a few thousands, and collecting and adding up the FID obtained after each pulse. As in ¹H-NMR, TMS is also used as internal standard. Carbons are affected by the shielding-deshielding effect of neighboring substituents in a way that parallels the effect on hydrogens. The typical chemical shift range for carbons is 0–210 ppm. This range is much wider than the range for hydrogens (0–12 ppm), showing that the chemical shift of carbons is more sensitive to changes in the environment.

The ^{13}C atoms are coupled to ^{1a}H directly attached to them (**C−H**), and also to ^{1}H on nearby carbons and separated by two or three bonds (**C−C−H** and **C−C−C−H**). Some of these coupling constants can be as large as 200 Hz. This number is comparable to the difference in resonance frequency between carbons, and thus, the carbon-hydrogen splitting results in clutter of the ^{13}C signals and produces very complex spectra. To avoid this problem, ^{13}C-NMR spectra are usually obtained by **decoupling** the hydrogens. This is achieved by irradiating the hydrogens in a wide frequency range that covers their whole chemical shift range. In this technique, called **hydrogen broadband decoupling**, hydrogens are continuously irradiated at their resonance frequency, making them flip spin orientation very rapidly. As a result, the ^{13}C atoms do not sense two separate spin states for the hydrogens, but rather "see" the hydrogens as nonmagnetic, and do not couple with them. In the hydrogen-decoupled ^{13}C-NMR, each carbon shows as a singlet regardless of the number of attached hydrogens. How hydrogen decoupling simplifies the ^{13}C-NMR spectrum is illustrated in Figure 33.47 with the ^{13}C-NMR spectra of camphor. It should be noticed

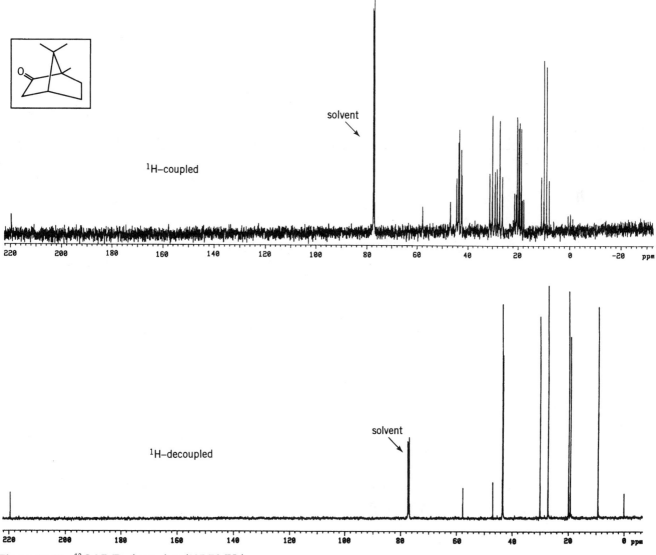

Figure 33.47 ^{13}C-NMR of camphor (125.7 MHz).

that coupling between carbons (^{13}C-^{13}C coupling) is not normally observed because the number of molecules with two adjacent ^{13}C atoms is very small.

^{13}C-NMR spectra are not usually integrated because the areas under the peaks are not directly proportional to the number of carbons that gave rise to them. This is in part due to the fact that the carbons do not have time to completely relax back to the lower spin state between pulses. Thus, when a new pulse is applied, a considerable number of nuclei are still in the higher energy state and do not absorb radiation. Also, not all carbons relax back at the same speed; some carbons relax faster than others. Methyl and methylene carbons relax faster than methine and quaternary carbons, leading to bigger peaks. To obtain a ^{13}C-NMR for integration purposes, long times between pulses should be used to allow for the relaxation of all carbons. This leads to very long experiments and thus the integration is normally not obtained. Other factors, such as the Nuclear Overhauser Effect (NOE) (which will not be discussed here; see Refs. 1, 2, 6, and 7) also contribute to make the integration of ^{13}C-NMR spectra meaningless.

It is possible to differentiate methyl, methylene, methine, and quaternary carbons by turning the hydrogen decoupler on and off at the right times during the pulse sequence. This is called the **attached proton test (APT)** experiment. In this experiment, methyl and methine groups give peaks pointing down, while methylene groups and carbons without attached hydrogens give peaks pointing up. The APT is a useful diagnostic tool to assign the peaks of the spectrum. This is illustrated with the spectra of *n*-propyl acetate, Figure 33.48.

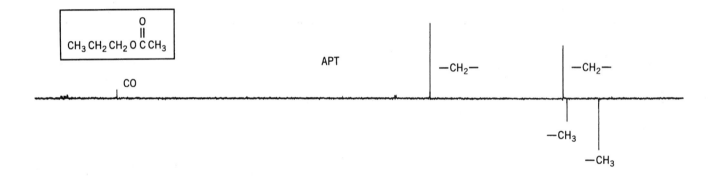

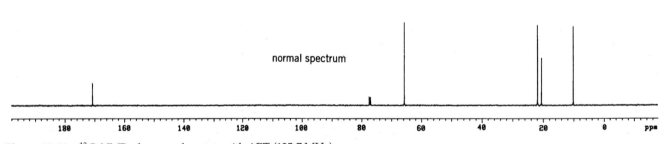

Figure 33.48 ^{13}C-NMR of *n*-propyl acetate with APT (125.7 MHz).

33.17 ^{13}C-NMR CHEMICAL SHIFTS

As in the case of hydrogens, the chemical shifts of carbons can be estimated using correlation tables. Carbon atoms are more susceptible than hydrogens to the effects of distant substituents, and the contribution of groups in the β and γ positions, which can be ignored for hydrogens, should be computed for carbons.

The chemical shifts of a substituted alkane can be estimated with the help of the chemical shifts for the unsubstituted alkane (Table 33.9), and the substituent constants (Table 33.10). Let's consider, for example, the case of *n*-butyl bromide shown below. To calculate its chemical shifts we take the chemical shifts of *n*-butane (Table 33.9) and add the effect of the bromine on each carbon (Table 33.10). We use the values in the "terminal" column because the bromine is attached to a terminal carbon. If the substituent is attached to an internal carbon, the "internal" values should be used instead. If two or more substituents are present, Table 33.10 can still be used but the errors are larger. The use of Tables 33.9 and 33.10 is illustrated below.

Table 33.9 ¹³C **Chemical Shifts for Selected Hydrocarbons**

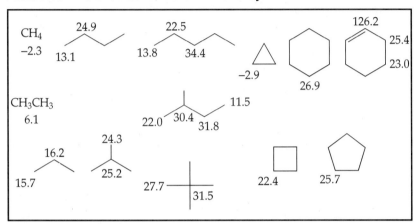

The chemical shifts of phenyl ring carbons can be estimated with the help of Table 33.11. To the base value of 128.5 ppm (chemical shift of benzene), the substituent constants are added. It is worth noticing that with the exception of −C≡C, −C≡N, −I and −Br, all substituents increase the chemical shift of the carbon to which they are directly attached. The error in the estimated chemical shift increases as the number of substituents increases.

The chemical shifts of alkenes can be estimated with Table 33.12. To the base value of 123.3 ppm, the substituent effects are added. The calculations do not distinguish between Z and E isomers.

The carbonyl group is easily distinguished by ¹³C-NMR. Carbonyls usually give small peaks in the range 160–210 ppm. Different subranges are observed, depending on the particular functional group; they are listed in Table 33.13.

Table 33.10 **^{13}C Shielding Constants for Groups Attached to Alkyl Chains**

	Substituent	Terminal α	Terminal β	Internal α	Internal β	Terminal or internal γ		Substituent	Terminal α	Terminal β	Internal α	Internal β	Terminal or internal γ
C	—CH$_3$	9	10	6	8	−2		—NH$_2$	29	11	24	10	−5
	—C=C	20	6			−1		—NHR	37	8	31	6	−4
	—C≡C	5	6			−4	**N**	—NR$_2$	42	6			−3
	—C$_6$H$_5$	23	9	17	7	−2		—NH$_3^+$	26	8	24	6	−5
C=O	—CO—NH$_2$	22	3			−1		—NR$_3^+$	31	5			−7
	—CO—OR	21	3	17	2	−2		—NO$_2$	63	4	57	4	
	—CO—OH	21	3	16	2	−2	**S**	—SH	11	12	11	11	−4
	—CO—O$^-$	24	4	20	3	−2		—SR	20	7			−3
	—CO—R	30	1	24	1	−2	**O**	—OH	48	10	41	8	−6
	—CO—H	31	0			−2		—OR	58	8	51	5	−4
	—C≡N	4	3	1	3	−3		—O—CO—R	51	6	45	5	−3
X	—I	−7	11	4	12	−1		—O—CO—Ar	53				
	—Br	20	11	25	10	−3							
	—Cl	31	11	32	10	−4							
	—F	68	9	63	6	−4							

Examples of the use of Table 33.9 and Table 33.10

C1:		**C3:**	
C1 n-butane	13.1 ppm	C3 n-butane	24.9 ppm
α Br (terminal)	20 ppm	γ Br	−3 ppm
total	33.1 ppm	total	21.9 ppm
	δ_{obs} = 33.6 ppm		δ_{obs} = 21.3 ppm

C2:		**C4:**	
C1 n-butane	24.9 ppm	C4 n-butane	13.1 ppm
β Br (terminal)	11 ppm	total	13.1 ppm
total	35.9 ppm		δ_{obs} = 13.2 ppm
	δ_{obs} = 34.8 ppm		

C1:		**C3:**	
C1 2-methylbutane	22.0 ppm	C3 2-methylbutane	31.8 ppm
β C1 (internal)	10 ppm	β Cl (internal)	10 ppm
total	32.1 ppm	total	41.8 ppm
	δ_{obs} = 31.9 ppm		δ_{obs} = 38.8 ppm

C2:		**C4:**	
C2 2-methylbutane	30.4 ppm	C4 2-methylbutane	11.5 ppm
α Cl (internal)	32 ppm	γ Cl	−4 ppm
total	62.4 ppm	total	7.5 ppm
	δ_{obs} = 71.6 ppm		δ_{obs} = 9.5 ppm

Table 33.11   **^{13}C Shielding Constants for Substituents Attached to Phenyl Rings**

$$\delta_{C1} = 128.5 \text{ ppm} + \sigma_{Zattach} \qquad \delta_{C3} = 128.5 \text{ ppm} + \sigma_{Zmeta}$$

$$\delta_{C2} = 128.5 \text{ ppm} + \sigma_{Zortho} \qquad \delta_{C4} = 128.5 \text{ ppm} + \sigma_{Zpara}$$

	Substituent	σ_{attach}	σ_{ortho}	σ_{meta}	σ_{para}		Substituent	σ_{attach}	σ_{ortho}	σ_{meta}	σ_{para}
	−H	0	0	0	0		−I	−32.3	9.9	2.6	−0.4
						X	−Br	−5.6	3.3	1.8	−1.3
	−CH$_3$	9.3	0.7	−0.1	−3.0		−Cl	6.4	0.3	1.2	−2.0
	−CH$_2$CH$_3$	15.7	−0.6	−0.1	−2.8		−F	35.0	−0.26	−13.6	−4.4
C	−CF$_3$	2.6	−2.8	−0.4	−3.3		−NH$_2$	18.7	−12.9	1.0	−9.5
	−C=C	9.5	−2.0	0.2	−0.5		−NR$_2$	22.4	−15.6	0.9	−11.7
	−C≡C	−6.2	3.7	0.4	−0.3		−NH−CO−R	11.1	−9.9	0.2	−5.6
	−C$_6$H$_5$	13.0	−1.1	0.5	−1.0	**N**	−N=N−C$_6$H$_5$	24.0	−5.8	0.3	2.2
	−CO−NH$_2$	5.5	−0.4	−1.0	5.0		−NO$_2$	19.8	−4.9	0.9	6.1
	−CO−OR	2.1	1.2	0	4.4						
C=O	−CO−OH	2.4	1.6	−0.1	5.0	**S**	−SH	2.2	0.7	0.4	3.1
	−CO−Cl	4.6	2.6	0.8	6.6		−SR	10.0	−1.8	0.2	−3.5
	−CO−CH$_3$	9.1	0.2	0.1	4.3		−OH	26.9	−12.8	1.4	−7.3
	−CO−H	8.7	1.2	0.6	5.8		−OCH$_3$	31.4	−14.6	1.0	−8.1
	−C≡N	−15.8	3.6	0.7	4.3	**O**	−O−C$_6$H$_5$	29.1	−9.4	0.3	−5.2
							−O−CO−R	23.0	−6.2	1.3	−2.3

R stands for any alkyl groups.

Example of the use of Table 33.11

C1			**C3**		
base	128.5 ppm		base	128.5 ppm	
—OH *attach*	26.9 ppm		—OH *meta*	1.4 ppm	
—NO$_2$ *para*	6.1 ppm		—NO$_2$ *ortho*	−4.9 ppm	
total	161.5 ppm		total	125.0 ppm	
	δ_{obs} = 163.9 ppm			δ_{obs} = 125.8 ppm	
C2			**C4**		
base	128.5 ppm		base	128.5 ppm	
—OH *ortho*	−12.8 ppm		—OH *para*	−7.3 ppm	
—NO$_2$ *meta*	0.9 ppm		—NO$_2$ *attach*	19.8 ppm	
total	116.6 ppm		total	141.0 ppm	
	δ_{obs} = 115.6 ppm			δ_{obs} = 139.7 ppm	

Table 33.12 ^{13}C Shielding Constants for Substituents Attached to Vinyl Carbons

$$\delta_A = 123.3 \text{ ppm} + \sigma_1$$
$$\delta_B = 123.3 \text{ ppm} + \sigma_2$$

	Substituent	σ_1	σ_2		Substituent	σ_1	σ_2
	$-H$	0	0		$-C\equiv N$	15.2	14.3
C	$-CH_3$	10.6	−9.8	**X**	$-I$	−38.1	7.3
	$-CH_2CH_3$	15.5	−9.7		$-Br$	−7.9	−1.0
	$-CH_2OR$	13.6	−8.5		$-Cl$	2.6	−5.7
	$-C=C$	13.6	−7.0		$-F$	24.9	−34.1
	$-C\equiv C$	12.0	−11.0	**N**	$-NR_2$	16.0	−29.0
	$-C_6H_5$	12.5	−11.0		$-NR_3^+$	19.8	−10.6
C=O	$-CO-OR$	6.2	7.0		$-NO_2$	22.3	−0.9
	$-CO-OH$	4.7	9.0	**S**	$-SR$	19.0	−16.0
	$-CO-R$	15.0	5.9	**O**	$-OR$	29.0	−39.0
	$-CO-H$	13.4	13.0		$-OCO-R$	18.0	−27.0

R stands for any alkyl group.

Example of the use of Table 33.12

A		B	
base	123.3 ppm	base	123.3 ppm
—COOR (σ_1)	6.2 ppm	—COOR (σ_2)	7.0 ppm
—CH$_3$ (σ_2)	−9.8 ppm	—CH$_3$ (σ_1)	10.6 ppm
total	119.7 ppm	total	140.9 ppm
	δ_{obs} = 122.3 ppm		δ_{obs} = 144.1 ppm

Table 33.13 ^{13}C Chemical Shifts of Carbonyls and Other Unsaturated Carbons

Compound	δ	Compound	δ
R−CO−H	200−204	H−CO−OR	161−163
Ar−CO−H	185−197	R−CO−OR	170−179
R−CO−R	206−212	Ar−CO−OR	159−167
Ar−CO−R; Ar−CO−Ar	196−206	H−CO−NR$_2$	162−166
H−CO−OH	166	R−CO−NR$_2$	168−180
R−CO−OH	178−185	Ar−CO−NR$_2$	168−171
Ar−CO−OH	168−178	R−CO−Cl	169−180
H−CO−O$^-$	171	Ar−CO−Cl	163−171
R−CO−O$^-$	182−189	R−C≡C−R; Ar−C≡C−R	65−89
Ar−CO−O$^-$	171−181	R−C≡N; Ar−C≡N	117−125

33.18 AROMATIC SYSTEMS

The study of substituted aromatic rings by ^{1}H- and ^{13}C-NMR often allows the unequivocal determination of the substituents' positions. When the total number of hydrogens in the molecule is known, the integration of the ^{1}H-NMR gives the number of hydrogens attached to the aromatic ring, and thus it indirectly reveals the number of substituents. For example, if the integration indicates that there are four aromatic hydrogens, it means that there are two substituents attached to the ring. The spin-spin splitting between the aromatic hydrogens can be used to determine the type of substitution. The number of peaks for aromatic carbons in the ^{13}C-NMR can also be used to study the substitution pattern.

In ^{1}H-NMR there are three types of coupling constants between aromatic hydrogens, J_{ortho}, J_{meta}, and J_{para}. J_{meta} and J_{para} are long-range coupling constants, and thus are smaller in magnitude than J_{ortho}. In most cases, J_{para} is very close to zero and the coupling between para hydrogens is not observed.

Monosubstituted benzene rings have three different types of aromatic hydrogens coupled to each other and often present complex ^{1}H-NMR spectra. However, if the substituent is an *alkyl group*, the hydrogens resonate at virtually identical frequencies, giving rise to a broad singlet (you may remember this from the case of benzyl acetate discussed in section 33.8). The ^{13}C-NMR of monosubstituted benzene rings presents four peaks for aromatic carbons.

Aromatic rings with two different substituents in position 1 and 4 give a typical coupling pattern in ^{1}H-NMR. There are two types of hydrogen and thus two signals are observed. H_A is coupled to the H_B next to it with a J_{ortho} (J_{ortho} = 7–9 Hz) and also to the other H_B in para with J_{para}, which is usually very small and can be ignored. Both H_A are also coupled to each other with a J_{meta}. In the first approximation, if we disregard the meta and para couplings of H_A and consider only its vicinal (ortho) coupling to H_B, we should expect to see a doublet for H_A. This doublet should integrate for two hydrogens because both H_A are equivalent. The same analysis can be made for H_B which should also show as a doublet. The spectrum of 4-bromoacetanilide is shown in Figure 33.49. As can be observed, the signals for the aromatic hydrogens are not perfect doublets; there are small peaks on the side of the big ones that arise from the long-range couplings that have been ignored in our analysis.

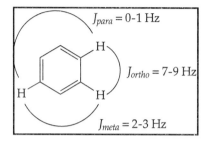

$$J_{para} = 0\text{-}1 \text{ Hz}$$
$$J_{ortho} = 7\text{-}9 \text{ Hz}$$
$$J_{meta} = 2\text{-}3 \text{ Hz}$$

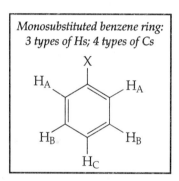

Monosubstituted benzene ring: 3 types of Hs; 4 types of Cs

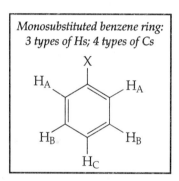

$$J_{AB} = J_{ortho} \approx 7\text{-}9 \text{ Hz}$$

The actual chemical shifts of H_A and H_B depend on Z and Y.

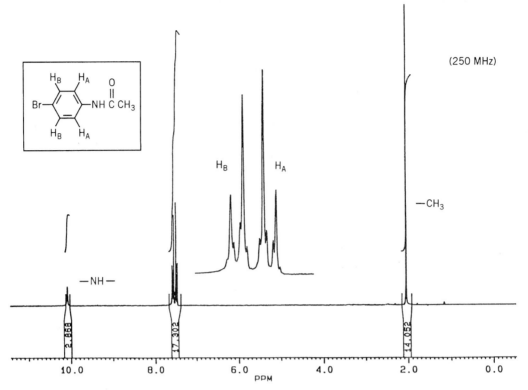

Figure 33.49 ¹H-NMR of 4-bromoacetanilide.

If the two substituents of a 1,4-disubstituted benzene ring are the same, the four hydrogens are equivalent and give a singlet in the ¹H-NMR.

The ¹³C-NMR of 1,4-disubstituted benzene rings gives four peaks for aromatic carbons if the substituents are different, and two peaks if they are the same (Fig. 33.50).

Ortho- and meta-disubstituted benzene rings with two different substituents have four different types of aromatic hydrogens and often give rise to complex ¹H-NMR spectra that cannot be simply interpreted by the $n + 1$ rule. The ¹³C-NMR spectrum shows six peaks for the aromatic carbons because they are all different. If the two substituents are the same, then the ¹H-NMR usually gets simplified and the ¹³C-NMR presents three peaks for the ortho-disubstituted benzene and four peaks for the meta-disubstituted benzene. A summary of the different types of hydrogens and carbons in substituted benzene rings is given in Figure 33.50.

The ¹H-NMR spectra of 1,2,4-trisubstituted benzene rings usually present a typical coupling pattern with three groups of peaks, one for each aromatic hydrogen. This is particularly true when the three hydrogens resonate at very different chemical shifts. They constitute what is called an **AMX system** and give a first-order spectrum (section 33.13) that can be easily studied. Let's call the three hydrogens A, M, and X.

$J_{AX} = J_{para} \approx 0\text{-}1$ Hz

$J_{AM} = J_{ortho} \approx 7\text{-}9$ Hz

$J_{MX} = J_{meta} \approx 2\text{-}3$ Hz

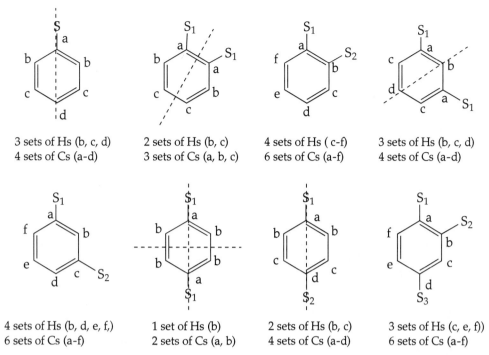

Figure 33.50 Number of different sets of hydrogens and carbons in substituted benzene rings The dashed lines represent symmetry planes. Some sets consist of only one atom.

H_A is coupled to H_M with a large coupling constant typical of hydrogens ortho to each other ($J_{AM} \approx 7\text{--}9$ Hz). H_A is also coupled to H_X with a smaller coupling constant, $J_{AX} \approx 0\text{--}1$ Hz, typical of para hydrogens. As a result of this double coupling, as you may recall from section 33.13, the signal for H_A is split into four lines of similar intensity called a **double doublet**.

Because the coupling constant between H_A and H_X is usually very small ($J_{para} \approx 0\text{--}1$ Hz), the coupling between these two hydrogens is not normally observed and the signal for H_A appears as a doublet with a coupling constant $J_{AM} = 7\text{--}9$ Hz, instead of a double doublet (Fig. 33.51).

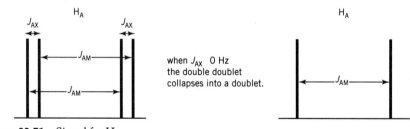

Figure 33.51 Signal for H_A.

Now let's consider H_M (Fig. 33.52). H_M is coupled to H_A and H_X with coupling constants J_{AM} (7–9 Hz) and J_{MX} (2–3 Hz), respectively. The signal for H_M is a double doublet.

Finally, H_X is coupled to H_A and H_M with coupling constants J_{AX} (0–1 Hz) and J_{MX} (2–3 Hz). In principle, its signal is also a double doublet, but the coupling between H_A and H_X (0–1 Hz), as explained before, is generally not observed and the signal for H_X appears as a doublet with a coupling of 2–3 Hz (J_{MX}) (see Figure 33.53).

Therefore, when the coupling between H_A and H_X is negligible, the ^{1}H-NMR of 1,2,4-trisubstituted benzenes shows, in the aromatic region, a doublet for H_A with

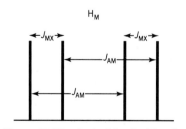

Figure 33.52 A double doublet for H_M.

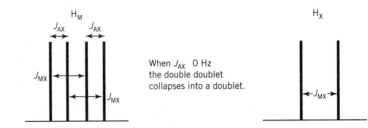

Figure 33.53   Signal for H_X.

a coupling of 7–9 Hz (J_{AM}), a double doublet for H_M with two coupling constants, 7–9 Hz (J_{AM}) and 2–3 Hz (J_{MX}), and a doublet for H_X with a coupling of 2–3 Hz (J_{MX}).

33.19   EFFECT OF CHIRALITY ON THE NMR

The ^1H- and ^{13}C-NMR spectra of enantiomers are indistinguishable in the absence of a chiral solvent or other chiral molecules in the medium. When a chiral solvent is used, enantiomers usually give rise to different spectra because of their stereospecific interactions with the solvent. (R)- and (S)-2-octanol and (R)- and (S)-1-(1-naphthyl)ethylamine are among the optically active solvents that can be used in NMR to distinguish enantiomers. Contrary to enantiomers, diastereomers normally give rise to different spectra even in ordinary solvents and can be easily distinguished by NMR (for an example, see Experiment 28).

Even though two enantiomers cannot be told apart by NMR in achiral media, the presence of a stereogenic center has far-reaching effects on the spectrum, as we will see in the next paragraphs.

Methylene Groups Attached to a Carbon with Three Different Substituents

The two hydrogens of a methylene group are chemically equivalent as long as the methylene group is free to rotate and is not bound to a carbon with three different substituents. Examples of chemically equivalent methylene hydrogens are shown below.

$$CH_3CH_2\underset{\underset{CH_3}{|}}{C}HCH_3 \qquad CH_3CH_2Br \qquad CH_3OCH_2\overset{\overset{O}{\|}}{C}CH_3$$

In each molecule the two hydrogens of the methylene group are chemically equivalent.

A different situation arises, however, when the methylene group is attached to a *carbon with three different substituents* X, Y, and Z. In this case, the two hydrogens are no longer chemically equivalent because they are now in different environments. To illustrate this point, let us consider Figure 33.54, which shows the Newman projections of the three staggered conformers of the molecule shown in the box. In our discussion we will assume the general case when the group R is different from —CXYZ; for a more detailed treatment of the subject, including the case when R is equal to —CXYZ, the reader can consult References 1 and 2.

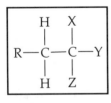

The environment that each hydrogen senses can be described by the order in which the hydrogen "sees" the other groups when looking around in either a clockwise or counterclockwise direction. For clarity, let us label the two methylene

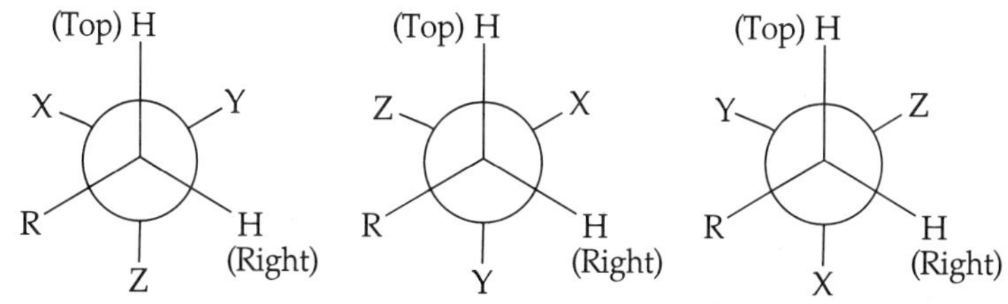

	1	**2**	**3**
Top H:	Y H Z R X	X H Y R Z	Z H X R Y
Right H:	Z R X H Y	Y R Z H X	X R Y H Z

Figure 33.54 Newman projections for a molecule $R-CH_2CXYZ$.

hydrogens as Top H and Right H. The order in which the Top H in conformer 1 sees the other groups, in a clockwise manner, is YHZRX. The environment seen by Right H in the same conformer is ZRXHY, and so on. If there is free rotation around the C—C bond, which is normally the case at ordinary temperatures, the environment seen by the Top H is an average of YHZRX, XHYRZ, and ZHXRY. This average environment is different from that seen by the Right H (an average of ZRXHY, YRZHX, and XRYHZ). As a result, the two methylene hydrogens resonate at different chemical shifts. These methylene hydrogens are called *diastereotopic*, because when each of them is replaced, in turn, by a deuterium atom, a different diastereoisomer results.

Aminoacids such as phenylalanine, tyrosine, and serine are just a few examples of systems where the nonequivalence of the methylene hydrogens can be observed. In each of these cases, the methylene is directly attached to a stereogenic carbon. (The ^{1}H-NMR of a derivative of tyrosine is studied in Experiment 16A.)

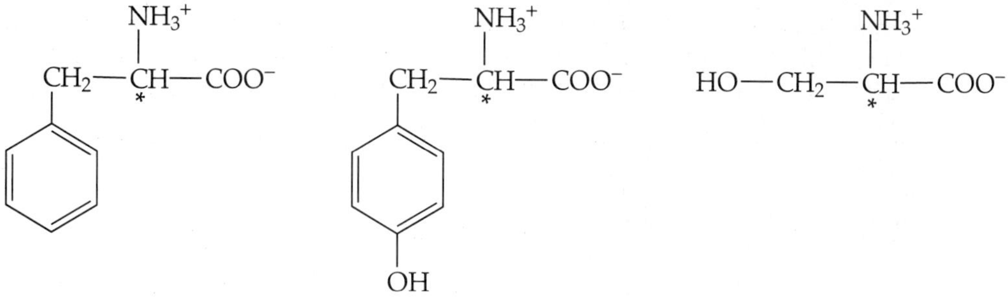

phenylalanine   tyrosine   serine

In each molecule, the two methylene hydrogens are nonequivalent because of the presence of the stereogenic carbon (*).

There are cases where nonequivalent methylene hydrogens resonate, by coincidence, at the same chemical shift. This fortuitous equivalence is highly dependent on experimental conditions and may disappear if the solvent is changed or if a more powerful spectrometer is used.

A stereogenic carbon attached to a methylene always has three different substituents (in fact, it has four different substituents) and therefore, the two methylene hydrogens are nonequivalent. However, it should be pointed out that the condition of three different substituents X, Y, and Z on the adjacent carbon to the methylene (RCH_2-CXYZ) does not necessarily make that carbon stereogenic. If one of the substituents X, Y, or Z is identical to $-CH_2R$, then the carbon is not stereogenic because it is bound to two identical groups ($RCH_2-CXYCH_2R$). Yet, in these cases the two methylene hydrogens are still nonequivalent. An example is shown below.

The two methylene hydrogens are nonequivalent because the adjacent carbon is attached to three different substituents ($-OH$, $-CH_3$, $-CH_2CH_3$), but the carbon is not stereogenic.

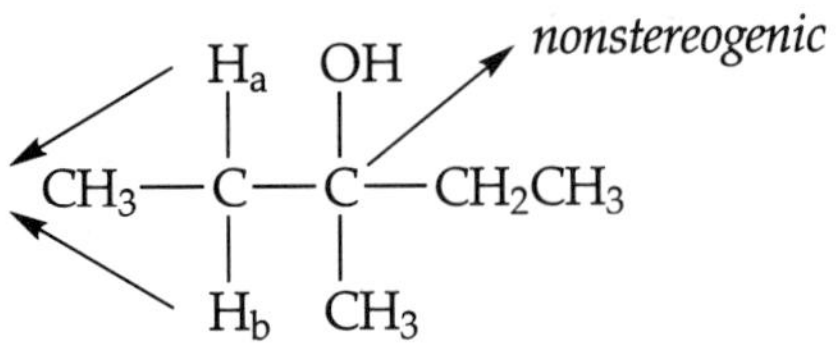

Now let us consider what happens if at least two of the groups X, Y, and Z are identical. In this case the methylene hydrogens become equivalent and resonate at the same chemical shift. Figure 33.55 illustrates this case when X = Y.

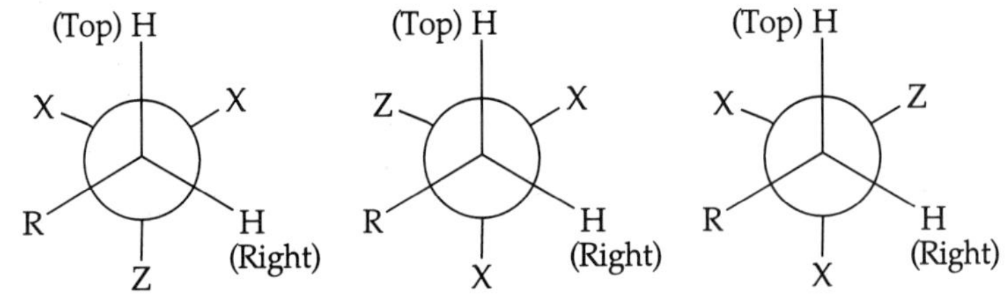

	1	**2**	**3**
Top H:	X H Z R X	X H X R Z	Z H X R X
Right H:	Z R X H X	X R Z H X	X R X H Z

Figure 33.55 Newman projections for a molecule RCH_2CXXZ.

The environment seen by the Top H in conformer 1 (XHZRX) is the mirror image of the environment seen by the Right H in conformer 2 (XRZHX). The environment seen by the Top H in conformer 2 (XHXRZ) is the mirror image of the environment seen by the Right H in conformer 1 (ZRXHX). Finally, the environments seen by both hydrogens in conformer 3 are mirror images of one another. As a result, both hydrogens see the same environments, except that one environment is the mirror image of the other. In the absence of a chiral solvent or other chiral molecules in the medium, one mirror image cannot be distinguished from the other and both methylene hydrogens resonate at the same chemical shift. Both methylene hydrogens become distinguishable, however, if a chiral solvent is used. These methylene hydrogens are called **enantiotopic** because when each of them is replaced, in turn, by a deuterium atom, a different enantiomer results.

If the carbon with the three different substituents is not directly attached to the methylene, but separated from it by more than one bond, the two methylene hydrogens are still nonequivalent. For example, the methylene hydrogens of the ethyl group in 2-ethoxytetrahydrofuran resonate at different chemical shifts because the molecule has a stereogenic center. In general, as the distance between the methylene and the stereogenic center increases, the difference in the environments seen by both methylene hydrogens becomes insignificant, and they resonate at the same chemical shift.

stereogenic carbon

The two hydrogens are nonequivalent and resonate at different chemical shifts (3.45 and 3.75 ppm in $CDCl_3$).

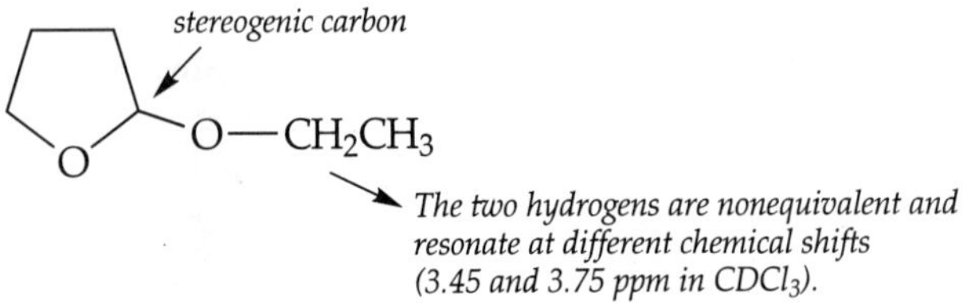

The arguments used to show that the two hydrogens of a methylene group are not always chemically equivalent can also be used to demonstrate that the methyl groups of an isopropyl group are nonequivalent when the isopropyl group is attached to a carbon with three different substituents. For example, the two methyl groups of the amino acid valine are not chemically equivalent and they resonate at different chemical shifts in ^{1}H-NMR as well as in ^{13}C-NMR.

The two methyl groups are nonequivalent because of the sterogenic carbon (*). They give different signals in both ^{1}H- and ^{13}C-NMR.

The ^{1}H-NMR of valine in D_2O shows two signals for the methyl groups at 0.97 and 1.02 ppm; each signal is a doublet because each methyl group is coupled to the neighboring H of the methine. The ^{13}C-NMR gives two signals for the methyl groups at 19.7 and 20.3 ppm.

33.20 RUNNING THE SPECTRUM

In this section we will discuss sample preparation and some important features in the use of NMR instruments. The operational details vary from instrument to instrument and will not be considered here. The student should check the instrument's instruction manual before obtaining the spectrum.

The amount of sample required to obtain a good spectrum depends on the power of the instrument. Spectrometers operating at 60 MHz for ^{1}H normally require between 10 and 30 mg of sample dissolved in about 0.7 mL of solvent. On the other hand, with 500 MHz instruments, ^{1}H-NMR spectra can be obtained with a few μg. The most common type of NMR sample holder is a slender tube 5 mm in diameter. Wider tubes for larger samples are also available.

The solvents used in ^{1}H-NMR should not contain ^{1}H; otherwise the solvent peak would be too intense and would interfere in the analysis. The use of deuterated solvents where the ^{1}H atoms have been replaced by D is very common in NMR. The number of D atoms replacing ^{1}H is specified after the solvent's name; for example, $CDCl_3$ (deuterochloroform) is represented as chloroform-d_1. Among the most commonly used solvents are: $CDCl_3$; D_3CCOCD_3 (acetone-d_6); D_3CSOCD_3 (dimethyl sulfoxide-d_6); CD_3OD (methanol-d_4); C_6D_6 (benzene-d_6); and D_2O (deuterium oxide).

Deuterated solvents are not 100% isotopically pure. They normally contain less than 0.1% of ^{1}H, which is enough to give a signal. For example, chloroform-d_1 has some residual $CHCl_3$ that gives a small peak at 7.27 ppm. The chemical shifts for some common solvents are given in Table 33.14. ^{1}H-NMR spectra often present a

Table 33.14 Common Solvents for NMR Experiments

Solvent	$\delta\,^1$H	$\delta\,^{13}$C	δ water	Solvent	$\delta\,^1$H	$\delta\,^{13}$C	δ water
acetone-d_6	2.05	206.68	2.8	dimethyl sulfoxide-d_6	2.50	39.51	3.3
		29.92					
acetonitrile	1.94	118.69	2.1	ethanol-d_6	5.29	56.96	5.3
		1.39			3.56	17.31	
					1.11		
benzene-d_6	7.16	128.4	0.4	methanol-d_4	4.87	4.9	7–9
					3.31		
chloroform-d_1	7.27	77.23	1.5	methylene chloride	5.32	54.00	1.5
deuterium oxide	4.65	—	—	trifluoroacetic acid-d_1	11.50	164.2	11.5
	(HDO)					116.6	

small peak for water that may arise from the solvent, a wet sample, or the glassware. The chemical shift of water depends on the solvent and is also listed in Table 33.14.

Solvent and concentration both play important roles in the spectrum. As discussed in section 33.14, the concentration especially affects the chemical shift of hydrogens attached to oxygen and nitrogen.

With modern spectrometers, it is no longer necessary to add TMS to the sample before obtaining its spectrum. If a deuterated solvent is employed, the peak from the residual ^{1}H, with its known chemical shift, is used to calibrate the spectrum.

To obtain good-quality spectra, the magnetic field should remain constant and homogeneous during the experiment. It is impossible to keep a magnetic field constant for long periods of time; it will inevitably fluctuate due to changes in temperature and the movement of metal objects in the room. To avoid this drift, the magnetic field is **locked** by feedback circuitry using the deuterium resonance of a deuterated solvent as a reference signal. As the magnetic field tends to drift, the lock circuitry readjusts it.

In addition to being stable, the magnetic field should be homogeneous. This means that the whole sample should experience the same magnetic field. This is in part achieved with the help of **shims** or **shim coils**, which are auxiliary magnets that create small magnetic fields around the sample and compensate for gradients in the applied magnetic field. Field inhomogeneities are further reduced by spinning the sample at a frequency of 15–40 revolutions per second. This fast spinning averages out residual field inhomogeneities in different parts of the sample.

33.21   TWO-DIMENSIONAL NMR

With the use of pulse FT-NMR we can extract detailed information about H—H, H—C, and C—C connectivities without analyzing complex coupling patterns. This is possible thanks to **two-dimensional NMR (2D NMR)**. In normal NMR one axis is the frequency (the chemical shift of ^{1}H or ^{13}C), and the other is the intensity. This type of NMR is called one-dimensional (even though there are two axes) because there is only one frequency axis. In 2D NMR, on the other hand, the intensity is plotted as a function of two frequency axes, resulting in a three-dimensional spectrum called the **stack plot** (Fig. 33.56). Normally, instead of showing the stack plot, which is difficult to analyze, a contour map of the spectrum is displayed. A **contour map** is obtained by intersecting the stack plot with parallel planes at different heights and overlaying them (Fig. 33.56).

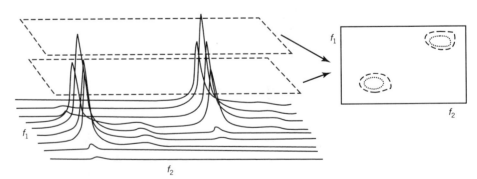

Figure 33.56   A stack plot of a 2D NMR.

Figure 33.57 shows a 2D spectrum of 2-(p-aminophenyl)-5-phenyloxazole. Both frequency axes in this case represent the ^{1}H chemical shift. To aid in the interpretation process, the normal ^{1}H-NMR of the sample is also shown along each axis. The 2D NMR is represented by the contour lines. As you can observe in Figure 33.57, the

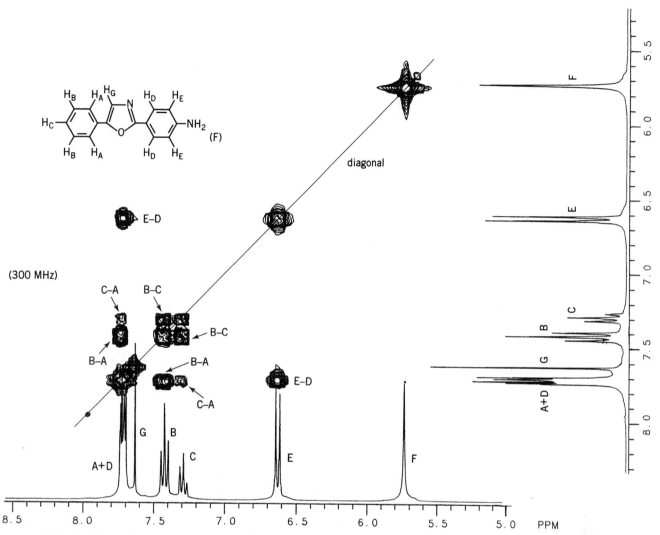

Figure 33.57 ^{1}H-^{1}H COSY of 2-(p-aminophenyl)-5-phenyloxazole.

contour map is symmetric about the diagonal. The contour lines on the diagonal represent the peaks observed in the one-dimensional spectrum and contain no new information. The off-diagonal contour lines, on the other hand, indicate hydrogen atoms coupled to each other. On each side of the diagonal, there is a set of contour lines for each pair of coupled hydrogens. Each set of contour lines is located at the intersection of the chemical shifts for both hydrogens.

Inspection of Figure 33.57 reveals that the hydrogen at 5.7 ppm (H$_F$) and the hydrogen at 7.6 ppm (H$_G$) are not coupled to any other hydrogens because they give singlets. Notice that these two hydrogens do not show off-diagonal contour lines, in agreement with the fact that they are not coupled to any other hydrogens. The analysis also reveals that the signal at 6.6 ppm (H$_E$) is coupled to the signal at 7.75 ppm (H$_D$) because there is a set of contour lines at the intersection of both chemical shifts. The signal at 7.3 ppm (H$_C$) is coupled to the signal at 7.4 ppm (H$_B$), and also to the signal at 7.75 ppm (H$_A$). The signal at 7.4 ppm (H$_B$) is coupled to the ones at 7.3 (H$_C$) and 7.75 (H$_A$). The coupling pattern of these hydrogens is in agreement with the expected couplings for this aromatic molecule.

This type of spectrum is called **^{1}H-^{1}H COSY (cor**related **s**pectro**s**cop**y**). It is also possible to correlate ^{1}H and ^{13}C chemical shifts, **^{1}H-^{13}C COSY** or **HETCOR** (**het**eronuclear **cor**relation), and even ^{13}C-^{13}C chemical shifts (**2D INADEQUATE**, **i**ncredible **n**atural **a**bundance **d**ouble **qua**ntum **t**ransfer **e**xperiment) among many other types of 2D techniques. 2D NMR spectra are obtained by a combination of pulses of different length and delay times between pulses. (A detailed discussion of them can be found in Refs. 1, 4, and 8).

33.22 INTERPRETING ^{1}H-NMR SPECTRA

The interpretation of NMR spectra can be learned only through extensive practice. There is no infallible recipe for obtaining the structure of an unknown from its spectra. Usually a combination of several methods (NMR, IR, mass spectrometry, and chemical characterization) is needed to elucidate the unknown's structure. We will outline here the steps followed in the interpretation of ^{1}H-NMR.

Let's consider the ^{1}H-NMR of an unknown sample Y that has a molecular formula C_8H_9Br and gives rise to the spectra shown in Figure 33.58 (solvent $CDCl_3$).

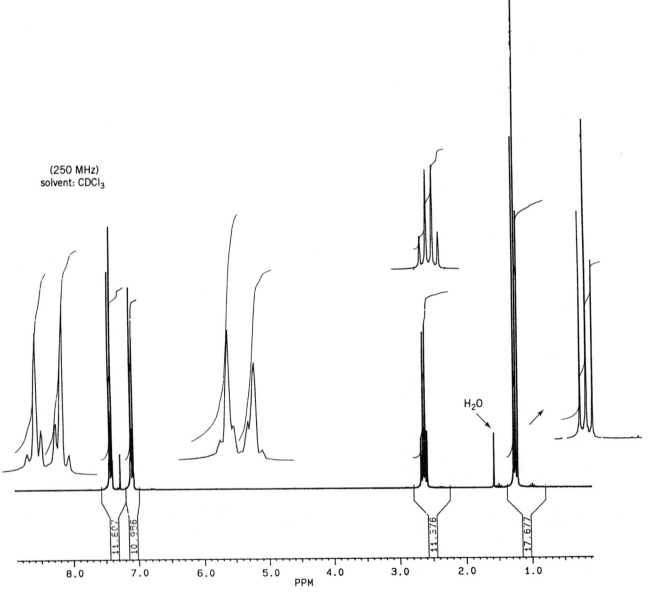

Figure 33.58 ^{1}H-NMR of an unknown sample Y of molecular formula C_8H_9Br in $CDCl_3$.

When the molecular formula is known, we should first calculate the **degree of unsaturation** with Equation 10:

$$\text{unsaturation} = \frac{2C + 2 - H - X + N}{2} \tag{10}$$

where C is the number of carbons, H is the number of hydrogens, X is the number of halogens, and N is the number of nitrogens. The degree of unsaturation calculated with this formula includes rings. Each ring increases this number by one. For compound Y, the degree of unsaturation (including rings) is 4.

The first step in the interpretation of ¹H-NMR is to **find out the number of signals** (singlets and multiplets). In doing so, let's keep in mind that doublets, triplets, quartets, quintets, and so forth are symmetrical, and the relative intensity of their lines follows Pascal's triangle (with minor deviations). Contrary to IR spectra, where not all the peaks are assignable, *all the NMR peaks should be accounted for*. If one fails to assign a peak of the ¹H-NMR, chances are that the proposed structure is wrong. In finding the peaks, one should **recognize the solvent's peaks** as such and should not try to assign them to the sample. See Table 33.14 to find out where the solvent resonates.

The spectrum of Y has a triplet around 1.23 ppm, a quartet around 2.61 ppm, and two distorted doublets at 7.08 and 7.41 ppm. The peaks at 1.5 ppm and 7.27 ppm belong to the water and $CHCl_3$ present in the solvent. They are in agreement with the values of Table 33.14.

The next step should be the **calculation of the number of hydrogens** under each signal using the integration numbers. In Figure 33.58 the integration numbers (the numbers in brackets) are 17.677 for the triplet at 1.23 ppm; 11.376 for the quartet at 2.61 ppm; 10.956 for the doublet at 7.08 ppm, and 11.807 for the doublet at 7.41 ppm. The total integration number is $I_T = 17.677 + 11.376 + 10.956 + 11.807 = 51.818$. Applying Equation 9 to each signal, the following numbers of hydrogen are obtained:

triplet at 1.23 ppm: $\quad \#H = \dfrac{I}{I_T} \times H_T$

$$= \frac{17.677}{51.818} \times 9 = 3.07, \text{ which is rounded off to 3H}$$

quartet at 2.61 ppm: $\quad \#H = \dfrac{11.376}{51.818} \times 9 = 1.98, \text{ which is rounded off to 2H}$

doublet at 7.08 ppm: $\quad \#H = \dfrac{10.956}{51.818} \times 9 = 1.90, \text{ which is rounded off to 2H}$

doublet at 7.41 ppm: $\quad \#H = \dfrac{11.807}{51.818} \times 9 = 2.05, \text{ which is rounded off to 2H}$

To double check that the integration is correct, one should add the number of hydrogens and compare the total number with the molecular formula. In our case: 3H + 2H + 2H + 2H = 9H, which agrees with the molecular formula.

The next step is to consider the **chemical shifts** and get a feeling for what types of hydrogens the sample has. Figure 33.15 is very useful in this respect. The NMR of Y shows peaks between 1 and 3 ppm and between 7 and 8 ppm. The former are in the chemical shift region of aliphatic hydrogens (methyl, methylene, methine), and the latter are in the region of aromatic hydrogens.

The next step is to find the **connectivities** using the multiplicities and the $n + 1$ rule. Compound Y gives a triplet that integrates for 3H and a quartet for 2H. These peaks are between 1 and 3 ppm and correspond to aliphatic hydrogens. A triplet for 3H implies a methyl group next to two hydrogens, according to the $n + 1$ rule ($n + 1 = 2 + 1$). A quartet for 2H implies a methylene next to three hydrogens ($n + 1 = 3 + 1$). This suggests a methylene and a methyl group next to each other, forming an ethyl group (CH_3CH_2-). Comparison with Figure 33.29 also confirms the presence of an ethyl group.

The peaks at 7.08 and 7.40 ppm correspond to aromatic hydrogens. Since they integrate for a total of 4H, the aromatic ring must have two substituents. It should be noticed that the presence of an aromatic ring is in agreement with the degree of unsaturation of 4. The two distorted doublets, each one integrating for 2H, is the typical pattern of a para-disubstituted phenyl ring, as discussed in section 33.18.

Now one can **put the molecule together**. One substituent on the ring is the ethyl group. The ethyl group (C_2H_5) plus the disubstituted phenyl ring (C_6H_4) account for all the hydrogens and carbons in the molecule (C_8H_9Br). Then, the bromine atom must be the other substituent. Compound Y is **4-bromoethylbenzene**.

To confirm the analysis, one should **calculate the chemical shifts** using tables. The calculation of the chemical shifts for the aromatic hydrogens is shown below Table 33.6. The expected chemical shift for the methylene, according to Table 33.4, is $\delta = 1.25$ ppm (base value) $+ 1.3$ ppm (phenyl) $= 2.55$ ppm, which agrees with the observed value of 2.61 ppm. The calculations are summarized below.

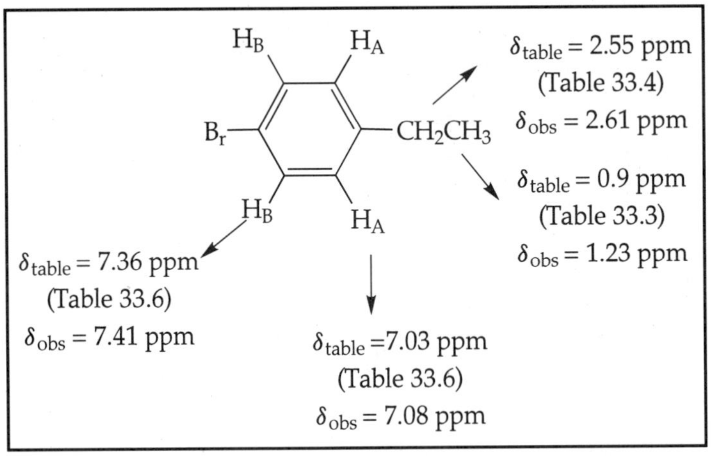

BIBLIOGRAPHY

1. Nuclear Magnetic Resonance Spectroscopy. F.A. Bovey, L. Jelinski, and P.A. Mirau. Academic Press, San Diego, 1988.

2. Spectroscopic Methods in Organic Chemistry. M. Hesse, H. Meier, and B. Zeeh. Translated by A. Linden and M. Murray. Thieme, New York, 1997.

3. Tables of Spectral Data for Structure Elucidation. E. Pretsch, T. Clerc, J. Seibl, and W. Simon. Springer-Verlag, New York, 1989.

4. NMR Spectroscopy. H. Günther. 2nd ed. Wiley. Chichester, New York, 1995.

5. Spectrometric Identification of Organic Compounds. R.M. Silverstein, G.C. Bassler, T.C. Morrill. 5th ed. Wiley, New York, 1991.

6. Organic Structure Analysis. P. Crews, J. Rodríguez, and M. Jaspars. Oxford University Press, New York, 1998.

7. NMR Spectroscopy Techniques. M.D. Bruch, ed. 2nd ed. Dekker, New York, 1996.

8. Modern NMR Spectroscopy. A Guide for Chemists. J.K.M. Sanders and B.K. Hunter. Oxford University Press. Oxford, 1988.

9. Structure Elucidation by Modern NMR. A Workbook,. H. Duddeck and W. Dietrich. Springer-Verlag. New York, 1989.

10. Spectroscopic Methods in Organic Chemistry. D.H. Williams and I. Fleming. 5th ed. McGraw-Hill. London, 1995.

11. NMR and Chemistry. An Introduction to Modern NMR Spectroscopy. J.W. Akitt. Chapman & Hall, London, 1992.

12. Spectroscopic Techniques for Organic Chemists. J.W. Cooper. Wiley, New York, 1980.

13. A Step-by-Step Picture of Pulsed (Time-Domain) NMR. L.J. Schwartz. *J. Chem Ed.*, **65**, 959–963 (1988).

14. Proton Magnetic Resonance Spectroscopy. D.A. McQuarrie. *J. Chem Ed.*, **65**, 427–433 (1988).

15. A General Approach for Calculating Proton Chemical Shifts for Methyl, Methylene, and Methine Protons When There Are One or More Substituents within Three Carbons. P.S. Beauchamp and R. Marquez. *J. Chem Ed.*, **74**, 1483–1485 (1997).

16. Varian Associates Technical Information Bulletin, Vol. 2, No 3. J.N. Shoolery. Varian Associates, Palo Alto, 1959.

17. The Aldrich Library of ¹³C and ¹H FT NMR Spectra. C.J. Pouchert and J. Behnke. Aldrich Chemical Company. Milwaukee, 1993.

18. The Aldrich Library of NMR Spectra. C.J. Pouchert. 2nd ed. Aldrich Chemical Company, Milwaukee, 1983.

EXERCISES

Problem 1

True or false. Circle the correct option.

(a) Spin-spin splitting arises because magnetically active nuclei interact with nearby magnetically active nuclei. T/F

(b) The value of the coupling constant in Hz depends on the operational frequency of the instrument. T/F

(c) Chemically equivalent hydrogens do not show spin-spin split. T/F

(d) Chemically equivalent hydrogens are not coupled. T/F

(e) ^{16}O is a magnetically active nucleus. T/F

(f) The Larmor frequency of an ^1H nucleus depends on its environment. T/F

(g) Electron-withdrawing substituents cause hydrogens to resonate at lower chemical shifts. T/F

(h) Hydrogens attached to oxygen and nitrogen usually do not show spin-spin splitting with nearby hydrogens. T/F

(i) Cyclohexane gives only one peak in ^1H-NMR regardless of the temperature. T/F

(j) Cyclohexane gives only one peak in ^{13}C-NMR regardless of the temperature. T/F

Problem 2

Complete the following paragraph (each line corresponds to one word):

Only nuclei with a nuclear spin quantum number different from _____ are magnetically active. TMS is used as _____ _____ in ^1H- and ^{13}C-NMR experiments. The chemical shift (in ppm) doesn't depend on the _____ _____ of the instrument. Hydrogens close to electron-withdrawing substituents are _____ . Pascal's triangle can be used to predict the relative _____ of multiplets in ^1H-NMR. In hydrogen-decoupled ^{13}C-NMR, the intensity of the peak is not directly _____ to the number of carbons that gave rise to the peak, and thus, the _____ of the spectrum is meaningless.

Problem 3

How many signals do you expect to see in the ^1H- and ^{13}C-NMR of these compounds? What would be the multiplicity of the ^1H-NMR signals according to the $n + 1$ rule?

(a) $CH_3CH_2OCH_3$	(b) $CH_3CH_2OCH_2CH_3$	(c) $\overset{\overset{\displaystyle O}{\|\|}}{CH_3CH_2CH}$	(d) $CH_3CH_2N(CH_3)_2$
(e) $\underset{\underset{\displaystyle CH_3}{\|}}{CH_3CHCH_2Br}$	(f) $\underset{\underset{\displaystyle CH_3}{\|}}{CH_3\overset{\overset{\displaystyle O}{\|\|}}{CH}CCH_3}$	(g) Cl—⬡—Cl	(h) Cl—⬡—NO₂

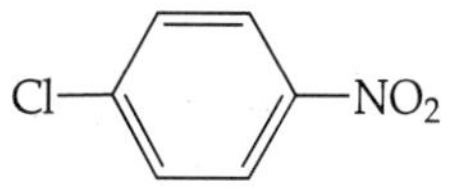

Problem 4

(a) Consider the ^{1}H-NMR shown in the figure below. Identify the multiplets as triplets or quartets. Pair them up according to the roof effect.

(b) Identify the compound that gives rise to the spectrum, knowing that the molecular formula is C_5H_9ClO.

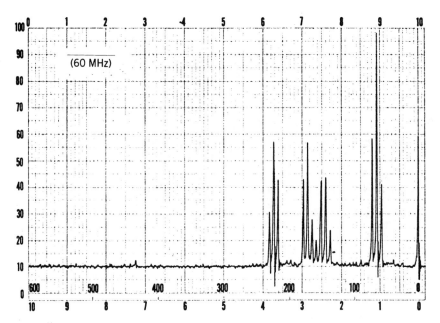

Problem 5

Give a structure consistent with the ^{1}H-NMR shown in the figure, for a compound of molecular formula C_3H_7Cl.

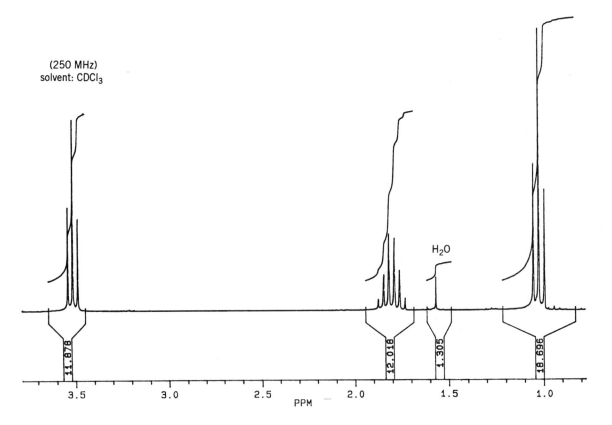

Problem 6

A compound $C_8H_6O_4$ gives rise to the ¹H- and ¹³C-NMR spectra shown in the figures below.

(a) Propose a structure in agreement with the data.

(b) Justify the chemical shifts using tables.

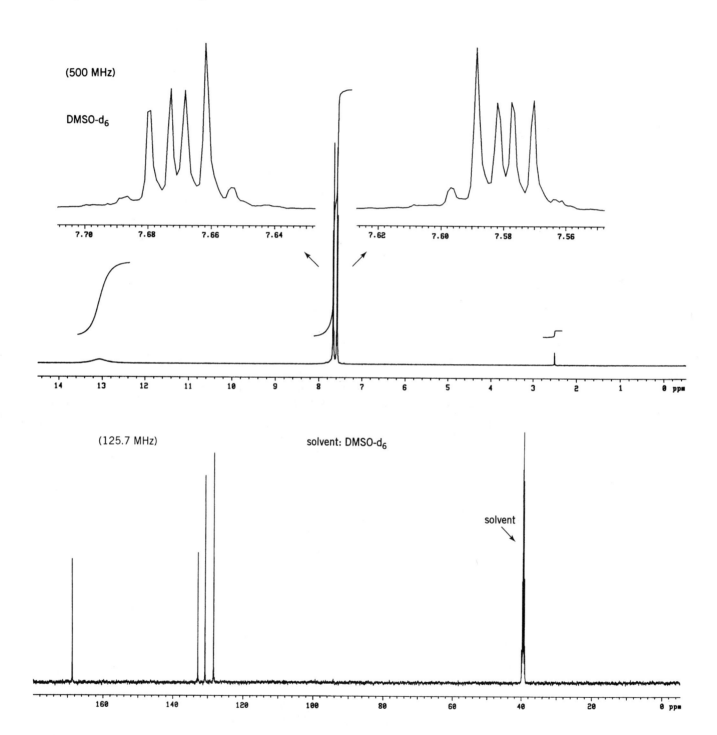

Problem 7

A compound of molecular formula $C_8H_8O_2$ gives rise to the ¹H-NMR shown in the figure.

(a) Propose a structure in agreement with the spectrum.

(b) Justify the chemical shifts using tables.

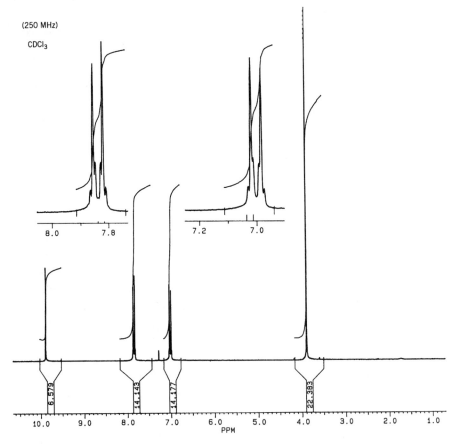

Problem 8

A compound $C_7H_7NO_2$ with analgesic properties gives rise to the ^{1}H-NMR shown in the figure. Propose a structure in agreement with the spectrum. Carefully analyze the splitting pattern in the 6.5–7.8 ppm region.

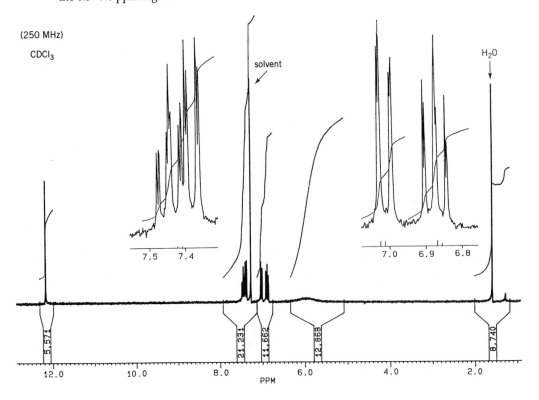

Problem 9

A compound $C_9H_{12}O_2$ gives rise to the ^{1}H-NMR shown in the figure.

(a) Give a structure in agreement with the spectrum.
(b) Justify the chemical shifts using tables.
(c) How many peaks do you expect to see in its ^{13}C-NMR?

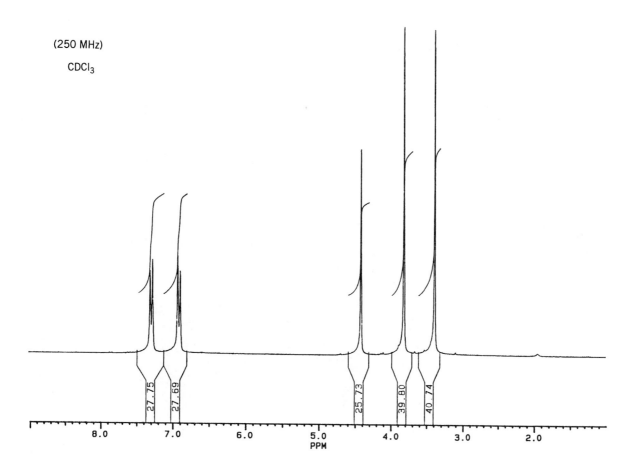

(250 MHz)

CDCl$_3$

8.0   7.0   6.0   5.0 PPM   4.0   3.0   2.0

27.75   27.69   25.73   39.80   40.74

Problem 10

A compound $C_6H_4N_2O_4$ gives rise to the spectra shown in the figure.

(a) Propose a structure in agreement with the spectroscopic information.
(b) Justify the chemical shifts using tables.

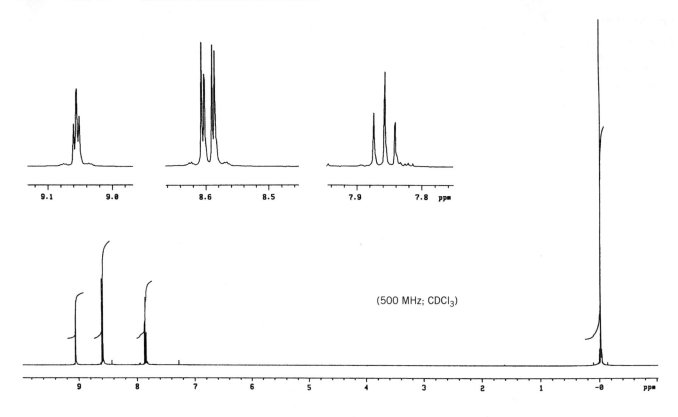

(500 MHz; CDCl$_3$)

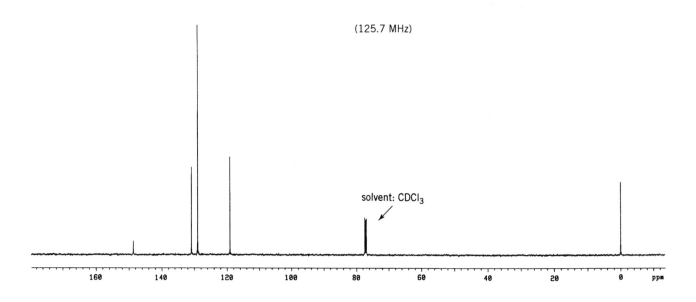

(125.7 MHz)

solvent: CDCl$_3$

Mass Spectrometry

34.1 OVERVIEW

Mass spectrometry, **MS**, unlike other spectroscopic methods, does not involve the interaction of light and matter. MS is based on the ancient principle of breaking something into pieces to see what it is made of. By analyzing the fragments the chemist can put the pieces together like a jigsaw puzzle.

In mass spectrometry, the sample in the vapor state is ionized by bombarding it with an electron beam of high energy (10–70 eV; 1–6.7 MJ/mol). The electrons collide with the sample and eject an electron from its valence shell. The molecule with one less electron becomes a **radical cation**: a radical because it has an odd number of electrons, and a cation because it has a positive charge. This is illustrated below with butanone.

$$\overset{\overset{\displaystyle ..}{\displaystyle \ddot{O}:}}{\underset{\displaystyle CH_3CH_2CCH_3}{\|}} + \ e^- \longrightarrow \overset{\overset{\displaystyle .+}{\displaystyle \dot{O}:}}{\underset{\displaystyle CH_3CH_2CCH_3}{\|}} + \ 2\,e^-$$

In general, the electrons most likely to be ejected are those in high-energy orbitals such as nonbonding and pi orbitals. The radical cation contains all the atoms of the original molecule, minus an electron, and is called the **molecular ion**. It is represented by $M^{+\cdot}$. As a result of the **electron impact** with the high-energy electrons, the molecular ion has an energy excess that is dissipated by breaking bonds and producing smaller cations, radicals, and neutral molecules. The cations generated in this fragmentation process are separated and analyzed by their **mass-to-charge ratio**, **m/z**, by using a combination of electric and magnetic fields. Most cations produced in MS have a charge of one, $z = 1$, and thus m/z directly gives the mass of the cation. Neutral fragments are not detected by the mass spectrometer.

$$ABC^{+\cdot} \xrightarrow{\textit{fragmentation}} AB^+ + C^{\cdot}$$
$$(M^{+\cdot})$$
$$\downarrow \textit{fragmentation}$$
$$A + B^+$$

The molecular ion of molecule ABC breaks down into cation AB^+ and radical $C^{\cdot}$. AB^+ then decomposes into cation B^+ and neutral fragment A. The mass spectrometer detects only $ABC^{+\cdot}$, AB^+, and B^+.

Molecules do not fragment randomly. Their fragmentation follows some basic principles and, thus, only certain fragments are produced from a given molecular ion. We will discuss these principles in section 34.5. A **mass spectrum** is a plot of a fragment's abundance versus its mass-to-charge ratio, m/z. The abundance of a fragment depends on the balance between its rate of formation and its rate of decomposition. Abundant fragments form easily and have a low tendency to undergo further fragmentation, or in other words, they are *relatively* stable. The most abundant fragment of a mass spectrum is assigned 100% **relative abundance (RA)** and is called the **base peak**. The abundances of the other fragments are given relative to the base peak. Because all the fragments are derived from the molecular ion, the molecular ion peak is also called the **parent peak** of the mass spectrum. The mass spectrum for butanone is shown in Figure 34.1. The molecular ion is at m/z 72, and the base peak at m/z 43.

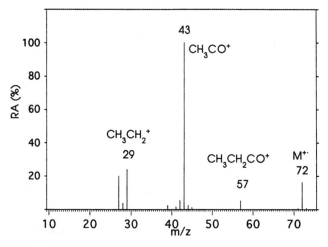

Figure 34.1   Mass spectrum of butanone.

By obtaining the mass spectrum of an unknown compound, a chemist can determine its molecular mass and often its molecular formula too. This information, along with a careful analysis of the fragments, is invaluable in the elucidation of the chemical's structure.

34.2 INSTRUMENTATION

A typical mass spectrometer consists of the following parts: a **vaporization chamber** where the sample is vaporized under vacuum, an **ionization chamber** where the sample is ionized, an **analyzer** that separates the fragments according to their mass-to-charge ratio, a **detector** that collects the charged fragments, an **amplifier** that increases the electrical current produced by the fragments, and a **recorder** that displays the data. Several types of mass spectrometers are commercially available; their main differences lie in how the sample is vaporized, ionized, and separated. One of the most common types of mass spectrometers, called a **magnetic sector mass spectrometer**, is shown in Figure 34.2. This instrument uses a magnetic field to separate the ions. The sample is ionized by the action of an electron beam. The sample ions are accelerated by an electric field and enter the analyzer tube where they are subjected to the action of a magnetic field. The magnetic field changes the trajectory of the ions. For a given magnetic field strength, only ions of a certain mass-to-charge ratio are focused into the detector; all other ions are deflected against the walls of the tube. By gradually increasing the strength of the magnetic field, ions of increasing m/z reach the detector in succession.

Ionization with an electron beam is called **hard ionization** because, as a result of the electron impact, the ions acquire a large energy excess that is used to break the molecule into smaller pieces. Hard ionization sometimes leads to mass spectra with a multitude of fragments but lacking the molecular ion peak. Without the molecular ion peak, these spectra are difficult to interpret. To avoid this problem, milder

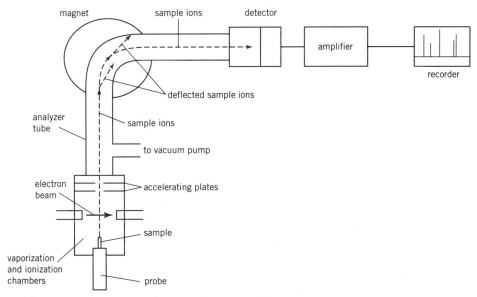

Figure 34.2 Diagram of a mass spectrometer with magnetic sector analyzer.

ionization methods (**soft ionization**) have been devised. For example, ionization by ionized gases (**chemical ionization**), ionization by beams of high energy argon or xenon atoms (**fast atom bombardment** or **FAB**), and ionization by lasers (**matrix assisted laser desorption/ionization** or **MALDI**) are commonly used. **Electrospray** is another soft ionization method that facilitates the study of nonvolatile compounds such as synthetic polymers, proteins, and nucleic acids. In this technique a solution of the sample is sprayed into an electric field, giving small droplets of electrically charged particles that are then separated and analyzed. Here we will focus mainly on electron impact (for a discussion of soft ionization methods, see Refs. 8 and 14).

Mass spectrometers can be used as detectors for gas chromatographs (**GC-MS**), and high-performance liquid chromatography systems (**LC-MS**). The versatility of these separating systems, coupled with the analytical power of a mass spectrometer, makes these techniques extremely powerful in all areas of chemical and biochemical research.

A mass spectrum can be obtained with very small amounts of material. With simple instruments less than a milligram is normally required, while modern instruments require as little as a few nanograms to produce quality spectra.

34.3 THE MOLECULAR ION

Isotopic Clusters

The first step in the interpretation of a mass spectrum is the identification of the molecular ion. Regardless of its relative abundance, the molecular ion peak is always the most important piece of information in the interpretation of the mass spectrum. The molecular ion contains all the atoms in the molecule, and thus it is the largest ion that the molecule can produce. The molecular ion never gives one single peak but rather a group of peaks, called an **isotopic cluster**, which arises because of the different isotopes of carbon, hydrogen, nitrogen, oxygen, and other elements present in organic molecules.

The atomic masses that we use to calculate molecular masses are the *average* atomic masses of the elements. These averages result from considering all natural isotopes of the elements and their abundances. For example, the average atomic mass of chlorine is 35.5 because there are two isotopes for this element, ^{35}Cl

and ^{37}Cl, with a natural abundance of approximately 76% and 24%, respectively ($35.5 = 0.76 \times 35 + 0.24 \times 37$). A list of the isotopes of the most common elements in organic chemistry is given in Table 34.1. Fluorine, phosphorus, and iodine have only one isotope, while all the other elements in the table have at least two natural isotopes.

Table 34.1 Isotopes of the Most Common Elements in Organic Compounds From *CRC Handbook of Physics and Chemistry*, 76th ed.

Element	Atomic mass	Isotope	Mass	Abundance (%)
hydrogen	1.00794	^1H	1.007825	99.985
		^2H (D)	2.014102	0.015
carbon	12.011	^{12}C	12.000000	98.90
		^{13}C	13.003355	1.10
nitrogen	14.00674	^{14}N	14.003074	99.635
		^{15}N	15.000109	0.367
oxygen	15.9994	^{16}O	15.994915	99.762
		^{17}O	16.999131	0.038
		^{18}O	17.999160	0.200
fluorine	18.998403	^{19}F	18.998403	100
silicon	28.086	^{28}Si	27.976927	92.23
		^{29}Si	28.976495	4.67
		^{30}Si	29.973771	3.10
phosphorus	30.973762	^{31}P	30.973762	100
sulfur	32.066	^{32}S	31.972071	95.03
		^{33}S	32.971459	0.75
		^{34}S	33.967867	4.21
		^{36}S	35.967081	0.02
chlorine	35.4527	^{35}Cl	34.968853	75.77
		^{37}Cl	36.965903	24.23
bromine	79.904	^{79}Br	78.91834	50.69
		^{81}Br	80.91629	49.31
iodine	126.90447	^{127}I	126.90447	100

Because there are two natural isotopes for carbon, ^{12}C and ^{13}C, and two natural isotopes for hydrogen, ^1H and D, a molecule like methane, for example, exists as several isotopic combinations.

$^{12}CH_4$	$^{12}CH_3D$	$^{12}CH_2D_2$	$^{12}CHD_3$	$^{12}CD_4$	
	$^{13}CH_4$	$^{13}CH_3D$	$^{13}CH_2D_2$	$^{13}CHD_3$	$^{13}CD_4$
mass: 16	17	18	19	20	21

These molecules have masses ranging from 16 to 21. The mass spectrometer is able to separate these molecules and measure their relative abundance. The mass spectrum of methane shows the molecular ion $M^{+\bullet}$ at $m/z = 16$ plus five other peaks at $m/z = 17, 18, 19, 20$, and 21, called the $M + 1$, $M + 2$, $M + 3$, $M + 4$, and $M + 5$ peaks, respectively. These peaks constitute an **isotopic cluster**. Although all these peaks represent a molecular ion of methane (they all have all the atoms), by convention *we reserve the term "molecular ion," $M^{+\bullet}$, to the molecular ion with the lowest mass*, 16 in the case of methane, and call the others $M + 1$, $M + 2$, and so on.

Of all these peaks found in the molecular ion region of methane, only the $M^{+\bullet}$ and the $M + 1$ peaks have a significant abundance (Fig. 34.3). Because ^{12}C accounts

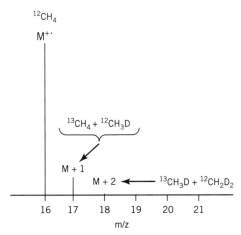

Figure 34.3   Isotopic cluster for methane. The $M+3$, $M+4$, and $M+5$ peaks are too small to show.

for almost 99% of all the carbon in nature and ^{13}C for only about 1%, the abundance of $^{13}CH_4$ (m/z = 17) is only about 1% of the abundance of $^{12}CH_4$ (m/z = 16). $^{12}CH_3D$ (m/z = 17), is present in an even smaller proportion because the natural abundance of deuterium is only 0.015% (Table 34.1). The abundance of species such as $^{13}CH_3D$ (m/z = 18), $^{12}CH_2D_2$ (m/z = 18), $^{13}CH_2D_2$ (m/z = 19), and so on, with two or more of the less abundant isotopes is negligible because the likelihood of two such isotopes being in the same molecule is extremely low.

Inspection of Table 34.1 indicates that the isotopes that contribute to the abundance of the $M+1$ peak are: D, ^{13}C, ^{15}N, ^{17}O, ^{29}Si, and ^{33}S. The contribution of D and ^{17}O can, in principle, be ignored because their abundance is very small (<0.1%). Thus, the abundance of the $M+1$ peak relative to the $M^{+\cdot}$ peak depends on the number of carbon, nitrogen, silicon, and sulfur atoms. If the molecule does not contain nitrogen, silicon, or sulfur, then we can estimate the number of carbon atoms by the relative intensities of the $M+1$ and $M^{+\cdot}$ peaks. Each carbon atom in a molecule increases the percentage of the $M+1$ peak relative to the $M^{+\cdot}$ peak by a factor of 1.1:

$$\frac{RA\ of\ (M+1)}{RA\ of\ M^{+\cdot}} \times 100 = 1.1 \times n^{\circ}C \tag{1}$$

Therefore, the number of carbon atoms can be calculated with the following expression:

$$n^{\circ}C = \frac{RA\ of\ (M+1)}{1.1 \times RA\ of\ M^{+\cdot}} \times 100 \tag{2}$$

When using this equation, we must make sure that the abundances of the $M^{+\cdot}$ and $M+1$ peaks are determined very carefully, otherwise the calculated number of carbon atoms may have an unacceptably large error.

When molecules contain Cl, Br, or S, the $M+2$ peak becomes an important tool in the analysis of the mass spectrum. Its relative abundance can shed light on the molecular formula. Inspection of Table 34.1 shows that the two isotopes of chlorine, ^{35}Cl and ^{37}Cl, are separated by two mass units and have approximate relative abundances of 76% and 24%. This means that molecules with one chlorine atom give two molecular ions, one containing ^{35}Cl ($M^{+\cdot}$), and another containing ^{37}Cl ($M+2$). The relative intensity of these two peaks is 76:24, or approximately 3:1 (Fig. 34.4).

Similarly, a compound with a bromine atom shows two peaks in the molecular ion region separated by two mass units; one for the molecular ion with ^{79}Br, $M^{+\cdot}$, and one for the molecular ion with ^{81}Br, $M+2$. Because the two isotopes of bromine have

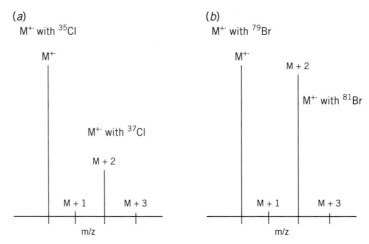

Figure 34.4 *a*) Isotopic cluster for a hypothetical molecule containing one chlorine atom (the M + 1 peak arises from the combination of ^{13}C and ^{35}Cl in the same molecule; the M + 3 arises from ^{13}C and ^{37}Cl); *b*) the same for one bromine.

approximately the same abundance ($\approx$50% each), both the $M^{+\bullet}$ and M + 2 peak have similar intensities (Fig. 34.4).

When the molecule contains two chlorine atoms, three prominent peaks separated by two mass units are observed in the molecular ion region, $M^{+\bullet}$, M + 2, and M + 4. $M^{+\bullet}$ belongs to the molecular ion with two ^{35}Cl, M + 2 has a ^{35}Cl and a ^{37}Cl, and M + 4 has two ^{37}Cl. The relative intensity of these peaks is approximately 10 : 6 : 1. In general, if a molecule contains m chlorine atoms, it gives $m + 1$ peaks separated by two mass units in the molecular ion region. If the molecule contains m chlorine atoms and n bromine atoms, the number of peaks separated by two mass units in the molecular ion region is $m + n + 1$. A chart with the relative intensities of isotopic clusters for different numbers of chlorine and bromine atoms is shown in Figure 34.5.

If a molecule has one sulfur atom (and no chlorine or bromine), the abundance of the M + 2 peak, due to ^{34}S, is approximately 4.4% the abundance of the $M^{+\bullet}$ peak.

Identifying the Molecular Ion

The molecular ion is the heaviest ion that a compound can produce, disregarding isotopic contributions. In principle, this fact should be enough to identify the molecular ion in the mass spectrum. However, there are instances when the identification of the $M^{+\bullet}$ peak is not trivial. Some samples, such as aliphatic alcohols and highly branched aliphatic compounds, give very unstable molecular ions that immediately decompose and do not show up in the mass spectrum. In such cases we should decrease the energy of the electron beam from a typical 70 eV to 10–20 eV. By bombarding the molecule with less-energetic electrons, the molecular ion gains less energy and has a lesser tendency to fragment.

Once a peak has been identified as the possible molecular ion, it must be subjected to some tests. Certain mass losses are virtually impossible from the molecular ion. The molecular ion does not normally lose fragments with masses in the range 4–14 and 21–25, because these masses do not correspond to any likely fragment. Thus, one should not observe peaks in the mass range between M − 4 and M − 14, and between M − 21 and M − 25. If such peaks are observed, it is an indication that the molecular ion was misidentified.

Another test for the molecular ion is the **nitrogen rule**. *When the molecular ion has an even mass, the molecule does not contain nitrogen, or it contains an even number of nitrogen atoms. If the molecular ion has an odd mass, the molecule contains an odd*

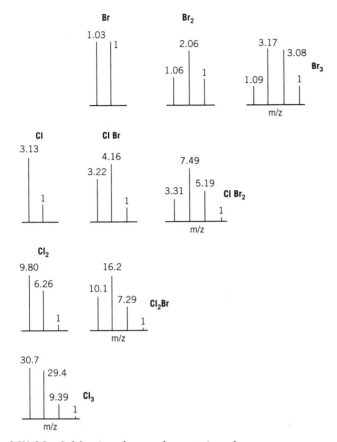

Figure 34.5 $M^{+\cdot}$, M + 2, M + 4, and so on, for organic molecules with different numbers of chlorine and bromine atoms (and no sulfur or silicon). The numbers above the peaks represent the relative abundances of the peaks (taking the abundance of the heaviest peak as 1).

number of nitrogen atoms. This rule arises because of the fortunate coincidence that the most common elements found in organic compounds, with the exception of nitrogen, either have even atomic masses and even valences (^{12}C, ^{16}O, ^{32}S), or odd atomic mass and odd valences (^{1}H, ^{19}F, ^{31}P, ^{35}Cl, ^{79}Br, ^{127}I). Nitrogen has an even atomic mass (^{14}N) and odd valence. A corollary of this rule, which is very useful in the recognition of the molecular ion, is that if we know that *a sample does not contain nitrogen, then we can rule out any odd mass peak as its molecular ion.*

34.4 CHARGE LOCALIZATION

The positive charge and the unpaired electron of the molecular ion drive its fragmentation. Thus, to understand the fragmentations of a mass spectrum, we must first consider the charge localization in the molecular ion. The electron ejected by electron impact belongs to the orbital with the lowest ionization potential, which is usually a nonbonding orbital with lone electron pairs. These electrons are ejected preferentially to those in pi bonds. In turn, pi bonds are ionized with preference to sigma bonds.

If the molecule contains several heteroatoms with lone electron pairs, the electron is preferentially, but not exclusively, ejected from the least electronegative heteroatom. For example, in a molecule containing nitrogen and oxygen, such as ethanolamine, nitrogen is more likely to get ionized than oxygen, and the fragmentations are driven by the presence of the radical cation on the nitrogen atom, as we will see later.

$$H_2\ddot{N}-CH_2CH_2\ddot{O}H + e^- \longrightarrow H_2\overset{\cdot+}{\ddot{N}}-CH_2CH_2\ddot{O}H + 2\,e^-$$

Carbon-carbon double bonds and aromatic rings are preferentially ionized in comparison to C—C single bonds. For isolated carbon-carbon double bonds, the radical cation is localized at the position of the double bond.

$$CH_3CH\overset{\cdot\cdot}{\underline{}}CH_2 + e^- \longrightarrow CH_3CH\overset{\cdot+}{\underline{}}CH_2 + 2\,e^-$$

Two different resonance forms are possible; $CH_3CH\overset{\cdot+}{\underline{}}CH_2$ can be represented as:

$$\left[CH_3\overset{+}{C}H-\overset{\cdot}{C}H_2 \longleftrightarrow CH_3\overset{\cdot}{C}H-\overset{+}{C}H_2\right]$$

In the case of aromatic rings and conjugated double bonds, the radical cation is delocalized in the entire pi system. For phenyl rings, this is usually indicated by the radical cation symbol inside a circle.

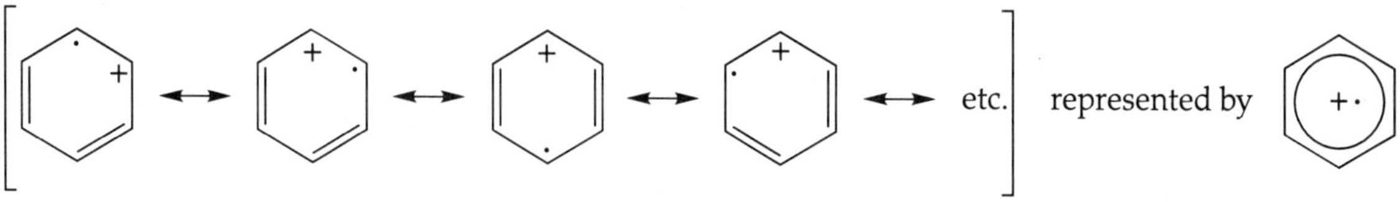

The ionization of molecules without heteroatoms, or double or triple bonds, such as alkanes, can take place in *any* of the sigma bonds. The radical cation in such cases is often represented by the +· symbol outside a bracket that embraces the whole molecule, as shown below for *n*-pentane.

$$[CH_3CH_2CH_2CH_2CH_3]^{+\cdot}$$

34.5 FRAGMENTATIONS

The peaks of a mass spectrum are generated by different types of fragmentation pathways. Most fragmentations can be classified as belonging to one of the following categories: **α-cleavage**, **C-heteroatom cleavage**, and **multibond cleavage**. Only the most important fragmentations will be discussed here (for a more complete treatment of the subject, see Refs. 1, 2, 12–14).

α-Cleavage

The fragmentation of the molecular ion can occur by either homolytic or heterolytic cleavage, or a combination of both. One of the most important mechanisms of fragmentation is the so-called **homolytic α-cleavage**, where *a sigma bond, one bond away from the atom bearing the radical cation, breaks homolytically.* For example, the molecular ion of diethyl ether undergoes the following cleavage:

$$\overset{\frown}{CH_3}\cdot\big\{\cdot CH_2\overset{\frown}{\underline{}}\overset{\cdot+}{O}CH_2CH_3 \longrightarrow CH_2=\overset{+}{O}CH_2CH_3 + CH_3\cdot$$

m/z = 74 (30%)   m/z = 59 (47%)

The carbon-carbon bond between the methyl and the methylene breaks homolytically with each carbon atom taking one electron. The electron on the methylene pairs up with the electron on the oxygen to form a new bond. This cleavage produces a methyl radical (which is not detected by the mass spectrometer because it has no charge) and a positively charged ion with mass 59. Homolytic cleavages are represented by "fishhook" half-arrows. To avoid clutter, usually only one arrow is drawn:

$$CH_3 \overset{\frown}{\text{---}} CH_2 \overset{\cdot+}{\text{---}} \overset{\cdot+}{O}CH_2CH_3 \longrightarrow CH_2 \overset{+}{=} \overset{+}{O}CH_2CH_3 + CH_3^{\cdot}$$

In the fragmentation of diethyl ether, a hydrogen atom instead of a methyl radical can also be lost in a homolytic α-cleavage. This reaction occurs to a lesser extent than the loss of a methyl radical, as indicated by the lower relative abundance of the product (3%):

$$CH_3CH \overset{\cdot+}{\text{---}} \overset{\cdot+}{O}CH_2CH_3 \longrightarrow CH_3CH \overset{+}{=} \overset{+}{O}CH_2CH_3 + H^{\cdot}$$

| |
H

$m/z = 74$ (30%) $\qquad\qquad\qquad m/z = 73$ (3%)

In general, when two different radicals can be lost in a homolytic α-cleavage, *the larger radical is lost preferentially*. This is illustrated in the mass spectrum of butanone (Fig. 34.1). The radical cation can lose either a methyl or an ethyl radical. Although both fragmentations occur, the loss of an ethyl radical is preferred. The cations obtained in these fragmentations of butanone are called **acylium ions**:

$$CH_3CH_2 \overset{\overset{\cdot+}{O}}{\underset{a\,\|\,b}{\overset{\|}{\text{---}}C\text{---}}} CH_3$$

$m/z = 72$ (16%)

$a \nearrow \quad CH_3C \overset{+}{\equiv} \overset{+}{O} + CH_3CH_2^{\cdot}$

an acylium ion
$m/z = 43$ (100%)

$b \searrow \quad CH_3CH_2C \overset{+}{\equiv} \overset{+}{O} + CH_3^{\cdot}$

an acylium ion
$m/z = 57$ (5%)

Another example of homolytic α-cleavage is provided by the mass spectrum of 2-butylamine, where three different such fragmentations occur:

$$CH_3CH_2 \overset{\overset{H}{|}}{\underset{\underset{\underset{+\cdot}{NH_2}}{|}}{\overset{|}{\underset{a}{\text{---}}}\overset{c}{\underset{\quad}{\text{C}}\overset{\quad}{\text{---}}}}} CH_3$$

$m/z = 73$ (1%)

$a \nearrow \quad CH_3CH \overset{+}{=} \overset{+}{N}H_2 + CH_3CH_2^{\cdot}$

$m/z = 44$ (100%)

$b \longrightarrow \quad CH_3CH_2CH \overset{+}{=} \overset{+}{N}H_2 + CH_3^{\cdot}$

$m/z = 58$ (11%)

$c \searrow \quad CH_3CH_2CCH_3 + H^{\cdot}$

$\quad\quad\quad\; \overset{||}{\underset{+}{N}H_2}$

$m/z = 72$ (1%)

The fragmentation of this molecule also illustrates that the largest radical ($CH_3CH_2^{\cdot}$ in this case) is lost preferentially.

Another example of homolytic α-cleavage is found in the fragmentation of toluene, which loses $H^{\cdot}$ to produce the benzyl cation:

$m/z = 92$ (65%)

benzyl cation
$m/z = 91$ (100%)

The benzyl cation undergoes a rearrangement reaction in the mass spectrometer, giving a seven-member ring, called the **tropilium ion**, which is fully aromatic. The mechanism of the rearrangement is complex and will not be discussed here. Being an aromatic species, the tropilium ion is relatively stable and usually gives rise to an intense m/z 91 peak.

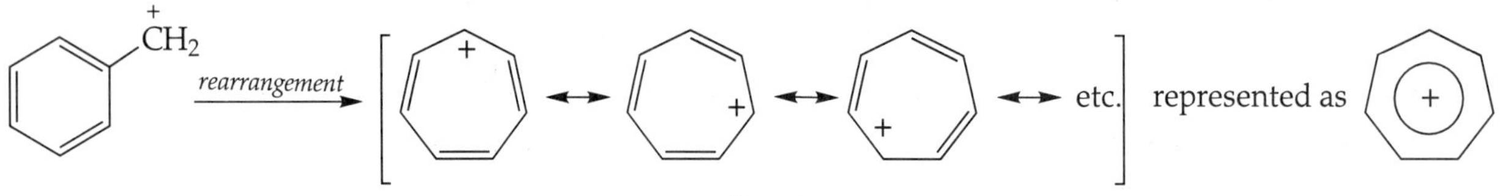

tropilium ion

Many other functional groups such as alkenes, alcohols, aldehydes, alkyl halides, carboxylic acids, carboxylic esters, and amides undergo homolytic α-cleavage reactions.

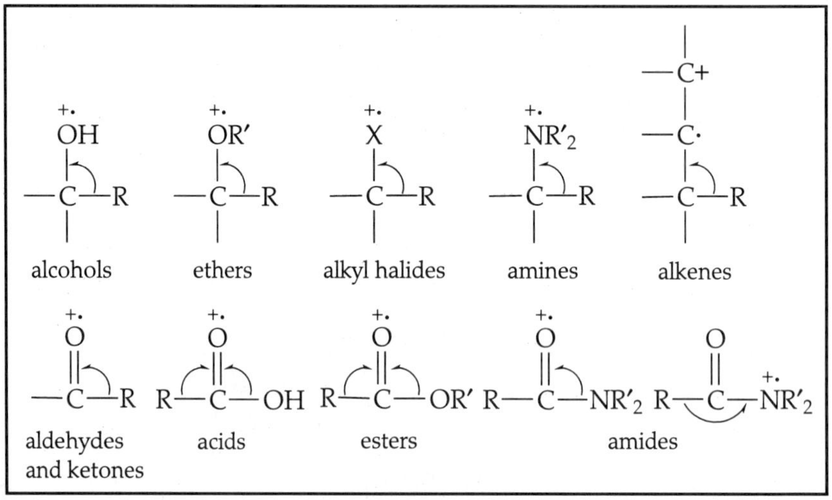

In the case of amides, the ionization takes place at either the oxygen or the nitrogen atom. The ionization of carboxylic acids and esters can, in principle, take place either at the carbonyl oxygen or the alcohol oxygen. The ionization of the carbonyl oxygen, however, is more prominent and drives the fragmentation of these compounds.

There is another type of α-cleavage: **heterolytic α-cleavage**. This fragmentation can take place directly from the molecular ion itself, or from a cation produced by the molecular ion. The heterolytic α-cleavage is an important fragmentation process in carbonyl compounds. It is illustrated below with the fragmentation of the molecular ion of *n*-butanal:

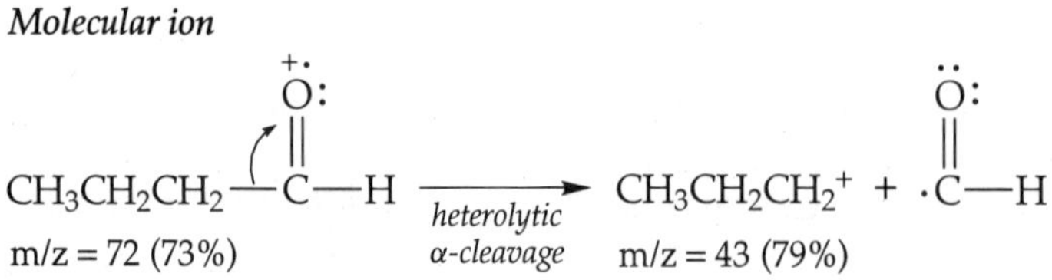

Molecular ion

To visualize this reaction and keep track of the electron flow, you can think that the carbon of the carbonyl takes the two electrons from the C—C bond that breaks; it keeps one electron and becomes a radical, and gives the other one to the oxygen, which completes its octet and becomes neutral.

The heterolytic α-cleavage also occurs from acylium ions produced in the fragmentation of carbonyl compounds. The driving force in this reaction is the elimination of a stable carbon monoxide molecule:

Molecular ion   *Acylium ion*

$$CH_3CH_2CH_2-\overset{\overset{+\cdot}{O}}{\underset{}{\underset{\|}{C}}}-H \xrightarrow[\substack{homolytic \\ \alpha\text{-}cleavage}]{-H^\cdot} CH_3CH_2CH_2-C\equiv\overset{+}{O} \xrightarrow[\substack{heterolytic \\ \alpha\text{-}cleavage}]{} CH_3CH_2CH_2^+ + CO$$

m/z = 72   m/z = 71 (7%)   m/z = 43
(73%)   (79%)

Carbocations formed in heterolytic α-cleavages can undergo a subsequent heterolytic α-cleavage. For example, an *n*-butyl carbocation breaks to give ethylene and ethyl carbocation. The driving force of this reaction is the elimination of the stable ethylene molecule:

$$CH_3CH_2-CH_2-\overset{+}{C}H_2 \longrightarrow CH_3CH_2^+ + CH_2=CH_2$$

m/z = 57   m/z = 29

C-Heteroatom Cleavage

Alkyl halides undergo a **heterolytic** cleavage between the halogen atom and the carbon atom. This heterolytic cleavage leaves the positive charge on the carbon atom, and produces a halogen atom (a radical).

$$CH_3\overset{\overset{CH_3}{|}}{\underset{\underset{CH_3}{|}}{C}}-\overset{+\cdot}{Cl} \longrightarrow CH_3\overset{\overset{CH_3}{|}}{\underset{\underset{CH_3}{|}}{C}}+ + Cl^\cdot$$

m/z = 92; 94   m/z = 57 (100%)
negligible abundance

A similar fragmentation occurs in ethers:

$$CH_3CH_2-\overset{+\cdot}{O}CH_2CH_3 \longrightarrow CH_3CH_2^+ + {}^\cdot OCH_2CH_3$$

m/z = 74 (30%)   m/z = 29 (40%)

It also occurs in alcohols and amines, although to a lesser extent.

$$CH_3\overset{\overset{CH_3}{|}}{\underset{\underset{CH_3}{|}}{C}}-\overset{+\cdot}{O}H \longrightarrow CH_3\overset{\overset{CH_3}{|}}{\underset{\underset{CH_3}{|}}{C}}+ + HO^\cdot$$

m/z = 74 (<1%)   m/z = 57 (9%)

$$CH_3\overset{\overset{CH_3}{|}}{\underset{\underset{CH_3}{|}}{C}}-\overset{+\cdot}{N}H_2 \longrightarrow CH_3\overset{\overset{CH_3}{|}}{\underset{\underset{CH_3}{|}}{C}}+ + H_2N^\cdot$$

m/z = 73 (<1%)   m/z = 57 (6%)

Multibond Cleavage

Some fragments are produced by the cleavage of two or more bonds. These multibond cleavages can occur with or without migration of atoms or **rearrangement**. Let's first consider **multibond cleavages without rearrangement**.

Cycloalkenes undergo a double homolytic cleavage of the allylic C—C bonds:

m/z = 82 (40%)   m/z = 54 (80%)

This fragmentation, which resembles a **retro Diels–Alder** reaction (Unit 19), produces an alkene (ethylene in this example) and the radical cation of a conjugated diene. This fragmentation also occurs in substituted cycloalkenes and can be used to determine the position of the substituent relative to the double bond. For example, the retro Diels–Alder fragmentation of the following substituted isomeric cycloalkenes leads to different ions:

m/z = 96 (10%)   m/z = 54 (59%)

m/z = 96 (100%)   m/z = 68 (45%)

Among the **multibond cleavages with rearrangement**, the **McLafferty rearrangement** is the most important. It takes place in carbonyl compounds where a hydrogen atom on the γ carbon with respect to a carbonyl migrates to the carbonyl, while the α, β carbon-carbon bond breaks homolytically. The McLafferty rearrangement is illustrated with n-butanal:

m/z = 72 (73%)   m/z = 44 (100%)

The McLafferty rearrangement takes place through a six-membered transition state. If there are substituents on the β and γ carbons, longer alkenes (instead of ethylene) are produced. For example, in the McLafferty rearrangement of methyl n-pentanoate, propene is eliminated:

m/z = 116 (1%)   m/z = 74 (100%)

The McLafferty rearrangement occurs in a variety of carbonyl compounds such as aldehydes, ketones, carboxylic esters, carboxylic acids, and amides. A similar rearrangement involving a six-membered transition state and the elimination of an alkene also occurs in phenylalkanes with a hydrogen atom on a γ carbon of the side chain:

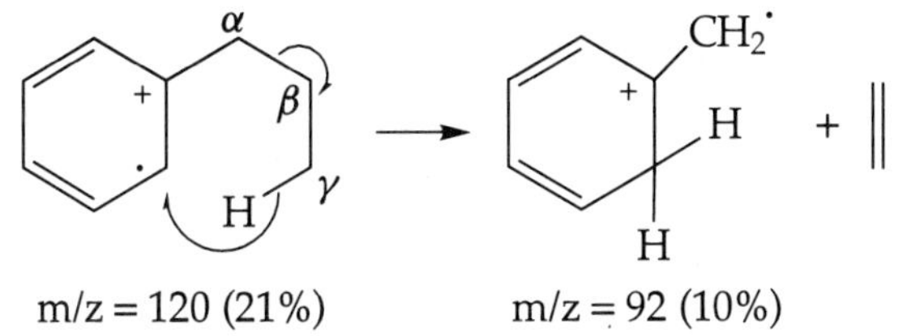

Another rearrangement mechanism with loss of an alkene takes place in amines and alcohols. After the homolytic α-cleavage from the molecular ion, the resulting cation undergoes, through a **four-center mechanism**, *a hydrogen migration and an alkene elimination*, giving a smaller cation:

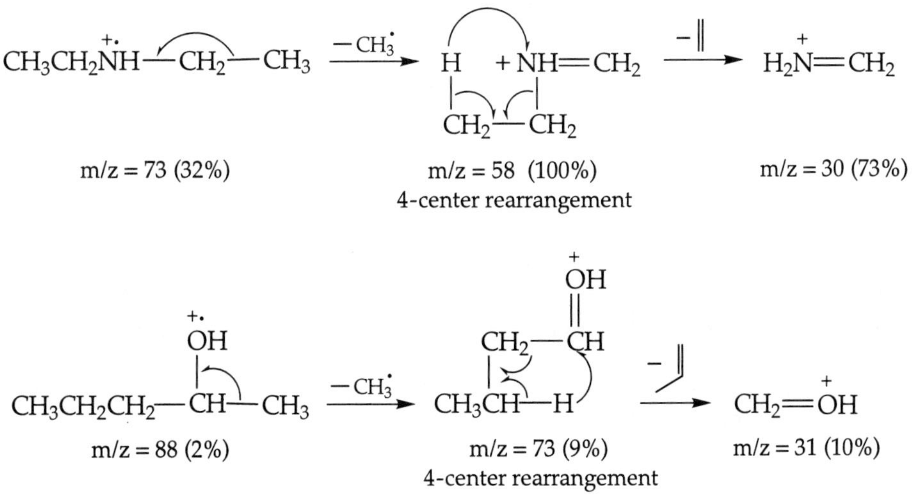

Neutral molecules can be lost from aromatic compounds by multibond cleavages. The tropilium ion and the phenyl cation lose acetylene (HC≡CH) to give noticeable peaks at m/z = 65 and 51, respectively:

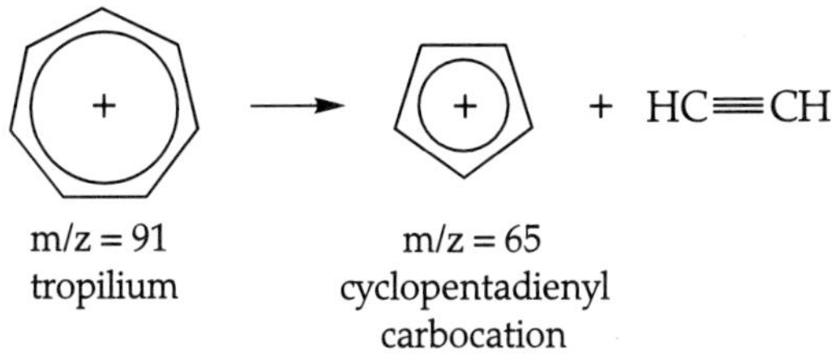

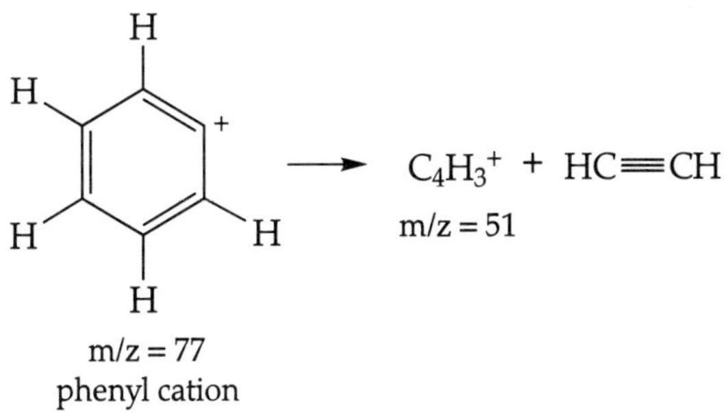

The presence of peaks at m/z = 91 and 65 or at m/z = 77 and 51 in the mass spectrum of an unknown compound is an indication of the presence of a phenyl ring.

Phenols lose carbon monoxide from their molecular ions. This loss is usually followed by a loss of hydrogen to give the cyclopentadienyl carbocation (m/z = 65):

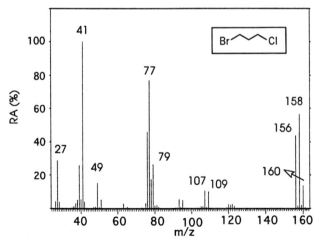

m/z = 94   m/z = 66   m/z = 65
(100%)   (28%)   (22%)

Aromatic nitrocompounds lose a molecule of NO (a radical) from the molecular ion in a reaction with oxygen rearrangement. This loss is illustrated with *p*-nitroaniline.

m/z = 138   m/z = 108
(100%)   (30%)

34.6 MS CASE STUDIES

Let's see how we can apply the concepts discussed in the previous sections to the interpretation of the mass spectra of known compounds.

Example 1

We will first consider the mass spectrum of 1-bromo-3-chloropropane shown in Figure 34.6. The isotopic cluster in the molecular ion region shows peaks at m/z 156 (44%; M$^{+\cdot}$), 158 (56%; M + 2), and 160 (14%; M + 4) in agreement with a chlorine

Figure 34.6   Mass spectrum of 1-bromo-3-chloropropane.

and a bromine in the molecule. The expected relative intensities for the $M^{+\cdot}$, $M + 2$ and $M + 4$ peaks for such a molecule, according to Figure 34.5, are $3.22 : 4.16 : 1$. This agrees with the observed values of $44\% : 56\% : 14\%$ or $3.14 : 4 : 1$.

The charge can be localized in either the chlorine or the bromine atom, and the fragmentation of this molecule is driven by both atoms. Two different *homolytic α-cleavages* occur. One leads to a fragment containing chlorine (m/z 49 and 51) and the other to a fragment with bromine (m/z 93 and 95). The relative abundance of the fragments m/z 49 and 51 is 15% and 5% (or 3:1), in agreement with the presence of one chlorine atom (Fig. 34.5). The relative abundance of the fragments m/z 93 and 95 is 5% and 5% (or 1:1), indicating the presence of one bromine.

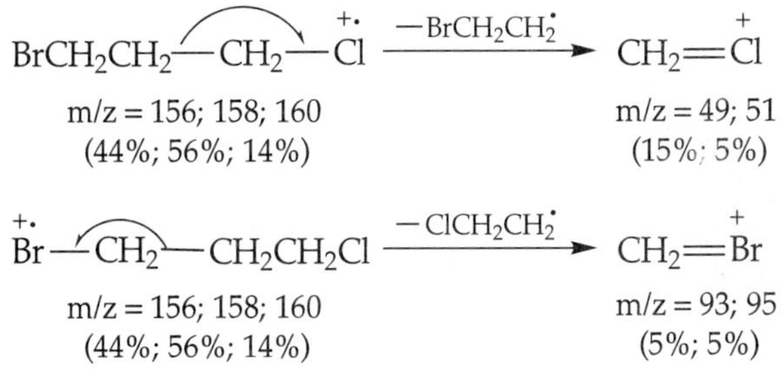

There are two possible *heterolytic α-cleavages* as shown below: the initial loss of the bromine from the molecular ion, or the initial loss of the chlorine atom. The loss of Br· results in the cations m/z 77 and 79. The loss of Cl· gives the fragments m/z 121 and 123. These cations then lose a molecule of either HCl or HBr to form the allyl carbocation (m/z 41), which is the base peak of the spectrum.

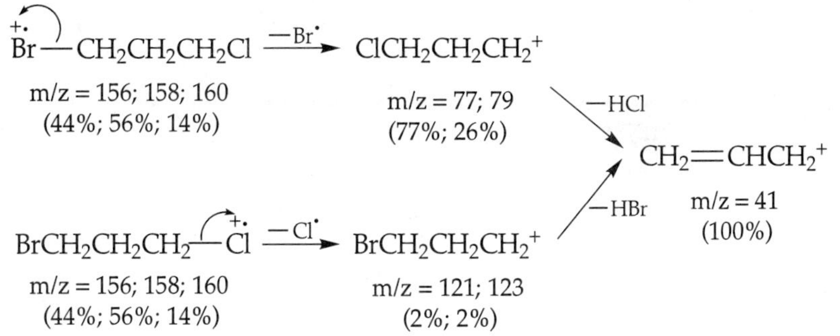

Example 2

The next example is the mass spectrum of *n*-butyrophenone (*n*-propyl phenyl ketone) (Fig. 34.7). The molecular ion (m/z 148) undergoes a *McLafferty rearrangement* with loss of ethylene and gives the fragment m/z 120:

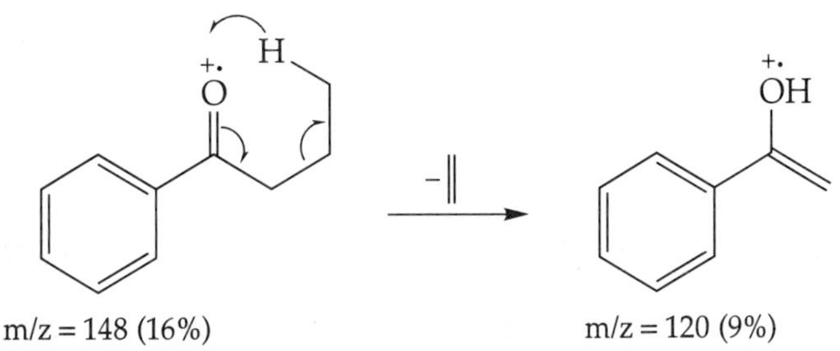

The molecular ion also loses a propyl radical by a *homolytic α-cleavage*, producing the benzoyl cation (m/z 105). This fragment then loses CO by a heterolytic

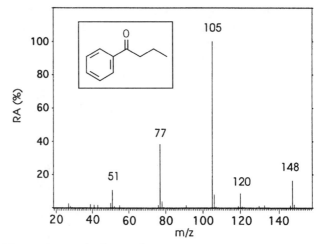

Figure 34.7   Mass spectrum of *n*-butyrophenone.

α-cleavage, giving phenyl cation (m/z 77), which loses acetylene to give the fragment at m/z 51:

m/z = 148	m/z = 105	m/z = 77	m/z = 51
(16%)	(100%)	(38%)	(11%)

$-CH_3CH_2CH_2^{\bullet}$ $-CO$ $-HC\equiv CH$ $C_4H_3^+$

34.7   INTERPRETING MASS SPECTRA

The first step in the interpretation of a mass spectrum is to find the molecular ion peak, if present. Aromatic compounds, alkenes, aldehydes, ketones, amines, esters, and ethers give intense, or at least recognizable, molecular ions. The molecular ion peak may be absent in the case of alcohols and highly branched aliphatic hydrocarbons.

If the compound in question is an unknown, the molecular formula can be calculated from the molecular ion. There are useful compilation tables that list all possible molecular formulas for a given molecular mass (Ref. 12). For example, the molecular formulas possible with a mass of 70 are: $C_2H_2N_2O$; $C_3H_2O_2$; $C_3H_6N_2$; C_4H_6O; and C_5H_{10}.

The use of the relative intensity of the M + 1 peak (Eq. 2), may also be helpful in calculating the molecular formula, or at least the number of carbon atoms. This calculation may not lead to an unequivocal molecular formula, but it will narrow the possibilities to a handful.

One should also consider the M + 2 peak. If the M + 2 peak is almost as intense as the $M^{+\bullet}$ peak, this is a clear indication of a bromine atom. If the intensity of the M + 2 peak is about one-third that of the $M^{+\bullet}$ peak, this is indicative of chlorine. If the abundance of the M + 2 peak is about 4% that of the $M^{+\bullet}$ peak, one should suspect the presence of one sulfur atom.

Making a list of the most important peaks of the spectrum is also very useful. The student may ask: How can we recognize the important peaks? Is there any particular value of the relative abundance above which a peak becomes important? The answer is no. Two peaks with the same intensity, but one at low m/z values and the other at high m/z values, do not necessarily have the same importance in the process of structure elucidation. *Peaks at high m/z values are usually more revealing than those at low masses.* These high-mass peaks arise from the loss of specific fragments from the molecular ion and often reveal the presence of certain functional groups. For example, a loss of 19 mass units from the molecular ion is indicative of fluorine; a loss of 35 mass units suggests chlorine, and a loss of 18 mass units corresponds to a loss of water, which is typical of alcohols. It is very helpful to calculate the mass difference between the molecular ion and the other peaks of the spectrum, especially those at high m/z values. Table 34.2 summarizes some of the most important fragments lost by the molecular ion.

Table 34.2 Some Common Fragmentations from the Molecular Ion

Fragment	Loss of	Typical compounds
M − 1	H·	hydrocarbons, aldehydes, general
M − 2	H_2	general
M − 15	CH_3·	general
M − 16	H_2N·	anilines
M − 17	HO·	acids
	NH_3	1° amines
M − 18	H_2O	alcohols, some acids
M − 19	F·	fluorocompounds
M − 20	HF	fluorocompounds
M − 26	HC≡CH	aromatic compounds
M − 27	HC≡N	nitriles
M − 28	CO	phenols
	$H_2C=CH_2$	retro Diels–Alder; McLafferty
M − 29	HCO·	phenols, aromatic aldehydes
	CH_3CH_2·	general
M − 30	$H_2C=O$	methyl ethers
	NO·	aromatic nitrocompounds
M − 31	CH_3O·	methyl esters; methyl ethers
M − 32	CH_3OH	methyl esters; methyl ethers
M − 33	CH_3· + H_2O	alcohols
	HS·	thiols
M − 34	H_2S	thiols
M − 35	Cl·	chlorocompounds
M − 36	HCl	chlorocompounds
M − 42	$CH_3CH=CH_2$	retro Diels–Alder; McLafferty
	$CH_2=C=O$	acetamides
M − 43	C_3H_7·	general
	CH_3CO·	methyl ketones
M − 44	CO_2	acid anhydrides
M − 45	CH_3CH_2O·	ethyl esters; ethyl ethers
M − 46	NO_2·	nitrocompounds

The actual m/z values of the most prominent peaks of the spectrum can also be very revealing. One should try to match the most abundant peaks of the spectrum with the typical fragments listed in Table 34.3. Fragments commonly observed in the mass spectrum can be characterized as belonging to one of the series shown in the table.

Table 34.3   **Typical MS Fragments**

Series	Mass	Example
$C_nH_{2n+1}^+$ (alkyl)	15; 29; 43; 57; 71; 85; etc.	CH_3^+; $CH_3CH_2^+$; etc.
$C_nH_{2n-1}^+$ (alkenyl)	27; 41; 55; 69; 83; etc.	$CH_2{=}CH^+$; $CH_2{=}CHCH_2^+$; etc.
$C_nH_{2n+1}O^+$ (oxygen)	31; 45; 59; 73; 87; etc.	$CH_2{=}\overset{+}{O}H$; $CH_3CH{=}\overset{+}{O}H$; etc.
$C_nH_{2n+1}CO^+$ (carbonyl)	29; 43; 57; 71; 85; etc.	$HC{\equiv}O^+$; $CH_3C{\equiv}O^+$; etc.
$C_6H_5C_nH_{2n}^+$ (aromatic)	77; 91; 105; 119; etc.	$C_6H_5^+$; $C_7H_7^+$; etc.

Inspection of Table 34.3 indicates that the alkyl and carbonyl series have fragments with the same mass. This coincidence arises because the mass of the carbonyl, 28, is equal to the mass of two methylene groups.

34.8   HIGH-RESOLUTION MASS SPECTRA

So far we have only considered round-off atomic masses: 1 for hydrogen, 14 for nitrogen, 16 for oxygen, and so on. With this level of approximation, a molecule of ethylene ($^{12}C_2{}^1H_4$) and a molecule of carbon monoxide ($^{12}C^{16}O$) have the same mass (28 mass units). However, when the decimal figures are considered these two molecules have slightly different masses. Using the values listed in Table 34.1, the following masses are obtained:

$$
\begin{array}{ll}
^{12}C_2{}^1H_4: & 2 \times {}^{12}C: \quad 24.0000 \\
 & 4 \times {}^1H: \quad \underline{4.0313} \\
 & \text{total:} \quad 28.0313
\end{array}
\qquad
\begin{array}{ll}
^{12}C^{16}O: & 1 \times {}^{12}C: \quad 12.0000 \\
 & 1 \times {}^{16}O: \quad \underline{15.9949} \\
 & \text{total:} \quad 27.9949
\end{array}
$$

Simple mass spectrometers are unable to detect such small mass differences, and give the mass of these two molecules as 28. These are called **low-resolution mass spectrometers**. More sophisticated spectrometers can distinguish up to the fourth decimal place. With these **high-resolution mass spectrometers** molecular formulas can be unequivocally determined. For example, a molecular ion was measured at m/z = 70.0535 ± 0.0005. There are five possible molecular formulas with C, H, O, and N and a mass of 70, as we have seen in the previous section; however, only one matches the measured molecular mass: $C_3H_6N_2$.

Molecular formula	Mass
$C_2H_2N_2O$	70.0167
$C_3H_2O_2$	70.0055
$C_3H_6N_2$	**70.0531**
C_4H_6O	70.0419
C_5H_{10}	70.0783

BIBLIOGRAPHY

1. Interpretation of Mass Spectra. F.W. McLafferty and F. Turecek. 4th ed. University Science Books, Mill Valley, Calif. (1993).

2. Mass Spectrometry of Organic Compounds. H. Budzikiewicz, C. Djerassi and D.H. Williams. Holden-Day, San Francisco (1967).

3. Practical Organic Mass Spectrometry. A Guide for Chemical and Biochemical Analysis. J.R. Chapman. 2nd ed. Wiley, Chichester (1993).

4. Mass Spectrometry. Principles and Applications. E. de Hoffmann, J. Charette, and V. Stroobant; translated by J. Trottier and the authors. Wiley, Chichester (1996).

5. Mass Spectrometry for Chemists and Biochemists. R.A.W. Johnstone and M.E. Rose. 2nd ed. Cambridge University Press, Cambridge (1996).

6. Introduction to Mass Spectrometry. J.T. Watson. 3rd ed. Lippincott-Raven, Philadelphia (1997).

7. Gas Chromatography and Mass Spectrometry. A Practical Guide. F.G. Kitson, B.S. Larsen, and C.N. McEwen. Academic Press, San Diego (1996).

8. Electrospray Ionization Mass Spectrometry. Fundamentals, Instrumentation, and Applications. Richard B. Cole, ed. Wiley, New York (1997).

9. Atlas of Mass Spectral Data. E. Stenhagen, S. Abrahamsson and F.W. McLafferty, ed. Interscience Publishers, New York (1969).

10. Registry of Mass Spectral Data. E. Stenhagen, S. Abrahamsson and F.W. McLafferty. Wiley, New York (1974).

11. Spectroscopic Methods in Organic Chemistry. D.H. Williams. I. Fleming. 5th ed. McGraw-Hill, London (1995).

12. Spectrometric Identification of Organic Compounds. R.M. Silverstein, G.C. Bassler and T.C. Morrill. 5th ed. Wiley, New York (1991).

13. Spectroscopic Methods in Organic Chemistry. M. Hesse, H. Meier, and B. Zeeh; translated by A. Linden and M. Murray. Thieme, New York (1997).

14. Organic Structure Analysis. P. Crews, J. Rodríguez, and M. Jaspars. Oxford University Press, New York (1998).

Problem 1

EXERCISES

Propose a structure in agreement with the following mass spectrum. Justify the most important peaks.

m/z	RA (%)	m/z	RA (%)	m/z	RA (%)
26	2.0	50	5.0	89	5.8
27	2.3	51	6.0	90	2.5
37	2.6	52	1.6	91	100
38	4.2	61	1.9	92	7.9
39	11.2	62	4.2	99	1.0
40	1.0	63	9.5	125	2.9
41	1.5	64	2.5	126	19.3
43	1.2	65	10.2	127	2.6
44	1.0	73	11.0	128	6.2
45	3.1	75	1.0	129	0.5
49	1.2	77	0.5		

Problem 2

A compound shows IR bands at 1715; 2720, and 2820 cm^{-1}. Its mass spectrum is shown below. Propose a structure for the unknown. Justify your answer.

m/z	RA (%)	m/z	RA (%)	m/z	RA (%)
25	1.2	38	5.4	53	1.0
26	8.7	39	25.7	55	1.5
27	69.3	40	3.1	56	0.5
28	14.8	41	69.0	57	2.5
29	45.1	42	8.8	71	1.0
30	1.0	43	100	72	35.5
31	1.5	44	4.4	73	1.8
37	3.5	45	1.6	74	0.1

Problem 3

Propose a structure in agreement with the following mass spectrum. Justify your answer.

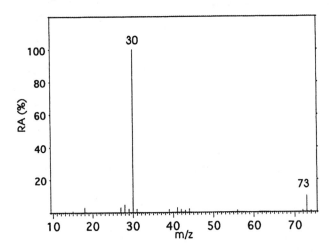

Problem 4

Propose a structure in agreement with the mass spectrum shown in the following figure. Justify your answer.

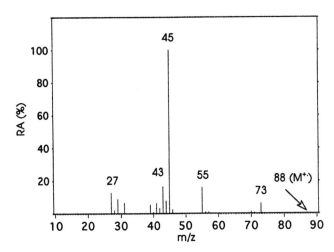

Problem 5

A compound with a smell that resembles pineapples gives the mass spectrum shown in the following figure. Propose a structure and justify your answer.

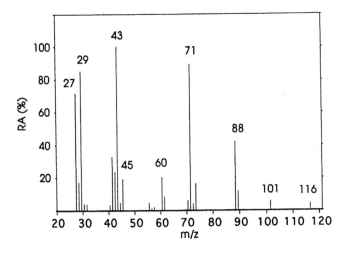

Answers to Odd-Numbered Exercises

UNIT 1. *Safety*

Problem 1

(a) Safety standards violations: Lab is in a mess. Nobody wears safety goggles. The alchemist wears slippers. Lab is obstructed with equipment lying on the floor. Children are playing in the lab. Most containers are not labeled. Most containers are open. Fume hood lacks a sash. Open flame is used indiscriminately.

(b) Instruments and apparatus: retort for distillation; crucibles; tongs; spatulas; measuring cups; hourglass; bellows; scales.

Operations being performed: distillation (the retort on the left side of the fume hood); fusion (in the crucible in the fume hood); evaporation (the big pot boiling in the fume hood).

Problem 3

1: c; 2: b; 3: b; 4: b; 5: c; 6: a; 7: c; 8: d.

Problem 5

(a) P.E.L.: permissible exposure limit; MSDS: material safety data sheet; LD_{50}: lethal dose 50; LC_{50}: lethal concentration 50.

(b) Carcinogenic $\longrightarrow$ malignant-tumor causing; teratogenic $\longrightarrow$ malformation of embryo; mutagenic $\longrightarrow$ heritable genetic changes.

UNIT 2. *Basic Concepts*

Problem 1

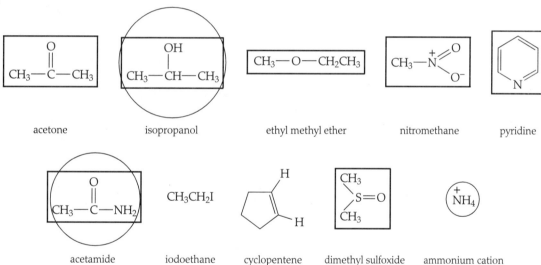

Problem 3

(a)

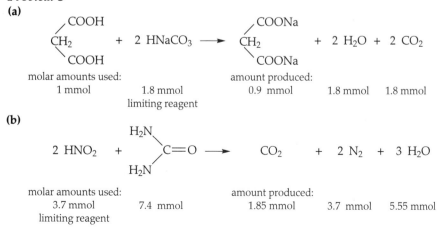

molar amounts used:

| 1 mmol | 1.8 mmol | 0.9 mmol | 1.8 mmol | 1.8 mmol |

limiting reagent

(b)

molar amounts used:

| 3.7 mmol | 7.4 mmol | 1.85 mmol | 3.7 mmol | 5.55 mmol |

limiting reagent

(c)

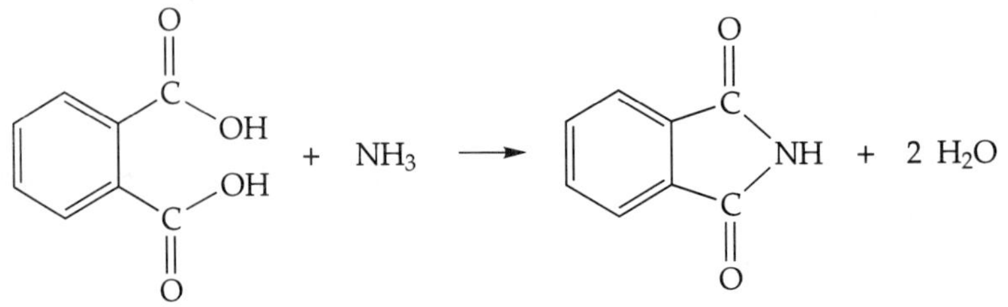

molar amounts used:
2.3 mmol   5.2 mmol
limiting reagent

amount produced:
2.3 mmol   4.6 mmol

(d)

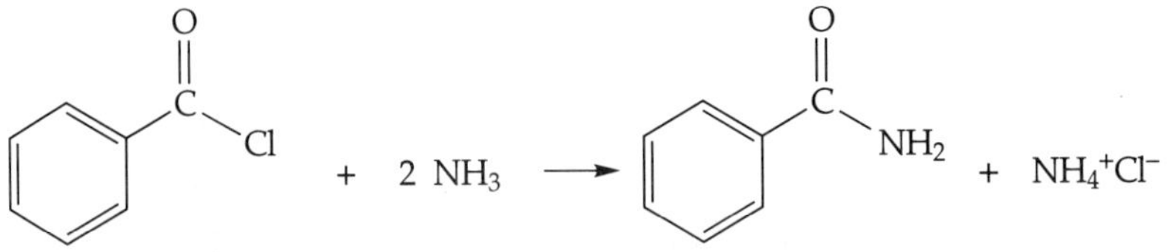

molar amounts used:
5.7 mmol   10.3 mmol
                  limiting reagent

amount produced:
5.15 mmol   5.15 mmol

(e)

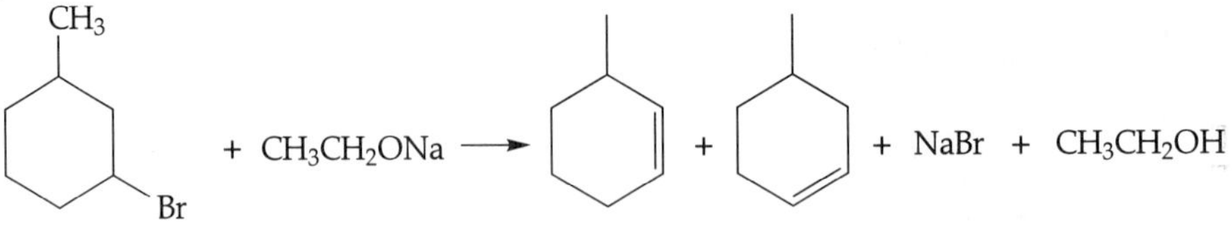

molar amounts used:

2.3 mmol   15.0 mmol
limiting reagent

amount produced:
2.3 mmol   2.3 mmol   2.3 mmol

Problem 5

The data are gathered in the table in the rows above the double line. The figures below the double line have been calculated.

	1-Methyl-cyclohexene	HBr (solution)	HBr (pure)	1-bromo-1-methyl-cyclohexane
volume (mL)	0.970	3.0	—	—
concentration	—	48% (w / w)	—	—
density (g / mL)	0.809	1.49	—	—
MM (g / mol)	96.17	—	80.92	177.09
mass (g)	0.785	4.47	2.15	1.10
mmoles	8.16	—	26.5	6.21

The limiting reagent is 1-methylcyclohexene. The theoretical yield of the synthesis is 8.16 mmol (1.445 g of product). The % yield is 76.1%.

UNIT 4. *Recrystallization and Melting Point*

Problem 1

Solvent III is the best choice to carry out the recrystallization. It displays the largest solubility difference between high and low temperature.

Problem 3

(a) Ethanol; **(b)** hexane; **(c)** water; **(d)** hexane or isopropanol.

Problem 5
Solubility = 0.29 g/100 mL.

Problem 7
Salt is separated in the cold filtration. Sand is removed in the hot filtration.

Problem 9
Y and Z are the same compound, X is different; 2:1 mixture of X and Z melts at 105–130°C.

UNIT 5. *Extraction*

Problem 1
Diethyl ether, petroleum ether, cyclohexane, toluene: top layers; methylene chloride and chloroform: bottom layers.

Problem 3
(a) $S_W = 0.0033$ g/mL; $S_C = 0.0588$ g/mL.

(b) $K_{C/W} \approx S_C/S_W = 17.8$.

(c) mass aspirin$_{original}$ = mass asprin$_{chlor}$ + mass aspirin$_{water}$ = 180 mg + 10.1 mg = 190.1 mg

Problem 5
1-Octanol is much less polar than water. As the length of the fluorocarbon chain increases, the polarity decreases, and since *like-dissolves-like*, the alcohols with longer fluorinated chains have higher affinity for the less polar solvent, 1-octanol, and less for water.

Problem 7

$$K_{C/W} \approx S_C/S_W = 18.1.$$

The partition coefficient for cholesterol under the same conditions is $K_{C/W} = 1.1\,10^5$ (Problem 6). Then cholic acid has more affinity for water and less for chloroform than cholesterol. Cholic acid is more polar than cholesterol (see structures in section 25.6).

Problem 9
$K_{C/W} = 50$; $K_{W/E} = 0.1$, therefore: $K_{E/W} = 10$; $K_{B/W} = 4$.
   The largest partition coefficient for acetanilide is between chloroform and water. So chloroform would be the best choice.

UNIT 6. *Distillation and Boiling Point*

Problem 1

atm	mm Hg	Torr	psi	pascal
1	760	760	14.6960	$1.01325\,10^5$
0.592	450	450	8.70	$0.600\,10^5$
$1.3\,10^{-7}$	10^{-4}	10^{-4}	$1.9\,10^{-6}$	$1.3\,10^{-2}$

Problem 3
(a)

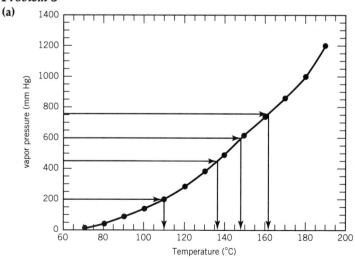

(b) 162°C. (c) 600 mm Hg: 148°C; 0.592 atm (450 mm Hg): 136°C; 200 Torr (200 mm Hg): 110°C.

Problem 5

(a) A schematic vapor-liquid phase diagram for the system is shown in the figure below.

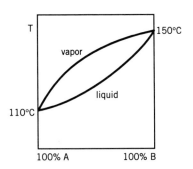

(b) A distills first (70 mL at 110°C) followed by B (50 mL at 150°C).

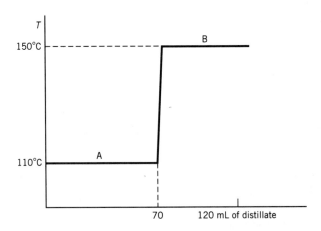

Problem 7

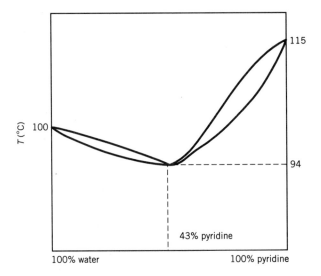

Distillation curve a, azeotrope distills first (with all of the water, 50 g), then pyridine.
Distillation curve b, azeotrope distills first (with all of the pyridine, 10 g), then water.
% Pyridine in azeotrope (from curve a) = 100 × (87.7 − 50)/87.7 = 43%. This agrees with the value calculated from curve b: % Pyridine in azeotrope (from curve b) = 100 × 10/23.2 = 43%.

Problem 9

The distillate contains 3.4 g of water per gram of oil (using Eq. 8).

UNIT 7. *Gas Chromatography*

Problem 1

Gas; solid; liquid; partition; adsorption; helium; nitrogen; Hamilton; flame ionization; thermal conductivity; sensitive; peak size; increase; length; diameter; flow rate; temperature; decrease.

Problem 3

Hydrocarbons of a homologous series come out of a column of medium polarity in order of increasing boiling points. The b.p. (°C) for benzene, toluene, ethylbenzene, *o*-xylene, *m*-xylene and *p*-xylene are: 80, 111, 136, 144, 139, and 138, respectively. The expected order of elution is: benzene, toluene, ethylbenzene, *p*-xylene, *m*-xylene, *o*-xylene. Depending on the experimental conditions, the resolution of *m*- and *p*-xylene peaks may be poor due to their similar b.p.

Problem 5

Menthone is likely to come out of a GC column before menthol for three reasons: 1) lower MM; 2) lower bp (207°C vs. 212°C); and 3) menthone is only an H-bond acceptor while menthol is an H-bond acceptor as well as an H-bond donor.

Problem 7

(a) $R_{1/IS} = 1.07$; $R_{2/IS} = 1.17$. **(b)** $C_1 = 0.84$ g (in 100 mL); $C_2 = 1.57$ g (in 100 mL).

UNIT 8. *TLC*

Problem 1

Adsorption; ion; dipole; H-bonds; stationary; silica gel; alumina; adsorptivity; high; low; mobile; development; ratio-to-the-front; lower; iodine; fluorescent indicators; analytical; number; preparative; isolate; cellulose; partition.

Problem 3

(a) The R_f values: at time zero (from bottom to top): 0.20 and 0.83; at 15 min: 0.20, 0.46, and 0.83; at 30 min: 0.44 and 0.83.

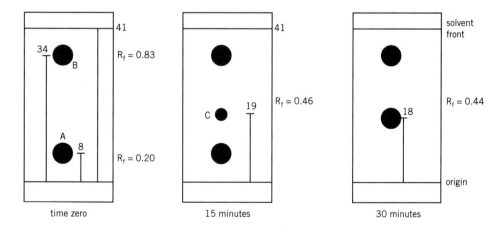

(b) and **(c)** The spots are identified in the figure. The R_f of pure A and B are 0.2 and 0.8, respectively; thus the spot in the middle ($R_f = 0.46$–0.44) must be C. After 30 min. there is no A left. A is the limiting reagent.

Problem 5

The R_f values are: gallic acid, 0.20; ethyl gallate, 0.41; resorcinol, 0.63; and salicylaldehyde, 0.84.

UNIT 9.   *Column Chromatography*

Problem 1

(a) Use cyclohexane first to remove A. After A is eluted, change to toluene to elute B. Then use methanol to elute C. **(b)** Cyclohexane.

Problem 3

(a) At least four. **(b)** Fractions 15–30 can be pooled together; also fractions 105–135. **(c)** All the other fractions should be reanalyzed to determine when the elution of each compound starts and finishes.

Problem 5

Each of a, b, and c is an elutropic series; c has the widest polarity range; d is not an elutropic series.

UNIT 10.   *HPLC*

Problem 1

C_{18} $\Longrightarrow$ reversed-phase chromatography; degassing $\Longrightarrow$ helium; deuterium lamp $\Longrightarrow$ multiwavelength detector; fast mass transfer $\Longrightarrow$ better resolution; guard column $\Longrightarrow$ column protection; injection $\Longrightarrow$ six-port valve; large particle size $\Longrightarrow$ stagnant solvent; low-pressure mixing $\Longrightarrow$ one pump; mercury lamp $\Longrightarrow$ fixed wavelength detector; outgassing $\Longrightarrow$ pressure drop; pore size $\Longrightarrow$ gel permeation; solvent mixture $\Longrightarrow$ miscibility; UV-cutoff $\Longrightarrow$ solvent absorbance.

Problem 3

In reversed-phase chromatography, C comes out followed by B, D, and A. C is the most polar and A the least.

Problem 5

The molecular mass of proteins with V_e 20.3 and 16.5 mL can be estimated directly from the plot V_e versus log MM. They are $10^{4.43} = 26{,}900$ and $10^{4.93} = 85{,}000$, respectively.

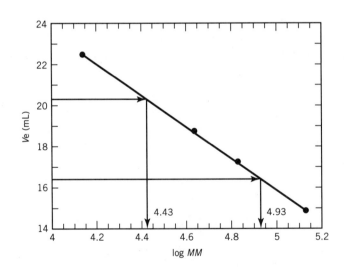

UNIT 11.   *Refractometry and Polarimetry*

Problem 1

The first step is to convert minutes and seconds from the sexagesimal to the decimal system. Crown glass $\alpha^c = 41°2'28'' = 41.041°$; n = 1.5230. Plexiglas, $\alpha^c = 42°9'19'' = 42.155°$; n = 1.4900. Diamond, $\alpha^c = 24°24'27'' = 24.4075°$; n = 2.4200.

Problem 3

The oil of cypress must contain some dextrorotatory compounds: *d*-pinene, *d*-camphene, and/or *d*-bornyl acetate. If the two oils are mixed, the optical rotation should decrease.

Problem 5

In the mixing process, both solutions get diluted. The final volume is 25 mL + 15 mL = 40 mL. Concentration of (−)-tartaric acid: 0.0125 g/mL; (+)-tartaric acid: 0.075 g/mL. The optical rotation of the solution is +1.5°.

UNIT 12. *Alcohols and Alkenes*

Problem 1

a)

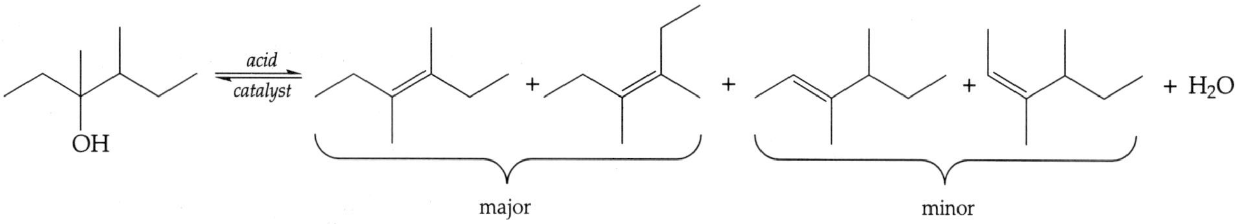

b)

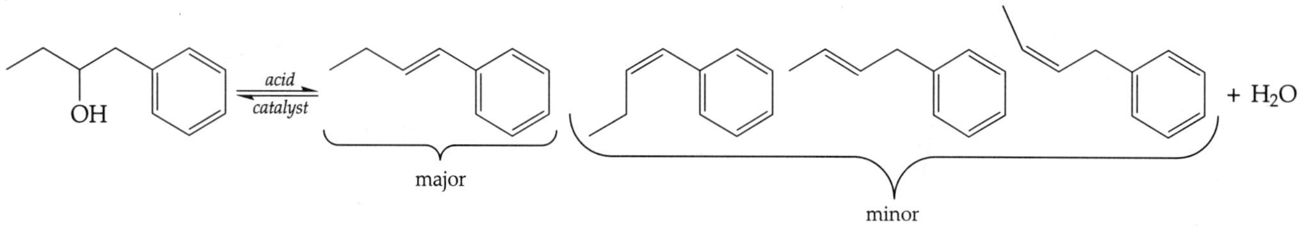

c)

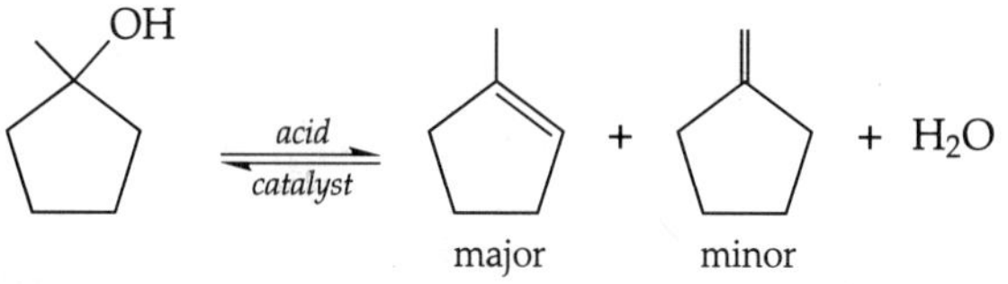

Problem 3

Synton *forgot to dry* the distillate with anhydrous magnesium sulfate to remove the water. He erroneously attributed the 3500 cm^{-1} band to the O−H stretching of the alcohol, when in fact this band was due to water.

Problem 5

1-Methylcyclohexene.

UNIT 13. *Alkyl Halides*

Problem 1

3-Chloro-3-methylpentane.

Problem 3

(a)

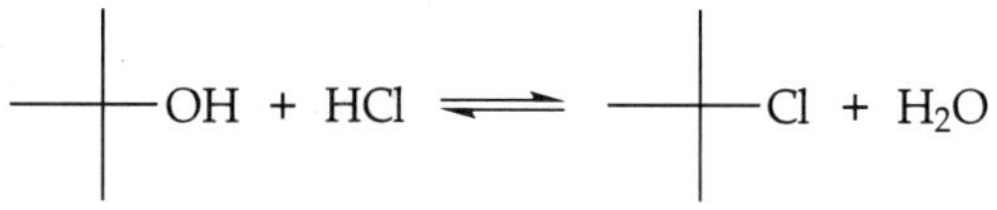

(b)

	Volume used or obtained in mL	Density g/mL	MM g/mole	Mass g	mmoles
t-butanol	1.56	0.786	74	1.226	16.6
t-butyl chloride	1.3	0.847	92.6	1.101	11.9
HCl	3.0	1.2	36.5	3.60/1.332	36.5

(c) *t*-Butanol is the limiting reagent.

(d) Theoretical yield is 16.6 mmoles; 1.537 g. % yield = 72%.

Problem 5
(a) i > iii > ii ≫ iv; (b) ii > iii ≫ i, iv.

UNIT 14. *Acid-Base Extraction*

Problem 1
(a) pH = 2.87; (b) pH = 9.12.

Problem 3
If the pK_a for phenylacetic acid is 4.31, the pK_a of *m*-phenyl benzoic acid (3-carboxybiphenyl) can be estimated as 4.16 from the graph.

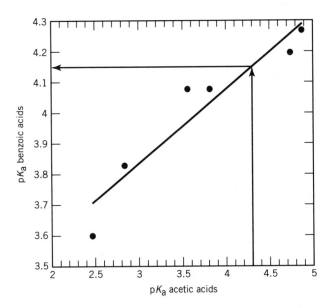

Problem 5
(a) Contrary to *N,N*-diethylaniline, where the amino group is free to rotate around the C–N bond that connects it with the aromatic ring, in benzoquinuclidine, the N atom is part of a rigid ring system and cannot rotate freely around the C–N bond. The lone electron pair on the nitrogen of benzoquinuclidine is in an orbital perpendicular to the p orbitals of the aromatic ring, and therefore it is unable to overlap with them. This makes benzoquinuclidine an unusually basic aromatic amine.

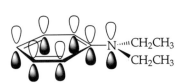

Diethylamino group can adopt a conformation almost coplanar with aromatic ring. Overlap between lone electron pair on N and the ring is possible.

Overlap between lone electron pair on N and aromatic ring is not possible.

(b) 2,5-Dihydroxybenzoic acid is more acidic than 3,5-dihydroxybenzoic acid because one of the −OH groups is in position *ortho* to the carboxylate group. Also, the formation of a strong intramolecular H-bond between the −OH and the −COO⁻ groups in *ortho* makes 2,5-dihydroxybenzoic acid a stronger acid.

Problem 7

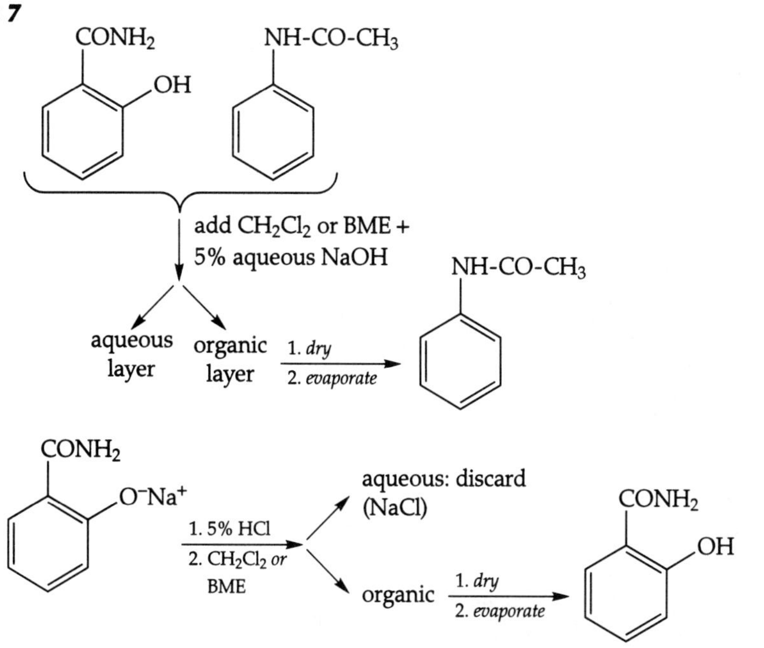

Problem 9

Dissolve the mixture in an organic solvent such as *t*-butyl methyl ether (BME), toluene, or methylene chloride. Extract the organic phase with 5% aqueous HCl. Dry and evaporate the organic solvent to obtain 4-carboxybiphenyl. Add NaOH to the aqueous layer and extract the amine with the organic solvent. Dry and evaporate the solvent to get the product. (Alternatively, extract the first solution in the organic solvent with a 5% aqueous NaOH solution instead of HCl. This extracts the carboxylic acid into the aqueous phase.)

UNIT 15. *Phenols and Ethers*

Problem 1

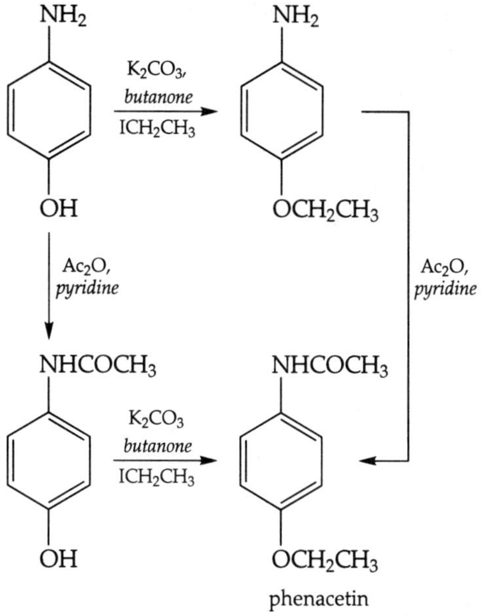

Problem 3

No. Both compounds are carboxylic acids and react similarly with solutions of NaOH or NaHCO₃. Because of their difference in polarity, they can be separated by column chromatography using silica gel as stationary phase.

Problem 5
The compound in question is methyl salicylate (methyl 2-hydroxybenzoate). This compound is also called wintergreen oil and is the main component of the betula and sweet birch oils.

UNIT 16. *Electrophilic Aromatic Substitution*

Problem 1

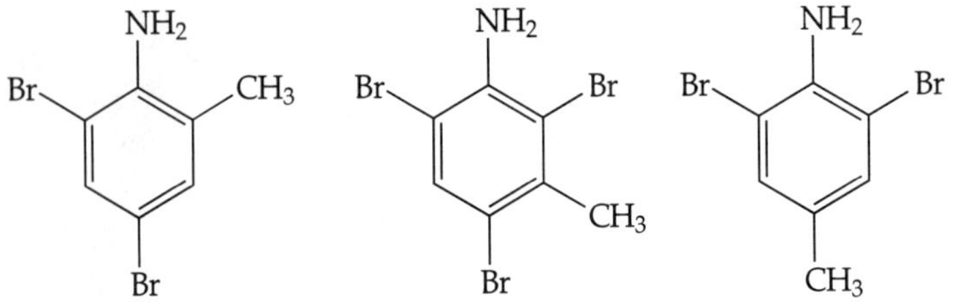

Problem 3

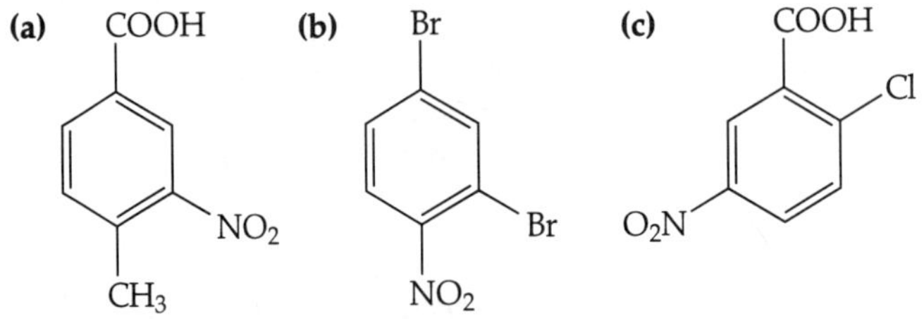

Problem 5
A: 2-Methyl-6-nitrophenol. It is more volatile than the other isomers because there is an intramolecular H-bond between OH and nitro in *ortho*. B: 2-methyl-6-nitroanisole.

Problem 7
4-Bromo-2-nitroanisole.

UNIT 17. *Nucleophilic Aromatic Substitution*

Problem 1
Deactivating: $-N(CH_3)_2$; $-OCH_3$, the rest are activating.

Problem 3
CH_3O^- forms in small amounts because of the following equilibrium (similar to the dissociation equilibrium of amines in water):

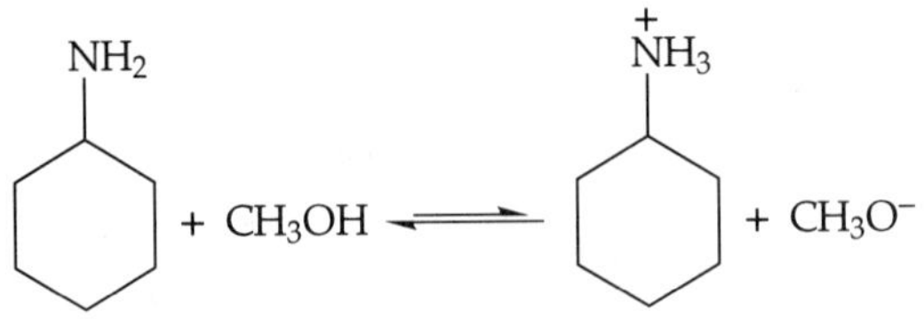

CH_3O^- reacts with 2,4-dinitrochlorobenzene, giving 2,4-dinitroanisole.

UNIT 18. *Chemical Kinetics*

Problem 1
(a) This is pseudo-first-order kinetics because a large excess of cyclohexylamine (0.402 M) is used in comparison to the initial concentration of anisole (1.64 10^{-4} M).

(b) A plot of ln[S] vs. time gives a good linear correlation:

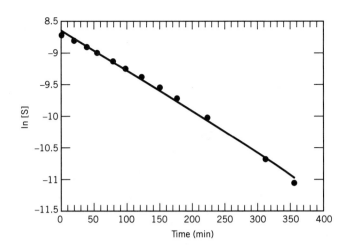

By the least-squares method: slope $= -0.00649$ min^{-1}, intercept $= -8.648$. Pseudo-first-order rate constant, $k_{obs} = 6.49\ 10^{-3}$ min$^{-1} = 1.08\ 10^{-5}$s^{-1}. The second-order rate constant, $k = k_{obs}/$[cyclohexylamine] $= 1.08\ 10^{-5}$ s$^{-1}/0.402$ M $= 2.69\ 10^{-4}$ M^{-1}s^{-1}.

(c) $\tau = 0.693/k_{obs} = 0.693/6.49\ 10^{-3}$ min$^{-1} = 107$ min.

(d) $\tau = 0.693/k_{obs} = 0.693/1.61\ 10^{-3}$ min$^{-1} = 430$ min.

Problem 3
23.3 min.

Problem 5
Plot of ln k vs. 1000/T. Slope: $-23.131\ 10^3$ K^{-1}; intercept: 32.949. $E_a = 193$ kJ/mol. The preexponential factor is: $A = e^{32.949} = 2.04\ 10^{14}$s^{-1}.

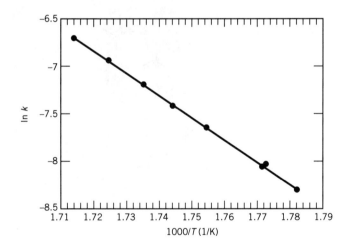

UNIT 19. *Diels–Alder*

Problem 1
In a and b only the major (endo) product is shown.

(a)   **(b)**   **(c)**

Problem 3
(a) exo; (b) endo; (c) exo.

Problem 5
Dienophile-diene-loving; hydrophobicity-water cavity; solvophobicity-solvent polarity; endo-*syn* to larger bridge; $4+2$ cycloaddition-decrease in conjugation; exo-*anti* to larger bridge; concerted — no intermediates.

UNIT 20. *Aldehydes and Ketones*

Problem 1
(a) Positive; (b) negative; positive; (c) derivatives; (d) aldehydes; unhindered; (e) yellow-orange; (f) negative.

Problem 3

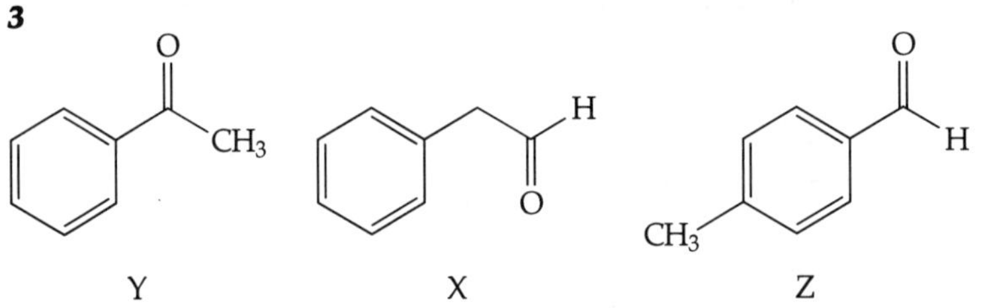

Y   X   Z

Problem 5

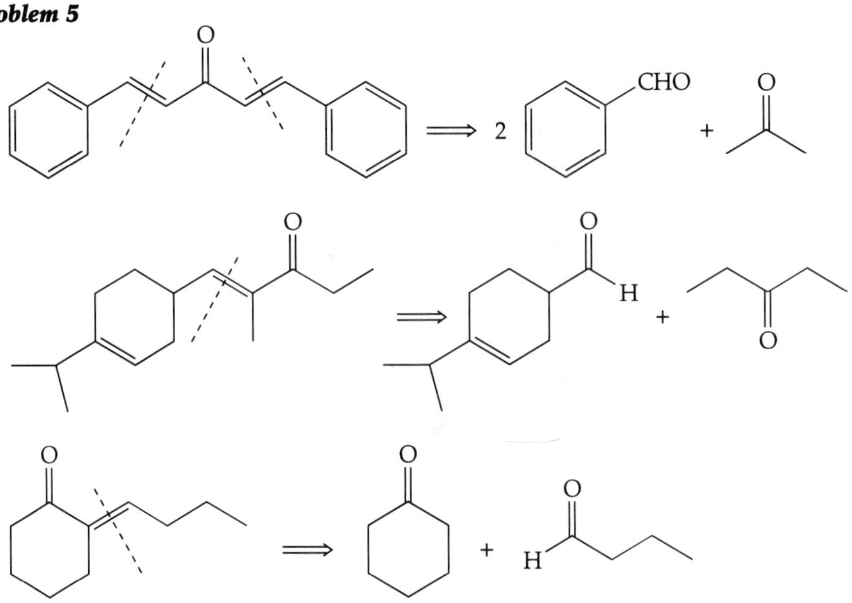

Problem 7

salicylaldehyde   coumarin

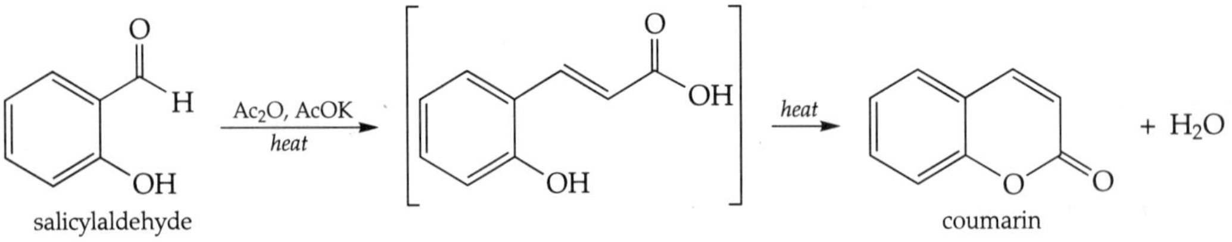

UNIT 21. *Oxidation-Reduction*

Problem 1
(a) $+1$; (b) $+3$; (c) $+1$; (d) 0; (e) 0; (f) -1.

Problem 3
(a) PCC; (b) MnO_2; (c) $KMnO_4$, H_3O^+, heat or CrO_3, H_2SO_4; (d) $Ag(NH_3)_2^+$; (e) Na_2S or Cu/HBr; (f) $NaBH_4$.

Problem 5

The aldehyde is contaminated with a carboxylic acid. Probably *trans*-cinnamic acid formed by air oxidation of *trans*-cinnamaldehyde. Simple distillation.

UNIT 22. *Esters*

Problem 1

(a) x = 0.339 mol of acetic acid and an equal amount of ethanol. **(b)** x = 0.965 mol of acetic acid and an equal amount of ethanol.

Problem 3

Methyl benzoate.

UNIT 23. *Multistep Synthesis*

Problem 1

Problem 3

A shorter pathway for the same transformation (with the help of Table 23.1): bromination of benzene, Grignard reagent of bromobenzene, treatment with CO_2.

Problem 5

UNIT 24. *Molecules of Life*

Problem 1

In general, the solubility decreases as the side chain of the amino acid becomes less polar. In going from Gly to Ile, the number of carbon atoms in the side chain increases and the solubility decreases. Ser is more soluble than Ala because of the polar HO— group present in Ser.

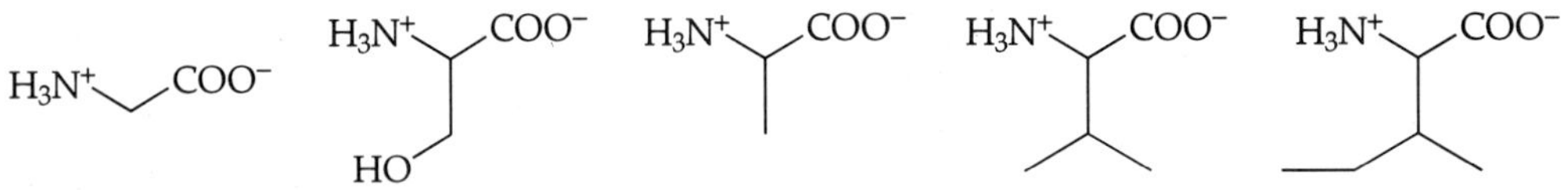

H$_3$N$^+$ COO$^-$

Gly
25.0

H$_3$N$^+$ COO$^-$
HO

Ser
25.0

H$_3$N$^+$ COO$^-$

Ala
16.7

H$_3$N$^+$ COO$^-$

Val
8.9

H$_3$N$^+$ COO$^-$

Ile
4.1

Problem 3
Tyrosine.

doublet at 3.2

NH$_3^+$

doublet at 7.2

CH$_2$—CH—COO$^-$

triplet at 4.4

doublet at 6.9   OH

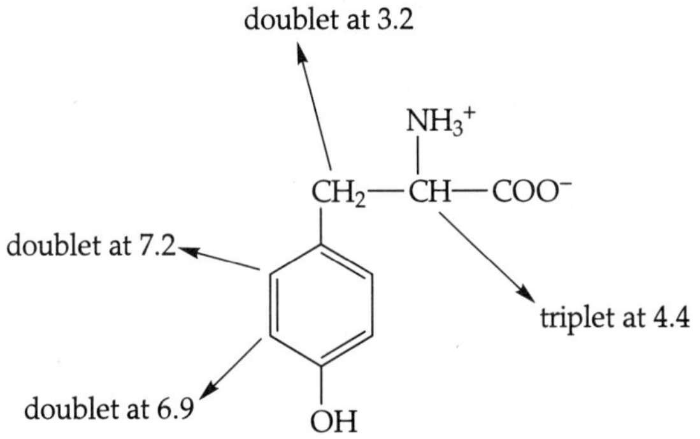

(The methylene hydrogens are diastereotopic [see section 33.19] and may give two different signals under different experimental conditions [different instrument, solvent, etc.]; see, for example, Experiment 16A.)

Problem 5
(a) Fehling's: maltose and galactose (reducing): positive. Sucrose (nonreducing): negative. Barfoed's: galactose (monosaccharide) reacts faster than maltose (disaccharide). Sucrose: negative.

(b) Alanine: blue-violet; proline: yellow-red; *N*-acetylglycine: negative.

UNIT 25. *Lipids*

Problem 1
The three constitutional isomers and their corresponding enantiomers. Notice that in each case the central carbon is stereogenic and thus two enantiomers are possible.

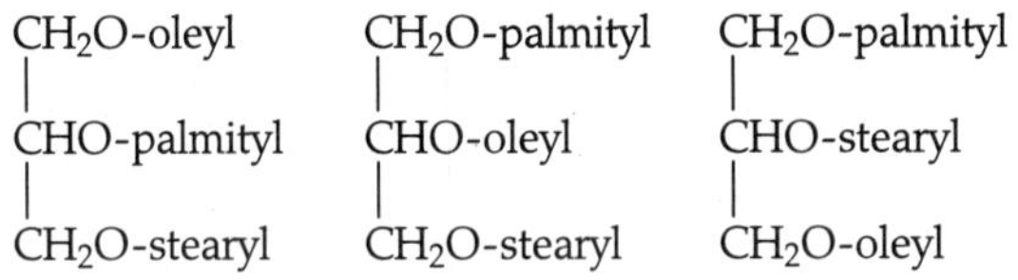

CH$_2$O-oleyl	CH$_2$O-palmityl	CH$_2$O-palmityl
CHO-palmityl	CHO-oleyl	CHO-stearyl
CH$_2$O-stearyl	CH$_2$O-stearyl	CH$_2$O-oleyl

Problem 3
The oil is likely to be sesame oil (mp: −6°C; iodine number: 104–116).

Problem 5
Peanut oil.

UNIT 26. *Polymers*

Problem 1
102.

Problem 3
They can be separated by flotation in water. Bottles are shredded. PE floats, PETE sinks.

Problem 5
Plastic cup: PMMA. Ruler: PS. Diskette box: HDPE.

UNIT 27. *Dyes and Pigments*

Problem 1

Intermediate resulting from attack at position 1 is more stable because in two of the resonance forms there is an intact benzene ring (structures 1 and 2):

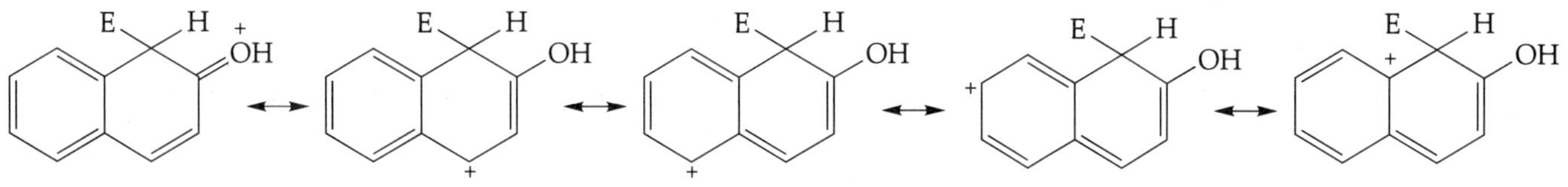

In the intermediate resulting from attack at position 3, only one resonance form has an intact benzene ring (structure 6):

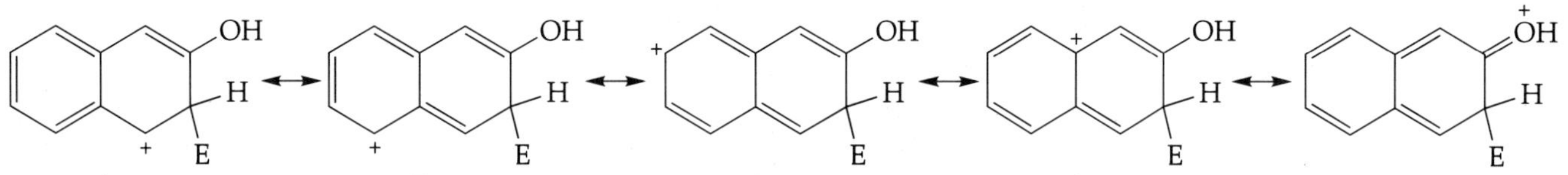

Problem 3

There is not enough hydrochloric acid to keep the aniline protonated. Aniline reacts with the diazonium salt, forming an azo dye:

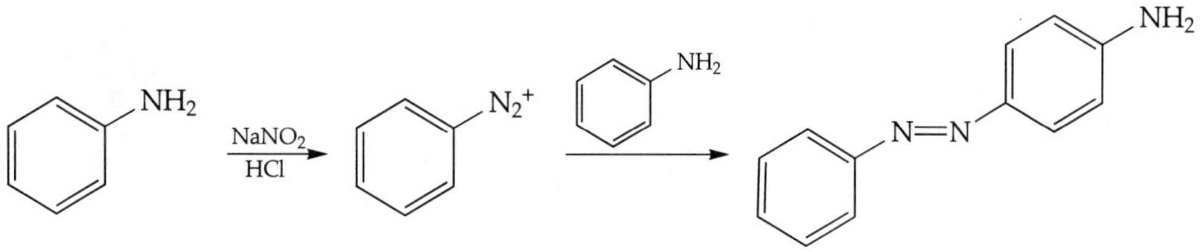

yellow solid

Problem 5

Compound X is triphenyl carbinol (triphenylmethanol).

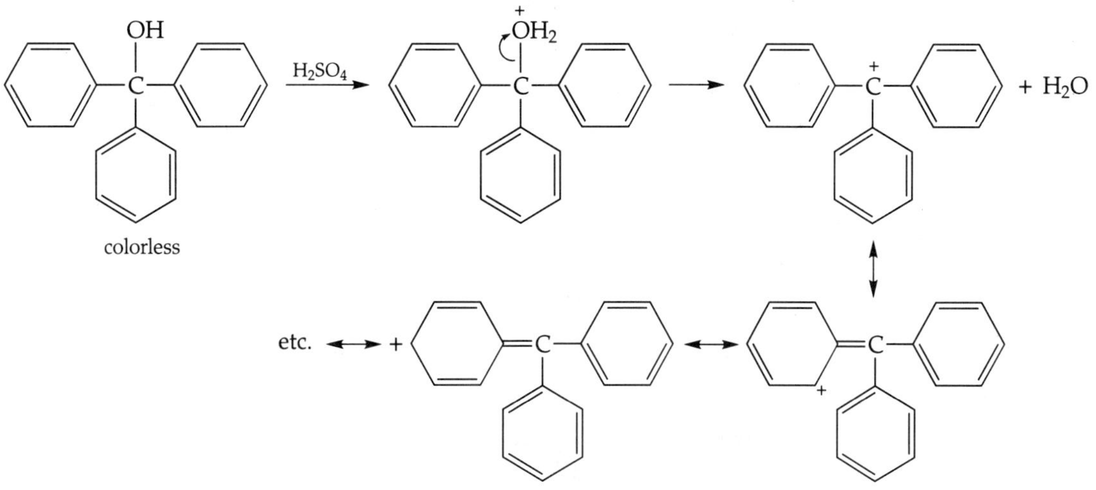

colorless

etc.

yellow

UNIT 28. *Bioorganic Chemistry*

Problem 1
(a) Both: si.

(b) **(c)**

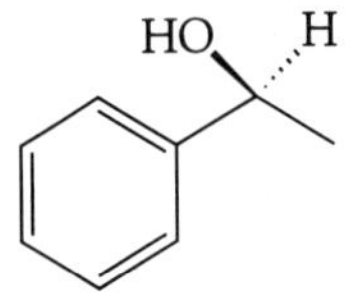

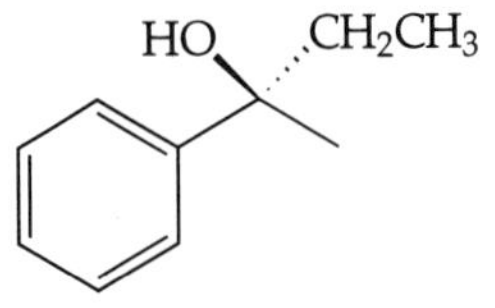

S   *R*

Problem 3
97.5% and 2.5%.

Problem 5
96%; 33%.

UNIT 29. *Molecules of Heredity*

Problem 1

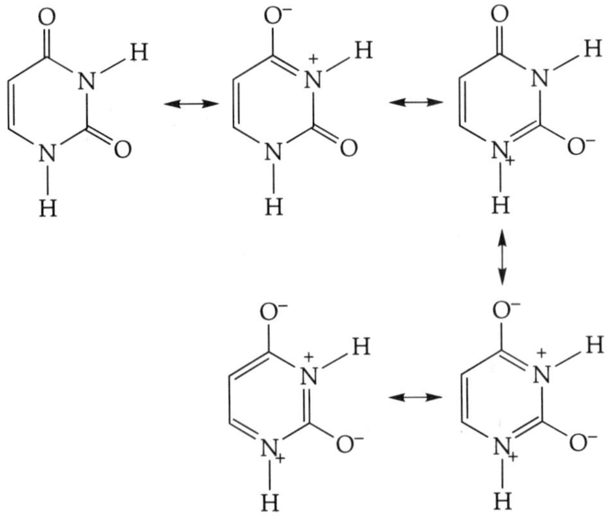

Problem 3
5′-ATTATACCAGGATGATAC-3′
3′-TAATATGGTCCTACTATG-5′

Problem 5
Either 260 nm, where uracil absorbs more strongly than cytosine, or 280 nm where the opposite is true.

UNIT 30. *Absorption Spectroscopy*

Problem 1

λ (nm)	λ (Å)	v (Hz)	$\bar{v}$ (cm^{-1})	E(kcal/mol)	Region
280	2800	$1.07\ 10^{15}$	$3.57\ 10^4$	102	UV
$5.9\ 10^3$	$5.9\ 10^4$	$5.1\ 10^{13}$	1700	4.9	IR
$1.0\ 10^7$	$1.0\ 10^8$	$3\ 10^{10}$	1	$2.9\ 10^{-3}$	microwave

Problem 3
$8.4\ 10^4\ M^{-1}\ cm^{-1}$.

UNIT 31. *IR Spectroscopy*

Problem 1
(a) 1.36; **(b)** 1.34.

Problem 3

Active: b; d; e; g. The rest inactive.

Problem 5

1–e; 2–d; 3–a; 4–c; 5–b.

Problem 7

2-methyl-1-butene: 886 cm^{-1}; 1-pentene: 992 and 911 cm^{-1}; *trans*-2-pentene: 964 cm^{-1}; *cis*-2-pentene: 695 cm^{-1}.

Problem 9

tert-Butylbenzene; 762 and 700 cm^{-1} typical C−H d out-of-plane of monosubstituted benzenes.

UNIT 32. *UV-Visible Spectroscopy*

Problem 1

p-Nitroaniline. It has a more extended conjugation.

Problem 3

In the order in which they are written: 313 nm; 267 nm; 316 nm; and 311 nm.

Problem 5

(a)

	Solution I			Solution II		
	Concentration (g/L)	Concentration (mol/L)	A (360 nm)	Concentration (g/L)	Concentration (mol/L)	A (360 nm)
stock solution	0.0836	5.46 10^{-4}	1.750	0.0642	4.19 10^{-4}	1.302
dilution factor 2/5	0.03344	2.18 10^{-4}	0.703	0.02568	1.68 10^{-4}	0.527
dilution factor 1/5	0.01672	1.09 10^{-4}	0.355	0.01284	8.38 10^{-4}	0.268
dilution factor 1/10	0.00836	5.46 10^{-5}	0.172	0.00642	4.19 10^{-5}	0.136

(b) Yes. **(c)** 3169 M^{-1} cm^{-1}.

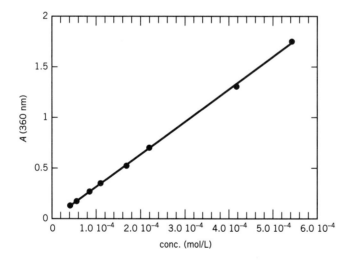

UNIT 33. *NMR*

Problem 1

(a) T; **(b)** F; **(c)** T; **(d)** F; **(e)** F; **(f)** T; **(g)** F; **(h)** T; **(i)** F; **(j)** T.

Problem 3

For ^{1}H-NMR: **(a)** Three. From left to right in the molecule: triplet; quartet; singlet; **(b)** two; triplet; quartet; **(c)** three; triplet; double quartet or quintet (depending on J values); triplet; **(d)** three; triplet; quartet; singlet; **(e)** three; doublet; multiplet (9); doublet; **(f)** three; doublet; septet; singlet; **(g)** one; singlet; **(h)** two; distorted doublet; distorted doublet.

For ^{13}C-NMR: **(a)** 3; **(b)** 2; **(c)** 3; **(d)** 3; **(e)** 3; **(f)** 4; **(g)** 2; **(h)** 4.

Problem 5
n-Propyl chloride.

Problem 7
p-Methoxybenzaldehyde (*p*-anisaldehyde).

Problem 9

CH_3O— (benzene ring) —CH_2OCH_3

Seven peaks in the ^{13}C-NMR.

UNIT 34. *Mass Spectrometry*

Problem 1
Benzyl chloride.

Problem 3
n-Butylamine or isobutylamine.

Problem 5
Ethyl butyrate.

Credits

Index

823

Index of Spectra

in alphabetical order by compound